GLOSSARY OF SYMBOLS

$\overline{A}$	Complement of set A
ANOVA	Analysis of variance
α(alpha)	Probability of a type I error
β(beta)	Probability of a type II error
$1 - \beta$	Power of a statistical test
β_0	y-intercept of the true linear relationship
β_1	Slope of the true linear relationship
b_0	y-intercept for the line of best fit for the sample data
b_1	Slope for the line of best fit for the sample data
$_nC_r$	Number of combinations of n things r at a time
c_j	Column total
c	Column number or class width
d	Difference in value between two paired pieces of data
$d(\ )$	Depth of
df or df($\ $)	Number of degrees of freedom
d_i	Difference in the rankings of the ith element
E	Expected value or maximum error of estimate
$e = y - \hat{y}$	Error (observed)
ε(epsilon)	Experimental error
ε_{ij}	Amount of experimental error in the value of the jth piece of data in the ith row
F	F distribution statistic
$F(\text{df}_n, \text{df}_d, \alpha)$	Critical value for the F distribution
f	Frequency
H	Value of the largest-valued piece of data in a sample
H_a	Alternative hypothesis
H_0	Null hypothesis
i	Index number when used with Σ notation
i	Position number for a particular data
k	Identifier for the kth percentile
k	Number of cells or variables
L	Value of the smallest-valued piece of data in a sample
LH	Lower hinge
m	Number of classes
MAD	Mean absolute deviation
MS($\ $)	Mean square
MSE	Mean square error
μ(mu)	Population mean
μ_d	Mean value of the paired differences

Elementary Statistics

SIXTH EDITION

Elementary Statistics

ROBERT JOHNSON
Monroe Community College

Duxbury Press
An Imprint of Wadsworth Publishing Company
Belmont, California

Duxbury Press
An Imprint of Wadsworth Publishing Company
A division of Wadsworth, Inc.

Reprinted with corrections, June 1992

Copyright © 1992, 1988 by PWS–KENT Publishing Company.
Copyright © 1984 by PWS Publishers, © 1980 by Wadsworth, Inc.

All rights reserved. No part of this book may be reproduced, stored in a retrieval system, or transcribed, in any form or by any means—electronic, mechanical, photocopying, recording, or otherwise—without the prior written permission of Wadsworth Publishing Company, Belmont, California 94002.

Library of Congress Cataloging-in-Publication Data

Johnson, Robert Russell, 1939–
  Elementary statistics / Robert Johnson. —6th ed.
    p. cm.
  Includes index.
  ISBN 0-534-92980-X
  1. Statistics. I. Title.
QA276.12.J64 1992   91-37750
519.5–dc20   CIP

International Student Edition ISBN: 0-534-97201-2

Printed in the United States of America
  94 95 96—10 9 8 7 6 5 4

Assistant Editor Marcia Cole
Production Coordinator Pamela Rockwell
Manufacturing Coordinator Marcia A. Locke
Interior Design Susan M. C. Caffey
Cover Design Jean Hammond
Cover Photo George Haling/Photo Researchers, Inc.
Composition Beacon Graphics
Cover Printer Henry N. Sawyer Co., Inc.
Text Printer/Binder R. R. Donnelley & Sons Company

 This book is printed on recycled, acid-free paper.

▼
To my mother
and to the memory
of my father
▲

Past and Present Reviewers for *Elementary Statistics*

Nancy Adcox
Mt. San Antonio College

Paul Alper
College of St. Thomas

William D. Bandes
San Diego Mesa College

Barbara Jeanne Blass
Oakland Community College

Austin Bonis
Rochester Institute of Technology

Louis F. Bush
San Diego City College

Rodney E. Chase
Oakland Community College

David M. Crystal
Rochester Institute of Technology

Joyce Curry and Frank C. Denny
Chabot College

Shirley Dowdy
West Virginia University

Joan Garfield
University of Minnesota General College

Carol Hall
New Mexico State University

Hank Harmeling
North Shore Community College

Bryan A. Haworth
California State College at Bakersfield

John C. Holahan
Xerox Corp.

James E. Holstein
University of Missouri

Robert Hoyt
Southwestern Montana State

T. Henry Jablonski, Jr.
East Tennessee State University

Sherry Johnson

Meyer M. Kaplan
The William Patterson College of New Jersey

Michael Karelius
American River College

Anand S. Katiyar
McNeese State University

Gayle S. Kent
Florida Southern College

Raymond Knodel
Bemidji State University

Pat Kuby
Monroe Community College

Robert O. Maier
El Camino College

Mark Anthony McComb
Mississippi College

Jeffrey Mock
Diablo Valley College

David Naccarato
University of New Haven

Harold Nemer
Riverside Community College

Janet M. Rich
Miami-Dade Junior College

Larry J. Ringer
Texas A & M University

John T. Ritschdorff
Marist College

John Rogers
California State Polytechnic Institute at San Luis Obispo

Barbara F. Ryan and Thomas A. Ryan
Pennsylvania State University

Robert J. Salhany
Rhode Island College

Howard Stratton
State University of New York at Albany

Larry Stephens
University of Nebraska — Omaha

Thomas Sturm
College of St. Thomas

Edward A. Sylvestre
Eastman Kodak Co.

William Tomhave
University of Minnesota

Richard Uschold
Canisius College

John C. Van Druff
Fort Steilacoom Community College

Philip A. Van Veldhuizen
University of Alaska

John Vincenzi
Saddleback College

Kenneth D. Wantling
Montgomery College

Mary Wheeler
Monroe Community College

CONTENTS

Preface x

PART ONE Descriptive Statistics 1

Chapter 1 STATISTICS 2

Miscellaneous Statistics 3
Chapter Objectives 4

1.1 What is Statistics? 4
1.2 Uses and Misuses of Statistics 6
  Case Study 1-1: Looking Around Campus 7
  Case Study 1-2: What Business Travelers Drink at Lunch 8
  Case Study 1-3: Where Motorcyclists Live 9
  Case Study 1-4: Home Prices Rise 10
1.3 Introduction To Basic Terms 11
  Case Study 1-5: What We Do in Bed 14
  Case Study 1-6: Average Outfield Distances 15
  Case Study 1-7: Eating Together: Still Important 17
1.4 Measurability and Variability 21
1.5 Data Collection 21
  Case Study 1-8: People have to tell the Truth or Statistics Don't Mean Anything 26
  Case Study 1-9: Money, Marriage: Tell Us 27
  Case Study 1-10: What Would Make Life Better? 28
1.6 Comparison of Probability and Statistics 29
1.7 Statistics and the Computer 30

In Retrospect 31
Chapter Exercises 32
Vocabulary List 33
Key Concepts 34
Quizzes 34

Chapter 2 DESCRIPTIVE ANALYSIS AND PRESENTATION OF SINGLE-VARIABLE DATA 36

Estimation of Flow for Evaluating and Classifying Pedestrian Facilities 37
Chapter Objectives 38

Graphic Presentation of Data 39
2.1 **Graphs and Stem-and-Leaf Displays** 39
2.2 **Frequency Distributions, Histograms, and Ogives** 51
  Case Study 2-1: Chalk It Up 60
  Case Study 2-2: Home Sweet Home 61
  Case Study 2-3: Transplant Movers, Inc. 62
Calculated Descriptive Statistics 68
2.3 **Measures of Central Tendency** 68
  Case Study 2-4: "Average" Means Different Things 74
2.4 **Measures of Dispersion** 81
  Case Study 2-5: Women in the State Houses 89
2.5 **Measures of Position** 94
  Case Study 2-6: Stock Market Mixed; Behind the Market 102
2.6 **Interpreting and Understanding Standard Deviation** 106
2.7 **The Art of Statistical Deception** 112
  Case Study 2-7: Bankruptcy Blues 116
  In Retrospect 117
  Chapter Exercises 117
  Vocabulary List 131
  Key Concepts 132
  Quizzes 132

Chapter 3 **DESCRIPTIVE ANALYSIS AND PRESENTATION OF BIVARIATE DATA** 136
  Changes of the Shinkansen Super Express Railway Noise Since the Opening in 1964 137
  Chapter Objectives 138
3.1 **Bivariate Data** 139
  Case Study 3-1: Exploratory Data Analysis for Possibly Censored Data from Skewed Distributions 146
  Case Study 3-2: Visual Preference as a Test of Infant Word Comprehension 148
3.2 **Linear Correlation** 158
3.3 **Linear Regression** 166
  Case Study 3-3: Thirty Years of Car Buying 176
  Case Study 3-4: Points and Fouls in Basketball 176
  In Retrospect 182
  Chapter Exercises 183
  Vocabulary List 190
  Key Concepts 190
  Quizzes 191

 WORKING WITH YOUR OWN DATA 194

PART TWO Probability 198

Chapter 4 PROBABILITY 200

Probability: An Important Tool in Police Science 201
Chapter Objectives 202
Concepts of Probability 202
4.1 **The Nature of Probability** 202
4.2 **Probability of Events** 204
4.3 **Simple Sample Spaces** 210
Case Study 4-1: Trying to Beat the Odds 220
4.4 **Rules of Probability** 217
Calculating Probabilities of Compound Events 222
4.5 **Mutually Exclusive Events and the Addition Rule** 222
4.6 **Independence, Multiplication Rule, and Conditional Probability** 230
Case Study 4-2: Save It for a Rainy Day 236
Case Study 4-3: Pick 10 237
4.7 **Combining the Rules of Probability** 240
Case Study 4-4: Ask Marilyn 245
4.8 **Bayes's Rule** 249
In Retrospect 253
Chapter Exercises 253
Vocabulary List 263
Key Concepts 263
Quizzes 263

Chapter 5 PROBABILITY DISTRIBUTIONS (DISCRETE VARIABLES) 266

An Analysis of Accidents at a Day Care Center 267
Chapter Objectives 268
5.1 **Random Variables** 268
5.2 **Probability Distributions of a Discrete Random Variable** 270
Case Study 5-1: Who Needs the Ambulance? 273
Case Study 5-2: Woes of the Weekend Jock 274
5.3 **Mean and Variance of a Discrete Probability Distribution** 278
5.4 **The Binomial Probability Distribution** 282
5.5 **Mean and Standard Deviation of the Binomial Distribution** 294
In Retrospect 297
Chapter Exercises 298
Vocabulary List 303
Key Concepts 303
Quizzes 303

▼ CONTENTS

Chapter 6 NORMAL PROBABILITY DISTRIBUTIONS 306

Teaching Breast and Testicular Self-Exams: Evaluation of a High School Curriculum Pilot Project 307

Chapter Objectives 308

6.1 **Normal Probability Distributions** 308
6.2 **The Standard Normal Distribution** 310
6.3 **Applications of Normal Distributions** 317

Case Study 6-1: The Adjustment of Birthweight for Very Early Gestational Ages: Two Related Problems in Statistical Analysis 322

6.4 **Notation** 325
6.5 **Normal Approximation of the Binomial** 331

In Retrospect 337
Chapter Exercises 337
Vocabulary List 341
Key Concepts 341
Quizzes 341

Chapter 7 SAMPLE VARIABILITY 344

Galluping Attitudes 345
Chapter Objectives 346

7.1 **Sampling Distributions** 346

Case Study 7-1: Average Aircraft Age 349
Case Study 7-2: Consumer Comfort Poll 350

7.2 **The Central Limit Theorem** 354
7.3 **Application of the Central Limit Theorem** 362

In Retrospect 368
Chapter Exercises 369
Vocabulary List 372
Key Concepts 372
Quizzes 373

▲▼ **WORKING WITH YOUR OWN DATA** 375

PART THREE Inferential Statistics 378

Chapter 8 INTRODUCTION TO STATISTICAL INFERENCES 380

Evaluation of Teaching Techniques for Introductory Accounting Courses 381

Chapter Objectives 382

8.1 **The Nature of Hypothesis Testing** 382
8.2 **The Hypothesis Test (A Classical Approach)** 392
8.3 **The Hypothesis Test (A Probability-Value Approach)** 407

8.4 Estimation 415
Case Study 8-1: How Much Water U.S. Troops in Saudi Arabia Drink 423
In Retrospect 426
Chapter Exercises 426
Vocabulary List 432
Key Concepts 433
Quizzes 433

Chapter 9 INFERENCES INVOLVING ONE POPULATION 436

The American Gender Evolution: Getting What We Want 437
Chapter Objectives 439

9.1 Inferences About the Population Mean 439
Case Study 9-1: Mother's Use of Personal Pronouns When Talking with Toddlers 446

9.2 Inferences About the Binomial Probability of Success 451
Case Study 9-2: Student Poll Says Alcohol No. 1 High School Problem 456
Case Study 9-3: Many Still Tan Despite Skin Cancer 456
Case Study 9-4: Gallup Report: Sampling Tolerances 457

9.3 Inferences About Variance and Standard Deviation 463
In Retrospect 472
Chapter Exercises 473
Vocabulary List 478
Key Concepts 479
Quizzes 479

Chapter 10 INFERENCES INVOLVING TWO POPULATIONS 482

We Like Our Work 483
Chapter Objectives 484

10.1 Independent and Dependent Samples 484
Case Study 10-1: Exploring the Traits of Twins 486
Case Study 10-2: Study Finds Unproven, Usual Cancer Treatments Equal 486

10.2 Inferences Concerning the Mean Difference Between Two Dependent Samples 488

10.3 Inferences Concerning the Difference Between the Means of Two Independent Samples (Variances Known or Large Samples) 496
Case Study 10-3: Who's Making What? 501

10.4 Inferences Concerning the Ratio of Variances Between Two Independent Samples 505
Case Study 10-4: Personality Characteristics of Police Applicants: Comparisons Across Subgroups and with Other Populations 511

10.5 Inferences Concerning the Difference Between the Means of Two Independent Samples (Variances Unknown and Small Samples) 515

Case Study 10-5: An Empirical Study of Faculty Evaluation Systems: Business Faculty Perceptions **520**

Case Study 10-6: Youthful Ideas About Old Age: An Analysis of Children's Drawings **521**

10.6 Inferences Concerning the Difference Between Proportions of Two Independent Samples 532

Case Study 10-7: Smokers Need More Time in Recovery Room **536**

Case Study 10-8: One Too Many **537**

Case Study 10-9: Gallup Report: Sampling Tolerances **538**

In Retrospect **542**

Chapter Exercises **543**

Vocabulary List **552**

Key Concepts **552**

Quizzes **552**

WORKING WITH YOUR OWN DATA 556

PART FOUR More Inferential Statistics 558

Chapter 11 ADDITIONAL APPLICATIONS OF CHI-SQUARE 560

Trial By Number **561**

Chapter Objectives **562**

11.1 Chi-Square Statistic 562

11.2 Inferences Concerning Multinomial Experiments 563

Case Study 11-1: Why We Rearrange Furniture **566**

Case Study 11-2: Methods Used to Fall Asleep **569**

11.3 Inferences Concerning Contingency Tables 573

Case Study 11-3: 50% Opposed to Building Nuclear Plants **580**

Case Study 11-4: Why We Don't Exercise **581**

Case Study 11-5: Percentages of Students Who Used Selected Drugs **582**

Case Study 11-6: Washer/Dryers by Region **582**

In Retrospect **587**

Chapter Exercises **588**

Vocabulary List **595**

Key Concepts **595**

Quizzes **595**

Chapter 12 ANALYSIS OF VARIANCE 598

The Behavioral Sciences and Management: An Evaluation of Relevant Journals **599**

Chapter Objectives **600**

12.1 Introduction to the Analysis of Variance Technique 601

	12.2	**The Logic Behind ANOVA** 606
		Case Study 12-1: Waiting for a Verdict 608
		Case Study 12-2: Quality of Faculty Life 609
		Case Study 12-3: Teens Are Big Spenders 610
	12.3	**Applications of Single-Factor ANOVA** 611
		Case Study 12-4: Tillage Test Plots Revisited 616
		Case Study 12-5: The Effect of the Class Evaluation Method on Learning in Certain Mathematics Courses 617
		In Retrospect 625
		Chapter Exercises 626
		Vocabulary List 632
		Key Concepts 632
		Quizzes 633
Chapter 13		**LINEAR CORRELATION AND REGRESSION ANALYSIS** 636
		Developmental Factors Associated with Self-Perceptions of Mentoring Competence and Mentoring Needs 637
		Chapter Objectives 638
	13.1	**Linear Correlation Analysis** 638
	13.2	**Inferences About the Linear Correlation Coefficient** 646
		Case Study 13-1: Personality Characteristics of Police Applicants 648
		Case Study 13-2: DNA Quantitation by Image Cytometry of Touch Preparations from Fresh and Frozen Tissue 649
	13.3	**Linear Regression Analysis** 652
	13.4	**Inferences Concerning the Slope of the Regression Line** 659
		Case Study 13-3: Reexamining the Use of Seriousness Weights in an Index of Crime 661
		Case Study 13-4: Charting the Pounds 663
	13.5	**Confidence Interval Estimates for Regression** 667
	13.6	**Understanding the Relationship Between Correlation and Regression** 676
		In Retrospect 677
		Chapter Exercises 678
		Vocabulary List 684
		Key Concepts 685
		Quizzes 685
Chapter 14		**ELEMENTS OF NONPARAMETRIC STATISTICS** 688
		Study Finds That Money Can't Buy Job Satisfaction 689
		Chapter Objectives 690
	14.1	**Nonparametric Statistics** 690
	14.2	**The Sign Test** 691
	14.3	**The Mann-Whitney U Test** 699

Case Study 14-1: Health Beliefs and Practices of Runners Versus Nonrunners 704
14.4 The Runs Test 708
14.5 Rank Correlation 713
Case Study 14-2: Effects of Therapeutic Touch on Tension Headache Pain 717
14.6 Comparing Statistical Tests 721
In Retrospect 723
Chapter Exercises 723
Vocabulary List 728
Key Concepts 728
Quizzes 728

▲▼ **WORKING WITH YOUR OWN DATA** 731

APPENDIXES

A **Summation Notation** A-1
B **Using the Random Number Table** B-1
C **Round-off Procedure** C-1
D **Basic Principles of Counting: Permutations, Combinations** D-1
E **Interpolation Procedure for F Distribution** E-1
F **Tables** F-1

1. Random Numbers F-1
2. Factorials F-3
3. Binomial Coefficients F-4
4. Binomial Probabilities F-5
5. Areas of the Standard Normal Distribution F-8
6. Critical Values of Student's t Distribution F-9
7. Critical Values of the χ^2 Distribution F-10
8a. Critical Values of the F Distribution ($\alpha = 0.05$) F-11
8b. Critical Values of the F Distribution ($\alpha = 0.025$) F-13
8c. Critical Values of the F Distribution ($\alpha = 0.01$) F-15
9. Critical Values of r when $\rho = 0$ F-17
10. Confidence Belts for the Correlation Coefficient F-18
11. Critical Values for the Sign Test F-19
12. Critical Values of U in the Mann-Whitney Test F-20
13. Critical Values for Total Number of Runs (V) F-21
14. Critical Values of Spearman's Rank Correlation Coefficient F-22

ANSWERS TO SELECTED EXERCISES ANS-1

ANSWERS TO CHAPTER QUIZZES ANS-24

INDEX I-1

PREFACE

The primary objective of *Elementary Statistics*, Sixth Edition is to present a truly readable introduction to statistics—one that is organized to promote learning and understanding and that will motivate students by presenting statistics in a context that relates to their personal experiences.

Statistics is a practical discipline that responds to the changing needs of our society. Today's student is a product of a particular cultural environment, and as such, is motivated differently from the student of a few years ago. Statistics is presented in this text as a very useful tool in learning about the world around us. While studying descriptive and inferential concepts, students will become aware of the practical application of these concepts to such fields as business, biology, engineering, industry, and the social sciences.

Statistics is different from other courses: (1) It has its own extensive technical vocabulary; (2) It is highly cumulative in that many of the concepts that you will be learning become the basis for other concepts learned throughout the rest of the course (failure to master each concept as presented will cause great difficulty later on); (3) It requires very precise measurements and calculations (a seemingly minor error will often lead to very wrong answers in some procedures); (4) It is an academic subject that is very real and touches each of us frequently in our everyday life.

This book was written for use in an introductory course for nonmathematics majors, students who need a working knowledge of statistics but do not have a strong mathematics background. Statistics requires the use of many formulas, so those students who have not had intermediate algebra should complete at least one semester of college mathematics.

▼ CHANGES IN THE SIXTH EDITION

The teaching objectives of this edition are the same as those of the previous editions. The following significant changes made in this revision should be helpful in attaining these objectives:

1. *Bibliographies* of *four statisticians* now open the four parts of the textbook.
2. *Dot plots* have been added.
3. The descriptive coverage of bivariate data has been extended to include *cross-tabulation* and formation of *contingency tables*.
4. More than 350 *real-life exercises* have been added; others have been extensively revised.

5. Exercises that emphasize use of the computer and statistical software have been added.
6. Conditional probability has been extended to include *Bayes's Rule*.
7. A new Appendix D covering the *Basic Principles of Counting (permutations* and *combinations)* has been added, as well as exercises covering this new material.
8. Chapter 10 has been reorganized by moving the section *Inferences Concerning the Mean Difference Between Two Dependent Samples* ahead for greater clarity and ease of understanding.
9. Each chapter now has *three levels of Quizzes* for self-examination by students.
10. *Answers to all odd-numbered* exercises are in the back of the textbook, as opposed to answers for selected exercises.

▼ TO THE INSTRUCTOR: THE TEXT AS A TEACHING TOOL

As stated earlier, one of the primary objectives in writing this book was to produce a truly readable presentation of elementary statistics. The chapters are designed to interest and involve students and to guide them step by step through the material in a logical manner. Each chapter includes the following main features:

- ▼ A *chapter outline* that shows students what to look for in the chapter.
- ▼ A *news article* that illustrates how statistics are actually applied to the real world and demonstrates the types of statistics to be studied in that chapter.
- ▼ *Chapter objectives* that tell students the specific information to be learned in that chapter.
- ▼ *Worked-out examples* with solutions to illustrate concepts as well as to demonstrate the applications of statistics in real-world situations.
- ▼ *Case studies* that are brief versions of actual newspaper and magazine articles specifically focusing on the use of statistics in everyday life.
- ▼ *End-of-section exercises* to facilitate practice of concepts as they are presented.
- ▼ An *in retrospect* section provides a summary of the material in the chapter and relates the material to the chapter objectives.
- ▼ *Chapter exercises* that give students further opportunity to master conceptual and computational skills.
- ▼ *Vocabulary* and *key concepts lists* to help students review key terms and concepts.
- ▼ Three *chapter quizzes* that help students evaluate their mastery of the material.

At the end of each part is a *working with your own data* section that has been included for student exploration. These data sections provide a more personalized learning experience by directing students to collect their own data and apply techniques they have been studying.

The first three chapters are introductory by nature. Chapter 3 is a descriptive (first-look) presentation of bivariate data. This material is presented at this point in the book because students often ask about the relationship between two sets of data (such as heights and weights).

In the chapters on probability (4 and 5), the concepts of permutations and combinations are deliberately avoided in Chapter 4. Instead, this material is contained in a new Appendix D entitled *The Basic Principles of Counting,* so that it may be included as the instructor wishes. The binomial coefficient is introduced in the probability chapters in connection with the binomial probability distribution.

The instructor has several options in the selection of topics to be studied in a given course. Chapters 1 through 9 are considered to be the basic core of a course (some sections of Chapters 2, 3, 4, and 6 may be omitted without affecting continuity). Following the completion of Chapter 9, any combination of Chapters 10 through 14 may be studied. However, there are two restrictions: Chapter 3 must be studied prior to Chapter 13, and Chapter 10 must precede Chapter 12.

The suggestions of instructors using the previous editions have been invaluable in helping me improve the text for the present revision. Should you or any of your students have comments or suggestions, I would be most grateful to receive them. Please address such communications to me at Monroe Community College, Rochester, New York 14623.

▼ TO THE STUDENT: THE TEXT AS A LEARNING TOOL

Plain talk and a stress on common sense are the book's main merits as a learning tool. Such a treatment should allow you, provided that you have the necessary basic mathematics skills, to work your way through the course with relative ease. Examples of this procedure are (1) Illustration 1-4 (p. 12), which is used to reemphasize the meaning of the eight basic definitions presented in Chapter 1, and (2) the use of real-life situations to introduce a new concept (see the introduction of hypothesis testing on page 382, and Illustration 13-4 on regression analysis, p. 653).

It is the aim of this book to motivate and involve you in the statistics that you are learning. The chapter format reflects these aims and can best promote learning if each part of each chapter is used as indicated:

1. Read the *annotated chapter outline* to gain an initial familiarity with several of the basic terms and concepts to be presented.
2. Read the *news article,* which puts to practical use some of the concepts to be learned in the chapter.
3. Use the *chapter objectives* as a guide to map out the direction and scope of the chapter.
4. Learn and practice using the concepts of each section by doing the *exercises* at the end of the respective section. Answers to the odd-numbered exercises are provided at the back of the book. The study illustrations and the odd-numbered answers complement one another, to enable you to

work independently. While working within a chapter, it will be helpful to save the results of the exercises, since some results will be used again in later exercises in the chapter. When this occurs, the later use has been cross-referenced.
5. Use the *in retrospect* section to reflect on the concepts you have just learned and the relationship of the material in this chapter to the material of previous chapters. At this point it would be meaningful to reread the news article.
6. The *chapter exercises* at the end of the chapter offer additional learning experiences, since in these exercises you must now identify the technique to be used and must be able to apply it. The exercises are graded—everyone will be able to complete the first exercises with reasonable ease, but succeeding exercises become more challenging.
7. The *vocabulary list* and *quizzes* are provided as self-testing devices. You are encouraged to use them. Use the *vocabulary list* as a guide; could you explain each of these terms to a friend who is not taking statistics? The quizzes represent three different levels of understanding: *Quiz A* is at a conceptual level, *Quiz B* is at a methods level, and *Quiz C* is intended to stretch your understanding. Correct responses for the quizzes may be found in the back of the book.
8. The *working with your own data* sections direct you to collect a set of data, often of your own interest, and to apply the techniques you have studied. This opportunity should (a) reinforce the concepts studied, and (b) result in an interesting and informative statistical experience with real data.

▼ SUPPLEMENTS

The complete instructional package that accompanies this book includes the following:

1. The *Study Guide with Self-Correcting Exercises* offers students an alternative approach to mastering difficult concepts. Each lesson in the study guide provides an alternative explanation to concepts that many students ask for a second explanation of (sort of a microscope view of a specific concept). The self-correcting exercises are graded from very easy to difficult, and you will find complete solutions for each in the back. The *Introduction* chapter presents ten review topics.
2. The *Solutions Manual* shows at least one complete solution to each exercise in the textbook. Occasionally some parenthetical comments have been added to aid the teacher in such areas as when to assign specific problems, and how some problems can be of greatest use.
3. A *Partial Solutions Manual* is available for students. It contains the complete solutions to all the odd-numbered exercises.
4. *Computerized Test Items*, providing a series of multiple-choice questions and problems, are available on disk for the Apple and IBM PC computers, along with the printed versions of the test items.

5. *Transparency Masters* highlighting important illustrations and pertinent examples are available for this edition.
6. The *Minitab Student Supplement* is provided for those interested in teaching or learning the course interactively with the computer. This supplement is a text-specific introduction to the MINITAB statistical analysis system and is keyed to text discussion and examples.

▼ ACKNOWLEDGMENTS

I owe a debt to many other books. Many of the ideas, principles, examples, and developments that appear in this text stem from thoughts provoked by these sources.

It is a pleasure to acknowledge the aid and encouragement I have received throughout the development of this text from my students and colleagues at Monroe Community College. A special thanks to those who read and offered suggestions about this and the previous editions, whose names are listed on page vi, before the Contents.

In addition, I would especially like to thank Larry Stephens of the University of Nebraska-Omaha for his invaluable assistance for helping me to develop exercises for this sixth edition.

I would also like to express my appreciation for the quality work that Marcia Cole and Pam Rockwell have put into the production of the book. To Michael Payne, former Senior Editor, I would like to say thanks for all your assistance and encouragement.

Thanks also to the many authors and publishers who so generously extended reproduction permissions for the news articles and tables used in the text. These acknowledgments are specified individually throughout the text.

The last and certainly the most significant of all—thank you to my loving wife, Barbara, for her assistance and for her incredible amount of patience through all my mood swings these last few months.

Robert Johnson

PART ONE

Descriptive Statistics

WHEN ONE EMBARKS on a statistical solution to a problem, a sequence of events must develop. The order in which these events occur should be: (1) the situation investigated is carefully and fully defined, (2) a sample of data is collected from the appropriate population following an established and appropriate procedure, (3) the sample data are converted into usable information (this usable information, either numerical or pictorial, is called the *descriptive statistics*), and (4) the theories of statistical inference are applied to the sample information in order to draw conclusions about the sampled population (these conclusions or answers are called *inferences*).

The first part of this textbook, Chapters 1 through 3, will concentrate on the first three of the four events identified above. The second part will deal with probability theory, the theory on which statistical inferences rely. The third and fourth parts will survey the various types of inferences that can be made from sample information.

Sir Francis Galton

SIR FRANCIS GALTON, English anthropologist and a pioneer of human intelligence studies, was born on February 16, 1822, in a village near Birmingham, England, to Samuel Tertiles and Anne (Violetta) Galton.

Galton's family included men and women of exceptional ability, one of whom was his cousin Charles Darwin. Although his family life was happy, and he was grateful to his parents for all they had done for him, he felt little use for the conventional religious and classical education he was given.

As a teen, Galton toured a number of medical institutions in Europe and began his medical training in hospitals in Birmingham and London. He continued his medical studies until the death of his father. Having been left a sizeable fortune, he decided to discontinue medical training and pursue his love of traveling. Several years of these travels led him to the exploration of primitive parts of southwestern Africa, where he gained valuable information that earned him recognition and a fellowship in the Royal Geographical Society. In 1853, at the age of only 31, Galton was awarded a gold medal from the Society in recognition of his hard work and many achievements. It was also in 1853 that Galton married Louisa Butler and they settled in London; their marriage remained childless.

Although Galton made important contributions to many fields of knowledge, he was best known for his work in eugenics; he spent most of the latter part of his life researching and promoting his belief that inheritance played a major role in the intelligence of man. Galton, a pioneer in the development of some of the refined statistical techniques that we use today, used the laws of probability to support his theory. His application of research techniques—curves of normal distribution, correlation coefficients, and percentile grading—to a large population revealed important facts about the intellectual and physical characteristics that are passed from one generation to the next and the ways in which children differ from their parents. He also discovered, with the use of the graph, that characteristics of two different generations could be plotted against one another, revealing important information.

Sir Francis Galton died in England on January 17, 1911, leaving behind a collection of 9 books, approximately 200 articles and lectures, and a valuable legacy of statistical techniques and knowledge.

1 STATISTICS

Chapter Outline

1.1 What Is Statistics?

1.2 Uses and Misuses of Statistics
Statistical analysis, like nuclear fission and firearms, can be **applied** either **responsibly** or **irresponsibly**. The way it is done is up to you.

1.3 Introduction to Basic Terms
Population, sample, variable, data, experiment, parameter, statistic, attribute data, and **numerical data.**

1.4 Measurability and Variability
Statistics is a study of the variability that takes place in the response variable.

1.5 Data Collection
The problem of selecting a **representative sample** from a defined population can be solved using the **random** method.

1.6 Comparison of Probability and Statistics
Probability is related to statistics as the phrase "**likelihood** of rain today" is related to the phrase "**actual amount** of rain that falls."

1.7 Statistics and the Computer
Today's state of the art.

WHAT AN IDEAL MAN IS . . .

Most Like

Receptive, responsive to the initiatives of others	89%
Strong intellectual, moral or physical presence	87%
Pays attention to diet, exercise, health	87%
Expresses feelings of sadness	86%
Stops often to wonder, appreciate, dream	82%
Follows inner authority	77%
Even-tempered, moderate	77%
Easy to be with	75%
Nonjudgmental	74%
Willingly accepts help	70%
A doer, takes charge	68%

Least Like

Basically ignores his body	2%
Never shows pain	6%
Has mood swings	10%
Critical	14%
Introverted	16%
Always where the action is	20%
Type A personality	20%
Suave, urbane	22%

Percentages listed reflect the proportion of survey respondents who consider each of these factors descriptive of "the beliefs, attitudes or behaviors of an ideal man."
Source: Reprinted with permission from *Psychology Today* (copyright © 1989, Sussex Publishers, Inc). August 1989.

4 out of 5 doctors recommend . . .

IOWA POLL SAYS 1 OF 9 IOWANS HAS A NAME FOR AUTOMOBILE

The Iowa Poll, known for probing respondents' attitude toward baldness and their favorite cold drinks, has done it again: It has found that one out of every nine Iowans names his or her car.

On Aug. 14, *The Des Moines Register* said 11 percent of Iowans name their cars, 86 percent do not, 2 percent have no cars and 1 percent were not sure whether they named them or not.

The names ranged from the negative Lemon, Leprosy, Clunker and Junker to the more cheerful Black Beauty.
Source: Copyright © 1989 by The New York Times Company. Reprinted by permission.

HAS IT BEEN GETTING EASIER OR TOUGHER FOR YOU TO LIVE IN NEW YORK?

Easier	11%
Tougher	72%
Hasn't changed	15%

Source: Copyright 1990 The Time Inc. Magazine Company. Reprinted by permission. Data from a telephone poll of 1,009 New York City residents taken for TIME/CNN on August 2 to 5 by Yankelovich Clancy Shulman. Sampling error is plus or minus 3%.

Average Gain on Today's Stockmarket Was 4.14

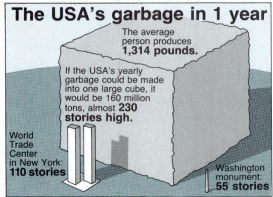

Source: Copyright 1990, USA TODAY. Reprinted with permission.

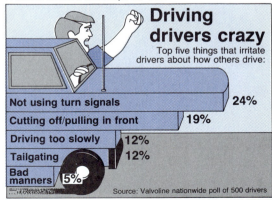

Source: Copyright 1990, USA TODAY. Reprinted with permission.

Chapter Objectives

The purpose of this introductory chapter is to present (1) an initial image of the field of statistics, (2) several of the basic vocabulary words used in studying statistics, and (3) the basic ideas and concerns about the processes used to obtain sample data.

1.1 ▼ What Is Statistics?

Statistics is the universal language of the sciences. Statistics is more than just a "kit of tools." As potential users of statistics, we need to master the "art" of using these tools correctly. Careful use of statistical methods enables us to (1) accurately describe the findings of scientific research, (2) make decisions, and (3) make estimations.

Statistics involves information, numbers to summarize this information, and their interpretation. The word *statistics* has different meanings to people of varied backgrounds and interests. To some people it is a field of "hocus-pocus" whereby a person in the know overwhelms the layperson. To other people it is a way of collecting and displaying large amounts of numerical information. And to still another group it is a way of "making decisions in the face of uncertainty." In the proper perspective, each of these points of view is correct.

descriptive and inferential statistics

The field of statistics can be roughly subdivided into two areas: descriptive statistics and inferential statistics. **Descriptive statistics** is what most people think of when they hear the word *statistics*. It includes the collection, presentation, and description of data. The term **inferential statistics** refers to the technique of interpreting the values resulting from the descriptive techniques and then using them to make decisions and draw conclusions about the population.

Statistics is more than just numbers — it is what is done to or with numbers. Let's use the following definition:

statistics

> **STATISTICS**
> The science of collecting, classifying, presenting, and interpreting data.

Before we begin our detailed study, let's look at a few examples of how and when statistics can be applied.

▼ ILLUSTRATION 1-1

State University is planning an expansion program of its physical facilities. To draw up an effective course of action, the board of trustees decides that it needs to answer this question: How many college students will we need to accommodate over the next ten years? This question can immediately be broken down into a number of smaller questions: How many college students will there be in the United States? How many will want to attend State U? and so on. To answer these questions the researcher will need to obtain data that will indicate the proportion of future high school graduates who will want to attend State U. Then he or she will somehow need to project, or predict, the number of high school graduates that will be available over the next ten years.

Let's consider the question of the proportion of high school graduates who will wish to attend State U. The best way to answer this question is to find out what proportion have attended in the past. Note that this "best" answer assumes that there is a relationship between the past and the future. This does not always hold true, however. Wars, economic depressions, and other events will alter the "natural" progression of events. ▲▲

As you can see, when accurate answers are desired, many problems must be overcome. One rather obvious problem is how to obtain historical data. Do we need to consider every student who has graduated from high school for the last several years? Do we need to account for all schools within a 500-mile radius? No, we do not; it would be impossible to research these questions completely. So we will obtain information about only some of this population — in other words, we will **sample** the population.

There are other considerations, too. How accurate are our results? What is the probability that there will be a larger proportion of students wanting to attend State U? and so on.

We have not begun to exhaust the questions that may be relevant. At this time I only wish to start you on your way to considering some of the problems involved in answering this type of question.

▼ ILLUSTRATION 1-2

"Everybody loves Pickadilly Pete!" Well, that is what Pickadilly Pete claims, so he has decided to run for mayor of Skunk Hollow (population 279). But his campaign manager is not sure that he understands exactly what Pickadilly Pete means. Does Pete mean that all 279 inhabitants of Skunk Hollow love him, or that the majority of the 279 (at least 140) love him, or that at least half of the inhabitants of voting age love him? Pete could have meant any or all of these, and what he means could make a big difference in the campaign. (Suppose, for example, that the 140 who love him are all children and can't vote.) If you were hired as an independent research agent, what would you do to test the "accuracy" of Pickadilly Pete's claim? ▲▲

▼ ILLUSTRATION 1-3

How tall are sports car drivers? This is the question posed by the owners of Custom Sport Coupe, a local manufacturer of the world's finest sports car. They want to design and build a new model that is truly comfortable for the driver. Their present model is designed for people between 5 feet 2 inches and 5 feet 8 inches tall. The manufacturers are concerned because they have heard rumors that their car is uncomfortable for a large proportion of sports car enthusiasts. (If you have ever ridden in a CSC, you will understand the rumor "built for short people with very short necks.") How might you go about obtaining an answer to the original question? What special considerations would you give to the process of obtaining sample information?

Each of these illustrations poses questions that should make you think about the situation and at the same time give you a feeling for statistics.

1.2 ▼ Uses and Misuses of Statistics

The uses of statistics are unlimited. It is much harder to name a field in which statistics is not used than it is to name one in which statistics plays an integral part. The following are a few examples of how and where statistics is used:

1. In education descriptive statistics are frequently used to describe test results.
2. In science the data resulting from experiments must be collected and analyzed.
3. In government many kinds of statistical data are collected all the time. In fact, the U.S. government is probably the world's greatest collector of statistical data.

Misuses of statistics are often colorful and sometimes troublesome. Many people are concerned about the neutrality of statistical descriptions; other people believe all statistics are lies. Most statistical lies are innocent, however, and result from using an inappropriate statistic; an open, nonspecific statement (such as Pickadilly Pete's claim); or data derived from a faulty sample. All these lead to a common result: misunderstanding of the information on the part of the consumer. Specific illustrations of the misuse of statistics will be given in Chapter 2.

▼▲ EXERCISES

 1.1 Refer to the *USA Snapshot* "Looking Around Campus" (Case Study 1-1).
   a. Who was surveyed?
   b. How many were surveyed?
   c. What question was asked of each college student? What information was collected?
   d. Explain the meaning of the "75% Punk." How many students answered "punk is out"?
   e. Why do the reported values (75%, 63%, 60%, . . .) add up to more than 100?
   f. The graphic shows four different "looks that are out." Do you believe these are the only four looks mentioned by the 1024 college students surveyed? Explain.

Section 1.2 ▼ USES AND MISUSES OF STATISTICS

Case Study 1-1

Looking Around Campus

The September 24, 1990 issue of USA Today *presented a graphic on the statistics on the "in" and "out" looks on today's campuses. How many college students were surveyed? What information was collected from each student? Why are there so many percentages given? Why do they not add up to 100? (See Exercise 1.1.)*

Source: Copyright 1990, USA TODAY. Reprinted with permission.

 1.2 Refer to the *USA Snapshot* "What Business Travelers Drink at Lunch" (Case Study 1-2, p. 8).
  a. What set of people were surveyed?
  b. How many were surveyed?
  c. What question was asked of each person?
  d. Explain the meaning of "Soft drinks–55.7%." (Does it mean that 55.7% of drinks were soft drinks? Or does it mean that 55.7% of the money spent on drinks was spent on soft drinks?)
  e. Do these percentages total 100?
  f. Does this mean that business travelers never drink more than one kind of drink for lunch? (Perhaps you may want to rethink your answer to (d).)

Case Study 1-2

What Business Travelers Drink at Lunch

The April 10, 1990 issue of USA Today *presented a graphic on the statistics about "what business travelers drink for lunch" (soft drink, coffee, and so on). The Snapshot reports that 55.7% drink soft drinks, 40.0% drink coffee or tea, while only 0.8% have a mixed drink. Do these percentages total 100? Does this mean that business travelers never drink more than one kind of drink for lunch? How were these percentages calculated? Are they percentages of drinks served to the business travelers? or are they percentages of money spent on drinks for lunch? or are they percentages of travelers who drank each type of drink? (See Exercise 1.2.)*

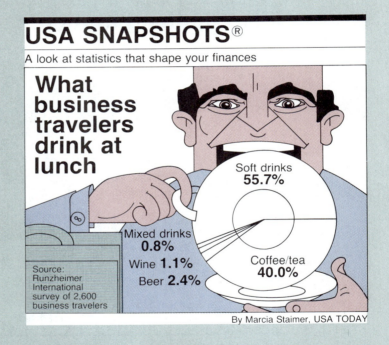

Source: Copyright 1990, USA TODAY. Reprinted with permission.

 1.3 Refer to the *USA Snapshot* "Where Motorcyclists Live" (Case Study 1-3).
  a. What population is represented by the statistics presented?
  b. The description of the average motorcyclist (32-year-old married male college graduate) results from data collected using four variables. Name the four variables.
  c. The graph presents percentage summaries of the information collected using another variable. Name this variable.

Section 1.2 ▼ USES AND MISUSES OF STATISTICS

Case Study 1-3

Where Motorcyclists Live

The March 15, 1991 issue of USA Today *presented a description of the average motorcyclist and showed a graph picturing the percentage of the motorcyclists who live in each of the four regions of the United States. What population is represented? What information (variables) was collected from each motorcyclist in order to describe the average motorcyclist? (See Exercise 1.3.)*

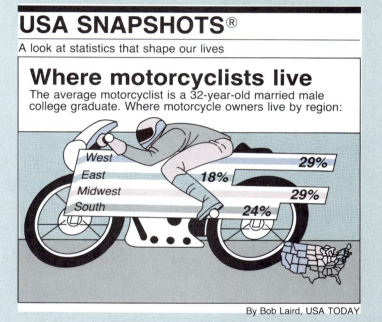

Source: Copyright 1991, USA TODAY. Reprinted with permission.

 1.4 Refer to the *USA Snapshot* "Home Prices Rise" (Case Study 1-4, p. 10).
  a. What was the source of the information?
  b. Verify (calculate) the 53% cited.
  c. The height of the graphic at 1990 looks to be approximately six times taller than the height at 1980, yet $80,825 is not six times larger than $52,900. Explain how this discrepancy occurred.

1.5 Determine which of the following statements are descriptive in nature and which are inferential. In reference to Case Study 1-1,
  a. 75% of the 1024 students answered "Punk look is out."
  b. 75% of all college students think "Punk look is out."

Case Study 1-4

Home Prices Rise

The February 21, 1991 issue of USA Today *presented the annual median price of homes sold to first-time buyers from 1980 to 1990. The median price rose from $52,900 to $80,825, a $27,925 increase over the ten years. The* Snapshot *summarizes this by saying "... rose 53%...." Yet when you look at the graph, the "height" of the house at 1990 seems to be many times the "height" at 1980. The graph seems to suggest, at first glance, that the 1990 median price is six or seven times the 1980 price. Does the graph show the prices fairly? (See Exercise 1.4.)*

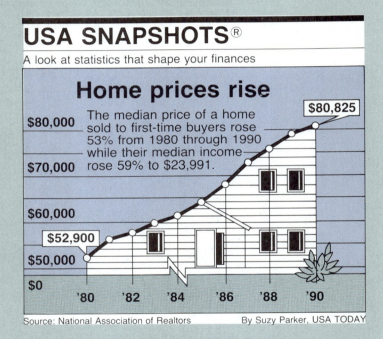

Source: Copyright 1991, USA TODAY. Reprinted with permission.

In reference to Case Study 1-2,
    c. 40% of the 2600 business travelers drank coffee or tea for lunch.
    d. 40% of all business travelers drink coffee or tea for lunch.

1.6 Determine which of the following statements are descriptive in nature and which are inferential. In reference to Case Study 1-3,
    a. The average American motorcyclist is 32 years old.
    b. The average age of the motorcyclists sampled was 32 years.
    c. 24% of the motorcyclists sampled live in the South.
    d. 29% of all motorcyclists in the United States live in the West.

1.3 ▼ Introduction to Basic Terms

To study statistics we need to be able to speak its language. Let's first define a few basic terms that will be used throughout this text. (These definitions are descriptive in nature and are not necessarily mathematically complete.)

population

> **POPULATION**
>
> A collection, or set, of individuals or objects whose properties are to be analyzed.

The population is the complete collection of individuals or objects that are of interest to the sample collector. The concept of a population is the most fundamental idea in statistics. The population of concern must be carefully defined and is considered fully defined only when its membership list of elements is specified. The set of "all students who have ever attended a U.S. college" is an example of a well-defined population.

Typically, we think of a population as a collection of people. However, in statistics the population could be a collection of animals, of manufactured objects, or whatever. For example, the set of all redwood trees in California could be a population.

There are two kinds of populations, finite and infinite. When the membership of a population can be (or could be) physically listed, the population is said to be *finite*. When the membership is unlimited, the population is *infinite*. The books in your college library are a finite population. The card catalog lists the exact membership. All registered voters in the United States are a very large finite population. If necessary, a composite of all voter lists from all voting precincts across the United States could be compiled. On the other hand, the population of all people who might use aspirin and the population of all 40-watt light bulbs to be produced by Sylvania are infinite.

finite
infinite

sample

> **SAMPLE**
>
> A subset of a population.

A sample consists of the individuals, objects, or measurements selected by the sample collector from the population.

response variable

> **RESPONSE VARIABLE (OR, SIMPLY, VARIABLE)**
>
> A characteristic of interest about each individual element of a population or sample.

A student's age at entrance into college, the color of her hair, her height, weight, and so on are all response variables.

data (singular)

> **DATA (SINGULAR)**
> The value of the response variable associated with one element of a population or sample.

For example, Bill Jones entered college at the age of "23," his hair is "brown," he is "6 feet 1 inch" tall, and he weighs "183 pounds." These four pieces of data are single values for the four response variables as applied to Bill Jones.

data (plural)

> **DATA (PLURAL)**
> The set of values collected for the response variable from each of the elements belonging to the sample.

The set of 25 heights collected from 25 students is an example of a set of data.

experiment

> **EXPERIMENT**
> A planned activity whose results yield a set of data.

parameter

> **PARAMETER**
> A numerical characteristic of an entire population.

The "average" age at time of admission of all students who have ever attended our college or the proportion of students who were over 21 years of age when they entered college are examples of two different population parameters. A parameter is a value that describes the entire population. It is a common practice in statistics to use a Greek letter to symbolize the names of the various parameters. These symbols will be assigned as we study individual parameters.

statistic

> **STATISTIC**
> A numerical characteristic of a sample.

The "average" height found by using the set of 25 heights is an example of a sample statistic. A statistic is a value that describes a sample. Most sample statistics are found with the aid of formulas and are assigned symbolic names using letters of the English alphabet (for example, $\bar{x}$, s, and r).

▼ **ILLUSTRATION 1-4**

A statistics student is interested in finding out something about the dollar value of the typical car owned by the faculty members of a college. Each of the eight terms just described can be identified in this situation.

1. The *population* is the collection of all cars owned by the faculty members.
2. A *sample* is any part of that population. For example, the cars owned by members of the mathematics department would be a sample.
3. The *response variable* is the "value" of each individual car.
4. One *data* would be the value of a particular car. Mr. Jones's car, for example, is valued at $9,400.
5. The *data* would be the set of values that correspond to the sample obtained (9,400; 8,700; 15,950, . . .).
6. An *experiment* would be the methods used to select the cars forming the sample and determining the value of each car in the sample. It could be carried out by questioning each member of the mathematics department or in other ways.
7. The *parameter* about which we are seeking information is the "average" value in the population.
8. The *statistic* that will be found is the "average" value in the sample. ▲▲

There are basically two kinds of data: (1) data obtained from qualitative information and (2) data obtained from quantitative information.

attribute or qualitative data

ATTRIBUTE OR QUALITATIVE DATA
Results from a process that categorizes or describes an element of a population.

For example, color is an attribute of a car. A sample of the colors of the cars in a nearby parking lot would result in such data as blue, red, yellow, and so on. In general, the resulting data are a collection of word responses (as opposed to number responses).

numerical or quantitative data

NUMERICAL OR QUANTITATIVE DATA
Results from a process that quantifies—that is, counts (of how many)— or measurements (length, weight, and so on).

discrete
continuous

Numerical data can be subdivided into two classifications: (1) **discrete** numerical data and (2) **continuous** numerical data. In most cases the two can be distinguished by deciding whether the data result from a count or from a measurement. A count will always yield discrete numerical data. The number of bugs on the right headlight of an automobile will be 0, 1, 2, . . . ,* but it cannot be 1.9 or 3.25. (The partial remains of any bug will be counted as a whole bug.) The idea of discontinuous numerical values is somewhat synonymous with that of discrete numerical values.

Case Study 1-5

What We Do in Bed

The February 21, 1991 issue of USA Today *presented a graphic representing the average number of hours people spend in bed doing various things. A sample of 1013 adults was selected and each was asked for the number of hours spent sleeping per week. The data resulting from this variable (1013 values) are described by the statistic, average number of hours 45. There are three other statistics cited. What are they? What three variables were used to determine these three statistics? (See Exercise 1.11, p. 19.)*

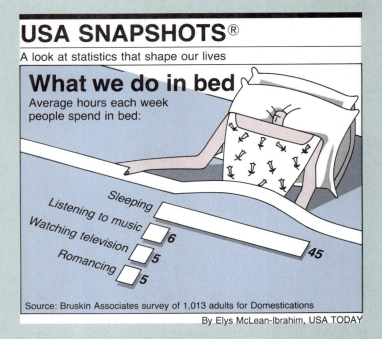

Source: Copyright 1991, USA TODAY. Reprinted with permission.

---

*Note: The notation "..." means that the listing of numbers continues indefinitely.

Case Study 1-6

Average Outfield Distances

The March 29, 1990 issue of USA Today *presented a pair of graphics showing the average distance from home plate to the outfield fence in the major league baseball parks. Let's focus our attention on the National League centerfield number of 404. If the set of all National League ballparks is the population, then the 404 is a parameter since 404 is the average value for the variable, distance to center field in all National League ballparks. The numbers on this graphic could be considered parameters or statistics. Under what circumstances are they parameters? statistics? (See Exercise 1.12, p. 19.)*

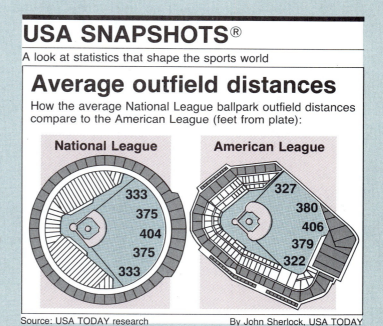

Source: Copyright 1990, USA TODAY. Reprinted with permission.

A measure of a quantity will usually be continuous. If John says, for example, that he weighs 162 lb to the nearest pound, all we can be sure of is that his weight is some value between 161.5 and 162.5 lb. He could actually weigh 162.000, 161.789, or any value in the interval 161.5–162.5. This is the basic concept of a continuous variable. There are many continuous variables that receive responses that appear to be discrete: a person's age, for example. When asked how old he was, Steve responded "19." He was 19 years old on his last birthday, but he is 19 plus some part of a year. The idea of continuous data is consistent with the concept of the number line having no gaps.

The use of fractions or decimals does not necessarily imply that data are continuous. The scoring of competitive diving is an example of a case in which decimals appear and the variable is actually discrete. A contestant can receive scores by halves only (5.5, 6.0, 6.5, and so on). These scores are actually discrete, because values *between* 6.0 and 6.5 cannot occur. There are other illustrations of this situation, but in this text a *count* or a *measurement* is about the only distinction that we will need to make.

There are situations in which data are collected in numerical form and reported and discussed in attribute form. Two such situations are (1) the measurement of air pollution, as commonly reported in the newspaper (measured in amount and reported as low, medium, or high), and (2) the Richter scale and the measurement of an earthquake.

Let's explore the difference between these terms and the types of data. If I were to go to a nearby parking lot right now and obtain some data, I would most likely think of it as a sample. However, the sample could be considered to be the population if I defined my population of concern to be the cars in that parking lot right now. It is more likely that the sample would be considered to be only part of a larger collection of cars, say of all cars that ever park in this lot or a sample of all cars parked in all parking lots at this time. (You must define your population and your sample carefully.) If I were to observe the make of each car, I would be collecting attribute data. If I were to record the number of people riding in each car as it left the parking lot, I would be collecting discrete data. (This clearly is a count of people.) If each car were to be weighed, continuous data would result, because all the various fractional parts of weight units could result from the weighing (see Section 1.4).

Don't let the appearance of the data fool you in regard to their type. For example, suppose that after surveying this parking lot, I summarized the sample by reporting five red, eight blue, six green, and two yellow cars. These data appear to be the count of cars; therefore, one would think them to be discrete. This is not the case, however; they are attribute data. *You must look at each individual source to determine the kind of variable being used.* One specific car was red; "red," then is the data for that one car. And red is an attribute. Thus this collection (five red, eight blue, and so on) is a *summary* of the *attribute* data.

Another example of a deceiving variable is the license plate number (number part only) of each car. The numbers appear to be a discrete variable since only whole number values occur; however, these numbers are merely identification numbers. Identification numbers are not variables because they do not measure anything; they serve only to identify. Consider the weights of the cars as recorded on the registration: 3485, 3860, 2091, 4175, and so on. They are all whole numbers, but that does not make the variable discrete. The variable is "weight," and these data are measured to the nearest pound. Thus the variable "weight" is continuous. As you can see, the appearance of the data *after* they are recorded can be misleading in respect to their type. Remember to inspect an individual piece of data and you should have little trouble in distinguishing among attribute, discrete, and continuous data.

Case Study 1-7

Eating Together: Still Important

The January 6, 1991 issue of the Democrat and Chronicle *presented the results of a New York Times/CBS News poll regarding today's attitudes about family togetherness. The graphic shows the results of four variables. Name the four variables. What kind of data results from each of these variables? (See Exercise 1.16, p. 20.)*

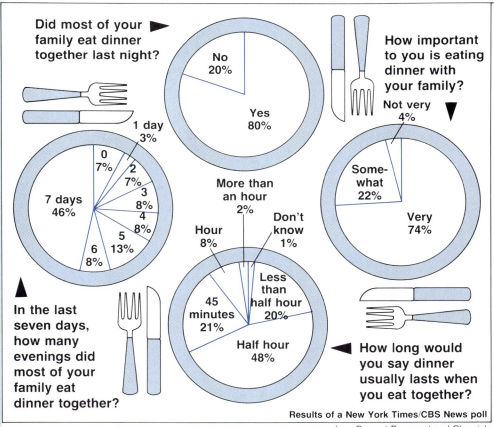

Source: Joan Bossert, "Eating Together: Still Important," *Democrat and Chronicle,* January 6, 1991, copyright 1991 by *Democrat and Chronicle.* Reprinted by permission.

▼▲ EXERCISES

1.7 A drug manufacturer is interested in the proportion of persons who have hypertension (elevated blood pressure) whose condition can be controlled by a new drug the company has developed. A study involving 5000 individuals with hypertension is conducted and it is found that 80% of the individuals are able to control their hypertension with the drug. Assuming that the 5000 individuals are representative of the group who have hypertension, answer the following questions:
  a. What is the population?
  b. What is the sample?
  c. Identify the parameter of interest.
  d. Identify the statistic and give its value.
  e. Do we know the value of the parameter?

1.8 A quality control technician selects assembled parts from an assembly line and records the following information concerning each part:

  A: defective or nondefective
  B: the employee number of individual who assembled part
  C: the weight of the part

  a. What is the population?
  b. Is the population finite or infinite?
  c. What is the sample?
  d. Classify the responses for each of the three variables as either attribute data, discrete data, or continuous data.

1.9 Perform the "first-ace" experiment five times (instructions follow) and observe the value of three different variables each time:

  Variable 1: the color of the first ace to appear
  Variable 2: the longest distance across the dealt pile of cards
  Variable 3: the count of the card on which the first ace appears

To perform the first-ace experiment, shuffle an ordinary deck of 52 playing cards containing 4 aces. Deal the cards one at a time onto a pile, stopping when the first ace appears. After the first ace has been placed on the pile, record its color. Now measure the longest distance across the scattered pile before it is disturbed and record it. Count and record the number of cards in the pile. Repeat the experiment four more times.
  a. What is the population?
  b. Is the population finite or infinite?
  c. What is the sample?
  d. Classify the responses for each of the three variables as either attribute data, discrete data, or continuous data.

Trial	x = Color of first ace	d = Distance across pile	y = Count of cards, including first ace
1	$x_1 =$	$d_1 =$	$y_1 =$
2	$x_2 =$	$d_2 =$	$y_2 =$
3	$x_3 =$	$d_3 =$	$y_3 =$
4	$x_4 =$	$d_4 =$	$y_4 =$
5	$x_5 =$	$d_5 =$	$y_5 =$
		$\sum_{i=1}^{5} d_i =$	$\sum_{i=1}^{5} y_i =$

Note: See Appendix A for information about the $\sum$ notation.

1.10 Select ten students currently enrolled at your college and collect data for these three variables:

X: number of courses enrolled in
Y: total cost of textbooks and supplies for courses
Z: method of payment used for textbooks and supplies

a. What is the population?
b. Is the population finite or infinite?
c. What is the sample?
d. Classify the responses for each of the three variables as either attribute data, discrete data, or continuous data.

1.11 Refer to the *USA Snapshot* "What We Do in Bed" (Case Study 1-5).
a. What is the population?
b. Is the population finite or infinite?
c. What is the sample?
d. What are the four statistics reported in this snapshot?
e. Name the four corresponding variables.

 1.12 Refer to the *USA Snapshot* "Average Outfield Distance" (Case Study 1-6).
a. Let's focus on the number 327, the average left field distance in American League ballparks. Think of 327 as a parameter.
  i. What is the population?
  ii. What is the variable?
b. Let's focus on the number 406, the average center field distance in American League ballparks. Think of 406 as a statistic.
  i. What is the sample?
  ii. What is the variable?
  iii. What is the population?

c. What conditions are necessary for the numbers reported to be parameters?
d. What conditions are necessary for the numbers reported to be statistics?

1.13 Identify each of the following as examples of (1) attribute, (2) discrete, or (3) continuous variables:
   a. the breaking strength of a given type of string
   b. the hair color of children auditioning for the musical *Annie*
   c. the number of stop signs in towns of less than 500 people
   d. whether or not a faucet is defective
   e. the number of questions answered correctly on a standardized test
   f. the length of time required to answer a telephone call at a certain real estate office

1.14 Identify each of the following as examples of (1) attribute, (2) discrete, or (3) continuous variables:
   a. a poll of registered voters as to which candidate they support
   b. length of time required for a wound to heal when using a new medicine
   c. the number of telephone calls arriving at a switchboard per ten-minute period
   d. the distance first-year college women can kick a football
   e. the number of pages per job coming off a computer printer
   f. the kind of tree used as a Christmas tree

1.15 The admissions office wants to estimate the cost of textbooks for students at our college. Let the variable x be the total cost of all textbooks purchased by a student this semester. The plan is to randomly identify 100 students and obtain their total textbook costs. The average cost for the 100 students will be used to estimate the average cost for all students.
   a. Describe the parameter the admissions office wishes to estimate.
   b. Describe the population.
   c. Describe the variable involved.
   d. Describe the sample.
   e. Describe the statistic and how you would use the 100 data collected to calculate the statistic.

1.16 Refer to the results of the *New York Times*/CBS News poll, "Eating Together: Still Important" (Case Study 1-7).
   a. Name the four variables.
   b. What kind of data results from each variable?

1.17 Suppose a 12-year-old asked you to explain the difference between a sample and a population.
   a. What information should your answer include?
   b. What reasons would you give him or her for why one would take a sample instead of surveying every member of the population?

1.18 Suppose a 12-year-old asked you to explain the difference between a statistic and a parameter.
   a. What information should your answer include?
   b. What reasons would you give him or her for why one would report the value of a statistic instead of a parameter?

1.4 ▼ Measurability and Variability

Within a set of experimental data, we *always* expect variation. If little or no variation is found, we would guess that the measuring device is not calibrated with a small enough unit. For example, we take a carton of a favorite candy bar (24) and weigh each bar individually. We observe that all 24 candy bars weigh $\frac{7}{8}$ of an ounce, to the nearest $\frac{1}{8}$ oz. Does this mean that they are all identical in weight? Not really. Suppose that we were to weigh them on an analytical balance that weighs to the nearest milligram. Now their weights will be variable.

variability

It does not matter what the response variable is; there will be **variability** in the numerical response if the tool of measurement is precise enough. A primary objective in statistical analysis is that of measuring variability. For example, in the study of quality control, measuring variability is an absolute essential. Controlling (or reducing) the variability in a manufacturing process is a field all its own (for example, statistical process control).

▼▲ EXERCISES

1.19 Suppose we measure the weights (in pounds) of the individuals in each of the following groups:

Group 1: cheerleaders for National Football League teams
Group 2: players for National Football League teams

For which group would you expect the data to have more variability? Explain why.

1.20 Suppose you were trying to decide which of two machines to purchase. Furthermore, suppose the length to which the machines cut a particular product part was important. If both machines produced parts that had the same length on the average, what other consideration regarding the lengths would be important? Why?

1.21 Teachers use examinations to measure a student's knowledge about their subject. Explain how "a lack of variability in the resulting scores might indicate that the exam was not a very effective measuring device."

1.22 A coin-operated coffee vending machine dispenses, on the average, 6 oz of coffee per cup. Can this statement be true of a vending machine that occasionally dispenses only enough to barely fill the cup half full (say 4 oz)? Explain.

1.5 ▼ Data Collection

One of the first problems a statistician faces is obtaining data. Data doesn't just happen. It must be collected. We must recognize the importance of good sampling techniques, because the inferences we ultimately make will be based on the statistics calculated from our sample data.

The collection of data for statistical analysis is an involved process and includes the following important steps:

1. Defining the objectives of the survey or experiment.
   Examples: (a) comparing the effectiveness of a new drug to the effectiveness of the standard drug, (b) estimating the average household income in our county.

2. Defining the variable and the population of interest.
   Examples: (a) length of recovery time for patients suffering from a particular disease, (b) total income for households in our county.
3. Defining the data-collection and data-measuring schemes.
   This includes sampling procedures, sample size, and the data-measuring device (questionnaire, telephone, and so on).
4. Determining the appropriate descriptive or inferential data-analysis techniques.

Very often an analyst is stuck with data already collected, possibly even data collected for other purposes, which makes it impossible to determine whether or not the data are "good." Collecting the data yourself using approved techniques is much preferred. Although this text will be chiefly concerned with various data-analysis techniques, you should be aware of the problems of data collection.

The following illustrations describe populations defined for specific investigations:

▼ ILLUSTRATION 1-5

The admissions office wishes to estimate the cost of college education at a local college. One of the components of total cost per semester is the cost of textbooks. Specifically, the admissions office would like to estimate the current "average" cost of textbooks per semester per student. The population of interest is the currently enrolled student body. ▲▲

▼ ILLUSTRATION 1-6

The cost of a one-minute television advertisement varies drastically according to the station, the day, and the time of day. This rate is determined by the proportion of potential viewers who are tuned to a particular station at a particular time. A national department store chain is opening a new store in a nearby city and is in the process of selecting time slots for advertising on WRQC TV. They want to choose the time slots that will be most cost effective in attracting new customers. The population of interest is the daily television viewing habits of all persons who live within the broadcasting area. ▲▲

The two methods used to collect data are *experiments* and *surveys*. In an experiment, the investigator controls or modifies the environment and observes the effect on the response variable. We often read about laboratory results obtained by using white rats to test some new product, such as different doses of a new medication and its effect on blood pressure. The experimental treatments were designed specifically to obtain the data needed to study the effect on the response variable. In a survey, data are obtained by sampling some population of interest. The investigator, however, does not modify the environment. The remainder of this section deals with some of the simpler methods that might be used in order to obtain sample data from surveys.

census

If every element in the population is listed, or enumerated, and observed, then a **census** is compiled. A census is a 100% survey. A census for the population in Illustration 1-5 could probably be obtained by contacting each student on the registrar's computer printout of all registered students. However, censuses are seldom used because they are difficult to compile, very time-consuming, and therefore expensive. Imagine the task of compiling a census of every person who lives within the broadcasting area of WRQC TV (Illustration 1-6). Instead of a census, a sample survey is usually conducted.

When selecting a sample for a survey, it is necessary to construct a *sampling frame*.

sampling frame

> **SAMPLING FRAME**
> A list of the elements belonging to the population from which the sample will be drawn.

representative sampling frame

The sampling frame should be identical to the population. Ideally, the frame will have every element of the population listed once and only once. However, this is not always possible because it may be impractical or impossible to select directly from the total population. Since only the elements in the frame have a chance to be selected as part of the sample, **it is important that the sampling frame be representative of the population.**

In Illustration 1-5, the registrar's computer list will serve as a sampling frame for the admissions office. In this case, the census becomes the sampling frame. In other situations a census may not be so easy to obtain, because a complete list is not available. Lists of registered voters or the telephone directory are sometimes used as sampling frames of the general public. One of these might be used in Illustration 1-6. Depending on the nature of the information sought, the list of registered voters or the telephone directory may or may not serve as a good sampling frame.

Once a representative sampling frame has been established, we are prepared to proceed with selecting the sample elements from the sampling frame. This selection process is defined by the *sample design*.

sample design, or sampling plan

> **SAMPLE DESIGN, OR SAMPLING PLAN**
> The procedures used to select the elements of the sample.

There are many different types of sample designs. However, all sample designs fit into two categories: *judgment samples* and *probability samples*.

judgment samples

> **JUDGMENT SAMPLES**
> Samples that are selected on the basis of being "typical."

When a judgment sample is drawn, the person selecting the sample chooses items that he or she thinks are representative of the population. The validity of the results from a judgment sample reflects the soundness of the collector's judgment.

probability samples

> ### PROBABILITY SAMPLES
> Samples in which the elements to be selected are drawn on the basis of probability. Each element in a population has a certain probability of being selected as part of the sample.

NOTE *Statistical inference* (tests of hypothesis and confidence interval estimates) *requires that the sample design be a probability sample.* Let's look at a few of the simpler and easier-to-use sample designs. One of the most common methods used to collect data for a probability sample is the *random sample*.

random sample

> ### RANDOM SAMPLE
> A sample selected in such a way that every element in the population has an equal probability of being chosen. Equivalently, all samples of size n have an equal chance of being selected. Random samples are obtained by either sampling with replacement from a finite population or by sampling without replacement from an infinite population.

When a random sample is drawn, an effort must be made to ensure that each element has an equal probability of being selected. Mistakes are frequently made because the term *random* (equal chance) is confused with *haphazard* (without pattern). The proper procedure for selecting a random sample is to use a random number generator or a table of random numbers.

To select a simple random sample, first assign a number to each element in the sampling frame. This is usually done sequentially using the same number of digits for each element. Then go to a table of random numbers and select as many numbers with that number of digits as are needed for the sample size desired. (See Appendix B for specific information regarding use of the random number table.) Each numbered element in the sampling frame that corresponds to a selected random number is chosen for the sample.

▼ ILLUSTRATION 1-7

Let's return to Illustration 1-5. Mr. Clar, who works in the admissions office, has obtained a computer list of this semester's full-time enrollment. There are 4265 student names on the list. He numbered the students 0001, 0002, 0003, and so on, up to 4265. Then, using four-digit random numbers, he identified a sample. (See Appendix B for a discussion of the use of the random number table.)

▲▲

One of the easiest-to-use methods for approximating a random sample is the *systematic sampling method*.

> **SYSTEMATIC SAMPLE**
>
> A sample in which every kth item in the sampling frame is selected after a random start among the first k elements.

systematic sample

To select a systematic sample (size equal to $c\%$ of the population), we use the random number table to find the starting point (first element sampled) from among the first $100/c$ elements listed on the sampling frame. The other elements then are located every $100/c$ positions along the sampling frame.

For example, if we desire a 3% systematic sample, we would determine the position of the first element by using the random number table to randomly select a number between 1 and 33 ($100/3 = 33$ rounded to the nearest integer). Suppose the random number table produced a 27. The first element in our sample would come from the source located in the 27th position on the sampling frame. Then starting at the 27th position, every 33rd element is selected until the sample is complete. The second piece of data comes from the element in the 60th position ($27 + 33 = 60$), the third is in the 93rd position ($60 + 33 = 93$), and so on.

When sampling very large populations, sometimes it is possible to divide the population into subpopulations on the basis of some characteristic. These subpopulations are called *strata*. These smaller, easier-to-work-with strata are sampled separately. One of the sample designs that starts by stratifying the sampling frame is the **stratified sampling method.**

strata

> **STRATIFIED SAMPLE**
>
> A sample obtained by stratifying the sampling frame and then selecting a fixed number of items from each of the strata by means of random sampling.

stratified sample

When a stratified sample is to be drawn, the population is subdivided into the various strata and then a subsample is drawn from each stratum. These subsamples may be drawn from each stratum randomly or systematically. Then the subsamples are summarized separately and this information is combined to draw conclusions about the whole population.

For example, a large university wants to poll its student body on some issue. The students are stratified by the program in which they have matriculated. A sample of 20 students taken from each of the 12 programs at the university would result in a stratified sample of 120 students.

Another sampling method that starts by stratifying results is a **cluster sample**.

> **CLUSTER SAMPLE**
>
> A sample obtained by stratifying the sampling frame and then selecting all of the items from some, but not all, of the strata.

cluster sample

The cluster sample is obtained by using either random numbers or a systematic method to identify the strata (clusters) to be sampled and then using all the items from within these strata. Then the subsamples are summarized separately and this information is combined.

For example, a political issue is to be resolved by the registered voters. A poll is to be taken before the actual voting. Each voting precinct can be used as a strata. If four precincts are randomly selected and then only registered voters from within those four precincts are polled, a cluster sample will result.

The sampling plan used in a particular situation depends on several factors: (1) the nature of the population and the variable, (2) the ease (or difficulty) of sampling, and (3) the cost of sampling. These and other factors must be weighed, with trade-offs usually occurring, before the exact sampling technique is determined. *In this text, all of the statistical methods presume that random sampling has been used to collect the data.*

Case Study 1-8

People Have to Tell the Truth or Statistics Don't Mean Anything

Statistical results are only as good as the sample information collected is accurate, as so well described by David Hoff in "People Have to Tell the Truth or Statistics Don't Mean Anything."

There is a scene in the farcical movie *Airplane* where Lloyd Bridges hands a printout of some sort of aviation data to one of his looney colleagues in the tower.

"What can you make of this?" Bridges frantically asks during a time of crisis.

"What can I make of this?" the pinhead replies as he hurriedly folds the paper into various shapes. "I can make a hat . . . I can make a brooch . . ."

That came to mind while reading a story in a local newspaper the other day from The Associated Press about a poll of married men and women.

In that poll, nine of 10 people surveyed claimed they had never been unfaithful, and four of five said they would marry the same person again.

The next day I read another story, this one in *The Wall Street Journal*, about the heavyweight of greeting cards, Hallmark.

The story said Hallmark was getting into divorce announcements in a big way in 1990 because, according to company literature, "about 50 percent of all marriages end in divorce—and people need a way to communicate about it."

A week later I read a story in *Parade* magazine that stated "two out of three married men today reportedly commit adultery."

The story also said that in a 1986 survey of married women age 25 to 50, "a whopping 41 percent of these wives said they were cheating or had cheated on their husbands in the past."

So, I ask, what can you make of this?

One story says nine out of 10 people claim they have never been unfaithful, another says 41 percent of the women and 66 percent of the men are cheating.

One story says four out of five spouses would marry the same mate again, and the other says divorce is so prevalent, it's a business opportunity. Whom do you believe?

I'll go with Hallmark.

A glaring problem with polls is that to be accurate or close to accurate, those surveyed must tell the truth.

In the poll of married people, three out of four respondents, said their spouse was their best friend. That's 75 percent.

But the divorce rate is 50 percent. Either 25 percent are lying, or they divorce their best friend.

Also, 64 percent said their marriage was "very happy," and 71 percent said "they're trying very hard to improve their marriage."

If Hallmark believed those numbers, it wouldn't be jumping into the divorce card business.

So credibility in polls—at least polls about our love life—is as stable as a teen-age romance, but that doesn't mean we'll stop reporting them.

They are not without some flicker of reality, and they do make fun reading.

Besides, you could at least use the story to make a hat . . . make a brooch . . .

Source: David Hoff, "People Have to Tell the Truth or Statistics Don't Mean Anything," *The New York Times.* Reprinted by permission.

Case Study 1-9

Money, Marriage: Tell Us

Samples are often collected by asking people to share information about themselves as in "Money, Marriage: Tell Us" in USA Today, *March 26, 1990. Will the resulting sample be a representative sample? a random sample? Explain. (See Exercise 1.29, p. 29.)*

Do you have a cash stash your spouse knows nothing about?

Do separate checking accounts mean you don't trust each other?

Couples fight about money more than they do about sex, experts say. Money is linked to power, independence and other complicated factors in a marriage.

Write us about your mate–money experience. You might consider:

▼ Prenuptial agreements.
▼ Who earns more—and what that means.
▼ How you deal with arguments over money.

Write a brief letter and include your name, age and a daytime telephone number if you are willing to be interviewed by a reporter.

We'll share your stories—plus some expert solutions—with USA TODAY readers. And we won't use names without permission. Write to: *Men, Women and Money, USA TODAY,* P.O. Box 7851, Washington, D.C. 20044

Source: Copyright 1990, USA TODAY. Reprinted with permission.

Case Study 1-10

What Would Make Life Better?

The Roper Organization conducted a poll of 3000 women in an attempt to find out what women think would make their lives better. Similar opinion polls appear in print frequently. These results appeared in McCall's *in September 1990. What information was collected from each woman in the survey? (See Exercise 1.30, p. 29.)*

When women were asked about what would make their lives better, money talked. Of the 3,000 women participating in the 1990 Virginia Slims Opinion Poll, 60 percent said more money topped their wish lists. A mere 6 percent told the Roper Organization, which conducted the poll, that staying home full time would improve their lives. Only 3 percent said a better sex life topped the list.

What Would Make Life Better?	
More money	60%
More control over the way things are going in life	28%
More help with household from spouse	26%
More leisure time	25%
More help with household from children	17%
Less stress at home	17%
Job or more interesting job	12%
A different relationship	8%
More flexible work schedule	7%
Less pressure on the job	7%
Being able to stay home full time	6%
A better sex life	3%

Source: Celia Slom, "What Would Make Life Better," *McCall's,* September 1990. Reprinted by permission from *McCall's* magazine. Copyright © 1990 by The New York Times Company.

▼▲ EXERCISES

1.23 a. What is a sampling frame?
b. What did Mr. Clar use for a sampling frame in Illustration 1-7?

1.24 Consider a simple population consisting of only the numbers 1, 2, and 3 (an unlimited number of each). There are nine different samples of size two that could be drawn from this population: (1, 1), (1, 2), (1, 3), (2, 1), (2, 2), (2, 3), (3, 1), (3, 2), (3, 3).
   a. If the population consists of the numbers 1, 2, 3, and 4, list all the samples of size two that could possibly be selected.
   b. If the population consists of the numbers 1, 2, and 3, list all the samples of size three that could possibly be selected.
   c. If the population consists of the numbers 1, 2, 3, and 4, list all the samples of size three that could possibly be selected.

1.25 The computer system for a large hospital contains the records for 30,000 patients. The records are sequentially numbered from 1 to 30,000. If we sampled the records by choosing patients numbered 100, 200, 300, . . . , 30,000, thus obtaining a sample of 300 patients, what type of sampling would this be called?

1.26 A wholesale food distributor in a large metropolitan area would like to test the demand for a new food product. He distributes food through five large supermarket chains. The food distributor selects a sample of stores from each chain and tests his new product in these stores. What type of sampling does this represent?

1.27 A random sample could be very difficult to obtain. Why?

1.28 Why is the random sample so important in statistics?

1.29 Refer to *USA Today*'s "Money, Marriage: Tell Us" (Case Study 1-9).
   a. Do you think the sample resulting from mail-in information will result in a sample representative of the population? Explain.
   b. Do you think this sample qualifies as a random sample?

1.30 Refer to *McCall's* "What Would Make Life Better" (Case Study 1-10).
   a. What population was sampled?
   b. What information was collected from each of the 3000 women in the survey?

1.6 ▼ Comparison of Probability and Statistics

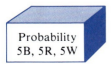

Probability and statistics are two separate but related fields of mathematics. It has been said that "probability is the vehicle of statistics." That is, if it were not for the laws of probability, the theory of statistics would not be possible.

Let's illustrate the relationship and the difference between these two branches of mathematics by looking at two boxes. We know the probability box contains five blue, five red, and five white poker chips. Probability tries to answer questions such as: If one chip is drawn from this box, what is the chance that it is blue? On the other hand, in the statistics box we don't know what the combination of chips is. We draw a sample and, based on the findings in the sample, we make conjectures about what we believe to be in the box. Note the difference: Probability asks about the chance that something specific (a sample) will happen when you know the possibilities (that is, you know the population). Statistics, on the other hand, asks you to draw a sample, describe the sample (descriptive statistics), and then make inferences about the population based on the information found in the sample (inferential statistics).

▼▲ EXERCISES

1.31 Classify each of the following as a probability or a statistics problem:
  a. determining whether a new drug shortens the recovery time from a certain illness
  b. determining the chance that a head will result when a coin is flipped
  c. determining the amount of waiting time required to check out at a certain grocery store
  d. determining the chance that you will be dealt a "blackjack"

1.32 Classify each of the following as a probability or a statistics problem:
  a. determining how long it takes to handle a typical telephone inquiry at a real estate office
  b. determining the length of life for the 100-watt light bulbs a company produces
  c. determining the chance that a blue ball will be drawn from a bowl that contains 15 balls of which 5 are blue
  d. determining the shearing strength of the rivets that your company just purchased for building airplanes

1.7 ▼ Statistics and the Computer

In recent years, the computer has had a tremendous effect on almost every aspect of life. The field of statistics is no exception. As you will see, the field of statistics uses many techniques that are repetitive in nature: formulas used to calculate descriptive statistics, procedures for constructing graphic displays of data, and procedures that are followed to formulate statistical inferences. The computer is very good at performing these repetitive operations. It is fairly common for someone with a set of sample data that needs to be analyzed to seek out someone who knows how to use and has access to a computer. If that computer has one of the standard statistical packages on line, it will make the analysis easy to complete. Your local computer center can provide you with a list of what is available. Some of the more readily available packaged programs are: MINITAB, Biomed (Biomedical Programs), SAS (Statistical Analysis System), IBM Scientific Subroutine Packages, and SPSS (Statistical Package for the Social Sciences). The examples of computer output that you will see throughout this textbook will all be MINITAB generated.

▼ ILLUSTRATION 1-8

Figure 1-1 shows two computer-generated drawings of stem-and-leaf diagrams. The two different diagrams depict the same set of data. Stem-and-leaf diagrams are studied in Section 2.1.

FIGURE 1-1

Weights of 50 College Students (lbs.)

09	8
10	8 1 8
11	2 8 8 5 5 0 6
12	0 8 0 0 9
13	5 7 2
14	8 3 5 2
15	0 8 7 4 5 4
16	2 7 8 2 1 5
17	0 7 6 0 6
18	6 8 4 3
19	1 5 0 5
20	5
21	5

Weights of 50 College Students (lbs.)

Female		Male
8	09	
8 1 8	10	
6 0 5 5 8 8 2	11	
9 0 0 8 0	12	
2 7 5	13	
2	14	8 3 5
	15	0 8 7 4 5 4
	16	2 7 8 2 1 5
	17	0 7 6 0 6
	18	6 8 4 3
	19	1 5 0 5
	20	5
	21	5

▼▲ EXERCISES

1.33 How have computers increased the usefulness of statistics to professionals such as researchers, government workers who analyze data, statistical consultants, and so on?

1.34 How might computers help you in statistics?

IN RETROSPECT

You should now have a general feeling of what statistics is about—an image that will grow and change as you work your way through this book. You know what a sample and a population are, the distinction between attribute and numerical data, and the difference between continuous and discrete variables. You even know the difference between statistics and probability (although we will not study probability in detail until Chapter 4). You should also have an appreciation for and a partial understanding of how sample data are obtained.

You should realize that the news articles at the beginning of the chapter represent various aspects of statistics. All of these articles, in one way or another, are the matter of statistics. The *USA Snapshots* picture a variety of information, for example, a serious ecology problem, "The USA's Garbage in 1 Year," and an everyday frustration for many, "Driving Drivers Crazy." Statistics can even be "recreational"; for example, "Iowa Poll Says 1 of 9 Iowans Has a Name for Automobile." Statistics are everywhere: on the news ("stockmarket was up . . ."), in commercials (four out of five doctors say . . .). The examples are endless. Look for some examples of statistics in your daily life. (See Exercise 1.45, p. 33.)

After some exercises using the material in this chapter, we will consider in detail in the next chapters the analysis of data.

CHAPTER EXERCISES

1.35 We want to describe the so-called typical student at your college. Describe a variable that measures some characteristic of a student and results in
    a. attribute data   **b.** discrete data   **c.** continuous data

1.36 A short survey consists of three questions:
    a. Is your religious preference Christian, Jewish, Moslem, or other?
    b. How many religious services do you attend per year?
    c. How much money did you contribute to religious organizations this last year?

Classify the responses to (a), (b), and (c) as attribute data, discrete variable data, or continuous variable data.

1.37 A candidate for a political office claims that he will win the election. A poll is conducted and 35 of 150 voters indicate that they will vote for the candidate, 100 voters indicate that they will vote for his opponent, and 15 voters are undecided.
    a. What is the population parameter of interest?
    b. What is the value of the sample statistic that might be used to estimate the population parameter?
    c. Would you tend to believe the candidate based on the results of the poll?

1.38 Refer to *USA Snapshot* "Driving Drivers Crazy" at the beginning of the chapter.
    a. What is the population under study?
    b. Describe the sample.
    c. Explain why 24% is a statistic.

1.39 A researcher studying consumer buying habits asks every 20th person entering Publix Supermarket how many times per week he or she goes grocery shopping. He then records the answer as T.
    a. Is $T = 3$ an example of (1) a sample, (2) a variable, (3) a statistic, (4) a parameter, or (5) a piece of data?

Suppose the researcher questions 427 shoppers during the survey.
    b. Give an example of a question that can be answered using the tools of descriptive statistics.

c. Give an example of a question that can be answered using the tools of inferential statistics.

1.40 A researcher studying the attitudes of parents of preschool children interviews a random sample of 50 mothers, each having one preschool child. She asks each mother, "How many times did you compliment your child yesterday?" She records the answer as C.
   a. Is C (1) a piece of data, (2) a statistic, (3) a parameter, (4) a variable, or (5) a sample?
   b. Give an example of a question that can be answered using the tools of descriptive statistics.
   c. Give an example of a question that can be answered using the tools of inferential statistics.

1.41 A study is designed to estimate the average income per household in a large city. The city is divided into blocks and a random sample of blocks is selected. For the blocks selected, each household is interviewed. What type of sampling does this represent?

1.42 The inspector on a bottle filling line at a local winery checks every 15th bottle for several properties. What type of sampling plan does this represent?

1.43 Describe, in your own words, and give an example of the following terms. Your example should not be one given in class or in the textbook.
   a. variable   b. data   c. sample
   d. population   e. statistic   f. parameter

1.44 Describe, in your own words, and give an example of the following terms:
   a. sampling frame   b. random sample   c. systematic sample
   d. stratified sample e. cluster sample

1.45 Find an article or an advertisement in a newspaper or a magazine that exemplifies the use of statistics.
   a. Identify and describe one statistic reported in the article.
   b. Identify and describe the variable related to the statistic in part (a).
   c. Identify and describe the sample related to the statistic in part (a).
   d. Identify and describe the population from which the sample in part (c) was taken.

VOCABULARY LIST

Be able to define each term. In addition, describe, in your own words, and give an example of each term. Your examples should not be ones given in class or in the textbook.

attribute data
census
cluster sample
continuous data
data
descriptive statistics
discrete data

experiment
finite population
inferential statistics
infinite population
judgment sample
numerical data
parameter

population
probability
probability sample
random
random sample
representative sampling frame
response variable
sample

sampling frame
sampling plan
statistic
statistics
strata
stratified sample
systematic sample
variability

KEY CONCEPTS

data
population
random sample

sample
statistic
variable

QUIZ A

Answer "True" if the statement is always true. If the statement is not always true, replace the words shown in bold with words that make the statement always true.

1.1 **Inferential** statistics is the study and description of data that result from an experiment.

1.2 **Descriptive** statistics is the study of a sample that enables us to make projections or estimates about the population from which the sample was drawn.

1.3 A **population** is typically a very large collection of individuals or objects about which we desire information.

1.4 A statistic is the calculated measure of some characteristic of a **population**.

1.5 A parameter is the measure of some characteristic of a **sample**.

1.6 As a result of surveying 50 freshmen, it was found that 16 had participated in interscholastic sports, 23 had served as officers of classes and clubs, and 18 had been in school plays during their high school years. This is an example of **discrete numerical data**.

1.7 The number of rotten apples in a shipping crate is an example of **continuous numerical** data.

1.8 The thickness of the sheet metal that a company uses in its manufacturing process is an illustration of **attribute** data.

1.9 A **representative** sample is a sample obtained in such a way that all individuals had an equal chance to be selected.

1.10 The basic objective of **statistics** is that of obtaining a sample, inspecting this sample, and then making inferences about the unknown characteristics of the population from which the sample was drawn.

QUIZ B

The owners of Corner Convenience Stores are concerned about the quality of service received by their customers. In order to study the service received, samples were collected for each of several variables.

1.1 Classify each of the following variables as being (A) attribute, (B) discrete, (C) continuous, or (D) not a variable:
   a. method of payment for purchases (cash, credit card, check)
   b. zip code for the customer's home mailing address
   c. amount of sales tax on purchase
   d. number of items purchased
   e. customer's driver's license number

1.2 The mean checkout time for all customers at Corner Convenience Stores is to be estimated by using the mean checkout time required by 75 randomly selected customers. Match the items in column II with the statistical term in column I.

I	II
___ data (one)	(a) the 75 customers
___ data (set)	(b) the mean time for all customers
___ experiment	(c) 2 minutes, one customer's checkout time
___ parameter	(d) the mean time for the 75 customers
___ population	(e) all customers at Convenient
___ sample	(f) the checkout time for one customer
___ statistic	(g) the 75 times
___ variable	(h) the process used to select the 75 customers and measure their times

QUIZ C

Write a brief paragraph in response to each question.

1.1 The *population* and the *sample* are both sets of objects. Describe the relationship between them and give an example.

1.2 The *variable* and the *data* for a specific situation are closely related. Explain this relationship and give an example.

1.3 The *data*, the *statistic*, and the *parameter* are all values used to describe a statistical situation. How does one distinguish among these three terms? Give an example.

1.4 What conditions are required in order for a sample to be a random sample?

2 DESCRIPTIVE ANALYSIS AND PRESENTATION OF SINGLE-VARIABLE DATA

Chapter Outline

GRAPHIC PRESENTATION OF DATA

2.1 Graphs and Stem-and-Leaf Displays
A **picture** is often worth a thousand words.

2.2 Frequency Distributions, Histograms, and Ogives
An **increase** in the amount of data requires us to modify our techniques.

CALCULATED DESCRIPTIVE STATISTICS

2.3 Measures of Central Tendency
The four measures of central tendency—mean, median, mode, and midrange—are **average values**.

2.4 Measures of Dispersion
Measures of dispersion—range, variance, and standard deviation—assign a numerical value to the **amount of spread** in a set of data.

2.5 Measures of Position
Measures of position allow us to **compare** one piece of data to the set of data.

2.6 Interpreting and Understanding Standard Deviation
A standard deviation is the length of a **standardized yardstick**.

2.7 The Art of Statistical Deception
How the unwitting or the unscrupulous can use **"tricky" graphs** and **insufficient information** to mislead the unwary.

Estimation of Flow for Evaluating and Classifying Pedestrian Facilities

Short-term manual counts are used in many pedestrian studies to overcome time, cost, and equipment constraints. Depending on the count period, several techniques are available for expanding short-term counts with the objective of estimating pedestrian volume over extended periods of time. When flow is highly variable and only a few short-term counts can be conducted, the expanded values can contain large margins of error. This paper presents the findings of two studies on the development and verification of models for expanding short-term (five-, 10-, 15-minute) counts.

. . . Two aspects were clearly evident from the above analyses. First, it was noted that the noon-hour peak is consistently the most pronounced of the three peaks as shown in Fig. 1. Moreover, over 80% of the time, the highest flow in any given hour occurs during the first or the last 15-minutes of the hour. Secondly, it was found that the pedestrians flows during all time intervals, at all sites, are normal random variables as illustrated in Fig. 2 (for one of the sites). . . .

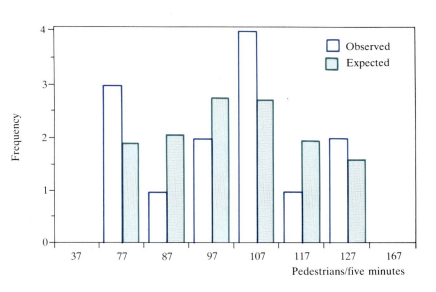

FIGURE 2 Frequency Distribution of Five-Minute Pedestrian Flow

Source: P. N. Seneviratne and J. F. Morrall, *Journal of Advanced Transportation,* Vol. 24, no. 1, 1990, pp. 31–46.

Chapter Objectives

In this chapter we will learn how to present and describe single-variable data. *Single variable* means that we are going to deal with only one numerical response value from a given element of the population. In Chapter 3 we will work with two values from a common element of the population.

The basic idea of descriptive statistics is to describe a set of data. Primarily we will try to describe a set of numerical values in a variety of abbreviated ways. Suppose your instructor just returned an exam paper to you. It carries a grade of 78, and you wish to interpret its value with respect to the grades in the rest of the class. Students usually are interested in three different things. What is the first thing you would ask your instructor?

Most students immediately ask: "What was the average exam grade?" The information being sought here is the location of the *middle* for the set of all the grades. *Measures of central tendency* are used to describe this concept. Your instructor replies that the class average was 68. So you know that your paper was above the average by 10 points. Now you would like an indication of the *spread* of the grades. Since 78 is 10 points above the average, you ask: "How close to the top is it?" Your instructor replies that the grades ranged from 42 to 87 points. Pictorially, the information we have so far looks like the accompanying figure.

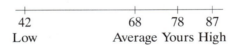

```
   42              68   78  87
   Low         Average Yours High
```

The third question is usually less important, since by now you have a fairly good idea of how your grade compares with those of your classmates. However, the third concept still has some importance. It answers the question "How are the data distributed?" You may have heard about the *normal curve,* but what is it? As you will see later, the idea of distribution describes the data by telling you whether the values are evenly distributed or clustered (bunched) around a certain value. For example, your instructor says that half the class had grades between 65 and 75.

When very large sets of data are involved, measures of *position* are useful. When college board exams are taken, for example, thousands of scores result. Your concern is the position of your score with respect to all the others. Averages, ranges, and so on, are not enough. We would like to say that your specific score is better than a certain percentage of scores. For example, your score is better than 75% of all the scores.

The four concepts necessary to describe sets of single-variable data are (1) measures of central tendency, (2) measures of dispersion (spread), (3) measures of position, and (4) types of distribution.

GRAPHIC PRESENTATION OF DATA

2.1 ▼ Graphs and Stem-and-Leaf Displays

Once we collect the sample of data, we must "get acquainted" with it. One of the most helpful ways to become acquainted is to use an initial exploratory technique that will result in a pictorial representation of the data. The resulting displays visually reveal patterns of behavior of the variable being studied. There are several graphic (pictorial) ways to describe data. The method used is determined by the type of data and the idea to be presented.

NOTE Realize from the start that there is no single correct answer when constructing a graphic display. The analyst's judgment and the circumstances surrounding the problem will play a major role in the development of the graphic.

CIRCLE AND BAR GRAPHS

Circle graphs (pie diagrams) and bar graphs are often used to summarize attribute data. You have seen many examples of these before.

Table 2-1 lists the number of cases of each type of operation performed at General Hospital last year. These data are displayed on a circle graph in Figure 2-1

TABLE 2-1 Operations Performed at General Hospital

Type of Operation	Number of Cases
Thoracic	20
Bones and joints	45
Eye, ear, nose, and throat	58
General	98
Abdominal	115
Urologic	74
Proctologic	65
Neurosurgery	23
	498

(p. 40), with each type being represented by its proportion and reported in percentages. Figure 2-2 (p. 40) displays the same set of data but in the form of a bar graph.

DOT PLOT DISPLAYS

dot plot

Dot plots display the data of a sample by representing each piece of data with a dot positioned along a scale. This scale can be either horizontal or vertical. The frequency of the values is represented along the other scale.

FIGURE 2-1
Circle Graph:
Operations Performed
at General Hospital
Last Year

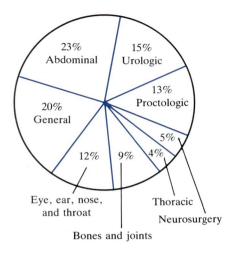

FIGURE 2-2
Bar Graph:
Operations Performed
at General Hospital
Last Year

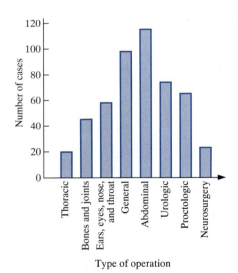

▼ ILLUSTRATION 2–1

A sample of 19 exam grades was randomly selected from a large class:

| 76 | 74 | 82 | 96 | 66 | 76 | 78 | 72 | 52 | 68 |
| 86 | 84 | 62 | 76 | 78 | 92 | 82 | 74 | 88 |

Figure 2-3 is a dot plot of the 19 exam scores. Notice how the data is "bunched" near the center and "spread out" near the extremes.

Section 2.1 ▼ GRAPHS AND STEM-AND-LEAF DISPLAYS

FIGURE 2-3
Dot Plot: 19 Exam Scores

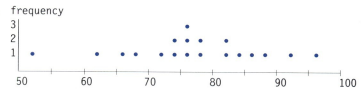

The dot plot display is a convenient technique to use as you first begin to analyze the data. It results in a picture of the data as well as sorts the data into numerical order. (To sort data is to make a listing of the data in rank order according to numerical value.)

STEM-AND-LEAF DISPLAYS

stem-and-leaf display

In recent years a technique known as the **stem-and-leaf display** has become very popular for summarizing numerical data. These displays are well-suited to computer application. This technique, very simple to create and use, is a combination of a graphic technique and a sorting technique. The data values themselves are used to do this sorting. The stem is the leading digit(s) of the data, and the leaf is the trailing digit(s). For example, the numerical data value 458 might be split 45–8 as shown:

Leading Digits	Trailing Digits
45	8
Used in sorting	Shown in display

▼ ILLUSTRATION 2-2

Let's construct a stem-and-leaf display for the 19 exam scores given in Illustration 2-1.

| 76 | 74 | 82 | 96 | 66 | 76 | 78 | 72 | 52 | 68 |
| 86 | 84 | 62 | 76 | 78 | 92 | 82 | 74 | 88 | |

At a quick glance we see that there are scores in the 50s, 60s, 70s, 80s, and 90s. Let's use the first digit of each score as the stem and the second digit as the leaf. Typically, the display is constructed in a vertical position. Draw a vertical line and place the stems, in order, to the left of it.

5 |
6 |
7 |
8 |
9 |

Next we place each leaf on its stem. This is done by placing the trailing digit on the right side of the vertical line opposite its corresponding leading digit. Our first data value is 76; 7 is the stem and 6 is the leaf. Thus we place a 6 opposite the stem 7.

7 | 6

The next data value is 74, so a leaf of 4 is placed on the 7 stem next to the 6.

7 | 6 4

The next data value is 82, so a leaf of 2 is placed on the 8 stem.

7 | 6 4
8 | 2

We continue until each of the other 16 leaves is placed on the display. Figure 2-4 shows the resulting stem-and-leaf display.

FIGURE 2-4

19 Exam Scores

5 | 2
6 | 6 2 8
7 | 4 4 8 6 2 6 6 8
8 | 2 8 4 6 2
9 | 6 2

Once the data have been entered into the computer, reworking it to create a display with different stems is usually easy to accomplish.

In Figure 2-4 all scores with the same tens digit are placed on the same branch, but this may not always be desired. Suppose we reconstruct the display; this time instead of grouping ten possible values on each stem, let's group the values so that only five possible values could fall on each stem. Do you notice a difference in the appearance of Figure 2-5? The general shape, approximately symmetrical about the 70s, is very much the same. It is fairly typical of many variables to display a distribution that is concentrated (mounded) about a central value and then in some manner dispersed in both directions. ▲▲

Often a graphic display reveals something that the analyst may or may not have anticipated. Illustration 2-3 demonstrates what generally occurs when two populations are sampled together.

Section 2.1 ▼ GRAPHS AND STEM-AND-LEAF DISPLAYS

FIGURE 2-5

```
          Exam Scores
(50–54)   5 | 2
(55–59)   5 |
(60–64)   6 | 2
(65–69)   6 | 6 8
(70–74)   7 | 4 4 2
(75–79)   7 | 8 6 6 6 8
(80–84)   8 | 2 4 2
(85–89)   8 | 8 6
(90–94)   9 | 2
(95–99)   9 | 6
```

▼ **ILLUSTRATION 2-3**

A random sample of 50 college students was selected. Their weights were obtained from their medical records. The resulting data are listed in Table 2-2.

TABLE 2-2

Student	1	2	3	4	5	6	7	8	9	10
Male/Female	F	M	F	M	M	F	F	M	M	F
Weight	98	150	108	158	162	112	118	167	170	120

Student	11	12	13	14	15	16	17	18	19	20
Male/Female	M	M	M	F	F	M	F	M	M	F
Weight	177	186	191	128	135	195	137	205	190	120

Student	21	22	23	24	25	26	27	28	29	30
Male/Female	M	M	F	M	F	F	M	M	M	M
Weight	188	176	118	168	115	115	162	157	154	148

Student	31	32	33	34	35	36	37	38	39	40
Male/Female	F	M	M	F	M	F	M	F	M	M
Weight	101	143	145	108	155	110	154	116	161	165

Student	41	42	43	44	45	46	47	48	49	50
Male/Female	F	M	F	M	M	F	F	M	M	M
Weight	142	184	120	170	195	132	129	215	176	183

Notice that the weights range from 98 to 215 lb. Let's group the weights on stems of ten units using the hundreds and the tens digits as stems and the units digit as the leaf. (See Figure 2-6, p. 44.) The leaves are in numerical order.

FIGURE 2-6

Weights of 50 College Students (lbs.)
Stem-and-Leaf of WEIGHT N = 50
Leaf Unit = 1.0

```
 9 | 8
10 | 1 8 8
11 | 0 2 5 5 6 8 8
12 | 0 0 0 8 9
13 | 2 5 7
14 | 2 3 5 8
15 | 0 4 4 5 7 8
16 | 1 2 2 5 7 8
17 | 0 0 6 6 7
18 | 3 4 6 8
19 | 0 1 5 5
20 | 5
21 | 5
```

Close inspection of Figure 2-6 suggests that there may be two overlapping distributions involved. That is exactly what we have: a distribution of female weights and a distribution of male weights. Figure 2-7, which shows a "back-to-back" stem-and-leaf display of this set of data, makes it quite obvious that we do in fact have two distinct distributions. Figure 2-8, a "side-by-side" dot plot of this same set of data, shows the same distinction between the two subsets of data. This situation involving more than one set of data is discussed further in Chapter 3.

FIGURE 2-7

Weights of 50 College Students (lbs.)

```
        Female        |    | Male
                    8 | 09 |
                1 8 8 | 10 |
      0 2 5 5 6 8 8   | 11 |
          0 0 0 8 9   | 12 |
              2 5 7   | 13 |
                  2   | 14 | 3 5 8
                      | 15 | 0 4 4 5 7 8
                      | 16 | 1 2 2 5 7 8
                      | 17 | 0 0 6 6 7
                      | 18 | 3 4 6 8
                      | 19 | 0 1 5 5
                      | 20 | 5
                      | 21 | 5
```

Section 2.1 ▼ GRAPHS AND STEM-AND-LEAF DISPLAYS

FIGURE 2-8

```
Female --- --:-- ---:..:::-- ----:---- ------:------ ------------ -- WEIGHTS
Male   --- ---- ---- ----:---:.::.:.- -:-.::-:- --:--- -:- ---- ---- WEIGHTS
         100        125       150       175        200        225
```

▼▲ EXERCISES

2.1 A computer-anxiety questionnaire was given to 200 students in a course in which computers are used. One of the questions was "I enjoy using computers." The responses to this particular question were

Response	Number
strongly agree	50
agree	75
slightly agree	25
slightly disagree	15
disagree	15
strongly disagree	20

a. Construct a bar graph that shows these responses.
b. Calculate the percentage for each response and construct a circle graph for these responses.
c. Compare the two graphs you constructed in (a) and (b); which one seems to be the most informative? Explain why.

2.2 Medical record practitioners are continually plagued with the problem of incomplete and delinquent medical files. Programs with positive incentives have been introduced to combat this problem. "The effects of positive incentive programs on physician chart completion" (*Topics in Health Record Management*, September 1990, pages 40–51) reported on the difficulties of implementing these programs in Florida.

a. Draw a circle graph showing the distribution of barriers.
b. Draw a bar graph for this data.
c. What would you consider the "average" barrier? Why?

Barriers	Number	Percentage
Lack of administrative support	10	43.5
Time	5	21.7
Money	4	17.4
Physician apathy	3	13.1
Rewards problem behavior	1	4.3
Total	23	100.0

2.3 The January 10, 1991 *USA Snapshots* "What's in U.S. Landfills" uses a circle graph to represent the percentages of each type of waste.

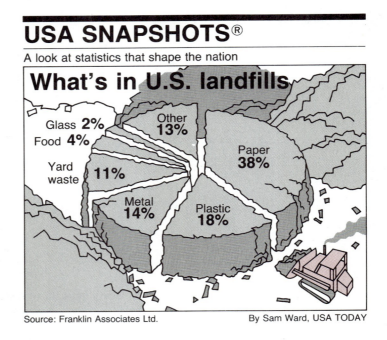

Source: Copyright 1991, USA TODAY. Reprinted with permission.

a. Draw a bar graph that shows this same information.
b. In your judgment, is the circle graph or the bar graph easier to read and understand? Explain.

 2.4 The November 30, 1990 *USA Snapshots* "Recycling Actions People in the USA Are Most Willing to Take" shows three recycling actions on a vertical bar graph (see p. 47).
a. The three percentages total 90. Explain the missing 10%.
b. Construct a circle graph showing these recycling actions. Be sure to include and label the 10%.
c. In your judgment, is the circle graph or the bar graph easier to read and understand? Explain.

 2.5 The *USA Snapshots* "Where Motorcyclists Live" (Case Study 1-3) presents a bar graph showing what percentage of America's motorcyclists live in each of four regions. What information was collected from each motorcyclist and used to construct the bar graph (name the variable)?

Section 2.1 ▼ GRAPHS AND STEM-AND-LEAF DISPLAYS

USA SNAPSHOTS®
A look at statistics that shape our lives

Recycling actions people in the USA are most willing to take:

- 51% Curbside pick-up
- 20% Take recyclables to centers for cash
- 19% Return containers to stores for deposit

Source: Glass Packaging Institute 1990 survey of 1,012 adults

By Bob Laird, USA TODAY

Source: Copyright 1990, USA TODAY. Reprinted with permission.

2.6 A sample of student-owned General Motors automobiles was identified and the make of each noted. The resulting sample follows (Ch = Chevrolet, P = Pontiac, O = Oldsmobile, B = Buick, Ca = Cadillac):

Ch	B	Ch	P	Ch	O	B	Ch	Ca	Ch
B	Ca	P	O	P	P	Ch	P	O	O
Ch	B	Ch	B	Ch	P	O	Ca	P	Ch
O	Ch	Ch	B	P	Ch	Ca	O	Ch	B
B	O	Ch	Ch	O	Ch	Ch	B	Ch	B

    a. Find the number of cars of each make in the sample.
    b. What percentage of these cars were Chevrolets? Pontiacs? Oldsmobiles? Buicks? Cadillacs?
    c. Draw a bar graph showing the percentage found in (b).

2.7 The following is the number of points scored during each game by a high school basketball team this past season:

56	54	67	53	61	71	46	61	55	56
68	60	66	54	61	52	36	64	51	52

Construct a dot plot of these data.

Chapter 2 ▼ DESCRIPTIVE ANALYSIS AND PRESENTATION OF SINGLE-VARIABLE DATA

 2.8 A city policewoman, using radar, checked the speed of cars as they were traveling down a city street:

27	23	22	38	43	24
25	23	22	52	31	30
29	28	27	25	29	28
26	33	25	27	25	
21	23	24	18	23	

Construct a dot plot of these data.

2.9 MINITAB was used to construct a dot plot of these data:

| 13.7 | 13.7 | 15.6 | 11.3 | 11.7 | 11.2 | 13.9 |
| 14.0 | 12.5 | 12.4 | 11.2 | 12.8 | 11.4 | 15.1 |

Verify the results.

```
MTB>       SET the data into C8
DATA>      13.7    13.7    15.6    11.3    11.7    11.2    13.9
DATA>      14.0    12.5    12.4    11.2    12.8    11.4    15.1
MTB>       NAME C8 'X'
MTB>       DOTPLOT C8;
SUBC>      INCREMENT =1.
        :..    .      ..      .       :  ..         .         .
     ---+------+------+------+------+------+------+------+---X
       11.0   12.0   13.0   14.0   15.0   16.0
```

 2.10 Delco Products, a division of General Motors, produces commutators designed to be 18.810 mm in overall length. (A commutator is a device used in the electrical system of an automobile.) The following sample of 35 commutators was taken while monitoring the manufacturing process:

The Overall Length of Commutators

18.802	18.810	18.780	18.757	18.824	18.827	18.825
18.809	18.794	18.787	18.844	18.824	18.829	18.817
18.785	18.747	18.802	18.826	18.810	18.802	18.780
18.830	18.874	18.836	18.758	18.813	18.844	18.861
18.824	18.835	18.794	18.853	18.823	18.863	18.808

Source: With permission of Delco Products Division, GMC.

Use a computer to construct a dot plot of this data. (If you use MINITAB, see Exercise 2.9 for a program.)

Section 2.1 ▼ GRAPHS AND STEM-AND-LEAF DISPLAYS

2.11 On the first day of class last semester, 30 students were asked for their one-way travel times from home to college (to the nearest 5 minutes). The resulting data were as follows:

20	20	30	25	20	25
35	25	15	25	25	40
25	30	15	20	45	25
15	20	20	20	20	25
5	20	20	10	5	20

Construct a stem-and-leaf display.

2.12 The following amounts are the fees charged by Quik Delivery for the 40 small packages they delivered last Thursday afternoon:

4.03	3.56	3.10	6.04	5.62	3.16	2.93	3.82
4.30	3.86	4.57	3.59	4.57	6.16	2.88	5.03
5.46	3.87	6.81	4.91	3.62	3.62	3.80	3.70
4.15	2.07	3.77	5.77	7.86	4.63	4.81	2.86
5.02	5.24	4.02	5.44	4.65	3.89	4.00	2.99

Construct a stem-and-leaf display.

2.13 MINITAB was used to construct a stem-and-leaf display of the following data:

5.97	6.01	6.08	6.04	6.12	6.19	6.10	6.43
6.24	6.20	6.35	6.38	6.16	6.16	6.22	6.27

Verify the results.

```
MTB>    SET data into C11
DATA>   (data as above)
MTB>    STEM C11;
SUBC>   INCREMENT =.1.
Stem-and-Leaf of C11          N =16
Leaf Unit = 0.010
    1    59   7
    4    60   148
   (5)   61   02669
    7    62   0247
    3    63   58
    1    64   3
```

2.14 Delco Products, a division of General Motors, produces commutators designed to be 18.810 mm in overall length. (A commutator is a device used in the electrical system of an automobile.) The following sample of 35 commutators was taken while monitoring the manufacturing process. (Same sample data as in Exercise 2.10.)

The Overall Length of Commutators

18.802	18.810	18.780	18.757	18.824	18.827	18.825
18.809	18.794	18.787	18.844	18.824	18.829	18.817
18.785	18.747	18.802	18.826	18.810	18.802	18.780
18.830	18.874	18.836	18.758	18.813	18.844	18.861
18.824	18.835	18.794	18.853	18.823	18.863	18.808

Source: With permission of Delco Products Division, GMC.

Use a computer to construct a stem-and-leaf display of this data. [If you use MINITAB, see Exercise 2.13 for a program and use a stem width (increment) of 0.01.]

 2.15 A term often used in solar energy research is *heating-degree-days.* This concept is related to the difference between an indoor temperature of 65°F and the average outside temperature for a given day. If the average outside temperature is 5°F, this would give 60 heating-degree-days. The annual heating-degree-days normals for several Nebraska locations follow:

6726	6796	6946	6197	6368	6437	6434	6740	6886	6197
6582	6297	6811	6102	6261	6919	6086	6320	6139	6420
6684	6070	6169	6955	6545					

Construct a stem-and-leaf display for these data. (Use the leading two digits as stems and the trailing two digits as leaves.)

MINITAB was used to construct a stem-and-leaf display of these data.

```
MTB>   SET the data into C12
DATA>  (data as above)
MTB>   STEM C12;
SUBC>  INCREMENT = 100.
Stem-and-leaf of C12           N = 25
Leaf Unit = 10
    2      60  78
    7      61  03699
    9      62  69
   11      63  26
   (3)     64  233
   11      65  48
    9      66  8
    8      67  249
    5      68  18
    3      69  145
```

Verify the results. Notice the effect that the stem width (increment) 100 and the resulting leaf unit of 10 had on the number of digits in each data value.

Section 2.2 ▼ FREQUENCY DISTRIBUTIONS, HISTOGRAMS, AND OGIVES

● 2.16 "Particle size" is an important property of latex paint and is monitored during production as part of the quality control process. Thirteen particle size measurements were taken using the Dwight P. Joyce Disc.

Particle Size Measurements						
4295	4271	4326	4530	4618	4779	4752
4744	3764	3797	4401	4339	4700	

Use a computer to construct a stem-and-leaf display of this data. [If you use MINITAB, see Exercise 2.15 for a program and use a stem width (increment) of 100.]

2.2 ▼ Frequency Distributions, Histograms, and Ogives

Listing a large set of data does not present much of a picture to the reader. Sometimes we want to condense the data into a more manageable form. This can be accomplished with the aid of a **frequency distribution**.

frequency distribution

FREQUENCY DISTRIBUTIONS

To demonstrate the concept of a frequency distribution, let's use the following set of data:

```
3   2   2   3   2
4   4   1   2   2
4   3   2   0   2
2   1   3   3   1
```

If we let x represent these data values, we can use a frequency distribution to represent this set of data by listing the x values with their frequencies. For example, the value 1 occurs in the sample three times; therefore, the frequency for $x = 1$ is 3. The set of data is represented by the frequency distribution shown in Table 2-3. The frequency f is the number of times the value x occurs in the sample. This is an **ungrouped frequency distribution**. We say "ungrouped" because each value of x in the distribution stands alone.

ungrouped frequency distribution

TABLE 2-3 Frequency Distribution

x	f
0	1
1	3
2	8
3	5
4	3

classes

When a large set of data has many different *x* values instead of a few repeated values, as in the previous example, we can group the values into a set of **classes** and construct a frequency distribution. The stem-and-leaf display in Figure 2-4 shows, in picture form, a grouped frequency distribution. Each stem represents a class. The number of leaves on each stem is the same as the frequency for that same class. The data represented in Figure 2-4 are listed as a frequency distribution in Table 2-4.

TABLE 2-4

Class	Frequency
50–59	1
60–69	3
70–79	8
80–89	5
90–99	2
	19

The stem-and-leaf process can be used to construct a frequency distribution; however, the stem representation is not compatible with all class widths. For example, class widths of 3, 4, or 7 are awkward to use. Thus, sometimes we will find it advantageous to have a separate procedure for constructing a grouped frequency distribution. To illustrate this grouping (or classifying) procedure, let's use a sample of 50 final exam scores. Table 2-5 lists the 50 scores.

The two basic guidelines that should be followed in constructing a grouped frequency distribution are

1. Each class should be of the same width.
2. Classes should be set up so that they do not overlap and so that each piece of data belongs to exactly one class.

TABLE 2-5
Statistics Exam Scores

60	47	82	95	88
72	67	66	68	98
90	77	86	58	64
95	74	72	88	74
77	39	90	63	68
97	70	64	70	70
58	78	89	44	55
85	82	83	72	77
72	86	50	94	92
80	91	75	76	78

Three additional helpful (but not necessary) guidelines are

3. For the exercises given in this textbook, 5 to 12 classes are most desirable since all samples contain less than 125 data.
4. When it is convenient, an odd class width is often advantageous.
5. Use a system that takes advantage of a number pattern, to guarantee accuracy. (This will be demonstrated in the following example.)

Section 2.2 ▼ FREQUENCY DISTRIBUTIONS, HISTOGRAMS, AND OGIVES

PROCEDURE

1. Identify the high and the low scores ($H = 98, L = 39$) and find the range. Range = $H - L = 98 - 39 = 59$.
2. Select a number of classes ($m = 9$) and a class width ($c = 7$) so that the product ($mc = 63$) is a bit larger than the range (range = 59).
3. Pick a starting point. This starting point should be a little smaller than the lowest score L. Suppose that we start at 35; counting from there by 7s (the class width), we get 35, 42, 49, 56, ..., 98. These are called the *lower class limits*. (They are all multiples of 7, an easily recognized pattern.)

lower and upper class limits

The **lower class limit** is the smallest value that can go into each class. The **upper class limits** are the largest values fitting into each class: 41, 48, 55, and so on, in our example. Our *classes* for this example are

35–41	70– 76
42–48	77– 83
49–55	84– 90
56–62	91– 97
63–69	98–104

NOTES

1. At a glance you can check the number pattern to determine whether the arithmetic used to form the classes was correct.

class width

2. The **class width** is the difference between a lower class limit and the next lower class limit. (It is *not* the difference between the lower and upper limit of the same class.)

class boundary

3. **Class boundaries** are numbers that do not occur in the sample data but are halfway between the upper limit of one class and the lower limit of the next class. In the previous example, the class boundaries are 34.5, 41.5, 48.5, 55.5, ..., 97.5, and 104.5. The difference between the upper and lower class boundaries will also give you the class width.

When classifying data it helps to use a standard chart (see Table 2-6).

TABLE 2-6 Standard Chart for Frequency Distribution

Class Number	Class Limits	Tallies	Frequency (f)
1	35– 41	|	1
2	42– 48	||	2
3	49– 55	||	2
4	56– 62	|||	3
5	63– 69	﹂﹉﹉ ||	7
6	70– 76	﹂﹉﹉ ﹂﹉﹉ |	11
7	77– 83	﹂﹉﹉ ||||	9
8	84– 90	﹂﹉﹉ |||	8
9	91– 97	﹂﹉﹉ |	6
10	98–104	|	1
			50

Once the classes are set up, we need to tally the data (Table 2-6).

NOTES:

1. If the data have been ranked, this tallying is unnecessary.
2. If the data are not ranked, be careful as you tally them.
3. The frequency f for each class is the number of pieces of data that belong in that class.
4. The sum of the frequencies should be exactly equal to the number of pieces of data n ($n = \Sigma f$). This summation serves as a good check.

summation

NOTE: See Appendix A for information about the Σ (read "**sum**") **notation**.

class mark

In Table 2-7, there is a column for the class mark, x. The **class mark** (sometimes called *class midpoint*) is the numerical value that is exactly in the middle of each class.

TABLE 2-7
Frequency Table with Class Marks, a Grouped Frequency Distribution

Class Number	Class Limits	f	Class Mark (x)
1	35– 41	1	38
2	42– 48	2	45
3	49– 55	2	52
4	56– 62	3	59
5	63– 69	7	66
6	70– 76	11	73
7	77– 83	9	80
8	84– 90	8	87
9	91– 97	6	94
10	98–104	1	101
		50	

In Table 2-7, the class marks are

$$x_1 = \frac{35 + 41}{2} = 38 \qquad x_2 = \frac{42 + 48}{2} = 45$$

and so on. As a check of your arithmetic, successive class marks should be a class width apart, which is 7 in this illustration.

NOTE Now you can see why it is helpful to have an odd class width: the class marks are whole numbers.

grouped frequency distribution

Once the class marks are determined, we have a **grouped frequency distribution**. Note that when we classify data in this fashion, we lose some information. Only when we have all the raw data before us do we know the exact values that were actually observed in each class. For example, we put two 58s and a 60 into class number 4, with class limits of 56 and 62. Once they are placed in the class,

Section 2.2 ▼ FREQUENCY DISTRIBUTIONS, HISTOGRAMS, AND OGIVES

their values are lost to us and we use the class mark, 59, as their representative value. This loss of identity costs some accuracy, but the computations are made easier.

HISTOGRAMS

histogram

The **histogram** is a type of bar graph representing an entire set of data. The distribution of frequencies from Table 2-7 appears in histogram form in Figure 2-9. A histogram is made up of the following components:

1. A title, which identifies the population of concern.
2. A vertical scale, which identifies the frequencies in the various classes. Figure 2-9 is a frequency histogram.
3. A horizontal scale, which identifies the variable x. Values for the class boundaries, class limits, or class marks may be labeled along the x-axis. Use whichever *one* of these sets of class numbers best presents the variable.

(*Note:* Be sure to identify both scales so that the histogram tells the complete story.)

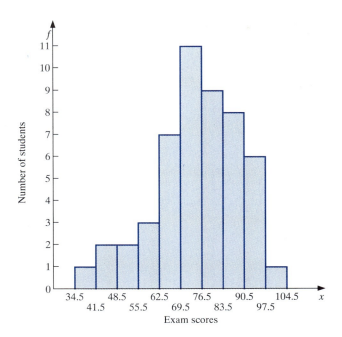

FIGURE 2-9
Frequency Histogram:
50 Final Exam Scores
in Elementary
Statistics

relative frequency histogram
relative frequency

By changing the vertical scale to $\frac{0}{50}, \frac{1}{50}, \frac{2}{50}$, and so on, the histogram would become a **relative frequency histogram**. The **relative frequency** is a proportional measure of the frequency of an occurrence. It is found by dividing the class frequency by the total number of observations. For example, in this illustration the frequency associated with the seventh class (77–83) is 9. The relative frequency is $\frac{9}{50}$ (or 0.18), since these values occurred 9 times in 50 observations. Relative frequency can often be useful in a presentation. They are particularly useful when comparing the frequency distributions of two different sets of data. Figure 2-10 is a relative frequency histogram of the sample of 50 scores. Compare and contrast Figures 2-9 and 2-10.

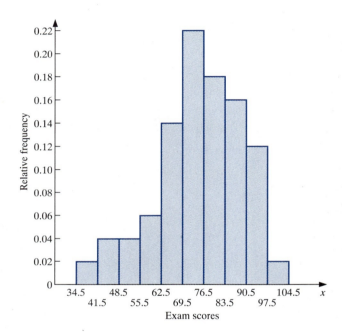

FIGURE 2-10
Relative Frequency Histogram: 50 Statistics Exam Scores

A stem-and-leaf display contains all the information in a histogram. Figure 2-4 shows the stem-and-leaf display constructed in Illustration 2-1. Figure 2-4 has been rotated 90° and labels have been added to form the histogram shown in Figure 2-11.

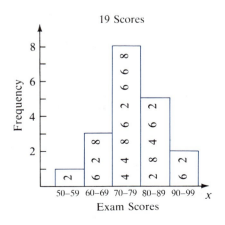

FIGURE 2-11
19 Exam Scores

Histograms are valuable tools. For example, the histogram of the sample should have a distribution shape that is very similar to that of the population from which the sample was drawn. If the reader of a histogram is at all familiar with the variable involved, he or she will usually be able to interpret several important facts. Figure 2-12 presents histograms with descriptive labels resulting from their geometric shape. Can you think of populations whose samples might yield histograms like these?

FIGURE 2-12
Shapes of Histograms

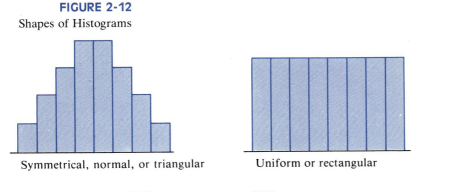
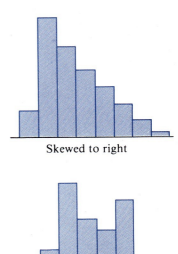

Symmetrical, normal, or triangular   Uniform or rectangular   Skewed to right

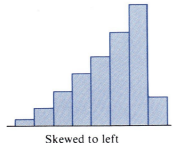
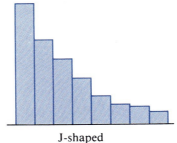
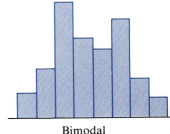

Skewed to left   J-shaped   Bimodal

Briefly, the terms used to describe histograms are as follows:

▼ **Symmetrical:** Both sides of this distribution are identical.
▼ **Uniform (rectangular):** Every value appears with equal frequency.
▼ **Skewed:** One tail is stretched out longer than the other. The direction of skewness is on the side of the longer tail.
▼ **J-shaped:** There is no tail on the side of the class with the highest frequency.
▼ **Bimodal:** The two most populous classes are separated by one or more classes. This situation often implies that two populations are being sampled.
▼ **Normal:** A symmetrical distribution that is mounded up about the mean and becomes sparse at the extremes. (Additional properties are discussed later.)

Chapter 2 ▼ DESCRIPTIVE ANALYSIS AND PRESENTATION OF SINGLE-VARIABLE DATA

NOTES

mode

modal class

bimodal frequency distribution

1. The **mode** is the value of the piece of data that occurs with the greatest frequency. (Mode will be discussed in Section 2.3, p. 73.)
2. The **modal class** is the class with the highest frequency.
3. A **bimodal frequency distribution** has two high-frequency classes separated by classes with lower frequencies.

OGIVE

cumulative and cumulative relative frequency distributions

A frequency distribution can very easily be converted to a **cumulative frequency distribution** by replacing the frequencies with cumulative frequencies. This is done by placing a subtotal of the frequencies next to each class, as shown in Table 2-8. The **cumulative frequency** for any given class is the sum of the frequency for that class and the frequencies of all classes of smaller values.

TABLE 2-8
Using a Frequency Distribution to Form a Cumulative Frequency Distribution

Frequency Distribution		Cumulative Frequency Distribution	
Class Limits	Frequencies	Class Boundaries	Cumulative Frequencies
35– 41	1	34.5– 41.5	1
42– 48	2	41.5– 48.5	3 (1 + 2)
49– 55	2	48.5– 55.5	5 (1 + 2 + 2)
56– 62	3	55.5– 62.5	8 (1 + 2 + 2 + 3)
63– 69	7	62.5– 69.5	15 (1 + 2 + 2 + 3 + 7)
70– 76	11	69.5– 76.5	26
77– 83	9	76.5– 83.5	35
84– 90	8	83.5– 90.5	43
91– 97	6	90.5– 97.5	49
98–104	1	97.5–104.5	50
	50		

The same information can be presented by using a **cumulative relative frequency distribution** (see Table 2-9). This combines the cumulative frequency idea and the relative frequency idea.

ogive

An **ogive** (pronounced o'jiv) is a cumulative frequency or cumulative relative frequency graph. An ogive has the following components:

1. A title, which identifies the population.
2. A vertical scale, which identifies either the cumulative frequencies or the cumulative relative frequencies. (Figure 2-13 shows an ogive with cumulative relative frequencies.)

TABLE 2-9 A Cumulative Relative Frequency Distribution

Class Boundaries	Cumulative Relative Frequency
34.5– 41.5	$\frac{1}{50} = 0.02$
41.5– 48.5	$\frac{3}{50} = 0.06$
48.5– 55.5	$\frac{5}{50} = 0.10$
55.5– 62.5	$\frac{8}{50} = 0.16$
62.5– 69.5	$\frac{15}{50} = 0.30$
69.5– 76.5	$\frac{26}{50} = 0.52$
76.5– 83.5	$\frac{35}{50} = 0.70$
83.5– 90.5	$\frac{43}{50} = 0.86$
90.5– 97.5	$\frac{49}{50} = 0.98$
97.5–104.5	$\frac{50}{50} = 1.00$

3. A horizontal scale, which identifies the upper class boundaries. Until the upper boundary of a class has been reached, you cannot be sure you have accumulated all the data in that class. Therefore, *the horizontal scale for an ogive is always based on the upper class boundaries.*

NOTE Every ogive starts on the left with a relative frequency of zero at the lower class boundary of the first class and ends on the right with a relative frequency of 100% at the upper class boundary of the last class.

All graphic representations of sets of data should be completely self-explanatory. That includes a descriptive and meaningful title and proper identification of the vertical and horizontal scales.

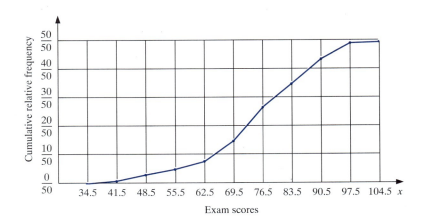

FIGURE 2-13 Ogive: 50 Final Exam Scores in Statistics

Case Study 2-1

Chalk It Up

A March 1990 issue of USA Today presented this graphic showing a relative frequency distribution resulting from a sample of 3960 teachers. Data on the amount of their own money spent to buy classroom supplies were collected. What class width is used in this histogram? (See Exercise 2.25, p. 66.)

USA SNAPSHOTS®
A look at statistics that shape our lives

Chalk it up
Some teachers use their own money to buy classroom supplies. Here's a breakdown of how much they spend per year:

- $1 to $50: 18%
- $51 to $100: 24%
- $101 to $250: 27%
- $251 or more: 26%
- Not sure/varies: 5%

Source: Editorial Projects in Education random sample of 3,960 *Teacher Magazine* subscribers

By Elys McLean-Ibrahim, USA TODAY

Source: Copyright 1990, USA TODAY. Reprinted with permission.

Case Study 2-2

Home Sweet Home

A March 1990 issue of USA Today presented this graphic which shows a relative frequency distribution in the form of a circle graph. Each section of the circle represents an interval of values for the variable, and the various percentages are relative frequencies for that class interval. This same information could be expressed as a grouped relative frequency distribution or a relative frequency histogram. What is the variable? What is the class width? Does this grouped distribution follow the guidelines for a grouped distribution? (See Exercise 2.26, p. 66.)

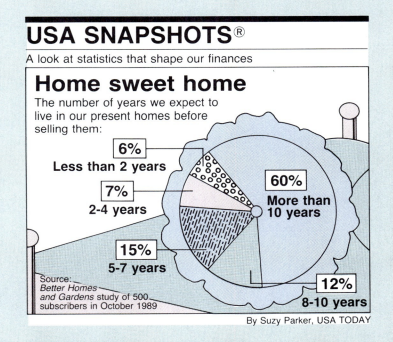

Source: Copyright 1990, USA TODAY. Reprinted with permission.

Case Study 2-3

The graph "Transplant Movers, Inc." appeared in the May/June issue of UTNE READER. The percentages shown appear to form a relative frequency distribution of the variable, age of person who moved. Does this graph represent a relative frequency distribution? Justify your answer. (See Exercise 2.30, p. 67.)

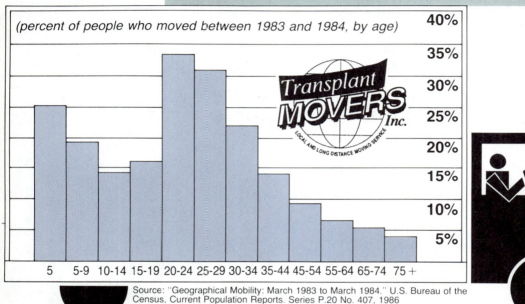

(percent of people who moved between 1983 and 1984, by age)

Source: "Geographical Mobility: March 1983 to March 1984." U.S. Bureau of the Census, Current Population Reports. Series P.20 No. 407, 1986

Source: UTNE READER, May/June, 1990. Reprinted with permission.

▼▲ EXERCISES

 2.17 In the May 1990 issue of the journal *Social Work,* the following ungrouped frequency distribution was reported:

Number of Children Living at Home	Mexican-American Women
0	23
1	22
2	17
3	7
4	1

Source: Copyright 1990, National Association of Social Workers, Inc. *Social Work*.

Construct a histogram of this of the above distribution.

2.18 An article in the *Therapeutic Recreation Journal* reported the following distribution for the variable "number of persistent disagreements between sixty-six patients and their therapeutic recreation specialist":

Number of Items	Frequency	Number of Items	Frequency
0	2	6	8
1	2	7	11
2	4	8	7
3	10	9	3
4	7	10	1
5	9	11	2

Source: Reprinted with permission of the National Recreation and Park Association, Alexandria, VA, from Pauline Petryshen and Diane Essex-Sorlie, "Persistent Disagreement Between Therapeutic Recreation Specialists and Patients in Psychiatric Hospitals," *Therapeutic Recreation Journal,* Vol. XXIV, Third Quarter, 1990.

Draw a histogram of this ungrouped frequency distribution.

2.19 The ages of 50 dancers who responded to a call to audition for a musical comedy are

21	19	22	19	18	20	23	19	19	20
19	20	21	22	21	20	22	20	21	20
21	19	21	21	19	19	20	19	19	19
20	20	19	21	21	22	19	19	21	19
18	21	19	18	22	21	24	20	24	17

a. Prepare an ungrouped frequency distribution of these ages.
b. Prepare an ungrouped relative frequency distribution of the same data.
c. Prepare a cumulative relative frequency distribution of the same data.
d. Prepare a relative frequency histogram of these data.
e. Prepare an ogive of these data.

2.20 The opening-round scores for the Ladies' Professional Golf Association tournament at Locust Hill Country Club were posted as

69	73	72	74	77	80	75	74	72	83	68	73
75	78	76	74	73	68	71	72	75	79	74	75
74	74	68	79	75	76	75	77	74	74	75	75
72	73	73	72	72	71	71	70	82	77	76	73
72	72	72	75	75	74	74	74	76	76	74	73
74	73	72	72	74	71	72	73	72	72	74	74
67	69	71	70	72	74	76	75	75	74	73	74
74	78	77	81	73	73	74	68	71	74	78	70
68	71	72	72	75	74	76	77	74	74	73	73
70	68	69	71	77	78	68	72	73	78	77	79
79	77	75	75	74	73	73	72	71	68	70	71
78	78	76	74	75	72	72	72	75	74	76	77
78	78										

a. Form an ungrouped frequency distribution of these scores.
b. Draw a histogram of the first-round golf scores. Use the frequency distribution from (a). (Retain these solutions for use in answering Exercises 2.52, p. 80, and 2.74, p. 93.)

2.21 The KSW computer-science aptitude test was given to 50 students. The following frequency distribution resulted from their scores:

KSW Test Score	Frequency
0– 3	4
4– 7	8
8–11	8
12–15	20
16–19	6
20–23	3
24–27	1

a. Give all class boundaries associated with this frequency distribution.
b. Give all class marks associated with this frequency distribution.
c. What is the class width?
d. Give the relative frequencies for the classes.

Section 2.2 ▼ FREQUENCY DISTRIBUTIONS, HISTOGRAMS, AND OGIVES

 2.22 The following distribution of commuting distances was obtained for a sample of Mutual of Nebraska employees:

Distance (Miles)	Frequency
1.0– 2.9	2
3.0– 4.9	6
5.0– 6.9	12
7.0– 8.9	50
9.0–10.9	35
11.0–12.9	15
13.0–14.9	5

a. Give all class boundaries associated with this frequency distribution.
b. Give all class marks associated with this frequency distribution.
c. What is the class width?
d. Find the relative frequencies for each class.

 2.23 In the March 1991 issue of the magazine *Club Industry* (page 20), the following distribution of salaries for club managers is given:

Range	Percentage
15k–20k	10
20k–25k	14
25k–30k	23
30k–35k	21
35k–40k	10
40k–50k	13
50k–70k	9

Suppose you conducted a similar survey and found the following results:

Range	Number
15k–24k	10
25k–34k	35
35k–44k	30
45k–54k	15
55k–64k	10
65k–74k	4

Draw a frequency histogram for your survey results. (Retain this solution for use in answering Exercise 2.54.)

2.24 The following relative frequency distribution for family income was reported in the May 1990 issue of *Social Work:*

Income	Anglo-American Women ($n = 72$)	
Less than $10,000	4%	
$10,000–19,999	26	Percentages do not sum to 100% due to round-off.
$20,000–29,999	29	
$30,000–39,999	9	
$40,000 and over	31	

Source: Copyright 1990, National Association of Social Workers, Inc. *Social Work.*

Construct a relative frequency histogram of this distribution, treating the class "$40,000 and over" as "$40,000–$49,999."

2.25 **a.** What class width is used in the histogram presented in Case Study 2-1 (p. 60)?
b. What are the difficulties with this graph as a histogram?

2.26 "Home Sweet Home" (Case Study 2-2 on p. 61) shows a relative frequency distribution in the form of a circle graph.
 a. Construct the relative frequency distribution shown.
 b. What is the variable?
 c. What is the class width?
 d. Does this grouped distribution follow the guidelines set forth in this textbook? Explain.

2.27 The speeds of 55 cars were measured by a radar device on a city street:

27	23	22	38	43	24	35	26	28	18	20
25	23	22	52	31	30	41	45	29	27	43
29	28	27	25	29	28	24	37	28	29	18
26	33	25	27	25	34	32	36	22	32	33
21	23	24	18	48	23	16	38	26	21	23

 a. Classify these data into a grouped frequency distribution by using classes of 15–19, 20–24, ..., 50–54. (Retain this solution for use in answering Exercises 2.47, p. 79, and 2.67, p. 92.)
 b. Find the class width.
 c. For the class 20–24, find (1) the class mark, (2) the lower class limit, (3) the upper class boundary.
 d. Construct a frequency histogram of these data.

2.28 The hemoglobin A_{1c} test, a blood test given to diabetics during their periodic checkups, indicates the level of control of blood sugar during the past two to three months. The following data were obtained for 40 different diabetics at a university clinic that treats diabetic patients.

6.5	5.0	5.6	7.6	4.8	8.0	7.5	7.9	8.0	9.2
6.4	6.0	5.6	6.0	5.7	9.2	8.1	8.0	6.5	6.6
5.0	8.0	6.5	6.1	6.4	6.6	7.2	5.9	4.0	5.7
7.9	6.0	5.6	6.0	6.2	7.7	6.7	7.7	8.2	9.0

   a. Classify these A_{1c} values into a grouped frequency distribution using the classes 3.7–4.6, 4.7–5.6, and so on.
   b. What are the class boundaries for these classes?
   c. What are the class marks for these classes?
   d. Construct a frequency histogram of these data.

2.29 The following 40 amounts are the fees that Fast Delivery charged for delivering small freight items last Thursday afternoon:

2.03	1.56	1.10	4.04	3.62	1.16	0.93	1.82
2.30	1.86	2.57	1.59	2.57	4.16	0.88	3.02
3.46	1.87	4.81	2.91	1.62	1.62	1.80	1.70
2.15	2.07	1.77	3.77	5.86	2.63	2.81	0.86
3.02	3.24	2.02	3.44	2.65	1.89	2.00	0.99

   a. Classify these data into a grouped frequency distribution by using classes of 0.01–1.00, 1.01–2.00, ..., 5.01–6.00.
   b. Find the class width.
   c. For the class 2.01–3.00, name the value of (1) the class mark, (2) the class limits, (3) the class boundaries.
   d. Construct a relative frequency histogram of these data.

2.30 Does the graph "Transplant Movers, Inc." presented in Case Study 2-3 (p. 62) represent a relative frequency distribution? Justify your answer.

2.31 MINITAB was used to form the frequency distribution and draw the histogram for the following data in which the size of high school graduating classes was being studied. The following list is a sample of class sizes taken from a list published in the *Democrat and Chronicle,* June 2, 1991, for school systems in western New York State. Verify the results.

80	160	66	58	30	30	32	235	26	138
35	71	188	160	156	72	19	275	300	72
71	54	235	72	315	77	141	30	59	240
172	75	228	36	477	62	260	74	348	123
3	200	287	311	61	126	270	77	72	149
222	76	138	53	70	306	131	74	220	161

Notice that by specifying the midpoint (class mark) of the first class, MINITAB will use your choice for classification.

```
MTB> SET C1
DATA> (data as above)
MTB> HIST C1;
SUBC> INCREMENT 50;
SUBC> START 25.5.
```

Histogram of C1   N = 60

Midpoint	Count	
25.5	9	*********
75.5	21	*********************
125.5	7	*******
175.5	7	*******
225.5	6	******
275.5	5	*****
325.5	4	****
375.5	0	
425.5	0	
475.5	1	*

2.32 The number of hits batters in the National League have produced through Friday's game were reported in the June 2, 1991 *Democrat and Chronicle*:

44	69	57	50	57	41	47	51	52	50	51	47	50	45
36	52	47	48	47	47	42	51	49	47	48	36	46	35
46	30	44	39	41	35	43	37	35	39	31	30	36	39
38	35	36	38	38	37	30	34	42	27	43	34	26	27
36	37	36	42	34	24	34	35	37	28	32	29	41	29
39	36	33	35	23	19	19							

Use a computer to construct a histogram of this data. (If you use MINITAB, see Exercise 2.31 for a program.)

CALCULATED DESCRIPTIVE STATISTICS

2.3 ▼ Measures of Central Tendency

measure of central tendency

Measures of central tendency are numerical values that tend to locate in some sense the *middle* of a set of data. The term *average* is often associated with these measures. Each of the several measures of central tendency can be called the *average* value.

Section 2.3 ▼ MEASURES OF CENTRAL TENDENCY

MEAN

mean

To find the **mean**, $\bar{x}$ (read "x bar"), the average with which you are probably most familiar, you will add all the values of the variable x (this sum of x values is symbolized by Σx) and divide by the number of these values, n. We express this in formula form as

$$\text{sample mean} = \bar{x} = \frac{\Sigma x}{n} \qquad (2\text{-}1)$$

NOTE See Appendix A for information about the Σ notation.

▼ **ILLUSTRATION 2-3**

A set of data consists of the five values 6, 3, 8, 5, and 3. Find the mean.

SOLUTION Using formula (2-1), we find

$$\bar{x} = \frac{6 + 3 + 8 + 5 + 3}{5} = \frac{25}{5} = 5$$

Therefore, the mean of this sample is 5. ▲▲

A physical representation of the mean can be constructed by thinking of a number line balanced on a fulcrum. A weight is placed on a number on the line corresponding to each number in the sample. In Figure 2-14 there is one weight each on the 6, 8, and 5 and two weights on the 3, since there were two 3s in the sample of Illustration 2-3. The mean is the value that balances the weights on the number line—in this case, 5.

FIGURE 2-14
Physical Representation of Mean

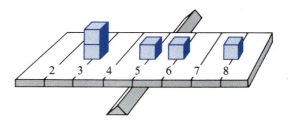

Don't be misled by Illustration 2-3 into believing that the mean value has to be a value in the data set. For example, consider the case in which a salesperson sells $130 worth of merchandise one week and $110 in another. The mean of these values is $120, which is not a value in the data set.

When the sample data has the form of a frequency distribution, we will need to make a slight adaptation in order to find the mean. Consider the frequency distribution of Table 2-10 on p. 70. This frequency distribution represents a

TABLE 2-10
Ungrouped Frequency Distribution

x	f
1	5
2	9
3	8
4	6
	28

sample of 28 values, five 1s, nine 2s, eight 3s, and six 4s. To calculate the mean $\bar{x}$ using formula (2-1), we need Σx, the sum of the 28 x values.

$$\Sigma x = \underbrace{1 + 1 + \cdots + 1}_{5 \text{ of them}} + \underbrace{2 + 2 + \cdots + 2}_{9 \text{ of them}} + \underbrace{3 + 3 + \cdots + 3}_{8 \text{ of them}} + \underbrace{4 + 4 + \cdots + 4}_{6 \text{ of them}}$$

$$\Sigma x = (5)(1) + (9)(2) + (8)(3) + (6)(4)$$

$$= 5 + 18 + 24 + 24 = \mathbf{71}$$

We can also obtain this sum by direct use of the frequency distribution. The extensions xf (see Table 2-11) are formed for each row by multiplying each x by its corresponding f and then totaled to obtain Σxf. The Σxf is the sum of the data. Therefore, the mean of a frequency distribution may be found by dividing the sum of the data, Σxf, by the sample size, Σf. We can rewrite formula (2-1) for use with a frequency distribution as

TABLE 2-11
Extension xf

x	f	xf
1	5	5
2	9	18
3	8	24
4	6	24
Total	28	71

$$\bar{x} = \frac{\Sigma xf}{\Sigma f} \qquad (2\text{-}2)$$

The mean for the example is found by using formula (2-2).

$$\text{Mean: } \bar{x} = \frac{\Sigma xf}{\Sigma f},$$

$$\bar{x} = \frac{71}{28} = 2.536 = \mathbf{2.5}$$

NOTES

1. The extensions in the *xf* column are the same subtotals as we found previously.
2. The two column totals, Σf and Σxf, are the same values as were previously known as n and Σx, respectively. That is, $\Sigma f = n$, the sum of the frequencies is the number of pieces of data. The Σxf is the sum of all the pieces of data.
3. The f in the summation expression Σxf indicates that the sum was obtained with the use of a frequency distribution.

Let's return now to the sample of 50 exam scores. In Table 2-7 (see p. 54) you will find a grouped frequency distribution for the 50 scores. We will calculate the mean for the sample by using the frequencies and the class marks in the same manner as in the last example. Table 2-12 shows the extensions and totals. (*Note:* The class marks are now being used as representative values for the observed data. Hence the sum of the class marks is meaningless and is therefore not found.)

TABLE 2-12 Calculations and Totals for Exam Scores

Class Number	Class Limits	f	Class Marks x	xf
1	35– 41	1	38	38
2	42– 48	2	45	90
3	49– 55	2	52	104
4	56– 62	3	59	177
5	63– 69	7	66	462
6	70– 76	11	73	803
7	77– 83	9	80	720
8	84– 90	8	87	696
9	91– 97	6	94	564
10	98–104	1	101	101
		50		3755

The mean: Using formula (2-2) we obtain

$$\bar{x} = \frac{\Sigma xf}{\Sigma f} = \frac{3755}{50} = \mathbf{75.1}$$

To illustrate again what the mean is, consider a balance beam (a real number line) with a cube placed on the value for each response (class mark). That is, nine cubes will be placed on 80, and so on. When the cubes are all placed, one cube for each piece of data, the beam will balance when the fulcrum is at 75.1, the mean. (See Figure 2-15, p. 72.) Formula (2-2) gives an approximation of $\bar{x}$ when working with a grouped frequency distribution. Recall that the value of the class mark is used to represent the value of each piece of data that fell into that class interval.

FIGURE 2-15
Mean Is 75.1

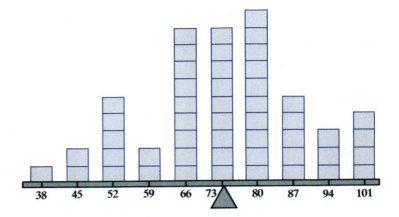

MEDIAN

median

The **median** $\tilde{x}$ (read "x tilde" or "median") is the value of the data that occupies the middle position when the data are ranked in order according to size. The data in Illustration 2-3, ranked in order of size, are 3, 3, 5, 6, and 8. The 5 is in the third, or middle, position of the five numbers. Thus, the median is 5. Notice that

$$\tilde{x} = 5$$

the median essentially "breaks" the ranked set of data into two subsets:

```
        5
    3       6
3   x̃       8
```

The *depth* (number of positions from either end), or *position,* of the median is determined by the formula

$$\text{depth of median} = d(\tilde{x}) = \frac{n+1}{2} \tag{2-3}$$

where 1 is the position of the smallest data value, and n is the number of pieces of data, or the position number, of the largest data value. The median's depth (or position) is then found by adding the position numbers of the smallest and largest data values and dividing by 2.

To find the median $\tilde{x}$ you must first have the data in ranked order. Next, determine the depth of the median and then count over the ranked data, starting from either end, finding the data in the $d(\tilde{x})$th position. The median will be the value of that data and will be the same regardless of which end of the ranked data (high or low) you count from. In fact, counting from both ends will serve as an excellent check.

When n is an odd number, the median will be the exact middle piece of data. In our example $n = 5$, and therefore the depth of the median is

$$d(\tilde{x}) = \frac{5 + 1}{2} = 3$$

That is, the median is the third number from either end in the ranked data, or $\tilde{x} = 5$.

However, if n is even, the depth of the median will always be a half-number. For example, let's look at a sample whose ranked data are 6, 7, 8, 9, 9, and 10. Here $n = 6$, and therefore the median's depth is

$$d(\tilde{x}) = \frac{6 + 1}{2} = 3.5$$

This says that the median is halfway between the third and fourth pieces of data. To find the number halfway between any two values, add the two values together and divide by 2. In this case, add the 8 and the 9, then divide by 2. **The median is 8.5**, a number halfway between the "middle" two numbers (see Figure 2-16). Notice that the median again "breaks" the ranked set of data into two subsets:

FIGURE 2-16
Median

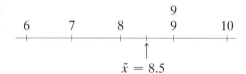

OTHER AVERAGES

mode

The **mode** is the value of x that occurs most frequently. In the set of data 3, 3, 5, 6, 8, the mode is 3. The mode in the sample 6, 7, 8, 9, 9, 10 is 9. In both cases these are the only numbers that occur more than once. If it happens that more than one of the values are tied for the highest frequency (number of occurrences), we say there is *no mode*. For example, in the sample 3, 3, 4, 5, 5, and 7, both the 3 and the 5 appear an equal number of times. There is no one value that appears most often. Thus this sample has no mode.

midrange

Another measure of central tendency is the **midrange**. A set of data will always have a lowest value L and a highest value H. The midrange is a number exactly midway between them. It is found by averaging the low and the high values:

$$\text{midrange} = \frac{L + H}{2} \tag{2-4}$$

For the sample 6, 7, 8, 9, 9, and 10, $L = 6$ and $H = 10$. Therefore, the midrange is

$$\frac{6 + 10}{2} = 8$$

The midrange is the numerical value halfway between the two extreme values, the low L and the high H.

The four measures of central tendency represent four different methods of describing the middle. These four values may be the same but more likely they will result in different values. For the sample data shown in Figure 2-**16**, the mean $\bar{x}$ is 8.2, the median $\tilde{x}$ is 8.5, the mode is 9, and the midrange is 8.

Case Study 2-4

"Average" Means Different Things

When it comes to convenience, few things can match that wonderful mathematical device called averaging.

How handy it is! With an average you can take a fistful of figures on any subject—temperatures, incomes, velocities, populations, light-years, hair-breadths, anything at all that can be measured—and compute one figure that will represent the whole fistful.

But there is one thing to remember. There are several kinds of measures ordinarily known as averages. And each gives a different picture of the figures it is called on to represent.

Take an example. Here are the annual incomes of ten families:

$54,000	$31,500
$39,000	$31,500
$37,500	$31,500
$36,750	$31,500
$35,250	$25,500

What would this group's "typical" income be? Averaging would provide the answer, so let's compute the typical income by the simpler and most frequently used kinds of averaging.

The arithmetic mean. When anyone cites an average without specifying which kind, you can probably assume that he has the arithmetic mean in mind.

It is the most common form of average, obtained by adding items in the series, then dividing by the number of items. In our example, the sum of the ten incomes divided by 10 is $35,400. The mean is representative of the series in the sense that the sum of the amounts by which the higher figures exceed the mean is exactly the same as the sum of the amounts by which the lower figures fall short of the mean.

The median. As you may have observed, six families earn less than the mean, four earn more. You might very well wish to represent this varied group by the income of the family that is right smack dab in the middle of the whole bunch. To do this, you need to find the median. It would be easy if there were 11 families in the group. The family sixth from highest (or sixth from lowest) would be in the middle and have the median income. But with ten families there is no middle family. So you add the two central incomes ($31,500 and $35,250 in this case) and divide by 2. The median works out to $33,375, less than the mean.

The midrange. The median, you will note, is the middle item in the series. Another number that might be used to represent the group is the midrange, computed by calculating the figure that lies halfway between the highest and lowest incomes. To find this figure, add the highest and lowest incomes ($54,000 and $25,500), divide by 2 and you have the amount that lies halfway between the extremes, $39,750.

The mode. So, three kinds of averages, and not one family actually has an income matching any of them. Say you want to represent the group by stating the income that occurs most frequently. That kind of representativeness is called a mode. In this example $31,500 would be the modal income. More families earn that income than any other. If no two families earned the same income, there would be no mode.

Four different averages, each valid, correct and informative in its way. But how they differ!

arithmetic mean	$35,400
median	$33,375
midrange	$39,750
mode	$31,500

And they would differ still more if just one family in the group were a millionaire—or one were jobless!

So there are three lessons to take away from today's class in averages. First, when you see or hear an average, find out which average it is. Then you'll know what kind of picture you are being given.

Second, think about the figures being averaged so you can judge whether the average used is appropriate.

And third, don't assume that a literal mathematical quantification is intended every time somebody says "average." It isn't. All of us often say "the average person" with no thought of implying a mean, median or mode. All we intend to convey is the idea of other people who are in many ways a great deal like the rest of us.

Source: Reprinted by permission from CHANGING TIMES Magazine (March 1980 issue). Copyright by The Kiplinger Washington Editors.

▼▲ EXERCISES

2.33 Consider the sample 2, 4, 7, 8, and 9. Find the following:
  a. the mean $\bar{x}$ **b.** the median $\tilde{x}$
  c. the mode   **d.** the midrange

2.34 Consider the sample 6, 8, 7, 5, 3, and 7. Find the following:
  a. the mean $\bar{x}$ **b.** the median $\tilde{x}$
  c. the mode   **d.** the midrange

2.35 Consider the sample 7, 6, 10, 7, 5, 9, 3, 7, 5, and 13. Find the following:
  a. the mean $\bar{x}$ **b.** the median $\tilde{x}$
  c. the mode   **d.** the midrange

2.36 Fifteen randomly selected college students were asked to state the number of hours they slept last night. The resulting data are 5, 6, 6, 8, 7, 7, 9, 5, 4, 8, 11, 6, 7, 8, 7. Find the following:
  a. the mean $\bar{x}$ **b.** the median $\tilde{x}$
  c. the mode   **d.** the midrange
(Retain these solutions for use in answering Exercise 2.58.)

2.37 Recruits for a police academy were required to undergo a test that measures their exercise capacity. The exercise capacity (measured in minutes) was obtained for each of 20 recruits:

25	27	30	33	30	32	30	34	30	27
26	25	29	31	31	32	34	32	33	30

  a. Find the mean, median, mode, and midrange.
  b. Construct a dot plot of these data and locate the mean, median, mode, and midrange on the graph.

2.38 Case Study 2-4 uses a sample of ten annual incomes to discuss the four averages. Calculate the mean, median, mode, and midrange for the ten incomes. Compare your results with those found in the article.

2.39 Say you are a budding entrepreneur. Here are the start-up costs for *Entrepreneur Magazine*'s 25 hottest franchises:

Subway	$ 32,400	McDonald's	$610,000
Jani-King	$ 6,500	Little Caesar's Pizza	$117,000
Hardee's	$694,000	Chem-Dry	$ 3,800
Arby's	$525,000	Electronic Realty Assoc.	$ 1,110
Kentucky Fried Chick.	$150,000	Jazzercise	$ 2,000
Service Master	$ 8,700	Intelligent Electronics	$150,000
Domino's Pizza	$ 76,500	Budget Rent A Car	$150,000
Dairy Queen	$375,000	Midas Int'l.	$182,000
Burger King	$333,600	H&R Block	$ 5,000
Coverall North America	$ 350	Choice Hotels Int'l.	$ 76,000
Nutri System	$ 60,000	Century 21 Real Estate	$ 15,000
Big Boy Restaurants	$450,000	Mail Boxes Etc.	$ 35,000
Re/Max International	$ 2,800		

Source: USA Today, May 6, 1991.

Section 2.3 ▼ MEASURES OF CENTRAL TENDENCY

Find:
    a. the mean start-up cost   **b.** the median start-up cost
    c. the midrange start-up cost

2.40 The May 13, 1991 *USA Snapshot* "How Much Governors Earn" summarized the annual salaries paid to the governors of the 50 states: New York pays the highest annual salary, $130,000, while Arkansas pays the lowest, $35,000, and the average annual salary is $79,800.
    a. What is the midrange?
    b. If the reported average of $79,800 is the mean salary, what can be concluded about the distribution of all 50 salaries since the midrange is so similar in value?

2.41 The following list of number of jobless individuals in each of the 11 most populous states was adapted from *USA Today* on May 6, 1991:

CA	1,096,000	FL	435,000	IL	388,000
MA	260,000	MI	464,000	NJ	261,000
NY	652,000	NC	196,000	OH	399,000
PA	423,000	TX	618,000		

Find the following statistics describing the number of jobless in these 11 states:
    a. total number   **b.** mean
    c. median   **d.** midrange

2.42 *The Atlantic Monthly* (November 1990, pages 78–106) contains an article entitled "The Case for More School Days." The number of days in the standard school year is given for several different countries and are as follows:

Country	n (days)/yr	Country	n (days)/year
Japan	243	New Zealand	190
West Germany	226–240	Nigeria	190
South Korea	220	British Columbia	185
Israel	216	France	185
Luxembourg	216	Ontario	185
Soviet Union	211	Ireland	184
Netherlands	200	New Brunswick	182
Scotland	200	Quebec	180
Thailand	200	Spain	180
Hong Kong	195	Sweden	180
England/Wales	192	United States	180
Hungary	192	French Belgium	175
Swaziland	191	Flemish Belgium	160
Finland	190		

Find the mean and median number of days per year of school for the countries listed. (Use the midpoint of the 226–240 interval for West Germany when computing your answers.)

2.43 a. Use formula (2-2) to find the mean for the following frequency distribution:

x	f
0	1
1	3
2	8
3	5
4	3

b. Find the median.   **c.** Find the mode.   **d.** Find the midrange.

2.44 The weight gains (in grams) for chicks fed on a high-protein diet were as follows:

Weight Gain	Frequency
12.5	2
12.7	6
13.0	22
13.1	29
13.2	12
13.8	4

a. Find the mean.   **b.** Find the median.
c. Find the mode.   **d.** Find the midrange.

2.45 a. Find the mean for the following grouped frequency distribution:

Class Limits	f
3– 5	2
6– 8	10
9–11	12
12–14	9
15–17	7

b. Find the median.   **c.** Find the mode.   **d.** Find the midrange.

2.46 a. Find the mean for the following grouped frequency distribution:

Class Limits	f
2– 5	7
6– 9	15
10–13	22
14–17	14
18–21	2

b. Find the median.   **c.** Find the mode.   **d.** Find the midrange.

Section 2.3 ▼ MEASURES OF CENTRAL TENDENCY

2.47 Find the mean for the set of speeds for the 55 cars given in Exercise 2.27. Use the frequency distribution found in answering Exercise 2.27. (Retain these solutions for use in answering Exercise 2.67, p. 92.)

2.48 A quality control technician selected 25 one-pound boxes from a production process and found the following distribution of weights (in ounces).

Weight	Frequency
15.95–15.97	2
15.98–16.00	4
16.01–16.03	15
16.04–16.06	3
16.07–16.09	1

Find the mean weight for this distribution. (Retain these solutions for use in answering Exercise 2.68, p. 92.)

2.49 The *News Edition* of *Computer Design* (January 1991, page 31) gives the distribution of salaries for Electrical Engineers having three different educational levels: BS/BS, MA/MS, and Ph.D. The average for a Ph.D. was $66,700; for a Master's, $59,000; and for a Bachelor's, $52,700.

Compute the mean for the following distribution of salaries for Electrical Engineers holding Bachelor's degrees:

Salary ($1000)	Number
20–34	5
35–49	20
50–64	45
65–79	20
80–94	10

2.50 The March/April 1991 issue of *Perspective on Aging* (published by the National Council on Aging) gives the following projection of the workforce for the year 2000:

Age	Millions
16–24	32.8
25–34	37.5
35–44	43.9
45–54	37.2
55–64	24.2
65+	34.9

Suppose the age group 16–24 had been 15–24, and the age group 65+ had been 65–74. Find the mean age of a worker in the year 2000 using this modified grouped frequency distribution.

2.51 The annual percentage rate charged for fixed 30-year mortgages varies considerably as reported in the *Democrat and Chronicle* of June 1, 1991:

9.86	9.81	9.26	9.00	9.05	9.40	9.53	9.74	9.38
9.73	9.50	9.50	9.85	9.13	9.48	9.00	9.67	9.51
9.45	9.37	9.93	9.33	9.73	9.57	9.99	9.10	9.45
9.51	9.13	9.46	9.47	9.40	9.46	9.55	9.30	9.50

The mean, median, and midrange were found using MINITAB. Verify the results.

```
MTB > SET C1
DATA> (data as above)
MTB > MEAN C1
   MEAN =    9.4750
MTB > MEDIAN C1
   MEDIAN =  9.4750
MTB > # K1= MIDRANGE
MTB > LET K1= (MAX(C1) + MIN(C1))/2
MTB > PRINT K1
K1     9.49500
MTB > STOP
```

2.52 Use a computer to find the mean, median, and midrange for the first-round golf scores given in Exercise 2.20. (If you use MINITAB, see Exercise 2.51 for a program.) (Retain these solutions for use in answering Exercise 2.74, p. 93.)

2.53 A survey was conducted to determine the number of radios per household. The results were summarized as follows:

x (n radios/household)	1	2	3	4	5	6	7
f (frequency)	20	35	100	90	65	40	5

MINITAB was used to find the mean of this frequency distribution. Verify the results.

```
MTB > READ C1 C2
      7 ROWS READ
MTB > NAME C1 'X'  C2 'F'
MTB > LET C3 = C1*C2
MTB > # K1 = MEAN
MTB > LET K1 = SUM(C3)/SUM(C2)
MTB > PRINT K1
K1    3.80282

MTB > STOP
```

2.54 Use a computer to calculate the mean salary for club managers discussed in Exercise 2.23. (If you use MINITAB, see Exercise 2.53 for a program.) (Retain this solution for use in answering Exercise 2.76, p. 93.)

2.4 ▼ Measures of Dispersion

measures of dispersion

Once the middle of a set of data has been determined, our search for information immediately turns to the **measures of dispersion** (spread). The measures of dispersion include the range, variance, and standard deviation. These numerical values describe the amount of spread, or variability, that is found among the data. *Closely grouped data will have relatively small values, and more widely spread-out data will have larger values for these various measures of dispersion.*

RANGE

range

The **range** is the simplest measure of dispersion. It is the difference between the highest- (largest) valued (H) and the lowest- (smallest) valued (L) pieces of data:

$$\text{range} = H - L \tag{2-5}$$

For the sample 3, 3, 5, 6, and 8, the range is $8 - 3 = 5$. The range tells us that the five pieces of data all fall within a distance of 5 units on the number line.

The other two measures of dispersion, variance and standard deviation, are actually measures of dispersion about the mean. To develop a measure of dispersion about the mean, let's look first at the concept of **deviation from the mean**. An individual value x deviates, or is located, from the mean by an amount equal to $(x - \bar{x})$. This deviation $(x - \bar{x})$ is zero when x is equal to the mean. The deviation $(x - \bar{x})$ is positive if x is larger than $\bar{x}$ and negative if x is less than $\bar{x}$.

deviation from the mean

Consider the sample 6, 3, 8, 5, and 3. Using formula (2-1), $\bar{x} = (\Sigma x)/n$, we find that the mean is 5. Each deviation is then found by subtracting 5 from each x value.

x	6	3	8	5	3
$x - \bar{x}$	1	-2	3	0	-2

Figure 2-17 shows the deviation from the mean for each value. We might suspect that the sum of all these deviations, $\Sigma(x - \bar{x})$, would serve as a measure of dispersion about the mean. However, that sum is exactly zero. As a matter

FIGURE 2-17
Deviations from the Mean

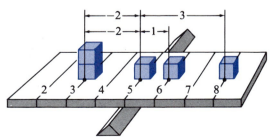

of fact, it will always be zero. Why? Think back to the definition of the mean (p. 69) and see if you can justify this statement.

If the sum of the deviations, $\Sigma(x - \bar{x})$, is always zero, it is not going to be of any value in describing a particular set of data. However, we still want to be able to use the idea of deviation about the mean, since the mean is the most common average used. Yet there is a neutralizing effect between the deviations of x values smaller than the mean (negative) and those values larger than the mean (positive). This neutralizing effect can be removed if we make all the deviations positive. By using the absolute value of $x - \bar{x}$, $|x - \bar{x}|$, we can accomplish this. For the previous illustration we now obtain the following absolute deviations:

x	6	3	8	5	3
$|x - \bar{x}|$	1	2	3	0	2

mean absolute deviation

The sum of these deviations is 8. Thus we define a value known as the **mean absolute deviation**:

$$\text{mean absolute deviation} = \frac{\Sigma |x - \bar{x}|}{n} \quad (2\text{-}6)$$

For our example the mean absolute deviation then becomes 8/5, or **1.6**. Although this particular measure of spread is not used too frequently, it is a measure of dispersion. It tells us the average distance that a piece of data is from the mean.

There is another way to eliminate the neutralizing effect. Squaring the deviations will cause all these values to be nonnegative (positive or zero). When these values are totaled, the result is positive. The sum of the squares of the deviations from the mean, $\Sigma(x - \bar{x})^2$, is used to define the variance.

VARIANCE

variance

The **variance**, s^2, of a sample is the numerical value found by applying the following formula to the data:

$$\text{sample variance} = s^2 = \frac{\Sigma(x - \bar{x})^2}{n - 1} \quad (2\text{-}7)$$

where n is the sample size, that is, the number of items in the sample. The variance of a sample is a measure of the spread of the data about the mean. The variance of our sample 6, 3, 8, 5, and 3 is found in Table 2-13.

TABLE 2-13 Computing s^2 for 6, 3, 8, 5, 3 Using Formula (2-7)

Step 1. x	Step 3. $x - \bar{x}$	Step 4. $(x - \bar{x})^2$
6	1	1
3	−2	4
8	3	9
5	0	0
3	−2	4
25	0	18

Step 2. $\bar{x} = \dfrac{\sum x}{n} = \dfrac{25}{5} = 5$

Step 5. $s^2 = \dfrac{\sum(x - \bar{x})^2}{n - 1} = \dfrac{18}{4} = \mathbf{4.5}$

Let's look at another sample: 1, 3, 5, 6, and 10. The calculations of the totals, the mean, and the variance are shown in Table 2-14.

TABLE 2-14 Computing s^2 for 1, 3, 5, 6, 10 Using Formula (2-7)

1. x	3. $x - \bar{x}$	4. $(x - \bar{x})^2$
1	−4	16
3	−2	4
5	0	0
6	1	1
10	5	25
25	0	46

2. $\bar{x} = \dfrac{\sum x}{n} = \dfrac{25}{5} = 5$

5. $s^2 = \dfrac{\sum(x - \bar{x})^2}{n - 1} = \dfrac{46}{4} = \mathbf{11.5}$

NOTES

1. The sum of all the xs is used to find $\bar{x}$.
2. The sum of the deviations, $\sum(x - \bar{x})$, is always zero, provided the exact value of $\bar{x}$ is used. Use this as a check in your calculations, as we did in Tables 2-13 and 2-14.
3. If a rounded value of $\bar{x}$ is used, then the $\sum(x - \bar{x})$ will not always be exactly zero. It will, however, be reasonably close to zero.

The last set of data is more dispersed than the previous set, and therefore its variance is larger. A comparison of these two samples is shown in Figure 2-18.

FIGURE 2-18 Comparison of Data

First sample: 3, 3, 5, 6, 8 $s^2 = 4.5$
Second sample: 1, 3, 5, 6, 10 $s^2 = 11.5$

STANDARD DEVIATION

standard deviation

The **standard deviation** of a sample, s, is the positive square root of the variance:

$$\text{sample standard deviation} = s = \sqrt{s^2} = \sqrt{\frac{\sum (x - \bar{x})^2}{n - 1}} \quad \text{(2-8)}$$

For the samples shown in Figure 2-16, the standard deviations are $\sqrt{4.5}$, or **2.1**, and $\sqrt{11.5}$, or **3.4**.*

NOTE The numerator for sample variance, $\sum (x - \bar{x})^2$, is often called the *sum of squares for x* and symbolized by $SS(x)$. Thus, formula (2-7) can be expressed

$$s^2 = \frac{SS(x)}{n - 1}, \quad \text{where } SS(x) = \sum (x - \bar{x})^2$$

The formulas for variance can be modified into other forms for easier use in various situations. For example, suppose that we have the sample 6, 3, 8, 5, and 2. The variance for this sample is computed in Table 2-15.

TABLE 2-15 Computing s^2 for 6, 3, 8, 5, 2 Using Formula (2-7)

1. x	3. $x - \bar{x}$	4. $(x - \bar{x})^2$
6	1.2	1.44
3	−1.8	3.24
8	3.2	10.24
5	0.2	0.04
2	−2.8	7.84
24	0	22.80

2. Using formula (2-1): $\bar{x} = \dfrac{\sum x}{n} = \dfrac{24}{5} = 4.8$

5. Using formula (2-7): $s^2 = \dfrac{\sum (x - \bar{x})^2}{n - 1} = \dfrac{22.80}{4} = 5.7$

*ROUND-OFF RULE

When rounding off an answer, let's agree to keep one more decimal place in our answer than was present in our original data. To avoid round-off buildup, round off only the final answer, not the intermediate steps. That is, do not use a rounded variance to obtain a standard deviation. In our previous examples, the data were composed of whole numbers; therefore, those answers that have decimal values should be rounded to the nearest tenth. See Appendix C for specific instructions on how to perform the rounding off.

Section 2.4 ▼ MEASURES OF DISPERSION

The arithmetic for this example has become more complicated because the mean contains nonzero digits to the right of the decimal point. However, the sum of squares for x can be rewritten:

$$\text{SS}(x) = \sum (x - \bar{x})^2 = \sum x^2 - \frac{(\sum x)^2}{n} \quad (2\text{-}9)$$

Combining formulas (2-7) and (2-9) yields the shortcut formula:

$$s^2 = \frac{\text{SS}(x)}{n-1} = \frac{\sum x^2 - \frac{(\sum x)^2}{n}}{n-1} \quad (2\text{-}10)$$

or equivalently,

$$s^2 = \frac{n(\sum x^2) - (\sum x)^2}{n(n-1)} \quad (2\text{-}10\text{a})$$

Formulas (2-10) and (2-10a) are called "shortcut" because they bypass the calculation of $\bar{x}$. Formulas (2-10) and (2-10a) can be used interchangeably. Use of formula (2-10) is preferred by many since SS(x) shows up in many other statistical formulas. Use of formula (2-10a) avoids the complex fraction format of formula (2-10). These shortcut formulas and the definition formula (2-7) can all be shown to be equivalent (see Exercises 2.77 and 2.78, p. 93).

The computations for s^2 using formulas (2-9) and (2-10) are performed as shown in Table 2-16.

TABLE 2-16 Computing s^2 with the Shortcut Formula

1. x	2. x^2
6	36
3	9
8	64
5	25
2	4
24	138

Using formula (2-9):

3. $\text{SS}(x) = 138 - \frac{(24)^2}{5} = 138 - 115.2 = 22.8$

$$s^2 = \frac{22.8}{4} = 5.7$$

$$s = \sqrt{5.7} = 2.4$$

The unit of measure for the standard deviation is the same as the unit of measure for the data. For example, if our data are in pounds, then the standard deviation s will also be in pounds. The unit of measure for variance might then be thought of as units squared. In our example of pounds, this would be *pounds squared*. As you can see, this unit has very little meaning.

The reason we have more than one formula to calculate variance is for convenience, not for confusion. As is so common in life, there is an easy way and a hard way to accomplish a goal. In statistics we have several formulas for variance, and if you use the appropriate formula your work will be kept to a minimum.

How do you decide which formula is applicable? With small samples there are only two formulas for variance,

$$\frac{\sum (x - \bar{x})^2}{n - 1} \quad \text{and} \quad \frac{\sum x^2 - \frac{(\sum x)^2}{n}}{n - 1}$$

Notice that the first formula involves the mean. If the mean is unknown or is a decimal (and thereby hard to work with), use the second formula. If the mean $\bar{x}$ is known and is a whole number (and thus convenient to work with), use the first formula. If you need to find both the mean and the variance of a sample, always find the mean first.

In order to use formula (2-10) to calculate the variance when the sample is in the form of a frequency distribution, we must determine the values of n, $\sum x$, and $\sum x^2$ using an extensions table. Recall that in Section 2.3 the values of n and $\sum x$ were found on an extensions table as the values $\sum f$ and $\sum xf$. To obtain $\sum x^2 f$ (it is the value of $\sum x^2$) we need only add a column to our extensions table. Let's look at Table 2-17, which is the same ungrouped frequency distribution used in Section 2-3 (Table 2-10).

TABLE 2-17
Ungrouped Frequency Distribution

x	f
1	5
2	9
3	8
4	6
	28

To calculate the sum, $\sum x^2$, we find the extension $x^2 f$ for each class. These extensions (see Tables 2-18 and 2-19) will have the same value as the sum of the squares of the like values of x. For example, $x = 3$ has a frequency of 8, meaning that the value 3 occurs 8 times in the sample. Thus, $x^2 f = (3^2)(8) = 72$, which is the same value obtained by adding the squares of eight 3s. These extensions serve as subtotals. When added, they result in the sum $\sum x^2 f$, the replacement for $\sum x^2$.

Section 2.4 ▼ MEASURES OF DISPERSION

TABLE 2-18 Σx and Σx^2

x	x²	x	x²
1	1	3	9
1	1	3	9
1	1	3	9
1	1	3	9
1	1	3	9
2	4	3	9
2	4	3	9
2	4	3	9
2	4	4	16
2	4	4	16
2	4	4	16
2	4	4	16
2	4	4	16
2	4	4	16
		71	209

TABLE 2-19 Extensions xf and x^2f

x	f	xf	x²f
1	5	5	5
2	9	18	36
3	8	24	72
4	6	24	96
Total	28	71	209

NOTES

1. The extension x^2f is found by multiplying x^2 by f or by multiplying xf by x.
2. The three column totals, Σf, Σxf, and Σx^2f, are the same values previously known as n, Σx, and Σx^2, respectively. That is, $\Sigma f = n$, the sum of the frequencies, is the number of pieces of data. The $\Sigma xf = \Sigma x$ and $\Sigma x^2 f = \Sigma x^2$. The f in the summation expression indicates only that the sum was obtained with the use of a frequency distribution. With these ideas in mind, the formula for the variance (2-10) becomes

$$s^2 = \frac{\Sigma x^2 f - \frac{(\Sigma xf)^2}{\Sigma f}}{\Sigma f - 1} \qquad (2\text{-}11)$$

The standard deviation is the positive square root of variance.

The variance and standard deviation for the example are found by using formula (2-11).

Variance:

$$SS(x) = 209 - \frac{(71)^2}{28}$$

$$= 209 - 180.036 = 28.964$$

$$s^2 = \frac{28.964}{27} = 1.073$$

$$= \mathbf{1.1}$$

Standard deviation:

$$s = \sqrt{s^2} = \sqrt{1.073} = 1.036$$

$$= \mathbf{1.0}$$

Let's return now to the sample of 50 exam scores. In Table 2-7 you will find a grouped frequency distribution for the 50 scores. We will calculate the variance and standard deviation for the sample by using the frequencies and the class marks in the same manner as in the preceding example. Table 2-20 shows the extensions and totals. (*Note:* The class marks are now being used as representative values for the observed data. Hence the sum of the class marks is meaningless and is therefore not found.)

TABLE 2-20 Calculations and Totals for Exam Scores

Class Number	Class Limits	f	Class Marks, x	xf	x^2f
1	35– 41	1	38	38	1,444
2	42– 48	2	45	90	4,050
3	49– 55	2	52	104	5,408
4	56– 62	3	59	177	10,443
5	63– 69	7	66	462	30,492
6	70– 76	11	73	803	58,619
7	77– 83	9	80	720	57,600
8	84– 90	8	87	696	60,552
9	91– 97	6	94	564	53,016
10	98–104	1	101	101	10,201
		50		3,755	291,825

The variance: Using formula (2-11) we have

$$s^2 = \frac{\sum x^2 f - \frac{(\sum xf)^2}{\sum f}}{\sum f - 1}$$

$$= \frac{291{,}825 - \frac{(3755)^2}{50}}{49} = \frac{9{,}824.5}{49} = \mathbf{200.5}$$

The standard deviation is
$$s = \sqrt{s^2} = \sqrt{200.5} = 14.1598 = \mathbf{14.2}$$

Case Study 2-5

Women in the State Houses

The percentage of women in state houses varies from state to state. The October USA Snapshot shows the extreme values 3 and 60 expressed as percentages, 2% and 33%. What is the population of interest for these statistics? What is the variable whose extreme values are 3 and 60? (See Exercise 2.84, p. 105.)

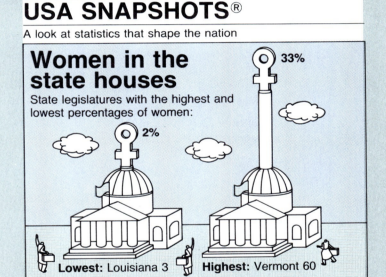

Source: Copyright 1990, USA TODAY. Reprinted with permission.

▼▲ EXERCISES

2.55 Consider the sample 2, 4, 7, 8, and 9. Find the following:
   a. range   **b.** variance s^2, using formula (2-7)   **c.** standard deviation s

2.56 Consider the sample 6, 8, 7, 5, 3, and 7. Find the following:
   a. range   **b.** variance s^2, using formula (2-7)   **c.** standard deviation s

2.57 Given the sample 7, 6, 10, 7, 5, 9, 3, 7, 5, 13, find the following:
   a. variance s^2, using formula (2-7)   **b.** variance s^2, using formula (2-10)
   c. standard deviation s

Chapter 2 ▼ DESCRIPTIVE ANALYSIS AND PRESENTATION OF SINGLE-VARIABLE DATA

2.58 Using the information from Exercise 2.36 concerning the sample of number of hours slept, find the following:
  a. variance s^2, using formula (2-7) b. variance s^2, using formula (2-10)
  c. standard deviation s

2.59 A company specializing in the manufacture of shafts is considering the purchase of a computer-controlled cutting machine. The company engineer tests machines from two different manufacturers. The diameters (in centimeters) of the shafts cut by the two machines were found to be

Manufacturer 1:   2.001, 2.000, 2.004, 1.998, and 1.997

Manufacturer 2:   2.002, 2.008, 1.995, 1.990, and 2.005

Compute the mean and standard deviation for each and comment on the results obtained from the two machines.

2.60 In an article entitled "Whatever Happened to Gold Prices?" (*Changing Times*, April 1991, page 69), it is pointed out that since mid-1981 gold prices have varied between $300 and $500 per troy ounce, but that recently the prices have varied between $360 and $420. Compare the variability in gold prices for the above-mentioned time periods by calculating their ranges.

2.61 Use formula (2-11) to find the variance and the standard deviation for the following frequency distribution:

x	f
0	1
1	3
2	8
3	5
4	3

2.62 A survey of medical doctors asked the number of children each had fathered. The results are summarized by this ungrouped frequency distribution:

Number of Children	0	1	2	3	4	6
Number of Doctors	15	12	26	14	4	2

Calculate the sample variance and standard deviation for the number of children in the doctors' families.

Section 2.4 ▼ MEASURES OF DISPERSION

2.63 Find the variance and the standard deviation for the following grouped frequency distribution:

Class Limits	f
3– 5	2
6– 8	10
9–11	12
12–14	9
15–17	7

2.64 Find the mean and variance for the following grouped frequency distribution:

Class Limits	f
2– 5	7
6– 9	15
10–13	22
14–17	14
18–21	2

2.65 The following distribution of commuting distances was obtained for a sample of Mutual of Nebraska employees.

Distance (Miles)	Frequency
1.0– 2.9	2
3.0– 4.9	6
5.0– 6.9	12
7.0– 8.9	50
9.0–10.9	35
11.0–12.9	15
13.0–14.9	5

Find the standard deviation for the commuting distances.

$ 2.66 A sheet metal firm employs several troubleshooters to make emergency repairs of furnaces. Typically, the troubleshooters take many short trips. For the purpose of estimating travel expenses for the coming year a sample of 20 travel-expense vouchers related to troubleshooting was taken. The following information resulted:

Dollar Amount on Voucher	Number of Vouchers
$0.01–10.00	2
10.01–20.00	8
20.01–30.00	7
30.01–40.00	2
40.01–50.00	1
Total in Sample	20

Calculate the mean and the standard deviation for these travel account dollar amounts.

 2.67 Find the standard deviation for the set of speeds for the 55 cars given in Exercise 2.27. Use the frequency distribution found in answering Exercise 2.27 and the results found in answering Exercise 2.47.

2.68 Find the standard deviation for the weights given in Exercise 2.48. (Use the results found in answering Exercise 2.48.)

2.69 Adding (or subtracting) the same number from each value in a set of data does not affect the measures of variability for that set of data.
   a. Find the variance of the following set of annual heating-degree-days data:

$$6017, 6173, 6275, 6350, 6001, \text{ and } 6300$$

   b. Find the variance of the following set of data [obtained by subtracting 6000 from each value in (a)]:

$$17, 173, 275, 350, 1, \text{ and } 300$$

 2.70 Calculate the standard deviation of the particle size for the Glidden data given in Exercise 2.16 by subtracting 3700 from each value.

2.71 Consider the following two sets of data:

| Set 1: | 46 | 55 | 50 | 47 | 52 |
| Set 2: | 30 | 55 | 65 | 47 | 53 |

Both sets have the same mean, which is 50. Compare these measures for both sets: $\Sigma(x - \bar{x})$, $\Sigma|x - \bar{x}|$, SS(x), and range. Comment on the meaning of these comparisons.

2.72 Comment on the following statement: "The mean loss for customers at First State Bank (which was not insured) was $150. The standard deviation of the losses was $-$125."

2.73 MINITAB was used to calculate the standard deviation for the mortgage rates given in Exercise 2.51. Verify the results.

```
MTB > SET  C1
DATA> (Data as above)
MTB > STDEV C1
   ST.DEV. =      0.26895
MTB > STOP
```

2.74 Use a computer to calculate the standard deviation for the first-round golf scores given in Exercise 2.20. (Exercise 2.52 used this same data.)

 2.75 MINITAB was used to calculate the standard deviation for the frequency distribution, number of radios per household, given in Exercise 2.53. Verify the results.

```
MTB > READ C1 C2
     7 ROWS READ
MTB > NAME C1 'X'  C2 'F'  C3 'XF'  C4 'XSQF'
MTB > LET C3 = C1*C2
MTB > LET C4 = C1*C1*C2
MTB > LET K1 = SQRT((SUM(C4)-(SUM(C3)**2/SUM(C2)))/(SUM(C2)-1))
MTB > PRINT K1
K1       1.38209
MTB > STOP
```

2.76 Use a computer to calculate the standard deviation for the frequency distribution of salaries for club managers given in Exercise 2.54.

2.77 Show that the shortcut formula for the sum of squares for x, $SS(x)$, and $\Sigma x^2 - (\Sigma x)^2/n$ [formula (2-9)] is equivalent to $\Sigma (x - \bar{x})^2$, which is the sum of squares by definition.

2.78 Show that the two forms of the shortcut formula for sample variance

$$\frac{\Sigma x^2 - \frac{(\Sigma x)^2}{n}}{n - 1}$$

formula (2-10), and

$$\frac{n(\Sigma x^2) - (\Sigma x)^2}{n(n - 1)}$$

formula (2-10a), are equivalent.

2.5 ▼ Measures of Position

measure of position

Measures of position are used to describe the location of a specific piece of data in relation to the rest of the sample. Quartiles and percentiles are two of the most popular measures of position.

QUARTILES

quartile

Quartiles are numbers that divide the ranked data into quarters; each set of data has three quartiles. The **first quartile**, Q_1, is a number such that at most one-fourth of the data are smaller in value than Q_1 and at most three-fourths are larger. The **second quartile** is the median. The **third quartile**, Q_3, is a number such that at most three-fourths of the data are smaller in value than Q_3 and at most one-fourth are larger (see Figure 2-19).

The procedure for determining the value of the quartiles is the same as that for percentiles and is shown in the following description of percentiles.

FIGURE 2-19
Quartiles

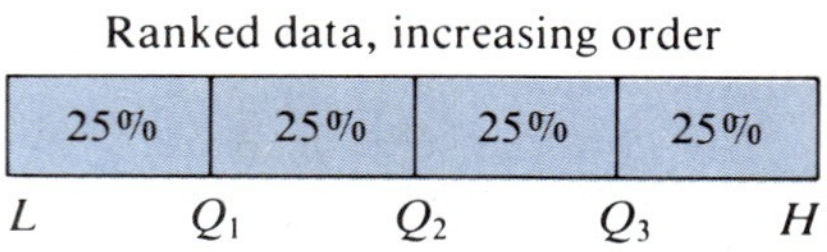

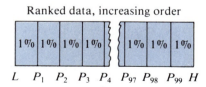

PERCENTILES

percentile

Percentiles are numbers that divide a set of ranked data into 100 equal parts; each set of data has 99 percentiles (see Figure 2-20). The kth percentile, P_k, is a number value such that at most k percent of the data are smaller in value than P_k and at most $(100 - k)$ percent of the data are larger (see Figure 2-21).

FIGURE 2-20
Percentiles

FIGURE 2-21
kth Percentile

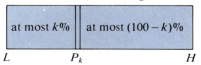

Section 2.5 ▼ MEASURES OF POSITION

NOTES

1. The 1st quartile and the 25th percentile are the same; that is, $Q_1 = P_{25}$. Also, $Q_3 = P_{75}$.
2. The median, the 2nd quartile, and the 50th percentile are all the same: $\tilde{x} = Q_2 = P_{50}$. Therefore, when asked to find P_{50} or Q_2, use the procedure for finding the median.

The procedure for determining the value of any kth percentile (or quartile) involves three basic steps.

Step 1: The data must be ranked: start at the lowest-valued piece of data and proceed toward the larger-valued data.

Step 2: Determine the depth of the kth percentile.

The depth of the kth percentile, $d(P_k)$, is determined by a two-step procedure. First, the value of $nk/100$ is calculated, n being the sample size and k being the identifying number for the percentile being found. Second, the value of $nk/100$ is then used to determine $d(P_k)$, the depth of P_k.

(a) If the calculated value of $nk/100$ is an integer, then $d(P_k) = nk/100 + 0.5$. For example, if $nk/100 = 23$, then $d(P_k) = 23.5$.

(b) If the calculated value of $nk/100$ is not an integer (that is, it contains a fraction), then $d(P_k)$ is equal to the next larger integer. For example, if $nk/100 = 17.2$, then $d(P_k) = 18$th.

Step 3: Determine the value of P_k.

The value of P_k is located by counting over the ranked data (found in step 1), starting at the lowest valued data, until the value in the $d(P_k)$th position is located.

(a) If $d(P_k)$ is an integer, the value of P_k will be the value of the data located.

(b) If $d(P_k)$ contains the fraction $\frac{1}{2}$ then the value of P_k is halfway between the $(nk/100)$th and the $[(nk/100) + 1]$th piece of data. That is, P_k is found by adding these two values and dividing by 2.

Optional technique when k is greater than 50: Subtract k from 100 and use $(100 - k)$ in place of k in step 2. For example, if $k = 80$, then use $100 - k$ $(100 - 80)$, which is 20. The depth of the 80th percentile is determined by using $k = 20$ and counting down from the largest-valued data to locate the value of P_k.

▼ **ILLUSTRATION 2-4**

A sample of 50 raw scores was taken from a population of raw scores for an elementary statistics final exam. These scores are listed in Table 2-21 in the order of collection. Find the first quartile Q_1, the 58th percentile P_{58}, and the third quartile Q_3.

TABLE 2-21
Raw Scores for Elementary Statistics Exam

60	47	82	95	88
72	67	66	68	98
90	77	86	58	64
95	74	72	88	74
77	39	90	63	68
97	70	64	70	70
58	78	89	44	55
85	82	83	72	77
72	86	50	94	92
80	91	75	76	78

SOLUTION The first step is to rank the data. A ranked list may be formulated, as shown in Table 2-22, or a graphic display showing the ranked data may be used. The stem-and-leaf is very handy for this purpose, especially if the leaf values are in numeric order (see Figure 2-22).

To find Q_1 we need to determine the depth of Q_1 by first finding $\frac{nk}{100}$.

$n = 50$ since there are 50 pieces of data

$k = 25$ since $Q_1 = P_{25}$

$$\frac{nk}{100} = \frac{(50)(25)}{100} = 12.5$$

Thus $d(Q_1) = 13$. So Q_1 is the thirteenth value, counting from L (see Table 2-22 or Figure 2-22). Thus $Q_1 = 67$.

To find P_{58} we need to determine its depth by first finding $\frac{nk}{100}$.

$n = 50$ since there are 50 pieces of data

$k = 58$ from P_{58}

$$\frac{nk}{100} = \frac{(50)(58)}{100} = 29.$$

TABLE 2-22
Ranked Data

39	64	72	78	89
44	66	72	80	90
47	67	74	82	90
50	68	74	82	91
55	68	75	83	92
58	70	76	85	94
58	70	77	86	95
60	70	77	86	95
63	72	77	88	97
64	72	78	88	98

13th position from L → 50 (row), 67 circled

13th position from H → 86 circled

29th and 30th positions from L → 77, 78

Section 2.5 ▼ MEASURES OF POSITION

FIGURE 2-22
Final Exam Scores

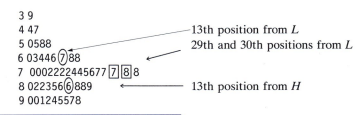

Thus $d(P_{58}) = 29.5$ from L. So P_{58} is the value halfway between the 29th and the 30th pieces of data counting from L (see Table 2-22 or Figure 2-22). Therefore,

$$P_{58} = \frac{77 + 78}{2} = \mathbf{77.5}$$

To find Q_3 we need to determine its location. Using the optional technique,

$$n = 50 \quad \text{and} \quad k = 25 \quad \text{since } Q_3 = P_{75}$$

(Notice that these calculations are the same ones that were completed for Q_1.) Thus $d(Q_3) = 13$. So Q_3 is the thirteenth value, counting from H (see Table 2-22 or Figure 2-22). Thus $Q_3 = \mathbf{86}$.

▲▲

midquartile

An additional measure of central tendency, the midquartile, can now be defined. The **midquartile** is the numerical value midway between the first quartile and the third quartile.

$$\text{midquartile} = \frac{Q_1 + Q_3}{2} \qquad (2\text{-}12)$$

▼ ILLUSTRATION 2-5

Find the midquartile for the set of 50 exam scores given in Illustration 2-4.

SOLUTION $Q_1 = 67$ and $Q_3 = 86$, as found in Illustration 2-4. Thus,

$$\text{midquartile} = \frac{67 + 86}{2} = \frac{153}{2} = \mathbf{76.5}$$

▲▲

NOTE The median, the midrange, and the midquartile are not necessarily the same value. They are each *middle* values, but by different definitions of the middle.

5-NUMBER SUMMARIES

A **5-number summary** is very effective in describing a set of data. It is easy information to obtain and is very informative to the reader. The 5-number summary is composed of

1. L, the smallest value in the data set
2. Q_1, the first quartile (also called P_{25}, 25th percentile)
3. $\tilde{x}$, the median
4. Q_3, the third quartile (also called P_{75}, 75th percentile)
5. H, the largest value in the data set

The 5-number summary for the set of 50 exam scores in Illustration 2-4 is

39	67	75.5	86	98
L	Q_1	$\tilde{x}$	Q_3	H

Notice that these five numerical values divide the set of data into four subsets, with one-quarter of the data in each subset. From this information we can observe how much the data is spread out in each of the fourths. An additional measure of dispersion can now be defined. The **interquartile range** is the difference between the first and third quartiles. It is *the range of the middle 50% of the data*.

This 5-number summary is even more informative when it is displayed on a diagram drawn to scale. One of the modern computer-generated graphic displays that accomplishes this is known as the *box-and-whisker* display.

BOX-AND-WHISKER DISPLAYS

The **box-and-whisker display** is a graphic 5-number summary of a set of data. Five numerical values (smallest, *lower hinge,* median, *upper hinge,* and largest) are located on a scale that can be either vertical or horizontal. The **box** is used to depict the middle half of the data that lies between the two hinges. The **whiskers** are line segments used to depict the other half of the data: one line segment represents the quarter of the data that is smaller in value than the lower hinge, and a second line segment represents the quarter of the data that is larger in value than the upper hinge. The hinges and the median break the ranked set of data into four subsets in much the same way that the median breaks the data into two subsets. The values used as **hinges** are often the same as the values of the quartiles, but, depending on the number of pieces of data, the value of a quartile and its corresponding hinge may be slightly different. This difference, when it occurs, results because the method used to determine the hinge is slightly different from that used for the quartiles.

The following procedure for finding the hinges is, however, quite similar to that for finding percentiles:

Step 1: Rank the data.

Step 2: The depth of the hinge, the number of positions from extreme values, is calculated by adding 1 to the *integer value* of the depth of the median and dividing by 2.

$$\text{depth of hinge} = d(H) = \frac{[d(\tilde{x})] + 1}{2}$$

integer value

NOTE The **integer value** of a number, symbolized by [ ], is the integer part only. Any fractional part is discarded. For example, 9.0 remains the value 9, and 9.5 becomes 9.

Let's consider the following set of 19 quiz scores (already ranked): 52, 62, 66, 68, 72, 74, 74, 76, 76, 76, 78, 78, 82, 82, 84, 86, 88, 92, 96

Depth of the median is 10 $\left(d(\tilde{x}) = \frac{n+1}{2} = \frac{19+1}{2} = 10 \right)$

Depth of the hinge is 5.5 $\left(d(\text{hinge}) = \frac{[d(\tilde{x})] + 1}{2} = \frac{10+1}{2} = 5.5 \right)$

lower hinge

upper hinge

The **lower hinge**, LH, is the value in the 5.5th position when you start counting from the L, the smallest value. (Recall that the half position number means that the value is halfway between two data values.) Thus $LH = 73$, the sum of the 5th and 6th values divided by 2. The **upper hinge**, UH, is the value in the 5.5th position when you start counting from the H, the largest value. Thus $UH = 83$, the sum of the 5th and 6th values in from H divided by 2.

The 5-number summary is $L = 52$, $LH = 73$, $\tilde{x} = 76$, $UH = 83$, $H = 96$. Notice the way the ranked set of data is "broken" into four subsets.

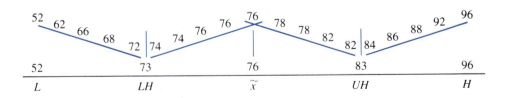

The box-and-whisker display for this set follows:

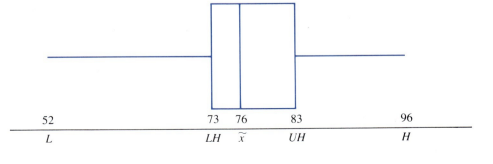

Examine the box-and-whisker display. Notice the amount of spread in each of the fourths: L to LH, LH to $\tilde{x}$, $\tilde{x}$ to UH, UH to H. One-fourth of the data is between 73 and 76 in value. One-half of the data is between 73 and 83.

Compare the box-and-whisker display on p. 99 to a dot plot and a histogram of the same data. Figures 2-23–2-25 contain the dot plot, the histogram, and the box-and-whisker diagram all drawn on the same horizontal scale.

FIGURE 2-23 Dot Plot: 19 Exam Scores

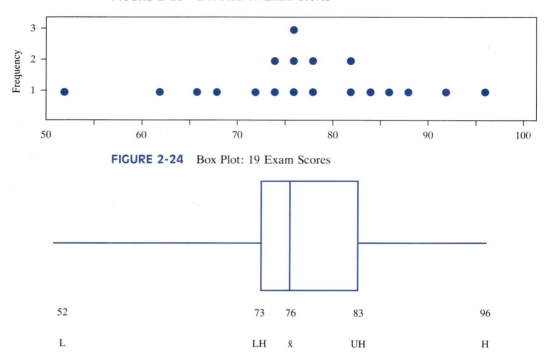

FIGURE 2-24 Box Plot: 19 Exam Scores

FIGURE 2-25 Histogram: 19 Exam Scores

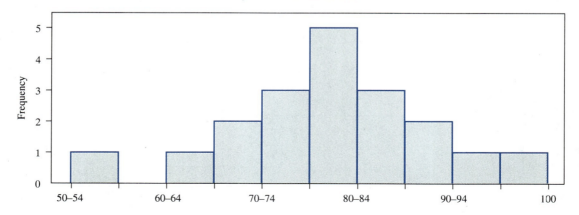

Section 2.5 ▼ MEASURES OF POSITION

Now let's consider a different set of 19 quiz scores:

52, 56, 58, 60, 64, 66, 68, 70, 72, 74, 76, 76, 78, 82, 84, 86, 86, 92, 96

The following figure shows this second set of quiz scores. Notice how the histogram is flatter than the histogram for the first set. Notice also the difference in appearance of the box-and-whisker display. The data in this second set are much more dispersed.

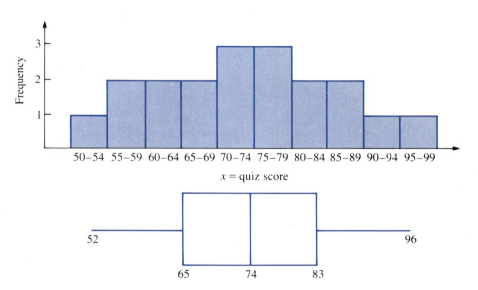

z-SCORE

z-, or standard, score

The *z-score*, also called the *standard score*, is the position a particular value of x has in terms of the number of standard deviations it is from the mean of the set of data to which it is being compared. The z-score is found by the formula

$$z = \frac{\text{piece of data} - \text{mean of the data}}{\text{standard deviation of the data}} = \frac{x - \bar{x}}{s} \tag{2-13}$$

▼ ILLUSTRATION 2-6

Find the standard scores for (a) 92 and (b) 72 with respect to a sample of exam grades that have a mean score of 75.9 and a standard deviation of 11.1.

SOLUTION

a. $x = 92$, $\bar{x} = 75.9$, and $s = 11.1$. Thus

$$z = \frac{92 - 75.9}{11.1} = \frac{16.1}{11.1} = \mathbf{1.45}$$

b. $x = 72, \bar{x} = 75.9$, and $s = 11.1$. Thus

$$z = \frac{72 - 75.9}{11.1} = \frac{-3.9}{11.1}$$

$$= -0.35$$

This means that the score 92 is approximately one and one-half standard deviations above the mean, while the score 72 is approximately one-third of a standard deviation below the mean. ▲▲

NOTES

1. Typically, the calculated value of z is rounded to the nearest hundredth.
2. z-scores typically range in value from approximately -3.00 to $+3.00$.

Since the z-score is a measure of relative position with respect to the mean, it can be used to help make a comparison of two raw scores that come from separate populations. For example, suppose that you want to compare a grade that you received on a test with a friend's grade on a comparable exam in her course. You received a raw score of 45 points; she obtained 72 points. Is her grade better? We need more information before we can draw such a conclusion. Suppose that the mean on the exam you took was 38 and the mean on her exam was 65. Your grades are both 7 points above the mean, so we still can't draw a definite conclusion. However, the standard deviation on the exam you took was 8 points, and it was 14 points on your friend's exam. This means that your score is $\frac{7}{8}$ of a standard deviation above the mean ($z = \frac{7}{8}$), whereas your friend's grade is only $\frac{1}{2}$ of a standard deviation above the mean ($z = \frac{1}{2}$). Since your score has the "better" relative position, we would conclude that your score was slightly higher than your friend's score. (Again, this is speaking from a relative point of view.)

Case Study 2-6

The stock market generates great quantities of statistical information every day. The daily Dow Jones Average statistics are shown in the March 1, 1991 Democrat and Chronicle *graph. The daily high, low, and closing values are shown graphically using a "3 value summary" with a "dot and whiskers" symbol. The dot represents the closing value for the day while the whiskers above and below show the variation in the day's average value. USA Today's "The Dow Wednesday" and "The Dow Thursday" trace the continuously changing average value during the market's hours for Wednesday, February 27, 1991, and Thursday, February 28, 1991. The two graphic presentations are quite different visually; however, they summarize the same information. Can you find the Wednesday and Thursday high, low, and closing on both the* Democrat and Chronicle *and the* USA Today *graphs? (See Exercise 2.83, p. 105.)*

STOCK MARKET MIXED

The Associated Press

The stock market was mixed yesterday in a busy but trendless session as traders tried to sort out prospects after the end of the Persion Gulf War.

The Dow Jones average of 30 industrials dropped 6.93 to 2,882.18, finishing February with a net gain of 145.79 points, or 5.33 percent.

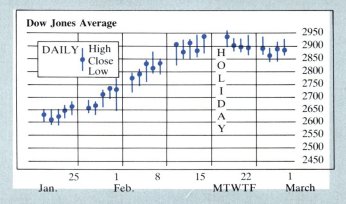

Source: Democrat and Chronicle, March 1, 1991 Rochester, N.Y., reprinted by permission.

BEHIND THE MARKET:

Stocks opened slightly higher in early trading, then the Dow surged nearly 30 points by noon on bargain hunting and program-related buying. (*Story, 1B.*)

Profit taking took hold as the Dow approached 2900, erasing most of the early gains. But a flurry of buy programs boosted blue chips in late trading.

Source: Copyright 1991, USA TODAY. Reprinted with permission.

BEHIND THE MARKET:

Stock prices climbed in early trading, the Dow rising more than 15 points on positive reaction to President Bush's declaration of a cease fire in the gulf war Wednesday evening.

But the investors still eager to lock in profits after the February rally soon sent stocks lower. There was widespread talk that some Kuwait and Saudi Arabian investors were selling stocks to raise cash to pay for war damage. By noon, the Dow was off nearly 16 points. It was unable to recover much by the end of the day.

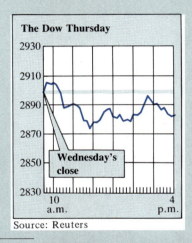

Source: Copyright 1991, USA TODAY. Reprinted with permission.

▼▲ EXERCISES

2.79 Find the following percentiles for the set of 50 final exam scores shown in Table 2-22.

    a. P_{35} **b.** P_{60} **c.** P_{95}

2.80 The following data are the yields, in pounds, of hops:

3.9	3.4	5.1	2.7	4.4
7.0	5.6	2.6	4.8	5.6
7.0	4.8	5.0	6.8	4.8
3.7	5.8	3.6	4.0	5.6

    a. Find the first and the third quartiles of the yield.
    b. Find the midquartile.
    c. Find the following percentiles: (1) P_{15}, (2) P_{33}, (3) P_{90}.

Section 2.5 ▼ MEASURES OF POSITION

2.81 A research study of manual dexterity involved determining the time required to complete a task. The time required for each of 40 handicapped individuals is as follows (data are ranked):

7.1	7.2	7.2	7.6	7.6	7.9	8.1	8.1	8.1	8.3
8.3	8.4	8.4	8.9	9.0	9.0	9.1	9.1	9.1	9.1
9.4	9.6	9.9	10.1	10.1	10.1	10.2	10.3	10.5	10.7
11.0	11.1	11.2	11.2	11.2	12.0	13.6	14.7	14.9	15.5

Find

    a. Q_1,   **b.** Q_2,   **c.** Q_3,   **d.** P_{95},   **e.** the 5-number summary.
    f. Draw the box-and-whisker display.

2.82 Consider the following set of ignition times that were recorded for a synthetic fabric:

30.1	30.1	30.2	30.5	31.0	31.1	31.2	31.3	31.3	31.4
31.5	31.6	31.6	32.0	32.4	32.5	33.0	33.0	33.0	33.5
34.0	34.5	34.5	35.0	35.0	35.6	36.0	36.5	36.9	37.0
37.5	37.5	37.6	38.0	39.5					

Find

    a. the median,   **b.** the midrange,   **c.** the midquartile,
    d. the 5-number summary.   **e.** Draw the box-and-whisker display.

2.83 The daily Dow Jones Average reports the daily high, low, and closing values. These are reported on the graphic presentations in Case Study 2-6 (page 103).
    a. "Stock Market Mixed" uses a dot-and-whiskers symbol to report the daily statistics. What does the dot represent? What do the whiskers represent? What are the values for Thursday, February 28, 1991?
    b. "Behind the Market" shows the Dow continuously throughout the day. What are the values for Thursday, February 28, 1991?

2.84 Case Study 2-5 (p. 89), "Women in the State Houses" shows the extreme values of 2% and 33%.
    a. What is the population of interest?
    b. What is the variable whose extreme values are 3 and 60?
    c. Louisiana has three women in its state legislature. How many legislators does Louisiana have?
    d. Why are the graphs drawn showing percentages?

2.85 A sample has a mean of 50 and a standard deviation of 4.0. Find the z-score for each value of x.
    a. $x = 54$ **b.** $x = 50$ **c.** $x = 59$ **d.** $x = 45$

2.86 An exam produced grades with a mean score of 74.2 and a standard deviation of 11.5. Find the z-score for each of the following test scores, x:
    a. $x = 54$ **b.** $x = 68$ **c.** $x = 79$ **d.** $x = 93$

2.87 A nationally administered test has a mean of 500 and a standard deviation of 100. If your standard score on this test was 1.8, what was your test score?

2.88 A sample has a mean of 120 and a standard deviation of 20.0. Find the value of x that corresponds to each of these standard scores:
  a. $z = 0.0$ **b.** $z = 1.2$ **c.** $z = -1.4$ **d.** $z = 2.05$

2.89 **a.** What does it mean to say that $x = 152$ has a standard score of $+1.5$?
  b. What does it mean to say that a particular value of x has a z-score of -2.1?
  c. In general, the standard score is a measure of what?

2.90 In a study involving mastery learning (*Research in Higher Education,* Vol. 20, No. 4, 1984, pages 491–498), 34 students took a pretest. The mean score was 11.04 and the standard deviation was 2.36. Find the z-score for scores of 9 and 15 on the twenty-question pretest.

2.91 Which x value has the higher value relative to the set of data from which it comes?

$$A: x = 85, \text{ where mean} = 72 \text{ and standard deviation} = 8$$

$$B: x = 93, \text{ where mean} = 87 \text{ and standard deviation} = 5$$

2.92 Which x value has the lower relative position with respect to the set of data from which it comes?

$$A: x = 28.1, \text{ where } \bar{x} = 25.7 \text{ and } s = 1.8$$

$$B: x = 39.2, \text{ where } \bar{x} = 34.1 \text{ and } s = 4.3$$

2.6 ▼ Interpreting and Understanding Standard Deviation

Standard deviation is a measure of fluctuation (dispersion) in the data. It has been defined as a value calculated with the use of formulas. But you may wonder what it really is. It is a kind of yardstick by which we can compare one set of data with another. This particular "measure" can be understood further by examining two statements, Chebyshev's theorem and the empirical rule.

Chebyshev's theorem

> **CHEBYSHEV'S THEOREM**
> The proportion of any distribution that lies within k standard deviations of the mean is at least $1 - (\frac{1}{k^2})$, where k is any positive number larger than 1. This theorem applies to any distribution of data.

This theorem says that within two standard deviations of the mean ($k = 2$) you will always find at least 75% (that is, 75% *or more*) of the data.

$$1 - \frac{1}{k^2} = 1 - \frac{1}{2^2} = 1 - \frac{1}{4} = \frac{3}{4} = 0.75$$

Figure 2-26 shows a mounded distribution that illustrates this.

If we consider the interval enclosed by three standard deviations on either side of the mean ($k = 3$), the theorem says that we will find at least $\frac{8}{9}$, or 89%, of the data $[1 - (\frac{1}{k^2}) = 1 - (\frac{1}{3^2}) = 1 - (\frac{1}{9}) = \frac{8}{9}]$, as shown in Figure 2-27.

FIGURE 2-26
Chebyshev's Theorem with $k = 2$

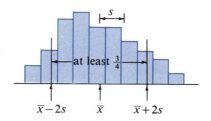

FIGURE 2-27
Chebyshev's Theorem with $k = 3$

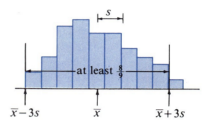

empirical rule

EMPIRICAL RULE

If a variable is normally distributed, then within one standard deviation of the mean there will be approximately 68% of the data. Within two standard deviations of the mean there will be approximately 95% of the data, and within three standard deviations of the mean there will be approximately 99.7% of the data. [This rule applies specifically to a normal (bell-shaped) distribution, but it is frequently applied as an interpretive guide to any mounded distribution.]

Figure 2-28, (p. 108) shows the intervals of one, two, and three standard deviations about the mean of an approximately normal distribution. Usually these proportions will not occur exactly in a sample, but your observed values will be close when a large sample is drawn from a normally distributed population.

The empirical rule can be used to determine whether or not a set of data is approximately normally distributed. Let's demonstrate this by working with the distribution of final exam scores that we have been using throughout this chapter. The mean, $\bar{x}$, was found to be 75.1, and the standard deviation, s, was 14.2. The interval from one standard deviation below the mean, $\bar{x} - s$, to one standard deviation above the mean, $\bar{x} + s$, is $75.1 - 14.2 = \mathbf{60.9}$ to $75.1 + 14.2 = \mathbf{89.3}$. This interval includes 61, 62, 63, ..., 88, 89. Upon inspection of the ranked data (Table 2-22), we see that 33 of the 50 data pieces, or 66%, lie within one standard deviation of the mean. Further, $\bar{x} - 2s = 75.1 - (2)(14.2) = 75.1 - 28.4 = 46.7$, $\bar{x} + 2s = 75.1 + 28.4 = 103.5$, and the interval from 46.7 to 103.5, two standard deviations about the mean, includes 48 of the 50 data pieces, or 96%. All 50 data, or 100%, are included within three standard deviations of the mean (from 32.5 to 117.7). This information can be placed in a table for comparison with the values given by the empirical rule (see Table 2-23).

FIGURE 2-28
Empirical Rule

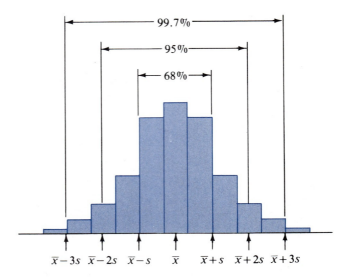

TABLE 2-23
Observed Percentages Versus the Empirical Rule

Interval	Empirical Rule Percentage	Percentage Found
$\bar{x} - s$ to $\bar{x} + s$	≈68	66
$\bar{x} - 2s$ to $\bar{x} + 2s$	≈95	96
$\bar{x} - 3s$ to $\bar{x} + 3s$	≈99.7	100

These percentages are reasonably close to those found by using the empirical rule. By combining this evidence with the shape of the histogram of the data, we can safely say that the sample data are approximately normally distributed.

If a distribution is approximately normal, it will be nearly symmetrical and the mean (the mean and the median are the same in a symmetrical distribution) will divide the distribution in half. This allows us to refine the empirical rule. Figure 2-29 shows this refinement.

FIGURE 2-29
Refinement of Empirical Rule

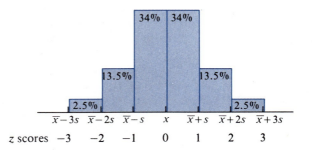

probability paper

There is also a graphic way to test for normality. This is accomplished by drawing a relative frequency ogive of the grouped data on **probability paper** (which can be purchased at your college bookstore). On this paper the vertical

Section 2.6 ▼ INTERPRETING AND UNDERSTANDING STANDARD DEVIATION

scale is measured in percentages and is placed on the right side of the graph paper. All the directions and guidelines given on pp. 58–59 for drawing an ogive must be followed. An ogive of the statistics exam scores is drawn and labeled on a piece of probability paper in Figure 2-30. To **test for normality**, first

test for normality

FIGURE 2-30
Ogive on Probability Paper: Final Exam Scores in Statistics

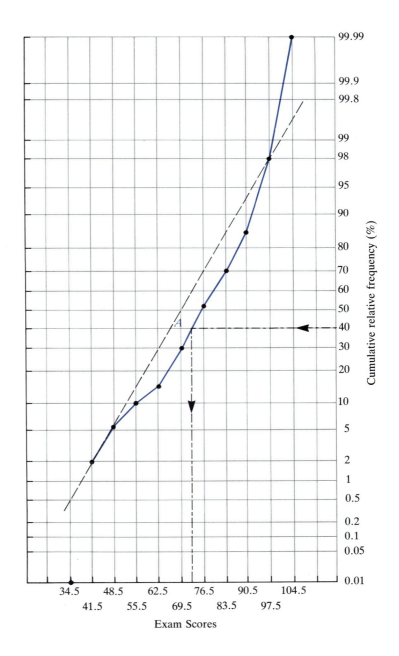

draw a straight line from the lower-left corner to the upper-right corner of the graph. Then, if the ogive lies close to this straight line, the distribution is said to be approximately normal. The ogive for an exactly normal distribution will trace the straight line.

The dashed line in Figure 2-30 is the straight-line test for normality. The ogive suggests that the distribution of exam scores is approximately normal. (*Warning:* This graphic technique is very sensitive to the scale used along the horizontal axis.)

The ogive is a very handy tool to use for finding percentiles of a frequency distribution. If the frequency distribution is drawn on probability paper, as in Figure 2-30, it is a very simple matter to determine any kth percentile. Locate the value k on the vertical scale along the right-hand side, and follow the horizontal line for k until it intersects the line of the ogive. Then follow the vertical line that passes through this point of intersection to the bottom of the graph. Read the value of x from the horizontal scale. This value of x is the value of the kth percentile. For example, let's find the value of P_{40} for the data shown in Figure 2-30. Then locate the value of k (40) along the right-hand side of the graph. Follow the horizontal line at k (40) until it intersects the graph of the ogive. Then determine the value of x that corresponds to this point of intersection (point A on Figure 2-30) by reading it from the scale on the x-axis: $P_{40} =$ **72**. This method can be used to find percentiles, quartiles, or the median. (*Note:* Because of the grouping into classes, the results may differ slightly from the answers obtained when using the ranked data.)

▼▲ EXERCISES

2.93 According to the empirical rule, practically all the data should lie between $(\bar{x} - 3s)$ and $(\bar{x} + 3s)$. The range also accounts for all the data.
   a. What relationship should hold (approximately) between the standard deviation and the range?
   b. How can you use the results of (a) to estimate the standard deviation in situations when the range is known?

2.94 The average clean-up time for a crew of a medium-size firm is 84.0 hours and the standard deviation is 6.8 hours. Assuming that the empirical rule is appropriate,
   a. what proportion of the time will it take the clean-up crew 97.6 or more hours to clean the plant?
   b. the total clean-up time will fall within what interval 95% of the time?

2.95 Chebyshev's theorem can be stated in a form equivalent to that given on p. 107. For example, to say that "at least 75% of the data fall within two standard deviations of the mean" is equivalent to stating that "at most, 25% will be more than two standard deviations away from the mean."
   a. At most, what percentage of a distribution will be three or more standard deviations from the mean?
   b. At most, what percentage of a distribution will be four or more standard deviations from the mean?

Section 2.6 ▼ INTERPRETING AND UNDERSTANDING STANDARD DEVIATION

2.96 The following information was generated from a study of stocks listed on the New York Stock Exchange and the American Stock Exchange:

	New York Exchange	American Exchange
Number of Stocks	24	25
Mean 52-week Price Difference	$ 5.57	$11.79
Standard Deviation	3.57	10.55

      a. At least 89% of the American Stock Exchange stocks had 52-week price differences within what range?
      b. At least 75% of the New York Stock Exchange stocks had 52-week price differences within what range?

2.97 An article entitled "Computer-enhanced algebra resources: Their effects on achievement and attitudes" (*International Journal of Math Education in Science and Technology,* 1980, Vol. 11, No. 4, pages 465–473) compared algebra courses that used computer-assisted instruction with courses that do not. The scores that the computer-assisted instruction group made on an achievement test consisting of 50 problems had the following summary statistics:

$$n = 57, \quad \bar{x} = 23.14, \quad s = 7.02$$

      a. Find the limits within which at least 75% of the scores fell.
      b. Find the standardized score for a raw score of 40.

2.98 An article entitled "Effects of Modifying Instruction in a College Classroom" (*Psychological Reports,* 1986, pages 965–966) describes an introductory psychology course in which 232 students received instruction consisting of lectures accompanied by unit testing to mastery and by student proctors. At the end of the course, students assigned ratings from 1 (significantly agree) to 5 (significantly disagree) to several statements. One such statement was: "Material was presented at an appropriate level and pace." The mean and standard deviation for the responses to this statement were 1.83 and 0.44. According to Chebyshev's theorem, at least what percentage of the ratings was a 1 or 2?

2.99 Sixty college freshmen were asked to give the number of children in their families (number of their brothers and sisters plus 1). The data collected follow:

1	6	3	5	5	3	4	1	2	7	3	2
3	4	5	3	1	3	2	1	4	4	2	2
3	9	4	3	3	5	3	5	7	3	1	1
3	5	2	6	4	3	3	3	3	3	2	3
4	3	5	7	3	2	1	2	3	2	4	3

      a. Construct an ungrouped frequency distribution of these data.
      b. Use the ungrouped frequency distribution found in (a) to find the mean and standard deviation of the data.

c. Find the values of $\bar{x} - s$ and $\bar{x} + s$.
d. How many of the 60 pieces of data have values between $\bar{x} - s$ and $\bar{x} + s$? What percentage of the sample is this?
e. Find the values of $\bar{x} - 2s$ and $\bar{x} + 2s$.
f. How many of the 60 pieces of data have values between $\bar{x} - 2s$ and $\bar{x} + 2s$? What percentage of the sample is this?
g. Find the values of $\bar{x} - 3s$ and $\bar{x} + 3s$.
h. What percentage of the sample has values between $\bar{x} - 3s$ and $\bar{x} + 3s$?
i. Compare the answers found in (f) and (h) to the results predicted by Chebyshev's theorem.
j. Compare the answers found in (d), (f), and (h) to the results predicted by the empirical rule. Does the result suggest an approximately normal distribution?

 2.100 On the first day of class last semester, 50 students were asked for the one-way distance from home to college (to the nearest mile). The resulting data were

6	5	3	24	15	15	6	2	1	3
5	10	9	21	8	10	9	14	16	16
10	21	20	15	9	4	12	27	10	10
3	9	17	6	11	10	12	5	7	11
5	8	22	20	13	1	8	13	4	18

a. Construct a grouped frequency distribution of the data by using 1–3 as the first class.
b. On probability paper, draw and label carefully the ogive depicting the distribution of one-way distances.
c. Does this distribution appear to have an approximately normal distribution? Explain.
d. Estimate the values of P_{10}, Q_1, $\tilde{x}$, Q_3, and P_{90} from the ogive drawn in (b).
e. Calculate the mean and standard deviation.
f. Determine the values of $\bar{x} \pm 2s$ and determine the percentage of data within two standard deviations of the mean using the ogive.

2.7 ▼ The Art of Statistical Deception

"There are three kinds of lies—lies, damned lies, and statistics." These remarkable words spoken by Disraeli (nineteenth-century British prime minister) represent the cynical view of statistics held by many people. Most people are on the consumer end of statistics and therefore have to "swallow" them.

GOOD ARITHMETIC, BAD STATISTICS

Let's explore an outright statistical lie. Suppose that a small business firm employs eight people who earn between $200 and $240 per week. The owner of the business pays himself $800 per week. He reports to the general public that the average wage paid to the employees of his firm is $280 per week. That may be an

Section 2.7 ▼ THE ART OF STATISTICAL DECEPTION

example of good arithmetic, but it is also an example of bad statistics. It is a misrepresentation of the situation, since only one employee, the owner, receives more than the mean salary. The public will think that most of the employees earn about $280 per week.

TRICKY GRAPHS

Graphic representation can be tricky and misleading. The frequency scale (which is usually the vertical axis) should start at zero in order to present a total picture. Usually, graphs that do not start at zero are used to save space. Nevertheless, this can be deceptive. Graphs in which the frequency scale starts at zero tend to emphasize the size of the numbers involved, whereas graphs that are *chopped off* may tend to emphasize the variation in the number without regard to the actual size of the number.

The graph in Figure 2-31 was presented in an annual report of a small business to show the amount of sales made by the company's three salespersons. Without careful study, a viewer of the graph might falsely conclude that B sold only a little more than half of what C sold. In reality, the graph should look like the one in Figure 2-32. The two graphs do not show any different information, they merely create different impressions. The variability among the three amounts of sales is the same in both figures. However, in Figure 2-31 the reader tends to *see* the variation, whereas in Figure 2-32 the reader tends to see that all the numbers are relatively large, and this de-emphasizes the variability that exists.

FIGURE 2-31
A Graph That Doesn't Start at Zero

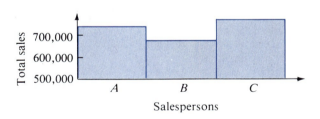

FIGURE 2-32
The Same Graph, but Starting at Zero

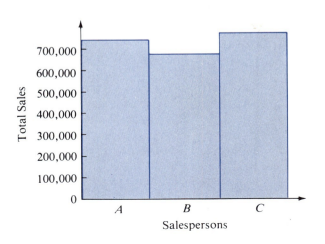

FIGURE 2-33
Air Pollution Chart:
Two-way Deception?

Air pollution

Reading represents average air-pollution levels for a 24-hour period ending at midnight this morning as measured at the Environmental Conservation Department's automatic monitoring station on Farmington Road near the Rochester-Irondequoit line.

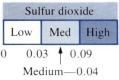

Sulfur dioxide

| Low | Med | High |
0   0.03 ↑ 0.09
Medium—0.04
(In parts per million): low, less than 0.03; medium, 0.03 to 0.08; high, 0.09 or above.

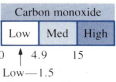

Carbon monoxide

| Low | Med | High |
0   ↑ 4.9   15
Low—1.5
(In parts per million): low, less than 4.9; medium, 5.0 to 14.9; high, 15 or above.

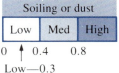

Soiling or dust

| Low | Med | High |
0   ↑ 0.4   0.8
Low—0.3
(In reflective units of dirt shade): low, less than 0.4; medium, 0.4 to 0.7; high, 0.8 or above.

From *The Times-Union*, Rochester, N.Y.; by permission

Figure 2-33 represents a daily report of air pollution as reported in a local newspaper. Notice that the attribute scales (Low, Medium, High) are approximately equal, but the numerical scale is not uniform. (The nonuniform scale is due to the pollutants' effect on a person's breathing.) Likewise, the arrow indicator points to the middle of the attribute level and is not in a position relative to the numerical scale. This seems to be a two-way deception.

The histogram in Figure 2-34 shows the number of drivers fatally injured in automobile accidents in New York State in 1968. Notice the "normal-appearing" distribution of ages of fatally injured drivers. The reason for grouping the ages in this manner is completely unclear. Certainly the age of fatally injured automobile drivers is not expected to be normally distributed. Notice too that the class width varies considerably. This histogram seems to indicate that 30- to 39-year-olds are the "poorest drivers." This may or may not be the case. Perhaps this age group is larger than any of the other groups. The group with the most drivers might be expected to have the highest number of driver fatalities. A vertical scale that showed the number of driver fatalities per thousand drivers might be more representative. An age grouping that created classifications with the same number of drivers in each class might also be more representative.

There are situations in which a change in the class width might be appropriate for the graphic representation. However, one must be extremely careful when calculating the values of the numerical measures. The annual salaries paid to the employees of a large firm is a situation for which a change in class width might be appropriate. A large firm has employees with salaries ranging from $8,000 to $80,000. Ten or 12 classes of equal width in this case would probably put 95% of the employees into the first two classes. To present a more accurate picture, one might consider a classification system similar to 8,000—9,000—10,000—12,000—15,000—18,000—24,000—30,000—50,000 and up. This system should

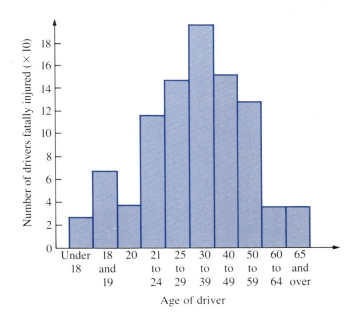

FIGURE 2-34
Histogram: Fatal Automobile Accidents

allow for more accurate description of the lower salaries and will also allow the higher salaries to be lumped together. There are, generally speaking, usually only a few very high salaries.

INSUFFICIENT INFORMATION

A realtor describes a mountainside piece of land as a perfect place for a nice summer home and tells you that the average high temperature during the months of July and August is 77 degrees. If the standard deviation is 5 degrees, this is not too bad. Assuming that these daily highs are approximately normally distributed, the temperature will range from three standard deviations below to three standard deviations above the mean temperature. The range would be from 62 to 92 degrees, with only a few days above 87 and a few days below 67. However, what she didn't tell you is that the standard deviation of temperature highs is 10 degrees, which means the daily high temperatures range from 47 degrees to 107 degrees. These extremes are not too pleasant!

The familiar advertisement that shows the "higher level of pain reliever" is a tricky graph with no statistics. Its statistical appearance is a result of the clever use of a graphic presentation. No variable, no units—just implications.

What it all comes down to is that statistics, like all languages, can be and is abused. In the hands of the careless, the unknowledgeable, or the unscrupulous, statistical information can be as false as "damned lies."

Case Study 2-7

Bankruptcy Blues

Filing for bankruptcy is becoming increasingly popular as a means of clearing debts. A quick glance at the bar graph in "Bankruptcy Blues" gives the impression that the number of filings in western New York has tripled in the last five years. However, when you look at the numbers, you see the number of filings has doubled. Clearly, doubling is bad enough; why draw a bar graph that implies tripling? How was this "deception" accomplished? (See Exercise 2.101.)

MORE PEOPLE FILE: BLAME RECESSION AND EASY CREDIT
By Janet Lively

Last Wednesday, a substance abuse counselor, a single mother on disability and the owner of a foreign automotive repair business gave up any hope of paying their bills.

They filed for bankruptcy, joining the more than 800 people and businesses who have asked for relief this year from the Western New York District of the U.S. Bankruptcy Court in Rochester. If filings continue at the same rate, 1991 will easily be another record year for the court.

Bankruptcies in Western New York*

- 1990/2,869
- 1989/2,489
- 1988/2,000
- 1987/1,860
- 1986/1,699
- 1985/1,461

*Monroe, Wayne, Livingston, Ontario, Steuben, Chemung, Schuyler, Yates and Seneca counties.

U.S. bankruptcies	
1990	782,960
1989	679,980
1988	613,606
1987	574,849
1986	530,008
1985	412,431

Source: U.S. Bankruptcy Court

David Cowies Democrat and Chronicle

Source: Democrat and Chronicle, Rochester, N.Y., April 1, 1991. Reprinted by permission.

EXERCISES

2.101 "Bankruptcy Blues" (Case Study 2-7) presents the number of bankruptcies filed in the Western New York District of the U.S. Bankruptcy Court. This "tricky" graph misrepresents the numbers reported. Explain why the graph is misrepresentative and why the graph was drawn this way.

2.102 Find an article or an advertisement that contains a graph that in some way misrepresents the information of statistics. Describe how this graph misrepresents the facts.

IN RETROSPECT

You have been introduced to some of the more common techniques of descriptive statistics. There are far too many specific types of statistics used in nearly every specialized field of study for us to review here. We have outlined the uses of only the most universal statistics. Specifically, you have seen several basic graphic techniques (circle and bar graphs, stem-and-leaf displays, histograms, and ogives) that are used to present sample data in picture form. You have also been introduced to some of the more common measures of central tendency (mean, median, mode, midrange, and midquartile), measures of dispersion (range, variance, and standard deviation), and measures of position (quartiles, percentiles, and z score).

You should now be aware that an average can be any one of five different statistics, and you should understand the distinctions among the different types of averages. The article "Average Means Different Things" (Case Study 2-4, p. 74) discusses four of the averages studied in this chapter. You might reread it now and find that it has more meaning and is of more interest. It will be time well spent!

You should also have a feeling for, and an understanding of, the concept of a standard deviation. You were introduced to Chebyshev's theorem and the empirical rule for this purpose.

The exercises in this chapter are extremely important; they will help you nail down the concepts studied before you go on to learn how to use these ideas in later chapters.

CHAPTER EXERCISES

2.103 Samples A and B are shown in the following table. Notice that the two samples are the same except that the 8 in A has been replaced by a 9 in B.

```
A:    2    4    5    5    7    8
B:    2    4    5    5    7    9
```

What effect does changing the 8 to a 9 have on each of the following statistics? Explain why.

    a. mean   **b.** median   **c.** mode   **d.** midrange
    e. range   **f.** variance   **g.** standard deviation

2.104 Samples C and D are shown in the following table. Notice that the two samples are alike except for two values.

C:	20	60	60	70	90
D:	20	30	70	90	90

What effect does changing the two 60s to 30 and 90 have on each of the following statistics?
  a. mean   b. median   c. mode   d. midrange
  e. range   f. variance   g. standard deviation

2.105 The cost of taxi fare from downtown to the airport varies greatly from city to city. As adapted from *USA Today* (May 7, 1991), the average cost of taxi fare for each of the 11 largest U.S. cities was

15   14   10   23   22.50   27   27.50   24   17   18   6.80

Find the following statistics describing the taxi fare from downtown to the airport in our largest cities:
  a. mean   b. median   c. standard deviation

2.106 The cost of staying overnight in a large city varies greatly from city to city. The following average cost figures (source Runzheimer International) are adapted from *USA Today,* May 7, 1991:

Atlanta	$138.24	Baltimore	$124.77
Boston	184.82	Chicago	185.83
Dallas	119.32	Detroit	133.27
Houston	129.27	Los Angeles	170.98
New York	267.33	Philadelphia	144.80
Washington	207.97		

Calculate the following statistics for the cost of an overnight stay in a big city:
  a. Construct a dot plot.
  b. Calculate the mean.
  c. Calculate the median.
  d. Calculate the standard deviation.
  e. On the dot plot in (a), identify the six cities on the East Coast (Atlanta, Baltimore, Boston, New York, Philadelphia, Washington).
  f. Calculate the mean for each group.
  g. Calculate the median for each group.
  h. Calculate the standard deviation for each group.
  i. How does the cost of an overnight stay in an East Coast city compare to that in the other cities?

2.107 The addition of a new accelerator is claimed to decrease the drying time of latex paint by more than 4%. Several test samples were conducted with the following percentage decreases in drying time:

5.2   6.4   3.8   6.3   4.1   2.8   3.2   4.7

CHAPTER EXERCISES

a. Find the sample mean.
b. Find the sample standard deviation.
c. Do you think these percentages average 4 or more? Explain.
(Retain these solutions for use in answering Exercise 9.69.)

2.108 Gasoline pumped from a supplier's pipeline is supposed to have an octane rating of 87.5. On 13 consecutive days a sample was taken and analyzed with the following results:

88.6	86.4	87.2	88.4	87.2	87.6	86.8
86.1	87.4	87.3	86.4	86.6	87.1	

a. Find the sample mean.
b. Find the sample standard deviation.
c. Do you think these readings seem to average 87.5? Explain.
(Retain these solutions for use in answering Exercise 9.14.)

2.109 Salaries for large groups of people are often summarized using the median. The median annual salaries for classroom teachers in each region of New York State, extracted from the *Democrat and Chronicle* (March 3, 1991), are

40,000	37,230	31,696	32,034	32,089	32,276	35,446
42,688	49,837	39,893	29,668	30,151	31,185	

a. Draw a dot plot of the median annual salaries.
b. Find the median value.
c. Find the first and third quartile values.
d. Draw a box plot of the median annual salaries.

2.110 What does room and board cost for full-time college students? A sample of New York State colleges and universities resulted in the following costs (adapted from the *Democrat and Chronicle,* September 28, 1990):

3,455	3,630	3,267	4,146
4,993	4,899	3,367	4,730
3,580	4,090	2,826	4,575
4,294	4,480	5,250	5,470

a. Draw a dot plot of the room and board costs. Find the following sample statistics:
b. mean
c. median
d. mode
e. midrange
f. first and third quartiles
g. Locate each of the sample statistics (b) through (f) on the dot plot drawn in (a).
h. Draw a box plot of the room and board costs.

2.111 "Light Fast Food Can Be Heavy on Percentage of Fat Calories" (*USA Today*, May 16, 1991) listed the percentage of calories derived from fat in commonly eaten fast food chicken items:

| 43 | 32 | 55 | 40 | 43 | 26 | 50 | 60 | 34 | 55 | 53 |

Find the following sample statistics for the variable, percentage of calories from fat:
  a. mean   b. standard deviation

2.112 The percentage earned on average shareholders' equity (the rate of return on common stock) for the Pacific Lighting Corporation for the last several years has been

| 8.6 | 9.8 | 10.9 | 11.8 | 13.7 | 13.9 | 14.6 |

Determine:
  a. the median rate of return for these years
  b. the mean rate of return for these years
  c. the mean absolute deviation for these data

2.113 The April unemployment rates for each of the last 12 years were (*USA Today*, May 6, 1991)

| 6.9 | 7.2 | 9.3 | 10.2 | 7.7 | 7.3 |
| 7.1 | 6.4 | 5.5 | 5.3 | 5.4 | 6.6 |

The August unemployment rates for each of the last 11 years were (*USA Today*, September 10, 1990)

| 7.7 | 7.4 | 9.8 | 9.5 | 7.5 | 7.1 |
| 6.9 | 6.0 | 5.6 | 5.3 | 5.6 |

  a. Calculate the mean and the standard deviation for both sets of unemployment rates.
  b. Compare the results found in (a).

2.114 In the May 1990 issue of the journal *Social Work*, Marlow reports the following results:

Number of Children Living at Home	Mexican-American Women	Anglo-American Women
0	23	38
1	22	9
2	17	15
3	7	9
4	1	1

Copyright 1990, National Association of Social Workers, Inc. *Social Work*.

a. Construct a histogram for each of the preceding distributions. Draw them on the same axis, using two different colors, so that you can compare their distributions.
b. Calculate the mean and standard deviation for the Mexican-American data.
c. Calculate the mean and standard deviation for the Anglo-American data.
d. Do these two distributions seem to be different? Cite specific reasons for your answer.

2.115 The numbers of pigs born per litter last year at Circle J Pig Farm are given in the following list:

11	8	13	14	11	8	14	7	11	13	10	8
7	4	9	11	10	12	6	12	5	10	9	10
12	10	3	6	9	10	10	13	12	9	12	7
9	5	12	7	11	9	4	8	13	12	11	
13	11	12	8	6	13	11	6	12	7	5	

a. Construct an ungrouped frequency distribution of these data.
b. How many litters of pigs were there last year?
c. What is the meaning and value of Σf?

2.116 The following set of data gives the ages of 118 known offenders who committed an auto theft last year in Garden City, Michigan:

11	14	15	15	16	16	17	18	19	21	25	36
12	14	15	15	16	16	17	18	19	21	25	39
13	14	15	15	16	17	17	18	20	22	26	43
13	14	15	15	16	17	17	18	20	22	26	46
13	14	15	16	16	17	17	18	20	22	27	50
13	14	15	16	16	17	17	19	20	23	27	54
13	14	15	16	16	17	18	19	20	23	29	59
13	15	15	16	16	17	18	19	20	23	30	67
14	15	15	16	16	17	18	19	21	24	31	
14	15	15	16	16	17	18	19	21	24	34	

a. Find the mean. b. Find the median. c. Find the mode.
d. Find Q_1 and Q_3. e. Find P_{10} and P_{95}.

2.117 A time-study analyst observed a packaging operation and collected the following times (in seconds) required for the operation to fill packages of a fixed volume:

| 10.8 | 14.4 | 19.6 | 18.0 | 8.4 | 15.2 | 11.0 | 13.3 | 23.1 | 17.2 |
| 17.1 | 16.1 | 12.6 | 14.6 | 9.1 | 12.0 | 11.6 | 14.7 | 12.4 | 17.3 |

a. Find the mean. b. Find the variance.
c. Find the standard deviation.

2.118 Ask one of your instructors for a list of exam grades (15 to 25 grades) from a class.
a. Find five measures of central tendency.
b. Find the three measures of dispersion.

c. Construct a stem-and-leaf display. Does this diagram suggest that the grades are normally distributed?
d. Find the following measures of location: (1) Q_1 and Q_3, (2) P_{15} and P_{60}, (3) the standard score z for the highest grade.

2.119 The stopping distance on a wet surface was determined for 25 cars each traveling at 30 miles per hour. The data (in feet) are shown on the following stem-and-leaf display:

```
 6 | 3 7 6 3 9
 7 | 4 2 0 1 1 2 0 5
 8 | 5 4 5 5 6
 9 | 4 1 0 0 5
10 | 5 4
```

Find the mean and standard deviation of these stopping distances.

2.120 Compute the mean and standard deviation for the following set of data. Then find the percentage of the data that is within two standard deviations of the mean.

```
1 | .4 .7 .1
2 | .4 .5
3 | .5 .0 .4 .1
4 | .4
5 | .5 .8 .7
6 | .8 .8 .2 .8 .6
7 | .5
8 |
9 | .4
```

2.121 The distribution of credit hours, per student, taken this semester at a certain college was found to be

Credit Hours	Frequency
3	75
6	150
8	30
9	50
12	70
14	300
15	400
16	1050
17	750
18	515
19	120
20	60

Find: **a.** mean   **b.** median   **c.** mode   **d.** midrange

CHAPTER EXERCISES

2.122 A survey of 32 workers at building 815 of Eastman Kodak Company was taken last May. Each worker was asked: "How many hours of television did you watch yesterday?" The results were as follows:

0	0	$\tfrac{1}{2}$	1	2	0	3	$2\tfrac{1}{2}$
0	0	1	$1\tfrac{1}{2}$	5	$2\tfrac{1}{2}$	0	2
$2\tfrac{1}{2}$	1	0	2	0	$2\tfrac{1}{2}$	4	0
6	$2\tfrac{1}{2}$	0	$\tfrac{1}{2}$	1	$1\tfrac{1}{2}$	0	2

a. Construct a stem-and-leaf display.   **b.** Find the mean.
c. Find the median   **d.** Find the mode.
e. Find the midrange.
f. Which one of the measures of central tendency would best represent the average viewer if you were trying to portray the typical television viewer? Explain.
g. Which measure of central tendency would best describe the amount of television watched? Explain.
h. Find the range.   **i.** Find the variance.
j. Find the standard deviation.

2.123 Given the following frequency distribution of ages, x, in years, of cars found in a parking lot,

x	1	2	3	4	5	6	7	8	9	10	11
f	16	20	18	12	9	7	6	3	5	3	1

a. draw a histogram of the data.
b. find the five measures of central tendency.
c. find Q_1 and Q_3.   **d.** find P_{15} and P_{12}.
e. find the three measures of dispersion (range, s^2, s).

2.124 An article in *Therapeutic Recreation Journal* reports a distribution for the variable, number of persistent disagreements. Sixty-six patients and their therapeutic recreation specialist each answered a checklist of problems with yes or no. Disagreement occurs when the specialist and the patient did not respond identically to an item on the checklist. It becomes a persistent disagreement if the item remains in disagreement after a second interview.
a. Draw a dot plot of this sample data.
b. Find the median number of persistent disagreements.
c. Find the mean number of persistent disagreements.
d. Find the standard deviation for the number of persistent disagreements.
e. Draw a vertical line on the dot plot at the mean.
f. Draw a horizontal line segment on the dot plot, whose length represents the standard deviation (start at the mean).

Number of Items	Frequency
0	2
1	2
2	4
3	10
4	7
5	9
6	8
7	11
8	7
9	3
10	1
11	2

Source: Data reprinted with permission of the National Recreation and Park Association, Alexandria, VA, from Pauline Petryshen and Diane Essex-Sorlie, "Persistent Disagreement Between Therapeutic Recreation Specialists and Patients in Psychiatric Hospitals," *Therapeutic Recreation Journal,* Vol. XXIV, Third Quarter, 1990.

2.125 Some of the 1991 Indianapolis 500 lineup of 33 race cars had the following qualifying speeds (*USA Today,* May 20, 1991):

224.1	222.4	221.8	221.4	220.9	219.8	219.1
218.9	218.3	218.1	217.6	216.8	224.5	223.9
223.1	222.8	219.5	219.0	218.6	218.2	216.7
214.9	218.7	218.4	218.0	215.6	214.8	214.0
217.4	215.3	214.6	214.0	213.8		

a. Form a grouped frequency distribution for the qualifying times. Use 213.0–214.9 for the limits of the first class.
b. Draw a frequency histogram for the distribution in (a).
c. Use the grouped frequency distribution in (a) to calculate the sample mean and standard deviation.

2.126 Earnings per share for 40 firms in the radio and transmitting equipment industry follow:

4.62	0.25	1.07	5.56	0.10	1.34	2.50	1.62
1.29	2.11	2.14	1.36	7.25	5.39	3.46	1.93
6.04	0.84	1.91	2.05	3.20	−0.19	7.05	2.75
9.56	3.72	5.10	3.58	4.90	2.27	1.80	0.44
4.22	2.08	0.91	3.15	3.71	1.12	0.50	1.93

a. Prepare a frequency distribution and a frequency histogram for these data.
b. Which class of your frequency distribution contains the median?

2.127 "Comparison of 'Light' Frozen Meals" (*The Boston Globe,* April 24, 1991) contained the following data on sodium content (mg) per frozen meal:

720	530	800	690	880	1050	340	810	760	300
400	680	780	390	950	520	500	630	480	940
450	990	910	420	850	390	600			

a. Construct a grouped frequency distribution.
b. Draw a horizontal frequency histogram of (a).
c. Calculate the mean and standard deviation using the frequency distribution in (a).

2.128 A chart, "Young Children Are Especially Vulnerable to the Consequences of Lead Poisoning" (*USA Today,* May 8, 1991), shows the number and percentage of children between the ages of 6 months and 5 years with unacceptable levels of lead in their blood. The data are from metro areas over 1 million. The number of affected children below has been extracted from this chart, rounded to the nearest thousand, and listed here in thousands, and the percentages have been rounded to the nearest whole percentage.

City	1	2	3	4	5	6	7	8	9	10
Number of Kids	53	82	99	123	54	372	66	95	49	124
Percent	33	47	58	69	62	62	55	65	51	43

City	11	12	13	14	15	16	17	18	19	20
Number of Kids	52	187	25	124	50	58	381	64	66	83
Percent	36	56	39	40	50	50	58	51	57	45

City	21	22	23	24	25	26	27	28	29	30
Number of Kids	61	65	491	95	222	39	87	44	64	34
Percent	36	58	75	71	62	32	59	45	41	41

City	31	32	33	34	35	36	37	38
Number of Kids	100	49	63	115	35	48	42	122
Percent	56	44	40	56	35	40	41	52

a. Construct a grouped frequency distribution for the variable x = number of affected kids per city.
b. Draw a frequency histogram for (a).
c. How many affected kids are there? In how many cities?
d. Calculate the mean and standard deviation for number of affected kids per city.
e. Construct a grouped frequency distribution for the variable x = percentage of affected kids per city.
f. Draw a frequency histogram for (e).
g. Calculate the mean and standard deviation for percentage of affected kids per city.

2.129 A survey was conducted to determine how many hours elementary school children spend viewing television per week. The results are as follows:

Hours Per Week	Frequency
5– 9	2
10–14	16
15–19	54
20–24	112
25–29	64
30–34	10

Find the mean and standard deviation for this distribution.

2.130 The grouped frequency distribution in the following table represents a sample of the ages of 120 randomly selected patients who were admitted to Memorial Hospital last June:

Class Limits	Frequency
0– 8	17
9–17	14
18–26	10
27–35	14
36–44	10
45–53	16
54–62	9
63–71	11
72–80	8
81–89	11

  a. Construct a relative frequency histogram that shows the ages of the 120 patients.
  b. Construct an ogive that shows the cumulative distribution of the ages.
  c. Calculate the mean age of the patients.
  d. Calculate the standard deviation of the ages.

2.131 Classify the data shown on the following stem-and-leaf diagram into a frequency distribution with classes of 3.0–5.1, 5.2–7.3, and so on:

```
 3 | .1  .6  .3
 4 | .0  .1  .0  .6
 5 | .7  .6  .5  .5  .8
 6 | .7  .0  .0  .5  .0  .5  .9
 9 | .1  .3  .1  .4
11 | .5  .7  .5
13 | .5
```

CHAPTER EXERCISES

Find: **a.** all class marks **b.** all relative frequencies
c. all cumulative frequencies

2.132 The lengths (in millimeters) of 100 brown trout in pond 2-B at Happy Acres Fish Hatchery on June 15 of last year were as follows:

15.0	15.3	14.4	10.4	10.2	11.5	15.4	11.7	15.0	10.9
13.6	10.5	13.8	15.0	13.8	14.5	13.7	13.9	12.5	15.2
10.7	13.1	10.6	12.1	14.9	14.1	12.7	14.0	10.1	14.1
10.3	15.2	15.0	12.9	10.7	10.3	10.8	15.3	14.9	14.8
14.9	11.8	10.4	11.0	11.4	14.3	15.1	11.5	10.2	10.1
14.7	15.1	12.8	14.8	15.0	10.4	13.5	14.5	14.9	13.9
10.1	14.8	13.7	10.9	10.6	12.4	14.5	10.5	15.1	15.8
12.0	15.5	10.8	14.4	15.4	14.8	11.4	15.1	10.3	15.4
15.0	14.0	15.0	15.1	13.7	14.7	10.7	14.5	13.9	11.7
15.1	10.9	11.3	10.5	15.3	14.0	14.6	12.6	15.3	10.4

a. Find the mean.
b. Find the median.
c. Find the mode.
d. Find the midrange.
e. Find the range.
f. Find Q_1 and Q_3.
g. Find the midquartile.
h. Find P_{35} and P_{64}.
i. Construct a grouped frequency distribution that uses 10.0–10.5 as the first class.
j. Construct a histogram of the frequency distribution.
k. Construct a cumulative relative frequency distribution.
l. Construct an ogive of the cumulative relative frequency distribution.
m. Find the mean of the frequency distribution.
n. Find the standard deviation of the frequency distribution.

2.133 The length of life of 220 incandescent 60-watt lamps was obtained and yielded the frequency distribution shown in the following table:

Class Limit	Frequency
500–599	3
600–699	7
700–799	14
800–899	28
900–999	64
1000–1099	57
1100–1199	23
1200–1299	13
1300–1399	7
1400–1499	4

a. Construct a histogram of these data using a vertical scale for relative frequencies.
b. Find the mean length of life.   **c.** Find the standard deviation.

2.134 The following table shows the age distribution of heads of families:

Age of Head of Family (years)	Number
20–24	23
25–29	38
30–34	51
35–39	55
40–44	53
45–49	50
50–54	48
55–59	39
60–64	31
65–69	26
70–74	20
75–79	16
	450

a. Find the mean age of the heads of families.
b. Find the standard deviation.

2.135 For a normal (or bell-shaped) distribution, find the percentile rank that corresponds to
a. $z = 2$ **b.** $z = -1$

2.136 For a normal (or bell-shaped) distribution, find the z-score that corresponds to the kth percentile:
a. $k = 20$ **b.** $k = 95$

2.137 An article entitled "The Use of Retesting as a Teaching Device in an Elementary Algebra Course" (*School Science and Mathematics*, 1973, pages 725–729) compares a control group to a retest group. The results on a standardized final may be displayed as follows:

Group	Mean	St. dev.
Control	25.48	6.33
Retest	26.46	5.74

a. Find the standardized score for a raw score of 30 for both groups.
b. Compare the test results of the two groups.

CHAPTER EXERCISES

2.138 Given the following information, compare an individual's achievement in relation to the rest of his class on a midterm exam and a final exam:

	His Score	Class Mean	Class Standard Deviation
Midterm	70	75	10
Final	50	55	15

On which exam did he perform better?

2.139 Bill and Bob are good friends, although they attend different high schools in their city. The city school system uses a battery of fitness tests to test all high school students. After completing the fitness tests, Bill and Bob are comparing their scores to see who did better in each event. They need help.

	Sit-ups	Pull-ups	Shuttle Run	50-Yard Dash	Softball Throw
Bill	$z = -1$	$z = -1.3$	$z = 0.0$	$z = 1.0$	$z = 0.5$
Bob	61	17	9.6	6.0	179 ft.
Mean	70	8	9.8	6.6	173 ft.
Standard Deviation	12	6	0.6	0.3	16 ft.

Bill received his test results in z-scores, whereas Bob was given raw scores. Since both boys understand raw scores, convert Bill's z-scores to raw scores in order to make an accurate comparison.

2.140 Mrs. Adam's twin daughters, Jean and Joan, are in fifth grade (different sections) and the class has been given a series of ability tests.

	Results	
Skill	Jean: z-Score	Joan: Percentile
Fitness	2.0	99
Posture	1.0	69
Agility	1.0	88
Flexibility	-1.0	35
Strength	0.0	50

If the scores for these ability tests are approximately normally distributed, which girl has the higher relative score on each of the skills listed? Explain your answers.

2.141 Chebyshev's theorem guarantees what proportion of a distribution will be included between the following?
  a. $\bar{x} - 2s$ and $\bar{x} + 2s$ **b.** $\bar{x} - 3s$ and $\bar{x} + 3s$

2.142 The empirical rule indicates that we can expect to find what proportion of the sample to be included between the following?
  a. $\bar{x} - s$ and $\bar{x} + s$ **b.** $\bar{x} - 2s$ and $\bar{x} + 2s$ **c.** $\bar{x} - 3s$ and $\bar{x} + 3s$

2.143 Why is it that the z-score for a value belonging to a normal distribution usually lies between −3 and 3?

2.144 The mean mileage per tire is 30,000 miles and the standard deviation is 2,500 miles for a certain tire.
  a. If we assume that the mileage is normally distributed, approximately what percentage of all such tires will give between 22,500 and 37,500 miles?
  b. If we assume nothing about the shape of the distribution, approximately what percentage of all such tires will give between 22,500 and 37,500 miles?

2.145 Manufacturing specifications are often based on the results of samples taken from satisfactory pilot runs. The following data resulted from just such a situation in which eight pilot batches were completed and sampled. The resulting particle sizes, in angstroms (where 1A = 10^{-8} cm.) were

| 3923 | 3807 | 3786 | 3710 | 4010 | 4230 | 4226 | 4133 |

  a. Find the sample mean.
  b. Find the sample standard deviation.
  c. Assuming that particle size has an approximately normal distribution, determine the manufacturing specs that bounds 95% of the particle sizes (that is, find the 95% interval, $\bar{x} \pm 2s$).

2.146 Delco Products, a division of General Motors, produces a bracket that is used as part of a power doorlock assembly. The length of this bracket is constantly being monitored. A sample of 30 power door brackets resulted in the following lengths (in millimeters):

11.86	11.88	11.88	11.91	11.88	11.88
11.88	11.88	11.88	11.86	11.88	11.88
11.88	11.88	11.86	11.83	11.86	11.86
11.88	11.88	11.88	11.83	11.86	11.86
11.86	11.88	11.88	11.86	11.88	11.83

Source: With permission of Delco Products Division, GMC.

  a. Without doing any calculations, what would you estimate for a sample mean?
  b. Construct an ungrouped frequency distribution.

c. Draw a histogram of this frequency distribution.
d. Use the frequency distribution and calculate the sample mean and standard deviation.
e. Determine the limits of the $\bar{x} + 3s$ interval and mark this interval on the histogram.
f. The product specification limits are 11.7–12.3. Does the sample indicate that production is within these requirements? Justify your answer.

VOCABULARY LIST

Be able to define each term. In addition, describe, in your own words, and give an example of each term. Your examples should not be ones given in class or in the textbook.

bell-shaped distribution
bimodal frequency distribution
box-and-whisker plot
Chebyshev's theorem
class
class boundary
class limits (lower and upper)
class mark
class width
cumulative frequency distribution
cumulative relative frequency distribution
depth
deviation from the mean
distribution
dot plot
empirical rule
5-number summary
frequency
frequency distribution
frequency histogram
grouped frequency distribution
histogram
integer value
interquartile range
lower hinge
mean
mean absolute deviation
measure of central tendency
measure of dispersion
measure of position

median
midquartile
midrange
modal class
mode
normal distribution
ogive
percentile
probability paper
quartile
range
rectangular distribution
relative frequency
relative frequency distribution
relative frequency histogram
sample
single-variable data
skewed distribution
standard deviation
standard score
stem-and-leaf display
summation
tally
test for normality
variability
variance
ungrouped frequency distribution
upper hinge
variance
x bar ($\bar{x}$)
z-score

KEY CONCEPTS

frequency
frequency distribution
histogram
mean
median

mode
percentile
standard deviation
standard score

QUIZ A

Answer "True" if the statement is always true. If the statement is not always true, replace the words in bold with words that make the statement always true.

2.1 The **mean** of a sample always divides the data into two equal halves—half larger and half smaller in value than itself.

2.2 A measure of **central tendency** is a quantitative value that describes how widely the data are dispersed about a central value.

2.3 The sum of the squares of the deviations from the mean, $\Sigma(x - \bar{x})^2$, will **sometimes** be negative.

2.4 For any distribution, the sum of the deviations from the mean equals **zero**.

2.5 The standard deviation for the set of values 2, 2, 2, 2, and 2 is **2**.

2.6 On a test John scored at the 50th percentile and Jim scored at the 25th percentile; therefore, John's test score was **twice** Jim's test score.

2.7 The frequency of a class is the number of pieces of data whose values fall within the **boundaries** of that class.

2.8 **Frequency distributions** are used in statistics to present large quantities of repeating values in a concise form.

2.9 The unit of measure for the standard score is always in **standard deviations**.

2.10 For a bell-shaped distribution, the range will be approximately equal to **six standard deviations.**

QUIZ B

The results of a consumer study completed at Corner Convenience Store were reported in the histogram on p. 133. Find the answer to each of the following.

2.1 **a.** What is the class width?
b. What is the class mark for the class 31–60?
c. What is the upper boundary for the class 61–90?
d. What is the frequency of the class 1–30?
e. What is the frequency of the class containing the largest observed value of x?
f. What is the lower class limit for the class 61–90?
g. How many pieces of data are shown in this histogram?

h. What is the value of the mode?
i. What is the value of the midrange?
j. Estimate the value of the ninetieth percentile, P_{90}.

The Amount of Time Needed to Check out at Corner Convenience Store

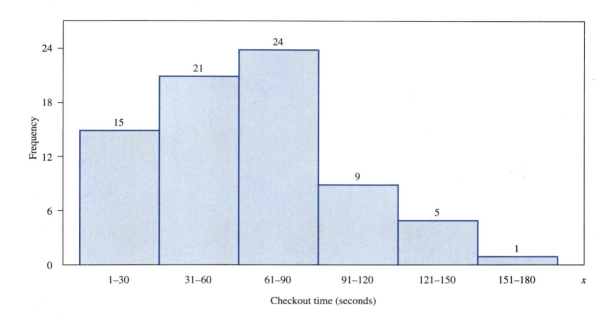

2.2 A sample of the purchases for several Corner Convenience Store customers resulted in the following sample data:

x = number of items purchased per customer

x	f
1	6
2	10
3	9
4	8
5	7

a. What does the "2" represent?
b. What does the "9" represent?
c. How many customers were used to form this sample?
d. How many items were purchased by the customers in this sample?
e. What is the largest number of items purchased by one customer?

Find each of the following (show formulas and work):
    f. mode   **g.** median   **h.** midrange
    i. mean   **j.** variance   **k.** standard deviation

2.3 Given the set of data: 4, 8, 9, 8, 6, 5, 7, 5, 8, find each of the following sample statistics:
   a. mean
   b. median
   c. mode
   d. midrange
   e. first quartile
   f. P_{40}
   g. variance
   h. standard deviation
   i. range

2.4 a. Find the standard score for the value $x = 452$ relative to its sample, where the sample mean is 500 and the standard deviation is 32.
   b. Find the value of x that corresponds to the standard score of 1.2, where the mean is 135 and the standard deviation is 15.

QUIZ C

Answer all questions.

2.1 The Corner Convenience Store kept track of the number of paying customers it had during the noon hour each day for 100 days. The following are the resulting statistics rounded to the nearest integer:

mean = 95   standard deviation = 12
median = 97   first quartile = 85
mode = 98   third quartile = 107
midrange = 93   range = 56

   a. The Corner Convenience Store served what number of paying customers during noon hour more often than any other number? Explain how you determined your answer.
   b. On how many days was there between 85 and 107 paying customers during the noon hour? Explain how you determined your answer.
   c. What was the greatest number of paying customers during any one noon hour? Explain how you determined your answer.
   d. For how many of the 100 days was the number of paying customers within three standard deviations of the mean ($\bar{x} \pm 3s$)? Explain how you determined your answer.

2.2 Mr. VanCott started his own machine shop several years ago. His business grew and has become very successful in recent years. Currently he employs 14 people, including himself, and pays the following annual salaries:

Owner, President	$80,000	Worker	$25,000
Business Manager	50,000	Worker	25,000
Production Manager	40,000	Worker	25,000
Shop Foreman	35,000	Worker	20,000
Worker	30,000	Worker	20,000
Worker	30,000	Worker	20,000
Worker	28,000	Worker	20,000

a. Calculate the four "averages": mean, median, mode, midrange.
   b. Draw a dot plot of these salaries and locate each of the four averages on it.
   c. Suppose you were the feature writer assigned to write this week's feature story on Mr. VanCott's machine shop, one of a series on local small businesses that are prospering. You plan to interview Mr. VanCott, his business manager, the shop foreman, and one of the newer workers. Which statistical average do you think each will give as their answer when asked, "What is the average annual salary paid to the employees here at VanCott's?" Explain why each has a different perspective and why this viewpoint will cause each to cite a different statistical average.
   d. What is there about the distribution of these annual salaries that causes the four "average values" to be so different from each other?

2.3 Create a set of data containing three or more values:
   a. where the mean is 12 and the standard deviation is zero.
   b. where the mean is 20 and the range is 10.
   c. where the mean, median, and mode are all equal.
   d. where the mean, median, and mode are all different.
   e. where the mean, median, and mode are all different and the median is the largest and the mode is the smallest of the three.
   f. where the mean, median, and mode are all different and the mean is the largest and the median is the smallest.

2.4 A set of test papers was machine scored. Later it was discovered that two points should be added to each score. Student A thought the mean score should also be increased by two points. Student B added that the standard deviation should also be increased by 2 points. Who's right? Justify your answer.

2.5 Student A stated that "both the standard deviation and the variance preserved the same unit of measurement as the observed data." Student B disagreed, arguing that "the unit of measurement for variance was a meaningless unit of measurement." Who's right? Justify your answer.

3 DESCRIPTIVE ANALYSIS AND PRESENTATION OF BIVARIATE DATA

Chapter Outline

3.1 Bivariate Data
Two variables are paired together for analysis.

3.2 Linear Correlation
Does an increase in the value of one variable indicate a **change** in the value of the other?

3.3 Linear Regression
The **line of best fit** is a mathematical expression for the relationship between two variables.

Changes of the Shinkansen Super Express Railway Noise since the Opening in 1964

Slow peak and sound exposure levels L_{Amax}, L_{AE} of the Shinkansen Super Express railway noise measured in 1976, 1981 and 1985 and the physical noise measure prescribed in the environmental quality standards were analyzed by a multivariate statistical method. The results of these surveys were compared. . . .

Figure 3 illustrates the scattergram between the measured L_{Amax} and the predicted values in this manner, where the correlation coefficient was 0.78. When predicting data in 1985 by applying the analytical results of data in 1976 in place of 1981, a coefficient of 0.75 was obtained. A conceivable reason for the decrease in the coefficient is a change in the contribution of each factor to L_{Amax} between 1976 and 1981. There might be a change in some factors or categories which was not explicitly included in this analysis. . . .

From category scores for distance it can be seen that attenuation of L_{AE} with distance is almost the same as that of L_{Amax}. The scores for train speed do not show a monotonically increasing tendency, unlike L_{Amax}. The scores for railway structure indicate, as in L_{Amax}, that an iron bridge increases L_{AE}, whereas a retaining wall decreases it. There is little difference between L_{Amax} and L_{AE} in the scores for train passby and noise barrier. The resemblance in analytical results of L_{Amax} and L_{AE} can be explained by the high correlation coefficient ($r = 0.97$) between them (Fig. 5).

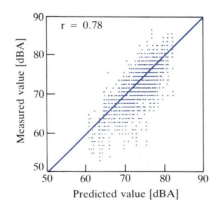

FIGURE 3. Scattergram between L_{Amax} measured in 1985 and predicted values from scores and average given by analysis of data measured in 1981.

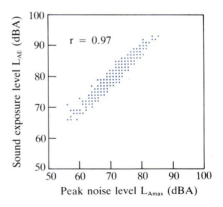

FIGURE 5. Scattergram of L_{Amax} vs L_{AE} measured in 1985.

Source: Yohzoh Okumura and Kazuhiro Kuno, *Applied Acoustics,* 30 (1990), 15–28.

Chapter Objectives

In the field of statistics there are many problems that require the analysis of more than one variable. In business, in education, and in many other fields, we often try to obtain answers to the following questions: Are these two variables related? If so, how are they related? Are these variables correlated? The relationships discussed here are not cause-and-effect relationships. They are only mathematical relationships that predict the behavior of one variable from knowledge about a second variable.

Let's look at a few specific illustrations.

▼ ILLUSTRATION 3-1

Are female voter opinions on the President's current position regarding tax increases related to political party affiliation? ▲▲

▼ ILLUSTRATION 3-2

As a person grows taller, he or she usually gains weight. Someone might ask, "Is there a relationship between height and weight?" ▲▲

▼ ILLUSTRATION 3-3

Students are constantly spending their time studying and taking examinations while going to school. We could ask, "Is it true that the more you study the higher your grade will be?" ▲▲

▼ ILLUSTRATION 3-4

Research doctors test new drugs (old ones, too) by prescribing different amounts and observing the responses of their patients. One question we could ask here is "Does the amount of drug prescribed determine the amount of recovery time needed by the patient?" ▲▲

▼ ILLUSTRATION 3-5

A high school guidance counselor would like to predict the academic success that students graduating from her school will have in college. In cases like this, the predicted value (grade point average at college) depends on many traits of the students: (1) how well they did in high school, (2) their intelligence, (3) their desire to succeed at college, and so on. ▲▲

These questions all require the analysis of bivariate data to obtain the answers. In this chapter we will take a first look at the techniques of tabling and graphing bivariate data and the descriptive aspects of correlation and regression analysis. The basic objectives of this chapter are (1) to be able to represent bivariate data in tabular and graphic form, (2) to gain an understanding of the distinction between the basic purposes of correlation analysis and regression analysis, and (3) to become familiar with the ideas of descriptive presentation. With these objectives in mind, we will restrict our discussion to the simplest and most basic form of correlation and regression analysis—the bivariate linear case.

3.1 ▼ Bivariate Data

bivariate data

Bivariate data consist of the values of two different response variables that are obtained from the same population element. Each of the two variables may be either qualitative or quantitative in nature. As a result, there are three combinations of variable types that can form bivariate data:

1. Both variables are qualitative (attribute).
2. One variable is qualitative (attribute) and the other is quantitative (discrete or continuous).
3. Both variables are quantitative (may be either discrete or continuous).

In this section we will be studying tabular and graphic methods for displaying each of these combinations of bivariate data.

TWO QUALITATIVE VARIABLES

When bivariate data result from two qualitative (attribute or categorical) variables, the data are often arranged on a cross-tabulation or contingency table. Let's look at an illustration.

▼ ILLUSTRATION 3-6

Thirty students from our college were identified and classified according to two variables: (1) gender (M/F) and (2) major (liberal arts, business administration, technology).

TABLE 3-1

Name	Gender	Major	Name	Gender	Major
Adams	M	LA	Kibby	M	BA
Argento	F	BA	Kleeberg	M	LA
Baker	M	LA	Light	M	BA
Bennett	F	LA	Linton	F	LA
Brock	M	BA	Lopez	M	T
Brand	M	T	McGowan	M	BA
Campbell	F	LA	Mowers	F	BA
Crain	M	T	Ornt	M	T
Cross	F	BA	Palmer	F	LA
Ellis	F	BA	Pullen	M	T
Feeney	M	T	Roffman	M	BA
Flanigan	M	LA	Sherman	F	LA
Hodge	F	LA	Small	F	T
Holmes	M	T	Tate	M	BA
Jopson	F	T	Vitale	M	LA

These 30 bivariate data can be summarized on a 2 × 3 cross-tabulation table (the 2 rows represent the two categories male and female, and the three columns represent the three major categories of liberal arts, business administration, and technology) by determining how many students fit categorically into each cell. Adams is male (M) and liberal arts (LA) and is classified in the cell in the first row, first column. See the tally mark in Table 3-2.

TABLE 3-2

		Major		
		Liberal Arts	Business Ad.	Technology
Gender	Male	|		
	Female			

The other 29 students are classified (tallied) in a similar fashion. The resulting 2 × 3 cross-tabulation (contingency table), Table 3-3, shows the frequency for each cross category of the two variables.

TABLE 3-3

		Major		
		Liberal Arts	Business Ad.	Technology
Gender	Male	⦀⦀ (5)	⦀⦀ | (6)	⦀⦀ || (7)
	Female	⦀⦀ | (6)	|||| (4)	|| (2)

The row and column totals shown in Table 3-4 are called *marginal totals* (or marginals). The total of the marginal totals is the grand total and is equal to n, the sample size.

TABLE 3-4

		Major			
		Liberal Arts	Business Ad.	Technology	Row Totals
Gender	Male	5	6	7	18
	Female	6	4	2	12
	Column totals	11	10	9	30

Contingency tables often show percentages (relative frequencies). These percentages can be based on the entire sample or on the subsample (row or column) classifications.

Section 3.1 ▼ BIVARIATE DATA

PERCENTAGES BASED ON THE GRAND TOTAL (ENTIRE SAMPLE) The contingency table shown in Table 3-4 can easily be converted to percentages of the grand total by dividing each frequency by the grand total and multiplying by 100. For example, 6 becomes 20% $[(\frac{6}{30}) \times 100 = 20]$. See Table 3-5.

TABLE 3-5

		Major			
		Liberal Arts	Business Ad.	Technology	Row Totals
Gender	Male	17%	20%	23%	60%
	Female	20	13	7	40
	Column totals	37%	33%	30%	100%

With the table expressed in percentages of the grand total you can easily see that 60% of the sample were male, 40% were female, 30% were technology majors and so on. These same statistics (numerical values describing sample results) can be shown in a bar graph (see Figure 3-1).

FIGURE 3-1 Bar Graph: Percentage Based on Grand Total

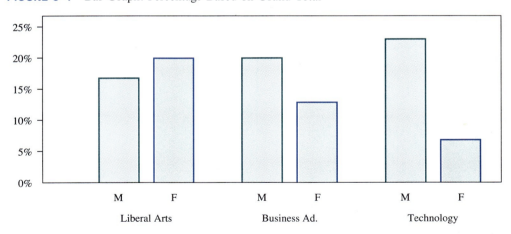

Table 3-5 and Figure 3-1 show the distribution of male liberal arts students, female liberal arts students, male business administration students, and so on relative to the entire sample.

PERCENTAGES BASED ON ROW TOTALS The entries in the same contingency table, Table 3-4, can be expressed as percentages of the row totals (or gender) by dividing each row entry by that row's total and multiplying by 100. Table 3-6 is based on row totals.

TABLE 3-6

		Major			Row Totals
		Liberal Arts	Business Ad.	Technology	
Gender	Male	28%	33%	39%	100%
	Female	50	33	17	100
	Column totals	37%	33%	30%	100%

From Table 3-6 we see that 28% of the male students were majoring in liberal arts while 50% of the female students were majoring in liberal arts.

FIGURE 3-2

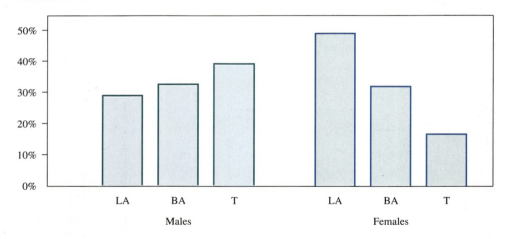

Table 3-6 and Figure 3-2 show the distribution of the three majors for male and female students separately.

PERCENTAGES BASED ON COLUMN TOTALS The entries in the contingency table in Table 3-4 can also be expressed as percentages of the column totals (or major) by dividing each column entry by that column's total and multiplying by 100. Table 3-7 is based on column totals.

TABLE 3-7

		Major			
		Liberal Arts	Business Ad.	Technology	Row Totals
Gender	Male	45%	60%	78%	60%
	Female	55	40	22	40
	Column totals	100%	100%	100%	100%

From Table 3-7 we see that 45% of the liberal arts students were male, while 55% of the liberal arts students were female.

FIGURE 3-3

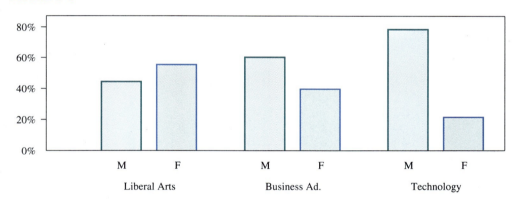

Table 3-7 and Figure 3-3 show the distribution of male and female students for each major separately.

ONE QUALITATIVE AND ONE QUANTITATIVE VARIABLE

When bivariate data result from one qualitative and one quantitative variable (the quantitative variable may be either discrete or continuous), they are often treated as separate sets of data viewed side by side.

▼ ILLUSTRATION 3-7

The distance required to stop a 3000-pound automobile on wet pavement was measured to compare the stopping capability of four different tread designs. Tires of each design were tested repeatedly on the same automobile on a controlled wet pavement.

TABLE 3-8 Stopping Distances for Four Tread Designs

Design A (n = 6)		Design B (n = 5)		Design C (n = 6)		Design D (n = 6)	
37	36	36	42	33	35	40	39
34	40	40	38	34	42	41	41
38	32	37		38	34	40	43

The design of the tread is a qualitative variable with four levels of response, and the stopping distance is a quantitative (continuous) variable. The distribution of the stopping distances for tread design A is to be compared with the distribution of stopping distances for each of the other tread designs. This comparison may be accomplished with both numerical and graphical techniques. Some of the available options are shown in Figure 3-4, Table 3-9, and Table 3-10.

FIGURE 3-4 Dot Plot and Box Plot of Stopping Distances

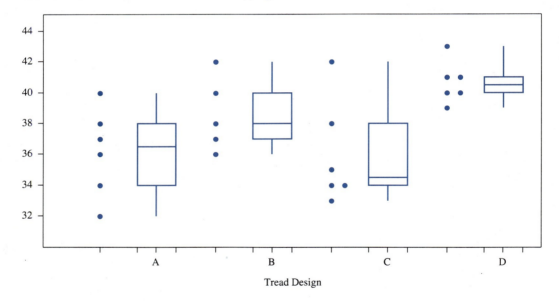

TABLE 3-9 5-Number Summary for Each Design

	Design A	Design B	Design C	Design D
High	40	42	42	43
Q_3	38	40	38	41
Median	36.5	38	35.5	40.5
Q_1	34	37	34	40
Low	32	36	33	39

TABLE 3-10 Mean and Standard Deviation for Each Design

	Design A	Design B	Design C	Design D
Mean	36.2	38.6	36.0	40.7
St. dev.	2.9	2.4	3.4	1.4

Much of the same kind of information described above can be demonstrated using many other statistical techniques such as stem-and-leaf displays, histograms, and other numerical statistics. When bivariate data is composed of one qualitative and one quantitative variable, the data is viewed as several samples, each set identified by the various levels of the qualitative variable. Each sample is described using the techniques from Chapter 2, and the results are displayed for side-by-side comparison.

TWO QUANTITATIVE VARIABLES

ordered pair

When the bivariate data is the result of two quantitative variables, it is customary to express the data mathematically as **ordered pairs**—let's call them x and y, where x is the value of the first variable and y is the value of the second variable. We write an ordered pair as (x, y). The data are said to be *ordered* because one value, x in this case, is always written first. They are said to be *paired* because for each x value there is a corresponding y value. For example, if x is height and y is weight, a height and corresponding weight are recorded for each person.

input and output variables

It is customary to call the **input variable** (sometimes called the *independent* variable) x and the **output variable** (sometimes called the *dependent* variable) y. The input variable x is measured or controlled in order to predict the output variable y. For example, in Illustration 3-4 the researcher can control the amount of drug prescribed. Therefore, the amount of drug would be referred to as x. In the case of height and weight, either variable could be treated as input, the other as output, depending on the question being asked. However, different results will be obtained from the regression analysis, depending on the choice made.

scatter diagram

In problems that deal with both correlation and regression analysis, we will present our sample data pictorially on a scatter diagram. A **scatter diagram** is a plot of all the ordered pairs of bivariate data on a coordinate axis system. The input variable x is plotted on the horizontal axis and the output variable y is plotted on the vertical axis.

NOTE When constructing a scatter diagram, it is convenient to construct scales so that the range of the y-values along the vertical axis is equal to or slightly shorter than the range of the x-values along the horizontal axis.

▼ ILLUSTRATION 3-8

In Mr. Chamberlain's physical fitness course, several fitness scores were taken. The following sample is the number of push-ups and sit-ups done by ten randomly selected students:

(27, 30), (22, 26), (15, 25), (35, 42), (30, 38),
(52, 40), (35, 32), (55, 54), (40, 50), (40, 43)

TABLE 3-11
Data for Push-ups and Sit-ups

	Student									
	1	2	3	4	5	6	7	8	9	10
Push-ups (x)	27	22	15	35	30	52	35	55	40	40
Sit-ups (y)	30	26	25	42	38	40	32	54	50	43

Table 3-11 shows these sample data, and Figure 3-5 shows a scatter diagram of these data.

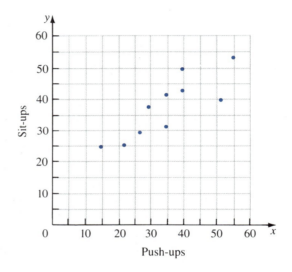

FIGURE 3-5
Scatter Diagram: Push-ups Versus Sit-ups for Ten Students

Case Study 3-1

Exploratory Data Analysis for Possibly Censored Data from Skewed Distributions

The data shown in Table 5 and summarized in Figure 1 is bivariate data; one variable is qualitative (batch) and the other is quantitative (lifetime). Essentially the data divide into ten separate sets of data being compared in a side-by-side manner. The box plots seem to demonstrate a shift in lifetime from batch to batch. Explain how this shift is shown (see Exercise 3.17, p. 157).

In recent years the ideas and techniques of exploratory data analysis (EDA) have received considerable attention in the statistics literature; see, for example, Tukey (1977) and the books edited by Hoaglin *et al.* (1983, 1985). One approach in EDA is based on selected order statistics from the data or from suitable subsets of the data. These can be used not only to investigate and display location, spread and skewness properties of the data, but also in

FIGURE 1
MINITAB Box Plot for Example 3

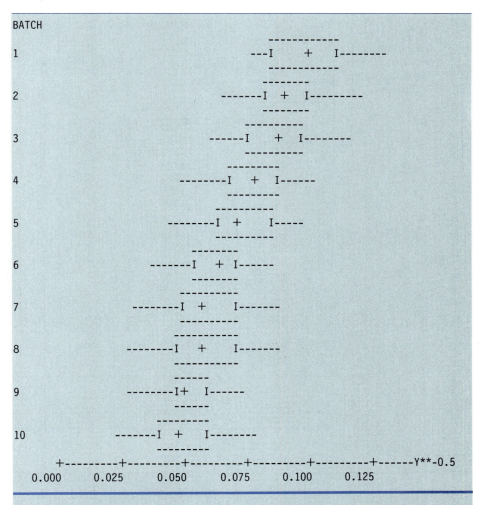

techniques such as preliminary screening for outliers and choosing appropriate transformations of the data. The emphasis is on resistant methods. . . .

EXAMPLE 3

Table 5 gives the lifetimes y in units of 1000 cycles of 200 steel specimens in 10 randomly assigned batches of 20 observations. Each batch has been subjected to a different stress amplitude. The source is Maennig (1967). Here an initial examination of the data is revealing. Box plots of the raw data show increasing medians and spreads with decreasing stress amplitude, . . . Fig. 1 shows the MINITAB Box plot for $1/\sqrt{y}$ against batch number. . . .

TABLE 5 Lifetimes y of 10 Batches of Steel Specimens Under Various Stress Amplitudes

y (× 1000 cycles) for the following batches and stress amplitudes: Batch									
1	2	3	4	5	6	7	8	9	10
Stress Amplitude									
38.5	38.0	37.5	37.0	36.5	36.0	35.5	35.0	34.5	34.0
51	59	65	83	89	117	112	115	140	146
57	66	71	98	105	127	125	129	155	159
60	69	78	100	108	141	133	143	169	168
67	80	84	104	118	151	156	169	174	224
68	87	89	110	119	162	166	177	218	246
69	90	93	111	121	173	168	178	248	253
75	97	98	122	130	181	173	218	265	291
76	98	103	125	133	186	202	230	293	326
82	99	105	132	152	192	227	271	321	358
83	100	109	136	164	198	247	280	326	385
87	107	113	141	170	203	253	285	348	397
95	109	118	143	181	209	261	305	350	425
106	117	124	146	182	218	285	326	364	449
109	118	131	155	192	255	286	342	374	498
111	125	147	165	199	262	309	381	397	532
119	128	156	194	211	288	365	431	426	610
122	132	171	200	238	295	442	495	461	714
128	158	182	201	273	309	559	568	504	763
132	177	199	251	324	394	702	734	738	987
140	186	220	318	398	585	852	1101	1063	1585

Source: A. C. Kimber, *Applied Statistics* 39:1 (1990), 21–30. Reprinted by permission.

Case Study 3-2

Visual Preference as a Test of Infant Word Comprehension

Table 3 displays two frequency distributions in the form of marginal totals. There is a frequency distribution for the variable, "14-month score" along the right side and another frequency distribution for the variable "20-month score" along the bottom of Table 3. Explain how the frequency numbers on Table 3 suggest "children with low comprehension scores at 14 months were most likely to be among those with low comprehension scores at 20 months." (See Exercise 3.18, p. 158.)

Most infants begin to understand words late in the first year and produce them early in the second year. Many theorists believe that early production vocabulary is a relatively small subset of comprehension vocabulary. . . . That is, a child who produces only 3 or 4 recognizable words may comprehend the meaning of 10 times that number.

The three experiments presented explore the usefulness of a visual preference technique for assessing word comprehension in infants. . . . In

> Experiment 2, word comprehension was assessed longitudinally, at 14 and 20 months, to see if individual differences at 14 months predicted word comprehension scores at 20 months.
>
> ### METHOD
>
> #### Subjects
>
> The original sample consisted of 100 infants assessed within 2 weeks of their 14th month. Ninety-one children returned to the laboratory for a second assessment at 20 months. . . .
>
> Table 3 displays the cross-tabulation of comprehension scores at 14 and 20 months. Word comprehension scores at 14 months ranged between 0 and 5, with an average of 2.31 words comprehended. At 20 months, scores ranged between 1 and 7, with an average of 3.17 words comprehended. . . . Table 3 suggests the nature of this relation: children with low comprehension scores at 14 months were more likely to be among those with low comprehension scores at 20 months.

TABLE 3
Distribution of Word Comprehension Scores at 14 and 20 Months

Words Comprehended at 14 Months	Words Comprehended at 20 Months							Total
	1	2	3	4	5	6	7	
0	1	0	1	2	0	0	0	4
1	4	8	8	1	1	1	0	23
2	5	5	5	3	3	3	1	25
3	2	1	6	2	3	0	2	16
4	1	3	6	2	2	0	0	14
5	0	1	1	0	1	1	1	5
Total	13	18	27	10	10	5	4	87

> *Source:* J. Steven Reznick, *Applied Psycholinguistics* 11 (1990), 145–166. Reprinted by permission.

▼▲ EXERCISES

 3.1 Vita Health Food Store conducted a market survey. The results are on the table on p. 150.
  a. Form a cross-tabulation of the variables, gender of customer and made a purchase (yes/no). Express the results in frequencies, showing marginal totals.
  b. Express the contingency table found in (a) in percentages based on the grand total.
  c. Express the contingency table found in (a) in percentages based on the marginal total for gender.
  d. Draw a bar graph showing the results from (c).

Vita Health Food Store Market Research Survey Findings

Person	Purchase	Gender	Age	Person	Purchase	Gender	Age
1	no	m	43	31	no	f	19
2	yes	m	58	32	yes	f	59
3	no	f	34	33	no	f	32
4	no	m	66	34	yes	m	46
5	yes	m	46	35	no	f	53
6	yes	f	52	36	no	m	40
7	no	f	18	37	yes	f	50
8	no	f	50	38	no	f	52
9	yes	f	39	39	yes	m	68
10	no	m	46	40	no	f	35
11	yes	m	62	41	yes	f	32
12	no	m	40	42	no	f	37
13	yes	m	47	43	no	f	39
14	no	m	64	44	no	m	28
15	no	m	21	45	no	m	68
16	no	f	42	46	yes	f	64
17	no	m	19	47	no	m	19
18	yes	f	69	48	yes	m	57
19	no	m	59	49	no	f	45
20	yes	m	54	50	no	f	22
21	no	f	48	51	yes	f	63
22	yes	f	49	52	no	m	35
23	yes	f	42	53	no	f	29
24	no	f	49	54	no	m	51
25	yes	f	61	55	yes	f	43
26	yes	m	56	56	no	m	30
27	no	f	44	57	yes	m	70
28	yes	f	60	58	no	f	21
29	yes	f	48	59	no	f	47
30	no	m	50	60	no	m	59

3.2 Using Vita's results on the table above,
   a. Form a cross-tabulation of the variables, age of customer (use categories: under 35, 35–50, over 50) and made a purchase (yes/no). Express the results in frequencies, showing marginal totals.
   b. Express the contingency table found in (a) in percentages based on the grand total.
   c. Express the contingency table found in (a) in percentages based on the marginal total for age.
   d. Draw a bar graph showing the results from (c).

3.3 A statewide survey was conducted to investigate the relationship between viewers preferences for ABC, CBS, NBC, or PBS for news information and their political party affiliations. The results are shown in tabular form:

	ABC	CBS	NBC	PBS
Democrat	200	200	250	150
Republican	450	350	500	200
Other	150	400	100	50

    a. How many viewers were surveyed?
    b. How many preferred to watch CBS?
    c. What percentage of the survey was Republican?
    d. What percentage of the Democrats preferred ABC?

3.4 Consider the accompanying contingency table, which presents the results of an advertising survey about the use of credit by Martan Oil Company customers:

Preferred Method of Payment	Number of Purchases at Gasoline Station Last Year					Sum
	0–4	5–9	10–14	15–19	20 and Over	
Cash	150	100	25	0	0	275
Oil company card	50	35	115	80	70	350
National or bank credit card	50	60	65	45	5	225
Sum	250	195	205	125	75	850

    a. How many customers were surveyed?
    b. How many customers preferred to use an oil company credit card?
    c. How many customers made 20 or more purchases last year?
    d. How many customers preferred to use an oil company credit card and made only between five and nine purchases last year?
    e. What does the 80 in the fourth cell in the second row mean?

3.5 The fixed rates on unsecured personal loans vary from bank to bank between banks and credit unions, and between credit unions. The following samples of rates were obtained from the *Democrat and Chronicle,* May 5, 1991. The data have been separated according to type of lending institution.

Banks	17.75	15.95	14.50	15.50	16.00	14.00	14.50
	15.00	14.25	17.80	15.95	14.95	14.00	14.75

Credit Unions	13.50	14.00	14.90	14.25	13.90	14.00
	13.95	12.95	14.40	10.50	18.00	

a. Draw a dot plot showing the rates charged by banks.
b. Draw a dot plot showing the rates charged by credit unions.
c. Find the median rate for both sets of rates.
d. Find the mean rate for both sets of rates.
e. Find the range for both sets of rates.
f. Find the standard deviation for both sets of rates.
g. In general, does it seem to make much difference which type of lending institution you borrow from? Compare the two sets of fixed rates.

3.6 An article entitled "Geographical Inquiry and Learning Reinforcement Theory," (*Journal of Geography,* May/June 1983, pages 121–122), compares a Personalized System of Instruction (PSI) with a formal mode of instruction. The following grade distributions are given for the two modes of instruction:

Mode of Instruction	n	Percent Grade Distribution					Pass-Fail
		A	B	C	D	F	
PSI	204	50	25	6	0	8	11
Formal	276	11	33	32	16	4	4

Code the grades A through F as A = 4, B = 3, C = 2, D = 1, F = 0. Remove the Pass-Fail scores and find the mean GPA for both modes of instruction.

3.7 MINITAB was used to construct stem-and-leaf displays that compare the heights of 32 randomly selected college students. Bivariate data: (weight, gender).

Code: 1 = male, 2 = female.

66, 2	67, 1	61, 2	68, 1	66, 1	70, 1	71, 1	71, 1
68, 1	73, 1	71, 1	73, 1	70, 1	62, 2	64, 2	65, 2
68, 1	69, 2	71, 1	65, 2	67, 1	63, 2	73, 1	69, 1
71, 1	75, 1	67, 2	63, 2	61, 2	65, 2	62, 2	60, 2

Verify the results.

```
MTB > Read data into C1 C2

     32 ROWS READ
MTB > # NOTE: GENDER CODE: 1 = MALE, 2 = FEMALE
MTB > NAME C1 'HEIGHT' C2 'GENDER'

MTB > STEM C1;
SUBC> INCREMENT = 1;
SUBC> BY C2.
```

Section 3.1 ▼ BIVARIATE DATA

```
Stem-and-leaf of HEIGHT    GENDER = 1     N = 18
Leaf Unit = 0.10

    1       66  0
    3       67  00
    6       68  000
    7       69  0
    9       70  00
    9       71  00000
    4       72
    4       73  000
    1       74
    1       75  0

Stem-and-leaf of HEIGHT    GENDER = 2     N = 14
Leaf Unit = 0.10

    1       60  0
    3       61  00
    5       62  00
    7       63  00
    7       64  0
    6       65  000
    3       66  0
    2       67  0
    1       68
    1       69  0
```

 3.8 The one-way distance from home to college for a sample of college commuters was obtained. Bivariate data: (distance, gender).

18, f	4, m	13, f	8, m	1, m	13, m	20, m	22, m
8, m	5, m	10, m	7, m	5, m	12, f	10, f	11, f
6, m	17, f	9, m	3, f	10, m	10, f	27, m	12, m
2, m	9, m	15, f	15, f	20, f	16, f	10, f	8, f

Use a computer to construct stem-and-leaf displays that compare the distances commuted by female and male students. (If you use MINITAB, see Exercise 3.7 for a program. Retain these solutions for use in answering Exercise 3.10.)

3.9 MINITAB was used to construct dot plots that compare the heights of 32 randomly selected college students. (Same data as in Exercise 3.7.)

Code: 1 = male, 2 = female.

66, 2	67, 1	61, 2	68, 1	66, 1	70, 1	71, 1	71, 1
68, 1	73, 1	71, 1	73, 1	70, 1	62, 2	64, 2	65, 2
68, 1	69, 2	71, 1	65, 2	67, 1	63, 2	73, 1	69, 1
71, 1	75, 1	67, 2	63, 2	61, 2	65, 2	62, 2	60, 2

Verify the results.

154   Chapter 3 ▼ DESCRIPTIVE ANALYSIS AND PRESENTATION OF BIVARIATE DATA

```
MTB > Read data into C1 C2

      32 ROWS READ
MTB > # NOTE: GENDER CODE: 1 = MALE, 2 = FEMALE
MTB > NAME C1 'HEIGHT' C2 'GENDER'

MTB > DOTPLOT C1;
SUBC> INCREMENT = 3;
SUBC> BY C2.
```

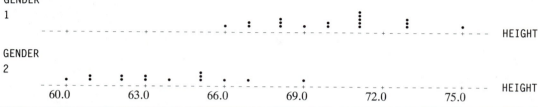

3.10 The one-way distance from home to college for a sample of college commuters was obtained. (Same data as in Exercise 3.8.)

18, f	4, m	13, f	8, m	1, m	13, m	20, m	22, m
8, m	5, m	10, m	7, m	5, m	12, f	10, f	11, f
6, m	17, f	9, m	3, f	10, m	10, f	27, m	12, m
2, m	9, m	15, f	15, f	20, f	16, f	10, f	8, f

Use a computer to construct dot plots that compare the distances commuted by female and male students. (If you use MINITAB, see Exercise 3.9 for a program.)

3.11 The accompanying data show the number of hours x studied for an exam and the grade y received on the exam. (y is measured in 10s; that is, $y = 8$ means that the grade, rounded to the nearest 10 points, is 80.) Draw the scatter diagram. (Retain this solution for use in answering Exercise 3.23, p. 164.)

x	2	3	3	4	4	5	5	6	6	6	7	7	7	8	8
y	5	5	7	5	7	7	8	6	9	8	7	9	10	8	9

3.12 An experimental psychologist asserts that the older a child is the fewer irrelevant answers he or she will give during a controlled experiment. To investigate this claim, the following data were collected. Draw a scatter diagram. (Retain this solution for use in answering Exercise 3.24, p. 164.)

Age (x)	2	4	5	6	6	7	9	9	10	12
Number of Irrelevant Answers (y)	12	13	9	7	12	8	6	9	7	5

Section 3.1 ▼ BIVARIATE DATA

3.13 A sample of 15 upperclassmen who commute to classes was selected at registration. They were asked to estimate the distance (x) and the time (y) required to commute each day to class (see the following table). Construct a scatter diagram depicting these data.

Distance, x (nearest mile)	Time, y (nearest 5 minutes)	Distance, x (nearest mile)	Time, y (nearest 5 minutes)
18	20	2	5
8	15	15	25
20	25	16	30
5	20	9	20
5	15	21	30
11	25	5	10
9	20	15	20
10	25		

3.14 A clinic that specializes in the treatment of hypertension (high blood pressure) designed an experiment to investigate the relationship between blood pressure readings obtained by medical personnel using the standard technique and readings obtained by patients using an electronic device. For a given patient, let x represent the diastolic blood pressure as determined by the standard technique and let y represent the diastolic blood pressure obtained using the electronic device.

x	72	84	96	90	85	105	98	102	87	80
y	73	80	95	90	88	101	96	100	90	82

Draw the scatter diagram for these data. Does there appear to be a consistent bias for either technique? (Retain this solution for use in answering Exercises 3.28 and 3.42.)

3.15 UNLV's basketball team had a fantastic 1990–91 season, winning all 30 of its games. Among the statistics reported (*USA Today*, March 11, 1991) were the following scores:

Game	1	2	3	4	5	6	7	8	9	10
UNLV	109	131	95	69	101	92	89	98	95	124
Opt.	68	81	75	35	69	72	65	67	63	93

Game	11	12	13	14	15	16	17	18	19	20
UNLV	117	117	114	88	97	126	88	115	113	112
Opt.	91	76	63	71	85	83	64	73	64	105

Game	21	22	23	24	25	26	27	28	29	30
UNLV	98	86	122	80	114	86	104	49	95	98
Opt.	71	74	75	59	86	74	83	29	67	74

a. Construct a scatter diagram using the opponents score as the independent variable, x.

b. Does there seem to be a pattern to this set of ordered pairs? Describe what you see in the relationship.

MINITAB was used to draw the scatter diagram. Verify the results.

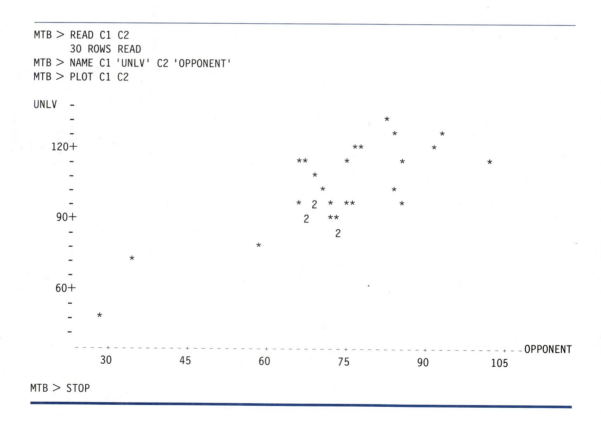

```
MTB > READ C1 C2
      30 ROWS READ
MTB > NAME C1 'UNLV' C2 'OPPONENT'
MTB > PLOT C1 C2
```

```
MTB > STOP
```

3.16 People are concerned about their weight: "I weigh 10 pounds too much for my height." Seldom do you hear, "I should gain 12 pounds." Occasionally you hear, "I'm too short for my weight!" The adjoining chart was printed in *USA Today*, November 6, 1990.

New Weight Guides

Hgt.	Weight (lbs.) Ages 19–34	Age 35+
5'0	97–128	108–138
5'1	101–132	111–143
5'2	104–137	115–148
5'3	107–141	119–152
5'4	111–146	122–157
5'5	114–150	126–162
5'6	118–155	130–167
5'7	121–160	134–172
5'8	125–164	138–178
5'9	129–169	142–183
5'10	132–174	146–188
5'11	136–179	151–194
6'0	140–184	155–199
6'1	144–189	159–205
6'2	148–195	164–210
6'3	152–200	168–216
6'4	156–205	173–222
6'5	160–211	177–228
6'6	164–216	182–234

Higher weights on table generally apply to men, who tend to have more muscle, bone. Lower weights more often apply to women, who tend to have less muscle, bone. Weight is without clothes; height without shoes.

Source: Nutrition and Your Health: Dietary Guidelines for Americans. "New Weight Guides," Copyright 1990, USA TODAY. Reprinted with permission.

a. Draw a scatter diagram for the "Ages 19–34" data. Plot a point for both the lower and upper interval values given in the chart. Draw a line connecting all of the lower values and all of the upper values separately.
b. What do the two lines just drawn on the scatter diagram represent?
c. Using the graph from (a), how much should a woman weigh who is 26 years old and has a medium build according to these new guides?
d. Repeat (a) for the "Age 35+" group.
e. Using the graph from (d), how much should a man weigh who is 46 years old and has a heavy build according to these new guides?

 3.17 The data shown on Table 5 and summarized in Figure 1 in Case Study 3-1, p. 146 is bivariate data: one variable is qualitative (batch) and the other is quantitative (lifetime). Essentially the data divide into ten separate sets of data being compared in a side-by-side manner. The box plots seem to demonstrate a shift in lifetime from one batch to the next. Explain how this shift is shown.

3.18 Table 3 in Case Study 3-2, p. 148, is a frequency contingency table. The marginal totals actually form frequency distributions for the two variables, "14 month score" and "20 month score."
   a. List these two frequency distributions.
   b. Explain how the frequency numbers on Table 3 suggest "children with low comprehension scores at 14 months were most likely to be among those with low comprehension scores at 20 months."

3.2 ▼ Linear Correlation

correlation analysis

independent and dependent variables

linear correlation

The primary purpose of linear **correlation analysis** *is to measure the strength of a linear relationship between two variables.* Let's examine some scatter diagrams demonstrating different relationships between input, or **independent variables**, x and output, or **dependent variables**, y. If as x increases there is no definite shift in the values of y, we say there is **no correlation,** or no relationship, between x and y. If as x increases there is a shift in the values of y, there is a **correlation**. The correlation is positive when y tends to increase and negative when y tends to decrease. If the values of x and y tend to follow a straight-line path, there is a **linear correlation**. The preciseness of the shift in y as x increases determines the strength of the linear correlation. The scatter diagrams in Figure 3-6 demonstrate these ideas.

FIGURE 3-6 Scatter Diagrams and Correlation

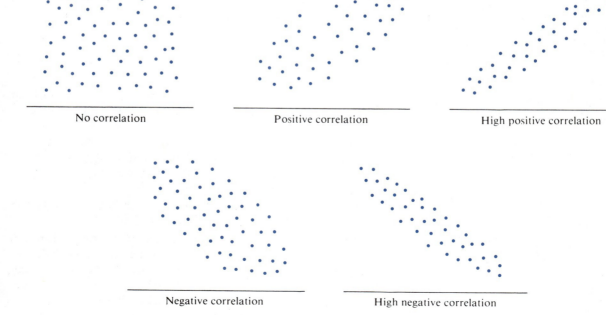

Perfect linear correlation occurs when all the points fall exactly along a straight line, as shown in Figure 3-7. This can be either positive or negative, depending on whether y increases or decreases as x increases. If the data form a straight horizontal or vertical line, there is no correlation, since one variable has no effect on the other.

FIGURE 3-7
Perfect Positive Correlation

Scatter diagrams do not always appear in one of the five forms shown in Figure 3-6. Sometimes they will suggest relationships other than linear. Such is the case in Figure 3-8. There appears to be a definite pattern; however, the two variables are not related linearly, and therefore they are not linearly correlated.

FIGURE 3-8
No Linear Correlation

coefficient of linear correlation

The **coefficient of linear correlation** r is the measure of the strength of the linear relationship between two variables. The coefficient reflects the consistency of the effect that a change in one variable has on the other. The value of the linear correlation coefficient helps us to answer the question "Is there a linear correlation between the two variables under consideration?" The linear correlation coefficient r always has a value between -1 and $+1$. A value of $+1$ signifies a perfect **positive correlation**, and a value of -1 shows a perfect **negative correlation**. If as x increases there is a general increase in the value of y, then r will indicate a positive linear correlation. For example, a positive value of r would be expected for age and height of children, because as children grow older, they grow taller. Also, consider the age x and relative value y of an automobile. As the car ages, its value decreases. Since as x increases, y decreases, their inverse relationship is denoted by a negative value.

positive, negative correlation

The value of r for a sample is obtained from the formula

$$r = \frac{\sum(x - \bar{x})(y - \bar{y})}{(n - 1)s_x s_y} \tag{3-1}$$

Pearson's product moment

NOTE s_x and s_y are the standard deviations of the x and y variables.

This formula is called **Pearson's product moment r**. The development of this formula is discussed in Chapter 13.

To calculate r, we will use an alternative formula that is equivalent to formula (3-1). We will calculate three separate sums of squares and then substitute them into formula (3-2) to obtain r.

$$r = \frac{SS(xy)}{\sqrt{SS(x)SS(y)}} \tag{3-2}$$

where $SS(x) = \Sigma x^2 - ((\Sigma x)^2/n)$, from formula (2-9) on p. 85,

$$SS(y) = \Sigma y^2 - \frac{(\Sigma y)^2}{n} \tag{3-3}$$

and

$$SS(xy) = \Sigma xy - \frac{(\Sigma x)(\Sigma y)}{n} \tag{3-4}$$

TABLE 3-12 Computations Needed to Calculate r

Student	Push-ups (x)	x^2	Sit-ups (y)	y^2	xy
1	27	729	30	900	810
2	22	484	26	676	572
3	15	225	25	625	375
4	35	1,225	42	1,764	1,470
5	30	900	38	1,444	1,140
6	52	2,704	40	1,600	2,080
7	35	1,225	32	1,024	1,120
8	55	3,025	54	2,916	2,970
9	40	1,600	50	2,500	2,000
10	40	1,600	43	1,849	1,720
Total	351	13,717	380	15,298	14,257

Let's return to Illustration 3-8, p. 145, and find the linear correlation coefficient. First we need to construct an extensions table (Table 3-12) listing all the pairs of values (x, y) and the extensions for x^2, xy, and y^2, and the column totals. Then we use the totals from the table and formulas (2-9), (3-3), (3-4), and (3-2) to calculate r.

Using formula (2-9),

$$SS(x) = 13{,}717 - \frac{(351)^2}{10} = 1396.9$$

Using formula (3-3),

$$SS(y) = 15{,}298 - \frac{(380)^2}{10} = 858.0$$

Using formula (3-4),

$$SS(xy) = 14{,}257 - \frac{(351)(380)}{10} = 919.0$$

Using formula (3-2), we find r to be

$$r = \frac{919.0}{\sqrt{(1396.9)(858.0)}} = 0.8394$$

$$= \mathbf{0.84}$$

NOTE Typically, r is rounded to the nearest hundredth.

The value of the calculated linear correlation coefficient is supposed to help us answer the question "Is there a linear correlation between the two variables under consideration?" When the calculated value of r is close to zero, we conclude that there is very little or no linear correlation. When the calculated value of r is close to +1 or −1, we suspect that there is a linear correlation between the two variables. Further discussion involving the interpretation of the calculated value of r is found in Chapter 13.

ESTIMATING THE LINEAR CORRELATION COEFFICIENT

With a formula as complex as formula (3-2), it would be very convenient to be able to inspect the scatter diagram of the data and estimate the calculated value of r. This would serve as a check on the calculations. The following **method for estimating r** is quick and generally yields a reasonable estimate.

1. Lay two pencils on your scatter diagram. Keeping them parallel, move them so that they are as close together as possible but yet have all the points on the scatter diagram between them. (See Figure 3-9).
2. Visualize a rectangular region that is bounded by the two pencils and that ends just beyond the points on the scatter diagram. (See the shaded portion of Figure 3-10.)

FIGURE 3-9

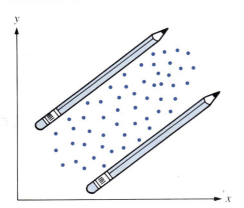

FIGURE 3-10

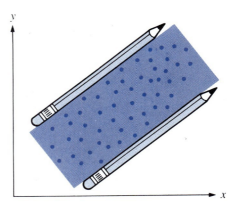

3. Estimate how many times longer the rectangle is than it is wide. An easy way to do this is to mentally mark off squares in the rectangle. (See Figure 3.11.) Call this number of multiples k.

FIGURE 3-11

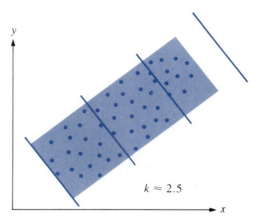

$k \approx 2.5$

4. The value of r may be estimated as

$$\pm \left(1 - \frac{1}{k}\right)$$

The sign assigned to r is determined by the general position of the length of the rectangular region. If it lies in an increasing position (see Figure 3-12), r will be positive; if it lies in a decreasing position, r will be negative. If the rectangle is in either a horizontal or a vertical position, then r will be zero, regardless of the length-width ratio.

FIGURE 3-12

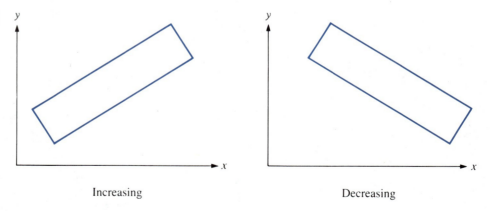

Increasing   Decreasing

NOTE *This estimating technique is very crude. It is very sensitive to the "spread" of the diagram.* However, if the range of the x values and the range of the y values are approximately equal, this approximation will be helpful. *This technique should be used only as a mental check.*

Section 3.2 ▼ LINEAR CORRELATION

Let's use this method to estimate the value of the linear correlation coefficient for the relationship between the number of push-ups and sit-ups. As shown in Figure 3-13, we find that the rectangle is approximately 3.5 times longer than it is wide; that is, $k = 3.5$, and the rectangle lies in an increasing position. Therefore, our estimate for r is $+0.7$ $[1 - (1/3.5) = 0.7]$.

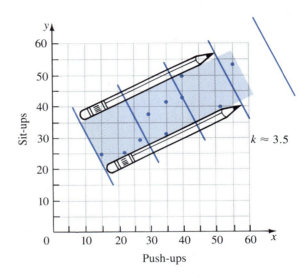

FIGURE 3-13
Push-ups Versus Sit-ups for Ten Students

▼▲ EXERCISES

3.19 How would you interpret the findings of a correlation study that reported a linear correlation coefficient of -1.34?

3.20 How would you interpret the findings of a correlation study that reported a linear correlation coefficient of $+0.03$?

3.21 Consider the following data, which give the weight (in thousands of pounds) x and gasoline mileage (miles per gallon) y for ten different automobiles.

x	2.5	3.0	4.0	3.5	2.7	4.5	3.8	2.9	5.0	2.2
y	40	43	30	35	42	19	32	39	15	44

Find **a.** SS(x) **b.** SS(y) **c.** SS(xy) **d.** Pearson's product moment r

  3.22 An article in the *Journal of Range Management* (September 1990, pages 413–415), states that Blue grama biomass was highly negatively related to canopy cover ($r^2 = 0.91$).
  a. Find the value of r.
  b. Explain what the statement above means.

3.23 **a.** Use the scatter diagram you drew in answering Exercise 3.11 to estimate r for the sample data relating the number of hours studied and exam grade.
b. Calculate r.

3.24 **a.** Use the scatter diagram you drew in answering Exercise 3.12 to estimate r for the sample data involving age and irrelevant answers.
b. Calculate r.

3.25 In the "News from the World of Medicine" section of *Reader's Digest* (April 1991, page 172), a study involving the television-viewing habits and levels of blood cholesterol of 1066 children and adolescents is discussed. Kids who watched more than two hours of TV per day had higher cholesterol levels than those who watched fewer than two hours per day. Find the linear correlation coefficient for the following data, where x = number of hours of TV per day, and y = cholesterol reading:

x	3.5	1.5	2.0	1.0	2.0	2.5	3.0	3.0
y	215	180	205	175	190	200	212	220

3.26 A paper in the Journal *Measurement and Evaluation in Counseling and Development* (October 1990, pages 121–127) discussed a survey instrument that is used to determine the level of mathematics anxiety of children (called the *Mathematics Anxiety Scale for Children,* MASC). One result reported was that the correlation coefficient between MASC scores and math grades was −0.37.

Suppose the MASC was administered to ten fifth-grade students with the following results:

Math Grade (%)	75	85	60	90	80	75	70	90	95	80
MASC Score	67	37	70	40	35	65	40	35	30	40

Calculate r, the linear correlation coefficient.

3.27 An experiment was conducted to study the relationship between the corn yield, y, and the amount of fertilizer applied, x, per plot.

x	1.3	2.5	1.8	3.1	2.7
y	60	72	61	70	68

Section 3.2 ▼ LINEAR CORRELATION

The Pearson product moment r was computed using the MINITAB statistical package:

```
MTB >READ DATA INTO C1, C2
DATA> 1.3, 60
DATA> 2.5, 72
DATA> 1.8, 61
DATA> 3.1, 70
DATA> 2.7, 68
DATA> END DATA
      5 ROWS READ
MTB > CORR C1, C2
Correlation of C1 and C2 = 0.879
MTB > STOP
```

The statement "Read data into C1, C2" is used to enter the data. The statement "Corr C1, C2" instructs the computer to calculate r. The value calculated was 0.879. Verify this value.

3.28 Is there any relationship between the fixed interest rate charged on an unsecured loan and the interest rate paid on money market savings accounts by lending institutions? The following set of data was adapted from the *Democrat and Chronicle*, May 6, 1991:

Institution	1	2	3	4	5	6
Loan Rate	17.75	15.95	14.50	15.50	16.00	14.00
Money Market	5.40	5.30	5.25	4.80	5.60	7.35

Institution	7	8	9	10	11	12	13
Loan Rate	14.50	15.00	14.25	17.80	15.95	14.95	14.00
Money Market	5.25	6.00	6.25	5.30	5.35	5.25	5.32

Institution	14	15	16	17	18	19	20
Loan Rate	14.75	14.00	14.90	14.25	13.90	14.00	13.95
Money Market	5.55	5.75	5.75	6.00	6.00	5.90	5.75

Institution	21	22	23	24
Loan Rate	12.95	14.40	10.50	18.00
Money Market	5.75	5.50	6.00	5.25

a. Draw a scatter diagram; let $x =$ loan rate, $y =$ money market rate.
b. Calculate the linear correlation coefficient, r.
c. Study your results from (a) and (b). What do you see about the relationship of the two rates? Be specific.
d. Can you justify the relationship you found?

3.29 A marketing firm wished to determine whether or not the number of television commercials broadcast were linearly correlated to the sales of its product. The data, obtained from each of several cities, are shown in the following table.

City	A	B	C	D	E	F	G	H	I	J
No. TV Commericals (x)	12	6	9	15	11	15	8	16	12	6
Sales Units (y)	7	5	10	14	12	9	6	11	11	8

a. Draw a scatter diagram.   b. Estimate r.   c. Calculate r.

3.30 A study was designed to investigate the relationship between the weight x and diastolic blood pressure y of adult males between the ages of 19 to 30.

x	173	178	145	146	157	175	173	137
y	76	76	74	70	80	68	90	70

x	199	131	152	171	163	170	135	159
y	96	80	90	72	76	80	68	72

a. Draw a scatter diagram.
b. Calculate r, the linear correlation between weight and diastolic blood pressure.

3.3 ▼ Linear Regression

Although correlation tells us the strength of a linear relationship, it does not tell us the exact numerical relationship. For example, the correlation coefficient calculated for the data in Table 3-12, p. 160, implies that there is a linear relationship between the number of push-ups and the number of sit-ups done by a student. That means that we should be able to use the number of push-ups to predict the number of sit-ups. But correlation analysis does not show us how to determine a y value given an x value. This is done by regression analysis. **Regression analysis** *calculates an equation that provides values of y for given values of x.* One of the primary objectives of regression analysis is to *make predictions*—for example, predicting the success a student will have in college based on high school results, or predicting the distance required to stop a car based on its speed.

regression analysis

Generally, the exact value of y is not predictable. We are usually satisfied if the predictions are reasonably close. The statistician seeks an equation to express the relationship between the two variables. The equation he or she chooses is the one that *best fits* the scatter diagram.

prediction equation

Here are some examples of various relationships, called ***prediction equations***:

$$y = b_0 + b_1 x \text{ (linear)}$$

$$y = a + bx + cx^2 \text{ (quadratic)}$$

$$y = a(b^x) \text{ (exponential)}$$

$$y = a \log_b x \text{ (logarithmic)}$$

Figures 3-14 through 3-17 illustrate some of these relationships. These graphs might result if we were to sample the following bivariate populations:

▼ the hourly wages earned by store clerks x and their weekly pay y

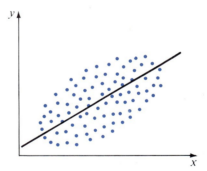

FIGURE 3-14
Linear Regression with Positive Slope

▼ the age of a used American economy car x and its resale value as a used car y

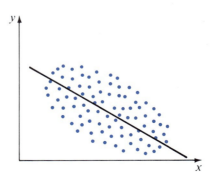

FIGURE 3-15
Linear Regression with Negative Slope

▼ the week of the semester x and the number of hours spent studying statistics y

FIGURE 3-16
Curvilinear Regression (looks quadratic)

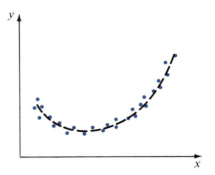

▼ the number of the row that a student sat in for a large lecture course x and the final grade received for the course y

FIGURE 3-17
No Relationship

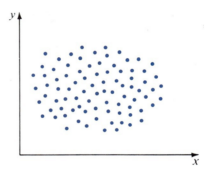

The relationship between these two variables will be an algebraic expression describing the mathematical relationship between x and y. In Figures 3-14, 3-15, and 3-16 there does appear to be a relationship. In Figure 3-17, however, the values of x and y do not seem to be related.

Given a set of bivariate (x, y) data, how do we find the line of best fit? If a straight-line relationship seems appropriate, the best-fitting straight line is found by using the **method of least squares**. Suppose that $\hat{y} = b_0 + b_1 x$ is the equation of a straight line, where $\hat{y}$ (read "y hat") represents the **predicted value** of y that corresponds to a particular value of x. The **least squares criterion** requires that we find the constants b_0 and b_1 such that the *sum* $\Sigma(y - \hat{y})^2$ *is as small as possible*.

Figure 3-18 shows the distance of an observed value of y from a predicted value of y. The length of this distance represents the value $(y - \hat{y})$. Note that $(y - \hat{y})$ is positive when the point (x, y) is above the line and negative when (x, y) is below the line. Figure 3-19 shows a scatter diagram with what might appear to be the line of best fit, along with ten individual $(y - \hat{y})$'s. The sum of the squares

method of least squares
predicted value
least squares criterion

of these differences is *minimized* (made as small as possible) if the line is indeed the line of best fit. Figure 3-20 shows the same data points as Figure 3-19 with the ten individual $(y - \hat{y})$'s associated with a line that is definitely *not* the line of best fit.

FIGURE 3-18
Observed and Predicted Values of y

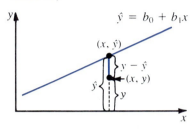

FIGURE 3-19
The Line of Best Fit

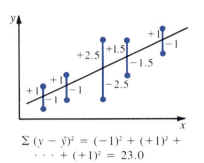

$\Sigma (y - \hat{y})^2 = (-1)^2 + (+1)^2 + \cdots + (+1)^2 = 23.0$

FIGURE 3-20
Not the Line of Best Fit

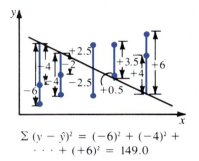

$\Sigma (y - \hat{y})^2 = (-6)^2 + (-4)^2 + \cdots + (+6)^2 = 149.0$

line of best fit

The **equation of the line of best fit** is determined by its slope (b_1) and its y-intercept (b_0). (See the Study Guide for a review of the concepts of slope and intercept of a straight line.) The values of the constants that satisfy the least squares criterion are found by using these formulas:

slope

$$\text{slope:} \quad b_1 = \frac{\Sigma (x - \bar{x})(y - \bar{y})}{\Sigma (x - \bar{x})^2} \quad (3\text{-}5)$$

intercept

$$\text{intercept:} \quad b_0 = \frac{1}{n} \left(\Sigma y - b_1 \cdot \Sigma x \right) \quad (3\text{-}6)$$

(The derivation of these formulas is beyond the scope of this text.)
We will use a mathematical equivalent of formula (3-5) that is easier to apply. We will calculate the slope b_1 by using the formula

$$b_1 = \frac{\text{SS}(xy)}{\text{SS}(x)} \quad (3\text{-}7)$$

and formulas (2-9), p. 85, and (3-4), p. 160. Notice that the numerator of formula (3-7) is the same as the numerator of formula (3-2) for the linear correlation coefficient. Notice also that the denominator is the first radicand (expression under a radical sign) in the denominator of formula (3-2). Thus if you calculate the linear correlation coefficient by using formula (3-2), you can easily find the slope of the line of best fit. If you did not calculate r, set up a table similar to Table 3-12, p. 160, and find the necessary values.

Now let's consider the data in Illustration 3-8 (pp. 145, 160) and the question of predicting a student's sit-ups based on the number of push-ups. We want to find the line of best fit, $\hat{y} = b_0 + b_1 x$. The necessary calculations and totals for determining b_1 have already been obtained in Table 3-12. Recall that $SS(xy) = 919.0$ and $SS(x) = 1396.9$ (see p. 160). Then, using formula (3-7), the slope is calculated to be

$$b_1 = \frac{919.0}{1396.9} = 0.6579 = \mathbf{0.66}$$

Using formula (3-6), the y-intercept is calculated to be

$$b_0 = \frac{1}{10}[380 - (0.6579)(351)] = \frac{1}{10}[380 - 230.9229] = 14.9077 = \mathbf{14.9}$$

Remember to keep the extra decimal places in the calculations to ensure an accurate answer. Thus the equation of the line of best fit is

$$\hat{y} = \mathbf{14.9 + 0.66}x$$

NOTE When rounding off the calculated values of b_0 and b_1, always keep at least two significant digits in the final answer, but *more* as calculations are being done.

Figure 3-21 shows a scatter diagram of the data and the line of best fit.

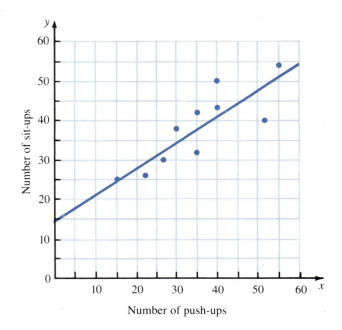

FIGURE 3-21
Line of Best Fit for Push-ups Versus Sit-ups

There are some additional facts about the least squares method that we need to discuss:

1. The slope b_1 represents the predicted change in y per unit increase in x. In our example, $b_1 = 0.66$; thus if a student had done an additional ten push-ups (x), we would predict that he or she would have done an additional seven (0.66×10) sit-ups (y).
2. The y-intercept is the value of y where the line of best fit intersects the y-axis, that is, where $x = 0$. However, in interpreting b_0 you first must consider whether $x = 0$ is a reasonable value before you conclude that you would predict $y = b_0$ if $x = 0$. To predict that if a student did no push-ups he or she would still do approximately 15 sit-ups ($b_0 = 14.9$) is probably incorrect. Second, the x value of zero is outside the domain of the data on which the regression line is based. *In predicting y based on an x value, you should check to be sure that the x value is within the domain of the x values observed.*
3. The line of best fit will always pass through the point $(\bar{x}, \bar{y})$. When drawing the line of best fit on your scatter diagram, you should use the point as a check. It must lie on the line of best fit.

▼ ILLUSTRATION 3-9

In a random sample of eight college women, each was asked for her height (to the nearest inch) and her weight (to the nearest 5 pounds). The data obtained are shown in Table 3-13. Find an equation to predict the weight of a college woman based on her height (the equation of the line of best fit) and draw it on the scatter diagram.

TABLE 3-13
Data for College Women's Heights and Weights

	Woman							
	1	2	3	4	5	6	7	8
Height (x)	65	65	62	67	69	65	61	67
Weight (y)	105	125	110	120	140	135	95	130

SOLUTION Before we start the process of finding the equation for the line of best fit, it is often helpful to draw the scatter diagram. The scatter diagram will give you visual insight about the relationship between the two variables. The scatter diagram for the data on the height and weight of college women is shown in Figure 3-22 (p. 172). The scatter diagram suggests that a linear line of best fit is appropriate.

FIGURE 3-22
Scatter Diagram:
College Women's
Heights Versus
Weights

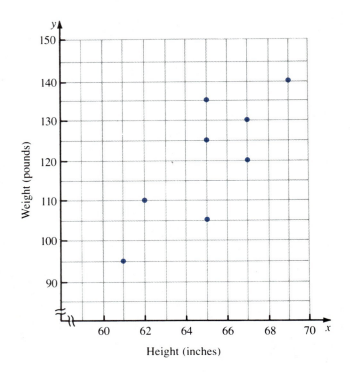

Let's calculate the equation for the line of best fit using the summations in Table 3-14.

TABLE 3-14
Calculations Needed to Find b_1 and b_0

Student	Height (x)	x^2	Weight (y)	xy
1	65	4,225	105	6,825
2	65	4,225	125	8,125
3	62	3,844	110	6,820
4	67	4,489	120	8,040
5	69	4,761	140	9,660
6	65	4,225	135	8,775
7	61	3,721	95	5,795
8	67	4,489	130	8,710
Total	521	33,979	960	62,750

By formulas (3-7), (3-4), and (2-9), we find

$$SS(xy) = 62{,}750 - \frac{(521)(960)}{8} = 230.0$$

$$SS(x) = 33{,}979 - \frac{(521)(521)}{8} = 48.875$$

$$b_1 = \frac{230.0}{48.875} = 4.706 = \mathbf{4.71}$$

By formula (3-6), we find

$$b_0 = \frac{1}{n}\left[\sum y - b_1 \cdot \sum x\right]$$

$$= \frac{1}{8}[960 - (4.706)(521)]$$

$$= -186.478 = \mathbf{-186.5}$$

Therefore,

$$\hat{y} = \mathbf{-186.5 + 4.71}x$$

To draw the line of best fit on the scatter diagram, we need to locate two points. Substitute two values for x, for example, 60 and 70, into the regression equation and obtain two corresponding values for y:

$$\hat{y} = -186.5 + (4.71)(60) = -186.5 + 282.6 = 96.1 = 96$$

and

$$\hat{y} = -186.5 + (4.71)(70) = -186.5 + 329.7 = 143.2 = 143$$

The values (60, 96) and (70, 143) represent two points (designated by a + in Figure 3-23) that enable us to draw the line of best fit.

FIGURE 3-23

Line of Best Fit for College Women's Heights Versus Weights

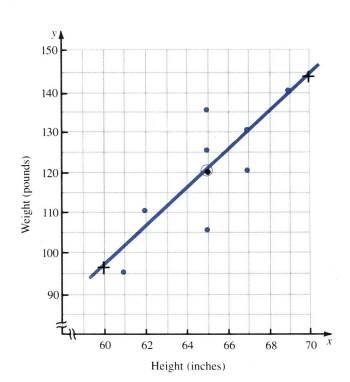

NOTE In Figure 3-23, $(\bar{x}, \bar{y}) = (65.1, 120)$ is also on the line of best fit. It is the point in the colored circle. Use $(\bar{x}, \bar{y})$ as a check on your work.

One of the main purposes for obtaining a regression equation is for making predictions. Once a linear relationship has been established and the value of the input variable x is known, we can predict a value of y ($\hat{y}$). For example, in the physical fitness illustration, the equation was found to be $\hat{y} = 14.9 + 0.66x$. If student A can do 25 push-ups, how many sit-ups do you therefore predict that A will be able to do? (This predicted value will be the average number of sit-ups that you would expect from a sample taken of students who could do exactly 25 push-ups.) The predicted value is

$$\hat{y} = 14.9 + (0.66)(25) = 14.9 + 16.5 = 31.4 = \mathbf{31}$$

When making predictions based on the line of best fit, you must observe the following restrictions:

1. The equation should be used to make predictions only about the population from which the sample was drawn. For example, using the relationship between the height and weight of college women to predict the weight of professional athletes given their height would be questionable.
2. The equation should be used only over the sample domain of the input variable. For example, for Illustration 3-9 the prediction that a college woman of height zero weighs -186.5 pounds is nonsense. You should not use a height outside the sample domain of 60 to 70 inches to predict weight.
3. If the sample was taken in 1987, do not expect the results to have been valid in 1929 or to hold in 1999. The women of today may be different from the women of 1929 and the women of 1999. On occasion you might wish to use the line of best fit to estimate values outside the domain interval of the sample. This can be done, but you should do it with caution and only for values close to the domain interval.

ESTIMATING THE LINE OF BEST FIT

It is possible to estimate the line of best fit and its equation. As with the approximation of r, it should be used mostly as a mental check of your work. On the scatter diagram of the data, draw the straight line that appears to be the line of best fit. [*Hint*: If you draw a line parallel to and halfway between your pencils (whose location was described in Section 3.2, p. 161, for estimating r) you will have a reasonable estimate for the line of best fit.] Then use this line to approximate the equation of the line of best fit.

Figure 3-24 shows the pencils and the resulting estimated line for Illustration 3-9. This line can now be used to approximate the equation. First, locate two points, (x_1, y_1) and (x_2, y_2), along the line and determine their coordinates. Two such points, circled in Figure 3-24, have the coordinates $(59, 85)$ and

(66, 125). These two pairs of coordinates can now be used in the following formula to obtain an estimate for the slope b_1:

$$\text{estimate for } b_1 = \frac{y_2 - y_1}{x_2 - x_1} = \frac{125 - 85}{66 - 59} = \frac{40}{7} = 5.7$$

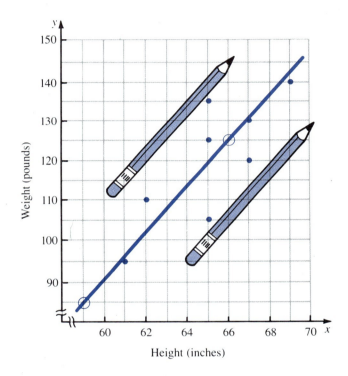

FIGURE 3-24

Estimating the Line of Best Fit for the College Women Data

Using this result, the coordinates of one of the points, and the following formula, we can determine an estimation for the y-intercept b_0:

$$\text{estimate for } b_0 = y - b_1 \cdot x = 85 - (5.7)(59)$$

$$= 85 - 336.3 = -251.3 \text{ or } -250$$

We now estimate the equation for the line of best fit to be $\hat{y} = -250 + 5.7x$. This should serve as a crude estimate. (*Note*: The graph's y-intercept is -250, not approximately 80, as might be read. Why?)

The techniques presented in this section presume that the scatter diagram suggests a linear relationship.

Case Study 3-3

Thirty Years of Car Buying

The following table lists the average price of a new car and the median annual family income for each of several years. The average price of a new car seems to be approximately one-half the annual median family income for that same year. Do you think there is a relationship between these two variables? Sketch a scatter diagram. What do you see? (See Exercise 3.37, p. 179.)

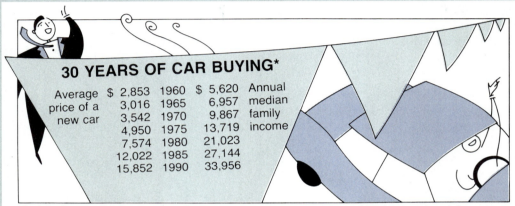

30 YEARS OF CAR BUYING*

Average price of a new car		Annual median family income
$ 2,853	1960	$ 5,620
3,016	1965	6,957
3,542	1970	9,867
4,950	1975	13,719
7,574	1980	21,023
12,022	1985	27,144
15,852	1990	33,956

*Statistics from the U.S. Department of Commerce, supplied by the Motor Vehicle Manufacturers Association.

Source: New Woman, November, 1990. Statistics from the U.S. Department of Commerce, supplied by the Motor Vehicles Manufacturers Association of the United States. Graphics by Laura Wallace.

Case Study 3-4

Points and Fouls in Basketball

Albert Shulte discusses the relationship between the number of personal fouls committed and total points scored during a season by the members of a junior-varsity basketball team. It appears, from Fig. 1, that the players score about 3 points for every personal foul committed. The data for the second year, Table 2, seem to indicate approximately the same relationship. Draw a scatter diagram for the data in Table 2. Does the relationship seem to be about the same as for the first-year data? (See Exercise 3.38, p. 179).

Sometimes you may think that two different sets of numbers are related in some way, although not perfectly. How can you decide if they are related? If they are related, is there any reason why they should be, or is it purely accidental? Does a change in one of the variables cause a change in the other?

The data come from the records of a junior-varsity basketball team . . . for two separate years. . . . A quick look at Table 2 makes one feel that a strong relationship exists between the number points . . . and the number of personal fouls. . . . In Fig. 1, the data for the first year have been plotted. . . .

Section 3.3 ▼ LINEAR REGRESSION

TABLE 2
Second Year

Player	Total Points	Personal Fouls	Player	Total Points	Personal Fouls
Brummett	2	1	McPartlin	35	5
Cooper	75	24	Pointer	46	20
Felice	0	1	Schuback	0	1
Hook	59	18	Wilson	2	3
Hurd	9	9	Zuelch	57	22
Kampsen	7	3			

Now let's think a bit more about the relationship that seems to exist.... The table and the figure both indicate that the more fouls he commits the more points he scores.... Surely no coach would coach a player to make lots of fouls in the hope that he would therefore score more points. The crucial word in the previous sentence is "therefore." Is the fact that a player commits more fouls the reason that he scores more points? Of course not. But maybe both of these... are the results of some other act. Perhaps... more game time.

A graph such as that in Fig. 1 can show that a high degree of correlation exists, but it does not tell us why.

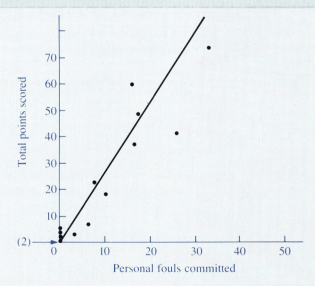

Source: Albert P. Shulte, "Points and Fouls in Basketball," in *Exploring Data,* from *Statistics by Example.* Edited and prepared by the Joint Committee on the Curriculum in Statistics and Probability of the American Statistical Association and the National Council of Teachers of Mathematics. Copyright 1973 by Addison-Wesley Publishing Co., Inc. Reprinted by permission.

EXERCISES

3.31 Draw a scatter diagram for these data:

x	2	12	4	6	9	4	11	3	10	11	3	1	13	12	14	7	2	8
y	4	8	10	9	10	8	8	5	10	9	8	3	9	8	8	11	6	9

Would you be justified in using the techniques of linear regression on these data to find the line of best fit? Explain.

3.32 Explain why the y-intercept is not 80 for the line of best fit in Figure 3-24.

3.33 A study was conducted to investigate the relationship between the cost y (in tens of thousands of dollars) per unit of equipment manufactured and the number of units produced per run x. The equation of the line of best fit was found to be $\hat{y} = 7.31 - 0.01x$, and x was observed for values between 10 and 200. If a production run is scheduled to produce 50 units, what would you predict the cost per unit to be?

3.34 A study was conducted to investigate the relationship between resale price y (in hundreds of dollars) and the age x (in years) of midsize luxury American-made automobiles. The equation of the line of best fit was $\hat{y} = 185.7 - 21.52x$.
  a. Find the resale value of such a car when it is 3 years old.
  b. Find the resale value of such a car when it is 6 years old.
  c. What is the average annual decrease in the resale price of these cars?

3.35 An article entitled "Microbioluminometry Assay (MBA): Determination of Erythromycin Activity in Plasma or Serum" (*Journal of Pharmaceutical Sciences,* December 1989, pages 979–985) compares the MBA assay with another method, Agar Diffusion Plate Assay, for determining the erythromycin activity in plasma or serum. The MBA technique is economical because it requires less sample and reagents.

If X represents the Agar Diffusion Plate assay in micrograms per milliliter and Y represents the MBA Assay in micrograms per milliliter, then the results may be expressed as

$$Y = 0.054128 + 0.92012X \qquad r = 0.9525 \qquad n = 206$$

If the Agar Diffusion Plate Assay determined the level of erythromycin to be 1.200, predict what the MBA assay would be.

3.36 In the article "Beyond Prediction: The Challenge of Minority Achievement in Higher Education" (Lunneborg and Lunneborg, *Journal of Multicultural Counseling and Development*), the relationship between high school GPA and first-year university GPA was investigated for the following groups: Asian-Americans, Black, Chicano, Native-American, and white. For the 43 Native-Americans, the correlation was found to be 0.26 and the equation of the line of best fit was found to be University GPA = 1.85 + 0.30 × high school GPA. Both GPAs were 4.0 scales.

Use the equation of the line of best fit to predict the mean first-year University GPA for all those Native-Americans who had a high school GPA equal to 3.0.

3.37 "Thirty Years of Car Buying," (Case Study 3-3, p. 176), lists the average price of a new car and the median annual family income for each of several years.
   a. Draw a scatter diagram; let x = annual median family income, y = annual average price of new car.
   b. Do you see a relationship between these two variables? Explain.

3.38 There seems to be a relationship between personal fouls committed and total points scored in basketball games. (Case Study 3-4, p. 176). The more fouls committed, the more points scored. Therefore, after reading this study, all basketball coaches are going to be encouraging their players to foul, since that means more points will be scored. Right? Let's investigate the second-year data.
   a. Construct the scatter diagram. Let x = number of fouls committed.
   b. Estimate the correlation coefficient, r.
   c. Estimate the line of best fit.
   d. Does there seem to be a relationship? Describe it.
   e. Does this mean that an increase in fouls will cause an increase in point production? Is there another variable involved that is not mentioned? Explain.

3.39 An article in the *Journal of Pharmaceutical Sciences* (September 1988, pages 796–798) was concerned with the distribution into human milk of 16 different drugs. Two variables were measured: (1) milk papp, which stands for milk lipid-ultrafiltrate partition coefficient; and (2) oct p'app, which stands for apparent octanol-water partition coefficient at ph 7.2. If y = log(milk papp) and x = log(oct p'app), then the article gives the equation for the line of best fit as $y = -0.88 + 1.29x$ for the following data. Verify the equation.

x	−0.47	0.45	0.38	1.06	1.23	1.33	1.26
y	−1.46	−0.28	−0.72	0.31	0.53	0.68	1.45

x	1.68	2.03	2.31	2.23	2.67	2.58	2.92
y	1.75	2.02	1.59	2.24	2.41	2.32	2.93

Note: Two drugs were omitted from the analysis because they did not fit the pattern of the other 14 drugs.

3.40 People not only live longer today but they are living longer independently. The May/June 1989 issue of *Public Health Reports* (pages 222–225) published an article entitled "A Multistate Analysis of Active Life Expectancy." Two of the variables studied were a person's current age and the expected number of years remaining.

age x	65	67	69	71	73	75	77	79	81	83
years remaining y	16.5	15.1	13.7	12.4	11.2	10.1	9.0	8.4	7.1	6.4

a. Draw a scatter diagram.
b. Calculate the equation for the equation of best fit.
c. Draw the line of best fit on the scatter diagram.
d. What is the expected years remaining for a person who is 70 years old? Find the answer two different ways: use the equation from (b) and use the line on the scatter diagram from (c).
e. Are you surprised that the data all lie so close to the line of best fit? Explain why the ordered pairs follow the line of best fit so closely.

3.41 Consider the following set of bivariate data:

x	8	4	2	5	6
y	6	4	3	5	4

The MINITAB statistical package was used to find the line of best fit. The MINITAB statements follow:

```
MTB > READ DATA INTO C1, C2
DATA> 6 8
DATA> 4 4
DATA> 3 2
DATA> 5 5
DATA> 4 6
DATA> END DATA
      5 ROWS READ
MTB > REGRESS Y IN C2 ON 1 PREDICTOR IN C1
The regression equation is
C2 = 2.15 + 0.450 C1
```

Verify that $\hat{y} = 2.15 + 0.45x$ is the line of best fit.

Section 3.3 ▼ LINEAR REGRESSION

3.42 A record of maintenance costs is kept for each of several cash registers throughout a department store chain. A sample of 14 registers gave the following data:

Age x (years)	Maintenance Cost, y (dollars)	Age x (years)	Maintenance Cost y (dollars)
6	92	2	49
7	181	1	64
1	23	9	141
3	40	3	110
6	126	8	105
4	82	9	181
5	117	8	152

a. Draw a scatter diagram that shows these data.
b. Calculate the equation of the line of best fit.
c. A particular cash register is eight years old. How much maintenance (cost) do you predict it will require this year?
d. Interpret your answer to (c).

3.43 An educational study was conducted to investigate the relationship between mathematical competency and computer-science aptitude. The Backmann-Beal mathematical competency test and the KSW computer-science aptitude test were used. The results for 20 students were as follows:

Math Competency (x)	28	35	42	41	44	42	36
Computer-science Aptitude (y)	4	16	20	13	22	21	15

x	44	39	36	40	40	33	27	32	45	41	31	41	43
y	20	19	16	18	17	8	6	5	20	18	11	19	22

a. Draw a scatter diagram.
b. Find the equation for the line of best fit.
c. Draw the line of best fit on the scatter diagram (a).
d. Predict the average KSW score for all students with a Backmann-Beal score of 36. [Do it two ways: use the equation from (b) and use the line drawn in (c).]

182 Chapter 3 ▼ DESCRIPTIVE ANALYSIS AND PRESENTATION OF BIVARIATE DATA

 3.44 The following data are the ages and the asking prices for 19 used foreign compact cars:

Age x (years)	Price y (× $100)	Age x (years)	Price y (× $100)
3	34	6	21
5	26	8	11
3	31.5	5	25
6	12	6	18
4	30	5	23
4	30	7	18
6	14	4	24
7	18	7	10
2	34	5	18
2	32		

  a. Draw a scatter diagram.
  b. Calculate the equation for the line of best fit.
  c. Graph the line of best fit on the scatter diagram.
  d. Predict the average asking price for all such foreign cars that are five years old. [Obtain this answer two ways: use the equation from (b) and use the line drawn in (c).]

IN RETROSPECT

To sum up what we have just learned, there is a distinct difference between the purpose of regression analysis and the purpose of correlation. In regression analysis, we seek a relationship between the variables. The equation that represents this relationship may be the answer desired or it may be the means to the prediction that is desired. In correlation analysis, on the other hand, we simply ask: Is there a significant linear relationship between the two variables?

The article at the beginning of this chapter and the articles that form the case studies show a wide variety of applications for the techniques of correlation and regression. These articles are worth reading again. When bivariate data appear to fall along a straight line on the scatter diagram, they suggest a linear relationship. But, as noted in two of the articles, this is not proof of cause and effect. Clearly, if a basketball player commits too many fouls, he will not be scoring more points. He will be in foul trouble and riding the bench. It also seems reasonable that the more game time he has, the more points he will score and the more fouls he will commit. Thus, a positive correlation and a positive regression relationship will exist between these two variables.

The linear methods we have studied thus far, the basic concepts, have been presented for the purpose of a first, descriptive look. More details must, by necessity, wait until additional developmental work has been completed. After completing this chapter you should have a basic understanding of bivariate data, how they are different from just two sets of data, how to present them, what correlation and regression analysis are, and how each is used. The evaluation and the interpretation of the meaning and the meaningfulness of these results are studied in Chapter 13. After a few more exercises to practice the methods of this chapter, we will study some basic probability.

CHAPTER EXERCISES

3.45 Gun laws, reasons for them, and demonstrations against them have all made the news frequently in recent years. How did people become so interested in gun-related sports? Is their interest related to their environment as a youth? The cross-tabulation table below summarizes the results when gun enthusiasts were asked the question, "Who introduced you to gun related sports?"

	Parents	Relatives	Friends	Combination	Others
Rural	75	46	24	37	2
Nonrural	30	32	44	57	7

  a. Find the marginal totals.
  b. Express the table as percentages of the grand total.
  c. Express the table as percentages of who influenced youth.
  d. Express the table as percentages of each type of environment.
  e. Draw a bar graph based on the type of environment as a youth.

3.46 "Fear of the dentist" (or the dentist's chair) is an emotion felt by many people of all age groups. A survey of 100 individuals in each of five age groups was conducted about this fear, and the results were as follows:

	Elementary	Jr. High	Sr. High	College	Adult
No. Who Fear	37	28	25	27	21
No. Who Do Not Fear	63	72	75	73	79

  a. Find the marginal totals.
  b. Express the table as percentages of the grand total.
  c. Express the table as percentages of each age group's marginal totals.
  d. Express the table as percentages of those who fear and those who do not fear.
  e. Draw a bar graph based on age groups.

Chapter 3 ▼ DESCRIPTIVE ANALYSIS AND PRESENTATION OF BIVARIATE DATA

3.47 When was the last time you saw your doctor? That was the question asked for the survey summarized below:

		Time Since Last Consultation with Your Physician		
		Less than 6 Months	6 Months to Less than 1 Year	1 Year or More
Age	Under 28 yrs.	413	192	295
	28–40	574	208	218
	Over 40	653	288	259

a. Find the marginal totals.
b. Express the table as percentages of the grand total.
c. Express the table as percentages of each age group's marginal totals.
d. Express the table as percentages of each time period.
e. Draw a bar graph based on the grand total.

3.48 Part of quality control is keeping track of what is occurring. The contingency table below shows the number of rejected castings that occurred last month.

	Causes for Rejection of Casting		
	1st Shift	2nd Shift	3rd Shift
Sand	87	110	72
Shift	16	17	4
Drop	12	17	16
Corebreak	18	16	33
Broken	17	12	20
Other	8	18	22

a. Find the marginal totals.
b. Express the table as percentages of the grand total.
c. Express the table as percentages of each shift's marginal totals.
d. Express the table as percentages of each type of rejection.
e. Draw a bar graph based on the shifts.

3.49 Determine whether each of the following questions requires correlation analysis or regression analysis to obtain an answer:
a. Is there a correlation between the grades a student attained in high school and the grades he or she attained in college?
b. What is the relationship between the weight of a package and the cost of mailing it first class?

CHAPTER EXERCISES

c. Is there a linear relationship between a person's height and shoe size?
d. What is the relationship between the number of worker hours and the number of units of production completed?
e. Is the score obtained on a certain aptitude test linearly related to a person's ability to perform a certain job?

3.50 An automobile owner records the number of gallons of gasoline, x, required to fill the gasoline tank and the number of miles traveled, y, between fill-ups.
  a. If he does a correlation analysis on the data, what would be his purpose and what would be the nature of his results?
  b. If he does a regression analysis on the data, what would be his purpose and what would be the nature of his results?

3.51 The following data were generated using the equation $y = 2x + 1$:

x	0	1	2	3	4
y	1	3	5	7	9

A scatter plot of these data results in five points that fall perfectly on a straight line. Find the correlation coefficient and the equation of the line of best fit.

3.52 Consider the set of bivariate data:

x	1	1	3	3
y	1	3	1	3

a. Draw a scatter diagram.
b. Calculate the correlation coefficient.
c. Calculate the line of best fit.

3.53 A biological study of a minnow called the blacknose dace was conducted. The length, x, in millimeters and the age, y, to the nearest year were recorded.

x	25	80	45	40	36	75	50	95	30	15
y	0	3	2	2	1	3	2	4	1	1

Calculate the correlation coefficient.

3.54 A survey was conducted at a large metropolitan university campus. Twenty-four students were interviewed. Two of the questions asked were: "How many hours per week are you employed?" and "How many credit hours are you currently registered for?"

Hours Employed (x)	20	40	35	15	40	20	20	0	20	40
Credit Hours (y)	6	3	6	9	6	6	3	15	6	9

x	10	20	30	40	15	0	0	0	10	40	0	0	30	25
y	9	3	6	6	3	12	15	18	6	6	21	12	6	9

Is there a correlation between x and y?

 3.55 Predicting tomorrow's weather is a favorite pastime of most people. Grandfather always said, "Predict tomorrow to be the same as today, and you will be right more often than the weatherman." Could he have been right? Let's look at some of the daily high temperatures from cities all around the United States as printed in the local newspaper daily (*Democrat and Chronicle,* May 1, 1991).

Tuesday's High	66	71	50	78	79	85	52	74	65	83
Wednesday's High	78	76	51	82	70	87	54	81	64	85

Tuesday's High	78	49	75	85	50	62	46	84	57	41	55	85	86
Wednesday's High	81	56	70	83	67	61	43	71	53	46	52	85	85

Tuesday's High	80	93	64	72	76	78	80	80	83	45	80	79	87
Wednesday's High	82	89	68	78	82	77	68	78	89	48	76	87	82

Tuesday's High	87	60	91	91	49	71	53	80	51	70	85	74	80
Wednesday's High	76	67	93	95	71	63	74	83	60	60	82	68	66

a. Draw a scatter diagram. Let x = Tuesday's high, and y = Wednesday's high.
b. Calculate the linear correlation coefficient, r.

CHAPTER EXERCISES

c. Study your results from (a) and (b). What do you see about the relationship of the two daily high temperatures? Be specific.
d. Comment on the wisdom of Grandfather.

3.56 "Meaning Construction in School Literacy Tasks: A Study of Bilingual Students," from the *American Educational Research Journal* (Fall 1990, pages 427–471) reported the following data for a sample of 12 fifth-graders of Mexican heritage and all from bilingual homes:

Number of Years in U.S.	Percentile Scores		English Envisionment*
	Reading	Language	
11	60	60	4.5
5	22	24	3.5
4	52	50	5.0
6	1	17	2.5
12	20	63	3.5
5	26	31	3.5
11	30	63	2.5
9	20	25	3.0
10	20	53	3.5
11	17	20	2.0
11	17	29	2.0
11	13	15	2.5

*Ability to unfold the meaning of what they are reading.

a. Construct a scatter diagram for reading, x, and language, y.
b. Construct a scatter diagram for reading, x, and English envisionment, y.
c. Construct a scatter diagram for years, x, and English envisionment, y.
d. Calculate the linear correlation coefficient for reading, x, and language, y.
e. Calculate the linear correlation coefficient for reading, x, and English envisionment, y.
f. Calculate the linear correlation coefficient for years, x, and English envisionment, y.

3.57 The following data resulted from a psychological experiment:

x	y	x	y
310	3.2	250	2.5
280	3.0	360	3.4
120	2.6	215	2.6
230	3.5	365	3.8
175	2.8	230	3.0
295	3.8	325	3.8
135	2.2	280	3.3

a. Draw a scatter diagram.
b. Calculate the linear correlation coefficient, r.
c. Is there evidence of correlation? Explain.
d. Calculate the equation for the line of best fit.
e. Graph the line of best fit on the scatter diagram.
f. If x equals 300, what value can be predicted for y?

3.58 The number of one-family building permits issued and the number of one-family housing starts for the last ten months in a certain town are recorded in the following table:

Month	1	2	3	4	5	6	7	8	9	10
Permits Issued	528	494	453	398	413	454	450	436	460	444
Starts	696	614	623	507	554	550	592	568	627	564

Determine whether there is a relationship between x, the number of permits issued, and y, the number of starts. The best relationship will probably be obtained by relating permits to starts one month later, since it takes time to get construction started after a permit is issued. [*Hint:* There will be only nine data points: (528, 614), (494, 623), ...]

3.59 *USA Today* (May 15, 1991) reported on major school systems with present or upcoming superintendent vacancies:

	Salary	Students
Baltimore	$125,000	110,000
Cincinnati	99,000	51,148
Dayton, OH	98,000	28,000
Houston	147,000	195,000
Memphis	109,000	104,000
Milwaukee	110,000	98,000
St. Louis	101,000	43,700
St. Paul, MN	93,000	35,730
Va. Beach, VA	100,000	70,000

a. Construct a scatter diagram. Let x = number of students, and y = salary offered.
b. Does there seem to be a pattern to this set of ordered pairs? Describe what you see in the relationship.
c. Calculate the linear correlation coefficient.
d. Calculate the equation for the line of best fit.
e. Do the results of (c) and (d) agree with your thoughts in (b)?

3.60 The following data on the April employment situation for the 11 most populous states were among those published in the May 6, 1991 issue of *USA Today* (data is in thousands):

CHAPTER EXERCISES   189

State	CA	FL	IL	MA	MI	NJ	NY	NC	OH	PA	TX
Employed, x	13,644	5,922	5,657	2,855	4,129	3,773	8,072	3,221	5,124	5,537	8,074
Jobless, y	1,096	435	388	260	464	261	652	196	399	423	618

    a. Construct a scatter diagram.
    b. Calculate r, the linear correlation coefficient.
    c. Calculate the equation of the line of best fit.

 3.61 Delco Products, a division of General Motors, produces engine cooling cases used for electric engine-cooling fans. A sample of 12 cases was taken to investigate the relationship between the depth of the case, y, and the depth of the pocket in the bottom of the case.

Depth of Pocket	6.886	6.891	6.893	6.894	6.894	6.877	6.883	6.896	6.894	6.890	6.873	6.875
Depth of Case	34.954	34.956	34.959	34.963	34.954	34.904	34.905	34.951	34.935	34.943	34.923	34.932

Source: With permission of Delco Products Division, GMC.

    a. Construct a scatter diagram of these data.
    b. Calculate r, the coefficient of linear correlation.
    c. Find the equation for the line of best fit.

3.62 Most of the 304 million acres of Western federal land is leased to ranchers for grazing livestock. Here are the number of ranchers in Western states, the number leasing federal land, and the number of grazing acres (*USA Today,* May 17, 1991):

State	Total Ranchers	Leasing Federal Land	Acres of Federal Grazing Land
Arizona	3,792	1,556	20,996,000
California	26,579	1,962	32,711,000
Colorado	16,127	3,750	22,917,000
Idaho	15,980	4,023	35,365,000
Kansas	47,008	11	108,000
Montana	15,822	5,340	25,773,000
Nebraska	39,555	153	471,000
Nevada	1,786	1,036	54,126,000
New Mexico	9,189	3,911	21,850,000
North Dakota	18,548	100	1,102,000
Oklahoma	58,236	39	48,000
Oregon	21,811	2,119	23,111,000
South Dakota	27,000	890	3,489,000
Utah	8,757	3,570	29,805,000
Washington	20,147	706	5,372,000
Wyoming	6,428	1,890	26,249,000

    a. Construct a scatter diagram; x = total number of ranchers per state, and y = number of ranchers leasing federal land.

b. Does there seem to be a pattern to this set of ordered pairs? Describe what you see in the relationship.
c. Construct a scatter diagram; x = number of ranchers leasing federal land, and y = number of acres of federal grazing land.
d. Does there seem to be pattern to this set of ordered pairs? Describe what you see in the relationship.
e. Calculate the equation for the line of best fit for the scatter diagram in (d).

3.63 a. Verify, algebraically, that formula (3-2) for calculating r is equivalent to the definition formula (3-1).
b. Verify, algebraically, that formula (3-7) is equivalent to formula (3-5).

3.64 The following equation gives a relationship that exists between b_1 and r:

$$r = b_1 \sqrt{\frac{SS(x)}{SS(y)}}$$

a. Verify this equation for the following data:

x	4	3	2	3	0
y	11	8	6	7	4

b. Verify this equation using formulas (3-2) and (3-7).

VOCABULARY LIST

Be able to define each term. In addition, describe, in your own words, and give an example of each term. Your examples should not be ones given in class or in the textbook.

bivariate data
coefficient of linear correlation
contingency table
correlation
correlation analysis
dependent variable
independent variable
input variable
least squares criterion
linear correlation
line of best fit
method of least squares
negative correlation

ordered pair
output variable
paired data
Pearson's product moment r
positive correlation
predicted value
prediction equation
regression
regression analysis
scatter diagram
slope, b_1
y-intercept, b_0
zero correlation

KEY CONCEPTS

contingency table
correlation
linear correlation coefficient
line of best fit

method of least squares
regression
scatter diagram

QUIZ A

Answer "True" if the statement is always true. If the statement is not always true, replace the words shown in bold with words that make the statement always true.

3.1 **Correlation** analysis is a method of obtaining the equation that represents the relationship between two variables.

3.2 The linear correlation coefficient is used to determine the **equation** that represents the relationship between two variables.

3.3 A correlation coefficient of **zero** means that the two variables are perfectly correlated.

3.4 Whenever the slope of the regression line is zero, the **correlation coefficient** will also be zero.

3.5 When r is positive, b_1 will always be **negative**.

3.6 The **slope** of the regression line represents the amount of change that is expected to take place in y when x increases by one unit.

3.7 When the calculated value of r is positive, the calculated value of b_1 will be **negative**.

3.8 Correlation coefficients range between **0 and +1**.

3.9 The value that is being predicted is called the **input variable**.

3.10 The line of best fit is used to predict the **average value of y** that can be expected to occur at a given value of x.

QUIZ B

Answer all questions.

3.1 The Horsepower and EPA Mileage Ratings for a Sample of 1988 American Automobiles

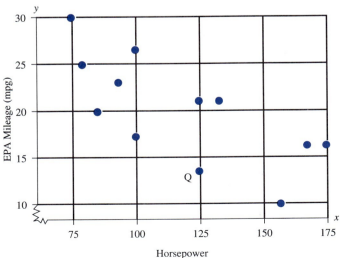

a. Match the items described in Column II with the term in Column I.

Column I	Column II
___ Population	A. The horsepower rating for an automobile
___ Sample	B. All 1988 American-made automobiles
___ Input Variable	C. The EPA mileage rating for an automobile
___ Output Variable	D. The 1988 automobiles whose ratings are shown on the scatter diagram

b. Find the sample size.
c. What is the smallest value reported for the output variable?
d. What is the largest value reported for the input variable?
e. Does the scatter diagram suggest a positive (P), negative (N) or zero (Z) linear correlation coefficient?
f. What are the coordinates of point Q?
g. Will the slope for the line of best fit be: positive (P), negative (N) or zero (Z)
h. Will the intercept for the line of best fit be: positive (P), negative (N) or zero (Z)

3.2 A research group reports the correlation coefficient for two variables to be 2.3. What can you conclude from this information?

3.3 For the bivariate data, the extensions, and the totals shown on the table, find

x	y	x^2	xy	y^2
2	6	4	12	36
3	5	9	15	25
3	7	9	21	49
4	7	16	28	49
5	7	25	35	49
5	9	25	45	81
6	8	36	48	64
28	49	124	204	353

a. SS(x)
b. SS(y)
c. SS(xy)
d. the linear correlation coefficient, r
e. the slope, b_1
f. the y-intercept, b_0
g. the equation of the line of best fit

QUIZ C

3.1 A test was administered to measure the mathematics ability of the people in a certain town. Some of the townspeople were totally surprised to find out that their test results and their shoe size correlated strongly. Explain why a strong positive correlation should not have been such a surprise.

3.2 Student A collected a set of bivariate data and calculated r, the linear correlation coefficient. The resulting value was -1.78. Student A proclaimed that this value indicated that there was no correlation between the two variables since the value of r was not between -1.0 and $+1.0$. Student B argued that -1.78 was impossible and that only values of r near zero implied no correlation. Who is correct? Justify your answer.

3.3 The linear correlation coefficient, r, is a numerical value that ranges from -1.0 to $+1.0$. Write a sentence or two describing the meaning of r for each of the following values:
    a. -0.93 **b.** $+0.89$ **c.** -0.03 **d.** $+0.08$ **e.** -2.3

3.4 Make up a set of three or more ordered pairs such that
    a. $r = 0.0$ **b.** $r = +1.0$ **c.** $r = -1.0$ **d.** $b_1 = 0.0$

WORKING WITH YOUR OWN DATA

Each semester, new students enter your college environment. You may have wondered, "What will the student body be like this semester?" As a beginning statistics student, you have just finished studying three chapters of basic descriptive statistical techniques. Let's use some of these techniques to describe some characteristic of your college's student body.

A ▼ Single Variable Data

1. Define the population to be studied.
2. Choose a variable to define. (You may define your own variable or you may use one of the variables in the accompanying table if you are not able to collect your own data. Ask your instructor for guidance.)
3. Collect 35 pieces of data for your variable.
4. Construct a stem-and-leaf display of your data. Be sure to label it.
5. Calculate the value of the measure of central tendency that you believe best answers the question "What is the average value of your variable?" Explain why you chose this measure.
6. Calculate the sample mean for your data (unless you used the mean in Question 5).
7. Calculate the sample standard deviation for your data.
8. Find the value of the 85th percentile, P_{85}.
9. Construct a graphic display (other than a stem-and-leaf) that you believe "best" displays your data. Explain why the graph best presents your data.

B ▼ Bivariate Data

1. Define the population to be studied.
2. Choose and define two variables that will produce bivariate data. (You may define your own variables or you may use two of the variables in the accompanying table if you are not able to collect your own data. Ask your instructor for guidance.)

3. Collect 15 ordered pairs of data.
4. Construct a scatter diagram of your data. (Be sure to label it completely.)
5. Using a table to assist with the organization, calculate the extensions x^2, xy, and y^2, and the summations of x, y, x^2, xy, and y^2.
6. Calculate the linear correlation coefficient r.
7. Calculate the equation of the line of best fit.
8. Draw the line of best fit on your scatter diagram.

The following table of data was collected on the first day of class last semester. You may use it as a source for your data if you are not able to collect your own.

Variable A = student's sex (male/female)

Variable B = student's age at last birthday

Variable C = number of completed credit hours toward degree

Variable D = "Do you have a job (full/part-time)?" (yes/no)

Variable E = number of hours worked last week, if D = yes

Variable F = wages (before taxes) earned last week, if D = yes

Student	A	B	C	D	E	F
1	M	21	16	No		
2	M	18	0	Yes	10	34
3	F	23	18	Yes	46	206
4	M	17	0	No		
5	M	17	0	Yes	40	157
6	M	40	17	No		
7	M	20	16	Yes	40	300
8	M	18	0	No		
9	F	18	0	Yes	20	70
10	M	29	9	Yes	8	32
11	M	20	22	Yes	38	146
12	M	34	0	Yes	40	340
13	M	19	31	Yes	29	105
14	M	18	0	No		
15	M	20	0	Yes	48	350
16	F	27	3	Yes	40	130
17	M	19	10	Yes	40	202
18	F	18	16	Yes	40	140
19	M	19	4	Yes	6	22
20	F	29	9	No		
21	F	21	0	Yes	20	80
22	F	39	6	No		
23	M	23	34	Yes	42	415

(*continued*)

Student	A	B	C	D	E	F
24	F	31	0	Yes	48	325
25	F	22	7	Yes	40	195
26	F	27	75	Yes	20	130
27	F	19	0	No		
28	M	22	20	Yes	40	470
29	F	60	0	Yes	40	390
30	M	25	14	No		
31	F	24	45	No		
32	M	34	4	No		
33	M	29	48	No		
34	M	22	80	Yes	40	336
35	M	21	12	Yes	26	143
36	F	18	0	No		
37	M	18	0	Yes	13	65
38	F	42	34	Yes	40	244
39	M	25	60	Yes	60	503
40	M	39	32	Yes	40	500
41	M	29	13	Yes	39	375
42	M	19	18	Yes	51	201
43	M	25	0	Yes	48	500
44	F	18	0	No		
45	M	32	68	Yes	44	473
46	F	21	0	No		
47	F	26	0	Yes	40	320
48	M	24	11	Yes	45	330
49	F	19	0	Yes	40	220
50	M	19	0	Yes	10	33
51	F	35	59	Yes	25	88
52	F	24	6	Yes	40	300
53	F	20	33	Yes	40	170
54	F	26	0	Yes	52	300
55	F	17	0	Yes	27	100
56	M	25	18	Yes	41	355
57	M	24	0	No		
58	M	21	0	Yes	30	150
59	M	30	12	Yes	48	555
60	F	19	0	Yes	38	169
61	M	32	45	Yes	40	385
62	M	26	90	Yes	40	340
63	M	20	64	Yes	10	45
64	M	24	0	Yes	30	150
65	M	20	14	No		
66	M	21	70	Yes	40	340
67	F	20	13	Yes	40	206
68	F	33	3	Yes	32	246
69	F	25	68	Yes	40	330
70	F	29	48	Yes	40	525

(*continued*)

Student	A	B	C	D	E	F
71	F	40	0	Yes	40	400
72	F	36	3	Yes	40	300
73	F	35	0	Yes	40	280
74	F	28	0	Yes	40	350
75	M	40	64	Yes	40	390
76	F	31	0	Yes	40	200
77	F	32	0	Yes	40	270
78	F	37	0	Yes	24	150
79	F	35	0	Yes	40	350
80	M	21	72	Yes	45	470
81	F	27	0	Yes	40	550
82	F	42	47	Yes	37	300
83	F	41	21	Yes	40	250
84	M	36	0	Yes	40	400
85	M	25	16	Yes	40	480
86	F	18	0	Yes	45	189
87	M	22	0	Yes	40	385
88	F	27	9	Yes	40	260
89	F	26	3	Yes	40	240
90	F	23	9	Yes	40	330
91	M	41	3	Yes	23	253
92	M	39	0	Yes	40	110
93	M	21	0	Yes	40	246
94	F	32	0	Yes	40	350
95	F	48	58	Yes	40	714
96	F	26	0	Yes	32	200
97	F	27	0	Yes	40	350
98	F	52	56	Yes	40	390
99	F	34	27	Yes	8	77
100	F	49	3	Yes	24	260

The required calculations are accomplished most easily with the assistance of an electronic calculator or a canned program on a computer. Your local computer center will have a list of the available canned programs and will also provide you with the necessary information and assistance to run your data on the computer.

PART TWO

Probability

BEFORE CONTINUING our study of statistics, we must make a slight detour and study some basic probability. Probability is often called the "vehicle" of statistics; that is, the probability associated with chance occurrences is the underlying theory for statistics. Recall that in Chapter 1 we described probability as the science of making statements about what will occur when samples are drawn from known populations. Statistics was described as the science of selecting a sample and making inferences about the unknown population from which it is drawn. To make these inferences, we need to study sample results in situations where the population is known so that we will be able to understand the behavior of chance occurrences.

In Part 2 we will study the basic theory of probability (Chapter 4), probability distributions of discrete variables (Chapter 5), and probability distributions for continuous random variables (Chapter 6). Following this brief study of probability, we will study the techniques of inferential statistics in Part 3.

Egon S. Pearson

EGON SHARPE PEARSON was born on August 11, 1895, in the Hampstead section of London. Egon, the second of three children, was the only son of Karl and Maria Sharpe Pearson. After being educated at the Oxford University and Winchester College, Pearson entered Trinity College in Cambridge, England. After taking time off to serve at the Admirality and Ministry of Shipping, Pearson returned to Cambridge where (in 1920) he received his B.A. and (in 1924) his M.A. Egon Pearson died in 1980 at the age of 85.

Pearson joined his father's department at the University College in London in 1924 and began giving lectures on statistics. It was here that he collaborated with Jerzy Neyman (a Polish mathematician who had been his student) in the development of the Neyman-Pearson Theory. The Neyman-Pearson theory of testing statistical hypotheses secured a recognized place in textbooks of statistical inference. The theory depends upon a compromise between the probability (a) of rejecting a true hypothesis and (b) that of accepting a false hypothesis. The Neyman-Pearson theory has had an immense impact on the world of statistics.

4 PROBABILITY

Chapter Outline

CONCEPTS OF PROBABILITY

4.1 The Nature of Probability
Probability can be thought of as the **relative frequency** with which an event occurs.

4.2 Probability of Events
There are three methods for assigning probabilities to an event: **empirical**, **theoretical**, or **subjective**.

4.3 Simple Sample Spaces
All **possible outcomes** of an experiment are listed.

4.4 Rules of Probability
The probabilities for the outcomes of a sample space **total exactly one**.

CALCULATING PROBABILITIES OF COMPOUND EVENTS

4.5 Mutually Exclusive Events and the Addition Rule
Events are mutually exclusive if they **cannot** both **occur** at the **same time**.

4.6 Independence, Multiplication Rule, and Conditional Probability
Events are independent if the **occurrence** of one event does **not change** the **probability** of the other event.

4.7 Combining the Rules of Probability
The addition and multiplication rules are often used together to calculate the probability of **compound events**.

4.8 Bayes's Rule
A method for finding **conditional probabilities**.

Probability: An Important Tool in Police Science

Larry J. Stephens discusses why what might seem to be unconvincing evidence is actually very convincing evidence. Probabilities are the key. Can you verify the probability stated by Professor Stephens? (See Exercise 4.59.)

Probability can be an extremely valuable tool to the police scientist in trying to access the likelihood of certain events or in trying to validate evidence....

The following is a burglary case description as supplied by Chief Richard R. Anderson, Omaha Police Department:

A man was shown a display of thirty-one (31) pieces of Meissen in an attempt to identify some as belonging to him. Within the display of thirty-one (31) pieces, there were eleven (11) that were accounted for; however, the man was unaware of this. Without the knowledge that eleven (11) were accounted for, the man selected nineteen (19) and stated these were his. Out of the nineteen selected, none were any of the eleven (11) that we had verified were not his. The one piece that was left could have been his; however he said he was unsure and would not select it just to say it was his.

All items on display are porcelain-type figurines, all painted differently with different musical instruments. The collection is referred to as a Meissen Monkey Band.

The fact that the man selected 19 pieces and none of them were from the accounted-for group provided strong evidence that the pieces, in fact, did belong to him.

Probability theory provided a way of measuring how strong the evidence, in fact, was....

If the man could not tell the accounted-for from the non-accounted-for pieces, then the odds of selecting 19 and all 19 being from the non-accounted-for set are 20 to 141,120,505 or 1 to 7,056,025. This provides strong evidence that the man could differentiate among the pieces in the collection....

"P[19 of type A]" is the probability that 19 were selected and that all 19 were from those not previously identified, that is, of type A....

$$P[19 \text{ of type A}] = \frac{20}{141{,}120{,}525}$$

Reproduced from the *Journal of Police Science and Administration,* Vol. 8, No. 3, pp. 353–354, with permission of the International Association of Chiefs of Police, P.O. Box 6010, 13 Firstfield Road, Gaithersburg, Maryland 20878.

Chapter Objectives

You may already be familiar with some ideas of probability, because probability is part of our everyday culture. We constantly hear people making probability-oriented statements such as

- ▼ "Our team will probably win the game tonight."
- ▼ "There is a 40% chance of rain this afternoon."
- ▼ "I will most likely have a date for the winter weekend."
- ▼ "If I park in the faculty parking area, I will probably get a ticket."
- ▼ "I have a 50-50 chance of passing today's chemistry exam."

Everyone has made or heard these kinds of statements. What exactly do they mean? Do they, in fact, mean what they say? Some statements may be based on scientific information and others on subjective prejudice. Whatever the case may be, they are probabilistic inferences—not fact but conjectures.

In this chapter we will study the basic concept of probability and the rules that apply to the probability of both simple and compound events.

CONCEPTS OF PROBABILITY

4.1 ▼ The Nature of Probability

Let's consider an experiment in which we toss two coins simultaneously and record the number of heads that occur. The only possible outcomes are 0H (zero heads), 1H (one head), and 2H (two heads). Let's toss the two coins 10 times and record our findings.

2H, 1H, 1H, 2H, 1H, 0H, 1H, 1H, 1H, 2H

Summary:

Outcome	Frequency
2H	3
1H	6
0H	1

Suppose that we repeat this experiment 19 times. Table 4-1 shows the totals for 20 sets of 10 tosses. (Trial 1 shows the totals from our first experiment.) The total of 200 tosses of the pair of coins resulted in 2H on 53 occasions, 1H on 104 occa-

TABLE 4-1
Experimental Results of Tossing Two Coins

	Trial																				
Outcome	1	2	3	4	5	6	7	8	9	10	11	12	13	14	15	16	17	18	19	20	Total
2H	3	3	5	1	4	2	4	3	1	1	2	5	6	3	1	4	1	0	3	1	53
1H	6	5	5	5	5	7	5	5	5	5	8	4	3	7	5	1	5	4	5	9	104
0H	1	2	0	4	1	1	1	2	4	4	0	1	1	0	4	5	4	6	2	0	43

sions, and 0H on 43 occasions. We can express these results in terms of relative frequencies and show the results using a histogram, as in Figure 4-1.

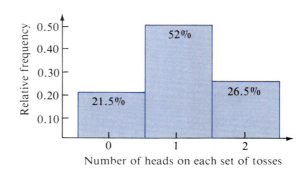

FIGURE 4-1
Relative Frequency Histogram for Coin-Tossing Experiment

What would happen if this experiment were repeated or continued? Would the relative frequencies change? If so, by how much? If we look at the individual sets of 10 tosses, we notice a large variation in the number of times each of the events (2H, 1H, and 0H) occurred. In both the 0H and the 2H categories, there were as many as 6 occurrences and as few as 0 occurrences in a set of 10 tosses. In the 1H category there were as few as 1 occurrence and as many as 9 occurrences.

If we were to continue this experiment for several hundred more tosses, what would you expect to happen in terms of the relative frequencies of these three events? It looks as if we have approximately a 1:2:1 ratio in the totals of Table 4-1. We might therefore expect to find the relative frequency for 0H to be approximately $\frac{1}{4}$, or 25%; the relative frequency for 1H to be approximately $\frac{1}{2}$, or 50%; and the relative frequency for 2H to be approximately $\frac{1}{4}$, or 25%. These relative frequencies accurately reflect the concept of probability.

▼▲ EXERCISES

4.1 Toss a single coin 10 times and record H (head) or T (tail) after each toss. Using your results, find the relative frequency of
    a. heads   **b.** tails

4.2 Roll a single die 20 times and record a 1, 2, 3, 4, 5, or 6 after each roll. Using your results, find the relative frequency of
    a. 1   **b.** 2   **c.** 3   **d.** 4   **e.** 5   **f.** 6

4.3 Place three coins in a cup, shake and dump them out, and observe the number of heads showing. Record 0H, 1H, 2H, or 3H after each trial. Repeat the process 25 times. Using your results, find the relative frequency of
    a. 0H   **b.** 1H   **c.** 2H   **d.** 3H

4.4 Place a pair of dice in a cup, shake and dump them out. Observe the sum of dots. Record 2, 3, 4, ..., 12. Repeat the process 25 times. Using your results, find the relative frequency for each of the values: 2, 3, 4, 5, ..., 12.

4.5 MINITAB can be used to simulate certain experiments such as flipping a coin or rolling a die. The following MINITAB output gives a simulated outcome for rolling a die 50 times and tossing a coin 100 times.

```
MTB > RANDOM 50 NUMBERS INTO C1;
SUBC> INTEGER 1 TO 6.
MTB > PRINT C1
C1
2  5  5  6  1  1  2  4  4  6  4  5  2  3  5
3  3  5  2  1  2  1  4  2  5  5  4  6  1  1
3  4  3  6  2  3  2  1  6  5  5  2  5  4  4
4  2  6  2  5
MTB > RANDOM 100 NUMBERS INTO C2;
SUBC> INTEGER 1 TO 2.
MTB > PRINT C2
C2
2  2  2  2  1  2  2  1  2  2  1  1  1  1  1
1  2  2  2  1  1  2  1  1  2  2  2  2  1  1
1  2  2  2  1  1  2  2  2  1  1  1  1  2  1
1  1  2  1  2  2  1  2  2  2  2  1  1  1  2
2  2  2  1  1  2  2  1  2  1  2  1  2  2  1
1  2  1  2  1  2  2  1  1  1  1  2  1  1  2
2  2  1  1  1  1  2  2  1  1
MTB > STOP
```

    a. Give the relative frequency for the outcomes 1, 2, 3, 4, 5, and 6 based on the MINITAB output.
    b. Give the relative frequency for a head (1) and a tail (2) based on the MINITAB output.

4.6 Use a computer to simulate the random selection of 100 single-digit numbers, 0 through 9. Have the computer list the 100 digits and print a histogram showing the frequency of each of the digits. (If you use MINITAB, see Exercise 4.5 for the necessary commands.)

4.2 ▼ Probability of Events

We are now ready to define what is meant by probability. Specifically, we talk about "the probability that a certain event will occur."

probability of an event

> **PROBABILITY THAT AN EVENT WILL OCCUR**
> The relative frequency with which that event can be expected to occur.

Section 4.2 ▼ PROBABILITY OF EVENTS

experimental or empirical probability

The probability of an event may be obtained in three different ways: (1) empirically, (2) theoretically, and (3) subjectively. The first method was illustrated in the experiment in Section 4-1 and might be called **experimental**, or **empirical probability**. This is nothing more than the *observed relative frequency with which an event occurs*. In our coin-tossing illustration we observed exactly one head (1H) on 104 of the 200 tosses of the pair of coins. The observed empirical probability for the occurrence of 1H was 104/200, or 0.52.

When the value assigned to the probability of an event results from experimental data, we will identify the probability of the event with the symbol $P'(\ )$. The name of the event is placed in parentheses:

$$P'(1H) = 0.52, \ P'(0H) = 0.215, \ \text{and} \ P'(2H) = 0.265$$

Thus the *prime notation* is used to denote empirical probabilities.

The value assigned to the probability of event A as a result of experimentation can be found by means of the formula

$$P'(A) = \frac{n(A)}{n} \tag{4-1}$$

where $n(A)$ is the number of times that event A actually is observed and n is the number of times the experiment is attempted. $P'(A)$ represents the relative frequency with which event A occurred. This may or may not be the relative frequency with which that event "can be expected" to occur. $P'(A)$, however, represents our best *estimate* of what "can be expected" to occur. But what is meant by the relative frequency that "can be expected" to occur?

Consider the rolling of a single die. Define event A as the occurrence of a 1. In a single roll of a die, there are six possible outcomes. Assuming that the die is symmetrical, each number should have an equal likelihood of occurring. Intuitively, we see that the probability of A, or the expected relative frequency of a 1, is $\frac{1}{6}$. (Later we will formalize this calculation.)

What does this mean? Does it mean that once in every six rolls a 1 will occur? No, it does not. Saying that the probability of a 1, $P(1)$, is $\frac{1}{6}$ means that in the long run the proportion of times that a 1 occurs is approximately $\frac{1}{6}$. How close to $\frac{1}{6}$ can we expect the observed relative frequency to be?

Table 4-2 (p. 206) shows the number of 1's observed in each set of six rolls of a die (Column 1), an observed relative frequency for each set of six rolls (Column 2), and a cumulative relative frequency (Column 3). Each trial is a set of six rolls. Figure 4-2a (p. 207) shows the fluctuation of the observed probability for event A on each of the 20 trials (Column 2, Table 4-2). Figure 4-2b shows the fluctuation of the cumulative relative frequency (Column 3, Table 4-2). Notice that the observed relative frequency on each trial of six rolls of a die tends to fluctuate about $\frac{1}{6}$. Notice also that the observed values on the cumulative graph seem to become more stable; in fact, they become relatively close to the expected $\frac{1}{6}$.

A cumulative graph such as Figure 4-2b demonstrates the idea of long-term average. When only a few rolls were observed (as on each trial), the probability $P'(A)$ fluctuated between 0 and $\frac{1}{2}$ (see the relative frequency of each trial, Column 2, Table 4-2). As the experiment was repeated, however, the cumulative

TABLE 4-2
Experimental Results of Rolling a Die Six Times in Each Trial

Trial	(1) Number of 1s Observed	(2) Relative Frequency	(3) Cumulative Relative Frequency
1	1	1/6	1/6 = 0.17
2	2	2/6	3/12 = 0.25
3	0	0/6	3/18 = 0.17
4	2	2/6	5/24 = 0.21
5	1	1/6	6/30 = 0.20
6	1	1/6	7/36 = 0.19
7	2	2/6	9/42 = 0.21
8	2	2/6	11/48 = 0.23
9	0	0/6	11/54 = 0.20
10	0	0/6	11/60 = 0.18
11	2	2/6	13/66 = 0.20
12	0	0/6	13/72 = 0.18
13	2	2/6	15/78 = 0.19
14	1	1/6	16/84 = 0.19
15	1	1/6	17/90 = 0.19
16	3	3/6	20/96 = 0.21
17	0	0/6	20/102 = 0.20
18	1	1/6	21/108 = 0.19
19	0	0/6	21/114 = 0.18
20	1	1/6	22/120 = 0.18

long-term average

graph suggests a stabilizing effect on the observed cumulative probability. This stabilizing effect, or **long-term average** value, is often referred to as the *law of large numbers*.

law of large numbers

> **LAW OF LARGE NUMBERS**
>
> If the number of times that an experiment is repeated is increased, the ratio of the number of successful occurrences to the number of trials will tend to approach the theoretical probability of the outcome for an individual trial.

The law of large numbers can be applied in a different manner. Consider the following experiment. A thumbtack is tossed and the result is observed and recorded as a point up (⌂) or a point down (⌐). What is the probability that the tack will land point up? That is, what is $P(⌂)$? When we worked with the die in the preceding example, we could easily see the true probability. With the tack, we have no idea. Is $P(⌂)$ equal to $\frac{1}{2}$? Since the tack is not symmetrical, we would suspect not. More than $\frac{1}{2}$? Less than $\frac{1}{2}$? About the only way we can tell is to conduct an experiment. And the law of large numbers can help us here. After many

FIGURE 4-2

Fluctuations Found in the Die-Tossing Experiment

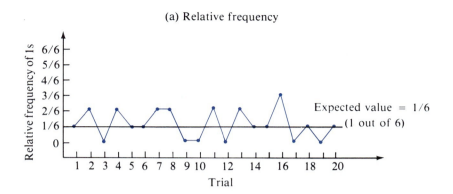

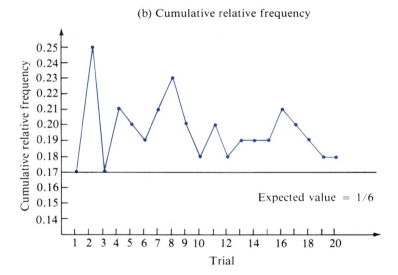

trials of the experiment, $P'(\text{tack})$ should indicate a value that is very close to the true value of $P(\text{tack})$, even though the true value cannot be proven mathematically.

Table 4-3 (p. 208) lists the data recorded after repeated tosses of a thumbtack. After each 10 tosses the number of times the tack landed point up was recorded. The cumulative value of the number of times it landed point up was computed and reported as a fraction of the total tosses. Data on 500 tosses are recorded. Figure 4-3 illustrates the value of $P(\text{tack})$ that we are attempting to estimate. The figure illustrates the tendency for the observed probability $P'(\text{tack})$ to be zeroing in on the value 0.605. We can conclude that $P(\text{tack})$ is very close to 0.605.

An important point is illustrated in Figure 4-3 and Table 4-3:

TABLE 4-3 Experimental Results of Tossing a Thumbtack

Multiples of 10 Tosses	$n(A)$ in Each Set of 10 Tosses	Cumulative $P'(A)$	Multiples of 10 Tosses	$n(A)$ in Each Set of 10 Tosses	Cumulative $P'(A)$
1	5	5/10 = 0.500	26	3	156/260 = 0.600
2	7	12/20 = 0.600	27	5	161/270 = 0.596
3	10	22/30 = 0.733	28	7	168/280 = 0.600
4	6	28/40 = 0.700	29	7	175/290 = 0.603
5	9	37/50 = 0.740	30	9	184/300 = 0.613
6	5	42/60 = 0.700	31	7	191/310 = 0.616
7	6	48/70 = 0.686	32	7	198/320 = 0.619
8	6	54/80 = 0.675	33	7	205/330 = 0.621
9	2	56/90 = 0.622	34	6	211/340 = 0.621
10	4	60/100 = 0.600	35	7	218/350 = 0.623
11	4	64/110 = 0.582	36	6	224/360 = 0.622
12	7	71/120 = 0.592	37	4	228/370 = 0.616
13	6	77/130 = 0.592	38	6	234/380 = 0.616
14	8	85/140 = 0.607	39	4	238/390 = 0.610
15	6	91/150 = 0.607	40	5	243/400 = 0.608
16	3	94/160 = 0.588	41	5	248/410 = 0.605
17	7	101/170 = 0.594	42	7	255/420 = 0.607
18	9	110/180 = 0.611	43	7	262/430 = 0.609
19	7	117/190 = 0.616	44	6	268/440 = 0.609
20	3	120/200 = 0.600	45	4	272/450 = 0.604
21	5	125/210 = 0.595	46	7	279/460 = 0.607
22	8	133/220 = 0.605	47	5	284/470 = 0.604
23	8	141/230 = 0.613	48	8	292/480 = 0.608
24	6	147/240 = 0.613	49	5	297/490 = 0.606
25	6	153/250 = 0.612	50	6	303/500 = 0.606

The larger the number of experimental trials n, the closer the experimental probability $P'(A)$ is expected to be to the true probability $P(A)$.

Many situations are analogous to the thumbtack experiment. We'll look at one example in Illustration 4-1.

▼ ILLUSTRATION 4-1

The key to establishing proper life insurance rates is using the probability that the insureds will live one, two, or three years, and so forth, from the time they purchase policies. These probabilities are derived from actual life and death statistics and hence are empirical probabilities. They are published by the government and are extremely important to the life insurance industry. ▲▲

FIGURE 4-3
Experimental Probability of Thumbtack Pointing Up

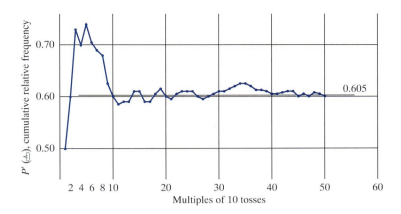

▼▲ EXERCISES

4.7 Explain what is meant by the statement, "The probability of a 1 when a single die is rolled is $\frac{1}{6}$."

4.8 **a.** In Exercise 4.1, you were asked to toss a single coin 10 times. What was the empirical probability of heads after 10 tosses?
b. Toss the coin 10 more times so that you have a total of 20 tosses. What is $P'(H)$ after 20 tosses?
c. Toss the coin 30 more times so that you have a total of 50 tosses. What is $P'(H)$ after 50 tosses?
d. Toss the coin 50 more times. What is $P'(H)$ after 100 tosses?

4.9 According to the *US News and World Report*, December 3, 1990 (page 33), 36 million Americans were victims of serious crimes in 1989, including nearly 19,000 who were murdered. If you assume that there were 200 million Americans in 1989 and that one of them is selected at random, what is the probability that the one selected was
a. a victim of a serious crime?
b. a murder victim?

4.10 Take two dice (one white and one black) and roll them 50 times, recording the results as ordered pairs [for example, (3.5) represents 3 on the white die and 5 on the black die]. Then calculate the following observed probabilities:
a. P'(white die is an odd number)
b. P'(sum is 6)
c. P'(both dice show odd numbers)
d. P'(number on black die is larger than number on white die)

4.11 Using a coin, perform the die and tack experiment discussed on pp. 205–208. Toss a single penny 10 times, observing the number of heads (or use a cup containing 10 pennies and use one toss for each block of 10), and record the results. Repeat until you have made 200 tosses. Chart and graph the data. Do your data tend to support the claim that $P(\text{head}) = \frac{1}{2}$? If not, what do you think caused the difference?

4.12 Use a computer to simulate the rolling of a pair of dice. Have the computer report the sum of the two dice by listing the 100 sums and by printing a histogram showing each possible sum as a separate class on the histogram. (If you use MINITAB, see Exercises 2.31, p. 68, and 4.5, p. 204, for help with the commands.)

4.3 ▼ Simple Sample Spaces

Let's return to an earlier question: What values might we expect to be assigned to the three events (0H, 1H, 2H) associated with our coin-tossing experiment? As we inspect these three events, we see that they do not tend to happen with the same relative frequency. Why? Suppose that the experiment of tossing two pennies and observing the number of heads had actually been carried out using a penny and a nickel—two distinct coins. Would this have changed our results? No, it would have had no effect on the experiment. However, it does show that there are more than three possible outcomes.

When a penny is tossed, it may land as heads or tails. When a nickel is tossed, it may also land as heads or tails. If we toss them simultaneously, we see that there are actually four different possible outcomes of each toss. Each observation would be one of the following possibilities: (1) heads on the penny and heads on the nickel, (2) heads on the penny and tails on the nickel, (3) tails on the penny and heads on the nickel, or (4) tails on the penny and tails on the nickel. Notice that the previous events (0H, 1H, and 2H) match up with these four in the following manner: event (1) is the same as 2H, event (4) is the same as 0H, and events (2) and (3) together make up what was previously called 1H.

In this experiment with the penny and the nickel, let's use an **ordered pair** notation. The first listing will correspond to the penny and the second will correspond to the nickel. Thus (H, T) represents the event that a head occurs on the penny and a tail occurs on the nickel. Our listing of events for the tossing of a penny and a nickel looks like this:

$$(H, H), (H, T), (T, H), (T, T)$$

What we have accomplished here is a listing of what is known as the *sample space* for this experiment.

ordered pair

experiment

EXPERIMENT
Any process that yields a result or an observation.

outcome

OUTCOME
A particular result of an experiment.

sample space

SAMPLE SPACE
The set of all possible outcomes of an experiment. The sample space is typically called S and may take any number of forms: a list, a tree dia-

Section 4.3 ▼ SIMPLE SAMPLE SPACES

gram, a lattice grid system, and so on. The individual outcomes in a sample space are called **sample points**. $n(S)$ is the number of sample points in sample space S.

> ### EVENT
> Any subset of the sample space. If A is an event, then $n(A)$ is the number of sample points that belong to event A.

Regardless of the form in which they are presented, the outcomes in a sample space can never overlap. Also, all possible outcomes must be represented. These characteristics are called *mutual exclusive* and *all inclusive*, respectively. A more detailed explanation of these characteristics will be presented later; for the moment, however, an intuitive grasp of their meaning is sufficient.

Now let's look at some illustrations of probability experiments and their associated sample spaces.

▼ EXPERIMENT 4-1

A single coin is tossed once and the outcome—a head (H) or a tail (T)—is recorded.

$$\text{Sample space: } S = \{H, T\}$$

▼ EXPERIMENT 4-2

Two coins, one penny and one nickel, are tossed simultaneously and the outcome for each coin is recorded using ordered pair notation: (penny, nickel). The sample space is shown here in two different ways:

TREE DIAGRAM REPRESENTATION*

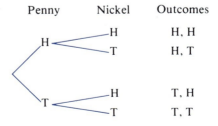

(four branches, each branch shows a possible outcome)

LISTING

$$S = \{(H, H), (H, T), (T, H), (T, T)\}$$

*See the *Study Guide* for information about tree diagrams.

Notice that both representations show the same four possible outcomes. For example, the top branch on the tree diagram shows heads on both coins, as does the first ordered pair in the listing.

▼ EXPERIMENT 4-3

A die is rolled one time and the number of spots on the top face observed. The sample space is

$$S = \{1, 2, 3, 4, 5, 6\}$$

▼ EXPERIMENT 4-4

A box contains three poker chips (one red, one blue, one white), and two are drawn *with replacement*. (This means that one chip is selected, its color is observed, and then the chip is replaced in the box.) The chips are scrambled before a second chip is selected and its color observed. The sample space is shown in two different ways:

TREE DIAGRAM REPRESENTATION

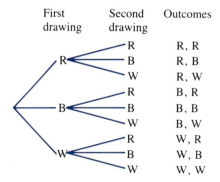

LISTING

$$S = \{(R, R), (R, B), (R, W), (B, R), (B, B), (B, W), (W, R), (W, B), (W, W)\}$$

▼ EXPERIMENT 4-5

This experiment is the same as Experiment 4-4 except that the first chip is not replaced before the second selection is made. The sample space is shown in two ways:

Section 4.3 ▼ SIMPLE SAMPLE SPACES

TREE DIAGRAM REPRESENTATION

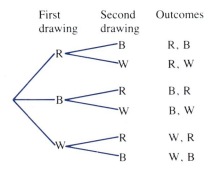

LISTING

$$S = \{(R, B), (R, W), (B, R), (B, W), (W, R), (W, B)\}$$ ▲▲

▼ EXPERIMENT 4-6

A white die and a black die are each rolled one time and the number of dots showing on each die is observed. The sample space is shown by a chart representation:

CHART REPRESENTATION

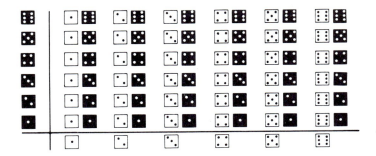

▲▲

▼ EXPERIMENT 4-7

Two dice are rolled and the sum of their dots is observed. The same space is

$$S = \{2, 3, 4, 5, 6, 7, 8, 9, 10, 11, 12\}$$

(or the 36-point sample space listed in Experiment 4-6). ▲▲

You will notice that two different sample spaces are suggested for Experiment 4-7. Both of these sets satisfy the definition of a sample space and thus either could be used. We will learn later why the 36-point sample space is more useful than the other.

▼ EXPERIMENT 4-8

A single marble is drawn from a box that contains 10 red and 30 blue marbles and its color is recorded. The sample space is

$$S = \{R, B\}$$

An alternative and more useful way to represent this sample space would be to repeat the elements proportionally to their possibilities. Let the sample space be

$$S = \{R, B, B, B\}$$

because there are three times as many blue marbles as there are red marbles. ▲▲

▼ EXPERIMENT 4-9

A thumbtack is tossed and the outcome is recorded [whether it lands point up (⚘) or point down (⚘)]. The sample space is

$$S = \{⚘, ⚘\}$$

▼ EXPERIMENT 4-10

A weather forecaster predicts that there will be a measurable amount of precipitation or no precipitation on a given day. The sample space is

$$S = \{\text{precipitation, no precipitation}\}$$

▼ EXPERIMENT 4-11

The 6024 students at a nearby college have been cross-tabulated according to gender and their college status:

	Full-time Student	Part-time Student
Female	2136	548
Male	2458	882

One student is to be picked at random from the student body.

▼ EXPERIMENT 4-12

After completing an inventory, a used-car dealer found the following information:

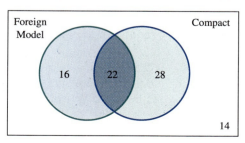

One lucky customer will randomly select one key from a barrel containing one ignition key to each of his 80 cars. ▲▲

Special attention should always be given to the sample space. Like the statistical population, the sample space must be well defined. Once the sample space is defined, you will find the remaining work much easier.

▼▲ EXERCISES

4.13 The face cards are removed from a regular deck and then 1 card is selected from this set of 12 face cards. List the sample space for this experiment.

4.14 An experiment consists of drawing one marble from a box that contains a mixture of red, yellow, and green marbles.
    a. List the sample space.
    b. Can we be sure that each outcome in the sample space is equally likely?
    c. If two marbles are drawn from the box, list the sample space.

4.15 In the July 23, 1990 issue of *Time*, it was reported that 600,000 new cases of skin malignancies would be diagnosed in the United States in 1990. Most of the malignancies would be either malignant melanoma, Basal-cell Carcinoma, or Squamous-cell carcinoma. The remaining malignancies were classified as "other." If two malignancy cases are randomly selected from the records of a dermatologist, give the sample space for the possible combinations of selected records.

4.16 **a.** A balanced coin is tossed twice. List a sample space showing the possible outcomes.
    b. A biased coin (it favors heads in a ratio of 3 to 1) is tossed twice. List a sample space showing the possible outcomes.

4.17 An experiment consists of two trials. The first is tossing a penny and observing heads or tails; the second is rolling a die and observing a 1, 2, 3, 4, 5, or 6. Construct the sample space.

4.18 A computer generates (in random fashion) pairs of integers. The first integer is between 1 and 5, inclusive, and the second is between 1 and 4, inclusive. Represent the sample space on a coordinate axis system where x is the first number and y is the second number.

4.19 A coin is tossed and a head or a tail observed. If a head results, the coin is tossed a second time. If a tail results on the first toss, a die is rolled.
    a. Construct the sample space for this experiment.
    b. What is the probability that a die is rolled in the second stage of this experiment?

4.20 A box stored in a warehouse contains 100 identical parts of which 10 are defective and 90 are nondefective. Three parts are selected without replacement. Construct a tree diagram representing the sample space.

4.21 MINITAB was used to simulate 100 trials of the experiment described in Exercise 4.17, the tossing of a penny and the rolling of a die (1 = H, 2 = T for the penny, and 1, 2, 3, 4, 5, 6 for the die).

```
MTB > #NOTE 1 = HEAD, 2 = TAIL
MTB > RANDOM 100 NUMBERS INTO C1;
SUBC> INTEGER 1 TO 2.

MTB > RANDOM 100 NUMBERS INTO C2;
SUBC> INTEGER 1 TO 6.

MTB > NAME C1 'COIN' C2 'DIE'
MTB > TABLE C1 C2
   ROWS: COIN      COLUMNS: DIE
              1       2       3       4       5       6      ALL

     1        5       9       6      12       8      10       50
     2        4       5      13       4      17       7       50
   ALL        9      14      19      16      25      17      100


    CELL CONTENTS --
                  COUNT

MTB > STOP
```

---

a. Find the relative frequency of heads.
b. Find the relative frequency of 3.
c. Find the relative frequency of (H, 3).

4.22 Use a computer to simulate the experiment described in Exercise 4.18; x is an integer 1 to 5, and y is an integer 1 to 4. Have the computer generate 100 random x-values and 100 y-values. Print a list and a graph (lattice grid or scatter diagram) of the 100 pairs of integers.

a. Find the relative frequency for $x = 2$.
b. Find the relative frequency for $y = 3$.
c. Find the relative frequency for the ordered pair (2, 3).

4.4 ▼ Rules of Probability

Let's return now to the concept of probability and relate the sample space to it. Recall that the probability of an event was defined as the relative frequency with which the event could be expected to occur.

In the sample space associated with Experiment 4-1, the tossing of one coin, we find two possible outcomes: heads (H) and tails (T). We have an "intuitive feeling" that these two events will occur with approximately the same frequency. The coin is a symmetrical object and therefore would not be expected to favor either of the two outcomes. We would expect heads to occur $\frac{1}{2}$ of the time. Thus the probability that a head will occur on a single toss of a coin is thought to be $\frac{1}{2}$.

This description is the basis for the second technique for assigning the probability of an event. In a sample space containing **sample points that are equally likely to occur**, the probability $P(A)$ of an event A is the ratio of the number $n(A)$ of points that satisfy the definition of event A to the number $n(S)$ of sample points in the entire sample space. That is,

equally likely events

$$P(A) = \frac{n(A)}{n(S)} \qquad (4\text{-}2)$$

theoretical probability

This formula gives a **theoretical probability** value of event A's occurrence. The *prime* symbol of formula (4-1) *is not used* with theoretical probabilities.

The probability of event A is still a relative frequency, but it is now based on all the possible outcomes regardless of whether they actually occur. We can now see that the probability of obtaining two heads, $P(2H)$, when the penny and nickel are tossed is $\frac{1}{4}$. The probability of exactly one head, $P(1H)$ [(H, T) or (T, H)], is $\frac{2}{4}$, or $\frac{1}{2}$. And $P(0H) = \frac{1}{4}$. These are the values that were indicated by the experiment.

The use of formula (4-2) requires the existence of a sample space in which each outcome is equally likely. Thus when dealing with experiments that have more than one possible sample space, it is helpful to construct a sample space in which the sample points are equally likely.

Consider Experiment 4-7, where two dice were rolled. If you list the sample space as the 11 sums, the sample points are not equally likely. If you use the 36-point sample space, all the sample points are equally likely. The 11 sums in the first sample space represent combinations of the 36 equally likely sample points. For example, the sum of 2 represents {(1, 1)}; the sum of 3 represents {(2, 1), (1, 2)}; and the sum of 4 represents {(1, 3), (3, 1), (2, 2)}. Thus we can use formula (4-2) and the 36-point sample space to obtain the probabilities for the 11 sums.

$$P(2) = \tfrac{1}{36}, \qquad P(3) = \tfrac{2}{36}, \qquad P(4) = \tfrac{3}{36}$$

and so forth.

In many cases the assumption of equally likely events does not make sense. Experiments 4-1 through 4-6 are examples in which the sample space elements are all equally likely. The sample points in Experiment 4-7 are not equally likely, and there is no reason to believe that the sample points in Experiments 4-8, 4-9, and 4-10 are equally likely.

What do we do when the sample space elements are not equally likely or not a combination of equally likely events? We could use empirical probabilities. But what do we do when no experiment has been done or can be performed?

subjective probability

Let's look again at Experiment 4-10. The weather forecaster often assigns a probability to the event "precipitation." For example, "there is a 20 percent chance of rain today," or "there is a 70 percent chance of snow tomorrow." In such cases the only method available for assigning probabilities is personal judgment. These probability assignments are called **subjective probabilities**. The accuracy of subjective probabilities depends on the individual's ability to correctly assess the situation.

Often, personal judgment of the probability of the possible outcomes of an experiment is expressed by comparing the likelihood among the various outcomes. For example, the weather forecaster's personal assessment might be that "it is five times more likely to rain (R) tomorrow than not rain (NR)"; $P(R) = 5 \cdot P(NR)$. If this is the case, what values should be assigned to $P(R)$ and $P(NR)$? To answer this question, we need to review some of the ideas about probability that we've already discussed.

1. Probability represents a relative frequency.
2. $P(A)$ is the ratio of the number of times an event can be expected to occur divided by the number of trials.
3. The numerator of the probability ratio must be a positive number or zero.
4. The denominator of the probability ratio must be a positive number (greater than zero).
5. The number of times an event can be expected to occur in n trials is always less than or equal to the total number of trials.

Thus it is reasonable to conclude that a probability is always a numerical value between zero and one.

Property 1 $0 \leq P(A) \leq 1$

NOTES

1. The probability is zero if the event cannot occur.
2. The probability is one if the event occurs every time.

Property 2 $\sum_{\text{all outcomes}} P(A) = 1$

Property 2 states that if we add up the probabilities of each of the sample points in the sample space, the total probability must equal one. This makes sense because when we sum up all the probabilities, we are really asking, "What is the probability the experiment will yield an outcome?" and this will happen every time.

Now we are ready to assign probabilities to $P(R)$ and $P(NR)$. The events R and NR cover the sample space, and the weather forecaster's personal judgment was

$$P(R) = 5 \cdot P(NR)$$

Section 4.4 ▼ RULES OF PROBABILITY

From Property 2, we know that

$$P(R) + P(NR) = 1$$

By substituting $5 \cdot P(NR)$ for $P(R)$, we get

$$5 \cdot P(NR) + P(NR) = 1$$
$$6 \cdot P(NR) = 1$$
$$P(NR) = \tfrac{1}{6}$$
$$P(R) = 1 - P(NR) = 1 - \tfrac{1}{6} = \tfrac{5}{6}$$

complementary event

COMPLEMENT OF AN EVENT
The set of all sample points in the sample space that do not belong to event A. The complement of event A is denoted by $\overline{A}$ (read "A complement").

For example, the complement of the event "success" is "failure"; the complement of "heads" is "tails" for the tossing of one coin; the complement of "at least one head" on 10 tosses of a coin is "no heads."

By combining the information in the definition of complement with Property 2 (p. 218), we can say that

$$P(A) + P(\overline{A}) = 1.0 \qquad \text{for any event A}$$

It then follows that

$$P(\overline{A}) = 1 - P(A) \tag{4-3}$$

NOTE Every event A has a complementary event $\overline{A}$. Complementary probabilities are very useful when the question asks for the probability of "at least one." Generally this represents a combination of several events, but the complementary event "none" is a single outcome. It is easier to solve for the complementary event and get the answer by using formula (4-3).

▼ **ILLUSTRATION 4-2**

Two coins are tossed. What is the probability that at least one head appears?

SOLUTION Let event A be the occurrence of no heads; then $\overline{A}$ represents the occurrence of one or more heads, that is, at least one head. The sample space is {(HH), (HT), (TH), (TT)}.

$$P(A) = \tfrac{1}{4} = \mathbf{0.25} \qquad \text{[using formula (4-2)]}$$
$$P(\overline{A}) = 1 - P(A) = 1 - \tfrac{1}{4} = \tfrac{3}{4} = \mathbf{0.75} \qquad \text{[using formula (4-3)]} \qquad ▲▲$$

Case Study 4-1

Trying to Beat the Odds

Many young men aspire to become professional athletes. Only a few make it to the big time as indicated in the following graph. For every 2400 college senior basketball players, only 64 make a professional team. That translates to a probability of 0.027 (64/2400). What is the probability that a high school senior basketball player makes a college team? Is he still playing by the time he is a college senior? What is the probability that a high school senior basketball player makes a professional team?

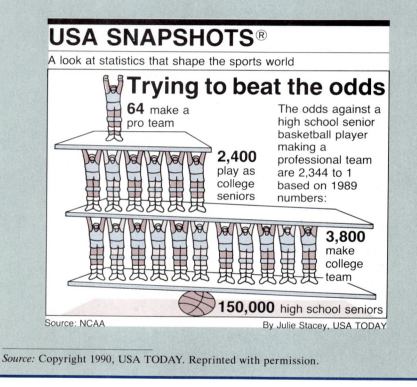

Source: Copyright 1990, USA TODAY. Reprinted with permission.

▼▲ EXERCISES

4.23 Suppose that a box of marbles contains an equal number of red marbles and yellow marbles but twice as many green marbles as red marbles. Draw one marble from the box and observe its color. Assign probabilities to the elements in the sample space.

 4.24 A transportation engineer in charge of a new traffic control system expresses the subjective probability that the system functions correctly 99 times as often as it malfunctions.

a. Based on this belief, what is the probability that the system functions properly?
b. Based on this belief, what is the probability that the system malfunctions?

4.25 Events A, B, and C are defined on sample space S. Their corresponding sets of sample points do not intersect and their union is S. Further, event B is twice as likely to occur as event A, and event C is twice as likely to occur as event B. Determine the probability of each of these three events.

4.26 Three coins are tossed and the number of heads observed is recorded. Find the probability for each of the possible results, 0H, 1H, 2H, and 3H.

4.27 Let x be the success rating of a new television show. The accompanying table lists the subjective probabilities assigned to each x for a particular new show as assigned by three different media people. Which of these sets of probabilities are inappropriate because they violate a basic rule of probability? Explain.

	Judge		
Success Rating (x)	A	B	C
Highly successful	0.5	0.6	0.3
Successful	0.4	0.5	0.3
Not successful	0.3	−0.1	0.3

4.28 Two dice are rolled (Experiment 4-6). Find the probabilities in parts (b) through (e). Use the sample space given on p. 213.
 a. Why is the set $\{2, 3, 4, \ldots, 12\}$ not a useful sample space?
 b. P(white die is an odd number)
 c. P(sum is 6)
 d. P(both dice show odd numbers)
 e. P(number on black die is larger than number on white die)

4.29 A group of files in a medical clinic classifies the patients by gender and by type of diabetes (I or II). The groupings may be shown as follows. The table gives the number in each classification.

		Type of Diabetes	
		I	II
Gender	Male	25	20
	Female	35	20

If one file is selected at random, find the probability that
  a. the selected individual is female.
  b. the selected individual is a Type II.

4.30 A lottery is conducted and 500 tickets are sold. The stubs to the tickets are well shuffled and the winner is chosen by randomly selecting one stub. If you bought 25 tickets, what is the probability that you will win?

4.31 According to an article in *Glamour* (April 1991, pages 65, 66), one out of every nine people diagnosed with AIDS during 1991 will be female. Based on this information, find the probability that an individual diagnosed with AIDS in 1991 will be male.

4.32 According to *Science News* (November 1990, page 333), sleep apnea affects 2 million individuals in the United States. The sleep disorder interrupts breathing and can awaken its sufferers as often as five times an hour. Many people do not recognize the condition even though it causes loud snoring. Assuming there are 200 million people in the United States, what is the probability that an individual chosen at random will not be affected by sleep apnea?

CALCULATING PROBABILITIES OF COMPOUND EVENTS

compound events

Compound events are formed by combining several simple events. We will study the following three compound events in the remainder of this chapter:

1. the probability that either event A or event B will occur, $P(A \text{ or } B)$
2. the probability that both events A and B will occur, $P(A \text{ and } B)$
3. the probability that event A will occur given that event B has occurred, $P(A|B)$

NOTE In determining which compound probability we are seeking, it is not enough to look for the words *either/or*, *and*, or *given*. We must carefully examine the question asked to determine what combination of events is called for.

4.5 ▼ Mutually Exclusive Events and the Addition Rule

MUTUALLY EXCLUSIVE EVENTS

mutually exclusive events

> **MUTUALLY EXCLUSIVE EVENTS**
> Events defined in such a way that the occurrence of one event precludes the occurrence of any of the other events. (In short, if one of them happens, the others cannot happen.)

The following illustrations give examples of events to help you understand the concept of mutually exclusive.

ILLUSTRATION 4-3

A group of 200 college students is known to consist of 140 full-time (80 female and 60 male) students and 60 part-time (40 female and 20 male) students. From this group one student is to be selected at random.

	200 College Students		
	Full-time	Part-time	Total
Female	80	40	120
Male	60	20	80
Total	140	60	200

Two events related to this selection are defined. Event A is "the student selected is full-time" and event B is "the student selected is a part-time male." Since no student is both "full-time" and "part-time male," the two events A and B are mutually exclusive events.

A third event, event C, is defined to be "the student selected is female." Now let's consider the two events A and C. Since there are 80 students that are "full-time" and "female," the two events A and C are not mutually exclusive events. This "intersection" of A and C can be seen on accompanying Venn diagram or in the table above.

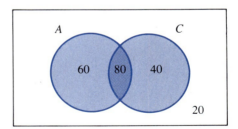

ILLUSTRATION 4-4

Consider an experiment in which two dice are rolled. Three events are defined:

A: The sum of the numbers on the two dice is 7.
B: The sum of the numbers on the two dice is 10.
C: Each of the two dice shows the same number.

Let's determine whether these three events are mutually exclusive.

Are events A and B mutually exclusive? Yes, they are, since the sum on the two dice cannot be both 7 and 10 at the same time. If a sum of 7 occurs, it is impossible for the sum to be 10.

FIGURE 4-4

Sample Space for the Roll of Two Dice

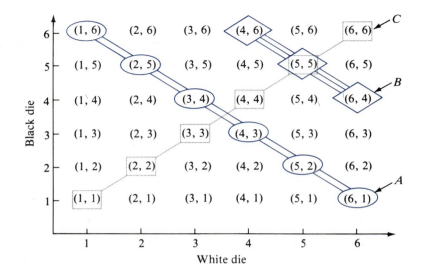

Figure 4-4 presents the sample space for this experiment. This is the same sample space shown in Experiment 4-6, except that ordered pairs are used in place of the pictures. The ovals, diamonds, and rectangles show the ordered pairs that are in events A, B, and C, respectively. We can see that events A and B do not **intersect** at a common sample point. Therefore, they are mutually exclusive. Point (5, 5) satisfies both events B and C. Therefore, B and C are not mutually exclusive. Two dice can each show a 5, which satisfies C; and the total of the two 5s satisfies B.

Figure 4-5 is a **Venn diagram** that shows the relationships between the events defined in Illustration 4-4. Venn diagrams are simple "area," or "region," diagrams. The basic concept of a Venn diagram is that those points (and only those

intersection

Venn diagram

FIGURE 4-5

Venn Diagram Events A, B, and C

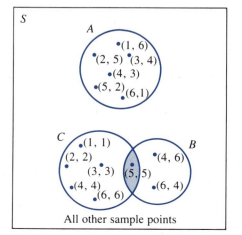

points) belonging to a given event are shown inside a circle. (For further background information, see the *Study Guide*.)

In Figure 4-5, events A and B are mutually exclusive. Six sample points belong to event A and there are three sample points in event B. The two events do not have a point in common; that is, they are mutually exclusive.

Similarly, events A and C do not intersect; therefore, they too are mutually exclusive. Events C and B, however, do share a common sample point, (5, 5). Notice that the point (5, 5) is in the intersection of the two Venn diagram circles. Consequently, events C and B are not mutually exclusive. ▲▲

addition rule

ADDITION RULE

Let us now consider the compound probability $P(A \text{ or } B)$, where A and B are mutually exclusive events.

▼ ILLUSTRATION 4-5

In Illustration 4-3 we considered an experiment in which one student was to be selected at random from a group of 200. Event A was "student selected is full-time," and event B was "student selected is part-time and male." The probability of event A, $P(A)$, is $\frac{140}{200}$, or 0.7, and the probability of event B, $P(B)$, is $\frac{20}{200}$, or 0.1. Let's find the probability of A or B, $P(A \text{ or } B)$. It seems reasonable to add the two probabilities 0.7 and 0.1 to obtain an answer of $P(A \text{ or } B) = 0.8$.

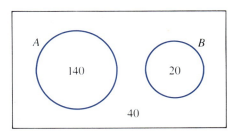

This is further justified by looking at the sample space. We see that there is a total of 160 that are either full-time or are male part-time students ($\frac{160}{200} = 0.8$). Recall that these two events, A and B, are mutually exclusive. ▲▲

When events are *not mutually exclusive*, we cannot find the probability that one or the other occurs by simply adding the individual probabilities, as shown in Illustration 4-5. Why not? Let's look at an illustration and see what happens when events are not mutually exclusive.

▼ ILLUSTRATION 4-6

Using the sample space and the events defined in Illustration 4-3, find the probability that the student selected is "full-time" or "female," $P(A \text{ or } C)$.

SOLUTION If we look at the sample space, we see that $P(A) = \frac{140}{200}$, or 0.7, and that $P(C) = \frac{120}{200}$, or 0.6. If we add the two numbers 0.7 and 0.6 together, we will get 1.3, a number larger than one. We also know, from the basic properties of probability, that probability numbers are never larger than one. So what happened? If we take another look at the sample space we will see that 80 of the 200 students have been counted twice if we add $\frac{140}{200}$ and $\frac{120}{200}$. There are only 180 students that are "full-time" or "female." Thus the probability of A or C is

$$P(A \text{ or } C) = \tfrac{180}{200} = 0.9$$

(See the figure in Illustration 4-3, p. 223.) ▲▲

We can add probabilities to find the probability of an "or" compound event, but we must make an adjustment in situations such as the previous example.

addition rule

> ### GENERAL ADDITION RULE
> Let A and B be two events defined in a sample space S.
>
> $$P(A \text{ or } B) = P(A) + P(B) - P(A \text{ and } B) \qquad (4\text{-}4a)$$
>
> ### SPECIAL ADDITION RULE
> Let A and B be two events defined in a sample space. If A and B are **mutually exclusive** events, then
>
> $$P(A \text{ or } B) = P(A) + P(B) \qquad (4\text{-}4b)$$
>
> This can be expanded to consider **more than two** mutually exclusive events:
>
> $$P(A \text{ or } B \text{ or } C \text{ or} \ldots \text{or } E) = P(A) + P(B) + P(C) + \ldots + P(E)$$
> $$(4\text{-}4c)$$

The key to this formula is the property "mutually exclusive." If two events are mutually exclusive, there is no double counting of sample points. If events are not mutually exclusive, then when probabilities are added, the double counting will occur. Let's look at some examples.

In Figure 4-6, events A and B are mutually exclusive. Simple addition is justified, since the total probability of the shaded regions is sought. [In a Venn diagram, probabilities are often represented by enclosed areas. In Figure 4-6, for example, we've shaded the areas representing the probabilities $P(A)$ and $P(B)$.]

In Figure 4-7, events A and B are not mutually exclusive. The probability of the event "A and B," $P(A \text{ and } B)$, is represented by the area contained in region II. The probability $P(A)$ is represented by the area of circle A. That is, $P(A) = P(\text{region I}) + P(\text{region II})$. Also, $P(B) = P(\text{region II}) + P(\text{region III})$. And $P(A \text{ or } B)$ is the sum of the probabilities associated with the three regions:

$$P(A \text{ or } B) = P(I) + P(II) + P(III)$$

FIGURE 4-6
Mutually Exclusive Events

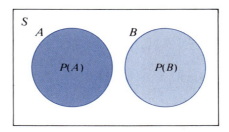

FIGURE 4-7
Nonmutually Exclusive Events

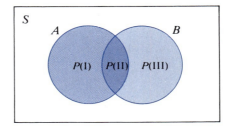

However, if $P(A)$ is added to $P(B)$, we have
$$P(A) + P(B) = [P(I) + P(II)] + [P(II) + P(III)]$$
$$= P(I) + 2P(II) + P(III)$$

This is the double count previously mentioned. However, if we subtract one measure of region II from this total, we will be left with the correct value.

The addition formula (4-4b) is a special case of the more general rule stated in formula (4-4a). If A and B are mutually exclusive events, $P(A \text{ and } B) = 0$. (They cannot both happen at the same time.) Thus the last term in formula (4-4a) is zero when events are mutually exclusive.

▼ **ILLUSTRATION 4-7**

One white die and one black die are rolled. Find the probability that the white die shows a number smaller than 3 or the sum of the dice is greater than 9.

SOLUTION 1   A = white die shows a 1 or a 2; B = sum of both dice is 10, 11, or 12.

$$P(A) = \frac{12}{36} = \frac{1}{3} \quad \text{and} \quad P(B) = \frac{6}{36} = \frac{1}{6}$$

$$P(A \text{ or } B) = P(A) + P(B) - P(A \text{ and } B)$$

$$= \frac{1}{3} + \frac{1}{6} - 0 = \mathbf{\frac{1}{2}}$$

$[P(A \text{ and } B) = 0$, since the events do not intersect.]

SOLUTION 2

$$P(A \text{ or } B) = \frac{n(A \text{ or } B)}{n(S)} = \frac{18}{36} = \frac{1}{2}$$

(Look at the sample space, Figure 4-4, p. 224, and count.)

▼ **ILLUSTRATION 4-8**

A pair of dice is rolled. Event T is defined as the occurrence of a "total of 10 or 11," and event D is the occurrence of "doubles." Find the probability $P(T \text{ or } D)$.

SOLUTION Look at the sample space of 36 ordered pairs for the rolling of two dice in Figure 4-4. Event T occurs if any one of 5 ordered pairs occurs; (4, 6), (5, 5), (6, 4), (5, 6), (6, 5). Therefore, $P(T) = \frac{5}{36}$. Event D occurs if any one of 6 ordered pairs occurs; (1, 1), (2, 2), (3, 3), (4, 4), (5, 5), (6, 6). Therefore, $P(D) = \frac{6}{36}$. Notice, however, that these two events are not mutually exclusive. The two events "share" the point (5, 5). Thus the probability $P(T \text{ and } D) = \frac{1}{36}$. As a result, the probability $P(T \text{ or } D)$ will be found using formula (4-4a).

$$P(T \text{ or } D) = P(T) + P(D) - P(T \text{ and } D)$$
$$= \tfrac{5}{36} + \tfrac{6}{36} - \tfrac{1}{36} = \tfrac{10}{36} = \tfrac{5}{18}.$$

(Look at the sample space in Figure 4-4).

▼▲ EXERCISES

4.33 Determine whether or not each of the following pairs of events is mutually exclusive:
   a. Five coins are tossed: "one head is observed," "at least one head is observed."
   b. A salesperson calls on a client and makes a sale: "the sale exceeds $100," "the sale exceeds $1000."
   c. One student is selected at random from a student body: the person selected is "male," the person selected is "over 21 years of age."
   d. Two dice are rolled: the total showing is "less than 7," the total showing is "more than 9."

4.34 Determine whether each of the following sets of events is mutually exclusive.
   a. Five coins are tossed: "no more than one head is observed," "two heads are observed," "three or more heads are observed."
   b. A salesperson calls on a client and makes a sale: the amount of the sale is "less than $100," is "between $100 and $1000," is "more than $500."
   c. One student is selected at random from the student body: the person selected is "female," is "male," or is "over 21."
   d. Two dice are rolled: the number of dots showing on each die are "both odd," are "both even," are "total seven," or "total eleven."

4.35 Explain why $P(A \text{ and } B) = 0$ when events A and B are mutually exclusive.

4.36 Explain why $P(A|B) = 0$ when events A and B are mutually exclusive.

4.37 If $P(A) = 0.3$ and $P(B) = 0.4$, and if A and B are mutually exclusive events, find the following:
    a. $P(\bar{A})$ **b.** $P(\bar{B})$ **c.** $P(A \text{ or } B)$ **d.** $P(A \text{ and } B)$

4.38 If $P(A) = 0.4$, $P(B) = 0.5$, and $P(A \text{ and } B) = 0.1$, find $P(A \text{ or } B)$.

4.39 One student is selected at random from a student body. Suppose the probability that this student is female is 0.5, and the probability that this student is working a part-time job is 0.6. Are the two events "female" and "working" mutually exclusive events? Explain.

4.40 The following table summarizes the teaching experience and educational background of the teachers in a public school:

Educational Background	Teaching Experience	
	Less Than 5 Years	5 Years or More
Less than a master's degree	75	40
Master's degree or more	55	30

Let A be the event that a teacher, selected at random, "has less than a master's degree" and let B represent "the teacher has less than five years experience." Find:
    a. $P(A \text{ and } B)$ **b.** $P(A \text{ or } B)$ **c.** $P(\bar{B})$

4.41 A parts store sells both new and used parts. Sixty percent of the parts in stock are used. Sixty-one percent are used or defective. If 5% of the store's parts are defective, what percentage are both used and defective?

4.42 Union officials report that 60% of the workers at a large factory belong to the union, 90% make over $5 per hour, and 40% belong to the union and make over $5 per hour. Do you believe these percentages? Explain.

4.43 The following survey results concerning the type of upgraded video boards purchased by 1126 different companies were presented in *PC Magazine* (May 29, 1990, page 64):

Type of Upgraded Video Board	Percent of Companies Purchasing
Hercules	7
8514 or compatible	9
EGA	20
VGA	45
Super VGA	58
Others	3

If each of the categories listed on the chart above is defined as an event, how do we know these events are not mutually exclusive?

4.44 In an article entitled "Trading Places" (*Newsweek*, July 16, 1990, pages 48–54), a survey of 7000 federal workers showed that nearly half said they cared for dependent adults. Of those approximately 3500, three quarters had missed some work. The hours of work missed could be presented as follows:

Hours of Work Missed in One Year	Percent
0	27
1–8	24
9–24	23
25–40	10
41–80	9
80 or more	7

For an individual randomly chosen from those in the survey who had cared for a dependent adult, find the probability that he/she
  a. missed more than a week of work in the year.
  b. missed a day or less of work in the year.

4.6 ▼ Independence, Multiplication Rule, and Conditional Probability

Next we will study the compound event "A and B." For example, what is the probability that two heads occur when two coins (one penny and one nickel) are tossed? Let A represent the occurrence of a head on the penny and B represent the occurrence of a head on the nickel. What is $P(A$ and $B)$? $P(A$ and $B)$ is found by using the definition $n(A$ and $B)/n(S)$. The sample space for Experiment 4-2 suggests that this value should be $\frac{1}{4}$. How can we obtain this value by using $P(A)$ and $P(B)$? Both $P(A)$ and $P(B)$ have values of $\frac{1}{2}$. If $P(A)$ is multiplied by $P(B)$, we obtain $\frac{1}{4}$. Thus we might suspect that $P(A$ and $B)$ equals $P(A) \cdot P(B)$.

Consider this example. The event that a 2 shows on a white die is A, and the event that a 2 shows on a black die is B. If both dice are rolled once, what is the probability that two 2s occur?

$$P(A) = \frac{1}{6} \quad \text{and} \quad P(B) = \frac{1}{6}$$

$$P(A \text{ and } B) = \frac{n(A \text{ and } B)}{n(S)} = \frac{1}{36}$$

However, note that $\frac{1}{6}$ multiplied by $\frac{1}{6}$ is also $\frac{1}{36}$. In this case, multiplication yields the correct answer. Multiplication does not always work, however. For example, P(sum of 7 and double) when two dice are rolled is zero (as seen in Figure 4-4). However, if $P(7)$ is multiplied by P(double), we obtain $\left(\frac{1}{6}\right)\left(\frac{1}{6}\right) = \frac{1}{36}$.

Multiplication does not work for P(sum of 10 and double), either. By definition and by inspection of the sample space, we know that $P(10$ and double$) = \frac{1}{36}$ (the point $(5,5)$ is the only element). However, if we multiply $P(10)$ by P(double), we obtain $\left(\frac{3}{36}\right)\left(\frac{6}{36}\right) = \frac{1}{72}$. The probability of this event cannot be both values.

independence

The property that is required for multiplying probabilities is **independence**. Multiplication worked in the two foregoing examples because the events were independent. In the other two cases, the events were not independent and multiplication gave us incorrect answers.

NOTE There are several situations that result in the compound event "and." Some of the more common ones are: (1) A followed by B, (2) A and B occurred simultaneously, (3) the intersection of A and B, (4) both A and B and (5) A but not B (equivalent to A and not B).

INDEPENDENCE AND CONDITIONAL PROBABILITIES

independent events

INDEPENDENT EVENTS

Two events A and B are independent events if the occurrence (or nonoccurrence) of one does not affect the probability assigned to the occurrence of the other.

Sometimes independence is quite easy to determine, for example, if the two events being considered have to do with unrelated trials, such as the tossing of a penny and a nickel. The results on the penny in no way affect the probability of heads or tails on the nickel. Similarly, the results on the nickel have no effect on the probability of heads or tails on the penny. Therefore, the results on the penny and the results on the nickel are *independent*. Also, if a coin and a die are tossed, either simultaneously or one following the other, the results on either one are independent of the results on the other. The coin and the die can each be thought of as a separate trial. However, if *events* are defined as combinations of outcomes from the separate trials, the independence of the events may or may not be so easy to determine. The separate results of each trial (dice in the next illustration) may be independent, but the compound events defined using both trials (both dice) may or may not be independent.

dependent events

Lack of independence, called dependence, is demonstrated by the following illustration. Reconsider the experiment of rolling two dice and observing the two events "sum of 10" and "double." As stated previously, $P(10) = \frac{3}{36} = \frac{1}{12}$ and P(double)$ = \frac{6}{36} = \frac{1}{6}$. Does the occurrence of 10 affect the probability of a double? Think of it this way. A sum of 10 has occurred; it must be one of the following: $\{(4,6), (5,5), (6,4)\}$. One of these three possibilities is a double. Therefore, we must conclude that the P (double, *knowing* 10 has occurred), written P (double $|$ 10), is $\frac{1}{3}$. Since $\frac{1}{3}$ does not equal the original probability of a double, $\frac{1}{6}$, we can conclude that the event "10" has an effect on the probability of a double. Therefore, "double" and "10" are dependent events.

Whether or not events are independent often becomes clear by examining the events in question. Rolling one die does not affect the outcome of a second roll. However, in many cases, independence is not self-evident, and the question of independence itself may be of special interest. Consider the events "having a checking account at a bank" and "having a loan account at the same bank." Having a checking account at a bank may increase the probability that the same person has a loan account. This has practical implications. For example, it would make sense to advertise loan programs to checking-account clients if they are more likely to apply for loans than are people who are not customers of the bank.

One approach to the problem is to *assume* independence or dependence. The correctness of the probability analysis depends on the truth of the assumption. In practice, we often assume independence and compare *calculated* probabilities with *actual frequencies* of outcomes in order to infer whether the assumption of independence is warranted.

conditional probability

> The symbol $P(A|B)$ represents the probability that A will occur given that B has occurred. This is called a **conditional probability**.

The previous definition of independent events can now be written in a more formal manner.

> **INDEPENDENT EVENTS**
>
> Two events A and B are independent events if
>
> $$P(A|B) = P(A) \text{ or if } P(B|A) = P(B) \qquad (4\text{-}5)$$

Let's consider conditional probability. Take, for example, the experiment in which a single die is rolled: $S = \{1, 2, 3, 4, 5, 6\}$. Two events that can be defined for this experiment are B = "an even number occurs" and A = "a 4 occurs." Then $P(A) = \frac{1}{6}$. Event A is satisfied by exactly one of the six equally likely sample points in S. The conditional probability of A given B, $P(A|B)$, is found in a similar manner, but S is no longer the sample space. Think of it this way. A die is rolled out of your sight, and you are told that the number showing is even, that is, event B has occurred. That is the given condition. Knowing this condition, you are asked to assign a probability to the event that the even number is a 4. There are only three possibilities in the new (or reduced) sample space, $\{2, 4, 6\}$. Each of the three outcomes is equally likely; thus $P(A|B) = \frac{1}{3}$. We can write this as

$$P(A|B) = \frac{P(A \text{ and } B)}{P(B)} \qquad (4\text{-}6)$$

Thus, for our example,

$$P(A|B) = \frac{\frac{1}{6}}{\frac{1}{2}} = \frac{1}{3}$$

Section 4.6 ▼ INDEPENDENCE, MULTIPLICATION RULE, AND CONDITIONAL PROBABILITY

▼ ILLUSTRATION 4-9

In a sample of 150 residents, each person was asked if he or she favored the concept of having a single countywide police agency. The county is composed of one large city and many suburban townships. The residence (city or outside the city) and the responses of the residents are summarized in Table 4-4. If one of these residents was to be selected at random, what is the probability that the person will (a) favor the concept? (b) favor the concept if the person selected is a city resident? (c) favor the concept if the person selected is a resident from outside the city? (d) Are the events F (favor the concept) and C (reside in city) independent?

TABLE 4-4
Sample Results for Illustration 4-9

Residence	Opinion		
	Favor (F)	Oppose ($\bar{F}$)	Total
In city (C)	80	40	120
Outside of city ($\bar{C}$)	20	10	30
Total	100	50	150

SOLUTION

(a) $P(F)$ is the proportion of the total sample that favor the concept. Therefore,

$$P(F) = \frac{n(F)}{n(S)} = \frac{100}{150} = \frac{2}{3}$$

(b) $P(F|C)$ is the probability that the person selected favors the concept given that he or she lives in the city. The sample space is reduced to the 120 city residents in the sample. Of these, 80 favored the concept; therefore,

$$P(F|C) = \frac{n(F \text{ and } C)}{n(C)} = \frac{80}{120} = \frac{2}{3}$$

(c) $P(F|\bar{C})$ is the probability that the person selected favors the concept, knowing that the person lives outside the city. The sample space is reduced to the 30 noncity residents; therefore,

$$P(F|\bar{C}) = \frac{n(F \text{ and } \bar{C})}{n(\bar{C})} = \frac{20}{30} = \frac{2}{3}$$

(d) All three probabilities have the same value, $\frac{2}{3}$. Therefore, we can say that the events F (favor) and C (reside in city) are independent. The location of residence did not affect $P(F)$. ▲▲

MULTIPLICATION RULE

general multiplication rule

GENERAL MULTIPLICATION RULE

Let A and B be two events defined in sample space S. Then

$$P(A \text{ and } B) = P(A) \cdot P(B|A) \quad \quad (4\text{-}7a)$$

or

$$P(A \text{ and } B) = P(B) \cdot P(A|B) \qquad (4\text{-}7b)$$

If events A and B are independent, then the general multiplication rule [formula (4-7)], reduces to the *special multiplication rule*, formula (4-8).

special multiplication rule

SPECIAL MULTIPLICATION RULE

Let A and B be two events defined in sample space S. If A and B are **independent events**, then

$$P(A \text{ and } B) = P(A) \cdot P(B) \qquad (4\text{-}8a)$$

This formula can be expanded. If A, B, C, ..., G are independent events, then

$$P(A \text{ and } B \text{ and } C \text{ and} \ldots \text{and } G) = P(A) \cdot P(B) \cdot P(C) \cdots P(G) \qquad (4\text{-}8b)$$

▼ ILLUSTRATION 4-10

One student is selected at random from a group of 200 known to consist of 140 full-time (80 female and 60 male) students and 60 part-time (40 female and 20 male) students. (See Illustration 4-3). Event A is "the student selected is full-time" and event C is "the student selected is female."

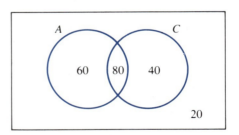

(a) Are events A and C independent? (b) Find the probability $P(A \text{ and } C)$ using the multiplication rule.

SOLUTION 1 (a) First find the probabilities $P(A)$, $P(C)$, and $P(A|C)$.

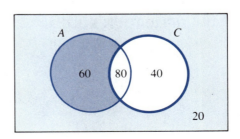

$$P(A) = \tfrac{140}{200} = 0.7$$
$$P(C) = \tfrac{120}{200} = 0.6$$
$$P(A|C) = \tfrac{80}{120} = 0.66$$

A and C are dependent events since $P(A) \ne P(A|C)$.

(b) $P(A \text{ and } C) = P(C) \times P(A|C)$
$$= \tfrac{120}{200} \times \tfrac{80}{120} = \tfrac{80}{200} = \mathbf{0.4}$$

SOLUTION 2 (a) First find the probabilities $P(A)$, $P(C)$, and $P(C|A)$.

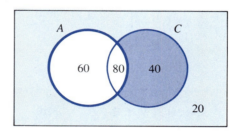

$$P(A) = \tfrac{140}{200} = 0.7$$
$$P(C) = \tfrac{120}{200} = 0.6$$
$$P(C|A) = \tfrac{80}{140} = 0.57$$

A and C are dependent events since $P(C) \ne P(C|A)$.

(b) $P(A \text{ and } C) = P(A) \times P(C|A)$
$$= \tfrac{140}{200} \times \tfrac{80}{140} = \tfrac{80}{200} = \mathbf{0.4}$$

▼ ILLUSTRATION 4-11

One white and one black die are rolled. Find the probability that the sum of their numbers is 7 and that the number on the black die is larger than the number on the white die.

SOLUTION A = "sum is 7"; B = "black number larger than white number." The "and" requires the use of the multiplication rule. However, we do not yet know whether events A and B are independent. (Refer to Figure 4-4 for the sample space of this experiment.) We see that $P(A) = \tfrac{6}{36} = \tfrac{1}{6}$. Also, $P(A|B)$ is obtained from the reduced sample space, which includes 15 points above the gray diagonal line. Of the 15 equally likely points, 3 of them—$(1,6)$, $(2,5)$, and $(3,4)$—satisfy event A. Therefore, $P(A|B) = \tfrac{3}{15} = \tfrac{1}{5}$. Since this is a different value than $P(A)$, the events are dependent. So we must use formula (4-7b) to obtain $P(A \text{ and } B)$.

$$P(A \text{ and } B) = P(B) \cdot P(A|B) = \tfrac{15}{36} \cdot \tfrac{3}{15} = \tfrac{3}{36} = \tfrac{1}{12}$$

NOTES

1. Independence and mutually exclusive are two very different concepts.
   a. Mutually exclusive says the two events cannot occur together; that is, they have no intersection.
   b. Independence says each event does not affect the other event's probability.
2. $P(A \text{ and } B) = P(A) \cdot P(B)$ when A and B are independent.
   a. Since $P(A)$ and $P(B)$ are not zero, $P(A \text{ and } B)$ is nonzero.
   b. Thus, independent events have an intersection.
3. Events cannot be both mutually exclusive and independent. Therefore,
   a. if two events are independent, then they are not mutually exclusive.
   b. if two events are mutually exclusive, then they are not independent.

Case Study 4-2

Save It for a Rainy Day

The bottom bar on the graph "Save It for a Rainy Day" seems to say a lot about the chances of winning a state lottery. According to this graph, you are 4 times more likely to be struck by lightning than you are to win Connecticut's lottery. Assuming you bought one ticket, what is the probability that you will win a state lottery? (See Exercise 4.61.)

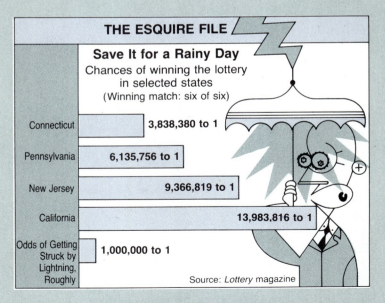

Source: "Save It for a Rainy Day," *Esquire* (February 1990). Reprinted with permission.

Section 4.6 ▼ INDEPENDENCE, MULTIPLICATION RULE, AND CONDITIONAL PROBABILITY   237

Case Study 4-3

New York State, like many other states and cities, runs several lottery games for the purpose of raising money to finance education. The PICK 10 game is currently one of several New York State games that a person can play. The game is explained and the chances of winning each of the various prizes is stated on the ticket. Did you know that you could win a lottery by being totally wrong? The rules say that if you pick ten numbers and zero of them match the numbers drawn, you win $4 on a $1 ticket with a 1 in 22 chance. Can you verify the probability of $\frac{1}{22}$? To have zero matches means that all ten numbers picked did not match the ten numbers picked by the lottery.

$$P(\text{zero matches}) = P(A \text{ and } B \text{ and } C \text{ and } D \ldots \text{ and } J),$$
$$(\text{all nonmatches})$$
$$= \left(\tfrac{60}{80}\right)\left(\tfrac{59}{79}\right)\left(\tfrac{58}{78}\right)\left(\tfrac{57}{77}\right)\left(\tfrac{56}{76}\right)\ldots\left(\tfrac{51}{71}\right)$$
$$= 0.04579$$
$$= \tfrac{1}{21.839} \text{ or approximately } \tfrac{1}{22}$$

Can you verify the other probabilities? (See Exercise 4.62.)

HOW TO PLAY PICK 10
(a) Each play card has 4 games, you may play 1, 2, 3 or all 4 games. Each game costs $1.00 to play
(b) Select 10 numbers in each game you want to play.
(c) Select the number of days you want to play these numbers.
(d) Use only blue or black ballpoint pen or pencil for marking. Red ink will not be accepted.
(e) Present your completed play card to your Pick 10 Agent for processing.
(f) If you can't think of 10 numbers, just ask for Quick Pick. The computer will choose your numbers for you.

HOW TO WIN
(a) Each night, the Lottery randomly selects 20 numbers from a field of 80; these 20 are the winning numbers.
(b) You may win if all, some or none of your selections match any of the winning numbers.

PRIZE LEVELS AND CHANCES OF WINNING

WINNING NUMBERS MATCHED PER GAME	PRIZE	CHANCES OF WINNING
10	$500,000	1 : 8,911,711
9	$ 6,000	1 : 163,381
8	$ 300	1 : 7,384
7	$ 40	1 : 621
6	$ 10	1 : 87
0	$ 4	1 : 22

Overall chances of winning: 1 in 17

Source: New York Pick 10 Play Card courtesy of NYS Division of the Lottery.

▼▲ EXERCISES

4.45 Determine whether or not each of the following pairs of events is independent:
  a. rolling a pair of dice and observing a "1" on the first die and a "1" on the second die
  b. drawing a "spade" from a regular deck of playing cards and then drawing another "spade" from the same deck without replacing the first card
  c. same as (b) except the first card is returned to the deck before the second drawing
  d. owning a red automobile and having blonde hair
  e. owning a red automobile and having a flat tire today
  f. studying for an exam and passing the exam

4.46 Determine whether or not the following pairs of events are independent:
  a. rolling a pair of dice and observing a "2" on one of the dice and having a "total of 10"
  b. drawing one card from a regular deck of playing cards and having a "red" card and having an "ace"
  c. raining today and passing today's exam
  d. raining today and playing golf today
  e. completing today's homework assignment and being on time for class

4.47 If $P(A) = 0.3$ and $P(B) = 0.4$ and A and B are independent events, what is the probability of each of the following?
  a. $P(A \text{ and } B)$ b. $P(B|A)$ c. $P(A|B)$

4.48 Suppose that $P(A) = 0.3$, $P(B) = 0.4$, and $P(A \text{ and } B) = 0.12$.
  a. What is $P(A|B)$?
  b. What is $P(B|A)$?
  c. Are A and B independent?

4.49 Suppose that $P(A) = 0.3$, $P(B) = 0.4$, and $P(A \text{ and } B) = 0.20$.
  a. What is $P(A|B)$?
  b. What is $P(B|A)$?
  c. Are A and B independent?

4.50 Suppose that A and B are events and that the following probabilities are known: $P(A) = 0.3$, $P(B) = 0.4$, and $P(A|B) = 0.2$. Find $P(A \text{ or } B)$.

4.51 A single card is drawn from a standard deck. Let A be the event that "the card is a face card" (a jack, a queen, or a king), B be the occurrence of a "red card," and C represent "the card is a heart." Check to determine whether the following pairs of events are independent or dependent:
  a. A and B   b. A and C   c. B and C

4.52 A box contains four red and three blue poker chips. What is the probability when three are selected randomly that all three will be red if we select each chip
  a. with replacement?   b. without replacement?

4.53 In a survey that questioned high school students about their attitudes toward business, 0.85 said, "Honesty is the best policy," and 0.28 stated, "Success in business requires some dishonesty." Assuming that the survey results both accurately re-

Section 4.6 ▼ INDEPENDENCE, MULTIPLICATION RULE, AND CONDITIONAL PROBABILITY   239

flect the attitudes of all high school students and that the answers to the two attitudes are independent of one another, what is the probability that
  a. a student did not say, "Honesty is the best policy," and agreed that "success in business requires some dishonesty?"
  b. a student said, "Honesty is the best policy," or agreed "success in business requires some dishonesty?

4.54 In the United States, the proportion of all births that involve Caesarean section deliveries is 15 out of 100. When Caesarean deliveries are made, 96 out of 100 babies survive. What is the probability that a randomly selected expectant woman will have a Caesarean section and that the baby will survive?

4.55 The owners of a two-person business make their decisions independently of each other, then compare their decisions. If they agree, the decision is made; if they do not agree, then further consideration is necessary before a decision is reached. If they each have a history of making the right decision 60% of the time, what is the probability that together they
  a. make the right decision on the first try?
  b. make the wrong decision on the first try?
  c. delay the decision for further study?

4.56 "Lying Is Just a Way of Life" is a report in the April 29, 1991 issue of *USA Today*. A new survey finds that 91% in the United States say they lie routinely, and 36% of those confess to dark, important lies. Based on these figures, if one person is picked at random, what is the probability that he or she routinely tells dark, important lies?

4.57 Consider the set of integers 1, 2, 3, 4, and 5.
  a. One integer is selected at random. What is the probability that it is odd?
  b. Two integers are selected at random (one at a time with replacement so that each of the five is available for a second selection).
Find the probability that (1) neither is odd; (2) exactly one of them is odd; (3) both are odd.

4.58 A box contains 25 parts, of which 3 are defective and 22 are nondefective. If 2 parts are selected without replacement, find the following probabilities:
  a. P(both are defective)   b. P(exactly one is defective)
  c. P(neither is defective)

4.59 In the article at the beginning of this chapter, a man was asked to identify the figurines he believed to be his from a set of 31 such figurines. The police knew that 11 of the figurines were not his. If the man had been guessing, what is the probability that he would be able to select 19 of the 20 unknown figurines? Does your answer agree with that of Professor Stephens? Why should this information convince the police that the man is telling the truth?

4.60 You are a contestant on a game show and given a choice of three doors. Behind one of them is a car and behind the other two are goats. You pick door number 3, and the host, knowing what's behind each door, has door number 2 opened, revealing a goat. He then asks you if you would like to keep door number 3 or switch to door number 1. Should you switch? What is the probability that door number 3 has the car? What is the probability that door number 1 has the car?

4.61 Case Study 4-2 shows the chances of winning lotteries in four different states and the chances of being struck by lightning. Using the numbers given in the Case Study, express each of the following probabilities in decimal form:
  a. The probability of being struck by lightning
  b. The probability of winning Connecticut's lottery
  c. The probability of winning California's lottery

4.62 Case Study 4-3 on p. 237, describes the rules for playing New York State's Pick 10 lottery. Many states have similar lottery games.
  a. If someone plays Pick 10, that person has a 1 in 8,911,711 chance of winning a half-million-dollar prize. Verify this probability.
  b. Verify that the chances of winning the $6,000 prize is 1 in 163,381.
  c. Verify that the overall chances of winning something are 1 in 17.

4.7 ▼ Combining the Rules of Probability

Many probability problems can be represented by tree diagrams. In these instances, the addition and multiplication rules can be applied quite readily. To illustrate the use of tree diagrams in solving probability problems, let's use Experiment 4-5. Two poker chips are drawn from a box containing one each of red, blue, and white chips. The tree diagram representing this experiment (Figure 4-8) shows a first drawing and then a second drawing. One chip was drawn on each drawing and not replaced.

After the tree has been drawn and labeled, we need to assign probabilities to each branch of the tree. If we assume that it is equally likely that any chip would be drawn at each stage, we can assign a probability to each branch segment of the tree, as shown in Figure 4-9. Notice that a set of branches that initiate from a

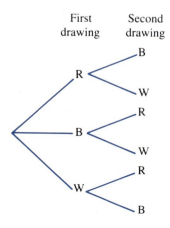

FIGURE 4-8 All Possible Combinations That Can Be Drawn

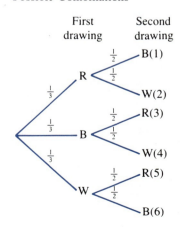

FIGURE 4-9 Probabilities of All Possible Combinations

single point has a total probability of 1. In this diagram there are four such sets of branch segments. The tree diagram shows six distinct outcomes. Reading down: branch (1) shows (R, B), branch (2) shows (R, W), and so on. (*Note:* Each outcome for the experiment is represented by a branch that begins at the common starting point and ends at the terminal points at the right.)

The probability associated with outcome (R, B), P(R on first drawing and B on second drawing), is found by multiplying P(R on first drawing) by P(B on second drawing | R on first drawing). These are the two probabilities $\frac{1}{3}$ and $\frac{1}{2}$ shown on the two branch segments of branch (1) in Figure 4-9. The $\frac{1}{2}$ is the conditional probability asked for by the multiplication rule. Thus we will multiply along the branches.

Some events will be made up of more than one outcome from our experiment. For example, suppose that we had asked for the probability that one red chip and one blue chip are drawn. You will find two outcomes that satisfy this event, branch (1) or branch (3). With "or" we will use the addition rule (4-4b). Since the branches of a tree diagram represent mutually exclusive events, we have

$$P(\text{one R and one B}) = \left(\frac{1}{3}\right)\left(\frac{1}{2}\right) + \left(\frac{1}{3}\right)\left(\frac{1}{2}\right) = \frac{1}{6} + \frac{1}{6} = \frac{1}{3}$$

NOTES

1. Multiply along the branches (Figure 4-10).
2. Add across the branches.

Now let's consider an example that places all the rules in perspective.

▼ ILLUSTRATION 4-12

A firm plans to test a new product in one randomly selected market area. The market areas can be categorized on the basis of location and population density. The number of markets in each category is presented in Table 4-5.

FIGURE 4-10
Multiply Along the Branches

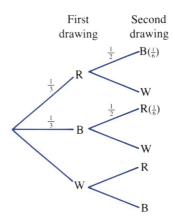

TABLE 4-5
Number of Markets by Location and by Population Density

Location	Population Density		Total
	Urban (U)	Rural (R)	
East (E)	25	50	75
West (W)	20	30	50
Total	45	80	125

What is the probability that the test market selected is in the East, $P(E)$? In the West, $P(W)$? What is the probability that the test market is in an urban area, $P(U)$? In a rural area, $P(R)$? What is the probability that the market is a western rural area, $P(W \text{ and } R)$? What is the probability it is an eastern or urban area, $P(E \text{ or } U)$? What is the probability that if it is in the East, it is an urban area, $P(U|E)$? Are "location" and "population density" independent? (What do we mean by independence or dependence in this situation?)

SOLUTION The first four probabilities, $P(E)$, $P(W)$, $P(U)$, and $P(R)$, represent "or" questions. For example, $P(E)$ means that the area is an eastern urban area or an eastern rural area. Since in this and the other three cases, the two components are mutually exclusive (an area can't be both urban and rural), the desired probabilities can be found by simply adding. In each case the probabilities are added across all the rows or columns of the table. Thus the totals are found in the total column or row.

$$P(E) = \frac{75}{125} \quad \text{(total for East divided by total number of markets)}$$

$$P(W) = \frac{50}{125} \quad \text{(total for West divided by total number of markets)}$$

$$P(U) = \frac{45}{125} \quad \text{(total for urban divided by total number of markets)}$$

$$P(R) = \frac{80}{125} \quad \text{(total for rural divided by total number of markets)}$$

Now we solve for $P(W \text{ and } R)$. There are 30 western rural markets and a total of 125 markets. Thus

$$P(W \text{ and } R) = \frac{30}{125}$$

Note that $P(W) \cdot P(R)$ does *not* give the right answer $[(\frac{50}{125})(\frac{80}{125}) = \frac{32}{125}.]$ Therefore, "location" and "population density" are dependent events.

$P(E \text{ or } U)$ can be solved in several different ways. The most direct way is to simply examine the table and count the number of markets that satisfy the condition that they are in the East or they are urban. We find 95, [25+50+20]. Thus

$$P(E \text{ or } U) = \frac{95}{125}$$

Note that the first 25 markets were both in the East and urban; thus E and U are not mutually exclusive events.

Another way to solve for $P(E \text{ or } U)$ is to use the addition formula:

$$P(E \text{ or } U) = P(E) + P(U) - P(E \text{ and } U)$$

which yields

$$\frac{75}{125} + \frac{45}{125} - \frac{25}{125} = \frac{95}{125}$$

A third way to solve the problem is to recognize that the complement of (E or U) is (W and R). Thus $P(E \text{ or } U) = 1 - P(W \text{ and } R)$. Using the previous calculation, we get $1 - \frac{30}{125} = \frac{95}{125}$.

Finally, we solve for $P(U|E)$. Looking at Table 4-5, we see that there are 75 markets in the East. Of the 75 eastern markets, 25 are urban. Thus

$$P(U|E) = \frac{25}{75}$$

The conditional probability formula could also be used:

$$P(U|E) = \frac{P(U \text{ and } E)}{P(E)}$$

$$= \frac{\frac{25}{125}}{\frac{75}{125}} = \frac{25}{75}$$

"Location" and "population density" are not independent events. They are dependent. This means that the probability of these events is affected by the occurrence of each other. ▲▲

Although each rule for computing compound probabilities has been discussed separately, you should not think they are only used separately. In many cases they are combined to solve problems. Consider the following two illustrations.

▼ **ILLUSTRATION 4-13**

A production process produces light bulbs. On the average, 20% of all bulbs produced are defective. Each item is inspected before being shipped. The inspector misclasses an item 10% of the time; that is,

$$P(\text{classified good}|\text{defective item}) = P(\text{classified defective}|\text{good item})$$

$$= 0.10$$

What proportion of the items will be "classified good"?

FIGURE 4-11

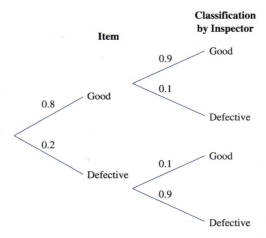

SOLUTION What do we mean by the event "classified good"?

$$G = \text{item good}$$
$$D = \text{item defective}$$
$$CG = \text{item called good by inspector}$$
$$CD = \text{item called defective by inspector}$$

CG consists of two possibilities: "the item is good and is correctly classified good" or "the item is defective and is misclassified good." Thus

$$P(CG) = P[(CG \text{ and } G) \text{ or } (CG \text{ and } D)]$$

Since the two possibilities are mutually exclusive, we can start by using the addition rule, formula (4-4b).

$$P(CG) = P(CG \text{ and } G) + P(CG \text{ and } D)$$

The condition of a bulb and its classification by the inspector are not independent. The multiplication rule for dependent events must be used. Therefore,

$$P(CG) = [P(G) \cdot P(CG|G)] + [P(D) \cdot P(CG|D)]$$

Substituting the known probabilities, we get

$$P(CG) = [(0.8)(0.9)] + [(0.2)(0.1)]$$
$$= 0.72 + 0.02$$
$$= \mathbf{0.74}$$

That is, 74% of the items are classified good.

▼ ILLUSTRATION 4-14

Reconsider Illustration 4-13. Suppose that only items that pass inspection are shipped. Items not classified good are scrapped. What is the quality of the shipped items? That is, what percentage of the items shipped are good, $P(G|CG)$?

SOLUTION Using the conditional probability formula (4-6),

$$P(G|CG) = \frac{P(G \text{ and } CG)}{P(CG)}$$

In Illustration 4-13 we found $P(CG)$; $P(G \text{ and } CG)$ was also found in the solution of Illustration 4-13. Thus,

$$P(G|CG) = \frac{P(G) \cdot P(CG|G)}{P(CG)}$$

$$= \frac{(0.8)(0.9)}{0.74} = 0.9729 = \mathbf{0.973}$$

In other words, 97.3% of all items shipped will be good. Inspection increases the quality of items sold from 80% good to 97.3% good. ▲▲

Case Study 4-4

Ask Marilyn

Marilyn presents a very simple and easy to understand solution to the game show problem, "Should you switch?" And the correct answer is...? Yes, you should switch. Marilyn puts at least three Ph.D.'s to shame. In this chapter you have learned to use tree diagrams to solve probability problems. Show that the strategy "switch" has a probability of $\frac{2}{3}$ while the strategy "do not switch" has only a $\frac{1}{3}$ probability of winning. Use tree diagrams. (See Exercise 4.70, p. 248.)

I'll come straight to the point. In the following question and answer, you blew it!

"You're on a game show and given a choice of three doors. Behind one is a car; behind the others are goats. You pick Door No. 1, and the host, who knows what's behind them, opens No. 3, which has a goat. He then asks if you want to pick No. 2. Should you switch?"

You answered, "Yes. The first door has a $\frac{1}{3}$ chance of winning, but the second has a $\frac{2}{3}$ chance."

Let me explain: If one door is shown to be a loser, that information changes the probability of either remaining choice to $\frac{1}{2}$. As a professional mathematician, I'm very concerned with the general public's lack of mathematical skills. Please help by confessing your error and, in the future, being more careful.

—Robert Sachs, Ph.D.,
George Mason University, Fairfax, Va.

You blew it, and you blew it big! Since you seem to have difficulty grasping the basic principle at work here, I'll explain: After the host reveals a goat, you now have a one-in-two chance of being correct. There is enough mathematical illiteracy in this country, and we don't need the world's highest IQ propagating more. Shame!

—Scott Smith, Ph.D.,
University of Florida

Your answer to the question is in error. But if it is any consolation, many of my colleagues have also been stumped by this problem.

—Barry Pasternack, Ph.D.,
California Faculty Association

Good heavens! With so much learned opposition, I'll bet this one is going to keep math classes all over the country busy on Monday.

My original answer is correct. But first, let me explain why your answer is wrong. The winning odds of $\frac{1}{3}$ on the first choice can't go up to $\frac{1}{2}$ just because the host opens a losing door. To illustrate this, let's say we play a shell game. You look away, and I put a pea under one of three shells. Then I ask you to put your finger on a shell. The odds that your choice contains a pea are $\frac{1}{3}$, agreed? Then I simply lift up an empty shell from the remaining two. As I can (and will) do this regardless of what you've chosen, we've learned nothing to allow us to revise the odds on the shell under your finger.

The benefits of switching are readily proved by playing through the six games that exhaust all the possibilities. For the first three games, you choose No. 1 and switch each time; for the second three games, you choose No. 1 and "stay" each time, and the host always opens a loser. Here are the results (each row is a game):

DOOR 1	DOOR 2	DOOR 3
AUTO	GOAT	GOAT

Switch and you lose.

GOAT	AUTO	GOAT

Switch and you win.

GOAT	GOAT	AUTO

Switch and you win.

AUTO	GOAT	GOAT

Stay and you win.

GOAT	AUTO	GOAT

Stay and you lose.

GOAT	GOAT	AUTO

Stay and you lose.

When you switch, you win two out of three times and lose one time in three; but when you don't switch, you only win one in three times and lose two in three. Try it yourself.

Section 4.7 ▼ COMBINING THE RULES OF PROBABILITY

Alternatively, you actually can play the game with another person acting as the host with three playing cards—two jokers for the goats and an ace for the auto. However, doing this a few hundred times to get statistically valid results can get a little tedious, so perhaps you can assign it for extra credit—or for punishment! (*That'll* get their goats!)

Source: Reprinted with permission from *Parade*, copyright © 1990.

▼▲ EXERCISES

4.63 If $P(A) = 0.4$ and $P(B) = 0.5$, and if A and B are independent events, find $P(A \text{ or } B)$.

4.64 $P(G) = 0.5$, $P(H) = 0.4$, and $P(G \text{ and } H) = 0.1$ (see the following diagram).
  a. Find $P(G|H)$. **b.** Find $P(H|G)$. **c.** Find $P(\overline{H})$.
  d. Find $P(G \text{ or } H)$. **e.** Find $P(G \text{ or } \overline{H})$.
  f. Are events G and H mutually exclusive? Explain.
  g. Are events G and H independent? Explain.

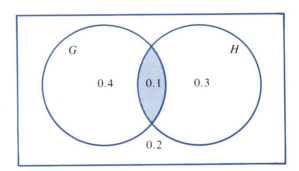

4.65 $P(R) = 0.5$, $P(S) = 0.3$, and events R and S are independent.
  a. Find $P(R \text{ and } S)$. **b.** Find $P(R \text{ or } S)$. **c.** Find $P(\overline{S})$.
  d. Find $P(R|S)$. **e.** Find $P(\overline{S}|R)$.
  f. Are events R and S mutually exclusive? Explain.

4.66 $P(M) = 0.3$, $P(N) = 0.4$, and events M and N are mutually exclusive.
  a. Find $P(M \text{ and } N)$. **b.** Find $P(M \text{ or } N)$. **c.** Find $P(M \text{ or } \overline{N})$.
  d. Find $P(M|N)$. **e.** Find $P(M|\overline{N})$.
  f. Are events M and N independent? Explain.

4.67 Two flower seeds are randomly selected from a package that contains five seeds for red flowers and three seeds for white flowers.
  a. What is the probability that both seeds will result in red flowers?
  b. What is the probability that one of each color is selected?
  c. What is the probability that both seeds are for white flowers?

4.68 The probability that a certain door is locked is 0.6. The key to the door is one of five unidentified keys hanging on a key rack. Two keys are randomly selected before approaching the door. What is the probability that the door may be opened without returning for another key? (*Hint*: Draw a tree diagram.)

4.69 A, B, and C each in turn toss a balanced coin. The first one to throw a head wins.
  a. What are their respective chances of winning if each tosses only one time?
  b. What are their respective chances of winning if they continue, given a maximum of two tosses each?

4.70 Marilyn presented a very clear and accurate explanation of the solution to the game show question, "Should I switch doors?" (Case Study 4-4, p. 245). Use a tree diagram, with probabilities, and show that switching does improve your chances of winning.
  a. Find the probability of winning if you play using the "switch doors" strategy.
  b. Find the probability of winning if you play using the "do not switch doors" strategy.

(You might also look back at your solution to Exercise 4.60, p. 239.)

4.71 Probabilities for events A, B, and C are distributed as shown on the following figure. Find
  a. $P(A \text{ and } B)$ **b.** $P(A \text{ or } C)$ **c.** $P(A|C)$

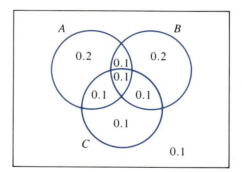

4.72 A coin is flipped three times.
  a. Draw a tree diagram that represents all possible outcomes.
  b. Identify all branches that represent the event "exactly one head occurred."

4.73 Box 1 contains two red balls and three green balls, and Box 2 contains four red balls and one green ball. One ball is randomly selected from Box 1 and placed in Box 2. Then one ball is randomly selected from Box 2. What is the probability that the ball selected from Box 2 is green?

4.74 A company that manufactures shoes has three factories. Factory 1 produces 25% of the company's shoes, Factory 2 produces 60%, and Factory 3 produces 15%. One percent of the shoes produced by Factory 1 are mislabeled, 0.5% of those produced by Factory 2 are mislabeled, and 2% of those produced by Factory 3 are mislabeled. If you purchase one pair of shoes manufactured by this company, what is the probability that the shoes are mislabeled?

4.75 An article entitled "A Puzzling Plague" found in the January 14, 1991 issue of *Time* (pages 48–52), stated that one out of every ten American women will get breast cancer. It also states that of those who do, one out of four will die of it. Use these probabilities to find the probability that a randomly selected American woman will

    a. never get breast cancer.
    b. get breast cancer and not die of it.
    c. get breast cancer and die from it.

4.76 An article in the March 25, 1991 issue of *US News and World Reports* (pages 69, 70) discussed the use of clotbusters to save the lives of Americans who have heart attacks. The following projections are given:

Treatment	Percent Surviving
A: with current clotbuster use	66
B: with optimum clotbuster use	70
C: with optimum clotbuster use plus prompt medical care	85

Suppose these projections are correct and that in a study 50% received treatment A, 30% received treatment B, and 20% received treatment C. What is the probability that a given individual in this study survived?

4.8 ▼ Bayes's Rule

The Reverend Thomas Bayes (1702–1761), an English Presbyterian minister and mathematician, developed an expanded form for conditional probabilities. This expanded rule, called Bayes's rule, allows us to revise (or adjust) the probabilities assigned to events in accordance with new information.

Bayes's rule

BAYES'S RULE

$$P(A_i | B) = \frac{P(A_i) \cdot P(B | A_i)}{\sum [P(A_i) \cdot P(B | A_i)]} \quad (4\text{-}9)$$

where $A_1, \ldots, A_n$ is an all-inclusive set of possible outcomes given B.

Although Bayes's formula looks difficult, if you use a tabular approach, you will find it easy to use.

▼ ILLUSTRATION 4-15

Let's solve Illustration 4-14 by using Bayes's rule.

SOLUTION The first step is to set up a table that shows all the possible outcomes given event B, that is, "shipped." These outcomes are listed in the first column of Table 4-6.

In the second column we list the probabilities for each of the A_i outcomes in the first column. In the third column we list the conditional probability that B happens for each A_i, $P(B|A_i)$. [For our illustration, $P(B|A_1) = P(CG|G)$ and $P(B|A_2) = P(CG|D)$.] These first three columns represent the information obtained from the problem.

TABLE 4-6 Tabular Presentation of Given Information

(1) A_i, Possible Outcomes	(2) $P(A_i)$	(3) $P(B\|A_i)$
A_1, item good	0.8	0.9
A_2, item defective	0.2	0.1
Total	1.0 ⓒⓚ	

To solve for the conditional probabilities $P(A_i|B)$, the first calculation is to multiply each number in the row of column (2) by the number in the same row of column (3). This product is placed in column (4) of the table (see Table 4-7). The column is labeled $P(A_i \text{ and } B)$. The values calculated represent the probability that both A_i and B will occur. Thus 72% of the items produced will be good and classified good; 2% of the items will be defective and classified good.

TABLE 4-7 Tabular Solution of Bayes's Rule

(1) A_i, Possible Outcomes	(2) $P(A_i)$	(3) $P(B\|A_i)$	(4) $P(A_i \text{ and } B)$ $= P(A_i)P(B\|A_i)$	(5) $P(A_i\|B)$
A_1, item good	0.8	0.9	0.72	$0.72/0.74 = \mathbf{0.973}$ $= P(G|\text{shipped})$
A_2, item defective	0.2	0.1	0.02	$0.02/0.74 = \mathbf{0.027}$ $= P(D|\text{shipped})$
Total	1.0 ⓒⓚ		$0.74 = P(B)$	1.000 ⓒⓚ

The second step is to add up column (4). The sum represents $P(B)$. Thus 74% of the items produced will be classified good.

Finally, the answers we are looking for, the conditional probabilities $P(A_i|B)$, are obtained by dividing each number in column (4) by the total of column (4). The results are placed in column (5) and are the answers. Thus 0.973 is the proportion of items classified good that are good. And 0.027 is the proportion of items classified good that are actually defective.

NOTE The total of columns (2) and (5) must equal 1. The total of column (3) need not equal 1. ▲▲

Bayes's rule is of special interest because it gives us a mechanism to revise initial probability estimates when new information is learned, as we see in the next illustration.

▼ **ILLUSTRATION 4-16**

prior

revised, or posterior

Consider the situation where we feel that the probability that a stock is a good buy is 0.4. That is, our **prior** (before new information) probabilities are $P(\text{good buy}) = 0.4$ and $P(\text{bad buy}) = 0.6$. Now we find out that an investment service that has a record of being right 80% of the time recommends the stock. What should be our **revised**, or **posterior** (after new information), probability that the stock is a good buy, that is, $P(\text{good buy}|\text{investment service recommends it})$, or $P(A_i|B)$?

SOLUTION Using the Bayesian tabular analysis, we find that the revised, or posterior, probability is 0.727 that the stock is a good buy and 0.273 that it is a bad buy (see Table 4-8).

TABLE 4-8
Tabular Analysis for Illustration 4-14

A_i	$P(A_i)$	$P(B\|A_i)$	$P(A_i)P(B\|A_i)$	$P(A_i\|B)$
A_1, good buy	0.4	0.8	0.32	$\dfrac{0.32}{0.44} = 0.727$
A_2, bad buy	0.6	0.2	0.12	$\dfrac{0.12}{0.44} = 0.273$
Total	1.0 ⓒⓚ		0.44 = P(B)	1.000 ⓒⓚ

▲▲

In analyzing the use of Bayes's rule to revise prior probability estimates in light of new information, we note the following relationship: The stronger the prior probability, the less the effect of the new information on changing the probability. Also, the more conclusive the new information, the greater the impact on the revised probability.

▼▲ EXERCISES

4.77

	Male		Female		
	Skilled	Unskilled	Skilled	Unskilled	Total
Satisfied	350	150	25	100	625
Unsatisfied	150	100	75	50	375
Total	500	250	100	150	1000

Use the accompanying table on worker satisfaction in the Russell Microprocessor Company.
  a. Find the probability that an unskilled employee is satisfied with the work.
  b. Find the probability that a skilled woman employee is satisfied with the work.
  c. Is satisfaction for women employees independent of their being skilled or unskilled?

4.78 In an article entitled "Why Quitting Means Gaining" (*Time*, March 25, 1991, page 55), it was reported that giving up cigarette smoking often results in gaining weight. In examining a group of quitters, the following data were found:

	Weight Gain			
	Major	Significant	Moderate	Slight
Men	9%	14%	22%	55%
Women*	12	11	26	50

*Due to rounding, numbers for women do not total 100%.

Suppose that 60% of the group were men and 40% were women. If a participant were randomly selected and found to have experienced
  a. a major weight gain, find the probability that it was a male.
  b. a slight weight gain, find the probability that it was a woman.

4.79 Given the information in the accompanying table, compute $P(A_1|UF)$ and $P(A_2|UF)$ by filling in the rest of the table. (UF = unfavorable survey results.)

	$P(A_i)$	$P[(UF)|A_i]$	$P(A_i \text{ and } UF)$	$P(A_i|UF)$
A_1, profitable	0.6	0.4		
A_2, not profitable	0.4	0.7		

4.80 Given the following:

$P(A_1) = 0.2$ $P(A_2) = 0.4$ $P(A_3) = 0.3$ $P(A_4) = 0.1$
$P(B|A_1) = 0.5$ $P(B|A_2) = 0.4$ $P(B|A_3) = 0.2$ $P(B|A_4) = 0.1$

Find $P(A_1|B)$, $P(A_2|B)$, $P(A_3|B)$, and $P(A_4|B)$.

IN RETROSPECT

You have now studied the basic concepts of probability. These fundamentals need to be understood to allow us to continue our study of statistics. Probability is the vehicle of statistics, and we have begun to see how probabilistic events occur. We have explored theoretical and experimental probabilities for the same event. Does the experimental probability turn out to have the same value as the theoretical? Not exactly, but over the long run we have seen that it does not have approximately the same value.

You must, of course, know and understand the basic definition of probability as well as understand the properties of mutual exclusiveness and independence as they apply to the concepts presented in this chapter.

From reading this chapter you should be able to relate the "and" and the "or" of compound events to the multiplication and addition rules. You should also be able to calculate conditional probabilities.

In the next three chapters we will look at distributions associated with probabilistic events. This will prepare us for the statistics that will follow. We must be able to predict the variability that the sample will show with respect to the population before we will be successful at "inferential statistics," in which we describe the population based on the sample statistics available.

CHAPTER EXERCISES

4.81 A fair coin is tossed 100 million times. Let B be the total number of heads observed. Identify each of the following statements as true or false. Explain your answer (no computations required).

    a. It is very probable that B is very close (within a few thousand) to 50 million.

    b. It is very probable that $\frac{B}{100}$ million is very close to $\frac{1}{2}$; maybe between 0.49 and 0.51.

    c. Since Hs and Ts fall with complete irregularity, one cannot say anything about what happens in 100 million tosses. It is just as likely that B is equal to one value as to any other value; for example, $P(B = 2) = P(B = 50 \text{ million})$.

    d. Hs and Ts fall about equally often. Therefore, if on the first ten tosses we get all Hs, it is more probable that the eleventh toss will yield a T than an H.

4.82 The article "Birth-Control Breakthrough" (*Newsweek*, December 24, 1990, page 68) stated that of America's 58 million women of childbearing age, 60% practice some form of contraception. A breakdown of the methods is as follows:

Method	Number (in millions)
Sterilization	13.7
The pill	10.7
Condom only	5.1
Diaphragm	2.0
Periodic abstinence	0.8
IUD	0.7
Spermicide only	0.6
Sponge	0.4
Other methods	0.9

What is the probability that an American woman of childbearing age
   a. uses the condom only method?
   b. uses the sterilization or the pill method?
   c. What assumption was necessary in order to obtain an answer in part (b)?
   d. uses one of the methods given above?
   e. What assumption was necessary in order to obtain an answer in part (d)? Since we do not have the mutually exclusive property, place an upper bound on the probability that one of the methods given was used.

4.83 Suppose a certain ophthalmic trait is associated with eye color. Three hundred randomly selected individuals are studied with results as follows:

Trait	Eye Color			Totals
	Blue	Brown	Other	
Yes	70	30	20	120
No	20	110	50	180
Totals	90	140	70	300

   a. What is the probability that a person selected at random has blue eyes?
   b. What is the probability that a person selected at random has the trait?
   c. Are events A (has blue eyes) and B (has the trait) independent? Justify your answer.
   d. How are the two events A (has blue eyes) and C (has brown eyes) related (independent, mutually exclusive, complementary, or all inclusive)? Explain why or why not each term applies.

CHAPTER EXERCISES

4.84 The employees at a large university were classified according to age as well as to whether they belonged to administration, faculty, or staff.

	Age Group			
	20–30	31–40	41–50	51 or Over
Administration	2	24	16	17
Faculty	1	40	36	28
Staff	16	20	14	2

For a randomly selected employee, find the probability that he or she
  a. was in the administration or was 51 or over.
  b. was not a faculty member.
  c. was a faculty member given that the individual was 41 or over.

4.85 Events R and S are defined on the same sample space. If $P(R) = 0.2$ and $P(S) = 0.5$, explain why each of the following statements is either true or false.
  a. If R and S are mutually exclusive, then $P(R \text{ or } S) = 0.10$.
  b. If R and S are independent, then $P(R \text{ or } S) = 0.6$.
  c. If R and S are mutually exclusive, then $P(R \text{ and } S) = 0.7$.
  d. If R and S are mutually exclusive, then $P(R \text{ or } S) = 0.6$.

4.86 Show that if event A is a subset of event B, then $P(A \text{ or } B) = P(B)$.

4.87 Let's assume that there are three traffic lights between your house and a friend's house. As you arrive at each light, it may be red (R) or green (G).
  a. List the sample space showing all possible sequences of red and green lights that could occur on a trip from your house to your friend's. (RGG represents red at the first light and green at the other two.)
     Assuming that each element of the sample space is equally likely to occur,
  b. what is the probability that on your next trip to your friend's house you will have to stop for exactly one red light?
  c. what is the probability that you will have to stop for at least one red light?

4.88 Assuming that a woman is equally likely to bear a boy as a girl, use a tree diagram to compute the probability that a four-child family consists of one boy and three girls.

4.89 Suppose that when a job candidate comes to interview for a job at RJB Enterprises, the probability that he or she will want the job (A) after the interview is 0.68. Also, the probability that RJB wants the candidate (B) is 0.36. The probability $P(A|B)$ is 0.88.
  a. Find $P(A \text{ and } B)$.   b. Find $P(B|A)$.
  c. Are events A and B independent? Explain.
  d. Are events A and B mutually exclusive? Explain.
  e. What would it mean to say A and B are mutually exclusive events in this exercise?

4.90 A traffic analysis at a busy traffic circle in Washington, D.C., showed that 0.8 of the autos using the circle entered from Connecticut Avenue. Of those entering the traffic circle from Connecticut Avenue, 0.7 continued on Connecticut Avenue at the opposite side of the circle. What is the probability that a randomly selected auto observed in the traffic circle entered from Connecticut and will continue on Connecticut?

4.91 A fair die is rolled five times. What is the probability that a 6 occurred for the first time on the fifth roll?

4.92 In the game of craps, you win on the first roll of a pair of dice if the sum of 7 or 11 occurs. You lose on the first roll if the sum 2, 3, or 12 occurs.
    a. What is the probability that you win on the first roll?
    b. What is the probability that you lose on the first roll?

4.93 The probabilities of women being employed in certain occupational groups are as follows:

Occupational Group	Probability
Professional/technical	0.16
Managerial/administrative	0.06
Sales	0.07
Clerical	0.34
Craft	0.02
Operatives, including transport	0.12
Non-farm laborers	0.01
Service, except private households	0.18
Private households	0.03
Farm	0.01
Total	1.00

    a. What is the probability that a randomly selected woman is either a professional or a managerial employee?
    b. What is the probability that two randomly selected women will both be clerical employees?
    c. What is the probability that a randomly selected woman will be both a sales and a clerical employee?

4.94 Patient's level of satisfaction concerning information they receive from doctors, nurses, and pharmacists about over the counter (OTC) medications is discussed in *Drug Topics* (special supplement, January 21, 1991, page 85). The results may be summarized as follows:

Patient's Level of Satisfaction with OTC Information

	From Pharmacist	From Doctor	From Nurse
Extremely Satisfied	66%	64%	46%
Somewhat Satisfied	32	34	51
Not Satisfied	2	2	3

Assuming the level of satisfaction is independent of the source of information, what is the probability of
    a. dissatisfaction with the information from doctors and nurses?
    b. satisfaction with the information from doctors or nurses?

4.95 In a health report found in the *US News and World Report* (February 18, 1991, page 66), the Center for Disease Control reports that the odds are between 1 in 42,000 and 1 in 417,000 of your picking up the human immunodeficiency virus (HIV) from an infected surgeon who uses proper precautions. The CDC puts the chance of contacting HIV from an infected dentist at between 1 in 263,000 and 1 in 2,632,000.
    a. Give a range for the probabilities of contacting the HIV from a surgeon. From a dentist.
    b. Assume that the probability of getting HIV from both is zero (i.e., mutually exclusive) and find a range for the probability of contacting the HIV from a surgeon or a dentist.

4.96 According to the National Cancer Data Base report for Hodgkin's disease (*CA—A Cancer Journal for Clinicians*, January/February 1991, page 16), the highest percentage of patients (31%) were 20 to 29 years of age, and they had a three-year observed survival rate of 91%. What is the probability that an individual who has been diagnosed with Hodgkin's disease is between 20 and 29 years of age *and* will survive for three years?

4.97 A store contains 12 aisles. Ten of the aisles are in the grocery section, labeled G1 to G10, and two are in the pharmacy section, labeled P1 and P2. If the store manager randomly selects an aisle to be used for a special display, find the probability that
    a. the aisle is an even-numbered aisle.
    b. the aisle is in the grocery section and has an odd number.
    c. the aisle is in the pharmacy section, given that it is an even-numbered aisle.

4.98 Tires salvaged from a train wreck are on sale at the Getrich Tire Company. Of the 15 tires offered in the sale, 5 tires have suffered internal damage and the remaining 10 are damage free. If you were to randomly select and purchase 2 of these tires,

    a. what is the probability that the tires you purchase are both damage free?
    b. what is the probability that exactly 1 of the tires you purchase is damage free?
    c. what is the probability that at least 1 of the tires you purchase is damage free?

4.99 One thousand persons screened for a certain disease are given a clinical exam. As a result of the exam, the sample of 1000 persons is distributed according to height and disease status.

	Disease Status				
Height	None	Mild	Moderate	Severe	Totals
Tall	122	78	139	61	400
Medium	74	51	90	35	250
Short	104	71	121	54	350
Totals	300	200	350	150	1000

Use this information to estimate the probability of being medium or short in height and of having moderate or severe disease status.

4.100 The following table shows the sentiments of 2500 wage-earning employees at the Spruce Company on a proposal to emphasize fringe benefits rather than wage increases during their impending contract discussions.

	Opinion			
Employee	Favor	Neutral	Opposed	Total
Male	800	200	500	1500
Female	400	100	500	1000
Total	1200	300	1000	2500

    a. Calculate the probability that an employee selected at random from this group will be opposed.
    b. Calculate the probability that an employee selected at random from this group will be female.
    c. Calculate the probability that an employee selected at random from this group will be opposed, given that the person is male.
    d. Are the events "opposed" and "female" independent? Explain.

4.101 A shipment of grapefruit arrived containing the following proportions of types: 10% pink seedless, 20% white seedless, 30% pink with seeds, 40% white with

seeds. A grapefruit is selected at random from the shipment. Find the probability that
- **a.** it is seedless.
- **b.** it is white.
- **c.** it is pink and seedless.
- **d.** it is pink or seedless.
- **e.** it is pink, given that it is seedless.
- **f.** it is seedless, given that it is pink.

4.102 The following information was generated from a recent survey of women employed by a predominantly female company when asked the question, "Is college more important for a man than a woman?"

Age	Is College More Important for a Man Than a Woman?		
	Yes	No	Total
18–24	0.17	0.43	0.60
Over 24	0.26	0.14	0.40
Total	0.43	0.57	1.00

Using this information,
- **a.** what is the probability that a randomly selected woman employee felt college was more important for a man than a woman?
- **b.** what is the probability that a randomly selected woman answered "Yes" or was "over 24"?
- **c.** what is the probability that a randomly selected woman was "over 24" given that she answered "No" to the question?

4.103 The probability that thunderstorms are in the vicinity of a particular midwestern airport on an August day is 0.70. When thunderstorms are in the vicinity, the probability that an airplane lands on time is 0.80. Find the probability that thunderstorms are in the vicinity and the plane lands on time.

4.104 According to automobile accident statistics, one out of every six accidents results in an insurance claim of $100 or less in property damage. Three cars insured by an insurance company are involved in different accidents. Consider the following two events:

A: The majority of claims exceed $100.
B: Exactly two claims are $100 or less.

- **a.** List the sample points for this experiment.
- **b.** Are the sample points equally likely?
- **c.** Find $P(A)$ and $P(B)$.
- **d.** Are A and B independent? Justify your answer.

4.105 A box contains ten ball-point pens, eight of which write properly and two of which do not write. You are going to randomly select two of these pens.
- **a.** Describe the sample space with a tree diagram.
- **b.** Assign probabilities to each branch of your tree diagram. Consider these three events: A, a good pen is drawn on the first drawing; B, a good pen

is drawn on the second drawing; C, at least one defective pen is selected. Find these probabilities:

  c. $P(A)$ d. $P(B)$ e. $P(C)$
  f. $P(\overline{A})$ g. $P(\overline{B})$ h. $P(\overline{C})$
  i. $P(A \text{ and } B)$ j. $P(A \text{ and } C)$ k. $P(B \text{ and } C)$
  l. $P(A \text{ or } B)$ m. $P(A \text{ or } C)$ n. $P(B \text{ or } C)$
  o. Are events A and B mutually exclusive? Explain.
  p. Are events A and C mutually exclusive? Explain.
  q. Are events A and B independent? Explain.
  r. Are events A and C independent? Explain.

4.106 Salesman Adams and saleswoman Jones call on three and four customers, respectively, on a given day. Adams could make 0, 1, 2, or 3 sales, whereas Jones could make 0, 1, 2, 3, or 4 sales. The sample space listing the number of possible sales for each person on a given day is given in the following table:

Adams	Jones				
	0	1	2	3	4
0	0, 0	1, 0	2, 0	3, 0	4, 0
1	0, 1	1, 1	2, 1	3, 1	4, 1
2	0, 2	1, 2	2, 2	3, 2	4, 2
3	0, 3	1, 3	2, 3	3, 3	4, 3

(3, 1 stands for 3 sales by Jones and 1 sale by Adams.) Assume that each sample point is equally likely. Let's define these events:

$$A = \text{at least one of the salespersons made no sales}$$
$$B = \text{together they made exactly three sales}$$
$$C = \text{each made the same number of sales}$$
$$D = \text{Adams made exactly one sale}$$

Find the following probabilities by *counting* sample points:
  a. $P(A)$ b. $P(B)$ c. $P(C)$
  d. $P(D)$ e. $P(A \text{ and } B)$ f. $P(B \text{ and } C)$
  g. $P(A \text{ or } B)$ h. $P(B \text{ or } C)$ i. $P(A|B)$
  j. $P(B|D)$ k. $P(C|B)$ l. $P(B|\overline{A})$
  m. $P(C|\overline{A})$ n. $P(A \text{ or } B \text{ or } C)$

Are the following pairs of events mutually exclusive? Explain.
  o. A and B   p. B and C   q. B and D

Are the following pairs of events independent? Explain.
  r. A and B   s. B and C   t. B and D

4.107 A testing organization wishes to rate a particular brand of television. Six TVs are selected at random from stock. If nothing is found wrong with any of the six, the brand is judged satisfactory.
    a. What is the probability that the brand will be rated satisfactory if 10% of the TVs actually are defective?
    b. What is the probability that the brand will be rated satisfactory if 20% of the TVs actually are defective?
    c. What is the probability that the brand will be rated satisfactory if 40% of the TVs actually are defective?

4.108 The town council has nine members. A proposal must have at least two-thirds of the votes to be accepted. A proposal to establish a new industry in this town has been tabled. If we know that two members of the town council are opposed and that the others randomly vote "in favor" and "against," what is the probability that the proposal will be accepted?

4.109 Coin A is loaded in such a way that $P(\text{heads})$ is 0.6. Coin B is a balanced coin. Both coins are tossed. Find the following:
    a. the sample space that represents this experiment; assign a probability measure to each outcome
    b. $P(\text{both show heads})$
    c. $P(\text{exactly one head shows})$
    d. $P(\text{neither coin shows a head})$
    e. $P(\text{both show heads} \mid \text{coin A shows a head})$
    f. $P(\text{both show heads} \mid \text{coin B shows a head})$
    g. $P(\text{heads on coin A} \mid \text{exactly one head shows})$

4.110 On a slot machine there are three reels with digits 0, 1, 2, 3, 4, and 5 and a flower on each reel. When a coin is inserted and the lever pulled, each of the three reels spins independently and comes to rest on one of the seven positions mentioned. Find these probabilities:
    a. a flower shows on all three reels
    b. a flower shows on exactly one reel
    c. a flower shows on exactly two reels
    d. no flowers show
    e. a three-digit sequence (in order) appears
    f. exactly one flower shows and the other two are odd integers

4.111 Professor French forgets to set his alarm with a probability of 0.3. If he sets the alarm it rings with a probability of 0.8. If the alarm rings, it will wake him on time to make his first class with a probability of 0.9. If the alarm does not ring, he wakes in time for his first class with a probability of 0.2. What is the probability that Professor French wakes in time to make his first class tomorrow?

4.112 A two-page typed report contains an error on one of the pages. Two proofreaders review the copy. Each has an 80% chance of catching the error. What is the probability that the error will be identified if
    a. each reads a different page?
    b. they each read both pages?

c. the first proofreader randomly selects a page to read, then the second proofreader randomly selects a page unaware of which page the first selected?

4.113 A computer program generates pairs of random integers. Each integer ranges between 0 and 5, inclusive, so that any one of 36 pairs is equally likely. Let A be the event "the first integer of the pair is a zero" and let B represent the event "the sum of the two integers is an even number." Are A and B independent or dependent events?

4.114 For a particular population, 30% are in the age group 0–20 years, 40% are in the age group 21–40 years, and the remainder are over 40 years old. Given that an individual from this population is in the 0–20 age group, the probability that the individual has an abnormal result on a glucose tolerance test is 0.05. The conditional probability for an abnormal glucose tolerance test is 0.04 for the 21–40 age group and is 0.10 for the over-40 age group. For a randomly selected individual from this population, what is the probability that he or she has an abnormal glucose tolerance test?

4.115 Solve this exercise by using Bayes's Rule in tabular form. The treasurer's initial opinion is that there is a 30% chance that an investment will exceed expectations, a 50% chance that it will equal expectations, and a 20% chance that it will return less than expected. A private investment consulting service reviews the investment and reports that it should equal expectations. In the past the consultants were correct 60% of the time, underestimated the return 10% of the time, and overestimated the return 30% of the time. What should be the treasurer's revised probabilities?

4.116 Ninety percent of the insulators produced by Superior Insulator Company are satisfactory. The firm hires an inspector. The inspector inspects all the insulators and correctly classifies an item 90% of the time; that is, P(classify good | good) = P(classify defective | defective) = 0.9. Items classified good are shipped and those classified defective are scrapped.
   a. What percentage of items shipped can be expected to be good?
   b. What percentage of items scrapped can be expected to be good?

4.117 The firm in exercise 4.116 hires a second inspector, who has the same accuracy record. The second inspector inspects all insulators independently of the first inspector. What percentage of items shipped and what percentage of items scrapped can be expected to be good if items are shipped only if
   a. both inspectors independently say they are good
   b. at least one inspector says they are good

4.118 In sports, championships are often decided by two teams playing each other in a championship series. Often the fans of the losing team claim they were unlucky and their team is actually the better team. Suppose team A is the better team, and the probability it will beat team B in any given game is 0.6. What is the probability that the better team A will loose the series if it is
   a. a one-game series?
   b. a best out of three series?

c. a best out of seven series?
d. Suppose the probability that A would beat B in any given game were actually 0.7. Recompute (a) through (c).
e. Suppose the probability that A would beat B in a given game were actually 0.9. Recompute (a) through (c).
f. What is the relationship between the "best" team winning and the number of games played? The best team winning and the probabilities that each will win?

VOCABULARY LIST

Be able to define each term. In addition, describe, in your own words, and give an example of each term. Your examples should not be ones given in class or in the textbook.

The bracketed numbers indicate the chapter in which the term first appeared, but you should define the terms again to show increased understanding of their meaning.

addition rule
all-inclusive events
Bayes's Rule
complementary event
compound event
conditional probability
dependent events
empirical probability
equally likely events
event
experiment [1]
experimental probability
general multiplication rule
independence
independent events
intersection
law of large numbers
listing

long-term average
multiplication rule
mutually exclusive events
odds
ordered pair
outcome
prior probability
posterior probability
probability of an event
relative frequency [2]
revised probability
sample point
sample space
special multiplication rule
subjective probability
theoretical probability
tree diagram
Venn diagram

KEY CONCEPTS

independent events
mutually exclusive events

probability
relative frequency

QUIZ A

Answer "True" if the statement is always true. If the statement is not always true, replace the words shown in bold with words that make the statement always true.

4.1 The probability of an event is a **whole number**.

4.2 The concepts of probability and relative frequency as related to an event are very **similar**.

4.3 The **sample space** is the theoretical population for probability problems.

4.4 The sample points of a sample space are **equally likely** events.

4.5 The value found for experimental probability will **always** be exactly equal to the theoretical probability assigned to the same event.

4.6 The probabilities of complementary events always **are equal**.

4.7 If two events are mutually exclusive, they are also **independent**.

4.8 If events A and B are **mutually exclusive**, the sum of their probabilities must be exactly one.

4.9 If the sets of sample points belonging to two different events do not intersect, the events are **independent**.

4.10 A compound event formed by use of the word *and* requires the use of the **addition rule**.

QUIZ B

4.1 A computer is programmed to generate the eight single-digit integers 1, 2, 3, 4, 5, 6, 7 and 8 with equal frequency. Consider the experiment—"the next integer generated." Define:

$$\text{Event A} = \text{"Odd number"} = \{1, 3, 5, 7\}$$

$$\text{Event B} = \text{"Number more than 4"} = \{5, 6, 7, 8\}$$

$$\text{Event C} = \text{"1 or 2"} = \{1, 2\}$$

Find:
- **a.** $P(A)$
- **b.** $P(B)$
- **c.** $P(C)$
- **d.** $P(\overline{C})$
- **e.** $P(A \text{ and } B)$
- **f.** $P(A \text{ or } B)$
- **g.** $P(B \text{ and } C)$
- **h.** $P(B \text{ or } C)$
- **i.** $P(A \text{ and } C)$
- **j.** $P(A \text{ or } C)$
- **k.** $P(A|B)$
- **l.** $P(B|C)$
- **m.** $P(A|C)$
- **n.** Are events A and B mutually exclusive? Explain.
- **o.** Are events B and C mutually exclusive? Explain.
- **p.** Are events A and C mutually exclusive? Explain.
- **q.** Are events A and B independent? Explain.
- **r.** Are events B and C independent? Explain.
- **s.** Are events A and C independent? Explain.

4.2 Given that events A and B are mutually exclusive and $P(A) = 0.4$ and $P(B) = 0.3$, find
- **a.** $P(A \text{ and } B)$
- **b.** $P(A \text{ or } B)$
- **c.** $P(A|B)$
- **d.** Are A and B independent? Explain.

4.3 Given that events C and D are independent and $P(C) = 0.2$ and $P(D) = 0.7$, find
    a. $P(C \text{ and } D)$ **b.** $P(C \text{ or } D)$
    c. $P(C|D)$ **d.** Are C and D mutually exclusive? Explain.

4.4 Given events E and F with probabilities $P(E) = 0.5$, $P(F) = 0.4$, and $P(E \text{ and } F) = 0.2$, find
    a. $P(E \text{ or } F)$ **b.** $P(E|F)$
    c. Are E and F mutually exclusive? Explain.
    d. Are E and F independent? Explain.

4.5 Given events G and H with probabilities $P(G) = 0.3$, $P(H) = 0.2$, and $P(G \text{ and } H) = 0.1$, find
    a. $P(G \text{ or } H)$ **b.** $P(G|H)$
    c. Are G and H mutually exclusive? Explain.
    d. Are G and H independent? Explain.

4.6 Janice wants to become a police officer. She must pass a physical exam and then a written exam. Records show the probability of passing the physical exam is 0.85, and that once the physical is passed the probability of passing the written exam is 0.60. What is the probability that Janice passes both exams?

QUIZ C

4.1 Explain briefly how you would decide which of the following two events is the more unusual:

    A: a 90-degree day in Vermont, or
    B: a 100-degree day in Florida

4.2 Student A says that "independence" and "mutually exclusive" are basically the same thing; namely, both mean "neither event has anything to do with the other one." Student B argues that although Student A's statement has some truth in it, Student A has missed the point of these two properties. Student B is correct. Carefully explain why.

4.3 Using complete sentences, describe in your own words
    a. mutually exclusive events   **b.** independent events
    c. the probability of an event   **d.** a conditional probability

5 PROBABILITY DISTRIBUTIONS (DISCRETE VARIABLES)

Chapter Outline

5.1 Random Variables
To study probability distributions, a **numerical value** will be assigned to each outcome in the sample space.

5.2 Probability Distributions of a Discrete Random Variable
The probability of a value of the random variable is expressed by a **probability function**.

5.3 Mean and Variance of a Discrete Probability Distribution
Population parameters are used to measure the probability distribution.

5.4 The Binomial Probability Distribution
Binomial probability distributions occur in situations where each trial of an experiment has **two possible outcomes**.

5.5 Mean and Standard Deviation of the Binomial Distribution
Two simple **formulas** are used to measure the binomial distribution.

An Analysis of Accidents at a Day Care Center

An analysis of 1324 accidents over a 42-month period at a university day care center revealed that toddlers had the highest average number of injuries, most of them self-induced, that accidents peaked in midmorning, and that September was the month with the highest accident rate. Although accidents were frequent, injuries were minor. Results are contrasted to those of earlier studies.

The dramatic increase in the number of working women with preschool age children has escalated the need for day care. Parents seek warm, caring, stimulating, affordable child care, but recent news reports of injuries from physical and sexual abuse in day care centers have raised questions of safety. Concern is so great that centers report insurance coverage is more difficult to obtain and, where available, rates have risen sharply. The true incidence of injuries from serious abuse and neglect in day care is not known, but injuries due to accidents are likely to be more common. According to Gratz,

> Accidents . . . are the leading cause of childhood mortality and rank second only to acute infections as the cause of morbidity and visits to the physician throughout childhood.

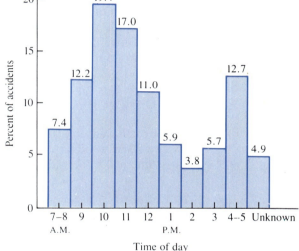

FIGURE 1 Percent of Accidents by Time of Day

Source: From Richard Elardo, Ph.D., Hope C. Solomons, Ed.D., Bill C. Snider, Ph.D., *American Journal of Orthopsychiatry,* 57(1), January 1987. Reprinted, with permission, from the *American Journal of Orthopsychiatry.* Copyright 1987 by the American Orthopsychiatric Association, Inc.

Chapter Objectives

Chapter 2 dealt with frequency distributions of data sets, and Chapter 4 dealt with the fundamentals of probability. Now we are ready to combine these ideas to form *probability distributions,* which are much like relative frequency distributions. The basic difference between probability and relative frequency distributions is that probability distributions are theoretical probabilities (populations) while relative frequency distributions are empirical probabilities (samples).

In this chapter we will investigate discrete probability distributions and study measures of central tendency and dispersion for such distributions. Special emphasis will be given to the binomial random variable and its probability distribution, since it is the most important discrete random variable encountered in most fields of application.

5.1 ▼ Random Variables

If each event in a probability experiment is assigned a numerical value, then as we observe the result of the experiment we are observing a random variable. This numerical value will be the value of the *random variable* under study.

random variable

> **RANDOM VARIABLE**
>
> A variable that assumes a unique numerical value for each of the outcomes in the sample space of a probability experiment.

In other words, a random variable is used to denote the outcome of a probability experiment. It can take on any numerical value that belongs to the set of all possible outcomes of the experiment. (It is called "random" because the value it assumes is the result of a chance, or random, event.)

Each event in a probability experiment must also be defined in such a way that only one value of the random variable is assigned to it, and each event must have a value assigned to it. Typically, the **discrete random variable** is a count of something.

discrete random variable

The following illustrations demonstrate what we mean by random variable.

▼ ILLUSTRATION 5-1

We toss five coins and observe the number of heads visible. The random variable x is the number of heads observed and may take on integer values from 0 to 5.

▲▲

▼ ILLUSTRATION 5-2

In Experiment 4-7 we rolled two dice and observed the sum of the dots on both dice. If a random variable had been assigned, it would have been the total number of dots showing. In this case, the random variable x could take on integer values from 2 to 12.

▼ ILLUSTRATION 5-3

Let the number of phone calls received per day by a company be a random variable. It could take on integer values ranging from zero to some very large number.

▼ ILLUSTRATION 5-4

The "model year" of automobiles owned by the faculty of a college could be used as a random variable. An observed probability distribution would result if the data were presented as a relative frequency distribution.

For example, suppose we find that 0.20 of the faculty-owned cars are 1991 models, 0.25 are 1990 models, 0.28 are 1989, 0.15 are 1988, and so on. (See Figure 5-1.) The random variable "model year of automobile" is a discrete random variable since it identifies the year in which the car was manufactured.

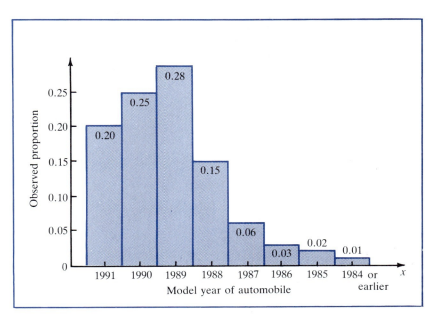

FIGURE 5-1
Model Year of Automobiles Owned by a College's Faculty

▼▲ EXERCISES

5.1 A social worker is involved in a study about family structure. She obtains information regarding the number of children per family for a certain community from the census data. Identify the random variable of interest and list its possible values.

5.2 An experiment involves the testing of a new on/off switch. The switch is flipped on and off until it breaks and the flip on which the break occurs is noted. Identify the random variable of interest and list its possible values.

5.3 An archer shoots eight arrows at a bull's-eye of one target. Identify the random variable most likely to be of interest and list its possible values.

5.4 Radios are packed in cartons of 12. A carton of radios is inspected and the number of defective radios found is recorded. Identify the random variable used and list its possible values.

5.2 ▼ Probability Distributions of a Discrete Random Variable

Recall the coin-tossing experiment we used at the beginning of Section 4-1 (p. 202). Two coins were tossed and no heads, one head, or two heads were observed. If we define the random variable x to be the number of heads observed when two coins are tossed, x can take on the values 0, 1, or 2. The probability of each of these three events is the same as we calculated in Chapter 4 (pp. 203, 210):

$$P(x = 0) = P(0H) = \tfrac{1}{4}$$
$$P(x = 1) = P(1H) = \tfrac{1}{2}$$
$$P(x = 2) = P(2H) = \tfrac{1}{4}$$

These probabilities can be listed in any number of ways, but they are best displayed in the form illustrated by Table 5-1. Can you see why the name "probability distribution" is used?

TABLE 5-1 Probability Distribution: Tossing Two Coins

x	P(x)
0	$\tfrac{1}{4}$
1	$\tfrac{1}{2}$
2	$\tfrac{1}{4}$

probability distribution

PROBABILITY DISTRIBUTION

A distribution of the probabilities associated with each of the values of a random variable. The probability distribution is a theoretical distribution; it is used to represent populations.

In the experiment in which a single die is rolled and the number of dots on the top surface is observed, the random variable is the number observed. The probability distribution for this random variable is shown in Table 5-2.

TABLE 5-2
Probability Distribution: Rolling a Die

x	P(x)
1	$\frac{1}{6}$
2	$\frac{1}{6}$
3	$\frac{1}{6}$
4	$\frac{1}{6}$
5	$\frac{1}{6}$
6	$\frac{1}{6}$

Sometimes it is convenient to write a rule that expresses the probability of an event in terms of the value of the random variable. This expression is typically written in formula form and is called a *probability function*.

probability function

> **PROBABILITY FUNCTION**
> A rule that assigns probabilities to the values of the random variables.

A probability function can be as simple as a list pairing the values of a random variable with their probabilities. Tables 5-1 and 5-2 show two such listings. However, a probability function is most often expressed in formula form.

Consider a die that has been modified so it has one face with one dot, two faces with two dots, and three faces with three dots. Let x be the number of dots observed when this die is rolled. The probability distribution for this experiment is presented in Table 5-3. Each of the probabilities can be represented by the value of x divided by 6. That is, each $P(x)$ is equal to the value of x divided by 6, where $x = 1$, 2, or 3. Thus $P(x) = \frac{x}{6}$ for $x = 1$, 2, and 3 is the formula expression of the probability function of this experiment.

TABLE 5-3
Probability Distribution: Rolling the Modified Die

x	P(x)
1	$\frac{1}{6}$
2	$\frac{2}{6}$
3	$\frac{3}{6}$

The probability function of the experiment of rolling one ordinary die is $P(x) = \frac{1}{6}$ for $x = 1, 2, 3, 4, 5,$ and 6. This particular function is called a **constant function** because the value of $P(x)$ does not change as x changes.

constant function

Every probability function must display the two basic properties of probability. These two properties are (1) the probability assigned to each value of the random variable must be between 0 and 1, inclusive, that is,

$$0 \leq \text{each } P(x) \leq 1$$

and (2) the sum of the probabilities assigned to all the values of the random variable must equal 1, that is,

$$\sum_{\text{all } x} P(x) = 1$$

Is $P(x) = \frac{x}{10}$ for $x = 1, 2, 3$, and 4 a probability function? To answer this question we need only test the function in terms of the two basic properties. The probability distribution is shown in Table 5-4. Property 1 is satisfied, since $\frac{1}{10}, \frac{2}{10}, \frac{3}{10}$, and $\frac{4}{10}$ are all numerical values between 0 and 1. Property 2 is also satisfied, since the sum of all four probabilities is exactly 1. Since both properties are satisfied, we can conclude that $P(x) = \frac{x}{10}$ for $x = 1, 2, 3, 4$ *is* a probability function.

TABLE 5-4 Probability Distribution for $P(x) = \frac{x}{10}$

x	P(x)
1	$\frac{1}{10}$
2	$\frac{2}{10}$
3	$\frac{3}{10}$
4	$\frac{4}{10}$
Total	$\frac{10}{10} = 1$

What about $x = 5$ (or any value other than 1, 2, 3, or 4) in the function $P(x) = \frac{x}{10}$ for $x = 1, 2, 3$, and 4? $P(x = 5)$ is considered to be zero. That is, the probability function provides a probability of zero for all values of x other than the values specified.

Figure 5-2 presents a probability distribution graphically. Regardless of the specific graphic representation used, the values of the random variable are plotted on the horizontal scale, and the probability associated with each value of the random variable is plotted on the vertical scale. A discrete random variable should really be presented by a **line histogram,** since the variable can assume only discrete values. Figure 5-2 shows the probability distribution of $P(x) = \frac{x}{10}$ for $x = 1, 2, 3$, and 4. This line representation makes sense because the variable x can

line histogram

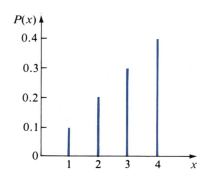

FIGURE 5-2 Line Histogram of the Probability Distribution for $P(x) = \frac{x}{10}$ for $x = 1, 2, 3, 4$

assume *only* the values 1, 2, 3, and 4. The length of the vertical line represents the value of the probability.

probability histogram

However, a regular **histogram** is more frequently used to present probability distributions. Figure 5-3 shows the same probability distribution of Figure 5-2 but in histogram form. This representation suggests that x can assume all numerical values (fractions, decimals, and so on), but this is not the case.

FIGURE 5-3

Histogram of the Probability Distribution of $P(x) = \frac{x}{10}$ for $x = 1, 2, 3, 4$

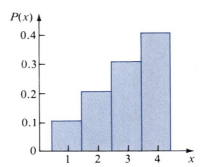

The histogram probability distribution uses the physical area of each bar to represent its assigned probability. The bar for $x = 2$ is 1 unit wide (from 1.5 to 2.5) and is 0.2 unit high. Therefore, its area is $(1)(0.2) = 0.2$, the probability assigned to $x = 2$. The areas of the other bars can be determined in similar fashion. This area representation will be an important concept in Chapter 6 when we begin to work with continuous random variables.

Case Study 5-1

Who Needs the Ambulance?

Robert Giordana used a relative frequency histogram to help him explain how his ambulance services are used. The histogram is of a discrete variable (number of trips per day) and shows, in percentages, the relative frequency of days with various numbers of service trips. Explain how this information satisfies the requirements of a discrete probability distribution. (See Exercise 5.11, p. 277.)

The Austin City Ambulance Company last week appealed to the City Council for additional municipal funding. Mr. Robert Giordana, the company's business manager, stated that while people see ambulances on the streets occasionally, they rarely have any real concept of the frequency with which an ambulance is called upon for assistance.

In surveying the company's records, the Council found that one ambulance responds to between one and six calls for help on a typical day. The records for a recent six-month period showed that an ambulance made three trips on 25% of the days and four trips on 21% of the days. It was further revealed that there was only one day in every three weeks when no trips were made. But on one day in the same three weeks, they made seven or more trips.

Mr. Giordana reminded the Council that the company was working hard for the community all year long.

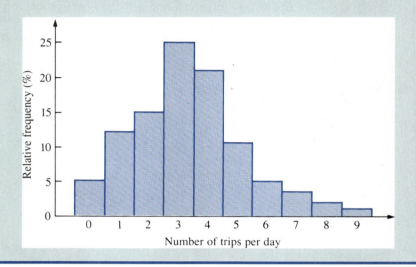

Case Study 5-2

Woes of the Weekend Jock

The information about injuries to the recreational athlete is summarized on a chart showing what appear to be many different relative frequency distributions. These numbers are relative frequencies, but they do not always form relative frequency distributions. Explain why. (See Exercise 5.12.)

THE PART-TIME ATHLETE IS NOT IMMUNE TO A HOST OF INJURIES

The pursuit of fitness, a phenomenon of the '70s and '80s, is responsible for generating another phenomenon: Both general practitioners and sports-medicine specialists are noticing a surge in sports-related injuries among weekend and after-work jocks. Basketball and soccer leagues, ski vacations, evening runs, dance classes, tennis games—all foster injuries once found chiefly among professional and college athletes.

For a comprehensive look at the problems of this new breed of recreational athlete, the Center for Sports Medicine at Saint Francis Memorial Hospital in San Francisco recently compiled statistics on over 10,000 injuries treated at the center. There were some interesting findings:

▼ Nine activities—basketball, dance, football, gymnastics, running, skiing, tennis, soccer and figure skating—account for nearly three-fourths of the injuries.

▼ More than two-thirds of the injuries are caused by overuse—problems such as shinsplints and tendinitis that develop from a repetitive

trauma to muscle and bone. Tennis, aerobic dance and running frequently cause such problems.
▼ The remaining injuries are acute ones, incidents that happen instantly, such as a sprained ankle. Not surprisingly, these tend to occur in skiing, football, basketball and soccer.
▼ Injuries to the knee occasion the most visits to the center, and skiers have the most knee problems.
▼ Aerobic dance causes more fractures than any other recreational activity.
▼ Many problems stem from acute injuries that occurred in the past.
▼ Unlike football injuries, most of the basketball injuries (62 percent) occur in participants over 25 years of age.

Injuries to the Recreational Athlete [age and sex, anatomic site of injury, and injury type (% of total) in 7,458 athletes]

Patients by sport (No.)	Basketball 377	Dance 2,242	Football 257	Running 3,004	Skiing 425	Soccer 271	Tennis 882
Age and Sex							
0–18	17.5	21.3	33.1	4.8	2.8	40.2	5.2
19–25	20.1	24.2	31.1	11.3	16.5	27.7	6.6
26–39	55.4	45.5	29.6	61.3	58.6	27.7	35.6
40 and over	6.9	8.8	6.2	22.6	22.1	4.4	52.6
Male	87.8	19.8	99.2	55.2	50.8	69.4	63.5
Female	12.3	80.2	0.8	44.8	49.2	30.6	36.5
Anatomic Site of Injury							
Shoulder	4.0	2.3	13.6	0.4	10.8	2.9	15.1
Upper extremity	3.4	1.4	8.2	0.3	14.3	2.2	22.2
Spine	4.8	9.6	9.3	4.9	3.3	4.8	7.9
Hip	4.0	8.7	1.5	7.7	1.2	3.7	3.9
Thigh	2.4	4.0	5.8	4.1	—	4.4	1.9
Knee	42.2	28.6	42.8	38.2	67.3	46.1	24.4
Leg	6.6	12.7	1.9	11.7	3.5	6.6	6.9
Ankle	21.2	12.1	8.6	8.5	4.7	15.5	6.6
Foot	8.5	14.5	0.8	20.7	2.3	9.6	7.9
Fracture	5.6	9.4	8.9	7.8	3.8	4.8	1.6
Dislocation	4.5	2.4	8.5	1.4	8.7	2.6	2.0
Sprain	36.3	9.5	29.2	5.9	42.6	26.9	10.1
Strain	12.2	23.9	14.0	18.0	11.8	16.6	27.5
Inflammation	15.9	18.6	5.8	36.3	7.1	15.5	31.2
Overuse	8.5	16.4	5.4	19.1	12.2	14.0	12.2
Other	13.2	11.7	18.3	5.8	8.8	14.4	10.0

All percentages don't total 100 because some categories weren't listed and there were multiple injuries in some sports

Source: Amy Wilbur, *Science Digest,* March 1986.

▼▲ EXERCISES

5.5 Census data are often used to obtain probability distributions for various random variables. Census data for families with a combined income of $50,000 or more in a particular state show that 20% have no children, 30% have one child, 40% have two children, and 10% have three children. From this information, construct the probability distribution for x, where x represents the number of children per family for this income group.

5.6 A company manufactures insulin needles and packages them in boxes of 100. Based on historical data obtained from sampling such boxes over a period of years, it is known that 90% of all such boxes contain no defective needles, 7% contain exactly one defective needle, and 3% contain exactly two defective needles. Based on this information, what is the probability distribution for x, where x represents the number of defective needles per box?

5.7 Test the following function to determine whether it is a probability function. If it is not, try to make it into a probability function. List the distribution of probabilities and sketch a histogram.

$$P(x) = \frac{5 - x}{10} \quad \text{for} \quad x = 1, 2, 3, 4$$

5.8 Test the following function to determine whether it is a probability function. If it is not, try to make it into a probability function. $R(x) = 0.2$ for $x = 0, 1, 2, 3, 4$.
  a. List the distribution of probabilities.
  b. Sketch a histogram.

5.9 Test the following function to determine whether it is a probability function. If it is not, try to make it into a probability function.

$$S(x) = \frac{6 - |x - 7|}{36} \quad \text{for} \quad x = 2, 3, 4, 5, 6, 7, \ldots, 11, 12$$

  a. List the distribution of probabilities and sketch a histogram.
  b. Do you recognize $S(x)$? If so, identify it.

5.10 In the November 1990 issue of *Fitness Management* (page 10), adult fitness participation percentages are given for several different income groups. For individuals with an annual dollar household income from $35,000 to $49,000, the fitness participation percentages are as follows:

Level of Participation	Percent
Frequent	22
Occasional	44
Nonparticipants	34

Define x to represent the level of participation. If the level is frequent, let $x = 3$; if the level is occasional, let $x = 2$; if nonparticipant, let $x = 1$. List the probability distribution for x.

5.11 a. List the information shown on the histogram in Case Study 5-1 (p. 273) as a probability distribution.
b. Are these theoretical probabilities? Explain.
c. Explain how this information forms a discrete probability distribution.

5.12 a. Why are the numbers reported on the chart in Case Study 5-2 (p. 274) relative frequencies?
b. Why does the set of numbers reported for skiing and age form a probability distribution?
c. Why does the set of numbers reported for skiing and anatomic site of injury not form a probability distribution?

5.13 MINITAB was used to generate a random sample of 25 observations drawn from the probability distribution

x	1	2	3	4	5
$P(x)$	0.2	0.3	0.3	0.1	0.1

Verify these results and draw a histogram of the sample.

```
MTB > READ C1 C2
     5 ROWS READ
MTB > RANDOM 25 OBSERVATIONS INTO C3;
SUBC> DISCRETE C1,C2.
MTB > PRINT C3

C3
    2   2   4   3   3   2   1   3   3   3   2   1   3
    1   2   2   5   5   3   2   5   4   2   2   3
MTB > HISTOGRAM C3;
SUBC> INCREMENT 1;
SUBC> START 1.
Histogram of C3    N = 25

Midpoint   Count
   1.00      3    ***
   2.00      9    *********
   3.00      8    ********
   4.00      2    **
   5.00      3    ***
```

5.14 Use a computer to generate a random sample of 100 observations drawn from the discrete probability population $P(x) = (5 - x)/10$, for $x = 1, 2, 3, 4$. List the resulting sample and print a histogram of the sample. (If you use MINITAB, see Exercise 5.13 for a program.)

5.3 ▼ Mean and Variance of a Discrete Probability Distribution

Recall that the mean of a frequency distribution (sample mean, p. 70) is

$$\bar{x} = \frac{\sum xf}{n}$$

which can also be written as

$$\bar{x} = \sum \left(x \cdot \frac{f}{n} \right)$$

We learned in our study of probability that the probability of an event is the expected relative frequency of its occurrence. Thus, if we replace f/n with $P(x)$, we can find the mean of a theoretical probability distribution:

$$\text{mean value of } x = \sum [x \cdot P(x)] \quad \quad (5\text{-}1)$$

NOTES

1. $\bar{x}$ is the mean of a sample.
2. s is the standard deviation of the individual elements of the sample.
3. $\bar{x}$ and s are called **sample statistics**.
4. μ (Greek letter mu) is the mean of the population under consideration.
5. σ (Greek letter sigma) is the standard deviation of the individual elements of the population under consideration.
6. μ and σ are called **population parameters**. (A parameter is a constant. μ and σ are typically unknown values.)

Recall that the "expected relative frequency" represents what will occur in the long run. The mean of x is therefore the mean of the entire population of experimental outcomes. The symbol for the **mean** value of x (in a probability distribution) is μ, and formula (5-1) can be written as

$$\mu = \sum [x \cdot P(x)] \quad \quad (5\text{-}2)$$

(Note: The summation is "over all x-values.")

▼ ILLUSTRATION 5-5

Let's return to the previous probability function: $P(x) = \frac{x}{10}$ for $x = 1, 2, 3$, and 4. We can find its mean by using formula (5-2) after we compile a probability distribution table (see Table 5-5). The mean μ, or the sum of the products of x times $P(x)$, is $\frac{30}{10}$, or **3.0**. [The sum of the probabilities, $\Sigma P(x)$, must be 1.0. Use this as a check.]

TABLE 5-5 Probability Distribution: $P(x) = \frac{x}{10}$, $x = 1, 2, 3, 4$

x	$P(x)$	$x \cdot P(x)$
1	$\frac{1}{10}$	$\frac{1}{10}$
2	$\frac{2}{10}$	$\frac{4}{10}$
3	$\frac{3}{10}$	$\frac{9}{10}$
4	$\frac{4}{10}$	$\frac{16}{10}$
Total	$\frac{10}{10} = 1.0$	$\mu = \frac{30}{10} = \mathbf{3.0}$

Section 5.3 ▼ MEAN AND VARIANCE OF A DISCRETE PROBABILITY DISTRIBUTION

variance of probability distribution

The **variance** of discrete probability distributions is defined in much the same way as the variance of sample data:

$$\sigma^2 = \sum [(x - \mu)^2 \cdot P(x)] \quad (5\text{-}3)$$

The variance of x for the probability distribution discussed in this illustration is found as shown in Table 5-6. The variance σ^2 is $\frac{10}{10}$, or **1.0**. ($\mu = 3$ was found in Table 5-5.)

TABLE 5-6
Finding the Variance for $P(x) = \frac{x}{10}$, $x = 1, 2, 3, 4$

x	$x - \mu$	$(x - \mu)^2$	$P(x)$	$(x - \mu)^2 P(x)$
1	−2	4	$\frac{1}{10}$	$\frac{4}{10}$
2	−1	1	$\frac{2}{10}$	$\frac{2}{10}$
3	0	0	$\frac{3}{10}$	$\frac{0}{10}$
4	1	1	$\frac{4}{10}$	$\frac{4}{10}$
Total				$\sigma^2 = \frac{10}{10}$

Formula (5-3) is often not convenient to use; fortunately it can be reworked to appear in the form

$$\sigma^2 = \sum [x^2 \cdot P(x)] - \left\{\sum [x \cdot P(x)]\right\}^2 \quad (5\text{-}4a)$$

or

$$\sigma^2 = \sum [x^2 P(x)] - \mu^2 \quad (5\text{-}4b)$$

It is left for you (Exercise 5.21) to verify that formulas (5-4a) and (5-4b) are equivalent to formula (5-3). ▲▲

▼ ILLUSTRATION 5-6

To find the variance of the probability distribution in Illustration 5-5 by using formula (5-4a), we will need to add two additional columns to Table 5-5. These columns are shown in Table 5-7.

TABLE 5-7
Calculations Needed to Find the Variance for $P(x) = \frac{x}{10}$

x	$P(x)$	$x \cdot P(x)$	x^2	$x^2 \cdot P(x)$
1	$\frac{1}{10}$	$\frac{1}{10}$	1	$\frac{1}{10}$
2	$\frac{2}{10}$	$\frac{4}{10}$	4	$\frac{8}{10}$
3	$\frac{3}{10}$	$\frac{9}{10}$	9	$\frac{27}{10}$
4	$\frac{4}{10}$	$\frac{16}{10}$	16	$\frac{64}{10}$
Total	$\frac{10}{10}$	$\frac{30}{10}$		$\frac{100}{10}$

The variance is

$$\sigma^2 = \sum [x^2 \cdot P(x)] - \left\{\sum [x \cdot P(x)]\right\}^2$$

$$= \frac{100}{10} - \left(\frac{30}{10}\right)^2 = 10 - (3)^2$$

$$= 10 - 9 = \mathbf{1.0}$$

standard deviation of probability distribution

The **standard deviation** is the square root of variance; therefore, $\boldsymbol{\sigma = 1.0}$.

▼ ILLUSTRATION 5-7

A coin is tossed three times. Let the number of heads occurring in those three tosses be the random variable x, which can take on the values 0, 1, 2, or 3. There are eight possible outcomes to this experiment: one results in $x = 0$, three in $x = 1$, three in $x = 2$, and one in $x = 3$. Therefore, the probabilities for this random variable are $\frac{1}{8}, \frac{3}{8}, \frac{3}{8}$, and $\frac{1}{8}$. The probability distribution associated with this experiment is shown in Figure 5-4 and in Table 5-8. The necessary extensions and summations for the calculation of its mean and standard deviation are also shown in Table 5-8.

The mean is found with the aid of formula (5-2):

$$\mu = \sum [x \cdot P(x)] = \mathbf{1.5}$$

This result, 1.5, is the mean number of heads expected per experiment.

The variance is found with the aid of formula (5-4a):

$$\sigma^2 = \sum [x^2 \cdot P(x)] - \left\{\sum [x \cdot P(x)]\right\}^2$$

$$= 3.0 - (1.5)^2 = 3.0 - 2.25 = \mathbf{0.75}$$

The standard deviation is the positive square root of the variance:

$$\sigma = \sqrt{0.75} = 0.866 = \mathbf{0.87}$$

FIGURE 5-4
Probability Distribution for Illustration 5-7

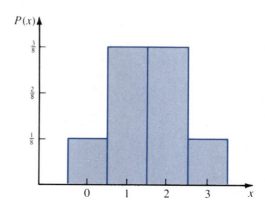

TABLE 5-8 Probability Distribution and Extensions, Illustration 5-7

x	P(x)	x · P(x)	x²	x² · P(x)
0	1/8	0/8	0	0/8
1	3/8	3/8	1	3/8
2	3/8	6/8	4	12/8
3	1/8	3/8	9	9/8
Total	8/8	$\mu = \frac{12}{8} = 1.5$		$\frac{24}{8} = 3.0$

That is, 0.87 is the standard deviation expected among the number of heads observed per experiment.

▼▲ EXERCISES

5.15 Given the probability function

$$P(x) = \frac{5 - x}{10} \quad \text{for} \quad x = 1, 2, 3, 4,$$

find the mean and standard deviation.

5.16 Given the probability function

$$R(x) = 0.2 \quad \text{for} \quad x = 0, 1, 2, 3, 4,$$

find the mean and standard deviation.

5.17 The number of calls x to arrive at a switchboard during any 1-minute period is a random variable and has the following probability distribution:

x	0	1	2	3	4
P(x)	0.1	0.2	0.4	0.2	0.1

a. Find the mean and standard deviation of x.
b. Draw the histogram of $P(x)$ and show the location of μ with a vertical line and the value of σ with a horizontal line segment drawn on the histogram.

5.18 Based on past history, the distribution of sales of 1-pound containers of cottage cheese at a convenience store is

Number of Cartons Sold per Day (x)	Probability
10	0.2
11	0.4
12	0.2
13	0.1
14	0.1
Total	1.0

a. Determine the mean and the standard deviation of x, the number of cartons sold per day.

b. Draw the histogram of this probability distribution and show the location of μ and the size of σ on the histogram.

5.19 A random variable x has the following probability distribution:

x	1	2	3	4	5
P(x)	0.6	0.1	0.1	0.1	0.1

How much of the probability distribution is within 2 standard deviations of the mean? That is, find the probability that x is between $\mu - 2\sigma$ and $\mu + 2\sigma$.

5.20 a. Draw a histogram of the probability distribution for the single-digit random numbers $(0, 1, 2, \ldots, 9)$.
b. Calculate the mean and standard deviation associated with the population of single-digit random numbers.
c. Represent (1) the location of the mean on the histogram with a vertical line, (2) the magnitude of the standard deviation with a line segment.
d. How much of this probability distribution is within 2 standard deviations of the mean?

5.21 Verify that formulas (5-4a) and (5-4b) are equivalent to formula (5-3).

5.22 Find the variance of the probability distribution

x	10	12	14	16
P(x)	0.1	0.2	0.3	0.4

a. Use formula (5-3).
b. Use formula (5-4b).
c. Compare the results of (a) and (b).

5.4 ▼ The Binomial Probability Distribution

Consider the following probability experiment. You are given a five-question multiple-choice quiz. You have not studied the material to be quizzed and therefore decide to answer the five questions by randomly guessing the answers without reading the question or the answers.

ANSWER PAGE TO QUIZ

Directions: Circle the best answer to each question.

1. A B C
2. A B C
3. A B C
4. A B C
5. A B C

Circle your answers before continuing.
Before we look at the correct answers to the quiz, let's think about some of the things we might consider when a quiz is answered in this way.

1. How many of the five questions do you think you answered correctly?
2. If an entire class answered the quiz by guessing, what do you think the "average" number of correct answers would be?
3. What is the probability that you selected the correct answers to all five questions?
4. What is the probability that you selected wrong answers for all five questions?

Let's construct the probability distribution associated with this experiment. Let x be the number of correct answers on your paper; x may then take on the values $0, 1, 2, \ldots, 5$. Since each individual question has three possible answers and since there is only one correct answer, the probability of selecting the correct answer to an individual question is $\frac{1}{3}$. The probability that a wrong answer is selected is $\frac{2}{3}$. $P(x = 0)$ is the probability that all questions were answered incorrectly.

$$P(x = 0) = \left(\frac{2}{3}\right)\left(\frac{2}{3}\right)\left(\frac{2}{3}\right)\left(\frac{2}{3}\right)\left(\frac{2}{3}\right) = \frac{32}{243} \approx \mathbf{0.132}$$

These events are independent because each question is a separate event. We can therefore multiply the probabilities according to formula (4-8b). Also,

$$P(5) = \left(\frac{1}{3}\right)^5 = \frac{1}{243} \approx \mathbf{0.004}$$

Before we find the other four probabilities, let's look at the events on a tree diagram, Figure 5-5 (p. 284). Notice that the event $x = 0$, "zero correct answers," is shown by the bottom branch, and that event $x = 5$, "five correct answers," is shown by the top branch. The other events, "one correct answer," "two correct answers," and so on, are represented by several branches of the tree. Figure 5-5 shows 32 branches representing five different values of x. We find that the event $x = 1$ occurs on 5 different branches, event $x = 2$ occurs on 10 branches, $x = 3$ occurs on 10 branches, and $x = 4$ on 5 branches. Thus the remaining probabilities can be calculated by finding the probability of a single branch representing the event and multiplying it by the number of branches that represent that event.

$$P(x = 1) = \left(\frac{1}{3}\right)\left(\frac{2}{3}\right)\left(\frac{2}{3}\right)\left(\frac{2}{3}\right)\left(\frac{2}{3}\right)(5) = \frac{80}{243} \approx \mathbf{0.329}$$

$$P(x = 2) = \left(\frac{1}{3}\right)^2\left(\frac{2}{3}\right)^3(10) = \frac{80}{243} \approx \mathbf{0.329}$$

$$P(x = 3) = \left(\frac{1}{3}\right)^3\left(\frac{2}{3}\right)^2(10) = \frac{40}{243} \approx \mathbf{0.165}$$

$$P(x = 4) = \left(\frac{1}{3}\right)^4\left(\frac{2}{3}\right)^1(5) = \frac{10}{243} \approx \mathbf{0.041}$$

The answer to question 1 above is suggested by the probability distribution in Table 5-9 (p. 284). The most likely occurrence would be to get one or two correct answers. One, two, or three correct answers are expected to result approximately 82% of the time (0.329 + 0.329 + 0.165 = 0.823). We might also argue that we

FIGURE 5-5

Tree Diagram for the Multiple-choice Quiz

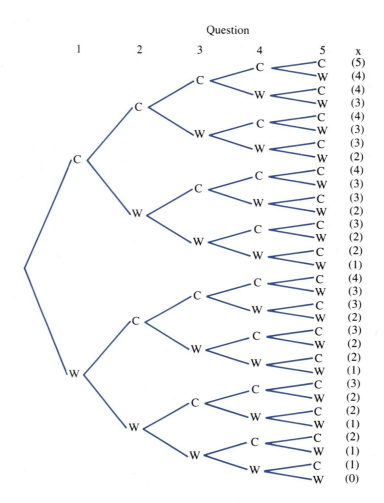

would expect $\frac{1}{3}$ of the questions to be answered correctly. This information provides a reason for answering both questions 1 and 2 on page 283 with the value of $\frac{5}{3}$. Questions 3 and 4 were asked with the expectation that you would feel the chance of "all correct" was very slight—and it is: 0.004. "All wrong" is not too likely, but it is not as rare an event as "all correct." It has a probability of 0.132.

TABLE 5-9

Probability Distribution for the Quiz Example

x	P(x)
0	0.132
1	0.329
2	0.329
3	0.165
4	0.041
5	0.004
Total	1.000

Section 5.4 ▼ THE BINOMIAL PROBABILITY DISTRIBUTION

The correct answers to the quiz are B, C, B, A, and C. How many correct answers did you have? You might ask several people to answer the quiz for you by guessing at the answers. Then construct an observed relative frequency distribution and compare it to the distribution shown in Table 5-9.

Many experiments result in outcomes that can be classified in one of two categories, **success or failure**. Examples of such experiments include the previous experiments of tossing coins or thumbtacks and other more practical experiments, such as determining whether a flashbulb flashes or does not flash, whether a product did its prescribed job or did not do it, whether quiz question answers are right or wrong. There are other experiments that have many outcomes that, under the right conditions, may fit this general description of being classified in one of two categories. For example, when we roll a single die, we usually consider six possible outcomes. However, if we are only interested in knowing whether an ace (a 1) shows or not, there are really only two outcomes: the ace shows or something else shows. The experiments described above are called *binomial probability experiments*.

> **BINOMIAL PROBABILITY EXPERIMENT**
>
> An experiment that is made up of repeated trials of a basic experimental event. The binomial experiment must possess the following properties:
>
> 1. Each **trial** has two possible outcomes (success, failure).
> 2. There are n repeated independent trials.
> 3. $P(\text{success}) = p$, $P(\text{failure}) = q$, and $p + q = 1$.
> 4. The **binomial random variable** x is the count of the number of successful trials that occur; x may take on any integer value from zero to n.
>
> [Don't forget: $p = P(\text{success})$ on *each* individual trial and $q = P(\text{failure})$ on *each* trial.]

The binomial experiment *must* demonstrate the two basic properties 1 and 2. Properties 3 and 4 concern notation and the identification of the variables involved.

NOTE It is of utmost importance that a probability p be assigned to the particular outcome that is considered to be the "success," since the binomial random variable indicates the number of "successes" that occur.

▼ ILLUSTRATION 5-8

Consider the roll of a die, where "ace" or "something else" occurs. In this experiment we would count the number of "aces." The random variable x would be assigned a value that is the number of times that an ace is observed in the n trials. Since "ace" is considered "success," $P(\text{ace}) = p$ and $P(\text{not ace}) = q$. ▲▲

▼ ILLUSTRATION 5-9

If you were an inspector on a production line in a plant where television sets were manufactured, you would be concerned with identifying the number of defective television sets. You would probably define "success" as the occurrence of a defective television. This is not what we normally think of as success, but if we count defective sets in a binomial experiment, we must define "success" as a "defective." The random variable x indicates the number of defective sets found per lot of n sets. p represents P(television is defective) and q is P(television is good).

▲▲

NOTE *Independent* trials mean that the result of one trial does not affect the probability of success of any other trial in the experiment. In other words, the probability of "success" remains constant throughout the entire experiment.

Each binomial probability experiment has its own specific probability function. However, all such functions have a common format and all such distributions display some similarities. Let's look at a relatively simple binomial experiment and discuss its probability function.

Let's reconsider Illustration 5-7 (p. 280); a coin is tossed three times and we observe the number of heads that occur in the three tosses. This is a binomial experiment because it displays all the properties of a binomial experiment:

1. Each trial (one toss of the coin) has two outcomes: success (heads, what we are counting) and failure (tails).
2. There are n (three) repeated trials that are independent (each is a separate toss and the outcome of any one trial has no effect on the outcome of another).
3. The probability of success (heads) is p ($\frac{1}{2}$) and the probability of failure (tails) is q ($\frac{1}{2}$), and $p + q = 1$.
4. The random variable x is the number of successes (heads) that occur in n (three) trials. x can assume the value 0, 1, 2, or 3.

Let's consider first the probability of $x = 1$.

$$P(x = 1) = P(\text{exactly one head is observed in three tosses})$$

We want to find this probability. What does it mean to say "exactly one head in three tosses"? The probability that one head appears on one toss is $\frac{1}{2}$. The probability that heads do not occur on the other two tosses is $\frac{1}{2} \times \frac{1}{2} = \frac{1}{4}$. The probability that exactly one head occurs in three tosses, then, must be related to $\frac{1}{2} \times \frac{1}{4} = \frac{1}{8}$ [from the multiplication rule, formula (4-8a), p. 234]. But there is one other consideration: In how many different ways can exactly one of the three tosses result in a head?

1. The first toss could be a head and the others tails.
2. The second toss could be a head and the first and last tails.
3. The first two tosses could be tails and the last toss a head.

These are the three ways in which three trials can result in exactly one success. This means that the $\frac{1}{8}$ found above can occur three different ways, resulting in $P(x = 1) = 3(\frac{1}{8}) = \frac{3}{8}$.

If we look at the case $x = 2$, we find a similar situation: there are three ways to obtain exactly two heads. Thus $P(x = 2) = \frac{3}{8}$, also. We can find only one way for $x = 0$ and $x = 3$ to occur—all tails or all heads. In both cases the probability

Section 5.4 ▼ THE BINOMIAL PROBABILITY DISTRIBUTION

is equal to $(\frac{1}{2})^3 = \frac{1}{8}$. The rest of the probability distribution is exactly as shown in Table 5-8 (p. 281).

If we take a careful look at the probability of each case shown in the tree diagram in Figure 5-5 (p. 284), we can argue that the probability that a random variable x takes on a particular value in a binomial experiment is always the product of three basic factors. These three factors are as follows:

1. the probability of exactly x successes, p^x
2. the probability that failure will occur on the remaining $(n - x)$ trials, q^{n-x}
3. the number of ways that exactly x successes can occur in n trials

The number of ways that exactly x successes can occur in a set of n trials is represented by the symbol

$$\binom{n}{x}$$

binomial coefficient

which must always be a positive integer. This is termed the **binomial coefficient**. The binomial coefficient is found by using the formula

$$\binom{n}{x} = \frac{n!}{x!(n-x)!} \quad (5\text{-}5)$$

NOTE $n!$ is an abbreviation for the product of the sequence of integers starting with n and ending with 1. For example, $3! = 3 \cdot 2 \cdot 1 = 6$, and $5! = 5 \cdot 4 \cdot 3 \cdot 2 \cdot 1 = 120$. There is one special case, $0!$, which is defined to be 1. For further information about factorial notation, see the *Study Guide*.

Also, see the *Study Guide* for general information on the binomial coefficient. The values for $n!$ when n is equal to or smaller than 20 are found in Table 2 of Appendix F. The values for $\binom{n}{x}$ when n is equal to or smaller than 20 are found in Table 3 of Appendix F. This information allows us to form a general *binomial probability function*.

binomial probability function

BINOMIAL PROBABILITY FUNCTION

If, for a binomial experiment, p is the probability of a success and q is the probability of a failure on a single trial, then the probability $P(x)$ that there will be exactly x successes in n trials is

$$P(x) = \binom{n}{x} \cdot p^x \cdot q^{n-x} \quad \text{for} \quad x = 0, 1, 2, \ldots, n \quad (5\text{-}6)$$

When this general binomial probability function is applied to the illustration with three coins (Illustration 5-7, p. 280), we find

$$P(x) = \binom{3}{x} \cdot \left(\frac{1}{2}\right)^x \cdot \left(\frac{1}{2}\right)^{3-x} \quad \text{for} \quad x = 0, 1, 2, 3$$

Let's calculate each of the probabilities and see whether they form a probability distribution:

$$P(x = 0) = \binom{3}{0} \cdot \left(\frac{1}{2}\right)^0 \cdot \left(\frac{1}{2}\right)^3 = 1 \cdot 1 \cdot \frac{1}{8} = \frac{1}{8}$$

$P(x = 0) = \frac{1}{8}$ is the probability of no heads occurring in three tosses of a coin. (*Note:* $\left(\frac{1}{2}\right)^0 = 1$.)

$$P(x = 1) = \binom{3}{1} \cdot \left(\frac{1}{2}\right)^1 \cdot \left(\frac{1}{2}\right)^2 = 3 \cdot \frac{1}{2} \cdot \frac{1}{4} = \frac{3}{8}$$

$P(x = 1) = \frac{3}{8}$ is the probability of one head occurring in three tosses of a coin.

$$P(x = 2) = \binom{3}{2} \cdot \left(\frac{1}{2}\right)^2 \cdot \left(\frac{1}{2}\right)^1 = 3 \cdot \frac{1}{4} \cdot \frac{1}{2} = \frac{3}{8}$$

$P(x = 2) = \frac{3}{8}$ is the probability of two heads occurring in three tosses of a coin.

$$P(x = 3) = \binom{3}{3} \cdot \left(\frac{1}{2}\right)^3 \cdot \left(\frac{1}{2}\right)^0 = 1 \cdot \frac{1}{8} \cdot 1 = \frac{1}{8}$$

$P(x = 3) = \frac{1}{8}$ is the probability of three heads occurring in three tosses of a coin. Since each of these probabilities is between 0 and 1, and the sum of all the probabilities is exactly 1, this is a probability distribution.

Let's look at another example. Consider an experiment that calls for drawing five cards, one at a time with replacement, from a well-shuffled deck of playing cards. The drawn card is identified as a spade or not a spade, returned to the deck, the deck reshuffled, and so on. The random variable x is the number of spades observed in five drawings. Is this a binomial experiment? Let's identify the various properties.

1. Each drawing is a trial and each drawing has two outcomes, "spade" or "not spade."
2. There are five repeated drawings, so $n = 5$. These individual trials are independent, since the drawn card is returned to the deck and the deck reshuffled before the next drawing.
3. $p = \frac{1}{4}$ and $q = \frac{3}{4}$.
4. x is the number of spades recorded in the five trials. The binomial probability function is

$$P(x) = \binom{5}{x} \cdot \left(\frac{1}{4}\right)^x \cdot \left(\frac{3}{4}\right)^{5-x}$$

for $x = 0, 1, 2, 3, 4,$ and 5. The probabilities are

$$P(0) = \binom{5}{0} \cdot \left(\frac{1}{4}\right)^0 \cdot \left(\frac{3}{4}\right)^5 = 1 \cdot 1 \cdot \left(\frac{3}{4}\right)^5 = \mathbf{0.2373}$$

$$P(1) = \binom{5}{1} \cdot \left(\frac{1}{4}\right)^1 \cdot \left(\frac{3}{4}\right)^4 = 5 \cdot \left(\frac{1}{4}\right) \cdot \left(\frac{3}{4}\right)^4 = \mathbf{0.3955}$$

$$P(2) = \binom{5}{2} \cdot \left(\frac{1}{4}\right)^2 \cdot \left(\frac{3}{4}\right)^3 = 10 \cdot \left(\frac{1}{4}\right)^2 \cdot \left(\frac{3}{4}\right)^3 = \mathbf{0.2637}$$

Section 5.4 ▼ THE BINOMIAL PROBABILITY DISTRIBUTION

$$P(3) = \binom{5}{3} \cdot \left(\frac{1}{4}\right)^3 \cdot \left(\frac{3}{4}\right)^2 = 10 \cdot \left(\frac{1}{4}\right)^3 \cdot \left(\frac{3}{4}\right)^2 = \mathbf{0.0879}$$

$$P(4) = \binom{5}{4} \cdot \left(\frac{1}{4}\right)^4 \cdot \left(\frac{3}{4}\right)^1 = 5 \cdot \left(\frac{1}{4}\right)^4 \cdot \left(\frac{3}{4}\right)^1 = \mathbf{0.0146}$$

$$P(5) = \binom{5}{5} \cdot \left(\frac{1}{4}\right)^5 \cdot \left(\frac{3}{4}\right)^0 = 1 \cdot \left(\frac{1}{4}\right)^5 \cdot \left(\frac{3}{4}\right)^0 = \mathbf{0.0010}$$

$$\Sigma P(x) = \overline{1.0000}$$

The preceding distribution of probabilities indicates that the single most likely value of x is 1, the event of observing exactly one spade in a hand of five cards. What is the least likely number of spades that would be observed?

Let's consider one more binomial probability problem.

▼ ILLUSTRATION 5-10

The manager of Steve's Food Market guarantees that none of his cartons of one dozen eggs will contain more than one bad egg. If a carton contains more than one bad egg, he will replace the whole dozen and allow the customer to keep the original eggs. The probability that an individual egg is bad is 0.05. What is the probability that Steve will have to replace a given carton of eggs?

SOLUTION Assuming that this is a binomial experiment, let x be the number of bad eggs found in a carton of a dozen eggs, $p = 0.05$, and let the inspection of each egg be a trial resulting in finding a "bad" or "not bad" egg. To find the probability that the manager will have to make good on his guarantee, we need the probability function associated with this experiment:

$$P(x) = \binom{12}{x} \cdot (0.05)^x \cdot (0.95)^{12-x} \quad \text{for} \quad x = 0, 1, 2, \ldots, 12$$

The probability that Steve will replace a dozen eggs is the probability that $x = 2, 3, 4, 5, \ldots,$ or 12. Recall that $\Sigma P(x) = 1$, that is,

$$P(x=0) + P(x=1) + P(x=2) + \cdots + P(x=12) = 1$$

Therefore,

$$P(x=2) + P(x=3) + \cdots + P(x=12) = 1 - [P(x=0) + P(x=1)]$$

Finding $P(x=0)$ and $P(x=1)$ and subtracting them from 1 is easier than finding each of the other probabilities.

$$P(x) = \binom{12}{x} \cdot (0.05)^x \cdot (0.95)^{12-x}$$

$$P(0) = \binom{12}{0} \cdot (0.05)^0 (0.95)^{12} = \mathbf{0.540}$$

$$P(1) = \binom{12}{1} \cdot (0.05)^1 (0.95)^{11} = \mathbf{0.341}$$

NOTE The value of many binomial probabilities, for small values of n and common values of p, are found in Table 4 of Appendix F. In this example, we have $n = 12$ and $p = 0.05$, and we want the probabilities for $x = 0$ and 1. We need to locate the section of Table 4 where $n = 12$, find the column marked $p = 0.05$, and read the numbers opposite $x = 0$ and 1. We find 540 and 341, as shown in Table 5-10. (Look these values up in Table 4.) The decimal point was left out of the table to save space, and we must replace it. It belongs at the front of each table entry. Therefore, our values are 0.540 and 0.341.

TABLE 5-10 Abbreviated Portion of Table 4 in Appendix F, Binomial Probabilities

n	x	0.01	0.05	0.10	0.20	0.30	0.40	P 0.50	0.60	0.70	0.80	0.90	0.95	0.99	x
	⋮	⋮													
12	0	886	540	282	069	014	002	0+	0+	0+	0+	0+	0+	0+	0
	1	107	341	377	206	071	017	003	0+	0+	0+	0+	0+	0+	1
	2	006	099	230	283	168	064	016	002	0+	0+	0+	0+	0+	2
	3	0+	017	085	236	240	142	054	012	001	0+	0+	0+	0+	3
	4	0+	002	021	133	231	213	121	042	008	001	0+	0+	0+	4
	⋮	⋮													⋮

Now let's return to our illustration:

$$P(\text{replacement}) = 1 - (0.540 + 0.341) = \mathbf{0.119}$$

If $p = 0.05$ is correct, Steve will be busy replacing cartons of eggs. If he replaces 11.9% of all the cartons of eggs he sells, he certainly will be giving away a substantial proportion of his eggs. This suggests that he should adjust his guarantee. For example, if he were to replace a carton of eggs only when four or more were found bad, he would expect to replace only 0.003 cartons [$1.0 - (0.540 + 0.341 + 0.099 + 0.017)$], or 0.3% of the cartons sold. Notice that he will be able to control his "risk" (probability of replacement) if he adjusts the value of the random variable stated in his guarantee.

▼▲ EXERCISES

5.23 Evaluate each of the following:

  a. 4!   **b.** 7!   **c.** 0!   **d.** $\dfrac{6!}{2!}$

  e. $\dfrac{5!}{3!\,2!}$ **f.** $\dfrac{6!}{4!\,(6-4)!}$ **g.** $(0.3)^4$ **h.** $\binom{7}{3}$

Section 5.4 ▼ THE BINOMIAL PROBABILITY DISTRIBUTION

i. $\binom{5}{2}$ j. $\binom{3}{0}$ k. $\binom{4}{1}(0.2)^1(0.8)^3$ l. $\binom{5}{0}(0.3)^0(0.7)^5$

5.24 Evaluate each of the following:

a. $5!$ b. $6!$ c. $1!$ d. $\dfrac{8!}{4!}$

e. $\dfrac{9!}{6!\,3!}$ f. $\dfrac{8!}{5!\,(8-5)!}$ g. $(0.9)^5$ h. $\binom{6}{4}$

i. $\binom{8}{1}$ j. $\binom{5}{0}$ k. $\binom{5}{2}(0.2)^2(0.8)^3$ l. $\binom{6}{0}(0.7)^0(0.3)^6$

5.25 A carton containing 100 T-shirts is inspected. Each T-shirt is rated "first quality" or "irregular." After all 100 T-shirts have been inspected, the number of irregulars is reported as a random variable. Explain why x is a binomial random variable.

5.26 A die is rolled 20 times and the number of "fives" that occurred is reported as being the random variable. Explain why x is a binomial random variable.

5.27 Four cards are selected, one at a time, from a standard deck of 52 cards. Let x represent the number of aces drawn in the set of 4 cards.
   a. If this experiment is completed without replacement, explain why x is not a binomial random variable.
   b. If this experiment is completed with replacement, explain why x is a binomial random variable.

5.28 The employees at a General Motors assembly plant are polled as they leave work. Each is asked, "What brand of automobile are you riding home in?" The random variable to be reported is the number of each brand mentioned. Is x a binomial random variable? Justify your answer.

5.29 The number of branches in a tree diagram representation of a binomial experiment having n trials is equal to 2^n.
   a. For a binomial experiment having two trials, draw the tree diagram and show that it has 2^2, or 4, branches.
   b. For a binomial experiment having three trials, draw the tree diagram and show that it has 2^3, or 8, branches.
   c. How many branches would a tree diagram representation for a binomial experiment have when five trials are made? Ten trials? (Do not draw the tree diagram.)

5.30 Draw a tree diagram picturing a binomial experiment of four trials.

5.31 Consider a binomial experiment made up of three trials with outcomes of success S and failure F, where $P(S) = p$ and $P(F) = q$.

a. Complete the accompanying tree diagram. Label all branches completely.

	Trial 1	Trial 2	Trial 3	(b) Probability	(c) x
			S	p^3	3
		S	F	p^2q	2
			⋮	⋮	⋮

b. In Column (b) of the tree diagram, express the probability of each outcome represented by the branches as a product of powers of p and q.

c. Let x be the random variable, the number of successes observed. In Column (c), identify the value of x for each branch of the tree diagram.

d. Notice that all the products in Column (b) are made up of three factors and that the value of the random variable is the same as the exponent for the number p.

e. Write the equation for the binomial probability function for this situation.

5.32 In reference to the five-question multiple-choice "quiz" on page 282, show that the guessing of the answers is actually a binomial experiment.

a. Specify exactly how this experiment satisfies the four properties of a binomial experiment.

b. The tree diagram (Figure 5-5) shows all possible outcomes for the quiz. There are a total of 32 outcomes shown. How many outcomes are there for each of the events "0 correct answers," "1 correct answer," "2 correct answers," …, "5 correct answers"?

c. Write the equation of the binomial probability function for this experiment.

5.33 If x is a binomial random variable, calculate the probability of x for each case.
  a. $n = 4, x = 1, p = 0.3$ **b.** $n = 3, x = 2, p = 0.8$
  c. $n = 2, x = 0, p = \frac{1}{4}$ **d.** $n = 5, x = 2, p = \frac{1}{3}$
  e. $n = 4, x = 2, p = 0.5$ **f.** $n = 3, x = 3, p = \frac{1}{6}$

5.34 If x is a binomial random variable, use Table 4, Appendix F, to determine the probability of x for each case.
  a. $n = 10, x = 8, p = 0.3$ **b.** $n = 8, x = 7, p = 0.95$
  c. $n = 15, x = 3, p = 0.05$ **d.** $n = 12, x = 12, p = 0.99$
  e. $n = 9, x = 0, p = 0.5$ **f.** $n = 6, x = 1, p = 0.01$
  g. Explain the meaning of the symbol 0+ that appears in Table 4.

5.35 Test the following function to determine whether or not it is a binomial probability function. List the distribution of probabilities and sketch a histogram.

$$T(x) = \binom{5}{x} \cdot \left(\frac{1}{2}\right)^x \cdot \left(\frac{1}{2}\right)^{5-x} \quad \text{for} \quad x = 0, 1, 2, 3, 4, 5$$

5.36 Let x be a random variable with the following probability distribution:

x	0	1	2	3
P(x)	0.4	0.3	0.2	0.1

Does x have a binomial distribution? Justify your answer.

5.37 Consider the experiment of tossing a coin five times and counting the number of heads that appear as the random variable (x).
   a. Is this a binomial experiment? Describe all the properties and assign the variable.
   b. Write the equation of the binomial probability function.
   c. Find the various binomial probabilities to form the probability distribution.

5.38 A manufacturer of matches knows that 0.1% of the matches produced by the company are defective.
   a. Write the equation of the binomial probability function.
   b. Find the probability that a package of 50 matches contains no defective matches.
   c. Find the probability that a package of 50 matches contains exactly one defective match.
   d. Find the probability that a package of 50 contains more than one defective match.

5.39 In the biathlon event of the Olympic Games, a participant skis cross-country and on four intermittent occasions stops at a rifle range and shoots a set of five shots. If the center of the target is hit, no penalty points are assessed. If a particular man has a history of hitting the center of the target with 90% of his shots, what is the probability that he will hit the center of the target with
   a. all five of his next set of five shots?
   b. at least four of his next set of five shots? (Assume independence.)

5.40 A basketball player has a history of making 80% of the foul shots taken during games. What is the probability that he will miss three of the next five foul shots he takes?

5.41 A machine produces parts of which 0.5% are defective. If a random sample of ten parts produced by this machine contains two or more defectives, the machine is shut down for repairs. Find the probability that the machine will be shut down for repairs based on this sampling plan.

5.42 The survival rate during a risky operation for patients with no other hope of survival is 80%. What is the probability that exactly four of the next five patients survive this operation?

5.43 According to an article in the February 1991 issue of *Reader's Digest* (page 144), Americans face a 1 in 20 chance of acquiring an infection while hospitalized. If the records of 15 randomly selected hospitalized patients are examined, find the probability that
  a. none of the 15 acquired an infection while hospitalized.
  b. one or more of the 15 acquired an infection while hospitalized.

5.44 According to a report in the journal *American Pharmacy* (November 1990, page 15), 34.4% of the population of Kentucky currently smokes cigarettes. Find the probability that in a random telephone survey of 20 individuals from Kentucky exactly 5 of the 20 currently smoke cigarettes.

5.45 State a very practical reason why the defective item in an industrial situation would be defined to be the "success" in a binomial experiment.

5.46 If the binomial $(q + p)$ is squared, the result is $(q + p)^2 = q^2 + 2qp + p^2$. For the binomial experiment with $n = 2$, the probability of no successes in two trials is q^2 (the first term in the expansion), the probability of one success in two trials is $2qp$ (the second term in the expansion), and the probability of two successes in two trials is p^2 (the third term). Find $(q + p)^3$ and compare its terms to the binomial probabilities for $n = 3$ trials.

5.5 ▼ Mean and Standard Deviation of the Binomial Distribution

mean and standard deviation of binomial distribution

The **mean** and **standard deviation** of a theoretical binomial probability distribution can be found by using these two formulas:

$$\mu = np \tag{5-7}$$

$$\sigma = \sqrt{npq} \tag{5-8}$$

The formula for the mean seems appropriate as the number of trials multiplied by the probability that a "success" will occur. (Recall that the mean number of correct answers on the binomial quiz was expected to be $\frac{1}{3}$ of 5, $5 \cdot \frac{1}{3}$, or np.) The formula for the standard deviation is not as easily understood. Thus at this point it is appropriate to look at an example that demonstrates that formulas (5-7) and (5-8) yield the same results as formulas (5-2) and (5-4).

Referring to the case of tossing three coins in Illustration 5-7, $n = 3$ and $p = \frac{1}{2}$. Using formulas (5-7) and (5-8), we find

$$\mu = np = (3)(\tfrac{1}{2}) = \mathbf{1.5}$$

$\mu = 1.5$ is the mean value of the random variable x.

$$\sigma = \sqrt{npq} = \sqrt{(3)(\tfrac{1}{2})(\tfrac{1}{2})} = \sqrt{\tfrac{3}{4}}$$
$$= \sqrt{0.75} = 0.866 = \mathbf{0.87}$$

Section 5.5 ▼ MEAN AND STANDARD DEVIATION OF THE BINOMIAL DISTRIBUTION

$\sigma = 0.87$ is the standard deviation of the random variable x. Look back at the solution for Illustration 5-7 (p. 280) and compare the use of formulas (5-2) and (5-4a) with formulas (5-7) and (5-8). Note that the results are the same, regardless of the formula you use. However, formulas (5-7) and (5-8) are much easier to use when x is a binomial random variable.

▼ ILLUSTRATION 5-11

Find the mean and standard deviation of the binomial distribution when $n = 20$ and $p = \frac{1}{5}$. Recall that the "binomial distribution where $n = 20$ and $p = \frac{1}{5}$" has a probability function

$$P(x) = \binom{20}{x} \cdot (0.2)^x \cdot (0.8)^{20-x}; \quad \text{for} \quad x = 0, 1, 2, \ldots, 20$$

and a corresponding distribution with 21 x-values and 21 probabilities as shown in the following distribution chart:

x	P(x)	x	P(x)
0	0.012	8	0.022
1	0.057	9	0.007
2	0.137	10	0.002
3	0.205	11	0.001
4	0.219	12	0+
5	0.174	13	0+
6	0.109	⋮	⋮
7	0.055	20	0+

Now let's find the mean and the standard deviation of this distribution of x using formulas (5-7) and (5-8).

$$\mu = np = (20)\left(\tfrac{1}{5}\right) = 4.0$$

$$\sigma = \sqrt{npq} = \sqrt{(20)\left(\tfrac{1}{5}\right)\left(\tfrac{4}{5}\right)} = \sqrt{\tfrac{80}{25}}$$

$$= \frac{(4\sqrt{5})}{5} = 1.79$$

▲▲

▼▲ EXERCISES

5.47 Given the binomial probability function

$$T(x) = \binom{5}{x} \cdot \left(\frac{1}{2}\right)^x \cdot \left(\frac{1}{2}\right)^{5-x} \quad \text{for} \quad x = 0, 1, 2, 3, 4, 5$$

    a. calculate the mean and standard deviation of the random variable by using formulas (5-2) and (5-4a).

b. calculate the mean and standard deviation using formulas (5-7) and (5-8).
c. compare the results of (a) and (b).

5.48 Answer the questions in Exercise 5.47 for the binomial probability function $P(x)$.
$$P(x) = \binom{6}{x} \cdot 0.3^x \cdot 0.7^{6-x} \quad \text{for} \quad x = 0, 1, 2, 3, 4, 5, 6$$

5.49 Find the mean and standard deviation of x for each of the following binomial random variables:
  a. the number of aces seen in 100 draws from a well-shuffled bridge deck (with replacement).
  b. the number of cars found to have unsafe tires among the 400 cars stopped at a road block for inspection. Assume that 6% of all cars have one or more unsafe tires.
  c. the number of melon seeds that germinate when a package of 50 seeds is planted. The package states that the probability of germination is 0.88.

5.50 Find the mean and standard deviation for each of the following binomial random variables:
  a. the number of tails seen in 50 tosses of a quarter.
  b. the number of defective televisions in a shipment of 125. The manufacturer claimed that 98% of the sets were operative.
  c. the number of operative televisions in a shipment of 125. The manufacturer claimed that 98% of the sets were operative.
  d. How are questions (b) and (c) related?

5.51 A binomial random variable has a mean equal to 200 and a standard deviation of 10. Find the values of n and p.

5.52 The probability of success on a single trial of a binomial experiment is known to be $\frac{1}{4}$. The random variable x, number of successes, has a mean value of 80. Find the number of trials involved in this experiment and the standard deviation of x.

5.53 Seventy-five percent of the foreign-made autos sold in the United States in 1984 are now falling apart.
  a. Determine the probability distribution of x, the number of these autos that are falling apart in a random sample of five cars.
  b. Draw a histogram of the distribution.
  c. Calculate the mean and the standard deviation of this distribution.

5.54 A binomial random variable x is based upon 15 trials with the probability of success equal to 0.3. Find the probability that this variable will take on a value more than two standard deviations from the mean.

5.55 For years, the manager of a certain company had sole responsibility for making decisions with regard to company policy. This manager has a history of making the correct decision with a probability of p. Recently company policy has changed and now all decisions are to be made by majority rule of a three-person committee.
  a. Each member makes a decision independently and each has a probability of p of making the correct decision. What is the probability that the committee's majority decision will be correct?
  b. If $p = 0.1$, what is the probability that the committee makes the correct decision?

c. If $p = 0.8$, what is the probability that the committee makes the correct decision?
d. For what values of p is the committee more likely to make the correct decision by majority rule than the former manager?
e. For what values (there are 3) of p is the probability of a correct decision the same for the manager and for the committee? Justify your answer.

$ 5.56 Suppose one member of the committee in Exercise 5.55 always makes his decision by rolling a die. If the die roll results in an even number, he votes for the proposal, and if an odd number occurs, he votes against it. The other two members still decide independently and have a probability of p of making the correct decision.
a. What is the probability that the committee's majority decision will be correct?
b. If $p = 0.1$, what is the probability that the committee makes the correct decision?
c. If $p = 0.8$, what is the probability that the committee makes the correct decision?
d. For what values of p is the committee more likely to make the correct decision by majority rule than the former manager?
e. For what values of p is the probability of a correct decision the same for the manager and for the committee? Justify your answer.
f. Why is the answer to (e) different than answer (e) to Exercise 5.55?

IN RETROSPECT

In this chapter we combined concepts of probability with some of the ideas presented in Chapter 2. We now are able to deal with distributions of probability values and find means, standard deviations, and so on.

In Chapter 4 we explored the concepts of mutually exclusive events and independent events. The addition and multiplication rules were used on several occasions in this chapter, but very little was said about mutual exclusiveness or independence. Recall that every time we add probabilities together, as we did in each of the probability distributions, we need to know that the associated events are mutually exclusive. If you look back over the chapter, you will notice that the random variable actually requires events to be mutually exclusive; therefore, no real emphasis was placed on this concept. The same basic comment can be made in reference to the multiplication of probabilities and the concept of independent events. Throughout this chapter, probabilities were multiplied together and occasionally independence was mentioned. Independence, of course, is necessary in order to be able to multiply probabilities together.

If we were to take a close look at some of the sets of data in Chapter 2, we would see that several of these problems could be reorganized to form probability distributions. For example: (1) Let x be the number of credit hours that a student is registered for in this semester, with the percentage of the student body being reported for each value of x. (2) Let x be the number of correct passageways through which an experimental laboratory animal passes before taking a wrong one. (3) Let x be the number of trips per day for the ambulance service (Case Study 5-1). (4) Let x be the age of the recreational athlete with an injury as discussed in Case Study 5-2. (This is an example of a continuous random variable.) The list of examples is endless.

We are now ready to extend these concepts to continuous random variables, which we will do in Chapter 6.

CHAPTER EXERCISES

5.57 What are the two basic properties of every probability distribution?

5.58 Explain the difference and the relationship between a *probability function* and a *probability distribution*.

5.59 Verify whether or not the following is a probability function. State your conclusion and explain.

$$f(x) = \frac{\frac{3}{4}}{x!(3-x)!} \quad \text{for} \quad x = 0, 1, 2, 3$$

5.60 Determine which of the following are probability functions:
   a. $f(x) = 0.25$ for $x = 9, 10, 11, 12$
   b. $f(x) = (3-x)/2$ for $x = 1, 2, 3, 4$
   c. $f(x) = (x^2 + x + 1)/25$ for $x = 0, 1, 2, 3$

5.61 The number of ships to arrive at a harbor on any given day is a random variable represented by x. The probability distribution for x is

x	10	11	12	13	14
P(x)	0.4	0.2	0.2	0.1	0.1

Find the probability that on a given day
   a. exactly 14 ships arrive.   **b.** at least 12 ships arrive.
   c. at most 11 ships arrive.

5.62 The number of data-entry mistakes per hour made by an individual entering data at a CRT terminal is a random variable represented by x and has the following probability distribution:

x	0	1	2	3
P(x)	0.40	0.30	0.25	0.05

   a. Find the mean number of mistakes for 1-hour sessions.
   b. Find the probability of at least one mistake during a particular 1-hour session.

5.63 A "wheel of fortune" in a professor's office produces uniformly distributed random integer values between zero and 100. Suppose that each student in this professor's class is assigned a grade by giving the wheel a whirl and that grades are determined by the following distribution:

    0 to   20 is F
    21 to 40 is D
    41 to 62 is C
    63 to 87 is B
    88 to 100 is A

    a. Find the probability associated with each letter grade.
    b. Construct a histogram depicting the letter-grade distribution.
    c. Construct a histogram depicting the random variable x, the random integer generated by the wheel.

5.64 A local elementary school holds four open houses annually. Records show that the probability that a child's parents (one or both) attend from 0 to 4 of the open houses is as shown in the accompanying table:

Number of Open Houses Attended (x)	0	1	2	3	4
Probability	0.12	0.38	0.30	0.12	0.08

    a. Is this a probability distribution? Explain.
    b. What is the probability that an individual child's parents attend at least one of these functions?
    c. Find the mean and the standard deviation for this distribution.

5.65 A coin has a 1 painted on the head and a 2 painted on the tail side. An experiment consists of flipping the coin and rolling a die. The random variable x is defined to be 0 if the coin shows a 1 and to be the number on the die if the coin shows a 2. Construct the probability distribution of x and calculate its mean.

5.66 The same coin and die from Exercise 5.65 are used again except the random variable x is defined to be the "sum" of the two outcomes. Construct the probability distribution of x and calculate its mean.

5.67 A discrete random variable has a standard deviation equal to 10 and a mean equal to 50. Find $\Sigma x^2 P(x)$.

5.68 A binomial random variable is based on $n = 20$ and $p = 0.4$. Find $\Sigma x^2 P(x)$.

5.69 If boys and girls are equally likely to be born, what is the probability that in a randomly selected family of six children, there will be boys? (Find the answer by using a formula.)

5.70 One-fourth of a certain breed of rabbits are born with long hair. What is the probability that in a litter of six rabbits, exactly three will have long hair? (Find the answer by using a formula.)

5.71 Suppose that you take a five-question multiple-choice quiz by guessing. Each question has exactly one correct answer of the four alternatives given. What is the probability that you guess more than one-half of the answers correctly? (Find the answer by using a formula.)

5.72 As a quality-control inspector for toy trucks, you have observed that wooden wheels that are bored off-center occur about 3% of the time. If six wooden wheels are used on each toy truck produced, what is the probability that a randomly selected set of wheels has no off-center wheels?

5.73 Ninety percent of the trees planted by a landscaping firm survive. What is the probability that eight or more of the ten trees they just planted will survive? (Find the answer by using a table.)

5.74 In California, 30% of the people have a certain blood type. What is the probability that exactly 5 out of a randomly selected group of 14 Californians will have that blood type? (Find the answer by using a table.)

5.75 On the average, 1 out of every 10 boards purchased by a cabinet manufacturer is unusable for building cabinets. What is the probability that 8, 9, or 10 of a set of 11 such boards are usable? (Find the answer by using a table.)

5.76 A local polling organization maintains that 90% of the eligible voters have never heard of John Anderson, who was a presidential candidate in 1980. If this is so, what is the probability that in a randomly selected sample of 12 eligible voters 2 or fewer have heard of John Anderson?

5.77 A doctor knows from experience that 10% of the patients to whom he gives a certain drug will have undesirable side effects. Find the probabilities that among the ten patients to whom he gives the drug,
   a. at most two will have undesirable side effects.
   b. at least two will have undesirable side effects.

5.78 In a recent survey of women, 90% admitted that they had never looked at a copy of *Vogue* magazine. Assuming that this is accurate information, what is the probability that a random sample of three women will show that fewer than two have looked at the magazine?

5.79 Of the Heisman Trophy winners in collegiate football, it is considered that only 0.35 have had successful professional football careers.
   a. What is the probability that none of the next five winners will have successful pro careers?
   b. What is the probability that more than three of the next five winners will have successful pro careers?

5.80 Seventy percent of those seeking a driver's license admitted that they would not report someone if he or she copied some answers during the written exam. You have just entered the room and see ten people waiting to take the written exam. What is the probability that, if the incident happened, five of the ten would not report what they saw?

5.81 The engines on an airliner operate independently. The probability that an individual engine operates for a given trip is 0.95. A plane will be able to complete a trip successfully if at least one-half of its engines operate for the entire trip.

Determine whether a four-engine or a two-engine plane has the higher probability of a successful trip.

5.82 CRT operators enter data for a large savings and loan institution. The probability that a character is incorrectly entered is 0.001. For a document containing 10,000 characters, compute the probability that at most one of the characters is incorrectly entered.

5.83 A box contains ten items of which three are defective and seven are nondefective. Two items are selected without replacement, and x is the number of defectives in the sample of two. Explain why x is not a binomial random variable.

5.84 A large shipment of radios is accepted upon delivery if an inspection of ten randomly selected radios yields no more than one defective radio.
  a. Find the probability that this shipment is accepted if 5% of the total shipment is defective.
  b. Find the probability that this shipment is not accepted if 20% of this shipment is defective.

5.85 In the education section of *Parents Magazine* (May 1990, pages 49, 50), it is stated that only 4% of all professional engineers are women. In a survey of professional engineers, 150 were asked to fill out a questionnaire concerning their educational background and experience.
  a. How many females would you expect to find, on the average, among the sample of 150?
  b. What is the probability of finding one or fewer females in a random sample of 150?

5.86 The "Health Update" section of *Better Homes and Garden* (July 1990, page 50) reported that patients who take long half-life tranquilizers are 70% more likely to suffer falls resulting in hip fractures than those taking similar drugs with a short half-life. It was also reported that in a Massachusetts study, 30% of nursing home patients who used tranquilizers used the long half-life ones. Suppose that in a survey of 15 nursing home patients in New York who used tranquilizers, it was found that 10 of the 15 used long half-life tranquilizers.
  a. If the 30% figure for Massachusetts also holds in New York, find the probability of finding 10 or more in a random sample of 15 who use long half-life tranquilizers.
  b. What might you infer from your answer in (a)?

5.87 The September 1990 issue of *Phi Delta Kappan* (pages 41–55) contains the 22nd Annual Gallup Poll of the public's attitudes toward the public schools. On page 54, the composition of the sample section contains the following age description:

Age	Percent
18–29 yrs	23
30–49 yrs	41
50 and over	36

If x represents the characteristic age and equals 1 if the age is between 18 and 29, 2 if the age is between 30 and 49, and 3 if 50 or over, give the probability distribution for x.

5.88 A business firm is considering two investments. It will choose the one that promises the greater payoff. Which of the investments should it accept? (Let the mean profit measure the payoff.)

Invest in Tool Shop		Invest in Book Store	
Profit	Probability	Profit	Probability
$100,000	0.10	$400,000	0.20
50,000	0.30	90,000	0.10
20,000	0.30	−20,000	0.40
−80,000	0.30	−250,000	0.30
Total	1.00	Total	1.00

5.89 A box contains eight identical balls numbered from 1 to 8. An experiment consists of selecting two of these balls, without replacement, and identifying their numbers:
    a. How many different sets of two balls could be selected?
    b. List the sample space for the experiment.
    c. If the balls are selected randomly, what is the probability of each of the possible sets in the sample space?
    d. Define the random variable x to be the total of the numbers on the two balls. Construct the probability distribution for x.
    e. Find the mean and standard deviation of x.

5.90 Bill has completed a ten-question multiple-choice test on which he answered seven questions correctly. Each question had one correct answer to be chosen from five alternatives. Bill says that he answered the test by randomly guessing the answers without reading the questions or answers.
    a. Define the random variable x to be the number of correct answers on this test and construct the probability distribution if the answers were obtained by random guessing.
    b. What is the probability that Bill guessed seven of the ten answers correctly?
    c. What is the probability that anybody can guess six or more answers correctly?
    d. Do you believe that Bill actually randomly guessed as he claims? Explain.

5.91 A random variable that can assume any one of the integer values $1, 2, \ldots, n$ with equal probabilities of $\frac{1}{n}$ is said to have a uniform distribution. The probability function is written $P(x) = \frac{1}{n}$, for $x = 1, 2, 3, \ldots, n$. Show that $\mu = (n+1)/2$. [*Hint*: $1 + 2 + 3 + \cdots + n = [n(n+1)]/2$.]

VOCABULARY LIST

Be able to define each term. In addition, describe, in your own words, and give an example of each term. Your examples should not be ones given in class or in the textbook. The bracketed numbers indicate the chapter in which the term first appeared, but you should define the terms again to show increased understanding of their meaning.

addition of probabilities [4]
binomial coefficient
binomial experiment
binomial probability function
binomial random variable
constant function
discrete random variable
experiment [1, 4]
failure
independent events [4]
line histogram
mean of probability distribution
multiplication of probabilities [4]
mutually exclusive events [4]

population parameter [1]
probability distribution
probability function
probability histogram
probability line histogram
random variable
relative frequency [2, 4]
sample statistic [1]
standard deviation of probability distribution
success
trial
variance of probability distribution

KEY CONCEPTS

binomial experiment
binomial probability function
discrete random variable
mean value of a probability distribution
population parameter

probability function
random variable
relative frequency
standard deviation of a probability distribution

QUIZ A

Answer "True" if the statement is always true. If the statement is not always true, replace the words shown in bold with words that make the statement always true.

5.1 The number of hours that you waited in line to register this semester is an example of a **discrete** random variable.

5.2 The number of automobile accidents that you were involved in as a driver last year is an example of a **discrete** random variable.

5.3 The sum of all the probabilities in any probability distribution is always exactly **two**.

5.4 The various values of a random variable form a list of **mutually exclusive events**.

5.5 A binomial experiment always has **three or more** possible outcomes to each trial.

5.6 The formula $\mu = np$ may be used to compute the mean of a **discrete** population.

5.7 The binomial parameter p is the probability of **one success occurring in n trials** when a binomial experiment is performed.

5.8 A parameter is a statistical measure of some aspect of a **sample**.

5.9 **Sample statistics** are represented by letters from the Greek alphabet.

5.10 The probability of events A or B is equal to the sum of the probability of event A and the probability of event B when A and B are **mutually exclusive events**.

QUIZ B

5.1 **a.** Show that the following is a probability distribution:

x	P(x)
1	0.2
3	0.3
4	0.4
5	0.1

  b. Find $P(x = 1)$.   **c.** Find $P(x = 2)$.
  d. Find $P(x > 2)$.   **e.** Find the mean of x.
  f. Find the standard deviation of x.

5.2 A T-shirt manufacturing company advertises the probability of an individual T-shirt being irregular is 0.1. A box of 12 such T-shirts is randomly selected and inspected.
  a. What is the probability that exactly 2 of these 12 T-shirts are irregular?
  b. What is the probability that exactly 9 of these 12 T-shirts are *not* irregular?
  Let x be the number of T-shirts that are irregular in all such boxes of 12 T-shirts.
  c. Find the mean of x.   **d.** Find the standard deviation of x.

QUIZ C

5.1 What properties must an experiment possess in order for it to be a binomial probability experiment?

5.2 Student A uses a relative frequency distribution for a set of sample data and calculates the mean and standard deviation using formulas learned in Chapter 5. Student A justified her choice of formulas by saying that since relative frequencies are empirical probabilities, her sample is represented by a probability distribution and therefore her choice of formulas was correct. Student B argues that since the distribution represented a sample, then the mean and the standard deviation involved are known as $\bar{x}$ and s and must be calculated using the corresponding frequency distribution and formulas learned in Chapter 2. Who is correct, A or B? Justify your choice.

5.3 Student A and Student B were discussing one entry in a probability distribution chart.

x	P(x)
−2	0.1

Student B felt that this entry was okay since the $P(x)$ was a value between 0.0 and 1.0. Student A argued that this entry was impossible for a probability distribution since x was a −2, and negatives are not possible. Who is correct, A or B? Justify your choice.

6 NORMAL PROBABILITY DISTRIBUTIONS

Chapter Outline

6.1 Normal Probability Distributions
The domain of **bell-shaped distributions** is the set of all real numbers.

6.2 The Standard Normal Distribution
To work with normal distributions we need the **standard score**.

6.3 Applications of Normal Distributions
Normal distribution can help us to determine **probabilities**.

6.4 Notation
The *z* **notation** is critical in the use of normal distributions.

6.5 Normal Approximation of the Binomial
Binomial probabilities can be **estimated** by using a normal distribution.

Teaching Breast and Testicular Self-Exams: Evaluation of a High School Curriculum Pilot Project

INTRODUCTION

This pilot project, a joint effort of the Cleveland Clinic Foundation and the American Cancer Society (Cuyahoga County Unit), promoted the concept of early detection of cancer to high school students by teaching the topics of breast and testicular self-examination (BSE and TSE). Although the concept for this project evolved from similar work done in Wisconsin, it represents the first comprehensive program in which classroom teachers have been used exclusively to teach students the topics.

Yearly in the United States, there are 115,000 new cases of breast cancer which lead to an estimated 37,300 deaths (American Cancer Society, 1983)....

EVALUATION METHODOLOGY

Two written questionnaires were used to evaluate the project: a teacher satisfaction questionnaire and a student questionnaire. The student questionnaire included questions about the students: use of self-exams (self-reported); knowledge about BSE and TSE (multiple choice items); and attitudes toward early cancer detection (Likert Scale items)....

Students were placed into one of two study groups, the "Pre-Test Post-Test Study Group" (given the student questionnaire before and after BSE/TSE teaching) and the "Post-Test Only" study group (given the questionnaire once, after BSE/TSE teaching).

TABLE 1 Mean Test Scores for Knowledge Questions

	N	BSE Score (possible = 13) Mean ± SD	TSE Score (possible = 13) Mean ± SD	Total Score (possible = 13) Mean ± SD
Pre-Test	532	5.4 ± 2.5	5.6 ± 2.7	11.0 ± 4.7
Post-Test	532	7.7 ± 2.5	8.0 ± 2.6	15.8 ± 4.2

FIGURE 1 Distribution of Difference Scores Obtained by Subtracting Each Student's Pre-test Knowledge Score from His or Her Post-test Knowledge Score

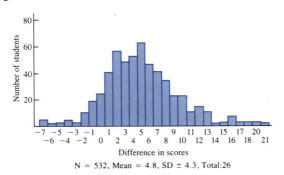

N = 532, Mean = 4.8, SD ± 4.3, Total:26

Source: This article by Stephen L. Luther, Stephen Sroka, Marlene Goormastic, and James E. Montie is reprinted with permission from *Health Education,* February/March 1985, pp. 40–43. *Health Education* is a publication of the American Alliance for Health, Physical Education, Recreation and Dance, 1900 Association Drive, Reston, VA 22091.

Chapter Objectives

Until now we have considered distributions of discrete variables only. In this chapter we will examine one particular family of probability distributions of major importance whose domain is the set of all real numbers. These distributions are called the *normal,* the *bell-shaped,* or the *Gaussian distributions.* "Normal" is simply the traditional title of this particular type of distribution and is not a descriptive name meaning "typical." Although there are many other types of continuous distributions (rectangular, triangular, skewed, and so on), many variables have an approximately normal distribution. For example, several of the histograms drawn in Chapter 2 suggested a normal distribution. A mounded histogram that is approximately symmetric is an indication of such a distribution.

In addition to learning what a normal distribution is, we will consider (1) how probabilities are found, (2) how they are represented, and (3) how normal distributions are used. Although there are other distributions of continuous variables, the normal distributions are the most important.

6.1 ▼ Normal Probability Distributions

As in all other probability distributions, there are formulas that give us information about the **normal distribution.** Previously, the probability function was the only function of interest. However, for the normal distribution there are two functions that define and describe it. Each formula uses the random variable as an independent variable. Formula (6-1) gives an ordinate (*y* value) for each point on the graph of the normal distribution for each given abscissa (*x* value). Formula (6-2) yields the probability associated with *x* when *x* is in the interval between the values *a* and *b.* Note that the random variable *x* is a continuous variable. For continuous variables we will discuss the probability that *x* has a value between the two extreme values of the interval, and we will depict the probability as an area under the graph of the probability distribution.

$$y = f(x) = \frac{1}{\sigma\sqrt{2\pi}} \cdot \exp\left[-\frac{1}{2}\left(\frac{x-\mu}{\sigma}\right)^2\right] \quad \text{for all real } x \quad \textbf{(6-1)}$$

$$P(a \leq x \leq b) = \int_a^b f(x)\, dx \quad \textbf{(6-2)}$$

If you don't understand these formulas, don't worry about it. We will not be using them in this book. However, a few comments about their meaning will be helpful if you ever use a probability table in any of the standard reference books. The directions for using these tables are often written with the aid of mathematical formulas such as (6-1) and (6-2).

Formula (6-1) and its associated table can be used to graph a normal distribution with a given mean and standard deviation. (The table associated with formula (6-1) is not included in this text, but it is available in many other textbooks and in all standard books of statistical tables.) In a more practical sense, we can draw a **normal,** or **bell-shaped, curve** and then label it to approximate scale, as indicated in Figure 6-1.

FIGURE 6-1
The Normal Distribution

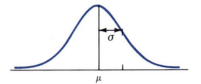

To talk intelligently about formula (6-2), we need to know what the symbols used in the formula represent. The $\int_a^b f(x)\,dx$ is called a "definite integral" and it comes from calculus. The definite integral is a number that is the measure of the area under the curve of $f(x)$. This area is bounded by a on the left, b on the right, $f(x)$ at the top, and the x-axis at the bottom. The equation of the normal curve in formula (6-2) gives the measure of the shaded area in Figure 6-2.

FIGURE 6-2
Shaded Area Under the Curve Is $P(a \leq x \leq b)$

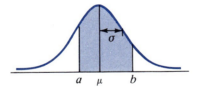

Since $P(a \leq x \leq b)$ (read "the probability that x is between a and b") is given by $\int_a^b f(x)\,dx$, the measure of this probability is also the measure of the probability that a continuous random variable will assume a value within a specified interval. As a student of elementary statistics, you are not expected to have studied calculus, so we will say nothing more about the definite integral in this book.

The values calculated by using formulas (6-1) and (6-2) are also found in a table, with the probabilities expressed to four decimal places. But, before we show you how to read the table, we must point out that the table is expressed in a "standardized" form. It is standardized so that it is not necessary to have a different table for every different pair of values of the mean and standard deviation. For example, the area under the normal curve of a distribution with a mean of 15 and a standard deviation of 3 must be somehow related to the area under the curve of a normal distribution with a mean of 113 and a standard deviation of 38.5. Recall that the empirical rule concerns the percentage of a distribution within a certain number of standard deviations of the mean (see page 107).

percentage—proportion

NOTE Percentage and probability are related concepts. **Percentage** is usually used when talking about a **proportion** of a population between certain values. **Probability** is usually used when talking about the **chance** that the next individual item will have a value between certain values.

The empirical rule is a fairly crude measuring device. With it we are able to find probabilities associated only with whole-number multiples of the standard deviation (within one standard deviation of the mean and so on). We will often be interested in the probabilities associated with fractional parts of the standard deviation. For example, we might want to know the probability that x is within 1.85 standard deviations of the mean. Therefore, we must refine the empirical rule so that we can deal with more precise measurements. This refinement is discussed in the next section.

6.2 ▼ The Standard Normal Distribution

The key to working with the normal distribution is the **standard score** z. The measures of probabilities associated with the random variable x are determined by the relative position of x with respect to the mean and the standard deviation of the distribution. The empirical rule tells us that approximately 68% of the data are within one standard deviation of the mean. The value of the mean and the size of the standard deviation do not change this fact.

Recall that z was defined in Chapter 2 to be

$$\frac{x - (\text{mean of } x)}{(\text{standard deviation of } x)}$$

Therefore, we can write the formula for z as

$$z = \frac{x - \mu}{\sigma} \tag{6-3}$$

(Note that if $x = \mu$, then $z = 0$.)

The z-score is known as a "standardized" variable because its units are standard deviations. The normal probability distribution associated with this standard score z is called the **standard normal distribution**. Table 5 in Appendix F lists the probabilities associated with the intervals from the mean for specific positive values of z. Other probabilities must be found by addition, subtraction, and so on, based on the concept of symmetry that exists in the normal distribution and the fact that the total area under the curve is 1.0.

standard normal distribution

area representation for probability

Let's look at an illustration to see how we read Table 5. The z-scores are in the margins: the left margin has the units digit and the tenths digit; the top has the hundredths digit. Let's look up a z-score of 1.52. We find 0.4357, as shown in Table 6-1. Now, exactly what is this 0.4357? It is the measure of the area under the standard normal curve between $z = 0$ (which locates the mean) and $z = 1.52$ (a number 1.52 standard deviations larger than the mean). See Figure 6-3. **This area is also the measure of the probability associated with the same interval**; that is,

Section 6.2 ▼ THE STANDARD NORMAL DISTRIBUTION

TABLE 6-1
A Portion of Table 5, Appendix F

z	0.00	0.01	0.02	...
⋮				
1.5			0.4357	
⋮				

FIGURE 6-3
Area Under Normal Curve from $z = 0$ to $z = 1.52$ Is 0.4357

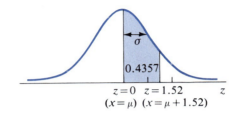

$$P(0 < z < 1.52) = 0.4357$$

Read this result as "the probability that a value picked at random falls between the mean and 1.52 standard deviations above the mean is 0.4357," or equivalently, "the probability that a z-score picked at random will fall between 0 and 1.52 is 0.4357."

Recall that one of the basic properties of probability is that the sum of all probabilities is exactly 1.0. Since the area under the normal curve represents the measure of probability, the total area under the bell-shaped curve is exactly 1 unit. This distribution is symmetric with respect to the vertical line drawn through $z = 0$, which cuts the area exactly in half at the mean. Can you verify this fact by inspecting formula (6-1)? That is, the area under the curve to the right of the mean is exactly one-half unit, 0.5, and the area to the left is also one-half unit, 0.5. Areas (probabilities) not given directly in the table can be found by relying on these facts.

Now let's look at some illustrations.

▼ **ILLUSTRATION 6-1**

Find the area under the normal curve to the right of $z = 1.52$; $P(z > 1.52)$.

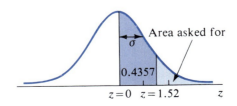

SOLUTION The area to the right of the mean (all the shading in the figure) is exactly 0.5000. The question asks for the shaded area that is not included in the 0.4357. Therefore, subtract 0.4357 from 0.5000.

$$P(z > 1.52) = 0.5000 - 0.4357 = \mathbf{0.0643}$$

SUGGESTION As we have done here, always draw and label a sketch. It is most helpful.

▼ ILLUSTRATION 6-2

Find the area to the left of $z = 1.52$; $P(z < 1.52)$.

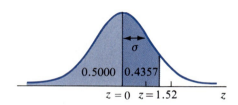

SOLUTION The total shaded area is made up of 0.4357 found in the table and the 0.5000 that is to the left of the mean. Therefore, add 0.4357 to 0.5000.

$$0.4357 + 0.5000 = \mathbf{0.9357}$$

NOTE The addition and subtraction done in Illustrations 6-1 and 6-2 are correct because the "areas" represent mutually exclusive events (discussed in Section 4-5).

The symmetry of the normal distribution is a key factor in determining probabilities associated with values below (to the left of) the mean. The area between the mean and $z = -1.52$ is exactly the same as the area between the mean and $z = +1.52$. This fact allows us to find values related to the left side of the distribution.

▼ ILLUSTRATION 6-3

The area between the mean ($z = 0$) and $z = -2.1$ is the same as the area between $z = 0$ and $z = +2.1$; that is, $P(-2.1 < z < 0) = P(0 < z < 2.1)$.

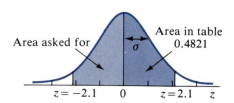

Thus we have

$$P(-2.1 < z < 0) = \mathbf{0.4821}$$

ILLUSTRATION 6-4

The area to the left of $z = -1.35$ is found by subtracting 0.4115 from 0.5000.

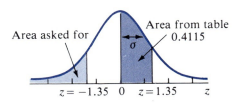

Therefore, we obtain

$$P(z < -1.35) = \mathbf{0.0885}$$

ILLUSTRATION 6-5

The area between $z = -1.5$ and $z = 2.1$, $P(-1.5 < z < 2.1)$, is found by adding the two areas together.

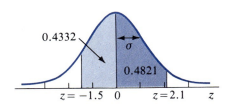

Therefore, we obtain

$$P(-1.5 < z < 0) + P(0 < z < 2.1) = 0.4332 + 0.4821 = \mathbf{0.9153}$$

ILLUSTRATION 6-6

The area between $z = 0.7$ and $z = 2.1$, $P(0.7 < z < 2.1)$, is found by subtracting. The area between $z = 0$ and $z = 2.1$ contains all the area between $z = 0$ and $z = 0.7$. The area between $z = 0$ and $z = 0.7$ is therefore subtracted from the area between $z = 0$ and $z = 2.1$.

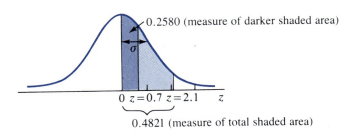

Thus we have

$$P(0.7 < z < 2.1) = 0.4821 - 0.2580 = \mathbf{0.2241}$$

The normal distribution table can also be used to determine a z-score if we are given an area. The next illustration considers this idea.

▼ ILLUSTRATION 6-7

What is the z-score associated with the 75th percentile? (Assume the distribution is normal.) See Figure 6-4.

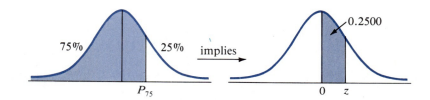

FIGURE 6-4
P_{75} and Its Associated z-Score

SOLUTION To find this z-score, look in Table 5, Appendix F, and find the "area" entry that is closest to 0.2500; this area entry is 0.2486. Now read the z-score that corresponds to this area. From the table the z-score is found to be $z = \mathbf{0.67}$. This says that the 75th percentile in a normal distribution is 0.67 (approximately $\frac{2}{3}$) standard deviation above the mean.

z	...	0.07	0.08	...
⋮				
0.6		0.2486	0.2517	
⋮				

▼ ILLUSTRATION 6-8

What z-scores bound the middle 95% of a normal distribution? See Figure 6-5.

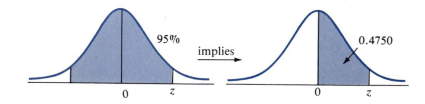

FIGURE 6-5
Middle 95% of Distribution and Its Associated z-Score

Section 6.2 ▼ THE STANDARD NORMAL DISTRIBUTION

SOLUTION The 95% is split into two equal parts by the mean; 0.4750 is the area (percentage) between $z = 0$, the mean, and the z-score at the right boundary. Since we have the area, we look for the entry in Table 5 closest to 0.4750 (it happens to be exactly 0.4750) and read the z-score in the margin. We obtain $z = 1.96$. Therefore, $z = -1.96$ and $z = 1.96$ bound the middle 95% of a normal distribution.

z	...	0.06	...
⋮			
1.9		0.4750	
⋮			

▲▲

▼▲ EXERCISES

6.1 Describe the distribution of the standard normal score z.

6.2 Why is this distribution called *standard normal*.

6.3 Find the area under the normal curve that lies between the following pairs of z values:
    a. $z = 0$ to $z = 1.30$ b. $z = 0$ to $z = 1.28$
    c. $z = 0$ to $z = -3.20$ d. $z = 0$ to $z = -1.98$

6.4 Find the probability that a piece of data picked at random from a normal population will have a standard score (z) that lies between the following pairs of z values:
    a. $z = 0$ to $z = 2.10$ b. $z = 0$ to $z = 2.57$
    c. $z = 0$ to $z = -1.20$ d. $z = 0$ to $z = -1.57$

6.5 Find the area under the standard normal curve that corresponds to the following z values:
    a. between 0 and 1.55   b. to the right of 1.55
    c. to the left of 1.55   d. between -1.55 and 1.55

6.6 Find the probability that a piece of data picked at random from a normal population will have a standard score (z) that lies
    a. between 0 and 0.84.   b. to the right of 0.84.
    c. to the left of 0.84.   d. between -0.84 and 0.84.

6.7 Find the area under the normal curve that lies between the following pairs of z values:
    a. $z = -1.20$ to $z = 1.22$ b. $z = -1.75$ to $z = 1.54$
    c. $z = -1.30$ to $z = 2.58$ d. $z = -3.5$ to $z = -0.35$

6.8 Find the probability that a piece of data picked at random from a normal population will have a standard score (z) that lies between the following pairs of z values:
    a. $z = -2.75$ to $z = 1.38$ b. $z = 0.67$ to $z = 2.95$
    c. $z = -2.95$ to $z = -1.18$

6.9 Find the following areas under the normal curve:
  a. to the right of $z = 0.00$
  b. to the right of $z = 1.05$
  c. to the right of $z = -2.30$
  d. to the left of $z = 1.60$
  e. to the left of $z = -1.60$

6.10 Find the probability that a piece of data picked at random from a normally distributed population will have a standard score that is
  a. less than 3.00.
  b. greater than -1.55.
  c. less than -0.75.
  d. less than 1.25.
  e. greater than -1.25.

6.11 Find the following:
  a. $P(0.00 < z < 2.35)$
  b. $P(-2.10 < z < 2.34)$
  c. $P(z > 0.13)$
  d. $P(z < 1.48)$

6.12 Find the following:
  a. $P(-2.05 < z < 0.00)$
  b. $P(-1.83 < z < 2.07)$
  c. $P(z > -1.52)$
  d. $P(z < -0.43)$

6.13 Find the z-score for the standard normal distribution shown on each of the following diagrams:

a.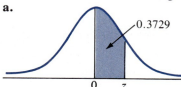
b.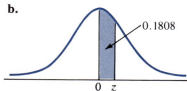
c.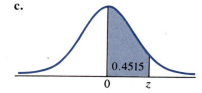
d.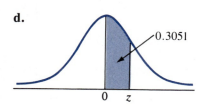
e.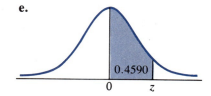
f.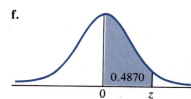

6.14 Find the z-score for the standard normal distribution shown in each of the following diagrams:

a.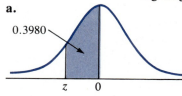
b.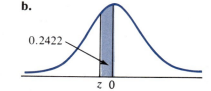
c.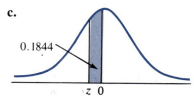
d.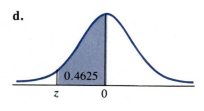
e.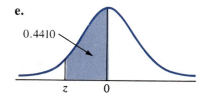
f.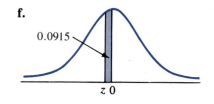

Section 6.3 ▼ APPLICATIONS OF NORMAL DISTRIBUTIONS   317

6.15 Find the standard score z shown on each of the following diagrams:

a.

b.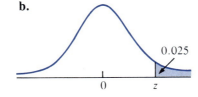

c.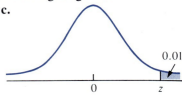

6.16 Find the standard score z shown on each of the following diagrams:

a.

b.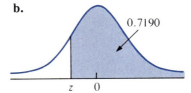

c.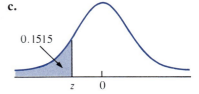

6.17 Find a value of z such that 40% of the distribution lies between it and the mean. (There are two possible answers.)

6.18 Find the standard score z such that
  a. 80% of the distribution is below (to the left of) this value.
  b. the area to the right of this value is 0.15.

6.19 Find the two z-scores that bound the middle 50% of a normal distribution.

6.20 Find the two standard scores z such that
  a. the middle 90% of a normal distribution is bounded by them.
  b. the middle 98% of a normal distribution is bounded by them.

6.21 Assuming a normal distribution, what is the z-score associated with the 90th percentile? the 95th percentile? the 99th percentile?

6.22 Assuming a normal distribution, what is the z-score associated with the 1 quartile? 2 quartile? 3 quartile?

6.3 ▼ Applications of Normal Distributions

The probabilities associated with any normal distribution can be found by applying the techniques discussed in Section 6-2. First, however, we must "standardize" the given information. When dealing with a normal distribution, we need to know its mean μ and its standard deviation σ. Once these values are known, any value of the random variable x can be easily converted to the standard score z by using formula (6-3):

$$z = \frac{x - \mu}{\sigma}$$

▼ ILLUSTRATION 6-9

Consider the intelligence quotient (IQ) scores for people. IQs are normally distributed with a mean of 100 and a standard deviation of 10. If a person is picked at random, what is the probability that his or her IQ is between 100 and 115; that is, what is $P(100 < x < 115)$?

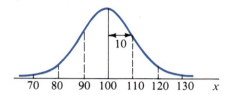

SOLUTION $P(100 < x < 115)$ is represented by the shaded area in the following figure:

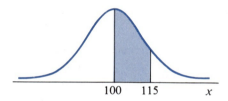

The variable x must be standardized by using formula (6-3). The z values are shown on the next figure.

$$z = \frac{x - \mu}{\sigma}$$

When $x = 100$: $\quad z = \dfrac{100 - 100}{10} = \mathbf{0.0}$

When $x = 115$: $\quad z = \dfrac{115 - 100}{10} = \mathbf{1.5}$

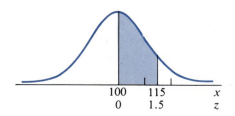

Section 6.3 ▼ APPLICATIONS OF NORMAL DISTRIBUTIONS 319

Therefore,

$$P(100 < x < 115) = P(0.0 < z < 1.5) = \mathbf{0.4332}$$

(The value 0.4332 is found by using Table 5, Appendix F.) Thus the probability is 0.4332 that a person picked at random has an IQ between 100 and 115. ▲▲

▼ ILLUSTRATION 6-10

Find the probability that a person selected at random will have an IQ greater than 95.

SOLUTION

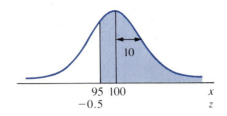

$$z = \frac{x - \mu}{\sigma} = \frac{95 - 100}{10} = \frac{-5}{10} = -0.5$$

$$P(x > 95) = P(z > -0.5)$$
$$= 0.1915 + 0.5000 = \mathbf{0.6915}$$

Thus the probability is 0.6915 that a person selected at random will have an IQ greater than 95. ▲▲

The normal table can be used to answer many kinds of questions that involve a normal distribution. Many times a problem will call for the location of a "cutoff point," that is, a particular value of x such that there is exactly a certain percentage in a specified area. The following illustrations concern some of these problems.

▼ ILLUSTRATION 6-11

In a large class, suppose that your instructor tells you that you need to obtain a grade in the top 10% of your class to get an A on a particular exam. From past experience she is able to estimate that the mean and standard deviation on this exam will be 72 and 13, respectively. What will be the minimum grade needed to obtain an A? (Assume that the grades will be approximately normally distributed.)

SOLUTION

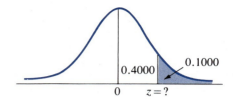

$$10\% = 0.1000 = 0.5000 - 0.4000$$

Look in Table 5 to find the value of z that is closest to 0.4000; it is $z = 1.28$. Thus,

$$P(z > 1.28) = 0.10 \quad \text{(information given)}$$

Now find the x value that corresponds to $z = 1.28$ by using formula (6-3):

$$z = \frac{x - \mu}{\sigma} \quad \text{or} \quad 1.28 = \frac{x - 72}{13}$$

$$x - 72 = (13)(1.28)$$

$$x = 72 + (13)(1.28) = 72 + 16.64$$

$$= 88.64, \text{ or } \mathbf{89}$$

Thus if you receive an 89 or higher, you can expect to be in the top 10% (which means an A). ▲▲

▼ **ILLUSTRATION 6-12**

Find the 33d percentile for IQ scores ($\mu = 100$ and $\sigma = 10$ from Illustration 6-9, p. 318).

SOLUTION

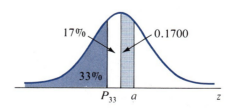

$$P(0 < z < a) = 0.17$$

$$a = 0.44 \quad \text{(cutoff value of } z \text{ from Table 5)}$$

33d percentile of $z = -0.44$ \quad (below mean)

Section 6.3 ▼ APPLICATIONS OF NORMAL DISTRIBUTIONS

Now we convert the 33d percentile of the z-scores, -0.44, to an x-score:

$$z = \frac{x - \mu}{\sigma} \quad \text{[formula (6-3)]}$$

$$-0.44 = \frac{x - 100}{10}$$

$$-4.4 = x - 100$$

$$100 - 4.4 = x$$

$$x = \mathbf{95.6}$$

Thus, 95.6 is the 33d percentile for IQ scores.

Illustration 6-13 concerns a situation where you are asked to find the mean μ when given the related information.

▼ ILLUSTRATION 6-13

The incomes of junior executives in a large corporation are normally distributed with a standard deviation of $1,200. A cutback is pending, at which time those who earn less than $28,000 will be discharged. If such a cut represents 10% of the junior executives, what is the current mean salary of the group of junior executives?

SOLUTION If 10% of the salaries are below $28,000, then 40% (or 0.4000) are between $28,000 and the mean μ. Table 5 indicates that $z = -1.28$ is the standard score that occurs at $x = \$28,000$. Using formula (6-3) we can find the value of μ:

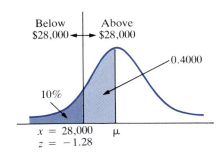

$$-1.28 = \frac{28,000 - \mu}{1,200}$$

$$-1,536 = 28,000 - \mu$$

$$\mu = \mathbf{\$29,536}$$

That is, the current mean salary of junior executives is $29,536.

Case Study 6-1

The Adjustment of Birthweight for Very Early Gestational Ages: Two Related Problems in Statistical Analysis

In this paper, we are concerned with two aspects of the relationship of birthweight to gestational age, both arising from the problem caused by babies born before term (i.e. before 37 completed weeks of gestation). The methods that we propose are statistically very simple. This is a great advantage when the users of the results of analysis, here mainly clinicians, are often familiar with only the simplest statistical methods.

An important problem in the creation of standards for birthweight is very early births. The relative growth of babies is usually determined using centile charts, showing the 3rd, 10th, 50th, 90th and 97th centiles of weight for each week of gestational age....

... Two data sets have been analyzed....

... The *Naples* set....

The *London* set consists of all births to Caucasian mothers booking for delivery at St. George's Hospital, London....

... However, the distribution of birthweight for fixed gestational age is symmetrical and follows an approximately normal distribution within gestational age (Altman and Coles, 1980). Fig. 2 shows the distribution of birthweight at 38 weeks and above from the London data, with the corresponding normal distribution curve.

FIGURE 2 Distribution of Birthweight for Gestational Age 38 weeks or More, London Data, with the Corresponding Normal Distribution Curve

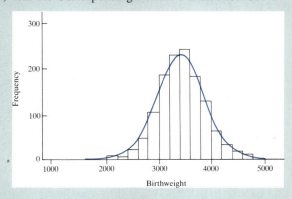

Source: J. M. Bland, J. L. Peacock, H. R. Anderson, O. G. Brooke, and M. De Curtis, *Applied Statistics* 39 (No. 2), 229–239. Reprinted by permission.

Referring again to the IQ scores, what is the probability that a person picked at random has an IQ of 125, $P(x = 125)$? (IQ scores are normally distributed with a mean of 100 and a standard deviation of 10.) This situation has two interpre-

Section 6.3 ▼ APPLICATIONS OF NORMAL DISTRIBUTIONS

tations: (1) theoretical and (2) practical. Let's look at the theoretical interpretation first. Recall that the probability associated with a continuous random variable is represented by the area under the curve. That is, $P(a \le x \le b)$ is equal to the area between a and b under the curve. $P(x = 125)$—that is, x is exactly 125—is then $P(125 \le x \le 125)$, or the area of the vertical line segment at $x = 125$. This area is zero. However, this is not the *practical* meaning of $x = 125$. It generally means 125 to the nearest integer value. Thus $P(x = 125)$ would most likely be interpreted as

$$P(124.5 < x < 125.5)$$

The interval from 124.5 to 125.5 under the curve has a measurable area and is then nonzero. In situations of this nature, you must be sure of the meaning being used.

▼▲ EXERCISES

6.23 Given that x is a normally distributed random variable with a mean of 60 and a standard deviation of 10, find the following probabilities:
  a. $P(x > 60)$ b. $P(60 < x < 72)$
  c. $P(57 < x < 83)$ d. $P(65 < x < 82)$
  e. $P(38 < x < 78)$ f. $P(x < 38)$

6.24 Let h be a normally distributed random variable with a mean of 26.7 and a standard deviation of 3.4. Find the probability that an individual value of h, selected at random, will fall in the following intervals:
  a. between 26.7 and 31.7   b. between 28.0 and 30.0
  c. between 20.0 and 24.0   d. between 20.0 and 30.0

6.25 For a particular age group of adult males the distribution of cholesterol readings, in mg/dl, is normally distributed with a mean of 210 and a standard deviation of 15.
  a. What percentage of this population would have readings exceeding 250?
  b. What percentage would have readings less than 150?

6.26 Computers are shut down for certain periods of time for routine maintenance, installation of new hardware, and so on. The down times for a particular computer are normally distributed with a mean of 1.5 hours and a standard deviation of 0.4 hours.
  a. What percentage of the down times exceed 3 hours?
  b. What percentage are between 1 and 2 hours?

6.27 Final averages are typically approximately normally distributed with a mean of 72 and a standard deviation of 12.5. Your professor says that the top 8% of the class will receive A; the next 20%, B; the next 42%, C; the next 18%, D; and the bottom 12%, F.
  a. What average must you exceed to obtain an A?
  b. What average must you exceed to receive a grade better than a C?
  c. What average must you obtain to pass the course? (You'll need a D or better.)

6.28 A study in the *Canadian Journal of Ophthalmology* (Vol. 25, No. 5, 1990, pages 229–233) reports the mean intraocular pressure (IOP) as 16.53 and the standard deviation as 2.84 for patients undergoing cataract and intraocular lens implant surgery. If we assume that the IOP measurements are normally distributed and use the mean and standard deviation quoted above as population parameters, what percentage of the patients undergoing this procedure would have an IOP exceeding 20.0?

6.29 The length of useful life of a fluorescent tube used for indoor gardening is normally distributed. The useful life has a mean of 600 hr and a standard deviation of 40 hr. Determine the probability that
  a. a tube chosen at random will last between 620 and 680 hr.
  b. such a tube will last more than 740 hr.

6.30 At Pacific Freight Lines, bonuses are given to billing clerks when they complete 300 or more freight bills during an eight-hour day. The number of bills completed per clerk per eight-hour day is approximately normally distributed with a mean of 270 and a standard deviation of 16. What proportion of the time should a randomly selected billing clerk expect to receive a bonus?

6.31 The waiting time x at a certain bank is approximately normally distributed with a mean of 3.7 min and a standard deviation of 1.4 min.
  a. Find the probability that a randomly selected customer has to wait less than 2.0 min.
  b. Find the probability that a randomly selected customer has to wait more than 6 min.
  c. Find the value of the 75th percentile for x.

6.32 A brewery filling machine is adjusted to fill quart bottles with a mean of 32 oz of ale, and a variance of 0.003. Periodically, a bottle is checked and the amount of ale is noted. Assuming that the amount of fill is normally distributed, what is the probability that the next randomly checked bottle contains more than 32.02 oz?

6.33 The weights of ripe watermelons grown at Mr. Smith's farm are normally distributed with a standard deviation of 2.8 lb. Find the mean weight of Mr. Smith's ripe watermelons if only 3% weigh less than 15 lb.

6.34 A machine fills containers with a mean weight per container of 16.0 oz. If no more than 5% of the containers are to weigh less than 15.8 oz, what must the standard deviation of the weights equal? (Assume normality.)

6.35 A radar unit is used to measure the speed of automobiles on an expressway during rush-hour traffic. The speeds of individual automobiles are normally distributed with a mean of 62 mph.
  a. Find the standard deviation of all speeds if 3% of the automobiles travel faster than 72 mph.
  b. Using the standard deviation found in (a), find the percentage of these cars that are traveling less than 55 mph.
  c. Using the standard deviation found in (a), find the 95th percentile for the variable "speed."

6.36 A study in the journal *Physical Therapy* (February 1991, pages 90–96) discusses range-of-motion measurements on patients with knee complaints. One measurement taken was the height (in cm) of the patients. The mean was 170.8 cm and the standard deviation was 10.5 cm. If the height of such patients is normally distributed and the mean and standard deviation are treated as population measurements, what percentage of such patients would exceed 200 cm in height?

6.37 Statistical computer packages are often used to generate simulated samples. Suppose, for example, we want to simulate a sample of adult male heights and that the distribution of these heights is normally distributed with a mean of 68 in. and a standard deviation of 2.5 in. MINITAB was used to generate a random sample of 25 heights and to find the mean and standard deviation of the simulated data. Verify the results.

```
MTB > # NOTE THIS MINITAB RUN SIMULATES A SAMPLE FROM
MTB > # NOTE A NORMAL DISTRIBUTION WITH MEAN = 68
MTB > # NOTE AND STANDARD DEVIATION = 2.5
MTB > RANDOM 25 OBSERVATIONS INTO C1;
SUBC> NORMAL WITH MU = 68 AND SIGMA = 2.5.
MTB > PRINT C1

C1
    68.0281   71.4545   66.0663   72.0609   66.4691   63.3665   68.2395
    72.7440   65.9620   64.9627   68.4147   75.4693   70.3609   66.5527
    67.8157   68.8205   66.5701   66.6655   72.5189   65.8856   69.3900
    65.6637   69.2138   65.6304   67.7742

MTB > MEAN C1
    MEAN    =       68.244
MTB > STDEV C1
    ST.DEV. =        2.8724
MTB > #
MTB > STOP
```

6.38 Use a computer to simulate a random sample of 40 from a normally distributed population with a mean equal to 35.2 and a standard deviation equal to 3.8. List the data, print a histogram, and find the sample mean and standard deviation.

6.4 ▼ Notation

When working with the standard score z, it is often helpful and necessary to identify z with an area under the normal curve. One standard procedure is to use the area under the curve and to the right of the z. We will write this area within parentheses following the z.

▼ ILLUSTRATION 6-14

$z(0.05)$ [read "z of 0.05"] is the value of z such that exactly 0.05 of the area under the curve lies to its right, as shown in Figure 6.6.

FIGURE 6-6
Area Associated with $z(0.05)$

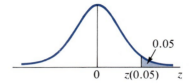

Now let's find the value of $z(0.05)$. We must convert this information into a value that can be read from Table 5; see the areas shown in Figure 6-7.

FIGURE 6-7
Finding the Value of $z(0.05)$

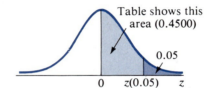

Now we look in Table 5 in Appendix F and find an area as close as possible to 0.4500.

z	...	0.04	0.05	...
⋮				
1.6	...	0.4495	0.4505	
⋮				

Therefore, $z(0.05) = $ **1.65** (*Note*: We will use the z that corresponds to the area closest in value. If the value happens to be exactly halfway between the table entries, always round up to the larger value of z.) ▲▲

$z(0.60)$ [read "z of 0.60"] is that value of z such that 0.60 of the area lies to its right, as shown in Figure 6-8.

FIGURE 6-8
Area Associated with $z(0.60)$

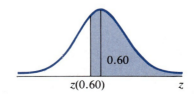

▼ ILLUSTRATION 6-15

Find the value of $z(0.60)$.

SOLUTION   The value 0.60 is related to Table 5 by use of the area 0.1000, as shown in the following diagram:

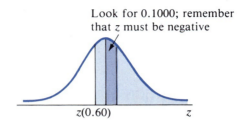

The closest values in Table 5 are 0.0987 and 0.1026.

z	...	0.05	0.06	...
⋮				
0.2	...	0.0987	0.1026	
⋮				

Therefore, $z(0.60)$ is related to 0.25. Since $z(0.60)$ is below the mean, we conclude that $z(0.60) = -0.25$. ▲▲

In later chapters we will use this notation on a regular basis. The two values of z that will be used regularly come from one of the following situations: (1) the z-score such that there is a specified area in one tail of the normal distribution, or (2) the z-scores that bound a specified middle proportion of the normal distribution.

Illustration 6-14 showed a commonly used one-tail situation; $z(0.05) = 1.65$ is located so that 0.05 of the area under the normal distribution curve is in the tail to the right.

▼ ILLUSTRATION 6-16

Find $z(0.95)$.

SOLUTION $z(0.95)$ is located on the left-hand side of the normal distribution since the area to the right is 0.95. The area in the tail to the left then contains the other 0.05, as shown in Figure 6-9 (p. 328). Because of the symmetrical nature of the normal distribution, $z(0.95)$ is $-z(0.05)$; that is, $z(0.05)$ with its sign changed. Thus $z(0.95) = -z(0.05) = -1.65$.

FIGURE 6-9
Area Associated with $z(0.95)$

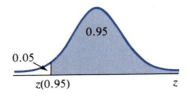

When the middle proportion of a normal distribution is specified, we can still use the "area to the right" notation to identify the specific z-score involved.

▼ **ILLUSTRATION 6-17**

Find the z-scores that bound the middle 0.95 of the normal distribution.

SOLUTION Given 0.95 as the area in the middle (Figure 6-10), the two tails must contain a total of 0.05. Therefore, each tail contains 0.025, as shown in Figure 6-11. In order to find $z(0.025)$ in Table 5, we must determine the area between the mean and $z(0.025)$. It is 0.4750, as shown in Figure 6-12.

FIGURE 6-10
Area Associated with Middle 0.95

FIGURE 6-11
Finding z-Scores for Middle 0.95

FIGURE 6-12
Finding the Value of $z(0.025)$

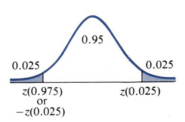
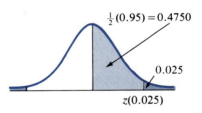

From Table 5,

	...	0.06	...
⋮			
1.9		0.4750	

Therefore, $z(0.025) = 1.96$ and $z(0.975) = -z(0.025) = -1.96$. The middle 0.95 of the normal distribution is bounded by **−1.96 and 1.96**.

▼▲ EXERCISES

6.39 Using the $z(\alpha)$ notation (identify the value of α used within the parentheses), name each of the standard normal variable z's shown in the following diagrams:

a. b. c.

(a) 0.03 (b) 0.14 (c) 0.75

d. e. f.

(d) 0.37 (e) 0.41 (f) 0.18

6.40 Using the $z(\alpha)$ notation (identify the value of α used within the parentheses), name each of the standard normal variable z's shown in the following diagrams:

a. b. c.

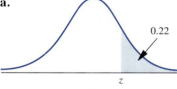  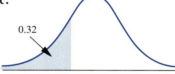

d. e. f.

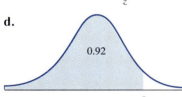 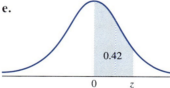 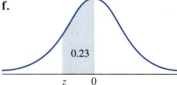

6.41 We are often interested in finding the value of z that bounds a given area in the right-hand tail of the normal distribution, as shown in the accompanying figure. The notation $z(\alpha)$ represents the value of z such that $P(z > z(\alpha)) = \alpha$. Find the following:

    a. $z(0.025)$ **b.** $z(0.05)$ **c.** $z(0.01)$

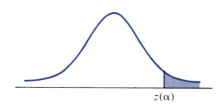

6.42 Use Table 5, Appendix F, to find the following values of z:
   a. $z(0.05)$ **b.** $z(0.01)$ **c.** $z(0.025)$ **d.** $z(0.975)$ **e.** $z(0.98)$

6.43 Complete the following charts of z-scores. The area A given in the tables is the area to the right under the normal distribution in the figures.
   a. z-scores associated with the right-hand tail: Given the area A, find $z(A)$.

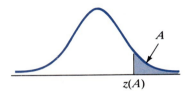

A	0.10	0.05	0.025	0.02	0.01	0.005
z(A)						

   b. z-scores associated with the left-hand tail: Given the area A, find $z(A)$.

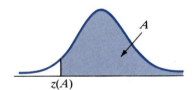

A	0.995	0.99	0.98	0.975	0.95	0.90
z(A)						

6.44 **a.** Find the area under the normal curve for z between $z(0.95)$ and $z(0.025)$.
   b. Find $z(0.025) - z(0.95)$.

6.45 The z notation, $z(\alpha)$, mixes two related concepts, the *z-score* and the *area to the right*, together into a mathematical symbol. Identify the letter in each of the following as being a *z-score* or being an *area*, and then with the aid of a diagram explain what both the given number and the letter represent on the standard normal curve.
   a. $z(A) = 0.10$
   b. $z(0.10) = B$
   c. $x(C) = -0.05$
   d. $-z(0.05) = D$

6.46 Understanding the z notation, $z(\alpha)$, requires us to know whether we have a *z-score* or the *area*. Each of the following expressions use the z notation in a variety of ways, some typical and some not so typical. Find the value asked for in each of the following and then with the aid of a diagram explain what your answer represents:
  a. $z(0.08)$
  b. the area between $z(0.98)$ and $z(0.02)$
  c. $z(1.00 - 0.01)$
  d. $z(0.025) - z(0.975)$

6.5 ▼ Normal Approximation of the Binomial

binomial probability

In Chapter 5 we introduced the binomial distribution. Recall that the binomial distribution is a probability distribution of the discrete random variable x, the number of successes observed in n repeated independent trials. We will now see how **binomial probabilities**, that is, probabilities associated with a binomial distribution, can be reasonably estimated by using the normal probability distribution.

Let's look first at a few specific binomial distributions. Figures 6-13a, 6-13b, and 6-13c show the probabilities of x for 0 to n for three situations: $n = 4$, $n = 8$, and $n = 24$. For each of these distributions, the probability of success for one trial is 0.5. Notice that as n becomes larger, the distribution appears more and more like the normal distribution.

FIGURE 6-13
Binomial Distributions

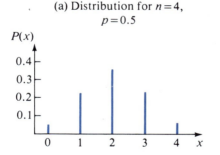
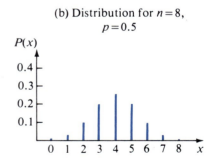
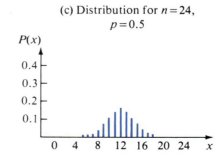

To make the desired approximation, we need to take into account one major difference between the binomial and the normal probability distributions. The binomial random variable is discrete, whereas the normal random variable is continuous. Recall that Chapter 5 demonstrated that the probability assigned to a particular value of x should be shown on a diagram by means of a straight-line segment whose length represents the probability (as in Figure 6-13). Chapter 5 suggested, however, that we can also use a histogram in which the area of each bar is equal to the probability of x.

Let's look at the distribution of the binomial variable x, where $n = 14$ and $p = 0.5$. The probabilities for each x value can be obtained from Table 4 in Appendix F. This distribution of x is shown in Figure 6-14. We see the very same distribution in Figure 6-15 in histogram form.

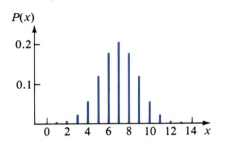

FIGURE 6-14 Line Histogram for the Distribution of x when $n = 14, p = 0.5$

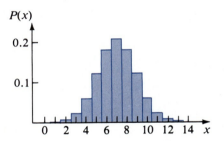

FIGURE 6-15 Histogram for the Distribution of x when $n = 14, p = 0.5$

Let's examine $P(x = 4)$ for $n = 14$ and $p = 0.5$ to study the approximation technique. $P(x = 4)$ is equal to 0.061 (*see* Table 4 of Appendix F), the area of the bar above $x = 4$ in Figure 6-16. Area is the product of width and height. In this case the height is 0.061 and the width is 1.0; thus the area is 0.061. Let's take a closer look at the width. For $x = 4$, the bar starts at 3.5 and ends at 4.5, so we are looking at an area bounded by $x = 3.5$ and $x = 4.5$. The addition and subtraction of 0.5 to the x value is commonly called the **continuity correction factor**. It is our method of converting a discrete variable into a continuous variable.

continuity correction factor

FIGURE 6-16 Area of Bar Above $x = 4$ Is 0.061 when $n = 14, p = 0.5$

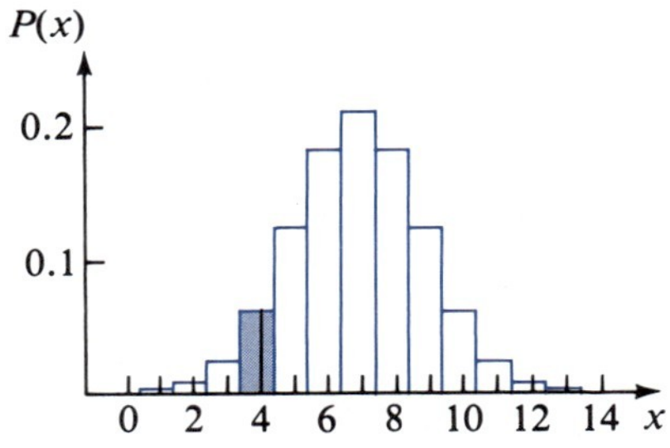

Section 6.5 ▼ NORMAL APPROXIMATION OF THE BINOMIAL

Now let's look at the normal distribution related to this situation. We will first need a normal distribution with a mean and a standard deviation equal to those of the binomial distribution we are discussing. Formulas (5-7) and (5-8) give us these values.

$$\mu = np = (14)(0.5) = 7.0$$
$$\sigma = \sqrt{npq} = \sqrt{(14)(0.5)(0.5)}$$
$$= \sqrt{3.5} = 1.87$$

The probability that $x = 4$ is approximated by the area under the normal curve between $x = 3.5$ and $x = 4.5$, as shown in Figure 6-17. Figure 6-18 shows the entire distribution of the binomial variable x with a normal distribution of the same mean and standard deviation superimposed. Notice that the bars and the interval areas under the curve cover nearly the same area.

FIGURE 6-17 Probability That $x = 4$ Is Approximated by Shaded Area

FIGURE 6-18 Normal Distribution Superimposed over Distribution for Binomial Variable x

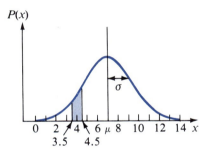
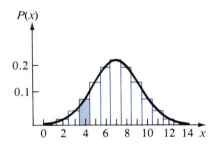

The probability that x is between 3.5 and 4.5 under this normal curve is found by using Table 5 and the methods outlined in Section 6.3.

$$P(3.5 < x < 4.5) = P\left(\frac{3.5 - 7.0}{1.87} < z < \frac{4.5 - 7.0}{1.87}\right)$$
$$= P(-1.87 < z < -1.34)$$
$$= 0.4693 - 0.4099 = \mathbf{0.0594}$$

Since the binomial probability of 0.061 and the normal probability of 0.0594 are reasonably close in value, the normal probability distribution seems to be a reasonable approximation of the binomial distribution.

The normal approximation of the binomial distribution is also useful for values of p that are not close to 0.5. The binomial probability distributions shown in Figures 6-19 and 6-20 suggest that binomial probabilities can be approximated by using the normal distribution. Notice that as n increases in size, the binomial

FIGURE 6-19 Binomial Distributions

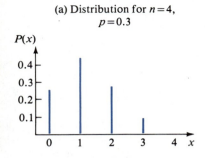
(a) Distribution for $n=4$, $p=0.3$

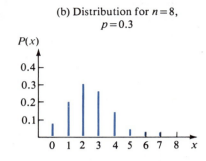
(b) Distribution for $n=8$, $p=0.3$

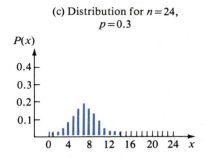
(c) Distribution for $n=24$, $p=0.3$

FIGURE 6-20 Binomial Distributions

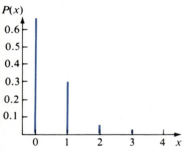
(a) Distribution for $n=4$, $p=0.1$

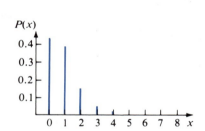
(b) Distribution for $n=8$, $p=0.1$

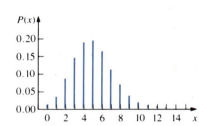
(c) Distribution for $n=50$, $p=0.1$

distribution begins to look like the normal distribution. As the value of p moves away from 0.5, a larger n will be needed in order for the normal approximation to be reasonable. The following "rule of thumb" is generally used as a guideline:

> **RULE**
>
> The normal distribution provides a reasonable approximation to a binomial probability distribution whenever the values of np and $n(1-p)$ both equal or exceed 5.

By now you may be thinking, "So what? I will just use the binomial table and find the probabilities directly and avoid all the extra work." But consider for a moment the situation presented in Illustration 6-18.

▼ **ILLUSTRATION 6-18**

An unnoticed mechanical failure has caused $\frac{1}{3}$ of a machine shop's production of 5000 rifle firing pins to be defective. What is the probability that an inspector will find no more than 3 defective firing pins in a random sample of 25?

Section 6.5 ▼ NORMAL APPROXIMATION OF THE BINOMIAL

SOLUTION In this illustration of a binomial experiment, x is the number of defectives found in the sample, $n = 25$, and $p = P(\text{defective}) = \frac{1}{3}$. To answer the question by using the binomial distribution, we will need to use the binomial probability function [formula (5-6)]:

$$P(x) = \binom{25}{x} \cdot \left(\frac{1}{3}\right)^x \cdot \left(\frac{2}{3}\right)^{25-x} \quad \text{for} \quad x = 0, 1, 2, \ldots, 25$$

We must calculate the values for $P(0)$, $P(1)$, $P(2)$, and $P(3)$, since they do not appear in Table 4. This is a very tedious job because of the size of the exponent. In situations such as this, we can use the normal approximation method.

Now let's find $P(x \leq 3)$ by using the normal approximation method. We first need to find the mean and standard deviation of x [formulas (5-7) and (5-8)]:

$$\mu = np = (25)(\tfrac{1}{3}) = \mathbf{8.333}$$

$$\sigma = \sqrt{npq} = \sqrt{(25)(\tfrac{1}{3})(\tfrac{2}{3})} = \mathbf{2.357}$$

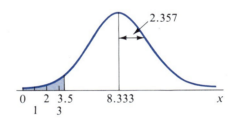

These values are shown in the figure above. The measure of the shaded area $(x < 3.5)$ represents the probability of $x = 0, 1, 2,$ or 3. Remember that $x = 3$, the discrete binomial variable, covers the continuous interval from 2.5 to 3.5.

$$P(x \text{ is no more than } 3) = P(x \leq 3) \quad [\text{for discrete variable } x]$$
$$= P(x < 3.5) \quad [\text{using a continuous variable } x]$$

$$P(x < 3.5) = P\left(z < \frac{3.5 - 8.333}{2.357}\right) = P(z < -2.05)$$

$$= 0.5000 - 0.4798 = \mathbf{0.0202}$$

Thus $P(\text{no more than 3 defectives})$ is approximately 0.02.

▼▲ EXERCISES

6.47 In which of the following binomial distributions does the normal distribution provide a reasonable approximation?
  a. $n = 10$, $p = 0.3$
  b. $n = 100$, $p = 0.005$
  c. $n = 500$, $p = 0.1$
  d. $n = 50$, $p = 0.2$

6.48 In order to see what happens when the normal approximation is improperly used, consider the binomial distribution with $n = 15$ and $p = 0.05$. Since $np = 0.75$, the rule of thumb ($np > 5$ and $nq > 5$) is not satisfied. Using the binomial tables, find the probability of one or fewer successes and compare this with the normal approximation.

6.49 Find the normal approximation for the binomial probability $P(x = 6)$, where $n = 12$ and $p = 0.6$. Compare this to the value of $P(x = 6)$ obtained from Table 4.

6.50 Find the normal approximation for the binomial probability $P(x = 4, 5)$, where $n = 14$ and $p = 0.5$. Compare this to the value of $P(x = 4, 5)$ obtained from Table 4.

6.51 Find the normal approximation for the binomial probability $P(x \leq 8)$, where $n = 14$ and $p = 0.4$. Compare this to the value of $P(x \leq 8)$ obtained from Table 4.

6.52 Find the normal approximation for the binomial probability $P(x \geq 9)$, where $n = 13$ and $p = 0.7$. Compare this to the value of $P(x \geq 9)$ obtained from Table 4.

6.53 A production run of T-shirts is expected to produce seconds 5% of the time. If you randomly select 100 T-shirts from a large run,
   a. what is the mean and the standard deviation of the binomial distribution when samples of 100 are taken?
   b. what is the probability that your sample will contain no seconds?
   c. what is the probability that your sample will contain no more than two seconds?

6.54 A drug manufacturer states that only 5% of the patients using a particular drug will experience side effects. Doctors at a large university hospital use the drug in treating 250 patients. What is the probability that 15 or fewer of the 250 patients experience side effects?

6.55 A company asserts that 80% of the customers who purchase its special lawn mower will have no repairs during the first two years of ownership. Your personal study has shown that only 70 of the 100 in your sample lasted the two years without repair expenses. What is the probability of your sample outcome or less if the actual expenses-free percentage is 80%?

6.56 Surveys have shown that 40% of the population of the United States have used Crest toothpaste at one time or another. A researcher, seeking to substantiate that 40% is correct, took a properly conducted random sample and found that 220 of the 600 individuals contacted had used Crest. If the 40% is really correct, what is the probability that a sample of 600 would have fewer than 221 individuals who had used Crest?

6.57 It is believed that the student body is equally split on a "new pub program" proposal. Assuming this to be the case, what is the probability that the Student Senate's straw poll of 100 student opinions shows at least 60% favoring the new proposal?

6.58 According to a report in *Time* (October 8, 1990, pages 47, 48), 12% of the children below the age of 18 suffer from some form of psychological illness. If this claim is correct, use the normal approximation to the binomial distribution to compute the probability of finding 40 or more children in a sample of 250 under age 18 with some form of psychological illness.

6.59 Consider a binomial distribution with $n = 500$ and $p = 0.2$.
  a. Set up, but do not evaluate, the probability expression for 90 or fewer successes in the 500 trials.
  b. Find the normal approximation to (a).

6.60 If 30% of all students entering a certain university drop out during or at the end of their first year, what is the probability that more than 600 of this year's entering class of 1800 will drop out during or at the end of their first year?

IN RETROSPECT

We now know what a normal distribution is, how to use it, and how it can help us. Let's again consider the article at the beginning of this chapter, which illustrated a normally distributed random variable. The random variable is the difference between the pre-test and post-test scores, and its distribution is mounded and symmetric about the mean value. We have seen many illustrations of normal distributions.

In the next chapter we will examine sampling distributions and learn how to apply the normal distribution to additional applications.

CHAPTER EXERCISES

6.61 According to Chebyshev's theorem, at least how much area is there under the standard normal distribution between $z = -2$ and $z = +2$? What is the actual area under the standard normal distribution between $z = -2$ and $z = +2$?

6.62 The middle 60% of a normally distributed population lies between what two standard scores?

6.63 The article at the beginning of this chapter includes a histogram of the variable, the difference between pre-test and post-test scores. This difference appears to be approximately normally distributed.
  a. Express this distribution as a frequency distribution.
  b. Using the mean of 4.8 and the standard deviation of 4.3, as reported in the article, find the proportion of this distribution that lies within one standard deviation of the mean, within two standard deviations of the mean, within three.
  c. Compare the results found in (b) to the proportions cited by the empirical rule.
  d. Determine what proportion of this distribution lies between the mean and 1.5 standard deviations above the mean. What proportion of a normal distribution lies between $z = 0$ and $z = 1.5$?

e. Do you think it is reasonable to conclude that this distribution is approximately normal?

6.64 Find the standard score z such that the area above the mean and below z under the normal curve is
    a. 0.3962   **b.** 0.4846   **c.** 0.3712

6.65 Find the standard score z such that the area below the mean and above z under the normal curve is
    a. 0.3212   **b.** 0.4788   **c.** 0.2700

6.66 Find the standard score of a normally distributed variable such that 42% of the distribution falls between the mean and this particular value.

6.67 Given that z is the standard normal variable, find the value of k such that
    a. $P(|z| > 1.68) = k$ **b.** $P(|z| < 2.15) = k$

6.68 Given that z is the standard normal variable, find the value of c such that
    a. $P(|z| > c) = 0.0384$ **b.** $P(|z| < c) = 0.8740$

6.69 Find the following values of z:
    a. $z(0.12)$ **b.** $z(0.28)$ **c.** $z(0.85)$ **d.** $z(0.99)$

6.70 Find the area under the normal curve that lies between the following pairs of z values:
    a. $z = -3.00$ and $z = 3.00$ **b.** $z = z(0.975)$ and $z(0.025)$
    c. $z = z(0.10)$ and $z(0.01)$

6.71 The test scores on a computer-science aptitude test are normally distributed with a mean of 15.0 and a standard deviation of 3.0. Find the 90th percentile for this distribution.

6.72 A soft drink vending machine can be regulated so as to dispense an average of μ oz of soft drink per glass. If the ounces dispensed per glass are normally distributed with a standard deviation of 0.2 oz, find the setting for μ that will allow a 6-oz glass to hold (without overflowing) the amount dispensed 99% of the time.

6.73 Suppose that a particular normal distribution has a mean of 70 and that the 90th percentile equals 84. Find the standard deviation.

6.74 An article in the April 4, 1991 issue of *USA Today* quotes a study involving 3365 people in Minneapolis–St. Paul between 1980 and 1982 and another 4545 between 1985 and 1987. It found that the average cholesterol level for males was 200. The authors of the study say the results of their study are probably similar nationwide. Assume that the cholesterol values for males in the United States are normally distributed with a mean equal to 200 and a standard deviation equal to 25.
    a. What percentage have readings between 150 and 225?
    b. What percentage have readings that exceed 250?

6.75 The length of life of a certain type of refrigerator is approximately normally distributed with a mean of 4.8 years and a standard deviation of 1.3 years.
    a. If this machine is guaranteed for 2 years, what is the probability that the machine you purchased will require replacement under the guarantee?
    b. What period of time should the manufacturer give as a guarantee if it is willing to replace only 0.5% of the machines?

6.76 The average length of time required for completing a certain academic achievement test is believed to be 150 min, and the standard deviation is 20 min. If we wish to allow sufficient time for only 80% to complete the test, when should the test be terminated? (Assume that the lengths of time required to complete this test are normally distributed.)

6.77 A machine is programmed to fill 10-oz containers with a cleanser. However, the variability inherent in any machine causes the actual amounts of fill to vary. The distribution is normal with a standard deviation of 0.02 oz. What must the mean amount μ be in order that only 5% of the containers receive less than 10 oz?

6.78 In a large industrial complex the maintenance department has been instructed to replace light bulbs before they burn out. It is known that the life of light bulbs is normally distributed with a mean life of 900 hours of use and a standard deviation of 75 hours. When should the light bulbs be replaced so that no more than 10% of them will burn out while in use?

6.79 Suppose that x has a binomial distribution with $n = 25$ and $p = 0.3$.
   a. Explain why the normal approximation is reasonable.
   b. Find the mean and standard deviation of the normal distribution that is used in the approximation.

6.80 Let x be a binomial random variable for $n = 30$ and $p = 0.1$.
   a. Explain why the normal approximation is not reasonable.
   b. Find the function used to calculate the probability of any x from $x = 0$ to $x = 30$.

6.81 The MINITAB statistical package was used to compute the binomial probabilities for $n = 50$ and $p = 0.1$. (The output stopped at $k = 14$ because the remaining probabilities are zero to four decimal places.)

```
MTB > PDF;
SUBC> BINOMIAL N = 50 AND P = .1.
     BINOMIAL WITH N = 50 P = 0.100000
           K              P( X = K)
           0               0.0052
           1               0.0286
           2               0.0779
           3               0.1386
           4               0.1809
           5               0.1849
           6               0.1541
           7               0.1076
           8               0.0643
           9               0.0333
          10               0.0152
          11               0.0061
          12               0.0022
          13               0.0007
          14               0.0002
          15               0.0001
MTB > STOP
```

Compute the normal approximation for $x \leq 6$ and compare with this output.

6.82 Use a computer to calculate the binomial probabilities for $n = 40$ and $p = 0.4$. (If you use MINITAB, see Exercise 6.81 for a program.)

6.83 A test-scoring machine is known to record an incorrect grade on 5% of the exams it grades. Find, by the appropriate method, the probability that the machine records
  a. 3 wrong grades in a set of 5 exams.
  b. no more than 3 wrong grades in a set of 5 exams.
  c. no more than 3 wrong grades in a set of 15 exams.
  d. no more than 3 wrong grades in a set of 150 exams.

6.84 It is believed that 58% of married couples with children agree on methods of disciplining their children. Assuming this to be the case, what is the probability that in a random survey of 200 married couples, we would find
  a. exactly 110 couples who agree?
  b. less than 110 couples who agree?
  c. more than 100 couples who agree?

6.85 A new drug is supposed to be 85% effective in treating a particular illness. (That is, 85% of the patients with this illness respond favorably to the drug.) Let x be the number of patients out of every group of 50 who respond favorably. Use the normal approximation method to find the following probabilities:
  a. $P(x > 45)$ b. $P(40 < x < 50)$ c. $P(x < 35)$

6.86 If 60% of the registered voters plan to vote for Ralph Brown for mayor of a large city, what is the probability that less than half of the voters, in a poll of 200 registered voters, plan to vote for Ralph Brown?

6.87 The following triangular distribution provides an approximation to the standard normal distribution. Line segment l_1 has the equation $y = x/9 + 1/3$ and segment l_2 has the equation $y = -x/9 + 1/3$.

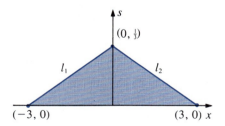

  a. Find the area under the entire triangular distribution.
  b. Find the area under the triangular distribution between 0 and 2.
  c. Find the area under the standard normal distribution between 0 and 2.

6.88 The grades on an examination whose mean score is 525 and whose standard deviation is 80 are normally distributed.
  a. Anyone who scores below 350 will be retested. What percentage does this represent?

b. The top 12% are to receive a special commendation. What score must be surpassed to receive this special commendation?
c. The interquartile range of a distribution is the difference between Q_1 and Q_3, $Q_3 - Q_1$. Find the interquartile range for the grades on this examination.
d. Find the grade such that only 1 out of 500 will score above it.

VOCABULARY LIST

Be able to define each term. In addition, describe, in your own words, and give an example of each term. Your examples should not be ones given in class or in the textbook. The bracketed numbers indicate the chapter in which the term first appeared, but you should define the terms again to show increased understanding of their meaning.

area representation for probability
bell-shaped curve
binomial distribution [5]
binomial probability
continuity correction factor
continuous random variable
discrete random variable [1, 5]
normal approximation of binomial
normal curve

normal distribution
percentage
probability [4]
proportion
random variable [5]
standard normal distribution
standard score [2]
z-score [2]

KEY CONCEPTS

continuous random variable
normal distribution
standard normal distribution

QUIZ A

Answer "True" if the statement is always true. If the statement is not always true replace the words shown in bold with words that make the statement always true.

6.1 The normal probability distribution is symmetric about **zero**.
6.2 The total area under the curve of any normal distribution is **1**.
6.3 The theoretical probability that a particular value of a **continuous** random variable will occur is exactly zero.
6.4 The unit of measure of the standard score is **the same as** the unit of measure of the data.
6.5 All **normal distributions** have the same general probability function and distribution.
6.6 When using the notation $z(0.05)$, the number in the parentheses is the measure of the area to the **left** of the z-score.

6.7 Standard normal scores have a mean of **1** and a standard deviation of **zero**.

6.8 Probability distributions of **all** continuous random variables are normal.

6.9 We are able to add and subtract the areas under the curve because these areas represent the probabilities of **independent events**.

6.10 The most common distribution of a continuous random variable is the **binomial probability**.

QUIZ B

6.1 Find the following probabilities for z, the standard normal score:
    a. $P(0 < z < 2.42)$
    b. $P(z < 1.38)$
    c. $P(z < -1.27)$
    d. $P(-1.35 < z < 2.72)$

6.2 Find the value of the z-score indicated.
    a. $P(z > 2) = 0.2643$
    b. $P(z < 2) = 0.17$
    c. $z(0.04)$

6.3 Using the symbolic notation $z(\alpha)$, give the symbolic name for the z-score shown.
    a.   b.

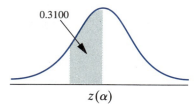 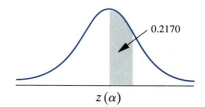

6.4 The lifetime of a flashlight battery is normally distributed about a mean of 35.6 hr with a standard deviation of 5.4 hr. Kevin selected one of these batteries at random and tested it. What is the probability that this one battery will last less than 40.0 hr?

6.5 The amount of time, x, spent commuting daily, one-way, to college by students is believed to have a mean of 22 min with a standard deviation of 9 min. If the length of time spent commuting is approximately normally distributed, find the time, x, that separates the 25% who spend the most time commuting from the rest of the commuters.

6.6 Thousands of high school students take the SAT exams each year. The scores attained by the students in a certain city are approximately normally distributed with a mean of 490 and a standard deviation of 70. Find
   a. the percentage of students who score between 600 and 700.
   b. the percentage of students who score less than 650.
   c. the 3d quartile.
   d. the 15th percentile, P_{15}.
   e. the 95th percentile, P_{95}.

QUIZ C

6.1 In 50 words or less, describe the standard normal distribution.

6.2 Describe the meaning of the symbol $z(\alpha)$.

6.3 Explain why the standard normal distribution, as computed in Appendix F, Table 5, can be used to find probabilities for all normal distributions.

7 SAMPLE VARIABILITY

Chapter Outline

7.1 Sampling Distributions
A distribution of values for a **sample statistic** is obtained by repeated sampling.

7.2 The Central Limit Theorem
The theorem describes the sampling distribution of **sample means**.

7.3 Application of the Central Limit Theorem
The behavior of sample means are **predicted**.

Galluping Attitudes

Since its founding in 1935, the American Institute of Public Opinion, better known as the Gallup Poll, has put approximately 20,000 questions to more than two million people.

One of the most interesting social changes to follow through the years in these pages is that of Americans' attitudes toward women at work and in politics. Herewith, an abbreviated reading of Gallup's progress report.

A WOMAN PRESIDENT

1937
Would you vote for a woman for President, if she qualified in every other respect?

		By Sex	Yes	No
Yes	34%	Men	27%	73%
No	66	Women	41	59

1955
If the party whose candidate you most often support nominated a woman for President of the United States, would you vote for her if she seemed best qualified for the job?

		By Sex	Yes	No
Yes	52%	Men	47%	48%
No	44	Women	57	40

1971
If your party nominated a woman for President, would you vote for her if she were qualified for the job?

		By Sex	Yes	No
Yes	66%	Men	65%	35%
No	29	Women	67	33

1978
If your party nominated a woman for president, would you vote for her if she were qualified for the job?

		By Sex	Yes	No
Yes	76%	Men	76%	19%
No	19	Women	77	18

Source: Copyright by the *Gallup Report*. Reprinted by permission.

Chapter Objectives

In Chapters 1 and 2 we discussed how to obtain and describe a sample. The description of the sample data is accomplished by using three basic concepts: (1) measures of central tendency (the mean is the most popularly used sample statistic), (2) measures of dispersion (the standard deviation is most commonly used), and (3) kind of distribution (normal, skewed, rectangular, and so on). The question that seems to follow is: What can be deduced about the statistical population from which a sample is taken?

To put this query at a more practical level, suppose that we have just taken a sample of 25 rivets made for the construction of airplanes. The rivets were tested for shearing strength, and the force required to break each rivet was the response variable. The various descriptive measures—mean, standard deviation, type of distribution—can be found for this sample. However, it is not the sample itself that we are interested in. The rivets that were tested were destroyed during the test, so they can no longer be used in the construction of airplanes. What we are trying to find out is information about the total population, and we certainly cannot test every rivet that is produced (there would be none left for construction). Therefore, we must somehow deduce information, or make inferences about, the population based on the results observed in the sample.

Suppose that we take another sample of 25 rivets and test them by the same procedure. Do you think that we would obtain the same sample mean from the second sample that we obtained from the first? the same standard deviation?

After considering these questions we might suspect that we would need to investigate the variability in the sample statistics obtained from **repeated sampling**. Thus we need to find (1) measures of central tendency for the sample statistics of importance, (2) measures of dispersion for the sample statistics, and (3) the pattern of variability (distribution) of the sample statistics. Once we have this information, we will be better able to predict the population parameters.

The objective of this chapter is to study the measures and the patterns of variability for the distribution formed by repeatedly observed values of a **sample mean**.

7.1 ▼ Sampling Distributions

To make inferences about a population, we need to discuss sample results a little more. A sample mean $\bar{x}$ is obtained from a sample. Do you expect that this value, $\bar{x}$, is exactly equal to the value of the population mean μ? Your answer should be "no." We do not expect that to happen, but we will be satisfied with our sample results if the sample mean is "close" to the value of the population mean. Let's consider a second question: If a second sample is taken, will the second sample have a mean equal to the population mean? Equal to the first sample mean? Again, no, we do not expect it to be equal to the population mean, nor do we expect the second sample mean to repeat the first one. We do, however, again expect the values to be "close." (This argument should hold for any other sample statistic and its corresponding population value.)

Section 7.1 ▼ SAMPLING DISTRIBUTIONS

The next questions should already have come to mind: What is "close"? How do we determine (and measure) this closeness? Just how would repeated sample statistics be distributed? To answer these questions we must take a look at a *sampling distribution*.

sampling distribution

> **SAMPLING DISTRIBUTION OF A SAMPLE STATISTIC**
> The distribution of values for that sample statistic obtained from all possible samples of a population. The samples must all be the same size, and the sample statistic could be any descriptive sample statistic.

▼ **ILLUSTRATION 7-1**

To illustrate the concept of a sampling distribution, let's consider the mean of each sample of size 2 that can be drawn with replacement from the set of even single-digit integers, (0, 2, 4, 6, 8). There are 25 possible samples of size 2:

(0,0)	(2,0)	(4,0)	(6,0)	(8,0)
(0,2)	(2,2)	(4,2)	(6,2)	(8,2)
(0,4)	(2,4)	(4,4)	(6,4)	(8,4)
(0,6)	(2,6)	(4,6)	(6,6)	(8,6)
(0,8)	(2,8)	(4,8)	(6,8)	(8,8)

Each of these samples has a mean $\bar{x}$. These means are, respectively,

0	1	2	3	4
1	2	3	4	5
2	3	4	5	6
3	4	5	6	7
4	5	6	7	8

Each of these samples is equally likely, and thus each of the 25 sample means can be assigned a probability of $\frac{1}{25} = 0.04$. (Why? See Exercise 7-1c, p. 351.) The sampling distribution for the sample mean then becomes as shown in Table 7-1. This is a probability distribution of $\bar{x}$ (see Figure 7-1).

TABLE 7-1 Sampling Distribution of Sample Means

$\bar{x}$	$P(\bar{x})$
0	0.04
1	0.08
2	0.12
3	0.16
4	0.20
5	0.16
6	0.12
7	0.08
8	0.04

FIGURE 7-1 Histogram: Sampling Distribution of Sample Means

For the same set of all possible samples of size 2, let's find the sampling distribution for sample ranges. Each sample has a range R. These ranges are, respectively,

0	2	4	6	8
2	0	2	4	6
4	2	0	2	4
6	4	2	0	2
8	6	4	2	0

Again, each of these 25 sample ranges has a probability of 0.04, and Table 7-2 shows the sampling distribution of sample ranges. This is a probability distribution of R (see Figure 7-2).

TABLE 7-2
Sampling Distribution of Sample Ranges

R	P(R)
0	0.20
2	0.32
4	0.24
6	0.16
8	0.08

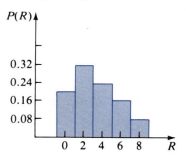

FIGURE 7-2 Histogram: Sampling Distribution of Sample Ranges

Most populations that are sampled are much larger than the one used in Illustration 7-1, and it would be a very tedious job to list all the possible samples. With this in mind, let's investigate a sampling distribution empirically (that is, by experimentation).

▼ ILLUSTRATION 7-2

Let's consider a portion of the sampling distribution of sample means for samples of size 5 obtained from the rolling of a single die. One sample will consist of 5 rolls of the die, and we will obtain a sample mean $\bar{x}$ from this sample. We repeat the experiment until 30 sample means have been obtained. Table 7-3 shows 30 such samples and their means. The resulting frequency distribution is shown in Figure 7-3. This distribution seems to display characteristics of a normal distribution; it is mounded and nearly symmetric about its mean (approximately 3.5).

TABLE 7-3
Sample Means for Rolling a Single Die Five Times

Trial	Sample	$\bar{x}$	Trial	Sample	$\bar{x}$
1	1, 2, 3, 2, 2	2.0	16	5, 2, 1, 3, 5	3.2
2	4, 5, 5, 4, 5	4.6	17	6, 1, 3, 3, 5	3.6
3	3, 1, 5, 2, 4	3.0	18	6, 5, 5, 2, 6	4.8
4	5, 6, 6, 4, 2	4.6	19	1, 3, 5, 5, 6	4.0
5	5, 4, 1, 6, 4	4.0	20	3, 1, 5, 3, 1	2.6
6	3, 5, 6, 1, 5	4.0	21	5, 1, 1, 4, 3	2.8
7	2, 3, 6, 3, 2	3.2	22	4, 6, 3, 1, 2	3.2
8	5, 3, 4, 6, 2	4.0	23	1, 5, 3, 4, 5	3.6
9	1, 5, 5, 3, 4	3.6	24	3, 4, 1, 3, 3	2.8
10	4, 1, 5, 2, 6	3.6	25	1, 2, 4, 1, 4	2.4
11	5, 1, 3, 3, 2	2.8	26	5, 2, 1, 6, 3	3.4
12	1, 5, 2, 3, 1	2.4	27	4, 2, 5, 6, 3	4.0
13	2, 1, 1, 5, 3	2.4	28	4, 3, 1, 3, 4	3.0
14	5, 1, 4, 4, 6	4.0	29	2, 6, 5, 3, 3	3.8
15	5, 5, 6, 3, 3	4.4	30	6, 3, 5, 1, 1	3.2

FIGURE 7-3
A Portion of a Sampling Distribution of Sample Means for Rolling a Die

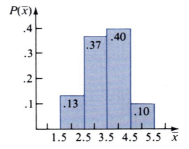

The theory involved with sampling distributions that will be described in the remainder of this chapter requires *random sampling*.

RANDOM SAMPLE

random sample

A sample obtained in such a way that each of the possible samples of fixed size n has an equal probability of being selected. (See p. 24.)

Case Study 7-1

Average Aircraft Age

USA TODAY's *"Average aircraft age"* shows the average age of the aircraft that make up the fleets of the 12 biggest airline companies. Each company reported their own fleet average. The information is presented in the form of a bar graph. This information could have been represented by a frequency distribution or a histogram using "average" as the variable. However, this distribution is not part of the sampling distribution of sample means. Why are these 12 averages not a portion of the sampling distribution of sample means? (See Exercise 7.5, p. 352.)

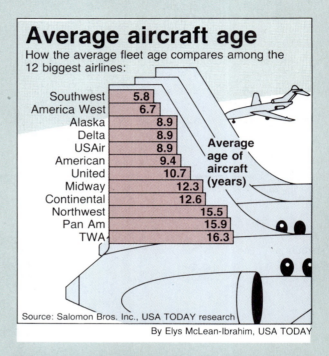

Source: Copyright 1991, USA TODAY. Reprinted with permission.

Case Study 7-2

Consumer Comfort Poll

Public opinion about a political question tends to fluctuate. The "Consumer Comfort Poll" from Money, *August 1990, shows the monthly results for repeated polls about the state of the nation's economy. The repeated statistics described here do not have the same intent as the repeated sampling and their related sampling distributions discussed in this chapter. Explain why this graph does not represent a sampling distribution. (See Exercise 7.6, p. 352.)*

−16	−21	−8
July	June	Year ago

Section 7.1 ▼ SAMPLING DISTRIBUTIONS

WOMEN OUTWORRY MEN

Since January, public concern has grown about the state of the U.S. economy. Women, however, have turned far more pessimistic than men, according to the MONEY/ABC Consumer Comfort poll (see graph). The economic-outlook index for women dropped from −32 in January to −50 in late May before inching up to −46. Men started the year marginally positive about the economy before becoming somewhat pessimistic. The overall index, still quite negative, rose in June from −21 to −16.

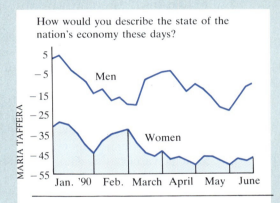

Index points are net negative and positive responses, with a potential range of +100 to −100. Based on a poll of 1,026 randomly chosen adults in the five weeks that ended July 1. Margin of error: plus or minus 3.5 points.

Source: Jordan E. Goodman, "Consumer Comfort Poll," *Money* (August 1990). The article above is reprinted from MONEY magazine by special permission; copyright 1990 The Time Inc. Magazine Company.

▼▲ EXERCISES

7.1 **a.** What is the sampling distribution of sample means?
 b. A sample of size 3 is taken from a population and the sample mean found. Describe how this sample mean is related to the sampling distribution of sample means.
 c. Why is the probability of 0.04 assigned to each of the sample mean values in Illustration 7-1, p. 347?

7.2 **a.** What is the sampling distribution of sample ranges?

b. A sample of size 12 is taken from a population and the sample range determined. Describe how this sample range is related to the sampling distribution of sample ranges.

c. Why is the probability of 0.20 assigned to the value $R = 0$ on Table 7-2, p. 348?

7.3 Consider the set of odd single-digit integers $\{1, 3, 5, 7, 9\}$.

a. Make a list of all samples of size 2 that can be drawn from this set of integers. (Sample with replacement; that is, the first number is drawn, observed, then replaced before the next drawing.)

b. Construct the sampling distribution of sample means for samples of size 2 selected from this set.

c. Construct the sampling distributions of sample ranges for samples of size 2.

7.4 Consider the set of even single-digit integers $\{0, 2, 4, 6, 8\}$.

a. Make a list of all the possible samples of size 3 that can be drawn from this set of integers. (Sample with replacement; that is, the first number is drawn, observed, then replaced before the next drawing.)

b. Construct the sampling distribution of the sample medians for samples of size 3.

c. Construct the sampling distribution of the sample means.

7.5 Refer to the information shown on the bar graph in Case Study 7-1 (p. 349).

a. What variable was reported in this information? (What did each company report?)

b. Construct a relative frequency histogram picturing this information. Describe the shape of the histogram. (Use classes 4.5 to 6.5, 6.5 to 8.5, etc.)

c. Explain why this distribution is not part of the sampling distribution of sample means.

7.6 Explain why the graph in Case Study 7-2 (p. 350) does not represent a sampling distribution.

7.7 Using the telephone numbers listed in your local directory as your population, obtain randomly 20 samples of size 3. From each number identified as a source, take the fourth, fifth, and sixth digits. (For example, for 245–826 8, you would take the 8, the 2, and the 6 as your sample of size 3.)

a. Calculate the mean of the 20 samples.

b. Draw a histogram showing the 20 sample means. (Use classes -0.5 to 0.5, 0.5 to 1.5, 1.5 to 2.5, and so on.)

7.8 Using a set of 5 dice, roll the dice and determine the mean number of dots showing on the five dice. Repeat the experiment until you have 25 sample means. Draw a dot plot showing the distribution of the 25 sample means. (See Illustration 7-2, p. 348.)

7.9 The following MINITAB program selects 100 samples (each of size 4) from a binomial distribution having $n = 4$ and $p = 0.1$. The means of these 100 samples are placed into column C8 and then a histogram of C8 is formed. Draw a histogram of the binomial distribution and compare it with the histogram of the sample means.

```
MTB > READ VALUE IN C1,PROB IN C2
DATA> 0 .656
DATA> 1 .292
DATA> 2 .049
DATA> 3 .004
DATA> 4 .0001
DATA> END DATA
      5 ROWS READ
MTB > RANDOM 100 OBSERVATIONS INTO C3;
SUBC> DISCRETE C1 C2.
MTB > RANDOM 100 OBSERVATIONS INTO C4;
SUBC> DISCRETE C1 C2.
MTB > RANDOM 100 OBSERVATIONS INTO C5;
SUBC> DISCRETE C1 C2.
MTB > RANDOM 100 OBSERVATIONS INTO C6;
SUBC> DISCRETE C1 C2.
MTB > LET C7 = (C3+C4+C5+C6)/4
MTB > HISTOGRAM C7 0 0.25

Histogram of C7    N = 100

Midpoint    Count
   0.000      20    ********************
   0.250      24    ************************
   0.500      32    ********************************
   0.750      17    *****************
   1.000       5    *****
   1.250       2    **

MTB > STOP
```

7.10 The following MINITAB program selects 100 samples from a uniform probability distribution for integers $1, 2, 3,$ and 4. Each sample is of size 4 and the means are put into column C6. A histogram of the 100 means is shown. Draw a probability histogram for the uniform distribution and compare it to the histogram of sample means.

```
MTB > RANDOM 100 OBSERVATIONS INTO C1;
SUBC> INTEGER  1 TO 4.
MTB > RANDOM 100 OBSERVATIONS INTO C2;
SUBC> INTEGER  1 TO 4.
MTB > RANDOM 100 OBSERVATIONS INTO C3;
SUBC> INTEGER  1 TO 4.
MTB > RANDOM 100 OBSERVATIONS INTO C4;
SUBC> INTEGER  1 TO 4.
MTB > LET C5 = (C1+C2+C3+C4)/4
MTB > HISTO C5;
SUBC> INCREMENT 0.25;
SUBC> START 1.
Histogram of C5    N = 100
Midpoint   Count
   1.000      1   *
   1.250      0
   1.500      7   *******
   1.750      3   ***
   2.000     14   **************
   2.250     20   ********************
   2.500     13   *************
   2.750     16   ****************
   3.000     13   *************
   3.250      8   ********
   3.500      2   **
   3.750      3   ***
MTB > STOP
```

7.2 ▼ The Central Limit Theorem

On the preceding pages we discussed two types of sampling distributions, for means and for ranges. There are many others that could be discussed; in fact, these two could themselves be discussed further. However, the only sampling distribution of concern to us here is the **sampling distribution of sample means**. The mean is the most commonly used sample statistic and thus is the most important.

sampling distribution of sample means

The central limit theorem (CLT) tells us about the sampling distribution of sample means of random samples of size n. Recall that there are basically three kinds of information that we want about a distribution: (1) where the center is, (2) how widely it is dispersed, and (3) how it is distributed. The central limit theorem tells us all three.

central limit theorem

CENTRAL LIMIT THEOREM

If all possible random samples, each of size n, are taken from any population with a mean μ and standard deviation σ, the sampling distribution of sample means will

1. have a mean $\mu_{\bar{x}}$ equal to μ.
2. have a standard deviation $\sigma_{\bar{x}}$ equal to $\sigma/\sqrt{n}$.

Section 7.2 ▼ THE CENTRAL LIMIT THEOREM

3. be normally distributed when the parent population is normally distributed or will be approximately normally distributed for samples of size 30 or more when the parent population is not normally distributed. The approximation to the normal distribution improves with samples of larger size.

In short, the central limit theorem states the following:
1. $\mu_{\bar{x}} = \mu$; the mean of the $\bar{x}$'s equals the mean of the x's.
2. $\sigma_{\bar{x}} = \sigma/\sqrt{n}$; the standard error of the mean (see the definition that follows) equals the standard deviation of the population divided by the square root of the sample size.
3. The sample means are approximately normally distributed (regardless of the shape of the parent population).

NOTE   The n referred to in the central limit theorem is the *size of each sample* in the sampling distribution.

standard error of the mean

STANDARD ERROR OF THE MEAN
The standard deviation of the sampling distribution of sample means

We are unable to prove the central limit theorem without using advanced mathematics. However, it is possible to check its validity by examining two illustrations. Let's consider a population for which we can construct the theoretical sampling distribution of all possible samples. For this example let's consider all possible samples of size 2 that could be drawn from a population that contains the three numbers 2, 4, and 6.

First let's look at the population itself. To calculate the mean μ and the standard deviation σ, we must use the formulas from Chapter 5 for discrete probability distributions:

$$\mu = \sum [x \cdot P(x)] \quad \text{and} \quad \sigma = \sqrt{\sum [x^2 \cdot P(x)] - \left\{\sum [x \cdot P(x)]\right\}^2}$$

These formulas are necessary because we are not drawing the samples but are discussing the theoretical possibilities (see Table 7-4).

TABLE 7-4
Probability Distribution and Extensions for $x = 2, 4, 6$

x	$P(x)$	$x \cdot P(x)$	$x^2 \cdot P(x)$
2	$\frac{1}{3}$	$\frac{2}{3}$	$\frac{4}{3}$
4	$\frac{1}{3}$	$\frac{4}{3}$	$\frac{16}{3}$
6	$\frac{1}{3}$	$\frac{6}{3}$	$\frac{36}{3}$
Total	$\frac{3}{3}$	$\frac{12}{3}$	$\frac{56}{3}$

$$\mu = \tfrac{12}{3} = \mathbf{4.0}$$
$$\sigma = \sqrt{56/3 - (12/3)^2} = \sqrt{18.66 - 16.0} = \sqrt{2.66} = \mathbf{1.63}$$

Table 7-5 lists all the possible samples that could be drawn if samples of size 2 were to be drawn from this population. (One number is drawn, observed, and then returned to the population before the second number is drawn.) Table 7-5 also lists the means of these samples. The probability distribution for these means and the extensions are given in Table 7-6. Thus we have

TABLE 7-5 All Possible Samples of Size 2 and Their Means

Possible Samples	$\bar{x}$
2, 2	2
2, 4	3
2, 6	4
4, 2	3
4, 4	4
4, 6	5
6, 2	4
6, 4	5
6, 6	6

TABLE 7-6 Probability Distribution for Means of All Possible Samples of Size 2 and the Extensions

$\bar{x}$	$P(\bar{x})$	$\bar{x} \cdot P(\bar{x})$	$\bar{x}^2 \cdot P(\bar{x})$
2	1/9	2/9	4/9
3	2/9	6/9	18/9
4	3/9	12/9	48/9
5	2/9	10/9	50/9
6	1/9	6/9	36/9
Total	9/9	36/9	156/9

$$\mu_{\bar{x}} = \frac{36}{9} = 4.0$$

$$\sigma_{\bar{x}} = \sqrt{\frac{156}{9} - \left(\frac{36}{9}\right)^2} = \sqrt{17.33 - 16} = \sqrt{1.33} = 1.15$$

The histogram for the distribution of possible $\bar{x}$'s is shown in Figure 7-4.

The CLT says that three things will occur in this sampling distribution:

1. It will be approximately normally distributed. The histogram (Figure 7-4) suggests this very strongly.
2. The mean $\mu_{\bar{x}}$ of the sampling distribution will equal the mean of the population. They both have the value 4.0.
3. The standard deviation $\sigma_{\bar{x}}$ of the sampling distribution will equal the standard deviation of the population divided by the square root of the sample size ($\sigma/\sqrt{n}$):

$$\sigma_{\bar{x}} = 1.15 \quad \text{and} \quad \frac{\sigma}{\sqrt{n}} = \frac{1.63}{\sqrt{2}} = \frac{1.63}{1.41} = 1.15$$

FIGURE 7-4
Histogram for Distribution of Table 7-6

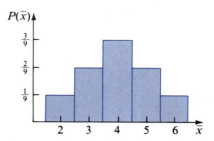

This illustration shows that the CLT is true for a **theoretical probability distribution**.

Now let's look at the empirical distribution that occurred in Illustration 7-2 (p. 348) and see whether it supports the three claims of the central limit theorem.

First, let's look at the theoretical probability distribution from which these samples were taken (Table 7-7). A histogram showing the probability distribution of the tossing of a die is shown in Figure 7-5. The population mean μ equals **3.5** (see Table 7-7). The population standard deviation σ equals $\sqrt{15.17 - (3.5)^2}$, which is $\sqrt{2.92} = $ **1.71**. (Note that this population has a uniform distribution.)

FIGURE 7-5
Probability Distribution for Rolling a Die

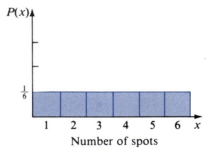

TABLE 7-7
Probability Distribution and Extensions for Rolling a Die

x	$P(x)$	$x \cdot P(x)$	$x^2 \cdot P(x)$
1	$\frac{1}{6}$	$\frac{1}{6}$	$\frac{1}{6}$
2	$\frac{1}{6}$	$\frac{2}{6}$	$\frac{4}{6}$
3	$\frac{1}{6}$	$\frac{3}{6}$	$\frac{9}{6}$
4	$\frac{1}{6}$	$\frac{4}{6}$	$\frac{16}{6}$
5	$\frac{1}{6}$	$\frac{5}{6}$	$\frac{25}{6}$
6	$\frac{1}{6}$	$\frac{6}{6}$	$\frac{36}{6}$
Total	$\frac{6}{6} = 1$	$\frac{21}{6} = 3.5$	$\frac{91}{6} = 15.17$

Now let's look at the empirical sampling distribution of the 30 sample means found earlier. Using the 30 values of $\bar{x}$ in Table 7-3, the observed mean of the $\bar{x}$'s turns out to be 3.43, and the observed standard deviation $s_{\bar{x}}$ turns out to be 0.73. The histogram appears in Figure 7-3 (p. 349).

The central limit theorem states that the $\bar{x}$'s should be approximately normally distributed, and the histogram certainly suggests this to be the case. The CLT also says that the mean $\mu_{\bar{x}}$ of the sampling distribution and the mean μ of the population are the same. The mean of the $\bar{x}$'s is 3.43 and $\mu = 3.5$; they seem to be reasonably close. Remember that we have taken only 30 samples, not all possible samples, of size 5.

The theorem says that $\sigma_{\bar{x}}$ should equal $\sigma/\sqrt{n}$. The observed standard deviation of $\bar{x}$'s is $s_{\bar{x}} = 0.73$ and the standard error of the mean is

$$\frac{\sigma}{\sqrt{n}} = \frac{1.71}{\sqrt{5}} = 0.76$$

These two values are very close.

The evidence seen in these two illustrations seems to suggest that the CLT is true, although this does not constitute a proof of the theorem, of course.

Having taken a look at these two specific illustrations, which support the CLT, let's now look at four graphic illustrations that present the same information in slightly different form. In each of these graphic illustrations there are four distributions. The first is a distribution of the parent population, the distribution of the individual x values. Each of the other three graphs shows a sampling distribution of sample means, using three different sample sizes. In Figure 7-6 we have a uniform distribution, much like Figure 7-5 for the die illustration, and the resulting distributions of sample means for samples of size 2, 5, and 30. Figure 7-7

FIGURE 7-6
Uniform Distribution

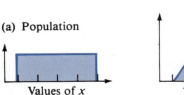

(a) Population

Values of x

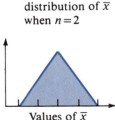

(b) Sampling distribution of $\bar{x}$ when $n = 2$

Values of $\bar{x}$

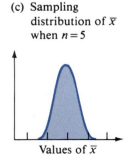

(c) Sampling distribution of $\bar{x}$ when $n = 5$

Values of $\bar{x}$

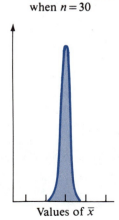

(d) Sampling distribution of $\bar{x}$ when $n = 30$

Values of $\bar{x}$

FIGURE 7-7
U-shaped Distribution

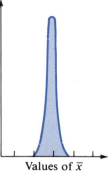

(d) Sampling distribution of $\bar{x}$ when $n=30$

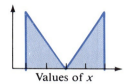

(a) Population

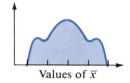

(b) Sampling distribution of $\bar{x}$ when $n=2$

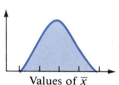

(c) Sampling distribution of $\bar{x}$ when $n=5$

FIGURE 7-8
J-shaped Distribution

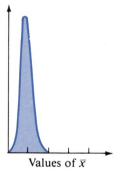

(d) Sampling distribution of $\bar{x}$ when $n=30$

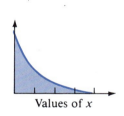

(a) Population

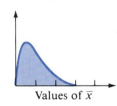

(b) Sampling distribution of $\bar{x}$ when $n=2$

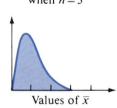

(c) Sampling distribution of $\bar{x}$ when $n=5$

shows a U-shaped population and the corresponding sampling distributions. Figure 7-8 shows a J-shaped population and the three corresponding distributions. Figure 7-9 (p. 360) shows a normally distributed population and the three sampling distributions.

All four illustrations seem to verify the CLT. Note that the sampling distribution of the three nonnormal distributions produced sample means with an approximately normal distribution for samples of size 30. In the normal population (Figure 7-9) the sampling distributions for all sample sizes appear to be normal. Thus you have seen an amazing phenomenon: no matter what the shape of a population, the sampling distribution of the mean becomes approximately normally distributed when n becomes sufficiently large.

FIGURE 7-9
Normal Distribution

(a) Population

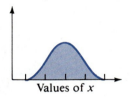
Values of x

(b) Sampling distribution of $\bar{x}$ when $n=2$

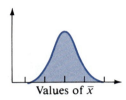
Values of $\bar{x}$

(c) Sampling distribution of $\bar{x}$ when $n=5$

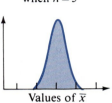
Values of $\bar{x}$

(d) Sampling distribution of $\bar{x}$ when $n=30$

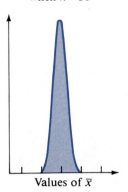
Values of $\bar{x}$

You should notice one other point: The sample mean becomes less variable as the sample size increases. Notice that as n increases from 2 to 30, all the distributions become narrower and taller. Can you explain how this implies less variability? How does the CLT state this? (See Exercise 7.11.) This point is discussed further in Section 7-3.

▼▲ EXERCISES

7.11 **a.** What is the measure of the total area for any probability distribution?
b. How does the CLT state that as n becomes larger, the sample mean becomes less variable?

7.12 If a normal population has a standard deviation σ of 25 units, what is the standard error of the mean ($\sigma_{\bar{x}}$) if samples of size 16 are used? of size 25? of size 50? of size 100?

7.13 A certain population has a mean of 500 and a standard deviation of 30. Many samples of size 36 are randomly selected and their mean calculated.
a. What value would you expect to find for the mean of all these sample means?
b. What value would you expect to find for the standard deviation of all these sample means?
c. What shape would you expect the distribution of all these sample means to have?

 7.14 Consider the experiment of taking a standardized mathematics test. The variable x is the raw score received. This exam has a mean of 720 and a standard deviation of 60. A group (sample) of 40 students takes the exam and the sample mean $\bar{x}$ is 725.6. A sampling distribution of means is formed from the means of all such groups of 40 students.

    a. Determine the mean of this sampling distribution.

    b. Determine the standard deviation for this sampling distribution.

7.15 The following MINITAB program simulates 100 samples, each of size 5, drawn from a normal distribution with mean equal to 5.5 and standard deviation equal to 2.5. The mean for each sample is placed in column C7.

    a. From the output, what is the mean of the 100 sample means? What is the standard deviation of the 100 sample means?

    b. According to the central limit theorem, if all possible sample means (not just 100 of them) were computed, what would be the mean of the sample means, $\mu_{\bar{x}}$? What would be the standard deviation of the sample means, $\sigma_{\bar{x}}$?

```
MTB > RANDOM 100 OBSERVATIONS INTO C1;
SUBC> NORMAL    MU = 5.5   SIGMA = 2.5.
MTB > RANDOM 100 OBSERVATIONS INTO C2;
SUBC> NORMAL    MU = 5.5   SIGMA = 2.5.
MTB > RANDOM 100 OBSERVATIONS INTO C3;
SUBC> NORMAL    MU = 5.5   SIGMA = 2.5.
MTB > RANDOM 100 OBSERVATIONS INTO C4;
SUBC> NORMAL    MU = 5.5   SIGMA = 2.5.
MTB > RANDOM 100 OBSERVATIONS INTO C5;
SUBC> NORMAL    MU = 5.5   SIGMA = 2.5.
MTB > LET C6 = (C1+C2+C3+C4+C5)/5
MTB > MEAN C6
    MEAN      = 5.5034
MTB > STDEV C6
    ST.DEV.   = 1.1104
MTB > STOP
```

7.16 A random variable that can take on the values $1, 2, \ldots, n$ (each with probability $1/n$) is called *uniform*. For such a variable the mean $\mu = (n + 1)/2$ and $\sigma = \sqrt{\frac{n^2-1}{12}}$. (*Hint*: Use these formulas in answering part (b).) The following MINITAB program gives 100 simulated samples, each of size 5, from a uniform distribution for integers from 1 to 10. The means of the 100 samples are found and put into column C7.

    a. From the output, what are the mean and standard deviation of the 100 sample means?

    b. For all samples of size 5, find $\mu_{\bar{x}}$ and $\sigma_{\bar{x}}$.

```
MTB > RANDOM 100 OBSERVATIONS INTO C1;
SUBC> INTEGER  1 TO 10.
MTB > RANDOM 100 OBSERVATIONS INTO C2;
SUBC> INTEGER  1 TO 10.
MTB > RANDOM 100 OBSERVATIONS INTO C3;
SUBC> INTEGER  1 TO 10.
MTB > RANDOM 100 OBSERVATIONS INTO C4;
SUBC> INTEGER  1 TO 10.
MTB > RANDOM 100 OBSERVATIONS INTO C5;
SUBC> INTEGER  1 TO 10.
MTB > LET C6 = (C1+C2+C3+C4+C5)/5
MTB > MEAN C6
    MEAN   = 5.6320
MTB > STDEV C6
    ST.DEV. = 1.2370
MTB > STOP
```

7.3 ▼ Application of the Central Limit Theorem

The central limit theorem tells us about the sampling distribution of sample means by describing the shape of the distribution of all possible sample means. It also specifies the relationship between the mean μ of the population and the mean $\mu_{\bar{x}}$ of the sampling distribution, and the relationship between the standard deviation σ of the population and the standard deviation $\sigma_{\bar{x}}$ of the sampling distribution. Since sample means are approximately normally distributed, we will be able to answer probability questions by using Table 5 of Appendix F.

▼ ILLUSTRATION 7-3

Consider a normal population with $\mu = 100$ and $\sigma = 20$. If a sample of size 16 is selected at random, what is the probability that this sample will have a mean value between 90 and 110? That is, what is $P(90 < \bar{x} < 110)$?

SOLUTION The CLT states that the distribution of $\bar{x}$'s is approximately normally distributed. To determine probabilities associated with a normal distribution, we will need to convert the statement $P(90 < \bar{x} < 110)$ to a probability statement concerning z in order to use Table 5, the standard normal distribution table. The sampling distribution is shown in the following figure, with $P(90 < \bar{x} < 110)$ represented by the shaded area:

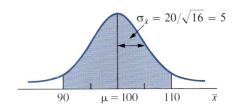

Section 7.3 ▼ APPLICATION OF THE CENTRAL LIMIT THEOREM

The formula for finding z if a value of $\bar{x}$ is known is

$$z = \frac{\bar{x} - \mu_{\bar{x}}}{\sigma_{\bar{x}}} \tag{7-1}$$

However, the CLT tells us that $\mu_{\bar{x}} = \mu$ and $\sigma_{\bar{x}} = \sigma/\sqrt{n}$. Therefore, we will rewrite formula (7-1) in terms of μ and σ:

$$z = \frac{\bar{x} - \mu}{\sigma/\sqrt{n}} \tag{7-2}$$

Using formula (7-2), we find that $\bar{x} = 90$ has a standard score of

$$z = \frac{90 - 100}{20/\sqrt{16}} = \frac{-10}{5}$$

$$= -2.0$$

$\bar{x} = 110$ has a standard score of

$$z = \frac{110 - 100}{20/\sqrt{16}} = \frac{10}{5} = 2.0$$

Therefore,

$$P(90 < \bar{x} < 110) = P(-2.0 < z < 2.0)$$

$$= 2(0.4772)$$

$$= \mathbf{0.9544}$$

▲▲

Before we look at more illustrations, let's consider for a moment what is implied by saying that $\sigma_{\bar{x}} = \sigma/\sqrt{n}$. To demonstrate, let's suppose that $\sigma = 20$ and let's use a sampling distribution of samples of size 4. Now $\sigma_{\bar{x}}$ would be $20/\sqrt{4}$, or 10, and approximately 95% (0.9544) of all such sample means should be within the interval from 20 below to 20 above the population mean (within two standard deviations of the population mean). However, if the sample size were increased to 16, $\sigma_{\bar{x}}$ would become

$$20/\sqrt{16} = 5$$

and approximately 95% of the sampling distribution would be within 10 units of the mean, and so on. As the sample size increases, the size of $\sigma_{\bar{x}}$ becomes smaller, so that the distribution of sample means becomes much narrower. Figure 7-10 (p. 364) illustrates what happens to the distribution of $\bar{x}$'s as the size of the individual samples increases. Recall that the area under the normal curve is always exactly one unit of area. So as the width of the curve narrows, the height will have to increase in order to maintain this area.

▼ ILLUSTRATION 7-4

Kindergarten children have heights that are approximately normally distributed about a mean of 39 in. and a standard deviation of 2 in. A random sample of

FIGURE 7-10
Distributions of Sample Means

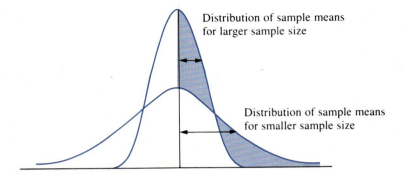

size 25 is taken and the mean $\bar{x}$ is calculated. What is the probability that this mean value will be between 38.5 and 40 in.?

SOLUTION We want to find $P(38.5 < \bar{x} < 40.0) = P(? < z < ?)$, where the z-scores are as follows:

$$\text{When } \bar{x} = 38.5: z = \frac{38.5 - 39.0}{2/\sqrt{25}} = \frac{-0.5}{0.4}$$

$$= -1.25$$

$$\text{When } \bar{x} = 40.0: z = \frac{40.0 - 39.0}{2/\sqrt{25}} = \frac{1.0}{0.4}$$

$$= 2.50$$

(See the following figure.)

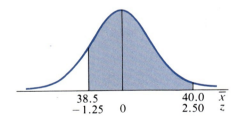

Therefore,

$$P(38.5 < \bar{x} < 40.0) = P(-1.25 < z < 2.50)$$

$$= 0.3944 + 0.4938 = \mathbf{0.8882} \qquad \blacktriangle\blacktriangle$$

▼ **ILLUSTRATION 7-5**

Referring to Illustration 7-4, within what limits would the middle 90% of the sampling distribution of sample means of sample size 100 fall?

Section 7.3 ▼ APPLICATION OF THE CENTRAL LIMIT THEOREM

SOLUTION The basic formula is

$$z = \frac{\bar{x} - \mu}{\sigma/\sqrt{n}}$$

and

$$\sigma_{\bar{x}} = \frac{\sigma}{\sqrt{n}} = \frac{2}{\sqrt{100}} = \frac{2}{10} = \mathbf{0.2}$$

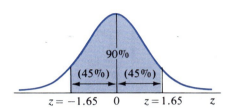

(Recall that the area to the right of the mean, 45% (0.4500), is related to the z-score of 1.65, according to Table 5.) We want to find the two values for $\bar{x}$.
If $z = -1.65$,

$$-1.65 = \frac{\bar{x} - 39}{0.2}$$

$$(-1.65)(0.2) = \bar{x} - 39$$

$$\bar{x} = 39 - 0.33$$

$$= \mathbf{38.67}$$

If $z = 1.65$,

$$1.65 = \frac{\bar{x} - 39}{0.2}$$

$$(1.65)(0.2) = \bar{x} - 39$$

$$\bar{x} = 39 + 0.33$$

$$= \mathbf{39.33}$$

Thus

$$P(\mathbf{38.67} < \bar{x} < \mathbf{39.33}) = 0.90$$ ▲▲

▼▲ EXERCISES

7.17 A random sample of size 36 is to be selected from a population that has a mean μ of 50 and a standard deviation σ of 10.
  a. This sample of 36 has a mean value of $\bar{x}$ which belongs to a sampling distribution. Find the shape of this sampling distribution.

b. Find the mean of this sampling distribution.
c. Find the standard error of this sampling distribution.
d. What is the probability that this sample mean will be between 45 and 55?
e. What is the probability that the sample mean will have a value greater than 48?
f. What is the probability that the sample mean will be within 3 units of the mean?

7.18 Consider the approximately normal population of heights of male students at a college. Assume that the individual heights have a mean of 69 in. and a standard deviation of 4 in. A random sample of 16 heights is obtained.
   a. Find the mean of this sampling distribution.
   b. Find the standard error of the mean.
   c. Find the shape of this sampling distribution.
   d. Find $P(\bar{x} > 70)$.
   e. Find $P(\bar{x} < 67)$.

7.19 The heights of the kindergarten children mentioned in Illustration 7-4 (p. 363) are approximately normally distributed with $\mu = 39$ and $\sigma = 2$.
   a. If an individual kindergarten child is selected at random, what is the probability that he or she has a height between 38 and 40 in.?
   b. A classroom of 30 of these children is used as a sample. What is the probability that the class mean $\bar{x}$ is between 38 and 40 in.?
   c. If an individual kindergarten child is selected at random, what is the probability that he or she is taller than 40 in.?
   d. A classroom of 30 of these kindergarten children is used as a sample. What is the probability that the class mean $\bar{x}$ is greater than 40 in.?

7.20 The amount of fill (weight of contents) put into a glass jar of spaghetti sauce is normally distributed. If the mean, μ, is 850 g and the standard deviation, σ, is 8 g,
   a. describe the distribution of x, the amount of fill per jar.
   b. find the probability that one jar selected at random contains between 848 and 855 g.
   c. describe the distribution of $\bar{x}$, the mean weight for a sample of 24 such jars of sauce.
   d. find the probability that a random sample of 24 jars has a mean weight between 848 and 855 g.

7.21 According to an article in *Pharmaceutical News* (January 1991, page 51), a person age 65 or older will spend, on the average, $300 on personal care products per year. If we assume that the amount spent on personal care products by individuals 65 or older is normally distributed and has a standard deviation equal to $75, what is the probability that the mean amount spent by 25 randomly selected such individuals will fall between $250 and $350?

7.22 The March 1, 1990 issue of *Criminal Justice Newsletter* (page 6) reported that the cost per prison bed for new prison construction programs ranged from $995 to $132,000. The average cost per bed was $52,000. Suppose we know σ, the standard deviation of all costs per bed, was equal to $25,000. If nine construction pro-

Section 7.3 ▼ APPLICATION OF THE CENTRAL LIMIT THEOREM

grams were randomly sampled, find the probability that the mean cost per bed would exceed $60,000. (Assume costs are normally distributed.)

7.23 The following MINITAB program selects 100 samples (each of size 4) from a normal distribution that has a mean of 5 and a standard deviation of 2. The sample means are put into column 6 and printed out.

   a. Find $P(4 < \bar{x} < 6)$.
   b. How many of the sample means computed fall between 4 and 6?
   c. Why do the answers in (a) and (b) differ?

```
MTB > RANDOM 100 OBSERVATIONS INTO C1;
SUBC> NORMAL   MU = 5  SIGMA = 2.
MTB > RANDOM 100 OBSERVATIONS INTO C2;
SUBC> NORMAL   MU = 5  SIGMA = 2.
MTB > RANDOM 100 OBSERVATIONS INTO C3;
SUBC> NORMAL   MU = 5  SIGMA = 2.
MTB > RANDOM 100 OBSERVATIONS INTO C4;
SUBC> NORMAL   MU = 5  SIGMA = 2.
MTB > LET C5 = (C1+C2+C3+C4)/4
MTB > PRINT C5

C5
   5.64967   4.98275   3.82674   6.04644   3.47323   4.66087   6.46622
   4.49978   5.68478   4.64910   5.38588   5.55961   5.64640   3.72777
   6.64830   5.80450   4.73553   4.32134   5.96149   6.17268   4.05841
   6.46568   4.43453   4.24916   5.68926   4.80518   5.15319   5.07253
   4.89557   5.28228   4.64055   6.54022   4.38253   4.76252   6.20739
   4.84407   4.09733   5.05982   4.48738   6.32345   6.98795   3.86917
   6.10219   5.01370   4.60655   6.16415   5.50360   6.23161   5.55847
   5.08329   5.93220   5.24714   5.15086   3.46194   5.28932   7.39262
   4.46692   3.45167   4.02496   5.66649   4.59504   5.76057   6.08165
   5.27776   5.25800   4.92345   4.45725   4.35168   5.82371   4.29552
   6.22617   5.25266   6.14570   4.91916   5.06789   5.18234   3.34981
   4.99196   5.68083   3.83586   4.86218   4.58404   5.95582   2.07919
   5.02754   6.23805   5.39304   5.35205   4.43152   5.81570   5.18657
   4.95501   4.48976   6.99998   5.37029   5.96970   5.65352   5.15685
   4.83735   4.43587
MTB > STOP
```

7.24 Use a computer to randomly select 200 samples, each of size 5, from a normal population with a mean equal to 50 and a standard deviation of 10. List the 200 sample means and print a histogram of their distribution.

   a. Find the theoretical probability $P(46 < \bar{x} < 55)$.
   b. What percentage of the sample means actually did have values between 46 and 55?
   c. Explain any difference that may have occurred between answers (a) and (b).

(If MINITAB is used, see Exercise 7.23 for a program.)

IN RETROSPECT

In Chapters 6 and 7 we used the standard normal probability distribution. There are now two formulas for calculating a *z*-score:

$$z = \frac{x - \mu}{\sigma} \quad \text{and} \quad z = \frac{\bar{x} - \mu}{\sigma/\sqrt{n}}$$

You must distinguish between these two formulas. The first gives the standard score when dealing with individual values from a normal population (*x* values). The second uses information provided by the central limit theorem. Sampling distributions of sample means are approximately normally distributed. Therefore, the standard scores and the probabilities in Table 5, Appendix F, may be used in connection with sample means ($\bar{x}$ values). The key to distinguishing between the formulas is to decide whether the problem deals with individual values of *x* from the population or sample means from the sampling distribution. If it deals with the individual values of *x*, we use the first formula, as presented in Chapter 6. If, on the other hand, the problem deals with the sample means, we use the second formula and proceed as illustrated in this chapter.

The basic purpose for considering what happens under repeated sampling, as discussed in this chapter, is to form sampling distributions. The sampling distribution is then used to describe the variability that occurs from one sample to the next. Once this pattern of variability is known and understood for a specific sample statistic, we will be able to make accurate predictions about the corresponding population parameters. The central limit theorem describes the distribution for sample means. We will begin to make inferences about population means in Chapter 8.

There are other reasons for repeated sampling. Repeated samples are commonly used in the field of production control, in which samples are taken to determine whether a product is of the proper size or quantity. When the sample is defective, a mechanical adjustment of the machinery is necessary. The adjustment is then followed by another sampling.

The "standard error of the _____" is the name used for the standard deviation of the sampling distribution of whatever statistic is named in the blank. In this chapter we have been concerned with the standard error of the mean. However, we could also work with the standard error of the range, median, or whatever.

The margin of error is related to the standard error of proportion (the standard deviation of the sampling distribution of sample proportion, binomial probability). This relationship between "margin of error" and "standard error" will be discussed in Chapter 9.

You should now be familiar with the concept of a sampling distribution and, in particular, with the sampling distribution of sample means. In Chapter 8 we will begin to make predictions about the values of the various population parameters.

CHAPTER EXERCISES

7.25 A population has a normal distribution with an unknown μ and a standard deviation $\sigma = 5$. Find the probability that $\bar{x}$ will be within one unit of μ if n equals the following values:
    a. $n = 25$ **b.** $n = 100$ **c.** $n = 225$

7.26 A population has a normal distribution with an unknown mean μ and a standard deviation $\sigma = 5$. A random sample of size $n = 25$ is taken from this population.
    a. Find the probability that $\bar{x}$ will be within one unit of the mean μ; that is, find $P(-1 < \bar{x} - \mu < 1)$.
    b. If the standard deviation were 2.5 instead of 5, find $P(-1 < \bar{x} - \mu < 1)$.
    c. If the standard deviation were 10 instead of 5, find $P(-1 < \bar{x} - \mu < 1)$.

7.27 Each of 25 students flips a coin 16 times. Let x represent the number of heads obtained by each student. What is the probability that the mean number of heads for the 25 students is between 7.5 and 8.5?

7.28 Consider a binomial distribution with $n = 400$ and $p = 0.5$. If a sample of size 25 is selected from this binomial distribution, find the probability that the mean of the sample is between 195 and 205.

7.29 Suppose that a box contains three identical blocks numbered 2, 4, and 6. A sample of three numbers is drawn with replacement (that is, the first number is drawn, observed, and returned; then the second number is drawn; then the third). The mean of the sample is determined.
    a. Make a list that shows all the possible samples that could result from the sampling. (*Hint*: There should be 27 samples.)
    b. Determine the mean of each of these samples and form a sampling distribution of these sample means. (Express as a probability of distribution.)
    c. Construct a probability histogram of this probability distribution.
    d. Find the mean of this sampling distribution, $\mu_{\bar{x}}$.
    e. Find the standard error of the mean $\sigma_{\bar{x}}$ for this sampling distribution.

7.30 Suppose that a box contains three identical blocks numbered 2, 4, and 8. A sample of three numbers is drawn with replacement. The mean of the sample is determined.
    a. Make a list that shows all the possible samples that could result from the sampling. (There are 27.)
    b. Determine the mean of each sample and form a sampling distribution of these means. (Express as a probability distribution.)
    c. Construct a probability histogram of this probability distribution.
    d. Find the mean of this sampling distribution, $\mu_{\bar{x}}$.
    e. Find the standard error of the mean, $\sigma_{\bar{x}}$, for this sampling distribution.

7.31 Consider the experiment of taking a standardized mathematics test. The variable x is the raw score received. The exam has a mean of 720 and a standard deviation of 60. Assume that the variable x is normally distributed.
    a. Consider the experiment of an individual student taking this exam. Describe the distribution (shape of distribution, mean, and standard deviation) for the experiment.

b. A group of 100 students takes the exam; the mean is reported as a result. Describe the sampling distribution (shape of distribution, mean, and standard deviation) for the experiment.
c. What is the probability that the student in (a) scored less than 725.6?
d. What is the probability that the mean of the group in (b) is less than 725.6?

7.32 The diameters of Red Delicious apples in a certain orchard are normally distributed with a mean of 2.63 in. and a standard deviation of 0.25 in.
a. What percentage of the apples in this orchard have diameters less than 2.25 in.?
b. What percentage of the apples in the orchard are larger than 2.56 in.?
A random sample of 100 apples is gathered and the mean diameter obtained is $\bar{x} = 2.56$.
c. If another sample of size 100 is taken, what is the probability that its sample mean will be greater than 2.56 in.?
d. Why is the z-score used in answering parts (a), (b), and (c)?
e. Why is the formula for z used in (c) different from that used in parts (a) and (b)?

7.33 Find a value for e such that 95% of the apples in Exercise 7.32 are within e units of the mean 2.63. That is, find e such that $P(2.63 - e < x < 2.63 + e) = 0.95$.

7.34 Find a value for E such that 95% of the samples of 100 apples taken from the orchard in Exercise 7.32 will have mean values within E units of the mean 2.63. That is, find E such that $P(2.63 - E < \bar{x} < 2.63 + E) = 0.95$.

7.35 According to an article in the January 1991 issue of *Health* magazine (page 41), root-canal therapy costs from $200 to $700. Suppose the mean cost for root-canal therapy is $450 and the standard deviation is $125. If a sample of 100 dentists was selected across the country, find the probability that the mean cost per root canal for the sample would fall between $425 and $475.

7.36 A report in *Newsweek* (November 12, 1990, page 6) stated that the day-care cost per week in Boston is $109. If this figure is taken as the mean cost per week and if the standard deviation were known to be $20, find the probability that a sample of 50 day-care centers would show a mean cost of $100 or less per week.

7.37 A shipment of steel bars will be accepted if the mean breaking strength of a random sample of 10 steel bars is greater than 250 pounds per square inch. In the past, the breaking strength of such bars has had a mean of 235 and a variance of 400.
a. What is the probability, assuming that the breaking strengths are normally distributed, that one randomly selected steel bar will have a breaking strength in the range from 245 to 255?
b. What is the probability that the shipment will be accepted?

7.38 From a sample of 50 employees taken in a random manner from all of the employees of a large firm, the mean weekly earnings of employed males was $259.30. Given the current wage structure, it has been estimated in labor negotiations that the standard deviation is $34.10. What is the probability of the sample outcome or less if the population mean is really $275.00 as the management negotiator insists?

7.39 The baggage weights for passengers using a particular airline are normally distributed with a mean of 20 lb and a standard deviation of 4 lb. If the limit on total luggage weight is 2125 lb, what is the probability that the limit will be exceeded for 100 passengers?

7.40 A trucking firm delivers appliances for a large retail operation. The packages (or crates) have a mean weight of 300 lb and a variance of 2500 lb.
  a. If a truck can carry 4000 lb and 25 appliances need to be picked up, what is the probability that the 25 appliances will have an aggregate weight greater than the truck's capacity? (Assume that the 25 appliances represent a random sample.)
  b. If the truck has a capacity of 8000 lb, what is the probability that it will be able to carry the entire lot of 25 appliances?

7.41 A manufacturer of light bulbs says that its light bulbs have a mean life of 700 hr and a standard deviation of 120 hr. You purchased 144 of these bulbs with the idea that you would purchase more if the mean life of your sample is more than 680 hr. What is the probability that you will not buy again from this manufacturer?

7.42 A tire manufacturer claims (based on years of experience with its tires) that the mean mileage is 35,000 mi and the standard deviation is 5,000 mi. A consumer agency randomly selects 100 of these tires and finds a sample mean of 31,000. Should the consumer agency doubt the manufacturer's claim?

7.43 As a result of several surveys during the last three years, it was concluded that people want pollution controlled and are willing to pay pollution taxes in order to do something about it. The amount that people are willing to pay has a mean of $22.80 per year and a standard deviation of $3.00. Given these values, what is the probability that a random sample of 200 individuals will show a mean differing from $22.80 by more than $0.75?

7.44 Every year, around Halloween, many street signs in a small city are defaced. The average repair cost per sign is $68.00 and the standard deviation is $12.40.
  a. If 300 signs are damaged this year, there is a 5% chance that the total repair costs for the 300 signs will exceed what value?
  b. You are about 68% certain that the total repair costs for 300 signs will fall within what interval centered around $20,400?

7.45 After several years of growth, Douglas fir trees being cultivated by a nursery currently have a mean height of 72 in. and a standard deviation of 10 in. The heights are approximately normally distributed.
  a. What proportion of the time will random samples of 100 trees show a mean height between 70 and 75 in.?
  b. What mean heights for samples of 100 will fall more than 3 standard errors from the mean of 72 in.?

7.46 A pop-music record firm wants the distribution of lengths of cuts on its records to have an average of 2 min and 15 sec (135 sec) and a standard deviation of

10 sec, so that disc jockeys will have plenty of time for commercials within each five-minute period. The population of times for cuts is approximately normally distributed with only a negligible skew to the right. You have just timed the cuts on a new release and have found that the 10 cuts average 140 sec.

    a. What percentage of the time will the average be 140 sec or longer, if the new release is randomly selected?

    b. If the music firm wants 10 cuts to average 140 sec less than 5% of the time, what must the population mean be given that the standard deviation remains at 10 sec?

7.47 A sample of 144 values is randomly selected from a population with a mean, μ, equal to 45 and a standard deviation, σ, equal to 18.

    a. Determine the interval (smallest value to largest value) within which you would expect a sample mean to lie.

    b. What is the amount of deviation from the mean for a sample mean of 46.3?

    c. What is the maximum deviation you have allowed for in your answer to part (a)?

    d. How is this maximum deviation related to the standard error of the mean?

$ 7.48 For large samples, the sample sum (Σx) has an approximately normal distribution. The mean of the sample sum is n and the standard deviation is $\sqrt{n} \cdot \sigma$. The distribution of savings per account for a savings and loan institution has a mean equal to $750 and a standard deviation equal to $25. For a sample of 50 such accounts, find the probability that the sum in the 50 accounts exceeds $38,000.

VOCABULARY LIST

Be able to define each term. In addition, describe, in your own words, and give an example of each term. Your examples should not be ones given in class or in the textbook.

The bracketed numbers indicate the chapter in which the term first appeared, but you should define the terms again to show increased understanding of their meaning.

central limit theorem
frequency distribution [2]
probability distribution [5]
random sample [2]
repeated sampling

sampling distribution of sample
    means
standard error of the mean
theoretical probability distribution
z-score [2, 6]

KEY CONCEPTS

central limit theorem
random sample [2]

sampling distribution
standard error of the mean

QUIZ A

Answer "True" if the statement is always true. If the statement is not always true, replace the words shown in bold with words that make the statement always true.

7.1 A sampling distribution **is** a distribution listing all the sample statistics that describe a particular sample.

7.2 The histograms of **all** sampling distributions are symmetrically shaped.

7.3 The mean of the sampling distribution of $\bar{x}$'s is equal to the mean of the **sample**.

7.4 The standard error of the mean is the standard deviation of the population **from which the samples have been taken**.

7.5 The standard error of the mean **increases** as the sample size increases.

7.6 The shape of the distribution of sample means is always that of a **normal** distribution.

7.7 A **probability** distribution of a sample statistic is a distribution of all the values of that statistic that were obtained from all possible samples.

7.8 The central limit theorem provides us with a description of the three characteristics of a sampling distribution of sample **medians**.

7.9 A **frequency** sample is obtained in such a way that all possible samples of a given size have an equal chance of being selected.

7.10 We **do not need** to take repeated samples in order to use the concept of the sampling distribution.

QUIZ B

7.1 The lengths of the lake trout in Conesus Lake are believed to be normally distributed about a mean length of 15.6 in. with a standard deviation of 3.8 in.
   a. Kevin is going fishing at Conesus Lake tomorrow. If he catches one lake trout, what is the probability that it is less than 15.0 in.?
   b. If Captain Brian's fishing boat takes ten people fishing on Conesus Lake tomorrow and they catch a random sample of 16 fish, what is the probability that the mean length of their total catch is less than 15 in.?

7.2 Cigarette lighters manufactured by Easyvice Company are claimed to have a mean life of 20 months with a standard deviation of 6 months. The money-back guarantee allows you to return the lighter if it does not last at least 12 months from the date of purchase.
   a. If the length of life of these lighters is normally distributed, what percentage of the lighters will be returned to the company?
   b. If a random sample of 25 lighters is tested, what is the probability the sample mean will be more than 18 months?

7.3 Aluminum rivets produced by Rivets Forever, Inc., are believed to have shearing strengths that are distributed about a mean of 13.75 and have a standard deviation of 2.4. If this information is true and a sample of 64 such rivets is tested for shear strength, what is the probability that the mean strength will be between 13.6 and 14.2?

QUIZ C

7.1 "Two heads are better than one." If that's true, then "how good would several heads be?" To find out, a statistics instructor drew a line across the chalkboard and asked her class to estimate its length to the nearest inch. She collected their estimates, which ranged from 33 to 61 in., and calculated the mean value. She reported that the mean was 42.25 in. She then measured the line and found it to be 41.75 in. long. Does this show that "several heads are better than one"? What statistical theory supports this occurrence? Explain how.

7.2 The sampling distribution of sample means is more than just a distribution of the mean values that occur from many repeated samples taken from the same population. Describe what other specific condition most be met in order to have a sampling distribution of sample means.

7.3 Student A stated that "a sampling distribution of the standard deviation tells you how the standard deviation varies from sample to sample." Student B argues that "a population distribution tells you that." Who is right? Justify your answer.

7.4 Student A says that it is the "size of each sample used" and Student B says that it is the "number of samples used" that determines the spread of an empirical sampling distribution. Who is right? Justify your choice.

WORKING WITH YOUR OWN DATA

The central limit theorem is very important to the development of the rest of this course. Its proof, which requires the use of calculus, is beyond the intended level of this course. However, the truth of the CLT can be demonstrated both theoretically and by experimentation. The following series of questions will help to verify the central limit theorem both ways:

A ▼ The Population

Consider the theoretical population that contains the three numbers 0, 3, and 6 in equal proportions.

1. a. Construct the theoretical probability distribution for the drawing of a single number, with replacement, from this population.
   b. Draw a histogram of this probability distribution.
   c. Calculate the mean μ and the standard deviation σ for this population.

B ▼ The Sampling Distribution, Theoretically

Let's study the theoretical sampling distribution formed by the means of all possible samples of size 3 that can be drawn from the given population.

2. Construct a list showing all the possible samples of size 3 that could be drawn from this population. (There are 27 possibilities.)
3. Find the mean for each of the 27 possible samples listed in answer to question 2.
4. Construct the probability distribution (the theoretical sampling distribution of sample means) for these 27 sample means.
5. Construct a histogram for this sampling distribution of sample means.
6. Calculate the mean $\mu_{\bar{x}}$ and the standard error of the mean $\sigma_{\bar{x}}$.
7. Show that the results found in answers 1(c), 5, and 6 support the three claims made by the central limit theorem. Cite specific values to support your conclusions.

C ▼ The Sampling Distribution, Empirically

Let's now see whether the central limit theorem can be verified empirically; that is, does it hold when the sampling distribution is formed by the sample means that result from several random samples?

8. Draw a random sample of size 3 from the given population. List your sample of three numbers and calculate the mean for this sample.

You may take three identical "tags" numbered 0, 3, and 6, put them in a "hat," and draw your sample using replacement between each drawing. Or you may use dice; let 0 be represented by 1 and 2, let 3 be represented by 3 and 4, and 6 by 5 and 6. You may also use random numbers to simulate the drawing of your samples. Or you may draw your sample from the list of random samples at the bottom of the page. Describe the method you decide to use. (Ask your instructor for guidance.)

9. Repeat question 8 forty-nine (49) more times so that you have a total of fifty (50) sample means that have resulted from samples of size 3.

10. Construct a frequency distribution of the 50 sample means found in answering questions 8 and 9.

11. Construct a histogram of the frequency distribution of observed sample means.

12. Calculate the mean $\bar{x}$ and standard deviation $s_{\bar{x}}$ of the frequency distribution formed by the 50 sample means.

13. Compare the observed values of $\bar{x}$ and $s_{\bar{x}}$ with the values of $\mu_{\bar{x}}$ and $\sigma_{\bar{x}}$. Do they agree? Does the empirical distribution of $\bar{x}$ look like the theoretical one?

The following table contains 100 samples of size 3 that were generated randomly by computer:

6 3 0	0 3 0	6 6 0	3 3 6	6 6 3	6 3 3
0 0 3	3 0 6	3 3 0	3 6 6	0 3 0	6 6 3
6 6 6	0 3 0	6 3 6	0 6 3	6 0 3	6 3 3
6 0 0	3 0 6	6 3 3	3 3 0	3 3 0	3 3 3
3 3 3	3 0 0	6 6 6	3 3 6	0 0 6	0 6 3
6 6 6	0 0 6	3 3 0	0 6 6	0 0 3	6 6 3
0 0 6	0 0 6	6 6 6	6 3 6	6 6 0	3 0 0
3 6 6	6 3 0	3 6 3	3 0 0	3 3 6	0 6 0
3 0 0	0 3 6	6 3 3	6 0 6	3 3 6	6 0 3
0 3 6	3 6 3	6 6 3	6 6 0	3 3 3	3 0 0
6 3 0	6 6 0	0 3 0	6 6 0	3 6 6	0 3 6
6 3 3	0 3 0	6 6 0	6 6 3	6 6 0	3 0 3
3 6 3	3 6 0	0 0 6	0 3 3	3 6 6	0 3 6
0 6 0	6 0 0	0 6 0	0 6 6	0 3 3	0 3 6
3 3 6	3 3 3	3 3 6	6 3 6	3 3 3	3 6 6
6 3 3	3 0 0	3 0 6	6 0 3	3 6 6	6 0 3
0 3 3	6 3 0	0 3 6	0 3 6		

The calculations required are accomplished most easily with the assistance of an electronic calculator or a canned program on a computer. Your local computer center can provide a list of the available canned programs and the necessary information and assistance to run your data on the computer.

PART THREE

Inferential Statistics

THE CENTRAL LIMIT THEOREM told us about the sampling distribution of sample means. Specifically, it stated that a distribution of sample means was normally or approximately normally distributed about the mean of the population; and it stated the relationship between the standard deviation of the population and the sampling distribution. With this information we are able to make probability statements about the likelihood of certain sample mean values occurring when samples are drawn from a population with a known mean and a known standard deviation. We are now ready to turn this situation around. We will draw one sample, calculate its mean value, and then make an inference about the value of the population mean based on the sample mean's value.

In this part of the textbook we will learn about making two types of inferences: (1) the decision-making process using a hypothesis test procedure, and (2) the procedures for estimating a population parameter. Specifically, we will learn about making these two types of inferences about the population mean μ and the population standard deviation for normally distributed populations, and for the probability parameter p of a binomial population.

William Gosset

WILLIAM GOSSET ("Student"), a British industrial statistician, was born in Canterbury, England, on June 13, 1876 to Frederick and Agnes (Vidal) Gosset. William's educational background included studies at Winchester College, New College, and Oxford University. In 1899, upon leaving Oxford University, Gosset was employed as a brewer by the Arthur Guinness & Son Brewing Company. In 1906, Gosset married Marjory Surtees Philpotts, and they became the parents of two daughters and a son. William Gosset died in Beaconsfield, England, on October 16, 1937.

Guinness liked their employees to use pen names if publishing papers, so in 1908 Gosset adopted the pen name "Student" under which he published what was probably his most noted contribution to statistics, "The Probable Error of a Mean." Guinness sent Gosset to work under Karl Pearson, at the University of London, and eventually he took charge of the new Guinness Brewery in London.

In his paper "The Probable Error of a Mean," "Student" set out to find the distribution of the amount of error in the sample mean, $(\bar{x} - \mu)$, when divided by s, where s was the estimate of σ from a sample of any known size. The probable error of a mean, $\bar{x}$, could be calculated for any size sample by using this distribution of $(x - \mu)/s$. Even though he was well aware of the insufficiency of a small sample to determine the form of the distribution of x, he chose the normal distribution for simplicity, stating his opinion: "It appears probable that the deviation from normality must be very severe to lead to serious error."

Student's t-distribution did not immediately gain popularity. In September 1922, even 14 years after its publication, Student wrote to Fisher: "I am sending you a copy of Student's Tables as you are the only man that's ever likely to use them!" Today, Student's t-distribution is widely used and respected in statistical research.

8 INTRODUCTION TO STATISTICAL INFERENCES

Chapter Outline

8.1 The Nature of Hypothesis Testing
A hypothesis test makes a **decision** about the value of a population parameter.

8.2 The Hypothesis Test (A Classical Approach)
To test a claim we must formulate a **null hypothesis** and an **alternative hypothesis**.

8.3 The Hypothesis Test (A Probability-Value Approach)
A probability-value approach is an **alternative** to the decision-making process.

8.4 Estimation
Another type of inference involves estimation, and we learn to make both **point estimates** and **interval estimates**.

Evaluation of Teaching Techniques for Introductory Accounting Courses

Abstract: This study tests the effect of homework collection and quizzes on exam scores. Expectancy theory as modified by Porter and Lawler [1968] suggests that performance (a student's exam score) is dependent upon effort, abilities and traits, and role perception (regarding the class). An accounting instructor is able to influence a student's effort through assigning different tasks (teaching techniques) which may include homework collection or quizzes....

The hypothesis for this study is that an instructor can improve a student's performance (exam scores) through influencing the student's perceived effort-reward probability. An instructor accomplishes this by assigning tasks (teaching techniques) which are a part of a student's grade and are perceived by the student as a means of improving his or her grade in the class. The student is motivated to increase effort to complete those tasks which should also improve understanding of course material. The expected final result is improved exam scores.

The null hypothesis for this study is:

H_0: Teaching techniques have no significant effect on student's exam scores. . . .

CONCLUSION

The results of this study were unable to provide evidence that collecting/grading accounting homework assignments and giving quizzes have an effect on students' exam scores. While students may perceive a benefit from homework and quizzes and expect introductory accounting courses to utilize them, their exam performance did not significantly improve in this study when they were rewarded for successfully completing those tasks.

Source: David R. Vruwink and Janon R. Otto, *The Accounting Review*, Vol. LXII, No. 2, April 1987. Reprinted by permission.

Chapter Objectives

A random sample of 36 pieces of data yields a mean of 4.64. What can we deduce about the population from which the sample was taken? We will be asked to answer two types of questions:

1. Is the sample mean significantly different in value from the hypothesized mean value of 4.5?
2. Based on the sample, what estimate can we make about the value of the population mean?

The first question requires us to make a *decision*, whereas the second question requires us to make an *estimation*.

In this chapter and in Chapters 9 and 10 we will find out how a hypothesis test is used to make a statistical decision about three basic parameters: the mean μ, the standard deviation σ, and the binomial probability of "success" p. We will also learn how an estimation of these three parameters is made. In this chapter we will concentrate on learning about the basic concepts of hypothesis testing and estimation. We will deal with questions about the population mean using two methods that assume that the population standard deviation is known. This assumption is seldom realized in real-life problems, but it will make our first look at inferences much simpler.

8.1 ▼ The Nature of Hypothesis Testing

Suppose that a certain airline company requires the manufacturer of its aircraft to use rivets whose mean shearing strength exceeds 120 lb. Each rivet manufacturer that wants to sell rivets to the aircraft manufacturer must demonstrate that its rivets meet the required specification, namely, that the mean shearing strength of all the manufacturer's rivets, μ, be greater than 120 lb.

NOTE 1 Each individual rivet has a shearing strength, which is determined by measuring the force required to shear (break) the rivet. The strength of each rivet is not in question but rather the mean shearing strength of all such rivets. How can this mean strength be determined? Clearly, not all the rivets can be tested. (If they were, only broken rivets would remain, and there would be none left to build airplanes.) Therefore, a sample of rivets will be tested and a decision about the mean strength of all the untested rivets will be based on the observed mean from those sampled.

NOTE 2 Throughout Chapter 8 we will treat the standard deviation σ as a known, or given, quantity and concentrate on learning the procedures for making inferences about the population mean μ. We will use $\sigma = 12$ for our rivet illustration. In this illustration the rivet supplier is interested in demonstrating that the mean shearing strength of its rivets is greater than 120 ($\mu > 120$). The statistical procedure used to make this determination is called a *hypothesis test*.

Section 8.1 ▼ THE NATURE OF HYPOTHESIS TESTING

The statistical hypothesis test is a well-organized four-step procedure. Each step of this procedure will be demonstrated and justified as we investigate the rivet illustration.

STEP 1 Formulate the null and alternative hypotheses.

hypothesis

> **HYPOTHESIS**
> A statement that something is true.

There are two hypotheses, the *null hypothesis* and the *alternative hypothesis*.

null hypothesis

> **NULL HYPOTHESIS, H_0**
> The hypothesis we will test. Generally this is a statement that a population parameter has a specific value. The null hypothesis is so named because it is the "starting point" for the investigation. The phrase "there is no difference" is often used in its interpretation.

alternative hypothesis

> **ALTERNATIVE HYPOTHESIS, H_a**
> This hypothesis, on which we focus our attention, is a statement about the same population parameter that is used in the null hypothesis. Generally this is a statement that specifies that the population parameter has a value different, in some way, from the value given in the null hypothesis. The rejection of the null hypothesis will imply the acceptance of this alternative hypothesis.

The null hypothesis and the alternative hypothesis are formulated by inspecting the problem or statement to be investigated and first formulating two opposing statements. For our illustration these two opposing statements are: (a) "The mean shearing strength is greater than 120 ($\mu > 120$), the airline company's requirement" versus (b) "The mean shearing strength is not greater than 120 ($\mu \leq 120$), the negation of the airline company's requirement."

NOTE The Trichotomy Law from algebra states that two numerical values must be related in exactly one of three possible relationships: $<$, $=$, or $>$. All three of these possibilities must be accounted for between the two opposing statements.

Statement (b) becomes the null hypothesis.

$$H_0: \mu = 120 \quad (\leq) \quad \text{(the mean is not greater than 120)}$$

The parameter of concern, the population mean μ, is assigned (is equal to) a specific value. Further, 120 is the value on which our attention has been focused. This hypothesis represents the opposition to the rivet supplier's desires and says that its rivets *do not meet* the required standards.

Statement (a) becomes the alternative hypothesis:

$$H_a: \mu > 120 \quad \text{(the mean is greater than 120)}$$

This statement says that the mean is greater than 120. "Greater than 120" is clearly different and separate from "equals 120." The statement represents the rivet supplier's desires and says that its rivets *do meet* the required standards.

From this point on in the hypothesis test procedure, we will work under the assumption that the null hypothesis is a true statement. This situation might be compared to a courtroom trial, where the accused is assumed to be innocent until sufficient evidence has been presented to show otherwise. At the conclusion of the hypothesis test, we will make one of two possible **decisions**. We will decide in agreement with the null hypothesis and say that we "fail to reject H_0" (this corresponds to "fail to convict" or an "acquittal" of the accused in a trial). Or we will decide in opposition to the null hypothesis and say that we "reject H_0" (this corresponds to "conviction" of the accused in a trial).

Before continuing with the illustration we need to look at the different possible situations regarding the truth of the null hypothesis and the correctness of the decision to be reached. There are four possible outcomes that could be reached as a result of the null hypothesis being either true or false and the decision being either "fail to reject" or "reject." Table 8-1 shows these four possible outcomes.

TABLE 8-1 Four Possible Outcomes in a Hypothesis Test

	Null Hypothesis	
Decision	True	False
Fail to reject H_0	Type A Correct decision	Type II Error
Reject H_0	Type I Error	Type B Correct decision

A **type A correct decision** occurs when the null hypothesis is true and we decide in its favor. A **type B correct decision** occurs when the null hypothesis is false and our decision is in opposition to the null hypothesis. A **type I error** will be committed when a true null hypothesis is rejected, that is, when the null hypothesis is true but we decided against it. A **type II error** is committed when we decide in favor of a null hypothesis that is actually false.

When we make a decision, it would be nice if it were always a correct decision. This, however, is statistically impossible, since we will be making our decision on the basis of sample information. The best we can hope for is to control the **risk**, or probability, with which an error occurs. The **probability** assigned to the type I error is called **alpha, α** (α is the first letter of the Greek alphabet). The **probability** of the type II error is called **beta, β** (β is the second letter of the Greek alphabet). See Table 8-2. To control these errors we will assign a small probability to them. The most frequently used probability values for α are 0.01 or

Section 8.1 ▼ THE NATURE OF HYPOTHESIS TESTING

0.05. The probability assigned to each error will depend on the seriousness of the error. The more serious the error, the less often we will be willing to allow it to occur, and therefore a smaller probability will be assigned to it. In this text we are going to devote our attention to α, the P(type I error). Further discussion of β, the P(type II error), is beyond the scope of this book.

TABLE 8-2 Probability with Which Error Occurs

Error	Type Error	Probability
Rejection of a true hypothesis	I	α
Failure to reject a false null hypothesis	II	β

Let's return to our illustration and proceed with Step 2 of the hypothesis test procedure.

STEP 2 Determine the test criteria.

test criteria

TEST CRITERIA

Consist of (1) specifying a level of significance α, (2) determining a test statistic, (3) determining the critical region(s), and (4) determining the critical value(s).

level of significance

LEVEL OF SIGNIFICANCE

The probability of committing the type I error, α.

This can be thought of as a "management decision." Typically, someone in charge determines the amount of risk (probability of) that seems reasonable in view of the seriousness of committing the type I error. If the type I error is a very serious or costly error, alpha should be assigned a very small value. If, on the other hand, the type I error is not that serious, then a larger value can be used.

NOTE There is a relationship between the probabilities of the type I and type II errors and the sample size. At this stage of development, α, the probability of the type I error, will be given in the statement of the problem, as will the sample size n. For now we will disregard the role of beta. For our illustration, let $\alpha = 0.05$.

test statistic

TEST STATISTIC

A random variable used to make the decision "fail to reject H_0" or "reject H_0."

For our illustration, we will want to compare the value of the sample mean to the hypothesized population mean (the value stated in the null hypothesis). In Chapter 7 we studied the central limit theorem and it told us about the distribution of means from all the possible samples of a given size. You will recall that the distribution of sample means, $\bar{x}$'s, is approximately normally distributed. Therefore, we will use the standard normal variable z as our test statistic for this illustration. Draw a sketch of the standard normal distribution and label it.

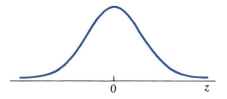

critical region

> **CRITICAL REGION**
> The set of values of the test statistic that will cause us to reject the null hypothesis.

Recall that we are working under the assumption that the null hypothesis is true. Thus, we are assuming that the mean shearing strength of all the rivets is actually 120. If this is the case, then when we select a random sample of 36 rivets, test them, and calculate the mean value for the sample, we can expect this sample mean, $\bar{x}$, to be part of a normal distribution that is centered at 120 and has a standard error of $12/\sqrt{36}$ or 2.0. (Recall that $\mu_{\bar{x}} = \mu$ and $\sigma_{\bar{x}} = \sigma/\sqrt{n}$.) Approximately 95% of the sample mean values will be within two standard deviations of 120, or between 116 and 124. Thus, if H_0 is true and $\mu = 120$, then we expect $\bar{x}$ to be between 116 and 124, approximately, 95% of the time.

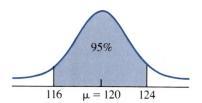

If, however, the value of $\bar{x}$ that we obtain from our sample is larger than 124, say 125, we will have to make a choice. It could be that such an $\bar{x}$ value (125) is either a member of the sampling distribution with mean 120 although it has a very low probability of occurrence (less than 0.05), or $\bar{x} = 125$ is a member of a sampling distribution whose mean is greater than 120, thereby being a value that is more likely to occur.

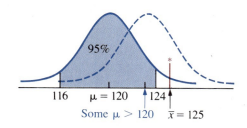

Section 8.1 ▼ THE NATURE OF HYPOTHESIS TESTING

In statistics, we "bet" on the "more probable to occur" and consider the second choice to be the right one. Thus, the right-hand tail of the z distribution becomes the critical region. And the value of alpha becomes the measure of its area, as pictured on the following figure:

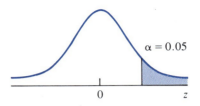

critical value

CRITICAL VALUE
The "first" (or "boundary") value in the critical region.

The critical value for our illustration is $z(0.05)$ and has the value of $+1.65$, as found in Table 5, Appendix F. The notation introduced in Section 6.4 (p. 325) is very handy for identifying critical values.

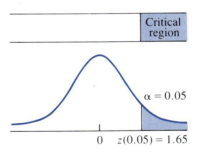

Having completed the ground rules for the test (Steps 1 and 2), we are now ready to obtain and present the sample evidence.

STEP 3 Obtain the sample data (evidence) and calculate the value of the test statistic.

At this point the sample data are collected and the value of the sample statistic that corresponds to the parameter being hypothesized about in the null hypothesis is calculated. The value of the test statistic is then determined from this sample statistic.

A sample of 36 rivets is tested and the resulting measurements yield a sample mean of 124.4. This sample statistic must now be converted to a z-score using formula (7-2). The resulting z-score will be our evidence.

$$z = \frac{\bar{x} - \mu}{\sigma/\sqrt{n}} = \frac{124.4 - 120}{12/\sqrt{36}} = \frac{4.4}{2.0} = \mathbf{2.20}$$

We now have a "calculated value" for the test statistic.

In fact, we now have two values of z. The first value of $z(0.05)$ is 1.65 and the second value is 2.20. Now we must compare these two values. To help keep track of which one is which, we will use an asterisk, *, to identify the calculated value of the test statistic. Thus, $z* = 2.20$.

STEP 4 Make a decision and interpret it.

Make the decision by comparing the value of the calculated test statistic found in Step 3 to the critical value of the test statistic found in Step 2. Graphically this comparison is made by locating an asterisk on the sketch of the z distribution at the location of the value for $z*$.

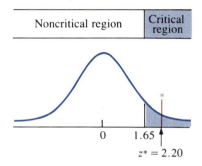

decision rule

DECISION RULE

If the test statistic falls within the critical region, we will reject H_0. If the test statistic does not fall within the critical region, we will fail to reject H_0. *Note*: The set of values that are not in the critical region is called the **noncritical region** or, sometimes, the **acceptance region**.

noncritical region
acceptance region

Since the calculated value of z, $z* = 2.20$, is within the critical region, the decision reached is "reject H_0."

We complete the hypothesis test by interpreting the decision (write the conclusion).

CONCLUSION RULE

If the decision is "reject H_0," then the **conclusion** should be worded something like "There is sufficient evidence at the α level of significance to show that . . . (the meaning of the alternative hypothesis)." If the decision is "fail to reject H_0," then the conclusion should be worded something like "There is *not* sufficient evidence at the α level of significance to show that . . . (the meaning of the alternative hypothesis)."

conclusion

For our illustration, the conclusion is: "There is sufficient evidence at the 0.05 level of significance to show that the population of rivets from which the sample was taken does meet the airline company's specifications."

Section 8.1 ▼ THE NATURE OF HYPOTHESIS TESTING

When writing the decision and the conclusion, remember that (1) the decision is about H_0, and (2) the conclusion is a statement about whether or not the contention of H_a was upheld. This is consistent with the "attitude" of the whole hypothesis test procedure. The null hypothesis is the statement that is "on trial" and therefore the decision must be about it. The contention of the alternative hypothesis is the thought (challenge or question) that brought about the need for a decision. Therefore, the question that led to the alternative hypothesis must be answered when the conclusion is written.

NOTE Some people use the phrase "accept H_0" instead of "fail to reject H_0." The difference between these two phrases is more substance than semantics. The key to understanding this difference lies in understanding the attitude of the statistician performing the hypothesis test. Perhaps this can best be illustrated using the courtroom trial analogy and looking at the prosecutor's viewpoint. The null hypothesis in court is "the accused is innocent." The trial takes place only if the prosecution is convinced that it has sufficient evidence of guilt. (Trials are not held to prove one's innocence.) That is, the prosecutor is convinced the null hypothesis is false. If the trial results in an acquittal, it seems very unlikely that the prosecution would change its convictions and conclude, "The accused is innocent!" A more accurate thought might be, "I was unable to get a conviction." The statistical hypothesis test follows much the same pattern. The hypothesis test exists for the sole purpose of disproving the null hypothesis. The statistician doing the testing (the prosecutor) does not believe the null hypothesis to be true. (That's why the test is being performed.) The statistician is not going to reverse position after seeing the results of one sample. "Acceptance of the null hypothesis" carries the wrong connotation.

We must also remember that when the decision is "fail to reject H_0," a type II error might have been committed. Lack of sufficient evidence has led to the acquittal of more than one guilty party.

▼▲ EXERCISES

8.1 As described in this section, the hypothesis testing procedure has many similarities to a courtroom procedure. The null hypothesis "the accused is innocent" is being tested.
  a. Describe the situation involved in each of the four possible outcomes shown in Table 8-1 (p. 384).
  b. If the accused is acquitted, does this "prove" him innocent? (Perhaps, in this respect, the phrase "fail to reject H_0" is more accurate in expressing the situation than is "accept H_0.")
  c. If the accused is found guilty, does this "prove" his guilt?

8.2 Consider the following nonmathematical situation as a hypothesis test. A medic at the scene of a serious accident tests the null hypothesis "this victim is alive."
  a. Carefully state the meaning of the four possible outcomes indicated in Table 8-1 (p. 384).
  b. Describe the seriousness of the type I error. the type II error.

c. If the type I error and type II error could be controlled statistically, which set of probabilities would you prefer be used if you were the victim?
 (1) $\alpha = 0.001$ and $\beta = 0.10$
 (2) $\alpha = 0.05$ and $\beta = 0.05$
 (3) $\alpha = 0.10$ and $\beta = 0.001$

8.3 Consider the following nonmathematical situation as a hypothesis test. You just received a parachute that was inspected by an inspector whose null hypothesis is "this parachute will open."
 a. Carefully state the meaning of the four possible outcomes indicated in Table 8-1.
 b. Describe the seriousness of the two possible errors.
 c. If the type I error and the type II error could be controlled statistically, which set of probabilities would you prefer be used if you were going to use the parachute?
  (1) $\alpha = 0.001$ and $\beta = 0.10$
  (2) $\alpha = 0.05$ and $\beta = 0.05$
  (3) $\alpha = 0.10$ and $\beta = 0.001$

8.4 Consider the following quality-control situation. You work for a company that builds airplanes and you are in charge of purchasing all the rivets used to assemble the airplanes. The rivets must be of a specified strength; otherwise the airplanes will fall apart. As a supplier of rivets used in the manufacturing of aircraft, I have assured you that the rivets I want to sell you are "strong enough." You take a sample of the rivets and test their strength in order to decide between the null hypothesis, "the rivets are strong enough," and the alternative hypothesis, "the rivets are not strong enough."
 a. Carefully state the meaning of the four possible outcomes indicated in Table 8-1.
 b. Describe the consequences and the seriousness of committing the type I error; the type II error.
 c. Which combination of probabilities would you prefer to use in making the decision? Explain.
  (1) $\alpha = 0.001$ and $\beta = 0.10$
  (2) $\alpha = 0.05$ and $\beta = 0.05$
  (3) $\alpha = 0.10$ and $\beta = 0.001$
  (4) $\alpha = 0.01$ and $\beta = 0.01$
  (5) $\alpha = 0.10$ and $\beta = 0.10$

8.5 A paint manufacturer wishes to test the hypothesis that "the addition of an additive will increase the average coverage of the company's paint." The average coverage has been 450 sq ft/gal. Let μ be the average coverage with the additive included. The null hypothesis is "the average coverage will not increase with the addition of the additive," ($\mu = 450$). The alternative hypothesis is "the average coverage will increase with the addition of the additive," ($\mu > 450$). Describe the meaning of the two possible types of errors that can occur in the decision when this test of the hypothesis is conducted.

Section 8.1 ▼ THE NATURE OF HYPOTHESIS TESTING

 8.6 A supplier of highway materials claims he can supply an asphalt mixture that will make roads paved with his materials less slippery when wet. A general contractor who builds roads wishes to test the supplier's claim. The null hypothesis is "roads paved with this asphalt mixture are no less slippery than roads paved with other asphalt." The alternative hypothesis is "roads paved with this asphalt mixture are less slippery than roads paved with other asphalt." Describe the meaning of the two possible types of errors that can occur in the decision when this hypothesis test is completed.

8.7   a. If the null hypothesis is true, what decision error could be made?
      b. If the null hypothesis is false, what decision error could be made?
      c. If the decision "reject H_0" is made, what decision error could have been made?
      d. If the decision "fail to reject H_0" is made, what decision error could have been made?

8.8   a. If the value of the test statistic falls in the critical region, what decision must we make?
      b. If the value of the test statistic does not fall in the critical region (that is, it falls in the noncritical region), what decision must we make?

8.9   a. If α is assigned the value 0.001, what are we saying about the type I error?
      b. If α is assigned the value 0.05, what are we saying about the type I error?
      c. If α is assigned the value 0.10, what are we saying about the type I error?

8.10 a. If β is assigned the value 0.001, what are we saying about the type II error?
      b. If β is assigned the value 0.05, what are we saying about the type II error?
      c. If β is assigned the value 0.10, what are we saying about the type II error?

 8.11 Refer to the article on p. 381. Discuss how the hypothesis for this study, "an instructor can improve a student's performance . . . ," and the null hypothesis as stated in the article fit the definitions of the null and alternative hypotheses as defined in this section.

8.12 Describe how the null hypothesis, as stated in Exercise 8.6, is a starting point for the decision to be made about the asphalt mixture.

8.13 The director of an advertising agency is concerned with the effectiveness of a television commercial.
      a. What null hypothesis is she testing if she commits a type I error when she erroneously says that the commercial is effective?
      b. What null hypothesis is she testing if she commits a type II error when she erroneously says that the commercial is effective?

8.14 a. If the null hypothesis is true, the probability of a decision error is identified by what name?
      b. If the null hypothesis is false, the probability of a decision error is identified by what name?
      c. If the test statistic falls in the critical region, what error could be made?
      d. If the test statistic falls in the noncritical region, what error might occur?

8.15 a. Suppose that a hypothesis test is to be carried out by using $\alpha = 0.05$. What is the probability of committing a type I error?
b. What proportion of the probability distribution is in the noncritical region provided the null hypothesis is correct?

8.16 a. What is the critical region?
b. What is the critical value?

8.17 There are only two possible decisions as a result of a hypothesis test.
  a. State the two possible decisions.
  b. Describe the conditions that will lead to each of the two decisions identified in (a).

8.18 The conclusion is the part of the hypothesis test that communicates the findings of the test to the reader. As such, it needs special attention so that the reader receives an accurate picture of the findings.
  a. Carefully describe the "attitude" and the statement of the conclusion when the decision is "reject H_0."
  b. Carefully describe the "attitude" and the statement of the conclusion when the decision is "fail to reject H_0."

8.19 Since the size of the type I error can always be made smaller by reducing the size of the critical region, why don't we always choose critical regions that make α extremely small?

8.20 The power of a statistical test is defined to be the probability of rejecting the null hypothesis when the null hypothesis is false. The probability of the type II error is the probability of failing to reject the null hypothesis when it is false. Therefore, the power of a test equals $1 - \beta$, since rejecting a false null hypothesis and failing to reject a false null hypothesis are complementary events. Find the power of a test when the probability of the type II error is
  a. 0.01   b. 0.05   c. 0.10

8.2 ▼ The Hypothesis Test (A Classical Approach)

In Section 8.1 we surveyed the steps for and some of the reasoning behind a hypothesis test while looking at a specific illustration. The approach we used is called the classical approach. In this section we are going to continue the study of the classical approach to the hypothesis test procedure as it applies to statements concerning the mean μ of a population. We will continue to impose the restriction that the population standard deviation is known. (This restriction will be removed in Chapter 9.) The first three illustrations deal with additional information about the procedures for formulating the null and alternative hypotheses.

▼ ILLUSTRATION 8-1

An ecologist would like to show that Rochester has an air pollution problem. Specifically, she would like to show that the mean level of carbon monoxide in downtown Rochester air is higher than 4.9 parts per million. State the null and alternative hypotheses.

SOLUTION To state the hypotheses we first need to identify the population parameter in question and the value to which it is being compared. The "mean level of carbon monoxide pollution" is the parameter μ and 4.9 parts per million is the specific value. Our ecologist is questioning the value of μ. It could be related to 4.9 in any one of three ways: (1) $\mu < 4.9$, (2) $\mu = 4.9$, or (3) $\mu > 4.9$. These three statements must be arranged to form two statements: one that states what the ecologist is trying to show and one that states the opposite. $\mu > 4.9$ represents the statement "the mean level is higher than 4.9," whereas $\mu < 4.9$ and $\mu = 4.9$ ($\mu \leq 4.9$) represent the opposite, "the mean level is not higher than 4.9." One of these two statements will become the null hypothesis H_0 and the other will become the alternative hypothesis H_a.

RECALL The null hypothesis states that the parameter in question has a specified value.

All this is suggesting that the statement containing the equals sign will become the null hypothesis; the other statement will become the alternative hypothesis. Thus we have

$$H_0: \mu = 4.9 \quad \text{and} \quad H_a: \mu > 4.9$$

Recall that once the null hypothesis is stated, we proceed with the hypothesis test under the assumption that the null hypothesis is true. Thus $\mu = 4.9$ locates the center of the sampling distribution of sample means. For this reason the null hypothesis will be written with an equals sign only. If $\mu = 4.9$ and $\mu < 4.9$ are stated together as the null hypothesis, the null hypothesis will be expressed as

$$H_0: \mu = 4.9(\leq)$$

where the ($\leq$) serves as a reminder of the grouping that took place when the hypothesis was formed. ▲▲

NOTE *The equals sign must be in the null hypothesis*, regardless of the statement of the original problem.

Illustration 8-1 expresses the viewpoint that an ecologist might take. Now let's consider how the Rochester Chamber of Commerce might view the same situation.

▼ **ILLUSTRATION 8-2**

In trying to promote the city, the Chamber of Commerce would be more likely to want to conclude that the mean level of carbon monoxide in downtown Rochester is less than 4.9 parts per million. State the null and alternative hypotheses related to this viewpoint.

SOLUTION Again, the parameter of interest is the mean level μ of carbon monoxide, and 4.9 is the specified value. $\mu < 4.9$ corresponds to "the mean level is less than 4.9," whereas $\mu \geq 4.9$ corresponds to "the mean level is not less than 4.9." Therefore, the hypotheses are

$$H_0: \mu = 4.9 \quad (\geq) \quad \text{and} \quad H_a: \mu < 4.9$$

▲▲

A more neutral point of view is suggested in Illustration 8-3.

▼ ILLUSTRATION 8-3

The "mean level of carbon monoxide in downtown Rochester is not 4.9 parts per million." State the null and alternative hypotheses that correspond to this statement.

SOLUTION The mean level of carbon monoxide is equal to 4.9 ($\mu = 4.9$) or the mean is not equal to 4.9 ($\mu \neq 4.9$). (*Note:* "Less than or greater than" is customarily expressed as "not equal to.") Therefore,

$$H_0: \mu = 4.9 \quad \text{and} \quad H_a: \mu \neq 4.9$$

▲▲

The viewpoint of the experimenter affects the way the hypotheses are formed, as we have seen in these three illustrations. Generally, the experimenter is trying to show that the parameter value is different from the value specified. Thus the experimenter is usually hoping to be able to reject the null hypothesis. Illustrations 8-1, 8-2, and 8-3 represent the three arrangements possible for the <, =, and > relationships between the parameter μ and the specified value 4.9.

The courtroom trial and the hypothesis test have many similarities, as already emphasized. Let's now look at another similarity. In the courtroom there is a trial only if the prosecutor has shown the court that there is sufficient reason to make the accused stand trial. If the prosecutor did not believe that the accused were guilty, there would be no trial. Much the same is true with the hypothesis test. If the experimenter believes that the null hypothesis is true, its truth will not be challenged and it will not be tested. The null hypothesis is tested only when the experimenter wishes to show that the alternative is correct. Thus the alternative hypothesis is a statement claiming that the value of the parameter is "smaller than" or "larger than" or "different from" the value claimed by the null hypothesis.

THE CLASSICAL HYPOTHESIS TEST: A FOUR-STEP MODEL

Step 1: State the null hypothesis (H_0) and the alternative hypothesis (H_a).
Step 2: Determine the test criteria:
   a. The level of significance, α, to be used
   b. The test statistic to be used
   c. The critical region(s)
   d. The critical value(s)

Step 3: Collect and present the sample evidence.
   a. Collect the sample information.
   b. Calculate the value of the observed test statistic.
Step 4: Determine the results.
   a. Compare the calculated value of the test statistic to the critical value(s) from Step 2.
   b. Make a decision about H_0.
   c. State the conclusion about H_a.

Illustration 8-4 demonstrates the complete hypothesis test procedure as it is used in questions dealing with the population mean.

▼ ILLUSTRATION 8-4

For many semesters an instructor has recorded students' grades, and the mean μ for all these students' grades is 72. The current class of 36 students seems to be better than average in ability and the instructor wants to show that according to their average "the current class is superior to previous classes." Does the class mean $\bar{x}$ of 75.2 present sufficient evidence to support the instructor's claim that the current class is superior? Use $\alpha = 0.05$ and $\sigma = 12.0$.

SOLUTION To be superior, this class must have a mean grade that is higher than the mean of all the previous classes. To be "equal to or less than" would not be superior.

STEP 1 H_0: $\mu = 72$ ($\leq$)   (class is not superior)
H_a: $\mu > 72$ (class is superior)

STEP 2 The level of significance $\alpha = 0.05$ is given in the statement of the problem. The standard score z is used as the test statistic when the null hypothesis is about a population mean and the standard deviation is known. Recall that the central limit theorem tells us that the sampling distribution of sample means is approximately normally distributed. Thus the normal probability distribution will be used to complete the hypothesis test. The critical region—values of the standard score z that will cause a rejection of the null hypothesis—has an area of 0.05 and is located at the extreme right of the distribution. The critical region is on the right because large values of the sample mean suggest "superior," while values near or below 72 support the null hypothesis. See the following figure:

If random samples of size 36 are taken from a population with a mean value equal to 72, then many of the sample means will have values near 72 (71, 72, 71.8, 72.5, 73, and so on). Only sample means that are considerably larger than 72 would cause us to reject the null hypothesis. The critical value, the cutoff between "not superior" and "superior," is determined by α, the probability of the type I error. $\alpha = 0.05$ was given. Thus the critical region (the shaded region of the following figure) has an area of 0.05 and a critical value of $+1.65$. (This value is obtained by using Table 5 of Appendix F.)

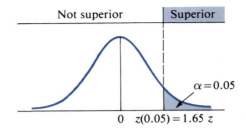

STEP 3 The calculated value of the test statistic z will be found by using formula (7-2) and the sample information

$$\bar{x} = 75.2 \quad \text{and} \quad n = 36$$

$$z = \frac{\bar{x} - \mu}{\sigma/\sqrt{n}}$$

RECALL We assumed that μ was equal to 72 in the null hypothesis and that $\sigma = 12.0$ was known.

$$z = \frac{75.2 - 72}{12.0/\sqrt{36}} = \frac{3.2}{2.0} = 1.60$$

$$z^* = \mathbf{1.60}$$

calculated value (*)

We will use an asterisk, *, to identify the **calculated value** of the test statistic; the asterisk will also be used to locate its value relative to the test criteria (Step 4).

STEP 4 We now compare the calculated test statistic, z^*, to the test criteria set up in Step 2 by locating the calculated value on the diagram and placing an asterisk at that value. Since the test statistic (calculated value) falls in the noncritical region (unshaded portion of the diagram), we must reach the following *decision*:

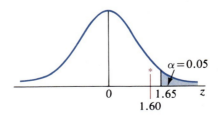

Decision Fail to reject H_0.

Section 8.2 ▼ THE HYPOTHESIS TEST (A CLASSICAL APPROACH)

Recall that the critical region was to be shaded in the diagram, and when the test statistic falls in the critical region, we must reject H_0. Step 4 is then completed by stating a **conclusion**.

Conclusion There is not sufficient evidence at the 0.05 level of significance to show that the current class is superior. ▲▲

Does this conclusion seem realistic? 75.2 is obviously larger than 72. Recall that 75.2 is a sample mean, and if a random sample of size 36 is drawn from a population whose mean is 72 and whose standard deviation is 12, then the probability that the sample mean is 75.2 or larger is greater than the risk, α, with which we are willing to make the type I error. [What is $P(\bar{x} > 75.2)$?]

▼ ILLUSTRATION 8-5

It has been claimed that the mean weight of women students at a college is 54.4 kg. Professor Schmidt does not believe the statement that the mean is 54.4 kg. To test the claim he collects a random sample of 100 weights from among the women students. A sample mean of 53.75 kg results. Is this sufficient evidence to reject the null hypothesis? Use $\alpha = 0.05$ and $\sigma = 5.4$ kg.

SOLUTION Professor Schmidt's statement suggests that the three possible relationships $(<, =, >)$ between the mean weight μ and the hypothesized 54.4 kg be split $(=)$ or $(<, >)$. Therefore, we have the following steps:

STEP 1 $H_0: \mu = 54.4$ (mean weight is 54.4 kg)

$H_a: \mu \neq 54.4$ (mean weight is not 54.4 kg)

The test statistic will be the standard score z, and the normal distribution is used since we are using the sampling distribution of sample means. The sample mean is our estimate for the population mean. Since the alternative hypothesis is "not equal to," a sample mean considerably larger than 54.4, as well as one considerably smaller than 54.4, will be in opposition to the null hypothesis. The values of the sample mean around 54.4 will support the null hypothesis. Therefore, the critical region will be split into two equal parts, one at each extreme of the normal distribution. The area of each region will be $\alpha/2$. Since $\alpha = 0.05$, each part of the critical region will have a probability of 0.025 (see the following figure).

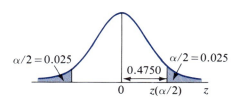

Look in Table 5 of Appendix F for the area 0.4750 (0.4750 = 0.5000 − 0.025). The z-score of 1.96 is found and becomes 1.96 on the right of the mean and −1.96 on the left.

STEP 2 The test criteria are shown in the following figure:

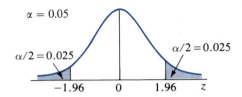

STEP 3 Calculate the test statistic z. The sample information is $\bar{x} = 53.75$ and $n = 100$. σ was given as 5.4, and the null hypothesis assumed that $\mu = 54.4$

$$z = \frac{\bar{x} - \mu}{\sigma/\sqrt{n}} = \frac{53.75 - 54.4}{5.4\sqrt{100}} = \frac{-0.65}{0.54} = -1.204$$

$z^* = -1.20$

Locate z^* on the diagram constructed in Step 2.

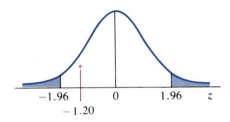

Now we are ready to identify the decision.

RECALL If the test statistic falls in the critical region, we must reject H_0. If the test statistic falls in the noncritical region, we must fail to reject H_0.

STEP 4 The calculated value of z, z^*, falls between the critical values cited in Step 2. Therefore, our decision is as follows:

Decision Fail to reject H_0.

The interpretation of our decision is the only thing left to do. This part of the hypothesis test may very well be the most significant part because the conclusions reached express the results found. We must be very careful to state precisely what is meant by the decision. It is often a temptation to overstate a conclusion. In general, a decision to "fail to reject H_0" may be interpreted to mean that the evidence found does not disagree with the null hypothesis. Note that this does not "prove" the truth of H_0. A decision to "reject H_0" will mean that the evidence found implies that null hypothesis is false and thus indicates the alternative hypothesis to be the case.

Conclusion The evidence (sample mean) found does not contradict the assumption that the population mean is 54.4 kg at the 0.05 level of significance. ▲▲

Before we look at another illustration, let's summarize briefly some of the details we have seen thus far:

1. The null hypothesis specifies a particular value of a population parameter.
2. The alternative hypothesis can take three forms. Each form dictates a specific location of the critical regions, as shown in the following table:

Sign in the Alternative Hypothesis	<	≠	>
Critical Region	One region Left side **One-tailed test**	Two regions One on each side **Two-tailed test**	One region Right side **One-tailed test**

one-tailed and two-tailed tests

3. For many hypothesis tests the sign in the alternative hypothesis "points" in the direction in which the critical region is located. [Think of the not equal to sign (≠) as being both less than (<) and greater than (>), thus pointing in both directions.]

The value assigned to α is called the significance level of the hypothesis test. Alpha cannot be interpreted to be anything other than the risk (or probability) of rejecting the null hypothesis when it is actually true. We will seldom be able to determine whether the null hypothesis is true or false; we will only decide to reject H_0 or to fail to reject H_0. The relative frequency with which we reject a true hypothesis is α, but we will never know the relative frequency with which we make an error in decision. The two ideas are actually quite different; that is, a type I error and an error in decision are two different things altogether (remember, there are two types of errors).

Let's look at some more illustrations of the hypothesis test.

▼ **ILLUSTRATION 8-6**

The student body at many community colleges is considered a "commuter population." The following question was asked of the Student Affairs Office: "How far (one way) does the average community college student commute to college daily?" The office answered: "No more than 9.0 mi." The inquirer was not convinced of the truth of this and decided to test the statement. He took a sample of 50 students and found a mean commuting distance of 10.22 mi. Test the hypothesis stated above at a significance level of $\alpha = 0.05$, using $\sigma = 5$ mi.

SOLUTION

STEP 1 $H_0: \mu = 9.0$ ($\leq$) (no more than 9.0 mi)

$H_a: \mu > 9.0$ (more than 9.0 mi)

The one-way distance traveled by the "average student" would be the same as the mean one-way distance traveled by all students. Therefore, the parameter of concern is μ. The claim "no more than 9.0 mi" implies that the three possible relationships should be grouped "no more than 9.0" ($\leq$) versus "more than 9.0" ($>$).

STEP 2 The critical region is on the right (since H_a contains the $>$ sign); H_0 will be rejected if it appears that the $\bar{x}$ observed is significantly greater than the 9.0 claimed. Notice that the sign in H_a ($>$) points toward the critical region. (See the following figure.)

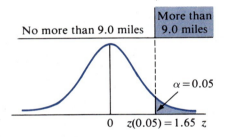

STEP 3

$$z = \frac{\bar{x} - \mu}{\sigma/\sqrt{n}} = \frac{10.22 - 9.0}{5/\sqrt{50}} = \frac{1.22}{0.707} = 1.73$$

$z^* = 1.73$

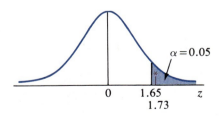

STEP 4 Decision Reject H_0. (z^* fell in the critical region.)

Conclusion At the 0.05 level of significance, we conclude that the average community college student probably travels more than 9.0 mi. ▲▲

▼ ILLUSTRATION 8-7

Draw a sample of 40 single-digit numbers from the random number table (Table 1, Appendix F) and test the null hypothesis $\mu = 4.5$. Use $\alpha = 0.10$ and $\sigma = 2.87$. The standard deviation of the random digits was found in Chapter 5. (See Exercise 5.20, p. 282.)

SOLUTION

STEP 1 H_0: $\mu = 4.5$ (mean is 4.5)

   H_a: $\mu \neq 4.5$ (mean is not 4.5)

STEP 2 $\alpha = 0.10$ (See the following figure.)

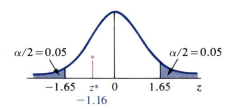

STEP 3   The following random sample was drawn from Table 1 of Appendix F.

2	8	2	1	5	5	4	0	9	1
0	4	6	1	5	1	1	3	8	0
3	6	8	4	8	6	8	9	5	0
1	4	1	2	1	7	1	7	9	3

$\sum x = 159$

$n = 40$

$\bar{x} = 3.975$

$$z = \frac{\bar{x} - \mu}{\sigma/\sqrt{n}} = \frac{3.975 - 4.50}{2.87/\sqrt{40}} = \frac{-0.525}{0.454} = -1.156$$

$z^* = -1.16$

z^* falls in the noncritical region, as shown in color on the diagram in Step 2.

STEP 4   Decision Fail to reject H_0.

Conclusion   The observed sample mean is not significantly different from 4.5 at the 0.10 level of significance.   ▲▲

Suppose that we were to take another sample of size 40 from the table of random digits. Would we obtain the same results? Suppose that we took a third sample or a fourth? What results might we expect? What is the level of significance α? Yes, its value is 0.10, but what does it measure? Table 8-3 lists the means obtained from 10 different random samples of size 40 that were taken from Table 1. The calculated value of z that corresponds to each $\bar{x}$ and the decision each would dictate are also listed. Each of the 10 calculated z-scores is shown in Figure 8-1. The sample number is used to identify each score. Note that one of the samples caused us to reject the null hypothesis, although we know that it is true for this situation. Why did this happen?

TABLE 8-3
Random Samples of Size 40 Taken from Table 1, Appendix F

Sample Number	Sample Mean ($\bar{x}$)	Calculated z (z^*)	Decision Reached
1	4.62	+0.26	Fail to reject H_0
2	4.55	+0.11	Fail to reject H_0
3	4.08	−0.93	Fail to reject H_0
4	5.00	+1.10	Fail to reject H_0
5	4.30	−0.44	Fail to reject H_0
6	3.65	−1.87	Reject H_0
7	4.60	+0.22	Fail to reject H_0
8	4.15	−0.77	Fail to reject H_0
9	5.05	+1.21	Fail to reject H_0
10	4.80	+0.66	Fail to reject H_0

REMEMBER α is the probability that we "reject H_0" when it is actually a true statement. Therefore, we can anticipate that a type I error will occur α of the time when testing a true null hypothesis.

FIGURE 8-1
z-Scores from Table 8-3; $\alpha = 0.10$

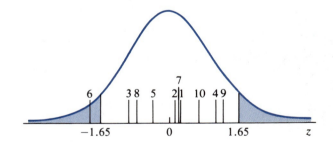

▼▲ EXERCISES

8.21 State the null hypothesis H_0 and the alternative hypothesis H_a that would be used for a hypothesis test related to each of the following statements:
   a. The mean age of the students enrolled in evening classes at a certain college is greater than 26 years.
   b. The mean weight of packages shipped on Air Express during the last month was less than 36.7 lb.
   c. The mean life of fluorescent light bulbs is at least 1600 hr.
   d. The mean weight of college football players is no more than 210 lb.

e. The mean distance from a home in suburban Chicago to the nearest fire station is less than 4.7 mi.

f. The mean daily high temperature during February at Daytona Beach, Florida, is not 80 degrees.

g. The mean grade (measure of steepness) for all ski runs at Greek Peak Ski Center is no less than 20 degrees.

h. The mean strength of welds by a new process is different from 570 lb per unit area, the mean strength of welds by the old process.

8.22 State the null hypothesis, H_0, and the alternative hypothesis, H_a, that would be used for a hypothesis test related to each of the following statements:

a. The mean age of the youths who hang out at the mall is less than 16 years.

b. The mean height of professional basketball players is more than 6 ft 6 in.

c. The mean elevation drop for ski trails at Eastern ski centers is at least 285 ft.

d. The mean diameter of the rivets is no more than 0.375 in.

e. The mean cholesterol level of male college students is different from 200 units.

8.23 Suppose that we want to test the hypothesis that the mean hourly charge for automobile repairs is at least $50 per hour at the repair shops in a nearby city. Explain the conditions that would exist if we make an error in decision by committing a

a. type I error   b. type II error

8.24 Suppose you want to test the hypothesis that the mean salt content of frozen "lite" dinners is more than 350 mg per serving.

a. What is the null hypothesis for your test?

b. Explain the conditions that would exist if your decision resulted in a type I error.

c. Explain the conditions that would exist if your decision resulted in a type II error.

8.25 Determine the test criteria (critical values and critical region for z) that would be used to test the null hypothesis at the given level of significance, as described in each of the following:

a. $H_0: \mu = 20 \quad (\alpha = 0.10)$
$H_a: \mu \neq 20$

b. $H_0: \mu = 24 \ (\leq) \quad (\alpha = 0.01)$
$H_a: \mu > 24$

c. $H_0: \mu = 10.5 \ (\geq) \quad (\alpha = 0.05)$
$H_a: \mu < 10.5$

d. $H_0: \mu = 35 \quad (\alpha = 0.01)$
$H_a: \mu \neq 35$

e. $H_0: \mu = 14.6 \ (\geq) \quad (\alpha = 0.02)$
$H_a: \mu < 14.6$

f. $H_0: \mu = 6.78 \ (\leq) \quad (\alpha = 0.10)$
$H_a: \mu > 6.78$

8.26 Determine the test criteria used to test the following null hypotheses:
  a. $H_0: \mu = 55 \;(\geq)\quad (\alpha = 0.02)$
  $H_a: \mu < 55$
  b. $H_0: \mu = -86 \;(\geq)\quad (\alpha = 0.01)$
  $H_a: \mu < -86$
  c. $H_0: \mu = 107 \quad (\alpha = 0.05)$
  $H_a: \mu \neq 107$
  d. $H_0: \mu = 17.4 \;(\leq)\quad (\alpha = 0.10)$
  $H_a: \mu > 17.4$

8.27 The calculated value of the test statistic is actually the number of standard errors that the sample mean differs from the hypothesized value of μ in the null hypothesis. Suppose that the null hypothesis is $H_0: \mu = 4.5$, σ is known to be 1.0, and a sample of size 100 results in $\bar{x} = 4.8$.
  a. How many standard errors is $\bar{x}$ above 4.5?
  b. If the alternative hypothesis is $H_a: \mu > 4.5$ and $\alpha = 0.01$, would you reject H_0?

8.28 Consider the hypothesis test where the hypotheses are

$$H_0: \mu = 26.4$$

$$H_a: \mu < 26.4$$

A sample of size 64 is randomly selected and yields a sample mean of 23.6.
  a. If it is known that $\sigma = 12$, how many standard errors below $\mu = 26.4$ is the sample mean, $\bar{x} = 23.6$?
  b. If $\alpha = 0.05$, would you reject H_0? Explain.

8.29 A machine cuts product parts that have a mean length equal to 15.0 cm with a standard deviation of 0.5 cm. It is known that the standard deviation has not changed, but a change in mean length is possible. A sample of 30 parts is taken with the following results:

15.1	15.6	15.8	16.0	15.0	15.5
16.2	15.0	15.5	16.0	15.9	14.8
16.0	15.5	15.6	15.5	16.4	15.2
15.8	16.0	15.5	14.9	15.9	16.2
15.9	15.8	15.0	16.1	15.7	15.6

A MINITAB analysis of the data follows.
  a. Verify the mean, standard deviation, standard error of the mean, and the z value shown.
  b. State H_0 and H_a.

Section 8.2 ▼ THE HYPOTHESIS TEST (A CLASSICAL APPROACH)

```
MTB > SET DATA IN C1
DATA> 15.1 15.6 15.8 16.0 15.0 15.5 16.2 15.0 15.5 16.0
DATA> 15.9 14.8 16.0 15.5 15.6 15.5 16.4 15.2 15.8 16.0
DATA> 15.5 14.9 15.9 16.2 15.9 15.8 15.0 16.1 15.7 15.6
DATA> END DATA
MTB > ZTEST OF MU = 15.0,SIGMA = .5,DATA IN C1
TEST OF MU = 15.0000 VS MU N.E. 15.0000
THE ASSUMED SIGMA = .0500
              N      MEAN     STDEV    SE MEAN      Z
C1           30    15.6333    0.4270    0.0913     6.94
MTB>STOP
```

8.30 "Obesity raises heart-attack risk" according to a study published in the March 1990 issue of the *New England Journal of Medicine*. "Those about 15 to 25 percent above desirable weight had twice the heart disease rate." Suppose the data listed below are the percentages above desired weight for a sample of patients involved in a similar study.

18.3	19.7	22.1	19.2	17.5	12.7	22.0
21.1	16.2	15.4	19.9	21.5	19.8	22.5
13.0	22.1	27.7	17.9	22.2	19.7	18.1
17.3	13.3	22.1	16.3	21.9	16.9	15.4

Use a computer to test the null hypothesis, $\mu = 18\%$, versus the alternative hypothesis, $\mu \neq 18\%$. Assume $\sigma = 4\%$ and use $\alpha = 0.05$.

8.31 A population has a standard deviation equal to 3.5 and the hypothesis H_0: $\mu = 20.0$ is to be tested against the hypothesis H_a: $\mu > 20.0$. α is specified to be 0.05.
  a. What is the critical value of the test statistic z?
  b. What sample mean value corresponds to the critical value found in part (a) for a sample of size 50?

8.32 A sample of 36 measurements was taken to test the hypothesis stated in Illustration 8-1. Does the sample mean of 5.3 provide sufficient evidence to reject the null hypothesis in the illustration? Complete the hypothesis test, using $\sigma = 1.8$ and $\alpha = 0.05$.

8.33 A machine produces ball bearings. The standard deviation remains constant at 0.010 cm. However, the mean diameter can change after the machine has been used for some time. In particular, the mean increases if it changes. A daily sample of 30 bearings is taken to test H_0: $\mu = 3.000$ cm versus H_a: $\mu > 3.000$ cm. For a given sample we find $\bar{x} = 3.003$. Test the hypothesis and state your conclusion for $\alpha = 0.01$.

8.34 An article entitled "The Trouble with Margarine" (*Consumer Reports,* March 1991, pages 196–197) discusses a Dutch research study which shows that trans-fatty acids tend to cause a rise in LDL cholesterol (which increases the risk of coronary heart disease) and tend to cause a lowering of HDL cholesterol (which lowers the risk of coronary heart disease). Margarine is much higher in trans-fatty acids than butter. The estimated U.S. intake of trans-fatty acids is 8 g per day. Consider a research project involving 150 individuals in which their daily intake of trans-fatty acids was measured. Suppose the sample mean intake was 12.5 g. Assuming that $\sigma = 8.0$, test the research hypothesis that $\mu > 8$ at $\alpha = 0.05$.

8.35 The manager at Air Express feels that the weights of packages shipped recently are less than in the past. Records show that in the past packages have had a mean weight of 36.7 lb and a standard deviation of 14.2 lb. A random sample of last month's shipping records yielded a mean weight of 32.1 lb for 64 packages. Is this sufficient evidence to reject the null hypothesis in favor of the manager's claim? Use $\alpha = 0.01$.

8.36 A fire insurance company felt that the mean distance from a home to the nearest fire department in a suburb of Chicago was at least 4.7 mi. It set its fire insurance rates accordingly. Members of the community set out to show that the mean distance was less than 4.7 mi. This, they felt, would convince the insurance company to lower its rates. They randomly identified 64 homes and measured the distance to the nearest fire department for each. The resulting sample mean was 4.4. If $\sigma = 2.4$ mi, does the sample show sufficient evidence to support the community's claim at the $\alpha = 0.05$ level of significance?

8.37 On a popular self-image test the mean score for public assistance recipients is 65 and the standard deviation is 5. A random sample of 42 public assistance recipients in Emerson County are given the test. They achieved a mean score of 60. Do the scores of Emerson County public assistance recipients differ from the average with respect to this variable? Use $\alpha = 0.01$.

8.38 A discussion in the drug research section of the magazine *Pharmacist of Tomorrow* (June 1991, page 39) concludes that calcium channel blockers are more effective than either beta blockers or angiotensin converting enzyme (ACE) inhibitors in treating African-Americans with mild to moderate hypertension. Suppose 350 African-Americans with mild to moderate hypertension were treated with calcium channel blockers and that the mean drop in blood pressure was 15.5 units. Test, at $\alpha = 0.05$, the following hypotheses:

$$H_0: \mu = 10 \text{ units vs. } H_a: \mu > 10 \text{ units}$$

(Assume that σ is 12.)

8.39 A manufacturer of automobile tires believes it has developed a new rubber compound that has superior wearing qualities. They produced a test run of tires made with this new compound and had them road tested. The data recorded was the amount of tread wear per 10,000 miles. In the past, the mean amount of tread wear per 10,000 miles, for tires of this quality, has been 0.0625 inches.

The null hypothesis to be tested here is "the mean amount of wear on the tires made with the new compound is the same mean amount of wear with the old compound, 0.0625 inches per 10,000 miles," H_0: $\mu = 0.0625$. There are three possible alternative hypotheses that could be used: (1) H_a: $\mu < 0.0625$, (2) H_a: $\mu \neq 0.0625$, (3) H_a: $\mu > 0.0625$.

  a. Explain the meaning of each of these three alternatives.
  b. Which one of the possible alternative hypotheses should the manufacturer use if it hopes to conclude that "use of the new compound does yield superior wear"?

8.40 All drugs must be approved by the Food and Drug Administration (FDA) before they can be marketed by a drug manufacturer. The FDA must weigh the error of marketing an ineffective drug, with the usual risks of side effects, against the consequences of not allowing an effective drug to be sold. Suppose, using standard medical treatment, that the mortality rate (r) of a certain disease is known to be a. A manufacturer submits for approval a drug that is supposed to treat this disease. The FDA sets up the hypothesis to test the mortality rate for the drug as (1) H_0: $r = a$, H_a: $r < a$, $\alpha = 0.005$; or (2) H_0: $r = a$, H_a: $r > a$, $\alpha = 0.005$.

  a. If $a = 0.95$, which test do you think the FDA would use?
  b. If $a = 0.05$, which test do you think the FDA would use? Explain.

8.3 ▼ The Hypothesis Test (A Probability-Value Approach)

The classical (or traditional) hypothesis testing procedure was described in Section 8.2. An alternative approach to the decision-making process in hypothesis testing has gained popularity in recent years, largely as a result of the convenience and the "number crunching" ability of computers. This alternative process is to calculate and report a *probability-value* related to the observed sample statistic. The value reported, *p*-value, is called the *probability-value*, the *prob-value*, or simply the *p-value*. The *p*-value is then compared to the level of significance α, the probability of the type I error, in order to make the decision to reject H_0 or fail to reject H_0.

prob-value

PROB-VALUE, *p*-VALUE

The prob-value, *p*-value, of a hypothesis test is the smallest level of significance for which the observed sample information becomes significant, provided the null hypothesis is true. The prob-value is computed by finding the probability that the test statistic could be the value it is or a more extreme value (in the direction of the alternative hypothesis) when the null hypothesis is true. (*Note*: The symbol **P** is often used to represent prob-value, especially in algebraic situations.)

THE PROB-VALUE HYPOTHESIS TEST: A FIVE-STEP MODEL

Step 1: State the null hypothesis (H_0) and the alternative hypothesis (H_a).
Step 2: Determine the level of significance, α, to be used.

Step 3: Collect and present the sample evidence.
   a. Collect the sample information.
   b. Calculate the value of the observed test statistic.
Step 4: Calculate the prob-value.
Step 5: Determine the results.
   a. Compare the calculated prob-value to the level of significance, α, from Step 2.
   b. Make a decision about H_0.
   c. State the conclusion about H_a.

Let's return to Illustration 8-4 and see how its solution would be different if solved using this "prob-value" method.

▼ ILLUSTRATION 8-8

For many semesters an instructor has recorded students' grades, and the mean μ for all these students' grades is 72. The current class of 36 students seems to be better than average in ability and the instructor wants to show that, according to their average, "the current class is superior to previous classes." Does the class mean $\bar{x}$ of 75.2 present sufficient evidence to support the instructor's claim that the current class is superior? Use $\alpha = 0.05$ and $\sigma = 12.0$.

SOLUTION To be superior, this class must have a mean grade that is higher than the mean of all the previous classes. To be "equal to or less than" would not be superior.

STEP 1 H_0: $\mu = 72$ ($\leq$)   (class is not superior)

H_a: $\mu > 72$ (class is superior)

Notice that Step 1 is exactly the same when using either the classical or the prob-value approach to hypothesis testing. (See Illustration 8-4, p. 395.)

STEP 2 Determine α, the probability of the type I error; $\alpha = 0.05$.

STEP 3 Using formula (7-2) and the sample information, we will obtain the calculated value of the test statistic z^*.

$$z = \frac{\bar{x} - \mu}{\sigma/\sqrt{n}} \quad \text{and} \quad \bar{x} = 75.2 \quad \text{and} \quad n = 36$$

RECALL We assume that μ was equal to 72 in the null hypothesis and that $\sigma = 12.0$ was known.

$$z = \frac{75.2 - 72.0}{12.0/\sqrt{36}} = \frac{3.2}{2.0} = 1.60$$

$$z^* = 1.60$$

(REMINDER We use an asterisk, *, to identify the calculated value of the test statistic; the asterisk will also be used to locate its value on the probability distribution in the next step.)

Section 8.3 ▼ THE HYPOTHESIS TEST (A PROBABILITY-VALUE APPROACH)

Notice that Step 3 in the prob-value approach is exactly the same as Step 3 in the classical approach.

STEP 4 Draw a sketch of the probability distribution for the test statistic — z in this case.

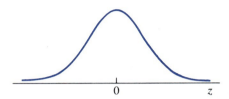

Next locate the value of z^* calculated in Step 3 on this distribution (see the following figure) and determine what part of the distribution relative to z^* represents the prob-value. Then calculate the p-value.

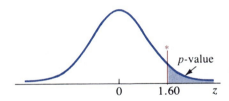

The alternative hypothesis indicates that we are interested in that part of the probability distribution that lies to the right of z^*, since the "greater than" sign was used.

p-value $= P(z > z^*) = P(z > 1.60)$

$\qquad = 0.5000 - 0.4452$ (use Table 5 in Appendix F to find the area under a normal curve)

p-value $= \mathbf{0.0548}$

The results of the statistical analysis are then reported.

STEP 5 The prob-value for this hypothesis test is 0.0548. If the decision is made using the 0.05 level of significance (or any α value less than or equal to 0.0548), then the decision will be to "fail to reject the null hypothesis" and the conclusion will be "there is no evidence to show the class as superior." If, however, we had chosen to use a level of significance of 0.10 (or any α value greater than 0.0548), our decision would be to "reject H_0," and our conclusion would be that "the current class is superior" at that level of significance. ▲▲

COMPUTER SOLUTION

MINITAB Printout for Illustration 8-8

Information given to the computer: z-test of H_0: $\mu = 72.0$, H_a: $\mu > 72.0$, $\sigma = 12.0$, and location of sample data

Sample statistics: $n = 36$, $\bar{x} = 75.2$, $s = 11.87$

The calculated z^*
The prob-value
The decision

```
MTB > ZTEST OF MU = 72 SIGMA = 12 DATA IN C1;
SUBC> ALTERNATIVE = 1.

TEST OF MU = 72.00 VS MU G.T. 72.00
THE ASSUMED SIGMA = 12.0

         N     MEAN    STDEV   SE MEAN     Z    P VALUE
C1      36     75.2    11.87     2.00    1.60    0.0548
```

Before looking at another illustration, let's summarize the details for this prob-value approach to hypothesis testing.

1. The null and alternative hypotheses are formulated in the same manner as that used in the classical approach.
2. Determine the level of significance, α, to be used.
3. The value of the test statistic is calculated in Step 3 in exactly the same manner as it was calculated in Step 3 in the classical approach.
4. The prob-value is the area that represents values of z that are more extreme than z^* under the curve of the probability distribution. There are three separate cases: two are one-tailed and one is two-tailed. The direction (or sign) in the alternative hypothesis is the key.
   ▼ Case I   If H_a is one-tailed to the right (that is, contains ">"), then p-value = $P(z > z^*)$, the area to the right of z^*.
   ▼ Case II   If H_a is one-tailed to the left (that is, contains "<"), then p-value = $P(z < z^*)$, the area to the left of z^*.
   ▼ Case III If H_a is two-tailed (that is, contains "$\neq$"), then p-value = $P(z < -|z^*|) + P(z > |z^*|)$, the sum of the area to the left of the negative value of z^* and the area to the right of the positive value of z^*. Since both areas are equal, you will probably find one and double it. Thus, p-value = $2 \times P(z > |z^*|)$.
5. The decision will be made by comparing the p-value to the previously established value of α.

   decision rule

   a. If the *calculated prob-value is less than or equal to the desired α*, then the decision must be *reject H_0*.
   b. If the *calculated prob-value is greater than the desired α*, then the decision must be *fail to reject H_0*.
6. Conclusions should be worded in the same manner as previously instructed.

Let's look at an illustration involving the two-tailed procedure.

▼ ILLUSTRATION 8-9

Many of the large companies in a certain city have for years used the Kelley Employment Agency for testing prospective employees. The employment selection test used has historically resulted in scores distributed about a mean of 82

Section 8.3 ▼ THE HYPOTHESIS TEST (A PROBABILITY-VALUE APPROACH)

and a standard deviation of 8. The Brown Agency has developed a new test that is quicker and easier to administer and therefore less expensive. Brown claims that their test results are the same as those obtained on the Kelley test. Many of the companies are considering a change from the Kelley Agency to the Brown Agency in order to cut costs. However, they are unwilling to make the change if the Brown test results have a different mean value.

An independent testing firm tested 36 prospective employees. A sample mean of 80 resulted. Determine the prob-value associated with this hypothesis test.

SOLUTION The Brown Agency's test results will be different if the mean test score is not equal to 82. They will be the same if the mean is 82. Therefore,

STEP 1 $H_0: \mu = 82$ (test results have same mean)

          $H_a: \mu \neq 82$ (test results have different mean)

STEP 2   Step 2 is omitted when a question asks for the prob-value and not a decision.

STEP 3   The sample information $n = 36$ and $\bar{x} = 80$ and formula (7-2) are used to calculate z^*:

$$z = \frac{\bar{x} - \mu}{\sigma/\sqrt{n}}$$

$$z = \frac{80 - 82}{8/\sqrt{36}} = \frac{-2}{8/6} = -1.50$$

$$z^* = -1.50$$

STEP 4   The value of z^* is located on the normal distribution. (See the accompanying figure.)

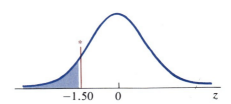

Since the alternative hypothesis indicates a two-tailed test, we must find the probability associated with two areas, namely, $P(z < -|z^*|)$ and $P(z > |z^*|)$. And since $z^* = -1.50$, the value of $|z^*| = 1.50$. Thus the p-value = $P(z < -1.50) + P(z > 1.50)$, as shown in Figure 8-2 (p. 412).

$$p\text{-value} = P(z < -1.50) + P(z > 1.50)$$
$$= (0.5000 - 0.4332) + (0.5000 - 0.4332)$$
$$= 0.0668 + 0.0668$$

p-value = **0.1336**

FIGURE 8-2

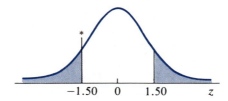

The prob-value for this hypothesis test is 0.1336. Each individual company now will make a decision whether to (a) continue to use Kelley's services or (b) change to the Brown Agency. Each will need to establish the level of significance that best fits its own situation and then make a decision using the decision rule described in this illustration. ▲▲

▼▲ EXERCISES

8.41 Calculate the prob-value for each of the following:
  a. $H_0: \mu = 10 \quad z^* = 1.48$
     $H_a: \mu > 10$
  b. $H_0: \mu = 105 \quad z^* = -0.85$
     $H_a: \mu < 105$
  c. $H_0: \mu = 13.4 \quad z^* = 1.17$
     $H_a: \mu \neq 13.4$
  d. $H_0: \mu = 8.56 \quad z^* = -2.11$
     $H_a: \mu < 8.56$
  e. $H_0: \mu = 110 \quad z^* = -0.93$
     $H_a: \mu \neq 110$
  f. $H_0: \mu = 54.2 \quad z^* = 0.46$
     $H_a: \mu > 54.2$

8.42 Calculate the prob-value for each of the following:
  a. $H_0: \mu = 20 \quad (\bar{x} = 17.8, \sigma = 9, n = 36)$
     $H_a: \mu < 20$
  b. $H_0: \mu = 78.5 \quad (\bar{x} = 79.8, \sigma = 15, n = 100)$
     $H_a: \mu > 78.5$
  c. $H_0: \mu = 1.587 \quad (\bar{x} = 1.602, \sigma = 0.15, n = 50)$
     $H_a: \mu \neq 1.587$

8.43 Find the value of z^* for each of the following:
  a. $H_0: \mu = 35$ versus $H_a: \mu > 35$ when p-value $= 0.0582$
  b. $H_0: \mu = 35$ versus $H_a: \mu < 35$ when p-value $= 0.0166$
  c. $H_0: \mu = 35$ versus $H_a: \mu \neq 35$ when p-value $= 0.0042$

8.44 The null hypothesis, $H_0: \mu = 48$, was tested against the alternative hypothesis, $H_a: \mu > 48$. A sample of 75 resulted in a calculated p-value of 0.102. If $\sigma = 3.5$, find the value of the sample mean, $\bar{x}$.

8.45 The calculated prob-value for a hypothesis test is p-value $= 0.084$. What decision about the null hypothesis would occur if
  a. the hypothesis test is completed at the 0.05 level of significance?
  b. the hypothesis test is completed at the 0.10 level of significance?

8.46 a. A one-tailed hypothesis test is to be completed at the 0.05 level of significance. What calculated values of p will cause a rejection of H_0?
b. A two-tailed hypothesis test is to be completed at the 0.02 level of significance. What calculated values of p will cause a "fail to reject H_0" decision?

8.47 A paint manufacturer knows from experience that the coverages for its brand of paint are normally distributed with a mean equal to 440 sq ft/gal and a standard deviation equal to 15 sq ft. The addition of an additive is known not to affect the variability of coverage, but may increase the mean coverage. Ten randomly selected gallons of the paint (with the additive) are tested and the following coverages obtained.

435 450 455 460 465 440 452 461 470 440

The following MINITAB output tests the hypothesis H_0: $\mu = 440$ versus H_a: $\mu > 440$.

a. Verify the sample mean, standard deviation, standard error of the mean, calculated z, z*, and p-value.
b. State H_0 and H_a.

```
MTB > SET DATA IN C1
DATA> 435 450 455 460 465 440 452 461 470 440
DATA> END DATA
MTB > ZTEST OF PAINT MU = 440   SIGMA = 15  DATA IN C1;
SUBC> ALTERNATIVE = 1.
TEST OF MU = 440.00 VS MU G.T. 440.00
THE ASSUMED SIGMA = 15.0
            N     MEAN    STDEV   SE MEAN    Z    P VALUE
C1         10    452.80   11.65    4.74    2.70   0.0035
MTB > STOP
```

8.48 For the last 11 years the August unemployment rates have been
7.7 7.4 9.8 9.5 7.5 7.1 6.9 6.0 5.6 5.3 5.6
Use a computer to test the hypothesis H_0: $\mu = 6.6$ against the alternative H_a: $\mu > 6.6$. Assume that unemployment rates are normally distributed with $\sigma = 1.5$.

8.49 The following MINITAB output was used to complete a hypothesis test:

```
MTB > ZTEST OF MU = 525  SIGMA = 60   DATA IN C1;
SUBC> ALTERNATIVE = -1.
TEST OF MU = 525.00 VS MU L.T. 525.00
THE ASSUMED SIGMA = 60.0
      N    MEAN    STDEV    SE MEAN    Z     P VALUE
     38   512.14   64.78    9.733    -1.32   0.093
```

a. State the null and alternative hypotheses.
b. If the test is completed using $\alpha = 0.05$, what decision and conclusion are reached?
c. Verify the value of the standard error of the mean.

8.50 The following MINITAB output was used to complete a hypothesis test.

```
MTB > ZTEST OF MU = 6.25  SIGMA = 1.4  DATA IN C1;
SUBC> ALTERNATIVE = 0.
TEST OF MU = 6.250 VS MU N.E. 6.250
THE ASSUMED SIGMA = 1.40
       N      MEAN    STDEV    SE MEAN    Z      P VALUE
      78     6.596    1.273     0.1585   2.18    0.029
```

a. State the null and alternative hypotheses.
b. If the test is completed using $\alpha = 0.05$, what decision and conclusion are reached?
c. Verify the value of the standard error of the mean.
d. Find the values for Σx and Σx^2.

8.51 An economist claims that when the Dow-Jones average increases, the volume of shares traded on the New York Stock Exchange tends to increase. Over the past two years the daily volume on the exchange has averaged 21.5 million shares and had a standard deviation of 2.5 million. A random sample of 64 days on which the Dow-Jones average increased was selected and the average daily volume was computed. The sample average was 22 million. Calculate the prob-value for this hypothesis test.

8.52 The marketing director of A & B Cola is worried that the product is not attracting enough young consumers. To test this hypothesis, she randomly surveys 100 A & B Cola consumers. The mean age of an individual in the community is 32 years and the standard deviation is 10 years. The surveyed consumers of A & B Cola have a mean age of 35. At the 0.01 level of significance, is this sufficient evidence to conclude that A & B Cola consumers are, on the average, older than the average person living in the community? Complete this hypothesis test using the prob-value approach.

8.53 An article entitled "Too Many Cesareans" appeared in *Consumer Reports* (February 1991, pages 120–126). Figures from 1988 indicated only 12.6 vaginal births per 100 women with a previous cesarean. Many experts believe that at least half of the women with prior cesareans could have a successful vaginal birth. Consider a study involving 125 hospitals. The average number per 100 women of vaginal births, following a previous cesarean, was 17.3. From prior experience, the standard deviation, σ, is believed to be 5. Calculate the value of the test statistic, z^*, and the prob-value for the hypothesis test of $H_0: \mu = 12.6$ vs. $H_a: \mu > 12.6$.

8.54 According to an article in *Good Housekeeping* (February 1991, page 98), a 128-lb woman who walks for 30 minutes four times a week at a steady, four-mi/hr pace can lose up to 10 pounds over a span of a year. Suppose 50 women with weights between 125 and 130 lb performed the four walks per week for a year and at the end of the year the average weight loss for the 50 was 9.1 lb. Assuming that the standard deviation, σ, is 5, calculate the value of the test statistic, z^*, and the prob-value for the hypothesis test of $H_0: \mu = 10.0$ vs. $H_a: \mu \neq 10.0$.

8.4 ▼ Estimation

estimation

Estimation is the second type of statistical inference. It is the procedure to use when answering a question that asks for the value of a population parameter. For example, "What is the mean one-way distance that students at our college commute daily?"

If you needed to answer this question, you might take a sample from the population and calculate the sample mean $\bar{x}$. Suppose that you did draw a random sample of 100 one-way distances, and a sample mean of 10.22 mi resulted. What is your estimate for the mean value of the population? If you report the sample mean $\bar{x}$ as your estimate, you will be making a point estimate.

point estimate

> **POINT ESTIMATE FOR A PARAMETER**
> The value of the corresponding sample statistic.

That is, the sample mean, $\bar{x} = 10.22$ mi, is the best point estimate for the mean one-way distance for this population. In other words, the mean one-way distance for all students is estimated to be 10.22 mi. We are not really implying that the population mean μ is exactly 10.22 mi. Rather, we intend this point estimate to be interpreted to say "μ is close to 10.22." From our study so far, you probably realize that when we say $\mu = 10.22$, there is very little chance that the statement is true. In previous chapters we drew samples from probability distributions where μ was known, and seldom did we obtain a sample mean exactly equal in value to μ. Then when should we expect a single sample to yield an $\bar{x}$ equal to μ when μ is unknown? We shouldn't.

What does it mean to say "μ is close to 10.22 mi"? Closeness is a relative term, but perhaps in this case "close" might be arbitrarily defined to be "within 1 mi" of μ. If 1 mi satisfies our intuitive idea of closeness, then to say "μ is close to 10.22" is comparable to saying "μ is between 9.22 (10.22 − 1.0) and 11.22 (10.22 + 1.0)." This suggests that we might make estimations by using intervals. The confidence interval estimate employs this interval concept and assigns a measure to the interval's reliability in estimating the parameter in question.

interval estimate

> **INTERVAL ESTIMATE**
> An interval bounded by two values and used to estimate the value of a population parameter. The values that bound this interval are statistics calculated from the sample that is being used as the basis for the estimation.

level of confidence

> **LEVEL OF CONFIDENCE 1 − α**
> The probability that the sample to be selected yields boundary values that lie on opposite sides of the parameter being estimated. The level of confidence is sometimes called the *confidence coefficient*.

confidence interval

> **CONFIDENCE INTERVAL**
> An interval estimate with a specified level of confidence.

The central limit theorem is the source for the information needed to construct confidence interval estimates for the mean. Our sample mean ($\bar{x} = 10.22$) is a member of the sampling distribution of sample means. The CLT describes this distribution as being approximately normally distributed with a mean $\mu_{\bar{x}} = \mu$ and a standard deviation $\sigma_{\bar{x}} = \sigma/\sqrt{n}$. The population mean is unknown; however, it exists and is constant. Let's assume that the standard deviation of the population is $\sigma = 6$. Figure 8-3 shows the sampling distribution of sample means to which our sample mean ($\bar{x} = 10.22$ and $n = 100$) belongs.

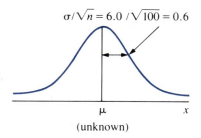

FIGURE 8-3
Sampling Distribution, $n = 100$

RECALL In this chapter we are studying the inferences about μ under the assumption that σ is known. This assumption is for convenience, and the restriction will be removed in Chapter 9.

When a sample of size 100 is randomly selected from a population whose mean is μ and whose standard deviation is 6, what is the probability that the sample mean is within one unit of μ? In other words, what is $P(\mu - 1 < \bar{x} < \mu + 1)$? See Figure 8-4. This probability may be found by using the normal probability distribution (Table 5, Appendix F) and formula (7-2):

$$z = \frac{\bar{x} - \mu}{\sigma/\sqrt{n}}$$

If $\bar{x} = \mu - 1$,

$$z = \frac{(\mu - 1) - \mu}{0.6} = \frac{-1.0}{0.6} = -1.67$$

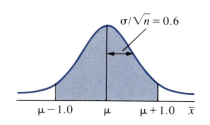

FIGURE 8-4
Probability That $\bar{x}$ Is Within 1 Unit of μ

Section 8.4 ▼ ESTIMATION

If $\bar{x} = \mu + 1$,

$$z = \frac{(\mu + 1) - \mu}{0.6} = \frac{+1.0}{0.6} = +1.67$$

Therefore,

$$P(\mu - 1 < \bar{x} < \mu + 1) = P(-1.67 < z < +1.67)$$
$$= 2 \cdot P(0 < z < 1.67)$$
$$= 2(0.4525) = \mathbf{0.9050}$$

The probability that the mean of a random sample is within one unit of this population mean is 0.9050. Thus the probability that the population mean is within one unit of the mean of a sample is also 0.9050. Therefore, the interval 9.22 to 11.22 is a 0.9050 confidence interval estimate for the mean one-way distance commuted by students at our college.

maximum error of estimate

> **MAXIMUM ERROR OF ESTIMATE, E**
>
> One-half the width of the confidence interval. In general, E is a multiple of the standard error.

The preceding illustration started with the maximum error being assigned the value of 1 mi. Typically, the maximum error of estimate is determined by the level of confidence that we want our confidence interval to have. That is, $1 - \alpha$ will determine the maximum error. The level of confidence will be split so that half of $1 - \alpha$ is above the mean and half below the mean (Figure 8-5). This will leave $\alpha/2$ as the probability in each of the two tails of the distribution. The z-score at the boundary of the confidence interval will be z of $z(\alpha/2)$. [Recall that $\alpha/2$ is the probability (or area) under the curve to the right of this point. See Chapter 6, pp. 325ff.] The standard score z is a number of standard deviations. Therefore, the maximum error of estimate E is

$$E = z(\alpha/2) \cdot \frac{\sigma}{\sqrt{n}} \tag{8-1}$$

See Figure 8-6 (p. 418).

FIGURE 8-5
Each Tail Contains $\alpha/2$

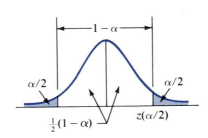

FIGURE 8-6
Maximum Error
of Estimate

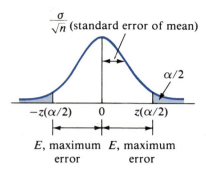

The **1 − α confidence interval** for μ is

$$\bar{x} - z(\alpha/2) \cdot \frac{\sigma}{\sqrt{n}} \quad \text{to} \quad \bar{x} + z(\alpha/2) \cdot \frac{\sigma}{\sqrt{n}} \tag{8-2}$$

lower confidence limit (LCL)

upper confidence limit (UCL)

$\bar{x} - z(\alpha/2) \cdot (\sigma/\sqrt{n})$ is called the **lower confidence limit (LCL)**, and $\bar{x} + z(\alpha/2) \cdot (\sigma/\sqrt{n})$ is called the **upper confidence limit (UCL)** for the confidence interval.

▼ **ILLUSTRATION 8-10**

Construct the 0.95 confidence interval for the estimate of the mean one-way distance that the students commute.

SOLUTION

$$1 - \alpha = 0.95$$
$$\alpha = 0.05$$
$$\frac{\alpha}{2} = 0.025 \quad \text{(see the accompanying figure)}$$

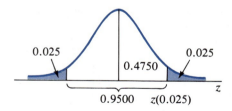

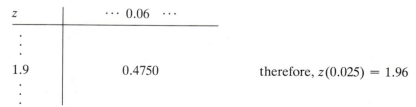

therefore, $z(0.025) = 1.96$

Section 8.4 ▼ ESTIMATION

$$\bar{x} \pm z(\alpha/2) \cdot \frac{\sigma}{\sqrt{n}}$$

$$10.22 \pm (1.96) \cdot \frac{6}{\sqrt{100}}$$

$$10.22 \pm (1.96)(0.6)$$

$$10.22 \pm 1.176$$

$$10.22 - 1.176 = 9.044 = \mathbf{9.04} \quad \text{and} \quad 10.22 + 1.176 = 11.396 = \mathbf{11.40}$$

Therefore, with 0.95 confidence we can say that the mean one-way distance is between 9.04 and 11.40, and we will write this as

(9.04, 11.40), the 0.95 confidence interval for μ   ▲▲

NOTE Many of the confidence interval formulas that you will be using will have a format similar to formula (8-2), namely, $\bar{x} - E$ to $\bar{x} + E$. However, perhaps the simplest way to handle the arithmetic involved is to think of the formula as $\bar{x} \pm E$, as was done in Illustration 8-10. Using this format you first calculate E. Then you calculate the lower bound by subtracting, and you calculate the upper bound by adding.

▼ ILLUSTRATION 8-11

To estimate the mean score on the first one-hour exam in statistics, we obtain a random sample of 38 exam scores. A sample mean of 74.3 is found. Construct the 0.98 confidence interval estimate for the mean of all first one-hour exam scores. Use $\sigma = 14$.

SOLUTION

$$1 - \alpha = 0.98$$

$$\alpha = 0.02$$

$$\frac{\alpha}{2} = 0.01 \text{ (see the accompanying figure)}$$

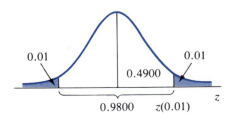

z	$\cdots$ 0.03 $\cdots$	
$\vdots$		
2.3	0.4901	(nearest table value to 0.4900); $z(0.01) = 2.33$
$\vdots$		

$$\bar{x} \pm z(\alpha/2) \cdot \frac{\sigma}{\sqrt{n}}$$

$$74.3 \pm (2.33) \cdot \frac{14}{\sqrt{38}}$$

$$74.3 \pm (2.33)(2.27)$$

$$74.3 \pm 5.29$$

$$74.3 - 5.29 = \mathbf{69.01} \qquad 74.3 + 5.29 = \mathbf{79.59}$$

(69.01, 79.59) is the 0.98 confidence interval for μ.

That is, with 0.98 confidence we can say that the mean exam score for all such exams is between 69.01 and 79.59. ▲▲

▼ ILLUSTRATION 8-12

Suppose that the sample in Illustration 8-7 had been obtained for the purpose of estimating the population mean μ. The sample results ($n = 40, \bar{x} = 3.975$) and $\sigma = 2.87$ are used in formula (8-2) to determine the 0.90 confidence interval.

$$\bar{x} \pm z(\alpha/2) \cdot \frac{\sigma}{\sqrt{n}}$$

$$3.975 \pm 1.65 \cdot \frac{2.87}{\sqrt{40}}$$

$$3.975 \pm (1.65)(0.454)$$

$$3.975 \pm 0.749$$

$$3.975 - 0.749 = 3.226 = \mathbf{3.23} \qquad 3.975 + 0.749 = 4.724 = \mathbf{4.72}$$

(3.23, 4.72) is the 0.90 confidence interval estimate for μ.

(See the accompanying figure.)

With 0.90 confidence, we think that μ is somewhere within this interval.

Section 8.4 ▼ ESTIMATION

Since this sample was taken from a known population—namely, the random digits—we can look back at our answer and say that the confidence interval is correct. That is, the true value of μ, 4.5, does fall within the confidence interval.

▲▲

Suppose that we were to take another sample of size 40; would we obtain the same results? Suppose we took a third and a fourth. What would happen? What is the level of confidence, $1 - \alpha$? Yes, it has the value 0.90, but what does that mean? Table 8-4 lists the means obtained from 10 different random samples of size 40 taken from Table 1. The 0.90 confidence interval for the estimate of μ based on each of these samples is also listed in Table 8-4. Since all the interval estimates were for the mean of the random digits ($\mu = 4.5$), we can inspect each confidence interval and see that 9 of the 10 contain μ and would therefore be considered correct estimates. One sample (sample 6) did not yield an interval that contains μ. This is shown in Figure 8-7.

TABLE 8-4
Samples of Size 40 Taken from Table 1, Appendix F

Sample Number	Sample Mean ($\bar{x}$)	0.90 Confidence Interval Estimate for μ
1	4.64	3.89 to 5.39
2	4.56	3.81 to 5.31
3	3.96	3.21 to 4.71
4	5.12	4.37 to 5.87
5	4.24	3.49 to 4.99
6	3.44	2.69 to 4.19
7	4.60	3.85 to 5.35
8	4.08	3.33 to 4.83
9	5.20	4.45 to 5.95
10	4.88	4.13 to 5.63

FIGURE 8-7
Interval Estimates from Table 8-4

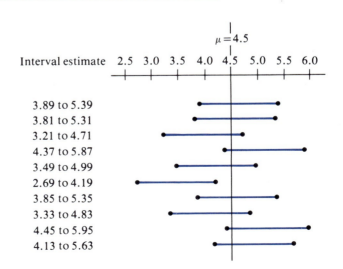

REMEMBER $1 - \alpha$ is the probability that we obtain a confidence interval such that μ is contained within it.

The maximum error formula can be used to determine the sample size required to obtain a $1 - \alpha$ confidence interval estimate with a specified maximum error.

▼ ILLUSTRATION 8-13

Determine the size of the sample that will be necessary to estimate the mean weight of second-grade boys if we want our estimate to be accurate to within 1 lb with 95% confidence. Assume that the standard deviation of such weights is 3 lb.

SOLUTION $1 - \alpha = 0.95$; therefore, $z(0.025) = 1.96$, as found by using Table 5, Appendix F. The maximum error $E = 1$ and $\sigma = 3$. Use formula (8-1).

$$E = z(\alpha/2) \cdot \frac{\sigma}{\sqrt{n}}$$

$$1.0 = 1.96 \cdot \frac{3}{\sqrt{n}}$$

$$1 = \frac{5.88}{\sqrt{n}}$$

$$\sqrt{n} = 5.88$$

$$n = 34.57$$

Therefore,

$$n = \mathbf{35}$$

NOTE When solving for the sample size n, all fractional (decimal) values are to be *rounded up* to the next larger integer. ▲▲

Does a sample size of 35 determine a maximum error of 1 pound in Illustration 8-13?

$$E = z(\alpha/2) \cdot \frac{\sigma}{\sqrt{n}} = 1.96 \cdot \frac{3.0}{\sqrt{35}} = (1.96)(0.51) = \mathbf{0.999}$$

This is a maximum error of just under 1.0.

The use of formula (8-1) can be made a little easier by rewriting the formula in a form that expresses n in terms of the other quantities:

$$n = \left[\frac{z(\alpha/2) \cdot \sigma}{E} \right]^2 \tag{8-3}$$

▼ ILLUSTRATION 8-14

A sample of what size would be needed to estimate the population mean to within $\frac{1}{5}$ of a standard deviation with 99% confidence?

SOLUTION $1 - \alpha = 0.99$; therefore, $\alpha/2 = 0.005$ and $z(0.005) = 2.58$, as found by using Table 5, Appendix F. The maximum error E is to be $\frac{1}{5}$ of σ; that is, $E = \sigma/5$. Using formula (8-3), we have

$$n = \left[\frac{z(\alpha/2) \cdot \sigma}{E}\right]^2 = \left[\frac{(2.58)(\sigma)}{\sigma/5}\right]^2$$
$$= \left[\frac{(2.58\sigma)(5)}{\sigma}\right]^2 = [(2.58)(5)]^2$$
$$= (12.90)^2 = 166.41 = \mathbf{167}$$

▲▲

Case Study 8-1

How Much Water U.S. Troops in Saudi Arabia Drink

Estimating how much water U.S. troops in Saudi Arabia drink has become a concern since the average soldier drinks much more than the average citizen at home. Supplying the soldiers with an adequate quantity of quality water is an even bigger problem. Based on the information in this graphic, what is the point estimate for the mean number of 8-oz glasses of water drunk per day per soldier? How much water will 1000 soldiers drink in a week? (See Exercise 8.56.)

Source: Copyright 1990, USA TODAY. Reprinted with permission.

▼▲ EXERCISES

8.55 Discuss the effect that each of the following have on the width of a confidence interval:
   a. level of confidence   b. sample size
   c. variability of the characteristic being measured

8.56 The graphics in Case Study 8-1 show that the soldiers are drinking much more than the average citizen at home.
   a. What is the point estimate for the average amount of water drunk by the citizens at home?
   b. What is the point estimate for the average amount of water drunk by soldiers in Saudi Arabia?
   c. How much water will 1000 soldiers drink in a week?
   d. Why are answers (a), (b), and (c) point estimates?

8.57 Adjustments to a machine sometimes change the mean length of the parts it makes. However, the adjustments do not affect the variability of lengths. The lengths are known to be normally distributed with a standard deviation equal to 0.5 mm. After a series of adjustments, the following lengths (in mm) were obtained for ten randomly selected parts:

   75.3   76.0   75.0   77.0   75.4   76.3   77.0   74.9   76.5   75.8

The following MINITAB output shows a 95% confidence interval estimate for μ. Verify that the given confidence interval is correct.

```
MTB > SET DATA IN C1
DATA> 75.3 76.0 75.0 77.0 75.4 76.3 77.0 74.9 76.5 75.8
DATA> END DATA
MTB > ZINTERVAL 95 PERCENT CONFIDENCE, SIGMA = .5, DATA IN C1
THE ASSUMED SIGMA = 0.500
              N     MEAN    STDEV   SE MEAN    95.0 PERCENT C.I.
C1           10    75.920   0.773    0.158    ( 75.610, 776.230)
MTB > STOP
```

8.58 The selling prices of mutual funds change daily. In order to estimate the mean daily change in price, a random sample of mutual funds was taken from the May 29, 1991 issue of the *Democrat and Chronicle*:

0.15	0.30	0.01	0.19	0.02	0.01	0.00	0.05
0.13	0.08	0.10	0.16	0.14	0.00	0.18	0.03
−0.01	0.16	−0.04	−0.08	0.12	0.12	0.05	−0.01
−0.03	0.13	0.17	0.24	0.01	0.00	0.13	0.00

Use a computer to find a 0.98 confidence interval estimate for μ, the mean change for all funds. Assume that sigma, σ, is equal to 0.20. (If you use MINITAB, see Exercise 8.57.)

8.59 A certain population of the annual incomes of unskilled laborers has a standard deviation of $1200. A random sample of 36 results in $\bar{x} = \$7280$.
  a. Give a point estimate for the population mean annual income.
  b. Estimate μ with a 95% confidence interval.
  c. Estimate μ with a 99% confidence interval.

8.60 In order to estimate the mean amount spent for textbooks by students during the fall semester at a large commuter university, a random sample of 75 students was surveyed. The mean of the sample was $85.30. Find the 90% confidence interval estimate for the mean cost for all the students. Based on experience with similar studies, it is reasonable to assume $\sigma = \$15.00$.

8.61 A sample of 60 night school students' ages is obtained in order to estimate the mean age of night school students. $\bar{x} = 25.3$ years. The population variance is 16.
  a. Give a point estimate for μ.
  b. Find the 95% confidence interval estimate for μ.
  c. Find the 99% confidence interval estimate for μ.

8.62 The lengths of 200 fish caught in Cayuga Lake had a mean of 14.3 in. The population standard deviation is 2.5 in.
  a. Find the 90% confidence interval for the population mean length.
  b. Find the 98% confidence interval for the population mean length.

8.63 How large a sample should be taken if the population mean is to be estimated with 99% confidence to within $75? The population has a standard deviation of $900.

8.64 A high-tech company wants to estimate the mean number of years of college education its employees have completed. A good estimate of the standard deviation for the number of years of college is 1.0. How large a sample needs to be taken to estimate μ to within 0.5 year with 99% confidence?

8.65 By measuring the amount of time it takes a component of a product to move from one work station to the next, an engineer has estimated that the standard deviation is 5 sec.
  a. How many measurements should be made in order to be 95% certain that the maximum error of estimation will not exceed 1 sec?
  b. What sample size is required for a maximum error of 2 sec?

8.66 A medical products firm has developed a new drug that requires very strict control of the amount of the drug placed in each capsule. It can be hazardous if too much drug is used and ineffective if not enough is used. The maximum error of estimate for the mean is 2 mg. The standard deviation is believed to be 7 mg. If 99% confidence is desired, what size sample should be taken to be certain that the mean amount of the drug per capsule is correct?

8.67 An article entitled "Capture Rates and Composition of Insect Prey of the Pitcher Plant *Sarracenia Purpurea*" (*The American Midland Naturalist,* January 1991, pages 1–9) discusses the carnivorous plant *Sarracenia Purpurea,* which captures insects in pitcher-shaped leaves that serve as traps. Suppose the biomass captured (in mg) was determined for 200 plants during a week in August. If the sample mean biomass was 40.7 mg per plant per week, estimate the population mean biomass per week in August with a 90% confidence interval (assume the standard deviation, σ, is 10.0 mg).

8.68 According to the publication *US Pharmacist* (February 1991, page 10), 70 poison control centers reported an average of 22,600 calls for 1989. Suppose we regard these 70 centers as representative of all such centers, and furthermore, suppose we know the standard deviation is 1500. Determine the 90% confidence interval on μ, the mean for all such poison control centers.

IN RETROSPECT

Two forms of inference were studied in this chapter: hypothesis testing and estimation. They may be, and often are, used separately. It would, however, seem quite natural for the rejection of a null hypothesis to be followed by a confidence interval estimate. (If the value claimed is wrong, we will usually want to know just what the true value is.)

These two forms of inference are quite different but they are related. There is a certain amount of crossover between the use of the two inferences. For example, suppose that you had sampled and calculated a 90% confidence interval for the mean of a population. The interval was 10.5 to 15.6. Following this, someone claims that the true mean is 15.2. Your confidence interval estimate can be compared to this claim. If the claimed value falls within your interval estimate, you would fail to reject the null hypothesis that $\mu = 15.2$ at a 10% level of significance in a two-tailed test. If the claimed value (say 16.0) falls outside the interval, you would then reject the null hypothesis that $\mu = 16.0$ at $\alpha = 0.10$ in a two-tailed test. If a one-tailed test is required, or if you prefer a different value of α, a separate hypothesis test must be used.

Many users of statistics (especially those marketing a product) will claim that their statistical results prove that their product is superior. But remember, the hypothesis test does not prove or disprove anything. The decision reached in a hypothesis test has probabilities associated with the four various situations. If "fail to reject H_0" is the decision, it is possible that an error has occurred. Further, if "reject H_0" is the decision reached, it is possible for this to be an error. Both errors have probabilities greater than zero.

In this chapter we have restricted our discussion of inferences to the mean of a population for which the standard deviation is known. In Chapters 9 and 10, we will discuss inferences about other population parameters and eliminate the restriction about the known value for standard deviation.

CHAPTER EXERCISES

8.69 The expected mean of a certain population is 100 and its standard deviation is 12. A sample of 50 measurements gives a sample mean of 96. Using a level of significance of 0.01, a test is to be made to see whether the population mean is really 100 or whether it is different from 100. State or calculate the desired answer in (a) through (j).

a. H_0
b. H_a
c. α
d. $z(\alpha/2)$
e. μ (based on H_0)
f. $\bar{x}$
g. σ
h. $\sigma_{\bar{x}}$
i. z-score for $\bar{x}$
j. decision
k. On the accompanying sketch, locate $\alpha/2$, $z(\alpha/2)$, the rejection region for H_0, and the z-score for $\bar{x}$.

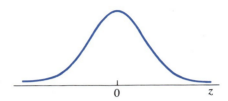

8.70 A sample of 64 measurements is taken from a continuous population and the sample mean is found to be 32.0. The standard deviation of the population is known to be 2.4. An interval estimation is to be made of the mean with a level of confidence of 0.90. State or calculate the following items:

a. $\bar{x}$
b. σ
c. n
d. $1 - \alpha$
e. $z(\alpha/2)$
f. $\sigma_{\bar{x}}$
g. E (maximum error of estimate)
h. upper confidence limit
i. lower confidence limit

8.71 A population has an unknown mean μ and a standard deviation σ equal to 5.0. Compare the concepts of hypothesis testing and confidence interval estimation by working the following:
   a. Calculate the value for z^* if the null hypothesis is H_0: $\mu = 40$, given that the sample size is 100 and the sample mean is 40.7. Determine the 95% confidence interval estimate for μ.
   b. Calculate the value for z^* if the null hypothesis is H_0: $\mu = 40$, given that the sample size is 100 and the sample mean is 42.5. Determine the 95% confidence interval estimate for μ.

8.72 The college bookstore tells prospective students that the average cost of its textbooks is $32 per book with a standard deviation of $4.50. The engineering science students think that the average cost of their books is higher than the average for all students. To test the bookstore's claim against their alternative, the engineering students collect a random sample of size 45.
   a. If they use $\alpha = 0.05$, what is the critical value of the test statistic?
   b. The engineering students' sample data are summarized by $n = 45$ and $\Sigma x = 1470.25$. Is this sufficient evidence to support their contention?

8.73 A rope manufacturer, after conducting a large number of tests over a long period of time, has found that the rope has a mean breaking strength of 300 lb and a standard deviation of 24 lb. Assume that these values are μ and σ. It is believed that by using a recently developed high-speed process, the mean breaking strength has been decreased.

    a. Design a null and alternative hypothesis such that rejection of the null hypothesis will imply that the mean breaking strength has decreased.

    b. If the decision rule in (a) is used with $\alpha = 0.01$, what is the critical value for the test statistic and what value of $\bar{x}$ corresponds to it if a sample of size 45 is used?

    c. Using the decision rule established in (a), what is the prob-value associated with rejecting the null hypothesis when 45 tests result in a sample mean of 295?

8.74 A manufacturing process produces ball bearings with diameters having a normal distribution and a standard deviation of $\sigma = 0.04$ cm. Ball bearings that have diameters that are too small or too large are undesirable. To test the null hypothesis that $\mu = 0.50$ cm, a sample of 25 is randomly selected and the sample mean is found to be 0.51.

    a. Design a null and alternative hypothesis such that rejection of the null hypothesis will imply that the ball bearings are undesirable.

    b. If the decision rule in (a) is used with $\alpha = 0.02$, what is the critical value for the test statistic?

    c. Using the decision rule established in (a), what is the prob-value for the sample results?

8.75 The admissions office at Memorial Hospital recently stated that the mean age of its patients was 42 years. A random sample of 120 ages was obtained from the admissions office records in an attempt to disprove the claim. Is a sample mean of 44.2 years significantly larger than the claimed 42 years at the $\alpha = 0.05$ level? Use $\sigma = 20$ years.

    a. Solve using the classical approach.
    b. Solve using the prob-value approach.

8.76 In a large supermarket the customer's waiting time to check out is approximately normally distributed with a standard deviation of 2.5 min. A sample of 24 customer waiting times produced a mean of 10.6 min. Is this evidence sufficient to reject the supermarket's claim that its customer checkout time averages no more than 9 min? Complete this hypothesis test using the 0.02 level of significance.

    a. Solve using the classical approach.
    b. Solve using the prob-value approach.

8.77 At a very large firm, the clerk-typists were sampled to see whether the salaries differed among departments for workers in similar categories. In a sample of 50 of the firm's accounting clerks, the average annual salary was $16,010. The firm's personnel office insists that the average salary paid to all clerk-typists in the firm is $15,650 and that the standard deviation is $1,800. At the 0.05 level of significance, can we conclude that the accounting clerks receive, on the average, a different salary from that of other clerk-typists?

    a. Solve using the classical approach.
    b. Solve using the prob-value approach.

CHAPTER EXERCISES

 8.78 A random sample of 280 rivets is selected from a very large shipment. The rivets are to have a mean diameter of no more than $\frac{5}{16}$ in. Does a sample mean of 0.3126 show sufficient reason to reject the null hypothesis that the mean diameter is no more than $\frac{5}{16}$ in. at the $\alpha = 0.01$ level of significance? Use $\sigma = 0.0006$.
   a. Solve using the classical approach.
   b. Solve using the prob-value approach.

8.79 Suppose that a hypothesis test is assigned a level of significance of $\alpha = 0.01$. How is the 0.01 used in completing the hypothesis test? If α were changed to 0.05, what effect would this have on the test procedure?

8.80 Suppose that a confidence interval is assigned a level of confidence of $1 - \alpha = 0.95$. How is the 0.95 used in constructing the confidence interval? If $1 - \alpha$ was changed to 0.90, what effect would this have on the confidence interval?

8.81 The following MINITAB output shows a simulated sample of size 28 drawn from a normal distribution with mean, μ, equal to 18 and standard deviation, σ, equal to 4. ZTEST is then used to complete a two-tailed test of the null hypothesis, $H_0: \mu = 18.0$. Verify the results.

```
MTB > RANDOM    28 OBSERVATIONS INTO C1;
SUBC> NORMAL    MU = 18  SIGMA = 4.
MTB > PRINT C1

C1
    18.7734    21.4352    15.5438    20.2764    23.2434    15.7222    13.9368
    14.4112    15.7403    19.0970    19.0032    20.0688    12.2466    10.4158
     8.9755    18.0094    20.0112    23.2721    16.6458    24.6146    17.8078
    16.5922    16.1385    12.3115    12.5674    18.9141    22.9315    13.3658

MTB > ZTEST   MU = 18 SIGMA = 4   C1

TEST OF MU = 18.000 VS MU N.E. 18.000
THE ASSUMED SIGMA = 4.00

               N       MEAN      STDEV     SE MEAN         Z     P VALUE
C1            28      17.217     4.053       0.756     -1.04        0.30

MTB > STOP
```

8.82 Use a computer to simulate a sample of size 45 being randomly drawn from a normal population with mean, μ, equal to 28 and standard deviation, σ, equal to 6. Have the computer use the sample mean and test $H_0: \mu = 30$ against $H_a: \mu < 30$. (If you use MINITAB, see Exercise 8.81.)

8.83 The following MINITAB output shows a simulated sample of size 25 from a normal distribution with $\mu = 130$ and $\sigma = 10$. ZINTERVAL is then used to set a 95% confidence interval for μ.

```
MTB > RANDOM   25 OBSERVATIONS INTO C1;
SUBC> NORMAL  MU = 130  SIGMA = 10.
MTB > PRINT C1
C1
   116.187    119.832   121.782   122.320   141.436   129.197   119.172
   120.713    135.765   131.153   122.307   126.155   137.545   141.154
   123.405    143.331   121.767   109.742   140.524   150.600   121.655
   127.992    136.434   139.768   125.594
MTB > ZINTERVAL 95 PERCENT CONFIDENCE, SIGMA = 10, DATA IN C1

THE ASSUMED SIGMA = 10.0

              N      MEAN     STDEV    SE MEAN    95.0 PERCENT C.I.
   C1        25    129.02     10.18      2.00     ( 125.10,  132.95)
```

Verify the last line of the MINITAB output; that is, verify the values reported for sample mean, sample standard deviation, standard error of the mean, and the 95% confidence interval.

8.84 Use a computer to simulate a sample of size 35 being randomly drawn from a normal population with mean, μ, equal to 42 and standard deviation, σ, equal to 8. Have the computer use the sample mean and find a 0.90 confidence interval estimate for the population mean. (If you use MINITAB, see Exercise 8.83.)

8.85 A random sample of the scores of 100 applicants for clerk-typist positions at a large insurance company showed a mean score of 72.6. The preparer of the test maintained that qualified applicants should average 75.0.
   a. Determine the 99% confidence interval estimate for the mean score of all applicants at the insurance company. Assume that the standard deviation of test scores is 10.5.
   b. Can the insurance company conclude that it is getting qualified applicants (as measured by this test)?

8.86 Waiting times (in hours) at a popular restaurant are believed to be approximately normally distributed with a variance of 2.25 hr during busy periods.
   a. A sample of 20 customers revealed a mean waiting time of 1.52 hr. Construct the 95% confidence interval for the estimate of the population mean.
   b. Suppose that the mean of 1.52 hr had resulted from a sample of 32 customers. Find the 95% confidence interval.
   c. What effect does a larger sample size have on the confidence interval?

CHAPTER EXERCISES

8.87 The weights of full boxes of a certain kind of cereal are normally distributed with a standard deviation of 0.27 oz. A sample of 18 randomly selected boxes produced a mean weight of 9.87 oz.
   a. Find the 95% confidence interval for the true mean weight of a box of this cereal.
   b. Find the 99% confidence interval for the true mean weight of a box of this cereal.
   c. What effect did the increase in the level of confidence have on the width of the confidence interval?

8.88 The standard deviation of a normally distributed population is equal to 10. A sample of size 25 is selected and its mean is found to be 95.
   a. Find an 80% confidence interval estimate for μ.
   b. If the sample size were 100, what would be the 80% confidence interval?
   c. If the sample size were 25 but the standard deviation were 5 (instead of 10), what would be the 80% confidence interval?

8.89 According to an article in *Health* magazine (March 1991, page 20), supplementation with potassium reduced the blood pressure readings in a group of mild hypertensive patients from an average of 158/100 to 143/84.5. Consider a study involving the use of potassium supplementation to reduce the systolic blood pressure for mild hypertensive patients. Suppose 75 patients with mild hypertension were placed on potassium for six weeks. The response measured was the systolic reading at the beginning of the study minus the systolic reading at the end of the study. The mean drop in the systolic reading was 12.5 units. Assume the population standard deviation, σ, to be 7.5 units. Calculate the value of the test statistic z^* and the prob-value for testing $H_0: \mu = 10.0$ vs. $H_a: \mu > 10.0$.

8.90 "Much Ado About Probability" (*Physical Therapy*, September 1990, pages 535, 536), the editor of the journal discusses the difference between statistical significance and practical significance. Suppose the mean value for a home in a residential area of a large city is listed as $150,000. A current sample of homes from the area finds a mean of $153,500 with a prob-value equal to 0.005. Does this statistical significance necessarily have practical significance?

8.91 The June 18, 1990 issue of *Insight* (page 52) describes the beneficial results of a Wellness Program begun in 1988 in Wellsburg, W. Va. One thousand of the town's 4000 residents participated in the program. Five hundred of the 1000 participants noted a reduction in their average cholesterol by 15 points. Consider another wellness program involving 500 participants. Suppose their cholesterol was lowered by 20 points on the average. If we regard these results as representative of what such a program can accomplish, estimate the mean reduction in cholesterol for all participants in such programs with a 0.95 confidence interval. (Assume the standard deviation, σ, is 7.5 units.)

8.92 A new paint has recently been developed by a research laboratory. Twenty-five gallons are tested and the mean coverage is found to be 515 sq ft/gal. Assuming the standard deviation to be 50 sq ft/gal, find a 99% confidence interval estimate for μ, the mean coverage per gallon of all such paint.

8.93 An automobile manufacturer wants to estimate the mean gasoline mileage that its customers will obtain with its new compact model. How many sample runs must be performed in order that the estimate be accurate to within 0.3 mpg at 95% confidence? (Assume that $\sigma = 1.5$.)

8.94 A fish hatchery manager wants to estimate the mean length of her three-year-old hatchery-raised trout. She wants to make a 99% confidence interval estimate accurate to within $\frac{1}{3}$ of a standard deviation. How large a sample does she need to take?

8.95 We are interested in estimating the mean life of a new product. How large a sample do we need to take in order to estimate the mean to within $\frac{1}{10}$ of a standard deviation with 0.90 confidence?

VOCABULARY LIST

Be able to define each term. In addition, describe, in your own words, and give an example of each term. Your examples should not be ones given in class or in the textbook.

The bracketed numbers indicate the chapters in which the terms previously appeared, but you should define the terms again to show increased understanding of their meaning.

acceptance region
alpha (α)
alternative hypothesis
beta (β)
calculated value (*)
conclusion
confidence interval estimate
critical region
critical value
decision
decision rule
estimation
hypothesis
hypothesis test
interval estimate
level of confidence
level of significance
lower confidence limit
maximum error of estimate
noncritical region

null hypothesis
one-tailed test
parameter [1]
point estimate
prob-value
risk
sample size
sample statistic [1, 2]
standard error [7]
standard error of mean [7]
statistic [1, 2]
test criteria
test statistic
two-tailed test
type A correct decision
type B correct decision
type I error
type II error
upper confidence limit
$x(\alpha)$ [6]

KEY CONCEPTS

alpha (α)
alternative hypothesis
confidence interval estimate
critical value
hypothesis test
level of confidence
level of significance

maximum error of estimate
null hypothesis
parameter [1]
point estimate
standard error of mean [7]
test statistic
type I error

QUIZ A

Answer "True" if the statement is always true. If the statement is not always true, replace the words shown in bold with words that make the statement always true.

8.1 **Beta** is the probability of a type I error.

8.2 $1 - \alpha$ is known as the level of significance of a hypothesis test.

8.3 The standard error of the mean is the standard deviation of the **sample selected**.

8.4 The maximum error of estimate is controlled by three factors: **level of confidence, sample size, and standard deviation**.

8.5 Alpha is the measure of the area under the curve of the standard score that lies in the **rejection region** for H_0.

8.6 The risk of making a **type I error** is directly controlled in a hypothesis test by establishing a level for α.

8.7 Failing to reject the null hypothesis when it is false is a **correct decision**.

8.8 If the acceptance region in a hypothesis test is made wider (assuming σ and n remain fixed), α becomes larger.

8.9 Rejection of a null hypothesis that is false is a **type II error**.

8.10 To conclude that the mean is higher (or lower) than a claimed value, the value of the test statistic must fall in the **acceptance region**.

QUIZ B

Answer all questions, showing all formulas, substitutions, and work.

8.1 State the null (H_0) and the alternative (H_a) hypotheses that would be used to test each of the following claims:
   a. The mean weight of professional football players is more than 245 lb.
   b. The mean monthly amount of rainfall in Monroe County is less than 4.5 in.
   c. The mean weight of the baseball bats used by major league players is not equal to 35 oz.

8.2 Determine the test criteria [level of significance, test statistic, critical region(s), and critical value(s)] that would be used in completing each hypothesis test using $\alpha = 0.05$.

    a. $H_0: \mu = 43$
        $H_a: \mu < 43$
        (given $\sigma = 6$)

    b. $H_0: \mu = 0.80$
        $H_a: \mu > 0.80$
        (given $\sigma = 0.13$)

    c. $H_0: \mu = 95$
        $H_a: \mu \neq 95$
        (given $\sigma = 12$)

8.3 Find each of the following:

    a. $z(0.05)$ **b.** $z(0.01)$ **c.** $z(0.12)$
    d. $z(0.95)$ **e.** $z(0.98)$ **f.** $z(0.75)$

8.4 A manufacturer of light bulbs claims that its light bulbs have a mean life of 1520 hr with a standard deviation of 85 hr. A random sample of 40 such bulbs is selected for testing. If the sample produces a mean value of 1498.3 hr, is there sufficient evidence to claim that the mean life is significantly less than the manufacturer claimed? Use $\alpha = 0.01$.

8.5 In the past, the grapefruits grown in a particular orchard have had a mean diameter of 5.50 in. and a standard deviation of 0.6 in. The owner believes this year's crop is larger than in the past. He collected a random sample of 100 and found a sample mean diameter of 5.65.

    a. Find the value of the test statistic, z^*, that corresponds to $\bar{x} = 5.65$.
    b. Calculate the prob-value for the owner's hypothesis.

8.6 An unhappy post office customer is disturbed with the waiting time to buy stamps. Upon registering his complaint he was told, "the average waiting time in the past has been about 4 min with a standard deviation of 2 min." The customer collected a sample of $n = 45$ customers and found the mean wait was 5.3 min. Find the 95% confidence interval estimate for the mean waiting time.

QUIZ C

8.1 The noise level in a hospital may be a critical factor influencing a patient's rate of recovery. Suppose for the sake of discussion that a research commission has recommended a maximum mean noise level of 30 db with a standard deviation of 10 db. The staff of a hospital intends to sample one of its wards to determine if the noise level is significantly higher than the recommended level. The following hypothesis test will be completed:

$$H_0: \mu = 30 \quad (\leq)$$

$$H_a: \mu > 30$$

$$\alpha = 0.05$$

    a. Identify the correct interpretation for the meaning of each hypothesis with regard to the recommendation and justify your choice.

    H_0: (a) noise level is not significantly higher than the recommended level or

    (b) noise level is significantly higher than the recommended level

H_a: (a) noise level is not significantly higher than the recommended level

or

    (b) noise level is significantly higher than the recommended level

  b. Which statement below best describes the type I error?
    (1) Decision reached was, noise level is within level recommended, when in fact it actually was within it.
    (2) Decision reached was, noise level is within level recommended, when in fact it actually exceeded it.
    (3) Decision reached was, noise level exceeds the level recommended, when in fact it actually exceeded it.
    (4) Decision reached was, noise level exceeds the level recommended, when in fact it actually was within it.
  c. Which statement above best describes the type II error?
  d. If alpha were changed from 0.05 to 0.01, identify and justify the effect (increases, decreases, or remains the same) on each of the following: P(type I error), P(type II error).

9 INFERENCES INVOLVING ONE POPULATION

Chapter Outline

9.1 Inferences About the Population Mean
When the standard deviation of the population is unknown, the **Student's distribution** is used to make inferences.

9.2 Inferences About the Binomial Probability of Success
The observed sample proportion p' is **approximately normally distributed** under certain conditions.

9.3 Inferences About Variance and Standard Deviation
The **chi-square distribution** is employed when testing hypotheses concerning variance or standard deviation.

The American Gender Evolution: Getting What We Want

THE *NEW WOMAN* SURVEY

The results reported here are based on a telephone survey of 1,201 adults supervised by the polling firm of Yankelovich Clancy Shulman and conducted during March 1990. It has a sampling error of ±3 percent and all statistics exclude "don't know" answers.

THE CHANGING FACE OF MARRIAGE

Marriage is as important as ever, but in the past 20 years, the roles of husband and wife—and their expectations of marriage—have changed dramatically. Today, couples often fall into one of two categories—Traditional and Egalitarian—which reflect opposing ideals of marriage. Other couples, struggling to integrate old and new ideals, fall somewhere in between, in marriages that could be called Transitional.

In the Traditional marriage, the wife generally derives her identity through her family and her role as wife and mother. The husband defines himself through his work and his role as breadwinner. The roles and rules for this marriage are clear, and the husband is usually the king of the castle.

Many older Americans—those over age 55—have Traditional marriages. They are most likely to say, for instance, that the ideal woman is a good homemaker (see Chart 1).

In an Egalitarian marriage, the couple views the union as an equal partnership. Both husband and wife gain a sense of identity through work and family; they share the burden of breadwinning as well as the pleasure of nurturing children.

Egalitarian marriages are most likely to be found among younger Americans—those under age 45. . . .

. . . Most American women and men under age 45 believe that *both* partners should be responsible for earning a living (see Chart 2). . . .

It's inevitable that over the next decade more couples will be hoping to find an Egalitarian marriage. This year, 65 percent of American wives under 65 are working outside the home, according to the U.S. Bureau of Labor Statistics. As more wives remain in the work force, fewer Americans will expect men to be the sole breadwinner. . . .

Perhaps that is why some men go even further and insist that wives are obligated to work. Almost half of American men today actually disapprove of a wife with no children who doesn't work; so do about one third of women (see Chart 3). . . .

THE WORK REVOLUTION

American women are discovering the importance of having both love and work. Eight in ten believe that the most satisfying life includes full- or part-time work (see Chart 4). Fifteen years ago, four in ten preferred to be supported by their mate. Today, about two thirds of women agree that raising children full time "cannot keep most women satisfied." . . .

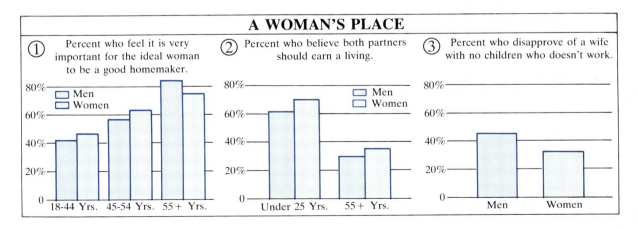

A WOMAN'S PLACE

① Percent who feel it is very important for the ideal woman to be a good homemaker.

② Percent who believe both partners should earn a living.

③ Percent who disapprove of a wife with no children who doesn't work.

Twenty years ago, barely one third of working women viewed their work as a career; today, 47 percent of working women do, as do 62 percent of men. Jacklin explains that "we use *career* to mean commitment to a job, and more and more women now see work as central to their existence."

Still, fewer women than men work full time: 44 percent of women in the *New Woman* survey do, compared with 78 percent of men. Another 23 percent of women work part time; however, only 9 percent of men do. The women who work full time tend to do so out of necessity (see Chart 5). . . .

Altered Visions of the Perfect Life

TABLE 4 "What Would Provide You with the Most Satisfying Life?"

Fifteen years ago, four in ten women wanted to be supported by their husbands; today, only two in ten do. Most American women now say that their vision of the ideal life includes some kind of paid work.

Most Satisfying To:	Men	Women
Work full time	89%	30%
Work part time	9%	49%
Be supported by mate	2%	21%

TABLE 5 Why Do Women Work?

Women who work full time are most likely to do so out of necessity. But those who work part time seem to have more choice in the matter.

Women Work To:	Full Time	Part Time
Support family	33%	14%
Support self	28%	12%
Earn extra money	21%	42%
Keep busy/be fulfilled	18%	32%

Source: "The American Gender Evolution: Getting What We Want," *New Woman* (November 1990). Reprinted by permission.

Chapter Objectives

In Chapter 8 we discussed two forms of statistical inference: (1) hypothesis tests and (2) confidence interval estimation. The study of these two inferences was restricted to the population parameter μ under the restriction that σ was known. However, the population standard deviation is generally not known. Thus the first section of this chapter deals with the inferences about μ when σ is unknown. In addition, we will learn how to perform hypothesis tests and confidence interval estimations about the population parameters σ (standard deviation) and p (binomial probability of success).

9.1 ▼ Inferences About the Population Mean

In Chapter 8, we learned about making inferences concerning the population mean when the population standard deviation σ was a known quantity. Now we will consider inferences about the population mean when σ is an *unknown* quantity.

The sample standard deviation, s, is a very effective point estimate to use for σ and will be used in place of σ in the various calculations. As a result of this substitution, the test statistic used is also changed. In place of z, the standard normal score, we will use the statistic t. The t-statistic is defined as

$$t = \frac{\bar{x} - \mu}{s/\sqrt{n}} \tag{9-1}$$

If the population being sampled is approximately normally distributed, then the calculated t-score belongs to the Student's t-distribution.

In 1908 W. S. Gosset, an Irish brewery employee, published a paper about this t-distribution under the pseudonym "Student." In deriving the t-distribution, Gosset assumed that the samples were taken from normal populations. Although this might seem to be quite restrictive, satisfactory results are also obtained when sampling from many nonnormal populations.

The t-distribution has the following properties (see also Figure 9-1):

PROPERTIES OF THE t-DISTRIBUTION

1. t is distributed with a mean of 0.
2. t is distributed symmetrically about its mean.
3. t is distributed with a variance greater than 1, but as the sample size n increases, the variance approaches 1.
4. t is distributed so as to be less peaked at the mean and thicker at the tails than the normal distribution.
5. t is distributed so as to form a family of distributions, a separate distribution for each sample size. The t-distribution approaches the normal distribution as the sample size increases.

FIGURE 9-1
Normal and Student's *t*-Distributions

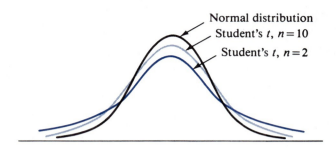

Although there is a separate *t*-distribution for each sample size, $n = 2$, $n = 3, \ldots, n = 20, \ldots, n = 40$, and so on, only certain key critical values of *t* will be necessary for our work. Consequently, the table for Student's *t*-distribution (Table 6 of Appendix F) is a table of critical values rather than a complete table, such as that for the standard normal distribution for *z*. As you look at Table 6, you will note that the left side of the table is identified by "df," which means "degrees of freedom." This left-hand column starts at 1 at the top and ranges to 29, then jumps to *z* at the bottom. As stated previously, as the sample size increases, the distribution approaches that of the normal. By inspecting Table 6, you will note that as you read down any one of the columns, the entry value approaches a familiar *z* value at the bottom. By now you should realize that as the sample size increases, so does the number of degrees of freedom, df.

degrees of freedom

DEGREES OF FREEDOM, df
A parameter in statistics that is very difficult to define completely. Perhaps it is best thought of as an *"index number" used for the purpose of identifying the correct distribution to be used*. In the methods presented in this chapter, the value of df for a given situation will be $n - 1$, the sample size minus 1.

As previously stated, as the sample size increases, the *t*-distribution approaches the characteristics of the standard normal *z*-distribution. Once *n* is larger than 30, the critical values of the *t*-distribution become very close in value to the corresponding critical values of the standard normal distribution. Thus for samples of size *n* greater than 30 (or degrees of freedom greater than 29), the critical values of the *t*-distribution are approximately the same as *z*. Thus, statisticians have generally agreed to abbreviate the Student's *t*-distribution table of critical values to include degrees of freedom from 1 to 29, and to use the corresponding critical value for *z* for all *t*'s with 30 or more degrees of freedom.

NOTE It is customary to divide samples into two categories according to size:

▼ *small sample*: a sample whose size *n* is 30 or less, or
▼ *large sample*: a sample whose size *n* is larger than 30.

Section 9.1 ▼ INFERENCES ABOUT THE POPULATION MEAN

The critical value of t to be used either in a hypothesis test or in constructing a confidence interval will be obtained from Table 6, Appendix F. To obtain the value of t, you will need to know two values: (1) df, the number of degrees of freedom, and (2) α, the area under the curve to the right of the right-hand critical value. A notation much like that used with z will be used to identify a critical value. $t(\text{df}, \alpha)$, read as "t of df, α," is the symbol for the value of t described in the previous sentence and shown in Figure 9-2.

FIGURE 9-2
t-Distribution Showing $t(\text{df}, \alpha)$

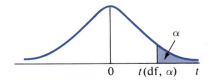

▼ ILLUSTRATION 9-1

Find the value of $t(10, 0.05)$ (see the following diagram).

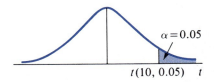

SOLUTION There are 10 degrees of freedom. So, in Table 6, Appendix F, we look for the column marked $\alpha = 0.05$ and come down to df = 10.

	Amount of α in One Tail		
df	...	0.05	...
.			
.			
.			
10		1.81	
.			
.			
.			

From the table we see that $t(10, 0.05) = \mathbf{1.81}$. ▲▲

For the values of t on the left-hand side of the mean, we can use one of two notations. The t value shown in Figure 9-3 (p. 442) could be $t(\text{df}, 0.95)$, since the area to the right of it is 0.95. Or it could be identified by $-t(\text{df}, 0.05)$, since the t-distribution is symmetric about its mean, zero.

FIGURE 9-3
t Value on Left Side of Mean

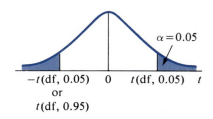

▼ ILLUSTRATION 9-2

Find the value of $t(15, 0.95)$.

SOLUTION There are 15 degrees of freedom. In Table 6 we look for the column marked $\alpha = 0.05$ and come down to df = 15. The table value is 1.75. Thus $t(15, 0.95) = -1.75$ (the value is negative because it is to the left of the mean; see the following figure).

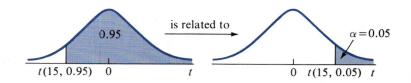

The *t*-statistic is used in problems concerned with μ in much the same manner as *z* was used in Chapter 8. In hypothesis-testing situations we will use formula (9-1) to calculate the test statistic value in Step 4 of our procedure.

▼ ILLUSTRATION 9-3

Let's return to the hypothesis of Illustration 8-1 (p. 392) concerning air pollution: "the mean carbon monoxide level of air pollution is no more than 4.9." Does a random sample of 25 readings (sample results: $\bar{x} = 5.1$ and $s = 2.1$) present sufficient evidence to cause us to reject this claim? Use $\alpha = 0.05$.

SOLUTION

STEP 1 $H_0: \mu = 4.9 (\leq)$

$H_a: \mu > 4.9$

STEP 2 $\alpha = 0.05$, df $= 25 - 1 = 24$, and $t(24, 0.05) = 1.71$, from Table 6, Appendix F. (See the accompanying figure.)

Section 9.1 ▼ INFERENCES ABOUT THE POPULATION MEAN

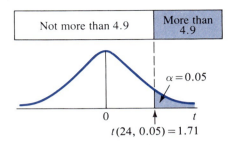

STEP 3 $\quad t = \dfrac{\bar{x} - \mu}{s/\sqrt{n}} = \dfrac{5.1 - 4.9}{2.1/\sqrt{25}} = \dfrac{0.2}{2.1/5} = \dfrac{0.20}{0.42} = 0.476$

$t^* = \mathbf{0.48}$

Comparing this value with the test criteria, we have the situation shown in the accompanying figure.

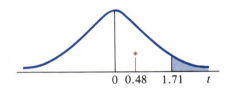

STEP 4 Decision Fail to reject H_0 (t^* is in the noncritical region).

Conclusion At the 0.05 level of significance, we do not have enough evidence to reject the claim that the mean carbon monoxide level is no more than 4.9. ▲▲

NOTE If the value of df (df $= n - 1$) in an exercise such as Illustration 9-3 is larger than 29, then the critical value for $t(\text{df}, \alpha)$ actually becomes nearly the same as $z(\alpha)$, the z-score listed at the bottom of Table 6. Therefore, $t(\text{df}, \alpha)$ is not given when df > 29; $z(\alpha)$ is used as the critical value.

Because Table 6 has critical values for the Student's t-distribution, the prob-value cannot be calculated for a hypothesis test that involves the use of t. However, the prob-value can be estimated.

▼ ILLUSTRATION 9-4

Let's return to Illustration 9-3. Note that $t^* = 0.48$, df $= 24$, and $H_a: \mu > 4.9$. Thus for Step 4 of the prob-value solution we have

$P = P(t > 0.48$, knowing df $= 24$; see the following figure)

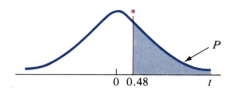

Portion of Table 6	
df	0.25
⋮	
24	0.685

By inspecting the df = 24 row of Table 6, you can determine that the prob-value is greater than 0.25. The 0.685 entry in the table tells us that the $P(t > 0.685) = 0.25$, as shown in the following figure:

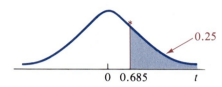

observed binomial probability (p')

By comparing $t^* = 0.48$ (see the following figure) we see that the prob-value p is greater than 0.25 and less than 0.50: **$0.25 < P < 0.50$**.

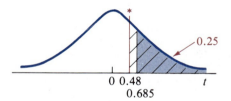

▲▲

▼ ILLUSTRATION 9-5

Determine the p-value for the following hypothesis test:

$$H_0: \mu = 55$$
$$H_a: \mu \neq 55$$
$$df = 15 \quad \text{and} \quad t^* = -1.84$$

Portion of Table 6		
df	0.05	0.025
⋮		
15	1.75	2.13

SOLUTION

$$\mathbf{P} = P(t < -1.84) + P(t > 1.84) = 2P(t > 1.84)$$

By inspecting the df = 15 row of Table 6 in Appendix F, we find that $P(t > 1.84)$ is between 0.025 and 0.05. Therefore, **P** is between the double of these amounts. Thus the prob-value is between 0.05 and 0.10, **$0.05 < P < 0.10$**.

▲▲

Section 9.1 ▼ INFERENCES ABOUT THE POPULATION MEAN

NOTE Since Table 6 has only critical values, we can only estimate the *p*-value when *t* is the test statistic.

The population mean may be estimated when σ is unknown in a manner similar to that used when σ is known. The difference is the use of Student's *t* in place of *z* and the use of *s*, the sample standard deviation, as an estimate of σ. The formula for the $1 - \alpha$ **confidence interval** of estimation then becomes

confidence interval formula

$$\bar{x} - t(\text{df}, \alpha/2) \cdot \frac{s}{\sqrt{n}} \quad \text{to} \quad \bar{x} + t(\text{df}, \alpha/2) \cdot \frac{s}{\sqrt{n}} \tag{9-2}$$

where df $= n - 1$.

▼ ILLUSTRATION 9-6

A random sample of size 20 is taken from the weights of babies born at Northside Hospital during the year 1982. A mean of 6.87 lb and a standard deviation of 1.76 lb were found for the sample. Estimate, with 95% confidence, the mean weight of all babies born in this hospital in 1982.

SOLUTION The information we are given is $\bar{x} = 6.87$, $s = 1.76$, and $n = 20$. $1 - \alpha = 0.95$ implies that $\alpha = 0.05$; $n = 20$ implies that df $= 19$. From Table 6, we get $t(19, 0.025) = 2.09$. See the following figure:

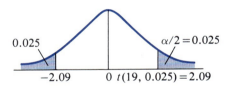

The interval is thus

$$\bar{x} \pm t(19, 0.025) \cdot \frac{s}{\sqrt{n}}$$

$$6.87 \pm 2.09 \cdot \frac{1.76}{\sqrt{20}}$$

$$6.87 \pm \frac{(2.09)(1.76)}{4.472}$$

$$6.87 \pm 0.82$$

$$6.87 - 0.82 = \mathbf{6.05} \quad \text{and} \quad 6.87 + 0.82 = \mathbf{7.69}$$

The 0.95 confidence interval for μ is $(\mathbf{6.05, 7.69})$.

That is, with 95% confidence we estimate the mean weight to be between 6.05 and 7.69 lb. ▲▲

Case Study 9-1

Mothers' Use of Personal Pronouns When Talking with Toddlers

The calculated t-value and the probability value for five different hypothesis tests are given in the following article. $t(44) = 1.92$ is $t^* = 1.92$ with $df = 44$ and is significant with $p < 0.05$. Can you verify the p-value? Explain. (See Exercise 9.23.)

ABSTRACT. The verbal interaction of 2-year-old children ($N = 46$; 16 girls, 30 boys) and their mothers was audiotaped, transcribed, and analyzed for the use of personal pronouns, the total number of utterances, the child's mean length of utterance, and the mother's responsiveness to her child's utterances. Mothers' use of the personal pronoun *we* was significantly related to their children's performance on the Stanford-Binet at age 5 and the Wechsler Intelligence Scale for Children at age 8. Mothers' use of *we* in social–vocal interchange, indicating a system for establishing a shared relationship with the child, was closely connected with their verbal responsiveness to their children. The total amount of maternal talking, the number of personal pronouns used by mothers, and their verbal responsiveness to their children were not related to mothers' social class or years of education.

Mothers tended to use more first person singular pronouns (*I* and *me*), $t(44) = 1.81$, $p < .10$, and used significantly more first person plural pronouns (*we*), $t(44) = 1.92$, $p < .05$, with female children than with male children. The mothers also were more verbally responsive to their female children, $t(44) = 2.0$, $p < .06$.

In general, mothers talked more to their first born children, $t(44) = 3.41$, $p < .001$, and were more responsive to their first born children, $t(44) = 3.71$, $p < .001$. Yet, the proportion of personal pronouns used when speaking to first born children was not different from that used when speaking to later born children.

Source: Dan R. Laks, Leila Beckwith, Sarale E. Cohen, THE JOURNAL OF GENETIC PSYCHOLOGY, 151(1), 25–32, 1990. Reprinted with permission of the Helen Dwight Reid Educational Foundation. Published by Heldref Publications, 4000 Albemarle St., N.W., Washington, D.C., 20016. Copyright © 1990.

▼▲ EXERCISES

9.1 State the null hypothesis, H_0, and the alternative hypothesis, H_a, that would be used to test each of the following claims:
   a. The mean weight of honey bees is at least 11 g.
   b. The mean age of patients at Memorial Hospital is no more than 54 years.
   c. The mean amount of salt in granola snack bars is different from 75 mg.

9.2 Determine the test criteria that would be used to test the null hypotheses below:
   a. $H_0: \mu = 10$ ($\alpha = 0.05, n = 15$)
   $H_a: \mu \neq 10$

b. $H_0: \mu = 37.2$ ($\alpha = 0.01, n = 25$)
$H_a: \mu > 37.2$
c. $H_0: \mu = -20.5$ ($\alpha = 0.05, n = 18$)
$H_a: \mu < -20.5$
d. $H_0: \mu = 32.0$ ($\alpha = 0.02, n = 42$)
$H_a: \mu > 32.0$

9.3 Find these critical values using Table 6 in Appendix F.
a. $t(25, 0.05)$ **b.** $t(10, 0.10)$ **c.** $t(15, 0.01)$ **d.** $t(21, 0.025)$
e. $t(21, 0.95)$ **f.** $t(26, 0.975)$ **g.** $t(27, 0.99)$ **h.** $t(60, 0.025)$

9.4 Using the notation of Exercise 9.1, name and find the following critical values of t:

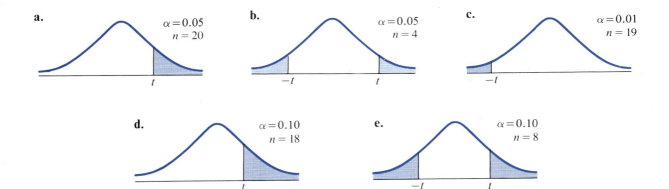

9.5 Ninety percent of Student's t-distribution lies between $t = -1.89$ and $t = 1.89$ for how many degrees of freedom?

9.6 Ninety percent of Student's t-distribution lies to the right of $t = -1.37$ for how many degrees of freedom?

9.7 **a.** Find the first percentile of Student's t-distribution with 24 degrees of freedom.
b. Find the 95th percentile of Student's t-distribution with 24 degrees of freedom.
c. Find the first quartile of Student's t-distribution with 24 degrees of freedom.

9.8 Find the percent of the Student's t-distribution that lies between the following values:
a. df = 12 and t ranges from -1.36 to 2.68
b. df = 15 and t ranges from -1.75 to 2.95

9.9 **a.** State two ways in which the standard normal distribution and the Student's t-distribution are alike.
b. State two ways in which they are different.

9.10 The variance for a Student's t-distribution is equal to df/(df $-$ 2). Find the standard deviation for a Student's t-distribution with each of the following degrees of freedom:
a. 10 **b.** 20 **c.** 30

9.11 A student group maintains that the average student must travel for at least 25 minutes in order to reach college each day. The college admissions office obtained a random sample of 36 one-way travel times from students. The sample had a mean of 19.4 min and a standard deviation of 9.6 min. Does the admissions office have sufficient evidence to reject the students' claim? Use $\alpha = 0.01$.
  a. Solve using the classical approach.
  b. Solve using the prob-value approach.

9.12 Homes in a nearby college town have a mean value of $88,950. It is assumed that homes in the vicinity of the college have a higher value. To test this theory, a random sample of 12 homes is chosen from the college area. Their mean valuation is $92,460 and the standard deviation is $5,200. Complete a hypothesis test using $\alpha = 0.05$.
  a. Solve using the classical approach.
  b. Solve using the prob-value approach.

9.13 It is claimed that the students at a certain university will score an average of 35 on a given test. Is the claim reasonable if a random sample of test scores from this university yields 33, 42, 38, 37, 30, 42. Complete a hypothesis test using $\alpha = 0.05$.
  a. Solve using the classical approach.
  b. Solve using the prob-value approach.

9.14 Gasoline pumped from a supplier's pipeline is supposed to have an octane rating of 87.5. On 13 consecutive days a sample was taken and analyzed with the following results:

| 88.6 | 86.4 | 87.2 | 88.4 | 87.2 | 87.6 | 86.8 |
| 86.1 | 87.4 | 87.3 | 86.4 | 86.6 | 87.1 | |

  a. Is there sufficient evidence to show that these octane readings were taken from gasoline with a mean octane significantly less than 87.5 at the 0.05 level? (The sample mean and standard deviation were found in answering Exercise 2.108.)
  b. Did the statistical decision reached in (a) result in the same conclusion as you expressed in answering part (c) of Exercise 2.108 for this same data?

9.15 In order to test the null hypothesis "the mean weight for adult males equals 160 lb" against the alternative "the mean weight for adult males exceeds 160 lb," the weights of 16 males were determined with the following results:

| 173 | 178 | 145 | 146 | 157 | 175 | 173 | 137 |
| 152 | 171 | 163 | 170 | 135 | 159 | 199 | 131 |

Verify the results shown in the following MINITAB analysis (*Note:* ALT = $-1, 0, +1$ represents lower-tail, two-tail, and upper-tail tests, respectively.)

Section 9.1 ▼ INFERENCES ABOUT THE POPULATION MEAN   449

```
MTB> SET THE FOLLOWING MALE WEIGHTS IN C1
DATA  173 178 145 146 157 175 173 137 199 131 152 171 163 170 135 159
DATA> END DATA
MTB > TTEST  OF MU = 160  DATA IN C1;
SUBC> ALTERNATIVE = 1.
TEST OF MU = 160.00 VS MU G.T. 160.00
              N       MEAN     STDEV    SE MEAN        T    P VALUE
C1           16     160.25     18.49       4.62     0.05       0.48
MTB > STOP
```

9.16 Use a computer to complete the calculations and the hypothesis test for this exercise. (If you use MINITAB, see Exercise 9.15 for program commands.) Delco Products, a division of General Motors, produces commutators designed to be 18.810 mm in overall length. (A commutator is a device used in the electrical system of an automobile.) The following sample of 35 commutators was taken while monitoring the manufacturing process:

18.802	18.810	18.780	18.757	18.824
18.827	18.825	18.809	18.794	18.787
18.844	18.824	18.829	18.817	18.785
18.747	18.802	18.826	18.810	18.802
18.780	18.830	18.874	18.836	18.758
18.813	18.844	18.861	18.824	18.835
18.794	18.853	18.823	18.863	18.808

Source: With permission of Delco Products Division, GMC.

Is there sufficient evidence to reject the claim that these parts meet the design requirements "mean length is 18.810" at the $\alpha = 0.01$ level of significance?

9.17 The data from a study reported in "White-Collar Crime and Criminal Careers: Some Preliminary Findings" (*Crime and Delinquency,* July 1990, pages 342–352) indicate that white-collar criminals are likely to be older and to show a lower frequency of offending than street criminals. For example, the mean age of onset of offending for those convicted of antitrust offenses was 54 years with $n = 27$. If the standard deviation is estimated to be 7.5 years, set a 90% confidence interval on the true mean age.

9.18 Taking a random sample of 25 individuals registering for classes at a particular college, we find that the mean waiting time in the registration line was 22.6 min and the standard deviation was 8.0 min. Using a 90% confidence interval, estimate the mean waiting time for all individuals registering.

9.19 Ten randomly selected shut-ins were each asked to list how many hours of television they watched per week. The results are

| 82 | 66 | 90 | 84 | 75 | 88 | 80 | 94 | 110 | 91 |

Determine the 90% confidence interval estimate for the mean number of hours of television watched per week by shut-ins.

9.20 While doing an article on the high cost of college education, a reporter took a random sample of the cost of textbooks for a semester. The random variable x is the cost of one book. Her sample data can be summarized by $n = 41$, $\Sigma x = 550.22$, and $\Sigma (x - \bar{x})^2 = 1617.984$.
  a. Find the sample mean, $\bar{x}$.
  b. Find the sample standard deviation, s.
  c. Find the 0.90 confidence interval to estimate the true mean textbook cost for the semester based on this sample.

9.21 The pulse rates for 13 adult women were found to be

83 58 70 56 76 64 80 76 70 97 68 78 108

Verify the results shown on the last line of the MINITAB output below.

```
MTB >  SET THE FOLLOWING FEMALE PULSE RATES IN C1
DATA>  83 58 70 56 76 64 80 76 70 97 68 78 108
DATA>  END DATA
MTB >  TINTERVAL 90 PERCENT CONFIDENCE INTERVAL FOR DATA IN C1
              N     MEAN    STDEV    SE MEAN    90.0 PERCENT C.I.
       C1    13    75.69    14.54     4.03      ( 68.50, 82.88)
MTB >  STOP
```

9.22 The weights of the drained fruit found in 21 randomly selected cans of peaches packed by Sunny Fruit Cannery were (in ounces)

11.0	11.6	10.9	12.0	11.5	12.0	11.2
10.5	12.2	11.8	12.1	11.6	11.7	11.6
11.2	12.0	11.4	10.8	11.8	10.9	11.4

Using a computer,
  a. Calculate the sample mean and standard deviation.
  b. Construct the 0.98 confidence interval for the estimate of the mean weight of drained peaches per can.
  (If you use MINITAB, see Exercise 9.21 for program commands.)

9.23 A calculated t, $t^* = 1.92$, with 44 degrees of freedom was reported in Case Study 9-1 (p. 446). Is it significant?
  a. Use Table 6 in Appendix F and place bounds on the p-value for $t^* = 1.92$ in a one-tailed test. Explain.
  b. Use Table 5 in Appendix F and calculate the p-value for $t^* = 1.92$ with $df = 44$ using z. Explain why this is acceptable.
  c. If $\alpha = 0.05$, is $t^* = 1.92$ significant?
  d. State the conclusion drawn as a result of $t^* = 1.92$ and the decision reached in (c).

9.24 Explain how the critical values of t are determined for a situation where $df > 30$.

9.2 ▼ Inferences About the Binomial Probability of Success

Recall that the binomial parameter p was defined to be the theoretical or population probability of success on a single trial in a binomial experiment. Also, the random variable x is the number of successes that occur in a set of n trials. By combining the definition of empirical probability, $P'(A) = n(A)/n$ [formula (4-1), p. 205], with the notation of the binomial experiment (p. 285), we define p', the **observed or sample binomial probability**, to be $p' = x/n$. Also recall that the mean and standard deviation of the binomial random variable x are found by using formulas (5-7) and (5-8): $\mu = np$ and $\sigma = \sqrt{npq}$, where $q = 1 - p$. This distribution of x is considered to be approximately normal if n is larger than 20 and if np and nq are both larger than 5. This commonly accepted rule of thumb allows us to use the normal distribution when making inferences concerning a binomial parameter p.

observed binomial probability (p')

Generally, it is easier to work with the distribution of p' rather than the distribution of x. Consequently, we will convert formulas (5-7) and (5-8) [p. 294] from the units of x to units of proportions. If we divide formulas (5-7) and (5-8) by n, we should change the units from those of x to those of proportion. The mean of x is np; thus the mean of p', $\mu_{p'}$, should be np divided by n, that is, np/n, or just p. (Does it seem reasonable that the mean of the distribution of observed values of p' should be p, the true proportion?) Further, the standard error of p' in this sampling distribution is

$$\sigma_{p'} = \sqrt{npq}/n = \sqrt{npq/n^2} = \sqrt{pq/n}$$

We summarize this information as follows:

An observed value of p' belongs to a sampling distribution that

1. is approximately normal
2. has a mean $\mu_{p'}$ equal to p
3. has a standard error $\sigma_{p'}$ equal to $\sqrt{pq/n}$

This approximation to the normal distribution is considered reasonable whenever n is greater than 20 and both np and nq are greater than 5.

RECALL The standard deviation of a sampling distribution is called the *standard error*.

As a result of these new definitions for μ and σ, the calculated **value of z** in Step 3 of a hypothesis test concerning p is obtained by using the formula

$$z = \frac{p' - p}{\sqrt{pq/n}} \quad \text{where} \quad p' = \frac{x}{n} \quad (9\text{-}3)$$

The value of p to be used in formula (9-3) will be the value stated in the null hypothesis.

▼ ILLUSTRATION 9-7

While talking about the cars that fellow students drive, Tom made the claim that at least 15% of the students drive convertibles. Bill found this hard to believe and decided to check the validity of Tom's claim, so he took a random sample. At a level of significance of 0.10, does Bill have sufficient evidence to reject Tom's claim if there were 17 convertibles in his sample of 200 cars?

SOLUTION

STEP 1 $H_0: p = 0.15$ ($\geq$) (at least 15%)

$H_a: p < 0.15$ (less than 15%)

STEP 2 $\alpha = 0.10$. z is found by using Table 5, Appendix F. See the accompanying figure:

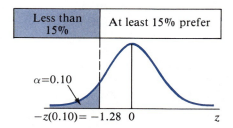

STEP 3
$$p' = \frac{17}{200} = 0.085$$

$$z = \frac{p' - p}{\sqrt{pq/n}} = \frac{0.085 - 0.150}{\sqrt{(0.15)(0.85)/200}} = \frac{-0.065}{\sqrt{0.00064}}$$

$$= \frac{-0.065}{0.025} = -2.6$$

$$z^* = -2.6$$

Compare this value with the test criteria, as shown in the accompanying figure:

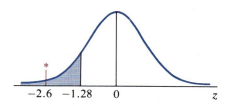

STEP 4 Decision Reject H_0 (z^* is in the critical region).

Section 9.2 ▼ INFERENCES ABOUT THE BINOMIAL PROBABILITY OF SUCCESS 453

Conclusion The evidence found contradicts Tom's claim. At the 0.10 level of significance, it appears that less than 15% of the students drive convertibles. ▲▲

The solution to Illustration 9-7 could have been carried out using the prob-value procedure. This alternative solution follows:

SOLUTION

STEP 1 $H_0: p = 0.15$ ($\geq$) (at least 15 percent)

$H_a: p < 0.15$ (less than 15 percent)

STEP 2 $\alpha = 0.10$

STEP 3 $z = \dfrac{p' - p}{\sqrt{pq/n}} = \dfrac{0.085 - 0.150}{\sqrt{(0.15)(0.85)/200}}$

$z^* = -2.60$

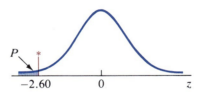

STEP 4 $P = P(z < z^*) = P(z < -2.60) = 0.5000 - 0.4953$

$P = 0.0047$ (see the accompanying figure)

STEP 5 At the 0.10 level of significance, the sample information is significant. That is, it appears that less than 15% of the students drive convertibles. ▲▲

When we estimate the true population proportion p, we will base our estimations on the observed value p'. The **confidence interval formula** is similar to the previous confidence interval formula.

confidence interval formula

$$p' - z(\alpha/2) \cdot \sqrt{\dfrac{p'q'}{n}} \quad \text{to} \quad p' + z(\alpha/2) \cdot \sqrt{\dfrac{p'q'}{n}} \qquad (9\text{-}4)$$

where $p' = x/n$ and $q' = 1 - p'$. Notice that the standard error, $\sqrt{pq/n}$, has been replaced by $\sqrt{p'q'/n}$. Since we do not know the value of p, we must use the best replacement available. That replacement is p', the observed value or the point estimate for p. This replacement will cause little change in the width of our confidence interval.

▼ ILLUSTRATION 9-8

Refer to Illustration 9-7. Suppose that Bill has taken his sample with the intention of estimating the value of p, the proportion of his fellow students who drive convertibles.

a. Find the best point estimate for p that he could use.
b. Determine the 90% confidence interval estimate for the true value of p by using formula (9-4).

SOLUTION

a. The best point estimate of p is 0.085, the observed value of p'.
b. The confidence interval is $p' \pm z(\alpha/2) \cdot \sqrt{p'q'/n}$, where $q' = 1 - p'$, and $1 - \alpha = 0.90$; therefore, $z(\alpha/2) = z(0.05) = 1.65$. Thus the confidence interval is

$$0.085 \pm (1.65) \cdot \sqrt{\frac{(0.085)(0.915)}{200}}$$

$$0.085 \pm (1.65)\sqrt{0.000389}$$

$$0.085 \pm (1.65)(0.020)$$

$$0.085 \pm 0.033$$

$$0.085 - 0.033 = \mathbf{0.052} \qquad 0.085 + 0.033 = \mathbf{0.118}$$

0.052 to **0.118, the 0.90 confidence interval for P**

That is, the true proportion of students who drive convertibles is between 0.052 and 0.118, with 90% confidence. ▲▲

By using the maximum-error part of the confidence interval formula, it is possible to determine the size of the sample that must be taken in order to estimate p with a desired accuracy. The **maximum error of estimate for a proportion** is

maximum error of estimate

$$E = z(\alpha/2) \cdot \sqrt{\frac{pq}{n}} \qquad (9\text{-}5)$$

When using this formula, we must decide how accurate our answer must be. (Remember that we are estimating p. Therefore, E will be expressed in hundredths.) We need to establish the level of confidence we wish to work with. If you have any indication of the value of p, use this value for p and $q = 1 - p$. If there is no indication of an approximate value for p, then by assigning p the value 0.5, you will obtain the largest possible sample size that is required.

Section 9.2 ▼ INFERENCES ABOUT THE BINOMIAL PROBABILITY OF SUCCESS

For ease of use, formula (9-5) can be expressed as

$$n = \frac{[z(\alpha/2)]^2 \cdot p \cdot q}{(E)^2} \tag{9-6}$$

▼ ILLUSTRATION 9-9

Determine the sample size that is required to estimate the true proportion of blue-eyed community college students if you want your estimate to be within 0.02 with 90% confidence.

SOLUTION

STEP 1 $1 - \alpha = 0.90$; therefore, $z(\alpha/2) = z(0.05) = 1.65$

STEP 2 $E = 0.02$

STEP 3 Use $p = 0.5$; therefore, $q = 1 - p = 0.5$.

STEP 4 Use formula (9-6) to find n:

$$n = \frac{(1.65)^2 \cdot (0.5) \cdot (0.5)}{(0.02)^2} = \frac{0.680625}{0.0004} = 1701.56$$

$$= 1702 \qquad \blacktriangle\blacktriangle$$

▼ ILLUSTRATION 9-10

An automobile manufacturer purchases bolts from a supplier who claims the bolts to be approximately 5% defective. Determine the sample size that will be required to estimate the true proportion of defective bolts if we want our estimate to be within 0.02 with 90% confidence.

SOLUTION

STEP 1 $1 - \alpha = 0.90$; therefore, $z(\alpha/2) = z(0.05) = 1.65$

STEP 2 $E = 0.02$

STEP 3 The supplier's claim is "5% defective"; thus $p = 0.05$; therefore, $q = 1 - p = 0.95$.

STEP 4 Use formula (9-6) to find n:

$$n = \frac{(1.65)^2 \cdot (0.05)(0.95)}{(0.02)^2} = \frac{0.12931875}{0.0004} = 323.3$$

$$= 324 \qquad \blacktriangle\blacktriangle$$

Notice the difference in the sample size required in Illustrations 9-9 and 9-10. The only real difference between the problems is the value that was used for p. In

Illustration 9-9 we used $p = 0.5$, and in Illustration 9-10 we used $p = 0.05$. Recall that $p = 0.5$ gives a sample of maximum size. Thus it will be of great advantage to have an indication of the value expected for p if p is much different from 0.5.

Case Study 9-2

Student Poll Says Alcohol No. 1 High School Problem

Forty-four percent of 1181 student leaders said alcohol was the biggest problem facing high schools. Why is 44% a statistic? For what parameter is 44% a point estimate? What is the variable? (See Exercise 9.39, p. 461.)

MILWAUKIE, ORE.—A lack of self-esteem, a desire to fit in and peer pressure drive high school students to drink, student leaders say.

A USA Today/Gannett News Service poll of 1,181 student leaders at the annual National Association of Student Councils convention here indicates alcohol is the biggest problem facing high schools as the 1990s begin.

Forty-four percent said drinking is the No. 1 problem facing their high schools.

Source: Copyright 1990, Gannett News Services. Reprinted with permission.

Case Study 9-3

Many Still Tan Despite Skin Cancer

The following article lists the results of a study involving more than 1000 people who underwent skin-cancer surgery. Twenty-four percent admitted they were still sun tanning one year later. Why is 24% a statistic? For what parameter is 24% a point estimate? (See Exercise 9.40, p. 461.)

CHICAGO—A study of more than 1,000 people who underwent skin-cancer surgery showed that nearly half refused to change their sunning habits.

Forty-four percent said they had not changed their habits concerning outdoor activities one year after removal of their cancers, despite repeated counseling from a physician that it would be beneficial to do so.

Thirty-eight percent said they still weren't using sunscreens, despite their doctor advising it, said the study, published in April's issue of the *American Medical Association's Archives of Dermatology*.

Though giving up deliberate sunbathing was the most common behavior among subjects after surgery, 24 percent admitted they were still suntanning a year later, the study reported.

"The attitude of these non-compliant individuals, who were usually women, was that skin cancer was not enough of a problem to give up a tan," said the author, Dr. June K. Robinson, a dermatologist at Northwestern University Medical School in Chicago.

The "non-compliant" patients, who did not listen to their physician's advice, said a suntan "made them feel good, and that the sunscreens have an objectionable sticky feeling," wrote Robinson, who [was] traveling yesterday and could not be reached for further comment.

All subjects had undergone surgery between May 1983 and May 1987 to remove non-melanoma cancers.

They were counseled before and after surgery by Robinson and a nurse, who advised them that they could reduce their risk for subsequent cancers by reducing sun exposure, wearing protective clothing and using sunscreens.

Source: "Many Still Tan Despite Skin Cancer," *Democrat and Chronicle*, April 19, 1990. Reprinted by permission.

Case Study 9-4

Gallup Report: Sampling Tolerances

The Gallup Report provided its readers with the accompanying information and table. The table gives the 0.95 confidence level sampling errors (maximum error of estimation) for various percentages and sample sizes. Can you verify the values on this table? (See Exercise 9.48, p. 462.)

SAMPLING ERROR

In interpreting survey results, it should be borne in mind that all sample surveys are subject to sampling error, that is, the extent to which the results may differ from those that would be obtained if the whole population surveyed had been interviewed. The size of such sampling errors depends largely on the number of interviews.

Table A shows the allowances that should be made for the sampling error of a percentage.

The table should be used as follows: Say a reported percentage is 33 for a group which includes 1500 respondents. Go to the row labeled "percentages near 30" and then to the column headed "1500." The number at this point is 3, which means that the 33 percent obtained in the sample is subject to a sampling error of plus or minus 3 points. Another way of saying it is that very probably (95 times out of 100) the average of repeated samplings would be somewhere between 30 and 36, with the most likely figure the 33 obtained.

TABLE A Sampling Tolerances

	Recommended Allowance for Sampling Error of a Percentage						
	In Percentage Points (at 95 in 100 confidence level) Size of Sample						
	1500	1000	750	600	400	200	100
Percentages near 10	2	2	3	4	4	5	7
Percentages near 20	2	3	4	4	5	7	9
Percentages near 30	3	4	4	5	6	8	10
Percentages near 40	3	4	4	5	6	9	11
Percentages near 50	3	4	4	5	6	9	11
Percentages near 60	3	4	4	5	6	9	11
Percentages near 70	3	4	4	5	6	8	10
Percentages near 80	2	3	4	4	5	7	9
Percentages near 90	2	2	3	4	4	5	7

Source: Copyright 1986 by the Gallup Report. Reprinted by permission.

▼▲ EXERCISES

9.25 State the null hypothesis, H_0, and the alternative hypothesis, H_a, that would be used to test the following claims:
  a. More than 60% of all students at our college work part-time jobs during the academic year.
  b. The probability of our team winning tonight is less than 0.50.
  c. No more than one-third of cigarette smokers are interested in quitting.
  d. At least 50% of all parents believe in spanking their children when appropriate.
  e. A majority of the voters will vote for the school budget this year.
  f. At least three-quarters of the trees in our county were seriously damaged by the storm.
  g. The results show the coin was not tossed fairly.
  h. The single-digit numbers generated by the computer do not seem to be random with regard to being odd or even.

9.26 Determine the test criteria that would be used to test the following hypotheses when z is used as the test statistic:
  a. $H_0: p = 0.5$ and $H_a: p > 0.5$, with $\alpha = 0.05$.
  b. $H_0: p = 0.5$ and $H_a: p \neq 0.5$, with $\alpha = 0.05$.
  c. $H_0: p = 0.4$ and $H_a: p < 0.4$, with $\alpha = 0.10$.
  d. $H_0: p = 0.7$ and $H_a: p > 0.7$, with $\alpha = 0.01$.

9.27 The binomial random variable, x, may be used as the test statistic when testing hypotheses about the binomial parameter, p. When n is small (say 15 or less), Table 4 in Appendix F provides the probabilities for each value of x separately, thereby making it unnecessary to estimate probabilities of the discrete binomial random variable with the continuous standard normal variable z. Use Table 4 and determine the value of α for each of the following:
  a. H_0: $p = 0.5$ and H_a: $p > 0.5$, where $n = 15$ and the critical region is $x = 12, 13, 14, 15$.
  b. H_0: $p = 0.3$ and H_a: $p < 0.3$, where $n = 12$ and the critical region is $x = 0, 1$.
  c. H_0: $p = 0.6$ and H_a: $p \neq 0.6$, where $n = 10$ and the critical region is $x = 0, 1, 2, 3, 9, 10$.
  d. H_0: $p = 0.05$ and H_a: $p > 0.05$, where $n = 14$ and the critical region is $x = 4, 5, 6, 7, \ldots, 14$.

9.28 Use Table 4 in Appendix F and determine the critical region used in testing each of the following hypotheses. (*Note*: Since x is discrete, choose critical regions that do not exceed the value of α given.)
  a. H_0: $p = 0.5$ and H_a: $p > 0.5$, where $n = 15$ and $\alpha = 0.05$.
  b. H_0: $p = 0.5$ and H_a: $p \neq 0.5$, where $n = 14$ and $\alpha = 0.05$.
  c. H_0: $p = 0.4$ and H_a: $p < 0.4$, where $n = 10$ and $\alpha = 0.10$.
  d. H_0: $p = 0.7$ and H_a: $p > 0.7$, where $n = 13$ and $\alpha = 0.01$.

9.29 The fairness (balance) of a coin is in question. It is believed that the probability of a head, p, is greater than 0.5. The null and alternative hypotheses are H_0: $p = 0.5$ and H_a: $p > 0.5$. The coin is flipped 14 times and the number of heads is observed. Calculate α for each of the following critical regions:
  a. $x = 10, 11, 12, 13, 14$ b. $x = 11, 12, 13, 14$ c. $x = 12, 13, 14$

9.30 We are testing H_0: $p = 0.2$ and decide to reject H_0 if after 15 trials we observe more than 5 successes.
  a. State an appropriate alternative hypothesis.
  b. What is the level of significance of this test?
  c. If we observe 5 successes, do we reject H_0?
  d. If we observe 6 successes, do we reject H_0?
  e. If H_0 is $p = 0.1$ and we use the same decision rule, what happens to the level of significance?

9.31 You are testing the hypothesis $p = 0.7$ and have decided to reject this hypothesis if after 15 trials you observe 14 or more successes.
  a. If the null hypothesis is true and you observe 13 successes, then which of the following will you do? (1) Correctly fail to reject H_0? (2) Correctly reject H_0? (3) Commit a type I error? (4) Commit a type II error?
  b. Find the significance level of your test.
  c. If the true probability of success is $\frac{1}{2}$ and you observe 13 successes, then which of the following will you do: (1) correctly fail to reject H_0? (2) correctly reject H_0? (3) commit a type I error? (4) commit a type II error?
  d. Calculate the prob-value for your hypothesis test after 13 successes are observed.

9.32 You are testing the null hypothesis $p = 0.4$ and will reject this hypothesis if z^* is less than -2.05.

    a. If the null hypothesis is true and you observe z^* equal to -2.12, then which of the following results: (1) correctly fail to reject H_0, (2) correctly reject H_0, (3) commit a type I error, (4) commit a type II error?

    b. What is the significance level for this test?

    c. What is the p-value for $z^* = -2.12$?

9.33 An insurance company states that 90% of its claims are settled within 30 days. A consumer group selected a random sample of 75 of the company's claims to test this statement. If the consumer group found that 55 of the claims were settled within 30 days, do they have sufficient reason to support their contention that fewer than 90% of the claims are settled within 30 days? Use $\alpha = 0.05$.

    a. Solve using the classical approach.

    b. Solve using the prob-value approach.

9.34 A county judge has agreed that he will give up his county judgeship and run for a state judgeship unless there is evidence that more than 25% of his party is in opposition. A random sample of 800 party members included 217 who opposed him. Does this sample suggest that he should run for the state judgeship in accordance with his agreement? Carry out this hypothesis test by using $\alpha = 0.10$.

    a. Solve using the classical approach.

    b. Solve using the prob-value approach.

9.35 A politician claims that she will receive 60% of the vote in an upcoming election. The results of a properly designed random sample of 100 voters showed that 50 of those sampled will vote for her. Is it likely that her assertion is correct at the 0.05 level of significance?

    a. Solve using the classical approach.

    b. Solve using the prob-value approach.

9.36 The full-time student body of a college is composed of 50% males and 50% females. Does a random sample of students (30 male, 20 female) from an introductory chemistry course show sufficient evidence to reject the hypothesis that the proportion of male and of female students who take this course is the same as that of the whole student body? Use $\alpha = 0.05$.

    a. Solve using the classical approach.

    b. Solve using the prob-value approach.

9.37 The January 14, 1991 issue of *Newsweek* (page 36) reported that in a telephone survey of 759 adults, 27% were worried about maintaining their mortgage payments. Suppose you conducted a similar survey and found that 150 out of 759 were worried about maintaining their mortgage payments. Does your evidence (150 of 759) show that the percentage of worriers is significantly less than the 27% reported in *Newsweek*? Test using $\alpha = 0.05$.

9.38 The article "Making Up for Lost Time" (*US News and World Report*, July 30, 1990, page 61) reported that more than half of the country's workers aged 45 to 64 want to quit work before they reach age 65. Suppose you conduct a survey of

Section 9.2 ▼ INFERENCES ABOUT THE BINOMIAL PROBABILITY OF SUCCESS   461

1000 randomly chosen workers in order to test $H_0: p = 0.5$ versus $H_a: p < 0.5$, where p represents the proportion who want to quit before they reach age 65. If 460 of the 1000 respond that they want to quit work before age 65,
  a. calculate the value of the test statistic and the prob-value.
  b. complete the hypothesis test using $\alpha = 0.01$.

9.39 Forty-four percent of 1181 student leaders said alcohol was biggest problem facing high schools. (See Case Study 9-2 on p. 456.)
  a. Why is 44% a statistic?
  b. What parameter is 44% a point estimate for?
  c. What population was sampled?
  d. What variable was used?

9.40 "Many Still Tan Despite Skin Cancer," (Case Study 9-3 on p. 456) lists the results of a study involving more than 1000 people who had previously undergone skin-cancer surgery.
  a. What population was sampled?
  b. Why is the 24% reported a statistic?
  c. What parameter is 24% a point estimate of?

9.41 A telephone survey was conducted to estimate the proportion of households with a personal computer. Of the 350 households surveyed, 75 had a personal computer.
  a. Give a point estimate for the proportion in the population who have a personal computer.
  b. Give the maximum error of estimate with 95% confidence.

9.42 A bank randomly selected 250 checking account customers and found that 110 of them also had savings accounts at this same bank. Construct a 0.95 confidence interval estimate for the true proportion of checking account customers who also have savings accounts.

9.43 In a sample of 60 randomly selected students, only 22 favored the amount being budgeted for next year's intramural and interscholastic sports. Construct the 0.99 confidence interval estimate for the proportion of all students who support the proposed budget amount.

9.44 According to an article in *Fortune* magazine (1990, page 82), 62% of American 13-year-olds spend less than one hour a week on mathematics homework. If this result was based on a sample of 300 13-year-olds, find a 95% confidence interval estimate for the true proportion of 13-year-olds who spend less than one hour per week on mathematics homework.

9.45 "Parents should spank children when they think it is necessary, said 51 percent of adult respondents to a survey—though most child-development experts say spanking is not appropriate. The survey of 7,225 adults . . . was co-sponsored by *Working Mother* magazine and Epcot Center at Walt Disney World." This statement appeared in the Rochester *Democrat & Chronicle* (December 20, 1990). Find the 0.99 confidence maximum error of estimate for the parameter p, P(should spank when necessary), for the adult population.

9.46 In a survey of 12,000 adults aged 19 to 74, National Cancer Institute researchers found that 9% in the survey ate at least the recommended two servings of fruit or juice and three servings of vegetables per day (*Ladies Home Journal,* April 1991). Use this information to determine a 95% confidence interval estimate for the true proportion in the population who follow the recommendation.

9.47 Construct 90% confidence intervals for the binomial parameter, p, for each of the following pairs of values. Write your answers on the chart.

Observed Proportion $p' = x/n$	Sample Size	Lower Limit	Upper Limit
a. $p' = 0.3$	$n = 30$		
b. $p' = 0.7$	$n = 30$		
c. $p' = 0.5$	$n = 10$		
d. $p' = 0.5$	$n = 100$		
e. $p' = 0.5$	$n = 1000$		

    f. Compare answers (a) and (b).
    g. Compare answers (c), (d), and (e).

9.48 Calculate the 0.95 confidence maximum error of estimate for p if
    a. p is near 0.1 and $n = 1000$.
    b. p is near 0.2 and $n = 100$.
    c. p is near 0.5 and $n = 50$.
    d. Compare these maximum errors to the sampling errors reported in Table A of Case Study 9-4, p. 457.

9.49 A bank believes that approximately $\frac{2}{5}$ of its checking account customers have used at least one other service provided by the bank within the last six months. How large a sample will be needed to estimate the true proportion to within 5% at the 0.98 level of confidence?

9.50 Members of the student senate are working on a proposal concerning a pub on campus. They feel that if 60% of the student body favor the proposal, it should be passed. How large a sampling needs to be taken to enable them to estimate the proportion of students who favor the proposal to within 4% with 99% confidence?

9.51 A hospital administrator wants to conduct a telephone survey to determine the proportion of people in a city who have been hospitalized for at least three days in the last five years. How large a sample must she take to be 95% confident that the sample proportion will be within 0.03 of the true proportion of the city?

9.52 According to the May 1990 issue of *Good Housekeeping* (p. 90), only about 14% of lung cancer patients survive for five years after diagnosis. Suppose you wanted to see if this survival rate were still true. How large a sample would you need to take to estimate the true proportion surviving for five years after diagnosis to within 1% with 95% confidence? (Use the 14% as the value of p.)

9.3 ▼ Inferences About Variance and Standard Deviation

Problems often arise that require us to make inferences about variability. For example, a soft drink bottling company has a machine that fills 32-oz bottles. It needs to control the variance σ^2 (or standard deviation σ) in the amount x of soft drink put in each bottle. The mean amount placed in each bottle is important, but a correct mean amount does not ensure that the filling machine is working correctly. If the variance is too large, there could be many bottles that are overfilled and many that are underfilled. Thus this bottling company will want to maintain as small a variance (or standard deviation) as possible.

We will study two kinds of inferences in this section: (1) the hypothesis test concerning the variance (or standard deviation) of one population and (2) the estimation of the variance (or standard deviation) of one population. In these two inferences it is customary to talk about variance instead of the standard deviation, because the techniques employ the sample variance rather than the standard deviation. However, remember that the standard deviation is the positive square root of the variance; thus to talk about the variance of a population is comparable to talking about the standard deviation.

Let's return to our problem. Suppose the bottling company wishes to detect when the variability in the amount of soft drink placed in each bottle gets out of control. A variance of 0.0004 is considered acceptable, and the company will want to adjust the bottle-filling machine when the variance becomes larger than this value. The decision will be made by using the hypothesis test procedure. The null hypothesis is "the variance is no larger than the specified value 0.0004"; the alternative hypothesis is "the variance is larger than 0.0004."

H_0: $\sigma^2 = 0.0004$ ($\leq$)   (variance is not larger then 0.0004)

H_a: $\sigma^2 > 0.0004$ (variance is larger than 0.0004)

The test statistic that will be used in making a decision about the null hypothesis is **chi-square**, χ^2 (χ is the Greek lowercase letter chi). The calculated value of chi-square will be obtained by using the formula

$$\chi^2 = \frac{(n-1)s^2}{\sigma^2} \qquad (9\text{-}7)$$

where s^2 is the sample variance, n is the sample size, and σ^2 is the value specified in the null hypothesis.

When random samples are drawn from a normal population of a known variance σ^2, the quantity $(n-1)s^2/\sigma^2$ possesses a probability distribution that is known as the ***chi-square distribution***. The equations that define the chi-square distribution are not given here; they are beyond the level of this book. However, to use the chi-square distribution, we must be aware of the following properties (see also Figure 9-4):

> **PROPERTIES OF THE CHI-SQUARE DISTRIBUTION**
> 1. χ^2 is nonnegative in value; it is zero or positively valued.
> 2. χ^2 is not symmetrical; it is skewed to the right.
> 3. There are many χ^2 distributions. As with the t-distribution, there is a different χ^2 distribution for each degree-of-freedom value.

FIGURE 9-4
Various Chi-square Distributions

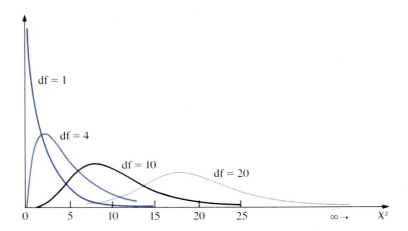

For the inferences discussed in this section, the number of degrees of freedom df is equal to $n - 1$.

The critical values for chi-square are obtained from Table 7 in Appendix F. The critical values will be identified by two values: degrees of freedom df and the area under the curve to the right of the critical value being sought. Thus $\chi^2(\text{df}, \alpha)$ is the symbol used to identify the critical value of chi-square with df degrees of freedom and with α being the area to the right, as shown in Figure 9-5. Since the chi-square distribution is not symmetrical, the critical values associated with both tails are given in Table 7.

FIGURE 9-5
Chi-square Distribution Showing $\chi^2(\text{df}, \alpha)$

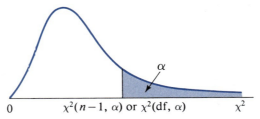

df $= (n-1)$, degrees of freedom;
α is the area under curve to the right of a particular value

ILLUSTRATION 9-11

Find $\chi^2(20, 0.05)$.

SOLUTION In Table 7 you will find the value shown in the following table. Therefore, $\chi^2(20, 0.05) =$ **31.4**.

df	Area Under Curve to the Right
	... 0.050 ...
⋮	
20	31.4
⋮	

ILLUSTRATION 9-12

Find $\chi^2(14, 0.90)$.

SOLUTION df = 14 and the area to the right of the critical value is 0.90, as shown in the following figure. Therefore, $\chi^2(14, 0.90) =$ **7.79**.

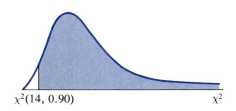

df	Area Under Curve to the Right
	... 0.90 ...
⋮	
14	7.79
⋮	

NOTE When df > 2, the mean value of the chi-square distribution is df. The mean is located to the right of the mode (the value where the curve reaches its high point). By locating the value of df on your sketch of the χ^2 distribution, you will establish an approximate scale so that the values can be located in their respective positions. See Figure 9-6.

FIGURE 9-6
Location of Mean and Mode for χ^2 Distribution

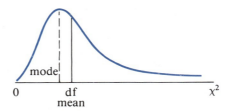

▼ **ILLUSTRATION 9-13**

Recall that the soft drink bottling company wanted to control the variance by not allowing the variance to exceed 0.0004. Does a sample of size 28 with a variance of 0.0007 indicate that the bottling process is out of control (with regard to variance) at the 0.05 level?

SOLUTION

STEP 1 H_0: $\sigma^2 = 0.0004$ ($\leq$)   (not out of control)

H_a: $\sigma^2 > 0.0004$ (out of control)

STEP 2 $\alpha = 0.05$ and $n = 28$; therefore, df = 27. The test statistic is χ^2, and the critical region is the right tail, with an area of 0.05. $\chi^2(27, 0.05)$ is found in Table 7. See the following figure:

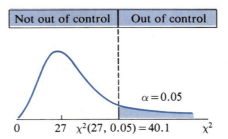

STEP 3

$$\chi^2 = \frac{(n-1)s^2}{\sigma^2} = \frac{(28-1)(0.0007)}{0.0004}$$

$$= \frac{(27)(0.0007)}{0.0004} = \frac{0.0189}{0.0004} = 47.25$$

$\chi^{2*} = \mathbf{47.25}$

See Figure 9-7.

FIGURE 9-7

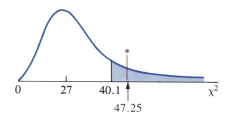

STEP 4 Decision Reject H_0 (χ^{2*} is in the critical region).

Conclusion At the 0.05 level of significance, we conclude that the bottling process is out of control with regard to the variance.

The prob-value can be estimated for hypothesis tests using the chi-square test statistic in much the same manner as when Student's t was used. ▲▲

▼ ILLUSTRATION 9-14

Find the prob-value for the hypothesis test in Illustration 9-13.

$$H_0: \sigma^2 = 0.0004$$
$$H_a: \sigma^2 > 0.0004$$
$$\text{df} = 27 \quad \text{and} \quad \chi^{2*} = 47.25$$

SOLUTION

$$P = P(\chi^2 > 47.25; \text{ see the accompanying figure})$$

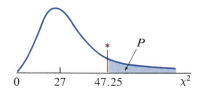

By inspecting the df = 27 row of Table 7, we find that 47.25 is between 47.0 and 49.7. Therefore the prob-value is between **0.005** and **0.010**. ▲▲

▼ ILLUSTRATION 9-15

One of the factors used in determining the usefulness of a particular exam as a measure of students' abilities is the amount of "spread" that occurs in the grades. A set of test results has little value if the range of the grades is very small. However, if the range of grades is quite large, there is a definite difference in the scores achieved by the "better" students and the scores achieved by the "poorer" students.

On an exam with a total of 100 points, it has been claimed that a standard deviation of 12 points is desirable. To determine whether or not the last one-hour exam was a good test, a professor tested this hypothesis at $\alpha = 0.05$ by using the exam scores of the class. There were 28 scores and the standard deviation of those 28 scores was found to be 10.5. Does the professor have evidence, at the 0.05 level of significance, that this exam does not have the specified standard deviation?

SOLUTION The information given is $n = 28$, $s = 10.5$, and $\alpha = 0.05$.

STEP 1 H_0: $\sigma = 12$

H_a: $\sigma \neq 12$

STEP 2 $\alpha = 0.05$; the critical values are $\chi^2(27, 0.975) = 14.6$ and $\chi^2(27, 0.025) = 43.2$. See the accompanying figure.

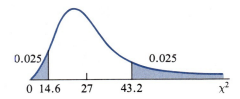

STEP 3

$$\chi^2 = \frac{(n-1)s^2}{\sigma^2} = \frac{(27)(10.5)^2}{(12)^2} = \frac{2976.75}{144} = 20.6719$$

$\chi^{2*} = \mathbf{20.67}$

The accompanying figure shows this value compared to the test criteria.

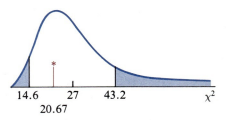

STEP 4 Decision Fail to reject H_0.

Conclusion At the 0.05 level of significance, the professor does not have sufficient evidence to claim that the standard deviation is different from 12. ▲▲

These tests for variance may be one-tailed or two-tailed tests, in accordance with the statement of the claim being tested.

Section 9.3 ▼ INFERENCES ABOUT VARIANCE AND STANDARD DEVIATION

The formula for chi-square may be reworked to give the values at the extremities of the confidence interval:

$$\chi^2 = \frac{(n-1) \cdot s^2}{\sigma^2} \quad \text{(solve for } \sigma^2\text{)}$$

$$\sigma^2 = \frac{(n-1) \cdot s^2}{\chi^2} \tag{9-8}$$

When constructing a $1 - \alpha$ confidence interval, the critical values of chi-square are separately substituted into formula (9-8) to obtain the two endpoints of the confidence interval of estimation. Note that $\chi^2(\mathrm{df}, 1 - \alpha/2)$ is less than $\chi^2(\mathrm{df}, \alpha/2)$. Therefore, after dividing, the numbers will be in the opposite order, yielding the following **confidence interval for variance**:

confidence interval for variance

$$\frac{(n-1)s^2}{\chi^2(\mathrm{df}, \alpha/2)} \quad \text{to} \quad \frac{(n-1)s^2}{\chi^2(\mathrm{df}, 1 - \alpha/2)} \tag{9-9}$$

If the confidence interval for the standard deviation is desired, we need only take the square root of each of the numbers in formula (9-9).

$$s \cdot \sqrt{\frac{(n-1)}{\chi^2(\mathrm{df}, \alpha/2)}} \quad \text{to} \quad s \cdot \sqrt{\frac{(n-1)}{\chi^2(\mathrm{df}, 1 - \alpha/2)}} \tag{9-10}$$

▼ ILLUSTRATION 9-16

Using the professor's results from Illustration 9-15 ($n = 28$, $s = 10.5$), calculate the 95% confidence interval for the estimate of the population variance and standard deviation.

SOLUTION The information given is $n = 28$ and $s = 10.5$. For a 95% confidence interval, $\alpha = 0.05$ and hence $\alpha/2 = 0.025$. The critical values for χ^2 are shown in the accompanying figure.

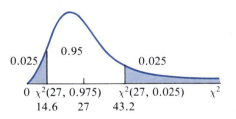

The confidence interval, using formula (9-9), is

$$\frac{(27)(10.5)^2}{43.2} \quad \text{to} \quad \frac{(27)(10.5)^2}{14.6}$$

$$\frac{2976.75}{43.2} \quad \text{to} \quad \frac{2976.75}{14.6}$$

68.9 to 203.9, the 0.95 confidence interval for σ^2

That is, with 95% confidence we estimate the population variance to be between 68.9 and 203.9.

The confidence interval for the standard deviation can be found by taking the square root of 68.9 and of 203.9. The 95% confidence interval estimate for the standard deviation is **8.3** to **14.3**.

▼▲ EXERCISES

9.53 State the null hypothesis, H_0, and the alternative hypothesis, H_a, that would be used to test the following claims:
  a. The standard deviation has increased from its previous value of 24.
  b. The standard deviation is no larger than 0.5 oz.
  c. The standard deviation is not equal to 10.
  d. The variance is no less than 18.
  e. The variance is different from the value of 0.025, the value called for in the specs.
  f. The variance has increased from 34.5.

9.54 Determine the test criteria that would be used to test the following:
  a. H_0: $\sigma = 0.5$ and H_a: $\sigma > 0.5$, with $n = 18$ and $\alpha = 0.05$.
  b. H_0: $\sigma^2 = 8.5$ and H_a: $\sigma^2 < 8.5$, with $n = 15$ and $\alpha = 0.01$.
  c. H_0: $\sigma = 20.3$ and H_a: $\sigma \neq 20.3$, with $n = 10$ and $\alpha = 0.10$.
  d. H_0: $\sigma^2 = 0.05$ and H_a: $\sigma^2 \neq 0.05$, with $n = 8$ and $\alpha = 0.02$.
  e. H_0: $\sigma = 0.5$ and H_a: $\sigma < 0.5$, with $n = 12$ and $\alpha = 0.10$.

9.55 Find these critical values by using Table 7 of Appendix F.
  a. $\chi^2(18, 0.01)$ b. $\chi^2(16, 0.025)$ c. $\chi^2(8, 0.10)$
  d. $\chi^2(28, 0.01)$ e. $\chi^2(22, 0.95)$ f. $\chi^2(10, 0.975)$
  g. $\chi^2(50, 0.90)$ h. $\chi^2(24, 0.99)$

9.56 Using the notation of Exercise 9.55, name and find the critical values of χ^2.

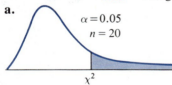
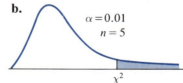
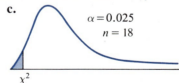
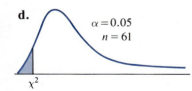
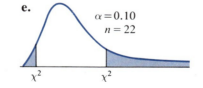
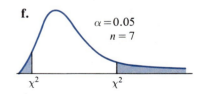
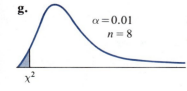
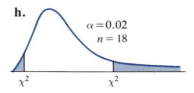

Section 9.3 ▼ INFERENCES ABOUT VARIANCE AND STANDARD DEVIATION

9.57 **a.** What value of chi-square for 5 degrees of freedom subdivides the area under the distribution curve such that 5% is to the right and 95% is to the left?
b. What is the value of the 95th percentile for the chi-square distribution with 5 degrees of freedom?
c. What is the value of the 90th percentile for the chi-square distribution with 5 degrees of freedom?

9.58 **a.** The central 90% of the chi-square distribution with 11 degrees of freedom lies between what values?
b. The central 95% of the chi-square distribution with 11 degrees of freedom lies between what values?
c. The central 99% of the chi-square distribution with 11 degrees of freedom lies between what values?

9.59 For a chi-square distribution having 12 degrees of freedom, find the area under the curve for chi-square values ranging from 3.57 to 21.0.

9.60 For a chi-square distribution having 35 degrees of freedom, find the area under the curve between $\chi^2(35, 0.96)$ and $\chi^2(35, 0.15)$.

9.61 A random sample of 51 observations was selected from a normally distributed population. The sample mean was $\bar{x} = 98.2$, and the sample variance was $s^2 = 37.5$. Does this sample show sufficient reason to conclude that the population standard deviation is not equal to 8 at the 0.05 level of significance?

9.62 In the past the standard deviation of weights of certain 32.0-oz packages filled by a machine was 0.25 oz. A random sample of 20 packages showed a standard deviation of 0.35 oz. Is the apparent increase in variability significant at the 0.10 level of significance?
a. Solve using the classical approach.
b. Solve using the prob-value approach.

9.63 A car manufacturer claims that the miles per gallon for a certain model has a mean equal to 40.5 mi with a standard deviation equal to 3.5 mi. Use the following data, obtained from a random sample of 15 such cars, to test the hypothesis that the standard deviation differs from 3.5. Use $\alpha = 0.05$.

| 37.0 | 38.0 | 42.5 | 45.0 | 34.0 | 32.0 | 36.0 | 35.5 |
| 38.0 | 42.5 | 40.0 | 42.5 | 36.0 | 30.0 | 37.5 | |

a. Solve using the classical approach.
b. Solve using the prob-value approach.

9.64 A commercial farmer harvests his entire field of a vegetable crop at one time. Therefore he would like to plant a variety of green beans that mature all at one time (small standard deviation between maturity times of individual plants). A seed company has developed a new hybrid strain of green beans that it believes to be better for the commercial farmer. The maturity time of the standard variety has an average of 50 days and a standard deviation of 2.1 days. A random sample of 30 plants of the new hybrid showed a standard deviation of 1.65 days. Does this sample show a significant lowering of the standard deviation at the 0.05 level of significance?
a. Solve using the classical approach.
b. Solve using the prob-value approach.

9.65 The variability of the scores made on the TOEFL exam was of interest. A random sample of ten scores made by foreign-born students on this exam were

495   525   580   605   552   490   590   505   551   600

Find a 95% confidence interval for the standard deviation of TOEFL exam scores.

9.66 A study involving the effect of clonidine on the cessation of cigarette smoking (*Clinical Pharmacological Therapy,* September 1988, pages 265–267) reported for 93 subjects using clonidine to help stop smoking that their average consumption at the beginning of the experiment was 1.57 packs/day with a standard deviation equal to 0.62. Set a 95% confidence interval on the standard deviation for the number of packs smoked per day for the population from which the sample was drawn. (Use chi-square value for df = 90.)

9.67 x is a random variable with a normal distribution. A sample of size 22 resulted in $\Sigma x = 397.3$ and $\Sigma x^2 = 7374.09$.
   a. What would be the point estimate for the population variance?
   b. What would be the 0.90 confidence interval estimate for the population variance?
   c. What would be the 0.90 confidence interval estimate for the population standard deviation?

9.68 In a study involving cooperative learning (*Journal of Educational Psychology,* Vol. 74, no. 4, 1982, pages 475–484), students were put into groups of three and participated in what is referred to as cooperative learning. Two different types of groups were used: Uniform Ability groups and Mixed Ability groups. The standard deviations in achievement scores were as follows:

Group	n	s
Uniform	14	6.44
Mixed	21	5.62

Find a 95% confidence interval estimate for the population standard deviation for each group.

IN RETROSPECT

We have studied inferences, both hypothesis testing and confidence interval estimation, for three basic population parameters: mean μ, proportion p, and standard deviation σ. When we make inferences about a single population, we are usually concerned with one of these three values. Table 9-1 identifies the formula that is used in each of the inferences for problems involving a single population.

CHAPTER EXERCISES

TABLE 9-1
Formulas to Use for Inferences Involving a Single Population

Situation	Test Statistic	Formula to Be Used	
		Hypothesis Test	Interval Estimate
One mean			
σ known	z	(7-2)	(8-2)
σ unknown	t	(9-1)	(9-2)
One proportion	z	(9-3)	(9-4)
One standard deviation	χ^2	(9-7)	(9-10)
One variance	χ^2	(9-7)	(9-9)

In this chapter we also used the maximum error of estimate term of formula (9-6) to determine the size of sample required to make estimates about the population proportion with the desired accuracy.

Case Study 9-4 reports several observed sample percentages as point estimates of population proportions. The article gives only the point estimates. However, the sampling error explanation that accompanies the article refers to the 95% confidence interval estimate. By combining a point estimate with its corresponding maximum error of estimate (sampling tolerance) we can construct an interval estimate. In Exercise 9.50 you did some calculating to verify the 95% level of confidence.

In the next chapter we will discuss inferences about two populations whose respective means, proportions, and standard deviations are to be compared.

CHAPTER EXERCISES

9.69 The addition of a new accelerator is claimed to decrease the drying time of latex paint by more than 4%. Several test samples were conducted with the following percentage decrease in drying time:

5.2   6.4   3.8   6.3   4.1   2.8   3.2   4.7

  a. Is there sufficient evidence to show that this accelerator has decreased the drying time by significantly more than 4% at the 0.01 level? (The sample mean and standard deviation were found in answering Exercise 2.107).
  b. Did the statistical decision reached in (a) result in the same conclusion as you expressed in answering part (c) of Exercise 2.107 for this same data?

9.70 It has been suggested that abnormal male children tend to occur more in children born to older-than-average parents. Case histories of 20 abnormal males were obtained and the ages of the 20 mothers were

31   21   29   28   34   45   21   41   27   31
43   21   39   38   32   28   37   28   16   39

The mean age at which mothers in the general population give birth is 28.0 years.
  a. Calculate the sample mean and standard deviation.
  b. Does the sample give sufficient evidence to support the claim that abnormal male children have older-than-average mothers? Use $\alpha = 0.05$.
     (1) Solve using the classical approach.
     (2) Solve using the prob-value approach.

9.71 The water pollution readings at State Park Beach seem to be lower than last year. A sample of 12 readings was randomly selected from the records of this year's daily readings:

3.5	3.9	2.8	3.1	3.1	3.4
4.8	3.2	2.5	3.5	4.4	3.1

Does this sample provide sufficient evidence to conclude that the mean of this year's pollution readings is significantly lower than last year's mean of 3.8 at the 0.05 level?

9.72 In a large cherry orchard the average yield has been 4.35 tons per acre for the last several years. A new fertilizer was tested on 15 randomly selected one-acre plots. The yields from these plots follow:

3.56	5.00	4.88	4.93	3.92
4.25	5.12	5.13	4.79	4.45
5.35	4.81	3.48	4.45	4.72

At the 0.05 level of significance, do we have sufficient evidence to claim that there was a significant increase in production?
  a. Solve using the classical approach.
  b. Solve using the prob-value approach.

9.73 The mean for a standardized reading test used in the state of Nebraska is 80. A school district wishes to test the hypothesis that its mean is different from that of the state. Twenty randomly selected students were tested and the results summarized by $\bar{x} = 77.5$ and $s = 2.5$. Complete the hypothesis test using $\alpha = 0.05$.
  a. Solve using the classical approach.
  b. Solve using the prob-value approach.

9.74 A manufacturer of television sets claims that the maintenance expenditures for its product will average no more than $50 during the first year following the expiration of the warranty. A consumer group has asked you to substantiate or discredit the claim. The results of a random sample of 50 owners of such television sets showed that the mean expenditure was $61.60 and the standard deviation was $32.46. At the 0.01 level of significance, should you conclude that the producer's claim is true or not likely to be true?
  a. Solve using the classical approach.
  b. Solve using the prob-value approach.

9.75 Determine the prob-value for the following hypothesis tests involving the Student's t-distribution with 10 degrees of freedom:
  a. $H_0: \mu = 15.5$ $H_a: \mu < 15.5$ $t^* = -2.01$
  b. $H_0: \mu = 15.5$ $H_a: \mu > 15.5$ $t^* = 2.01$
  c. $H_0: \mu = 15.5$ $H_a: \mu \neq 15.5$ $t^* = 2.01$

d. $H_0: \mu = 15.5$ $H_a: \mu \neq 15.5$ $t^* = -2.01$

9.76 According to an article in *Changing Times* (March 1991, page 27), financial assets in the United States fell by $6700 per household between the end of 1989 and the end of 1990. Suppose that in order to check this statement a survey of 500 households is conducted and that the decrease in financial assets was $6300 with a standard deviation equal to $3250. Calculate the value of the test statistic and the *p*-value for the research hypothesis $\mu < \$6700$.

9.77 One of the objectives of a large medical study was to estimate the mean physician fee for cataract removal. For 25 randomly selected cases the mean fee was found to be $1550 with a standard deviation of $125. Set a 99% confidence interval on μ, the mean fee for all physicians.

9.78 A natural gas utility is considering a contract for purchasing tires for its fleet of service trucks. The decision will be based on expected mileage. For a sample of 100 tires tested, the mean mileage was 36,000 and the standard deviation was 2,000 miles. Estimate the mean mileage that the utility should expect from these tires using a 96% confidence interval.

9.79 According to the January 28, 1991 issue of the *Christian Science Monitor* (page 10), the cost of replacing light bulbs shot out by drug dealers was $7200 per public housing complex last year. Suppose a study was conducted to determine the average maintenance costs for public housing complexes. For a sample of 20 such complexes in the Northeast, the mean maintenance cost was $75,000 with a standard deviation of $15,500. Determine a 95% confidence interval estimate for μ, the mean, cost per complex in the Northeast.

9.80 Oranges are selected at random from a large shipment that just arrived. A sample is taken to estimate the size (circumference, in inches) of the oranges. The sample data are summarized as follows: $n = 100$, $\Sigma x = 878.2$, and $\Sigma (x - \bar{x})^2 = 49.91$.
  a. Determine the mean and standard deviation for this sample.
  b. What is the point estimate for μ, the mean circumference of oranges in this shipment?
  c. What is the 0.95 confidence interval estimate for μ?
  d. What is the point estimate for the standard deviation σ of the circumference?
  e. What is the 0.95 confidence interval for σ?

9.81 A manufacturer wishes to estimate the mean life of a new line of automobile batteries. How large a sample should it take in order to estimate the mean life to within two months at 99% confidence, if the lifetimes of batteries of this type typically have a standard deviation of six months?

9.82 A tobacco company advertises its leading brand of cigarettes as "containing no more than 4 milligrams of tar." You have been asked to test a sample of these cigarettes and estimate the mean tar content to within 0.1 mg with 99% confidence. If past testing suggests that the standard deviation will be approximately 0.35 mg, how many cigarettes will you need to test for your sample?

9.83 A West Coast radio station is promoting a popular music group named Warren Peace and his Atom Bombs. In the past, 60% of the listeners of the station have

liked the music groups that the station promotes. Out of a randomly selected sample of 200 listeners, 102 of them like the Warren Peace group. At the 0.02 level of significance, test the hypothesis that there is no difference between the attitude of the listeners in this sample and listeners in the past.
    a. Solve using the classical approach.
    b. Solve using the prob-value approach.

9.84 A machine is considered to be operating in an acceptable manner if it produces 0.5% or fewer defective parts. It is not performing in an acceptable manner if more than 0.5% of its production is defective. The hypothesis $H_0: p = 0.005$ is tested against the hypothesis $H_a: p > 0.005$ by taking a random sample of 50 parts produced by the machine. The null hypothesis is rejected if two or more defective parts are found in the sample. Find the probability of the type I error.

9.85 You are interested in comparing the hypothesis $p = 0.8$ against the alternative $p < 0.8$. In 100 trials you observe 73 successes. Calculate the prob-value associated with this result.

9.86 An instructor asks each of the 54 members of his class to write down "at random" one of the numbers $1, 2, 3, \ldots, 13, 14, 15$. Since the instructor believes that students like gambling, he considers that 7 and 11 are lucky numbers. He counts the number of students, x, who selected 7 or 11. How large must x be before the hypothesis of randomness can be rejected at the 0.05 level?

9.87 *The LEXIS*, a national law journal, found from a survey conducted on April 6–7, 1991 that nearly two-thirds of the 800 people surveyed said doctors should not be prosecuted for helping people with terminal illnesses commit suicide. The poll carries a margin of error of plus or minus 3.5%.
    a. Describe how this survey of 800 people fits the properties of a binomial experiment. Specifically identify: n, a trial, success, p, and x.
    b. Exactly what is the "two-thirds" reported? How was it obtained? Is it a parameter or a statistic?
    c. Calculate the 0.95 confidence maximum error of estimate for the population proportion of all people who believe doctors should not be prosecuted.
    d. How is the maximum error, found in (c), related to the 3.5% mentioned in the survey report?

9.88 The marketing research department of an instant-coffee firm conducted a survey of married men to determine the proportion of married men who preferred their brand. Twenty of the 100 in the random sample preferred the company's brand. Use a 95% confidence interval to estimate the proportion of all married men who prefer this company's brand of instant coffee. Interpret your answer.

9.89 A company is drafting an advertising campaign that will involve endorsements by noted athletes. In order for the campaign to succeed, the endorser must be both highly respected and easily recognized. A random sample of 100 prospective customers are shown photos of various athletes. If the customer recognizes an athlete, then the customer is asked whether he or she respects the athlete. In the case of a top woman golfer, 16 of the 100 respondents recognized her picture and dicated that they also respected her. At the 95% level of confidence, what is the true proportion with which this woman golfer is both recognized and respected?

CHAPTER EXERCISES

9.90 A local auto dealership advertises that 90% of customers whose autos were serviced by their service department are pleased with the results. As a researcher, you take exception to this statement because you are aware that many people are reluctant to express dissatisfaction even if they are not pleased. A research experiment was set up in which those in the sample had received service by this dealer within the last two weeks. During the interview, the individuals were led to believe that the interviewer was new in town and was considering taking his car to this dealer's service department. Of the 60 sampled, 14 said that they were dissatisfied and would not recommend the department.
  a. Estimate the proportion of dissatisfied customers using a 95% confidence interval.
  b. Given your answer to (a), what can be concluded about the dealer's claim?

9.91 In obtaining the sample size to estimate a proportion, the formula $n = [z(\alpha/2)]^2 pq/E^2$ is used. If a reasonable estimate of p is not available, then it is suggested that $p = 0.5$ be used because this will give the maximum value for n. Calculate the value of $pq = p(1 - p)$ for $p = 0.1, 0.2, 0.3, \ldots, 0.8, 0.9$ in order to obtain some idea about the behavior of the quantity pq.

9.92 A consumer group was interested in determining the proportion of dentists who would accept the patient's insurance payment as the full payment for a routine exam. Determine the sample size needed to estimate the true proportion to within 0.01 with 95% confidence.

9.93 *Prevention* magazine reported in its latest survey that 64% of adult Americans, or 98 million people, were overweight. The telephone survey of 1254 randomly selected adults was conducted November 8–29, 1990 and had a margin of error of three percentage points.
  a. Calculate the maximum error of estimate for 0.95 confidence with $p' = 0.64$.
  b. How is the margin of error of three percentage points related to answer (a)?
  c. How large a sample would be needed to reduce the maximum error to 0.02 with 0.95 confidence?

9.94 "Two of five Americans believe the country should rely on nuclear power more than other energy sources for energy in the 1990s, according to a poll released yesterday.... The telephone poll, taken April 10–11, has a margin of error of plus or minus 3 points." This statement appeared in the Rochester *Democrat & Chronicle* on April 21, 1991. Forty percent plus or minus three points sounds like a confidence interval.
  a. What is another name for the "margin of error of plus or minus 3 points"?
  b. If we assume a 95% level of confidence, how large a sample is needed for a maximum error of 0.03?

9.95 In order to test the hypothesis that the standard deviation on a standard test is 12, a sample of 40 randomly selected students was tested. The sample variance was found to be 155. Does this sample provide sufficient evidence to show that the standard deviation differs from 12 at the 0.05 level of significance?

9.96 Bright-lite claims that its 60-watt light bulb burns with a length of life that is approximately normally distributed with a standard deviation of 81 hours. A sample of 101 bulbs had a variance of .8075. Is this sufficient evidence to reject Bright-

lite's claim in favor of the alternative, "the standard deviation is larger than 81 hours," at the 0.05 level of significance?
  a. Solve using the classical approach.
  b. Solve using the prob-value approach.

9.97 A drug manufacturer is concerned not only about the mean potency of its 500-mg tablets but also about the variability in potency. A random sample of 50 tablets was tested and the sample variance was determined to be 0.9 mg. Find a 95% confidence interval estimate for the population standard deviation.

9.98 The case histories for the 20 children in Exercise 9.70 showed their fathers to have ages of

| 32 | 48 | 33 | 22 | 29 | 30 | 45 | 25 | 26 | 43 |
| 36 | 27 | 34 | 20 | 35 | 55 | 52 | 38 | 34 | 37 |

  a. Calculate the sample mean and standard deviation.
  b. Estimate the mean age of all fathers of abnormal male children using a 0.95 confidence interval.
  c. Estimate the standard deviation of the ages of all fathers for these children using a 0.95 confidence interval.

9.99 A production process is considered to be out of control if the produced parts have a mean length different from 27.5 mm or a standard deviation that is greater than 0.5 mm. A sample of 30 parts yields a sample mean of 27.63 mm and a sample standard deviation of 0.87 mm.
  a. At the 0.05 level of significance, does this sample indicate that the process should be adjusted in order to correct the standard deviation of the product?
  b. At the 0.05 level of significance, does this sample indicate that the process should be adjusted in order to correct the mean value of the product?

9.100 Suppose that a sample of size 30 was used to test $H_0: \sigma^2 = 17$ versus $H_a: \sigma^2 > 17$ at $\alpha = 0.05$. How large would s^2 need to be before the null hypothesis would be rejected?

9.101 Verify algebraically that formula (9-6) is equivalent to formula (9-5).

9.102 A set of 75 measurements has a mean equal to 40 and a variance equal to 100. Find the Σx and Σx^2.

VOCABULARY LIST

Be able to define each term. In addition, describe, in your own words, and give an example of each term. Your examples should not be ones given in class or in the textbook.

  The bracketed numbers indicate the chapters in which the term previously appeared, but you should define the terms again to show increased understanding of their meaning.

acceptance region [8]
calculated value [8]

chi-square
chi-square distribution

conclusion [8]
confidence interval estimate
confidence interval formula
critical region [8]
critical value [8]
decision [8]
degrees of freedom
hypothesis test
inference [8]
level of confidence [8]
level of significance [8]
maximum error of estimate [8]
observed binomial probability (p')
one-tailed test [8]
parameter [1, 8]
proportion [6]
random variable [5, 6]
response variable [1]
sample size [8]
sample statistic [1, 2]
σ known
σ unknown
standard error [7, 8]
standard normal, z [2, 6, 8]
Student's t
test statistic [8]
two-tailed test [8]

KEY CONCEPTS

chi-square
confidence interval estimate
critical value
degrees of freedom
hypothesis test
σ unknown
standard normal, z [2, 6, 8]
Student's t
test statistic

QUIZ A

Answer "True" if the statement is always true. If the statement is not always true, replace the words shown in bold with words that make the statement always true.

9.1 The Student's t-distribution is an approximately normal distribution but is more **dispersed** than the normal distribution.

9.2 The **chi-square** distribution is used for inferences about the mean when σ is unknown.

9.3 The **Student's t**-distribution is used for all inferences about a population's variance.

9.4 If the test statistic falls in the critical region, the null hypothesis has **been proved true**.

9.5 When the test statistic is t and the number of degrees of freedom exceeds 30, the critical value of t is very close to that of z.

9.6 When making inferences about one mean when the value of σ is not known, the **z-score** is the test statistic.

9.7 The chi-square distribution is a skewed distribution whose mean is always **2**.

9.8 Often the concern with testing the variance (or standard deviation) is to keep its size under control or relatively small. Therefore, many of the hypothesis tests with chi-square will be **one-tailed**.

9.9 $\sqrt{npq}$ is the standard error of proportion.

9.10 The sampling distribution of p' is approximately distributed as **chi-square**.

QUIZ B

Answer all questions, showing all formulas, substitutions, and work.

9.1 State the null (H_0) and the alternative (H_a) hypotheses that would be used to test each of the following claims:
   a. The mean weight of professional basketball players is no more than 225 lb.
   b. The standard deviation for the monthly amounts of rainfall in Monroe County is less than 3.7 in.
   c. Approximately 40% of MCC's daytime students own their own car.

9.2 Determine the test criteria [level of significance, test statistic, critical region(s), and critical value(s)] that would be used in completing each hypothesis test using $\alpha = 0.05$.

   a. $H_0: \mu = 43$
      $H_a: \mu < 43$
      (given $\sigma = 6$)

   b. $H_0: p = 0.80$
      $H_a: p > 0.80$

   c. $H_0: \mu = 95$
      $H_a: \mu \neq 95$
      (σ unknown, $n = 22$)

   d. $H_0: \sigma = 12$
      $H_a: \sigma \neq 12$
      ($n = 28$)

9.3 Find each of the following:
   a. $z(0.02)$
   b. $t(18, 0.95)$
   c. $\chi^2(25, 0.95)$

9.4 A random sample of 25 was selected from a normally distributed population for the purpose of estimating the population mean, μ. The sample statistics are $n = 25, \bar{x} = 28.6, s = 3.50$.
   a. Find the point estimate for μ.
   b. Find the maximum error of estimate for the 0.95 confidence interval estimate.
   c. Find the lower confidence limit (LCL) and the upper confidence limit (UCL) for the 0.95 confidence interval estimate for μ.

9.5 Thousands of area elementary school students were recently given a nationwide standardized exam testing their composition skills. If 64 of a random sample of 100 students passed this exam, construct the 0.98 confidence interval estimate for the true proportion of all area students who passed the exam.

9.6 The manufacturer of a new model car, called Orion, claims the typical Orion will average 26 mpg of gasoline. An independent consumer group is somewhat skeptical of this claim and thinks the mean gas mileage is less than the 26 claimed. A sample of 24 randomly selected Orions produced the following results:

$$\text{Sample statistics: mean} = 24.15, \quad \text{st. dev.} = 4.87$$

At the 0.05 level of significance, does the consumer group have sufficient evidence to refute the manufacturer's claim?

9.7 A coffee machine is suppose to dispense 6 fluid ounces of coffee into a paper cup. In reality, the amount dispensed varies from cup to cup. However, if the machine is operating properly, the standard deviation of the amounts dispensed should be 0.1 or less ounces. A random sample of 15 cups produced a standard deviation of 0.13 oz. Does this represent sufficient evidence, at the 0.10 level of significance, to conclude that "the machine is not operating properly"?

9.8 An unhappy customer is disturbed with the waiting time at the post office when buying stamps. Upon registering his complaint, he was told, "You wait more than one minute for service no more than half of the time when only buying stamps." Not believing this to be the case, the customer collected some data from people who had just purchased stamps only.

$$\text{Sample statistics: } n = 60, \quad x = n(\text{less than one minute}) = 35$$

At the 0.02 level of significance, does our unhappy customer have sufficient evidence to refute the post office's claim?

QUIZ C

9.1 Student B says the range of a set of data may be used to obtain a crude estimate for the standard deviation of the population. Student A is not sure. How will Student B correctly explain how and under what circumstances his statement is true?

9.2 Is it (a) the null hypothesis or (b) the alternative hypothesis that the researcher usually believes to be true? Explain.

9.3 When you reject a null hypothesis, student A says that you are expressing disbelief in the value of the parameter as claimed in the null hypothesis, while student B says that you are expressing the belief that the sample statistic came from a population other than one related to the parameter claimed in the null hypothesis. Who is correct? Explain.

9.4 "The Student t-distribution must be used when making inferences about the population mean, μ, when the population standard deviation, σ, is not known" is a true statement. Student A states that the z-score plays a role sometimes when using the t-distribution. Explain the conditions that exist and the role played by z that make student A's statement correct.

9.5 Student A says that the percentage of the sample means that fall outside the critical values of the sampling distribution determined by a true null hypothesis is the p-value for the test. Student B says that the percentage that student A is describing is the level of significance. Who is correct? Explain.

9.6 Student A carries out a study in which he is willing to run a 1% risk of making a type I error. He rejects the null hypothesis and claims that his statistic is significant at the 0.99 level of confidence. Student B argues that student A's claim is not properly worded. Who is right? Explain.

9.7 Student A claims that when you employ a 95% interval estimation to determine a confidence interval, you do not know for sure whether your inference is correct (the parameter is contained within the interval) or not. Student B claims that you do; you have shown that the parameter cannot be less than the lower limit or greater than the upper limit of the interval. Who is right? Explain.

9.8 Student A says that the best way to improve a confidence interval estimate is to increase the level of confidence. Student B argues that using a high confidence level does not really improve the desirability of the resulting interval estimate. Who is right? Explain.

10 INFERENCES INVOLVING TWO POPULATIONS

Chapter Outline

10.1 Independent and Dependent Samples
Independent samples are obtained by using **unrelated** sets of **subjects**; **dependent** samples result from using **paired subjects**.

10.2 Inferences Concerning the Mean Difference Between Two Dependent Samples
The use of dependent samples helps **control** otherwise **untested factors**.

10.3 Inferences Concerning the Difference Between the Means of Two Independent Samples (Variances Known or Large Samples)
The comparison of the mean values of **two populations** is a common objective.

10.4 Inferences Concerning the Ratio of Variances Between Two Independent Samples
To investigate the relationship between the variances of two populations, we need the concept of the *F* distribution.

10.5 Inferences Concerning the Difference Between Means of Two Independent Samples (Variances Unknown and Small Samples)
We distinguish between cases in which the **variances** are **equal** and cases in which they are **not equal**.

10.6 Inferences Concerning the Difference Between Proportions of Two Independent Samples
Questions about the percentages or proportions of **two populations** are answered by means of hypothesis testing or a confidence interval estimate.

We Like Our Work

"Work isn't as bad as we thought it was going to be; we're satisfied." Living to work or working to live, jobs determine our standard of living and, often, our sense of self-worth. Here's a sampling, from a USA TODAY poll of 802 people of today's attitudes about the workplace.

WHAT WE THINK ABOUT WORK

Looking Forward to a Good Day's Work

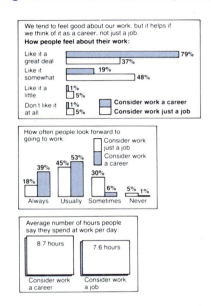

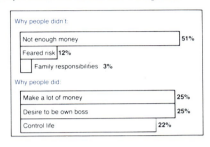

When You Do Your Job Well

The way you're rewarded for good work depends greatly on your field of work. Here's a breakdown of how workers in different fields say their bosses respond to a job well done:

	Professional	Sales	Assembly Line	Clerical
Verbal compliment	57%	70%	40%	60%
Nothing	24%	11%	45%	28%
Bonus	10%	21%	12%	7%
Written note	10%	6%	2%	6%

Starting Your Own Business

52% of people have thought about going into business for themselves. Some reasons why people did and why people didn't.

Source: Copyright 1987, USA TODAY. Reprinted with permission.

Chapter Objectives

In Chapters 8 and 9 we introduced the basic concepts of hypothesis testing and confidence interval estimation in connection with inferences about one population and the parameters: mean, proportion, and variance (or standard deviation). In this chapter we continue to investigate the inferences about these same parameters, but we will now use these parameters as a basis for *comparing two populations*.

First we will learn the difference between independent and dependent samples (Section 10.1). Throughout the remainder of this chapter, each section will present the techniques for using two means, or two proportions, or two variances (or two standard deviations) for comparing two populations. We will be studying several difference situations; the following chart is offered to help you initially organize the relationship of these various inference techniques:

Parameter	Kind of Samples		Section
Difference of Two Means	Dependent Samples		10.2
	Independent Samples	σ's known or large samples	10.3
		σ's unknown and small samples	10.5
Difference of Two Proportions	Independent Samples		10.6
Ratio of Two Variances (standard deviations)	Independent Samples		10.4

10.1 ▼ Independent and Dependent Samples

In this chapter we are going to study the procedures for making inferences about two populations. When comparing two populations we need two samples, one from each. Two basic kinds of samples can be used: independent and dependent. The dependence or independence of a sample is determined by the sources used for the data. A **source** can be a person, an object, or anything that yields a piece of data. If the same set of sources or related sets are used to obtain the data representing both populations, we have **dependent sampling**. If two unrelated sets of sources are used, one set from each population, we have **independent sampling**. The following illustrations should amplify these ideas.

source

dependent and independent samples

▼ ILLUSTRATION 10-1

A test will be conducted to see whether the participants in a physical fitness class actually improve in their level of fitness. It is anticipated that approximately

500 people will sign up for this course, and the instructor decides that she will give 50 of the participants a set of tests before the course actually begins (pretest) and then will give another set of tests to 50 participants at the end of the course (posttest). The following two sampling procedures are proposed:

▼ *Plan A*: Randomly select 50 participants from the list of those enrolled and give them the pretest. At the end of the course, select another random sample of size 50 and give them the posttest.
▼ *Plan B*: Randomly select 50 participants and give them the pretest; give the same set of 50 the posttest upon completion of the course.

Plan A illustrates independent sampling—the sources (class participants) used for each sample (pretest and posttest) were selected separately. Plan B illustrates dependent sampling—the sources used for both samples (pretest and posttest) are the same. ▲▲

Typically, when both a pretest and posttest are used, the same subjects will be used in the study. Thus pretest versus posttest (before versus after) studies are usually dependent samples.

▼ ILLUSTRATION 10-2

A test is being designed to compare the wearing quality of two brands of tires. The automobiles will be selected and equipped with the new tires and then driven under "normal" conditions for one month. Then a measurement will be taken to determine how much wear took place. Two plans are proposed:

▼ *Plan C*: n cars will be selected randomly and equipped with brand A and driven for the month, and n other cars will be selected and equipped with brand B and driven for the month.
▼ *Plan D*: n cars will be selected randomly, equipped with one tire of brand A and one tire of brand B (the other two tires are not part of the test), and driven for the month.

In this illustration we might suspect that many other factors must be taken into account when testing automobile tires—such as age, weight, and mechanical condition of the car; driving habits of drivers; location of the tire on the car; and where and how much the car is driven. However, at this time we are only trying to illustrate dependent and independent samples. Plan C is independent (unrelated sources) and plan D is dependent (common sources). ▲▲

Independent and dependent cases each have their advantages; these will be emphasized later. Both methods of sampling are often used.

Case Study 10-1

Exploring the Traits of Twins

Studies that involve identical twins are a natural for the dependent sampling technique discussed in this section.

A NEW STUDY SHOWS THAT KEY CHARACTERISTICS MAY BE INHERITED

Like many identical twins reared apart, Jim Lewis and Jim Springer found they had been leading eerily similar lives. Separated four weeks after birth in 1940, the Jim twins grew up 45 miles apart in Ohio and were reunited in 1979. Eventually they discovered that both drove the same model blue Chevrolet, chain-smoked Salems, chewed their fingernails and owned dogs named Toy. Each had spent a good deal of time vacationing at the same three-block strip of beach in Florida. More important, when tested for such personality traits as flexibility, self-control and sociability, the twins responded almost exactly alike.

The two Jims were the first of 348 pairs of twins studied at the University of Minnesota, home of the Minnesota Center for Twin and Adoption Research. Much of the investigation concerns the obvious question raised by siblings like Springer and Lewis: How much of any individual's personality is due to heredity?

The project, summed up in a scholarly paper that has been submitted to the *Journal of Personality and Social Psychology,* is considered the most comprehensive of its kind. The Minnesota researchers report the results of six-day tests of their subjects, including 44 pairs of identical twins who were brought up apart. Well-being, alienation, aggression and the shunning of risk or danger were found to owe as much or more to nature as to nurture. Of eleven key traits or clusters of traits analyzed in the study, researchers estimated that a high of 61 percent of what they call "social potency" (a tendency toward leadership or dominance) is inherited, while "social closeness" (the need for intimacy, comfort and help) was lowest, at 33 percent.

All the twins took several personality tests, answering more than 15,000 questions on subjects ranging from personal interests and values to phobias, aesthetic judgment and television and reading habits. Twins reared separately also took medical exams and intelligence tests and were queried on life history and stresses. Not all pairs matched up as well as the two Jims.

Source: Copyright 1987 by Time Inc. All rights reserved. Reprinted by permission of TIME.

Case Study 10-2

Study Finds Unproven, Usual Cancer Treatments Equal

Dependent samples of data can be thought of as "paired data." The pairs of data occur in an obvious fashion with the identical twins in Case Study 10-1; however, the pairing can occur as a result of matching two subjects to form the pairs. The article below demonstrates how two subjects might be paired off so that the resulting data can be analyzed using dependent sample techniques. Sometimes these dependent samples are referred to as sets of matched pairs.

BOSTON—Terminally ill cancer patients who get one unproven treatment fare no better—and no worse—than people who get chemotherapy and other standard care, a study concludes.

"Our study showed that regardless of whether they receive unorthodox or conventional treatment, these patients will die just as soon," said Dr. Barrie R. Cassileth, a psychologist from the University of Pennsylvania who directed the research. Median survival for both groups was 15 months.

The study, published in today's *New England Journal of Medicine,* compared 78 pairs of patients matched for age, sex, race, diagnosis and duration of disease. One of each pair was treated at the Livingston-Wheeler Medical Clinic in San Diego, while the other got routine care at the University of Pennsylvania Cancer Center.

Patients at the Livingston-Wheeler clinic are treated with vaccines and coffee enemas and put on strict diets consisting mostly of raw vegetables. The regimen is based on the unproven theory that cancer is caused by a microbe.

The researchers said they were surprised to find that the quality of patients' remaining months of life was no better at the San Diego clinic, even though some avoided the rigors of chemotherapy.

They measured pain, appetite, breathing difficulties, psychological state and other aspects of living, and found that those entering the clinic were already worse off than people at the university cancer center. In comparisons every two months until their deaths, clinic patients continued to have a worse quality of life.

At the Livingston-Wheeler clinic, a woman said clinic officials would not comment on the study.

Source: *Democrat and Chronicle*, April 25, 1991. Reprinted by permission.

▼▲ EXERCISES

 10.1 In trying to estimate the amount of growth that took place in the trees planted by the County Parks Commission recently, 36 trees were randomly selected from the 4000 planted. The heights of these trees were measured and recorded. One year later another set of 42 trees was randomly selected and measured. Do the two sets of data (36 heights, 42 heights) represent dependent or independent samples? Explain.

10.2 Twenty people were selected to participate in a psychology experiment. They answered a short multiple-choice quiz about their attitudes on a particular subject and then viewed a 45-minute film. The following day the same 20 people were asked to answer a follow-up questionnaire about their attitudes. At the completion of the experiment, the experimenter will have two sets of scores. Do these two samples represent dependent or independent samples? Explain.

10.3 An experiment is designed to study the effect diet has on the uric acid level. Twenty white rats are used for the study. Ten rats are randomly selected and given a junk-food diet. The other ten receive a high-fiber, low-fat diet. Uric acid levels of the two groups are determined at the beginning and at the end of the study. Do the resulting sets of data represent dependent or independent samples? Explain.

10.4 Two different types of disc centrifuges are used to measure the particle size in latex paint. A gallon of paint is randomly selected and 10 specimens are taken from it for testing on each of the centrifuges. There will be two sets of data, 10 data each, as a result of the testing. Do the two sets of data represent dependent or independent samples? Explain.

10.5 An insurance company is concerned that garage A charges more for repair work than garage B charges. It plans to send 25 cars to each garage and obtain separate estimates for the repairs needed for each car.
   a. How can the company do this and obtain independent samples? Explain in detail.
   b. How can the company do this and obtain dependent samples? Explain in detail.

10.6 A study is being designed to determine the reasons why adults choose to follow a healthy diet plan. One thousand men and 1000 women will be surveyed. Upon completion, the reasons that men choose a healthy diet will be compared to the reasons that women choose a healthy diet.
   a. How can the data be collected if independent samples are to be obtained? Explain in detail.
   b. How can the data be collected if dependent samples are to be obtained? Explain in detail.

10.7 Suppose that 400 students in a certain college are taking elementary statistics this semester. Describe how you would obtain two independent samples of size 25 from these 400 students in order to test some precourse skill against the same skill after the students complete the course.

10.8 Describe how you would obtain your samples in Exercise 10.7 if you were to use dependent samples.

10.2 ▼ Inferences Concerning the Mean Difference Between Two Dependent Samples

The procedure for comparing the means of two dependent samples is quite different from that of comparing the means of two independent samples. When comparing the means of two independent samples we make inferences about the difference between the two population means by using the difference between the two observed sample means. However, because of the relationship between two dependent samples, we will actually compare each data value of the first

Section 10.2 ▼ INFERENCES BETWEEN TWO DEPENDENT SAMPLES 489

paired difference

sample with the data value in the second sample that came from the same source. The two data values, one from each set, that come from the same source are called *paired* data. These pairs of data are compared by using the difference in their numerical values. This difference is called a *paired difference*. The mean difference of the two dependent populations is then tested by using the observed mean of the resulting paired differences. The concept of using paired data for this purpose has a built-in ability to remove many otherwise uncontrollable factors. The tire-wear problem (Illustration 10-2) is an excellent example of such additional factors. The wearing ability of a tire is greatly affected by a multitude of factors: the size and weight of the car, the age and condition of the car, the driving habits of the driver, the number of miles driven, the condition of and types of roads driven on, the quality of the material used to make the tire, and so on. By pairing the tires and considering differences, all of these extraneous causes of variability are removed.

▼ **ILLUSTRATION 10-3**

If we were to test the wearing quality of two tire brands by plan D, as described in Illustration 10-2, all the aforementioned factors will have an equal effect on both tire brands. Table 10-1 gives the amount of wear (in thousandths of an inch) that took place in such a test. One tire of each brand was placed on each of the six test cars. The position (left or right side, front or back) was determined with the aid of random number table.

TABLE 10-1
Sample Results for Illustration 10-3

Tire Brand	Car					
	1	2	3	4	5	6
A	125	64	94	38	90	106
B	133	65	103	37	102	115

Since the various cars, drivers, and conditions are the same for each paired set of data, it would make sense to introduce a new measure, the paired difference d, that was observed in each pair of related data. Therefore, we add a third row to the data of Table 10-1, d = brand B − brand A.

Car	1	2	3	4	5	6
d = B − A	8	1	9	−1	12	9

Do the sample data provide sufficient evidence for us to conclude that the two brands show unequal wear, at the 0.05 level of significance? Before we can solve the problem posed in this illustration, we need some new formulas. Our two sets of sample data have been combined into one set, a set of n values of d, $d = x_1 - x_2$. The sample statistics that we will use are $\bar{d}$, the observed *mean* value of d,

$$\bar{d} = \frac{\sum d}{n} \tag{10-1}$$

and s_d, the observed *standard deviation* of the d's,

$$S_d = \sqrt{\frac{\sum d^2 - \frac{(\sum d)^2}{n}}{n-1}} \qquad (10\text{-}2)$$

The observed values of d in our sample are from the population of *paired differences*, whose distribution is assumed to be approximately normal with a mean value of μ_d and a standard deviation of σ_d. Since σ_d is unknown, it will be estimated by s_d. The inferences about μ_d are completed in the same manner as the inferences about μ in Chapter 8 and in Section 9.1 by using $\bar{d}$ as the point estimate of μ_d. The inferences are completed by using the t distribution and $s_d/\sqrt{n}$ as the approximate measure for the standard error,

$$t = \frac{\bar{d} - \mu_d}{s_d/\sqrt{n}} \qquad (10\text{-}3)$$

with $n - 1$ degrees of freedom. Now let's complete Illustration 10-3.

SOLUTION

mean difference

STEP 1 H_0: $\mu_d = 0$ (μ_d is **mean difference**)

H_a: $\mu_d \neq 0$

STEP 2 $\alpha = 0.05$. The test criteria are shown in the accompanying figure. The critical values are $\pm t(5, 0.025) = \pm 2.57$.

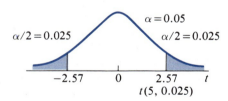

NOTE n is the number of paired differences (d). df $= n - 1$, and $t(5, 0.025)$ is obtained from Table 6, Appendix F.

STEP 3

$$t = \frac{\bar{d} - \mu_d}{s_d/\sqrt{n}}$$

First we must find $\bar{d}$ and s_d; some preliminary calculations are shown in Table 10-2.

TABLE 10-2 Preliminary Calculations

d	d²
8	64
1	1
9	81
−1	1
12	144
9	81
Total 38	372

$$\bar{d} = \frac{\sum d}{n} = \frac{38}{6} = 6.333 = \mathbf{6.3}$$

$$s_d = \sqrt{\frac{(372) - \frac{(38)^2}{6}}{5}} = \sqrt{26.27} = 5.13 = \mathbf{5.1}$$

Therefore,

$$t = \frac{6.3 - 0}{5.1/\sqrt{6}} = \frac{(6.3)\sqrt{6}}{5.1}$$

$$= \frac{(6.3)(2.45)}{5.1} = 3.026$$

$$t^* = \mathbf{3.03}$$

See the following figure:

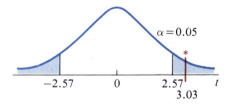

STEP 4 **Decision:** Reject H_0 (t^* is in the critical region).

Conclusion There is a significant difference in the mean amount of wear.

Since the calculated value of t fell in the critical region on the right, we might want to conclude that one brand of these tires is better than the other. This is possible, but we must be extremely careful to interpret the conclusion correctly. Recall that the recorded data concerned the amount of wear. Therefore, the tire that shows less wear should be the better one. Our hypothesis test showed that tire brand B had significantly more wear, thereby implying that brand A is the longer-wearing tire.

It is also possible to estimate the mean difference in paired data. The formula used for this **confidence interval estimate** is

$$\bar{d} - t(\text{df}, \alpha/2) \cdot \frac{s_d}{\sqrt{n}} \quad \text{to} \quad \bar{d} + t(\text{df}, \alpha/2) \cdot \frac{s_d}{\sqrt{n}} \tag{10-4}$$

▼ ILLUSTRATION 10-4

Construct the 95% confidence interval for the estimation of the mean difference in the paired data on tire wear, as found in Illustration 10-3. The given information is $n = 6$ pieces of paired data, $\bar{d} = 6.33$, and $s_d = 5.1$.

SOLUTION

$$\bar{d} \pm t(\text{df}, \alpha/2) \cdot \frac{s_d}{\sqrt{n}}$$

$$6.3 \pm 2.57 \cdot \frac{5.1}{\sqrt{6}}$$

$$6.3 \pm 5.4$$

0.9 to **11.7**, the 0.95 confidence interval for $\mu_{d=\text{B}-\text{A}}$.

That is, with 95% confidence we can say that the mean difference in the amount of wear is between 0.9 and 11.7.

NOTE This confidence interval is quite wide. This is due, in part, to the small sample size. Recall from the central limit theorem that as the sample size increases, the standard error (estimated by $s_d/\sqrt{n}$) decreases. ▲▲

▼▲ EXERCISES

10.9 State the null hypothesis, H_0, and the alternative hypothesis, H_a, that would be used to test the following claims:
   a. The mean of the differences between the posttest and the pretest scores is greater than 10.
   b. The mean weight gain, due to the change in diet for the laboratory animals, is at least 10 oz.
   c. The mean weight loss experienced by people on a new diet plan was no less than 12 lb.
   d. As a result of a special training session, it is believed that the mean of the difference in performance scores will not be zero.

10.10 Determine the test criteria that would be used to test the following hypotheses when t is used as the test statistic:

a. H_0: $\mu_d = 0.5$ and H_a: $\mu_d > 0.5$, with $n = 15$ and $\alpha = 0.05$.
b. H_0: $\mu_d = 4.5$ and H_a: $\mu_d \neq 4.5$, with $n = 25$ and $\alpha = 0.05$.
c. H_0: $\mu_d = 1.45$ and H_a: $\mu_d < 1.45$, with $n = 12$ and $\alpha = 0.10$.
d. H_0: $\mu_d = 0.75$ and H_a: $\mu_d > 0.75$, with $n = 18$ and $\alpha = 0.01$.

10.11 The corrosive effects of various soils on coated and uncoated steel pipe were tested by using a dependent sampling plan. The data collected are summarized by

$$n = 40 \qquad \Sigma d = 220 \qquad \Sigma d^2 = 6222$$

where d is the amount of corrosion on the coated portion subtracted from the amount of corrosion on the uncoated portion. Does this sample provide sufficient reason to conclude that the coating is beneficial? Use $\alpha = 0.01$.
   a. Solve using the classical approach.
   b. Solve using the prob-value approach.

10.12 To test the effect of a physical fitness course on one's physical ability, the number of sit-ups that a person could do in one minute, both before and after the course, was recorded. Ten randomly selected participants scored as shown in the following table. Can you conclude that a significant amount of improvement took place? Use $\alpha = 0.01$.

Before	29	22	25	29	26	24	31	46	34	28
After	30	26	25	35	33	36	32	54	50	43

   a. Solve using the classical approach.
   b. Solve using the prob-value approach.

10.13 A group of ten recently diagnosed diabetics were tested to determine whether an educational program was effective in increasing their knowledge of diabetes. They were given a test, before and after the educational program, concerning self-care aspects of diabetes. The scores on the test were as follows:

Patient	1	2	3	4	5	6	7	8	9	10
Before	75	62	67	70	55	59	60	64	72	59
After	77	65	68	72	62	61	60	67	75	68

The following MINITAB output may be used to determine whether the scores improved as a result of the program. Verify the following as shown on the output: mean difference (MEAN), standard deviation (STDEV), standard error of the difference (SE MEAN), t^* (T), and p-value.

```
MTB > READ BEFORE IN C1,AFTER IN C2
DATA> 75 77
DATA> 62 65
DATA> 67 68
DATA> 70 72
DATA> 55 62
DATA> 59 61
DATA> 60 60
DATA> 64 67
DATA> 72 75
DATA> 59 68
DATA> END DATA
     10 ROWS READ
MTB > LET C3 = C2 - C1
MTB > TTEST  OF MU = 0  DIFFERENCES IN C3;
SUBC> ALTERNATIVE = 1.

TEST OF MU = 0.000 VS MU G.T. 0.000
          N     MEAN    STDEV    SE MEAN      T    P VALUE
   C3    10    3.200    2.741     0.867     3.69    0.0025

MTB > STOP
```

10.14 As metal parts experience wear the metal is displaced. The table lists displacement measurements (mm) on metal parts that have undergone durability cycling for the equivalent of 100,000 plus miles. The first column is the serial number for the part, the second column lists the before test (BT) displacement measurement of the new part, the third column lists the end of test (EOT) measurements, the fourth column lists the change (i.e., wear) in the parts.

Serial Number	Displacement (mm)		
	BT	EOT	Difference
1	4.609	4.604	−0.005
2	5.227	5.208	−0.019
3	5.255	5.193	−0.062
4	4.622	4.601	−0.021
5	4.630	4.589	−0.041
6	5.207	5.188	−0.019
7	5.239	5.198	−0.041
8	4.605	4.596	−0.009
9	4.622	4.576	−0.046
10	4.753	4.736	−0.017
11	5.226	5.218	−0.008
12	5.094	5.057	−0.037
13	4.702	4.683	−0.019
14	5.152	5.111	−0.041

Source: Problem data provided by AC Rochester Division, General Motors, Rochester, NY.

a. Does it seem right that all the difference values are negative? Explain.
Use a computer to complete the following:
b. Find the sample mean, variance, and standard deviation for the before test (BT) data.
c. Find the sample mean, variance, and standard deviation for the end of test (EOT) data.
d. Find the sample mean, variance, and standard deviation for the difference data.
e. How is the sample mean difference related to the means of BT and EOT? Are the variances and standard deviations related in the same way?
f. At the 0.01 level, do the data show that a significant amount of wear took place?

10.15 We want to know which of two types of filters should be used over an oscilloscope to help the operator pick out the image on the screen of the cathode-ray tube. A test was designed in which the strength of a signal could be varied from zero to the point where the operator first detects the image. At this point, the intensity setting is read. The lower the setting, the better the filter. Twenty operators were asked to make one reading for each filter. Do the data in the following table show a significant difference in the filters at the 0.10 level?

| | Filter | | | Filter | |
Operator	1	2	Operator	1	2
1	96	92	11	88	88
2	83	84	12	89	89
3	97	92	13	85	86
4	93	90	14	94	91
5	99	93	15	90	89
6	95	91	16	92	90
7	97	92	17	91	90
8	91	90	18	78	80
9	100	93	19	77	80
10	92	90	20	93	90

a. Solve using the classical approach.
b. Solve using the prob-value approach.

10.16 Two men, A and B, who usually commute to work together, decide to conduct an experiment to see whether one route is faster than the other. The men feel that their driving habits are approximately the same, and therefore they decide on the following procedure. Each morning for two weeks A will drive to work on one route and B will use the second route. On the first morning, A will toss a coin. If heads appear, he will use route I; if tails appear, he will use route II. On the second morning, B will toss the coin: heads, route I; tails, route II. The times, recorded to the nearest minute, are shown in the following table. Do these data show that one of the routes is significantly faster than the other at $\alpha = 0.05$?

	Day								
Route	M	Tu	W	Th	F	M	Tu	W	Th	F
I	29	26	25	25	25	24	26	26	30	31
II	25	26	25	25	24	23	27	25	29	30

a. Solve using the classical approach.
b. Solve using the prob-value approach.

10.17 An experiment was designed to estimate the mean difference in weight gain for pigs fed ration A as compared to those fed ration B. Eight pairs of pigs were used. The pigs within each pair were littermates. The rations were assigned at random to the two animals within each pair. The gains (in pounds) after 45 days are shown in the following table:

Litter	1	2	3	4	5	6	7	8
Ration A	65	37	40	47	49	65	53	59
Ration B	58	39	31	45	47	55	59	51

Find the 95% confidence interval estimate for the mean of the differences μ_d, where d = ration A − ration B.

10.18 A sociologist is studying the effects of a certain motion picture film on the attitudes of black men toward white men. Twelve black men were randomly selected and asked to fill out a questionnaire before and after viewing the film. The scores received by the 12 men are shown in the following table. Construct a 95% confidence interval estimate for the mean shift in score that takes place when this film is viewed.

Before	10	13	18	12	9	8	14	12	17	20	7	11
After	5	9	13	17	4	5	11	14	13	18	7	12

10.3 ▼ Inferences Concerning the Difference Between the Means of Two Independent Samples (Variances Known or Large Samples)

When comparing the means of two populations, we typically consider the difference between their means, $\mu_1 - \mu_2$. The inferences we will make about $\mu_1 - \mu_2$ will be based on the difference between the observed sample means, $\bar{x}_1 - \bar{x}_2$.

Section 10.3 ▼ INFERENCES BETWEEN THE MEANS OF TWO INDEPENDENT SAMPLES

This observed difference belongs to a sampling distribution, the characteristics of which are described in the following statement:

independent means

If independent samples of sizes n_1 and n_2 are drawn randomly from large populations with means μ_1 and μ_2 and variances σ_1^2 and σ_2^2, respectively, the sampling distribution of $\bar{x}_1 - \bar{x}_2$, the difference between the means,

1. is approximately normally distributed.
2. has a mean of $\mu_{\bar{x}_1 - \bar{x}_2} = \mu_1 - \mu_2$.
3. has a standard error of $\sigma_{\bar{x}_1 - \bar{x}_2} = \sqrt{(\sigma_1^2/n_1) + (\sigma_2^2/n_2)}$.

This normal approximation is good for all sample sizes given that the populations involved are approximately normal and the population variances σ_1^2 and σ_2^2 are known quantities.

Since the sampling distribution is approximately normal, we will use the *z*-statistic in our inferences. In the **hypothesis tests**, *z* will be determined by

z-statistic
hypothesis test

$$z = \frac{(\bar{x}_1 - \bar{x}_2) - (\mu_1 - \mu_2)}{\sqrt{(\sigma_1^2/n_1) + (\sigma_2^2/n_2)}} \tag{10-5}$$

if both σ_1 and σ_2 are known quantities.

▼ ILLUSTRATION 10-5

Suppose that we are interested in comparing the academic success of college students who belong to fraternal organizations with the academic success of those who do not belong to fraternal organizations. These two populations are clearly separate, and we should take independent samples from each of the two populations. The "Greeks" claim that fraternity members achieve academically at a level no lower than that of nonmembers. (Cumulative grade point average is the measure of academic success.) Samples of size 40 are taken from each population. The means obtained are 2.03 for the 40 fraternity members (*g*) and 2.21 for the 40 nonmembers (*n*). Assume that the standard deviation of both populations is $\sigma = 0.6$. Complete a hypothesis test of the Greeks' claim, using $\alpha = 0.05$.

SOLUTION The hypotheses for this test are as follows:
STEP 1 $H_0: \mu_g = \mu_n \ (\geq)$ or $\mu_g - \mu_n = 0 \ (\geq)$ (no lower)

$H_a: \mu_g < \mu_n$ or $\mu_g - \mu_n < 0$ (lower)

The null hypothesis is usually interpreted as "there is no difference between the means," and, therefore, it is customary to express it by $\mu_1 - \mu_2 = 0$.

STEP 2 The test statistic used will be *z*. The test criteria for $\alpha = 0.05$ will be as shown in the accompanying figure.

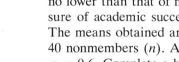

STEP 3 The formula for the test statistic z is formula (10-5):

$$z = \frac{(\bar{x}_g - \bar{x}_n) - (\mu_g - \mu_n)}{\sqrt{(\sigma_g^2/n_g) + (\sigma_n^2/n_n)}}$$

$$= \frac{(2.03 - 2.21) - 0}{\sqrt{[(0.6)^2/40] + [(0.6)^2/40]}} = \frac{-0.18}{\sqrt{(0.36/40) + (0.36/40)}}$$

$$= \frac{-0.18}{\sqrt{0.009 + 0.009}} = \frac{-0.18}{\sqrt{0.018}} = \frac{-0.180}{0.134} = -1.343$$

$$z^* = -1.34$$

This value is compared to the test criteria in the accompanying figure.

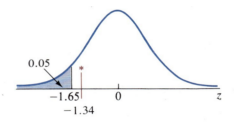

STEP 4 **Decision:** Fail to reject H_0 (z^* is in the noncritical region).

Conclusion At the 0.05 level of significance, the claim that the fraternity members achieve at a level no lower than that of nonmembers cannot be rejected.

▲▲

We often wish to estimate the difference between the means of two different populations. When independent samples are involved, we will use the information about the sampling distribution of $\bar{x}_1 - \bar{x}_2$ and the z statistic to construct our **confidence interval estimate for $\mu_1 - \mu_2$**.

$$(\bar{x}_1 - \bar{x}_2) - z(\alpha/2) \cdot \sqrt{\frac{\sigma_1^2}{n_1} + \frac{\sigma_2^2}{n_2}} \quad \text{to} \quad (\bar{x}_1 - \bar{x}_2) + z(\alpha/2) \cdot \sqrt{\frac{\sigma_1^2}{n_1} + \frac{\sigma_2^2}{n_2}} \quad \text{(10-6)}$$

▼ **ILLUSTRATION 10-6**

Construct the 95% confidence interval estimate for the difference between the two independent means of Illustration 10-5 (mean grade point average of fraternity members and mean grade point average of nonmembers). Use the sample values found in Illustration 10-5.

SOLUTION The information given is $\bar{x}_g = 2.03$, $\sigma_g = 0.6$, $n_g = 40$, $\bar{x}_n = 2.21$, $\sigma_n = 0.6$, $n_n = 40$, and $1 - \alpha = 0.95$. The interval is

$$(\bar{x}_g - \bar{x}_n) \pm z(0.025) \cdot \sqrt{\frac{\sigma_g^2}{n_g} + \frac{\sigma_n^2}{n_n}}$$

$$(2.03 - 2.21) \pm (1.96) \cdot \sqrt{\frac{(0.6)^2}{40} + \frac{(0.6)^2}{40}}$$

(See Illustration 10-5 for calculations.)

$$(-0.18) \pm (1.96)(0.134)$$

$$-0.18 \pm 0.26$$

−0.44 to **0.08**, the 0.95 confidence interval for $\mu_g - \mu_n$.

That is, with 95% confidence we estimate the difference between the means to be between -0.44 and $+0.08$. ▲▲

As we saw in previous chapters, the variance of a population is generally unknown when we wish to make an inference about the mean. Therefore, it is necessary to replace σ_1 and σ_2 with the best estimates available, namely, s_1 and s_2. If both samples have sizes that exceed 30, we may replace σ_1 and σ_2 in formulas (10-5) and (10-6) with s_1 and s_2, respectively, without appreciably affecting our level of significance or confidence. Thus for inferences about the difference between two population means, based on independent samples where the σ's are *unknown* and both $n_1 > 30$ and $n_2 > 30$, we will use

$$z = \frac{(\bar{x}_1 - \bar{x}_2) - (\mu_1 - \mu_2)}{\sqrt{(s_1^2/n_1) + (s_2^2/n_2)}} \tag{10-7}$$

test statistic for the calculation of the **test statistic** *in the hypothesis test*. We will use

$$(\bar{x}_1 - \bar{x}_2) - z(\alpha/2) \cdot \sqrt{\frac{s_1^2}{n_1} + \frac{s_2^2}{n_2}} \quad \text{to} \quad (\bar{x}_1 - \bar{x}_2) + z(\alpha/2) \cdot \sqrt{\frac{s_1^2}{n_1} + \frac{s_2^2}{n_2}} \tag{10-8}$$

confidence interval estimate for calculating the endpoints of the $1 - \alpha$ **confidence interval estimate**.

▼ **ILLUSTRATION 10-7**

Two independent samples are taken to compare the means of two populations. The sample statistics are given in Table 10-3. Can we conclude that the mean of population A is greater than the mean of population B, at the 0.02 level of significance?

TABLE 10-3
Sample Statistics for Illustration 10-7

Sample	n	$\bar{x}$	s
A	50	57.5	6.2
B	60	54.4	10.6

Chapter 10 ▼ INFERENCES INVOLVING TWO POPULATIONS

SOLUTION This problem calls for a hypothesis test for the difference of two independent means. Both n's are larger than 30; therefore, formula (10-7) will be used to calculate z.

STEP 1 H_0: $\mu_A - \mu_B = 0$ ($\leq$)

H_a: $\mu_A - \mu_B > 0$

STEP 2 $\alpha = 0.02$. The test criteria are shown in the following figure:

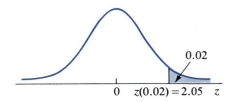

STEP 3 Using formula (10-7),

$$z = \frac{(57.5 - 54.4) - 0}{\sqrt{[(6.2)^2/50] + [(10.6)^2/60]}}$$

$$= \frac{3.1}{\sqrt{0.7688 + 1.8727}}$$

$$= \frac{3.1}{1.625}$$

$$= 1.908$$

$z^* = \mathbf{1.91}$

STEP 4 **Decision:** Fail to reject H_0 (z^* is in the noncritical region).

Conclusion At the 0.02 level of significance, we do not have sufficient evidence to conclude that $\mu_A > \mu_B$.

NOTE The null hypothesis can claim a difference; for example, the mean of A is 10 larger than the mean of B. In this situation the null hypothesis is written $\mu_A - \mu_B = 10$, and the value 10 is used in Step 3 when z is calculated. ▲▲

▼ ILLUSTRATION 10-8

Suppose that the samples given in Illustration 10-7 were taken for the purpose of estimating the difference between the two population means. Construct the 0.99 confidence interval for the estimation of this difference.

SOLUTION This estimation calls for the difference between the means of two independent samples whose sizes are both greater than 30. Therefore, we will use formula (10-8).

$$(57.5 - 54.4) \pm 2.58 \sqrt{\frac{(6.2)^2}{50} + \frac{(10.6)^2}{60}}$$

$$3.1 \pm (2.58)(1.625)$$

$$3.1 \pm 4.19$$

-1.09 to 7.29, the 0.99 confidence interval for $\mu_A - \mu_B$.

That is, our 99% confidence interval estimate for the difference between the two population means is -1.09 to 7.29.

Case Study 10-3

Who's Making What?

Traffic Management's Salary Survey reports average salaries for various subgroups from the population of readers of their magazine. Since only averages are given, and no measures of variation, we can determine only point estimates for the differences. Estimate the difference in average salary for respondents with and without a college education. Does this difference seem large enough to indicate that a college education has an effect on average salary? Explain. (See Exercise 10.21, p. 502.)

We haven't hit the $40,000 mark yet, but we're knocking on the door.

The results of *Traffic Management*'s second annual salary survey reveal that the average reader now makes $38,350 a year, an increase of $1,800 over last year. If pay increases follow past patterns, next year's average should easily surpass $40,000.

The survey findings tell an interesting story. They depict a mature, experienced, well-educated work force. The average respondent is a 43-year-old male with a college degree who has been working in transportation or distribution for more than 15 years. He's been with his company for 12 years and in his current job for the past six.

RESULTS AT A GLANCE

- ▼ **Average salary is $38,350.** The 4,100 plus readers who responded to the survey earn an average of $38,350. Men, who make up 88 percent of the total, average $39,930. Women earn $25,980.
- ▼ **VPs and GMs make the most.** Readers with vice president or general management titles average almost $62,000. At the other end of the salary spectrum, traffic analysts and specialists earn around $30,000.
- ▼ **Education pays.** Respondents with a college degree average $42,690. That's 24 percent higher than readers with some college and 33 percent more than high-school-only respondents. MBAs average $48,310.
- ▼ **Big companies pay best.** In companies with sales of $1 billion or more, six out of 10 readers earn more than $40,000. By contrast, only 25 percent of readers in $25-million-and-under companies earn $40K.

Source: Traffic Management, April 1986.

▼▲ EXERCISES

10.19 State the null hypothesis, H_0, and the alternative hypothesis, H_a, that would be used to test the following claims:
  a. The means of these two populations are not equal.
  b. The difference between the two population means is at least 25.
  c. The two population means differ by more than 10.
  d. The mean of population A is no less than 50 more than the mean of population B.

10.20 Determine the test criteria that would be used to test the following hypotheses when z is used as the test statistic:
  a. $H_0: \mu_A = \mu_B$ and $H_a: \mu_A > \mu_B$, with both σ's $= 15$ and $\alpha = 0.05$.
  b. $H_0: \mu_A = \mu_B$ and $H_a: \mu_A \neq \mu_B$, with $\sigma_A = 25$, $\sigma_B = 30$ and $\alpha = 0.05$.
  c. $H_0: \mu_A - \mu_B = 0$ and $H_a: \mu_A - \mu_B < 0$, with $n_A = 42$, $n_B = 54$ and $\alpha = 0.02$.
  d. $H_0: \mu_A - \mu_B = 7.5$ and $H_a: \mu_A - \mu_B > 7.5$, with $n_A = 108$, $n_B = 90$ and $\alpha = 0.01$.

$ 10.21 Refer to the information about average salaries of respondents to *Traffic Management*'s survey (Case Study 10-3).
  a. Make a point estimate for how much more money respondents make, on the average, when they have a college education.
  b. Does the difference (the answer in (a)) in average salaries seem to suggest that a college education has a significant effect on average salary? Explain. What other information is needed to determine significance?

10.22 Determine the p-value for the following hypothesis test (assume that both sample sizes exceed 30):
  a. $H_0: \mu_1 - \mu_2 = 10$, $H_a: \mu_1 - \mu_2 > 10$, $z^* = 1.85$
  b. $H_0: \mu_1 - \mu_2 = 0$, $H_a: \mu_1 - \mu_2 \neq 10$, $z^* = -2.33$
  c. $H_0: \mu_1 - \mu_2 = 0$, $H_a: \mu_1 - \mu_2 < 0$, $z^* = -2.76$

10.23 Independent samples are taken from normal populations I and II with variances of 200 and 700, respectively. Do the sample data shown in the following table provide sufficient evidence to reject the hypothesis that the populations have equal means at the 0.10 level of significance?

Sample	n	$\bar{x}$
I	32	104.5
II	40	110.9

  a. Solve using the classical approach.
  b. Solve using the prob-value approach.

$ 10.24 The purchasing department for a regional supermarket chain is considering two sources from which to purchase 10-lb bags of potatoes. A random sample taken from each source shows the following results:

Section 10.3 ▼ INFERENCES BETWEEN THE MEANS OF TWO INDEPENDENT SAMPLES

	Idaho Supers	Idaho Best
Number of Bags Weighed	100	100
Mean Weight	10.2 lb	10.4 lb
Sample Variance	0.36	0.25

At the 0.05 level of significance, is there a difference between the mean weights of the 10-lb bags of potatoes?
    a. Solve using the classical approach.
    b. Solve using the prob-value approach.

10.25 A study was designed to investigate the effect of a calcium-deficient diet on lead consumption in rats. One hundred rats were randomly divided into 2 groups of 50 each. One group served as a control group and the other was the experimental, or calcium-deficient, group. The response recorded was the amount of lead consumed per rat. The results were summarized by:

$$\text{Control group:} \quad n = 50 \quad \bar{x} = 5.2 \quad s = 1.1$$
$$\text{Experimental:} \quad n = 50 \quad \bar{x} = 7.6 \quad s = 1.3$$

Test $H_0: \mu_C - \mu_E = 0$ versus $H_a: \mu_C - \mu_E < 0$ at $\alpha = 0.05$.
    a. Solve using the classical approach.
    b. Solve using the prob-value approach.

10.26 The February 10, 1990 issue of *Science News* (page 92) reported on a study involving two groups of women from seven hospitals: 2817 pregnant women and 1818 infertile women (defined as not conceiving after one year of unprotected intercourse). The women were asked about their daily intake of regular or decaffeinated coffee and tea, as well as weekly consumption of cola. Suppose the caffeine consumption summary statistics were as follows:

	n	$\bar{x}$	s
Pregnant Women	2817	520.5	120.5
Infertile Women	1818	535.2	176.3

    a. Calculate the z-value for testing the research hypothesis, "there is a difference in caffeine consumption."
    b. Find the prob-value.

10.27 A sample of the heights of female and male students participating in the college intermural program was obtained, with the results as shown in the following table. Do these data provide sufficient evidence to reject the hypothesis that the average male participant is no more than 2.5 inches taller than the average female participant at the 0.02 level of significance?

Sample	n	Σx	$\Sigma (x - \bar{x})^2$
Female	30	1952	74.2
Male	40	2757	284.3

   a. Solve using the classical approach.
   b. Solve using the prob-value approach.

10.28 "Is the length of a steel bar affected by the heat treatment technique used?" This was the question being tested when the following data were collected:

Heat Treatment	Lengths (to nearest inch)							
I	156	159	151	153	157	159	155	155
	151	152	158	154	156	156	157	155
	156	159	153	157	157	159	158	155
	159	152	150	154	156	156	157	160
II	154	156	150	151	156	155	153	154
	149	150	150	151	154	155	155	154
	154	156	150	151	156	154	153	154
	149	150	150	151	154	148	155	158

Is there a significant difference in the lengths of these two sets of data at the 0.05 level?

10.29 An experiment was conducted to compare the mean absorptions of two drugs in specimens of muscle tissue. Seventy-two tissue specimens were randomly divided into two equal groups. Each group was tested with one of the two drugs. Assume that the variance of absorption is 0.10 for this type of drug. The means found were $\bar{x}_A = 7.9$ and $\bar{x}_B = 8.5$. Construct the 98% confidence interval for the difference in the mean absorption rates.

10.30 Two groups of university students were compared with respect to their computer-science aptitude. Group I was composed of students who had taken at least one computer-science course in high school, and group 2 was composed of students who had never had a computer-science course. None of the students had had any other computer-related experience. Both groups were given the KSW computer-science aptitude test. The results were

group 1: $n = 125$ $\bar{x} = 15.5$ $s = 2.7$
group 2: $n = 115$ $\bar{x} = 14.3$ $s = 3.0$

Find a 90% confidence interval estimate for $\mu_1 - \mu_2$.

10.31 In a certain species of plant the white-flowered plants appear to have smaller and more abundant flowers than do the red-flowered plants. The following data were collected when the numbers of flowers on several plants were counted. Use these sample data to construct the 95% confidence interval estimate for the difference between the mean number of flowers on the white- and the red-flowered plants ($\mu_W - \mu_R$).

Flower	n	$\sum x$	$\sum (x - \bar{x})^2$
Red	32	2253	32,462.0
White	35	5157	58,600.0

10.32 Experimentation with a new rocket nozzle has led to two slightly different designs. The following data summaries resulted from testing these two designs:

	n	$\sum x$	$\sum x^2$
Design 1	36	278.4	2163.76
Design 2	42	310.8	2332.26

Determine the 0.99 confidence interval estimate for the difference in the means for these two rocket nozzles.

10.4 ▼ Inferences Concerning the Ratio of Variances Between Two Independent Samples

When comparing two populations, it is quite natural that we compare their variances, or standard deviations. The inferences concerning two population variances (or standard deviations) are much like those comparing means. We will study two kinds of inferences about the comparison of the variances of two populations: (1) *the hypothesis test for the equality of the two variances* and (2) *the estimation of the ratio of the two population variances σ_1^2 / σ_2^2.*

The soft drink bottling company discussed in Section 9.3 (pp. 463–466) is trying to decide whether to install a modern high-speed bottling machine. There are, of course, many concerns in making this decision. The variance in the amount of fill per bottle is one. In this respect the manufacturer of the new system contends that the variance in fills is no larger with the new machine than with the old. A hypothesis test for the equality of the two variances can be used to make a decision in this situation. The null hypothesis will be that the variance of the high-speed machine (m) is no larger than the variance of the present machine (p); that is, $\sigma_m^2 \leq \sigma_p^2$. The alternative hypothesis will then be $\sigma_m^2 > \sigma_p^2$.

$$H_0: \sigma_m^2 = \sigma_p^2 \ (\leq) \quad \text{or} \quad \frac{\sigma_m^2}{\sigma_p^2} = 1 \ (\leq)$$

$$H_a: \sigma_m^2 > \sigma_p^2 \quad \text{or} \quad \frac{\sigma_m^2}{\sigma_p^2} > 1$$

The test statistic that will be used in making a decision about the null hypothesis is F. The calculated value of F will be obtained by using the following formula:

$$F = \frac{s_1^2}{s_2^2} \tag{10-9}$$

where s_1^2 and s_2^2 are the variances of two independent samples of sizes n_1 and n_2, respectively.

When independent random samples are drawn from normal populations with equal variances, the ratio of the sample variances, s_1^2/s_2^2, will have a probability distribution known as the ***F* distribution** (see Figure 10-1 below).

F distribution

PROPERTIES OF THE *F* DISTRIBUTION

1. F is nonnegative in value; it is zero or positively valued.
2. F is nonsymmetrical; it is skewed to the right.
3. There are many F distributions, much like the t and χ^2 distributions.

There is a distribution for each pair of degree-of-freedom values.

For the inferences discussed in this section, the degrees of freedom for each of the samples are $df_1 = n_1 - 1$ and $df_2 = n_2 - 1$.

The critical values for the F distribution may be obtained from Tables 8a, 8b, and 8c in Appendix F. Each critical value will be determined by three identification values: df_n, df_d, and the area under the curve to the right of the critical value being sought. Each F distribution has two degrees-of-freedom values: df_n, the degrees of freedom associated with the sample whose variance is in the numerator of the calculated F; and df_d, the degrees of freedom associated with the sample whose variance is in the denominator. Therefore, the symbolic name for a critical value of F will be $F(df_n, df_d, \alpha)$, as shown in Figure 10-2.

FIGURE 10-1
F Distribution

FIGURE 10-2
F Distribution Showing $F(df_n, df_d, \alpha)$

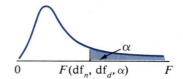

Section 10.4 ▼ INFERENCES CONCERNING THE RATIO OF VARIANCES

Table 8a in Appendix F shows the critical values for $F(\text{df}_n, \text{df}_d, \alpha)$, where α is equal to 0.05; Table 8b gives the critical values when $\alpha = 0.025$; Table 8c gives values when $\alpha = 0.01$.

▼ **ILLUSTRATION 10-9**

Find $F(5, 8, 0.05)$, the critical F value for samples of size 6 and 9 with 5% of the area in the right tail.

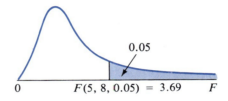

SOLUTION From Table 8a ($\alpha = 0.05$), we obtain the value shown in the accompanying table. Therefore, $F(5, 8, 0.05) = \mathbf{3.69}$.

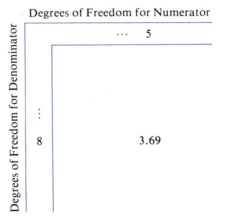

▲▲

Notice that $F(8, 5, 0.05)$ is 4.82. The degrees of freedom associated with the numerator and with the denominator must be kept in the correct order. (3.69 is quite different from 4.82. Check some other pairs to verify this fact.)

Tables showing the critical values for the F distribution give only the right-hand critical value. If the critical value for the left-hand tail is needed, we obtain it by calculating the reciprocal of the related critical value obtained from the table. Expressed as a formula, this is

$$F(\text{df}_1, \text{df}_2, 1 - \alpha) = \frac{1}{F(\text{df}_2, \text{df}_1, \alpha)} \tag{10-10}$$

FIGURE 10-3
Finding the Critical Value for the Left Tail of the F Distribution

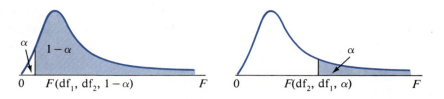

See Figure 10-3. Notice that when the reciprocal is taken, the degrees of freedom for the numerator and the denominator are switched also. (Why is this switch necessary?)

▼ **ILLUSTRATION 10-10**

Find the value for $F(10, 15, 0.99)$.

SOLUTION Using formula (10-10),

$$F(10, 15, 0.99) = \frac{1}{F(15, 10, 0.01)} = \frac{1}{4.56} = 0.219$$

$$= \mathbf{0.22}$$ ▲▲

▼ **ILLUSTRATION 10-11**

Recall that our soft drink bottling company was to make a decision about the equality of the variances of amounts of fill between its present machine and a modern high-speed outfit. Does the sample information in Table 10-4 present sufficient evidence to reject the manufacturer's claim that the modern, high-speed, bottle-filling machine fills bottles with no more variance than the company's present machine? Use $\alpha = 0.01$.

TABLE 10-4
Sample Results for Illustration 10-11

Sample	n	s²
Present machine (p)	22	0.0008
Modern high-speed machine (m)	25	0.0018

SOLUTION

STEP 1 H_0: $\sigma_m^2 = \sigma_p^2$ or $\sigma_m^2/\sigma_p^2 = 1$ ($\leq$) (no more variance)
H_a: $\sigma_m^2 > \sigma_p^2$ or $\sigma_m^2/\sigma_p^2 > 1$ (more variance)

STEP 2 $\alpha = 0.01$. The test statistic to be used is F, since the null hypothesis is about the equality of the variances of two populations. The critical region is one-tailed and on the right because the alternative hypothesis says "greater than." $F(24, 21, 0.01)$ is the critical value. The number of degrees of freedom for the numerator is 24 (25 − 1) because the sample from the modern high-speed machine is associated with the numerator, as specified by the null hypothesis. $df_d = 21$ because the sample associated with the denominator has size 22. The critical value is found in Table 8c and is 2.80. See the accompanying figure.

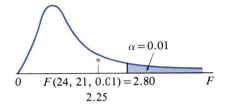

STEP 3

$$F = \frac{s_m^2}{s_p^2} = \frac{0.0018}{0.0008} = 2.25$$

$$F^* = 2.25$$

This value is shown in the figure.

STEP 4   Decision: Fail to reject H_0 (F^* is in the noncritical region).

Conclusion   At the 0.01 level of significance, the samples do not present sufficient evidence to reject the manufacturer's claim. ▲▲

OPTIONAL TECHNIQUE

When completing a hypothesis test about the equality of two population variances, it would be convenient if we could always use a right-hand critical value for F without having to calculate the left-hand critical value. This can be accomplished by minor adjustments in the null hypothesis and in the calculation of F in Step 3. The two cases we would like to change are (1) the two-tailed test and (2) the one-tailed test where the critical region is on the left. The one-tailed test with the critical region on the right already meets our criterion.

CASE 1   When a two-tailed test is to be completed, we will state the hypotheses in the normal way. The calculated value of F, F^*, will be the larger of s_1^2/s_2^2 or s_2^2/s_1^2. (One of these will have a value between zero and 1; the other will be larger than 1.) The critical value of F will be $F(df_n, df_d, \alpha/2)$, where df_n is the number of degrees of freedom for the sample whose variance is used in the numerator; df_d represents the degrees of freedom used in the denominator. Only the right-tail critical value will be needed. The test is completed in the usual fashion.

Consider the following hypothesis test situation:

$$H_0: \sigma_1^2/\sigma_2^2 = 1 \quad \text{versus} \quad H_a: \sigma_1^2/\sigma_2^2 \neq 1$$

$$\alpha = 0.05, \, n_1 = 5, \text{ and } n_2 = 11, \text{ with } F^* = \frac{5.0}{7.0} = 0.714$$

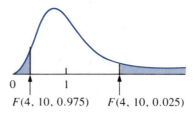

In order to complete the hypothesis test, we will need to determine the left-hand critical value, $F(4, 10, 0.025)$. This would require the use of formula (10-10), but notice that with a few simple changes, we can avoid using formula (10-10). Let's simply reverse the order of things.

$$H_0: \sigma_2^2/\sigma_1^2 = 1 \quad \text{versus} \quad H_a: \sigma_2^2/\sigma_1^2 \neq 1$$

$$\alpha = 0.05, n_2 = 11, \text{ and } n_1 = 5, \text{ with } F^* = \frac{7.0}{5.0} = 1.40$$

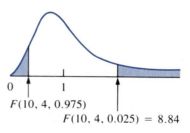

We also need $F(10, 4, 0.025)$ from Table 8b. There were only two changes that were made: the calculated F^* value was inverted and the degrees of freedom for the critical value was reversed. Thus we avoided using formula (10-10).

CASE 2 When a one-tailed test where the critical region is on the left is to be completed, we will interchange the position of the two variances in the statement of the hypotheses. This will reverse the direction of the alternative hypothesis and put the critical region on the right. From this point on, the procedure is the same as for the one-tailed test with a critical region on the right.

Even though the hypothesis test is the most commonly used inference about two variances, you may occasionally be asked to estimate the **ratio of two variances**, σ_A^2/σ_B^2. The best point estimate is s_A^2/s_B^2. The $1 - \alpha$ confidence interval for the estimate is constructed by using the following formula:

ratio of variances

$$\frac{s_A^2/s_B^2}{F(\text{df}_A, \text{df}_B, \alpha/2)} \quad \text{to} \quad \frac{s_A^2/s_B^2}{F(\text{df}_A, \text{df}_B, 1 - \alpha/2)} \quad (10\text{-}11)$$

Recall that left-tail values of F are not in the tables. These left-tail values are found by using formula (10-10). Using formula (10-10), we can rewrite formula (10-11) as

$$\frac{s_A^2/s_B^2}{F(\text{df}_A, \text{df}_B, \alpha/2)} \quad \text{to} \quad \frac{s_A^2}{s_B^2} F(\text{df}_B, \text{df}_A, \alpha/2) \quad (10\text{-}12)$$

ratio of standard deviation

(Notice that when the reciprocal is used, the degrees of freedom for the numerator and denominator are switched.)

If the **ratio of standard deviations** is desired, you need only take the positive square root of each of the bounds of the interval found by using formula (10-12).

NOTE The formula for estimating the ratio of population variances, or population standard deviations, requires the use of the ratio of sample variances. If the standard deviation of the sample is given, it must be squared to obtain the variance.

Case Study 10-4

Personality Characteristics of Police Applicants: Comparisons Across Subgroups and with Other Populations

Bruce N. Carpenter and Susan M. Raza concluded that "police applicants are somewhat more like each other than are those in the normative population" when the F test of homogeneity of variance resulted in a p-value of less than 0.005. Homogeneity means that the group's scores are less variable than the scores for the normative population. What null and alternative hypotheses did Carpenter and Raza test? What was the critical value for the test? What does "p < .005" mean? (See Exercise 10.39, p. 513.)

To determine whether police applicants are a more homogeneous group than the normative population, the F test of homogeneity of variance was used. With the exception of scales F, K, and 6, where the differences are nonsignificant, the results indicate that the police applicants form a somewhat more homogeneous group than the normative population (mean $F(237, 305) = 1.36$, $p < .005$, range of F from 1.10 to 1.65). Thus, police applicants are somewhat more like each other than are those in the normative population.

Source: Reproduced from the *Journal of Police Science and Administration*, Vol. 15, no. 1, pp. 10–17, with permission of the International Association of Chiefs of Police, P.O. Box 6010, 13 Firstfield Road, Gaithersburg, Maryland 20878.

▼▲ EXERCISES

10.33 State the null hypothesis, H_0, and the alternative hypothesis, H_a, that would be used to test the following claims:
   a. The variances of populations A and B are not equal.
   b. The standard deviation of population I is larger than the standard deviation of population II.
   c. The ratio of the variances for populations A and B is different from 1.
   d. The variability within population C is less than the variability within population D.

10.34 Determine the test criteria that would be used to test the following hypotheses when F is used as the test statistic:

a. $H_0: \sigma_1^2 = \sigma_2^2$ and $H_a: \sigma_1^2 > \sigma_2^2$, with $n_1 = 10$, $n_2 = 16$ and $\alpha = 0.05$.
b. $H_0: \sigma_1^2/\sigma_2^2 = 1$ and $H_a: \sigma_1^2/\sigma_2^2 \neq 1$, with $n_1 = 25$, $n_2 = 31$ and $\alpha = 0.05$.
c. $H_0: \sigma_1^2/\sigma_2^2 = 1$ and $H_a: \sigma_1^2/\sigma_2^2 > 1$, with $n_1 = 10$, $n_2 = 10$ and $\alpha = 0.01$.
d. $H_0: \sigma_1 = \sigma_2$ and $H_a: \sigma_1 < \sigma_2$, with $n_1 = 16$, $n_2 = 16$ and $\alpha = 0.01$.

10.35 Using the $F(\mathrm{df}_1, \mathrm{df}_2, \alpha)$ notation, name each of the critical values shown on the following figures:

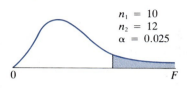
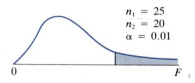
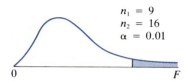
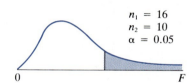

10.36 Using the $F(\mathrm{df}_1, \mathrm{df}_2, \alpha)$ notation, name each of the critical values shown on the following figures. (For two-tail cases, use $\alpha/2$ in each tail.)

(a)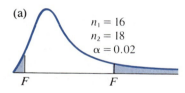
(b)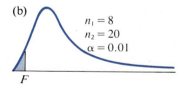
(c)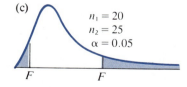
(d)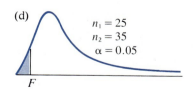

10.37 Find the following critical values for F from Tables 8a, 8b, and 8c in Appendix F.
    a. $F(24, 12, 0.05)$ **b.** $F(30, 40, 0.01)$ **c.** $F(12, 10, 0.05)$
    d. $F(5, 20, 0.01)$ **e.** $F(15, 18, 0.025)$ **f.** $F(15, 9, 0.025)$
    g. $F(40, 30, 0.05)$ **h.** $F(8, 40, 0.01)$

Section 10.4 ▼ INFERENCES CONCERNING THE RATIO OF VARIANCES

10.38 Find the following critical values for F. [*Hint*: Use formula (10-10).]
    a. $F(12, 20, 0.95)$ **b.** $F(5, 15, 0.975)$
    c. $F(20, 15, 0.99)$ **d.** $F(17, 20, 0.95)$

10.39 Referring to the article in Case Study 10-4, p. 511,
    a. what null and alternative hypotheses did Carpenter and Raza test?
    b. what was the critical value for the test?
    c. what does "$p < .005$" mean?

10.40 What formula is used to calculate the value of F? Explain it.

10.41 A bakery is considering buying one of two gas ovens. The bakery requires that the temperature remain constant during a baking operation. A study was conducted to measure the variance in temperature of the ovens during the baking process. The variance in temperature before the thermostat restarted the flame for the Monarch oven was 2.4 for 16 measurements. The variance for the Kraft oven was 3.2 for 12 measurements. Does this information provide sufficient reason to conclude that there is a difference in the variances for the two ovens? Use a 0.01 level of significance.
    a. Solve using the classical approach.
    b. Solve using the prob-value approach.

10.42 "Body Composition in Paraplegic Male Athletes" (*Medicine and Science in Sports and Exercise*, 1987, Vol. 19, no. 3, pages 195–201) compares the body composition and anthropometric characteristics of male paraplegic athletes with able-bodied athletes. One comparison was the circumference of the neck. The data may be summarized as follows:

	n	Mean	St. Dev.
Paraplegic	22	40.3 cm	2.1 cm
Able-bodied	30	41.5 cm	3.1 cm

Test the null hypothesis of equal variances against the alternative of unequal variances. Use $\alpha = 0.05$.

10.43 A study was conducted to determine whether or not there was equal variability in male and female systolic blood pressures. Random samples of 16 men and 13 women were used to test the experimenter's claim that the variances were unequal. MINITAB was used to calculate the observed value of F, F^*. Verify the results.

Men	120	120	118	112	120	114	130	114
	124	125	130	100	120	108	112	122
Women	122	102	118	126	108	130	104	
	116	102	122	120	118	130		

```
MTB > SET C1
      16 ROWS READ
MTB > SET C2
      13 ROWS READ
MTB > STDEV C1 INTO K1
     ST.DEV. =     7.89
MTB > STDEV C2 INTO K2
     ST.DEV. =     9.92
MTB > LET K3 = (K2**2)/(K1**2)
MTB > #  K3 = F*
MTB > PRINT K3
K3         1.58
MTB > STOP
```

10.44 The quality of the end product is somewhat determined by the quality of the materials used. Textile mills monitor the tensile strength of the fibers used in weaving their yard goods. The following data are tensile strengths of cotton fibers from two suppliers:

Supplier A	78	82	85	83	77	84	90	82	93	82
	80	82	77	80	80					
Supplier B	76	79	83	78	72	73	69	80	74	77
	78	78	73	76	78	79				

Calculate the observed value of F, F^*, for comparing the variances of these two sets of data.

10.45 A consumer agency wishes to compare the variability in potency of a comparable drug manufactured by companies 1 and 2. Both drugs are distributed in the form of 250-mg tablets. The potency was determined for 25 tablets from each company, and it was found that $s_1^2 = 1.25$ and $s_2^2 = 1.18$. Find a 95% confidence interval for σ_1^2/σ_2^2.

10.46 Assume that the data in Exercise 10.31 were collected for the purpose of estimation.
  a. What would be your point estimate for the ratio of the two variances?
  b. Construct the 95% confidence interval for the estimation of the ratio of the two variances.

10.47 Two independent samples, each of size 3, are drawn from a normally distributed population. Find the probability that one of the sample variances is at least 19 times larger than the other one.

10.48 Two independent samples, each of size 6, are drawn from a normally distributed population. Find the probability that one of the sample variances is at least 11 times larger than the other one.

10.5 ▼ Inferences Concerning the Difference Between the Means of Two Independent Samples (Variances Unknown and Small Samples)

In Section 10.3 we treated the cases for inferences about the difference between the means of two independent samples where the samples were both large. We will now investigate the inference procedures to be used in situations where two independent small samples (that is, one or both samples are of size less than or equal to 30) are taken from approximately normal populations for the purpose of comparing their means. For these inferences we must use Student's t-distribution. However, we must distinguish between two possible cases: (1) the variances of the two populations are equal, $\sigma_1^2 = \sigma_2^2$; or (2) the variances of the two populations are unequal, $\sigma_1^2 \neq \sigma_2^2$.

The two population variances are unknown; therefore, we will use the F test studied in Section 10.4 to determine whether we have case 1 or case 2. The two sample variances will be used in a two-tailed test of the null hypothesis H_0: $\sigma_1^2 = \sigma_2^2$, or $\sigma_1^2/\sigma_2^2 = 1$. If we fail to reject H_0, we will proceed with the methods for case 1. If we reject H_0, we will proceed with case 2. Typically, the same α is used for the F test as is given for the difference between the two means.

CASE 1 The procedures here are very similar to those used when the Student's t-distribution was employed. The standard error of estimate must be estimated by

$$s_p \sqrt{\frac{1}{n_1} + \frac{1}{n_2}}$$

pooled estimate for the standard deviation

where s_p symbolizes the **pooled estimate for the standard deviation**. (*Pooled* means that the information from both samples is combined so as to give the best possible estimate.) The formula for s_p is

$$s_p = \sqrt{\frac{(n_1 - 1)s_1^2 + (n_2 - 1)s_2^2}{n_1 + n_2 - 2}} \qquad (10\text{-}13)$$

The number of degrees of freedom, df, is the sum of the number of degrees of freedom for the two samples, $(n_1 - 1) + (n_2 - 1)$; that is,

$$df = n_1 + n_2 - 2 \qquad (10\text{-}14)$$

With this information we can now write the formula for the *test statistic t* that will be used in a *hypothesis test*,

$$t = \frac{(\bar{x}_1 - \bar{x}_2) - (\mu_1 - \mu_2)}{s_p \sqrt{(1/n_1) + (1/n_2)}} \qquad (10\text{-}15)$$

with $n_1 + n_2 - 2$ degrees of freedom.

▼ ILLUSTRATION 10-12

In studying the nature of the student body at a community college, the following question was raised: Do male and female students tend to drive the same distance (one way) to college daily? To answer this question a random sample of size 25 was taken from each segment of the student body. The results are shown in Table 10-5. Does this evidence contradict the null hypothesis that the two groups travel the same mean distances, at $\alpha = 0.10$?

TABLE 10-5
Sample Results for Illustration 10-12

Sample	n	$\bar{x}$	s^2
Male (m)	25	10.22	33.95
Female (f)	25	10.55	24.47

SOLUTION Since the σ's are unknown, we must first test the variances to determine whether we have case 1 or case 2.

STEP 1 $H_0: \sigma_m^2 = \sigma_f^2$ or $\sigma_m^2/\sigma_f^2 = 1$

$H_a: \sigma_m^2 \neq \sigma_f^2$

STEP 2 $\alpha = 0.10$. See the following figure for the test criteria:

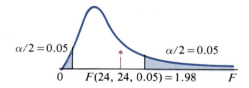

STEP 3 $F^* = 33.95/24.47 = \mathbf{1.387}$ (F^* is in the noncritical region).

STEP 4 **Decision:** Fail to reject H_0.

Conclusion Assume $\sigma_m^2 = \sigma_f^2$, at the 0.10 level of significance, and the test for the difference between means will be completed according to case 1 procedures, using t and formula (10-15).

STEP 1 $H_0: \mu_m = \mu_f$ or $\mu_m - \mu_f = 0$

$H_a: \mu_m \neq \mu_f$

STEP 2 $\alpha = 0.10$; df $= 25 + 25 - 2 = 48$; $t(48, 0.05) = 1.65$. See the accompanying figure.

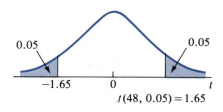

Section 10.5 ▼ INFERENCES BETWEEN THE MEANS OF TWO INDEPENDENT SAMPLES

STEP 3

$$t = \frac{(\bar{x}_m - \bar{x}_f) - (\mu_m - \mu_f)}{s_p \sqrt{(1/n_m) + (1/n_f)}}$$

$$= \frac{(10.22 - 10.55) - (0)}{\sqrt{[(24)(33.95) + (24)(24.47)]/48} \cdot \sqrt{(1/25) + (1/25)}}$$

$$= \frac{-0.33}{\sqrt{29.21} \cdot \sqrt{0.08}} = \frac{-0.33}{\sqrt{2.3368}}$$

$$= \frac{-0.33}{1.52866} = -0.2159$$

$$t^* = -0.22$$

STEP 4 **Decision:** Fail to reject H_0 (t^* is in the noncritical region).

Conclusion There appears to be no significant difference between the mean distances traveled to college daily by male and female students. ▲▲

Having completed our discussion of case 1, let's continue on to case 2.

unequal variance

CASE 2 If we must assume that the two populations have **unequal variances**, then the hypothesis test for the difference between two independent means is completed by using the test statistic t. The **calculated value of t** is obtained by using

calculated value of t

$$t = \frac{(\bar{x}_1 - \bar{x}_2) - (\mu_1 - \mu_2)}{\sqrt{(s_1^2/n_1) + (s_2^2/n_2)}} \tag{10-16}$$

with the number of degrees of freedom for the critical value being given by the smaller of $n_1 - 1$ or $n_2 - 1$.

▼ ILLUSTRATION 10-13

Many students have complained that the soft drink vending machine A (in the student recreation room) dispenses less drink than machine B (in the faculty lounge). To test this belief several samples were taken and weighed carefully, with the results as shown in Table 10-6. Does this evidence support the hypothesis that the mean amount dispensed by A is less than the mean amount dispensed by B, at the 5% level of significance?

TABLE 10-6
Sample Results for Illustration 10-13

Machine	n	$\bar{x}$	s
A	10	5.38	1.59
B	12	5.92	0.83

SOLUTION First we test for the equality of the variances.

STEP 1 $H_0: \sigma_A^2 = \sigma_B^2$ or $\sigma_A^2/\sigma_B^2 = 1$

$H_a: \sigma_A^2 \neq \sigma_B^2$

STEP 2 $\alpha = 0.05$. See the following figure:

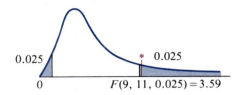

STEP 3 $F^* = (1.59)^2/(0.83)^2 = \mathbf{3.67}$

STEP 4 **Decision:** Reject H_0.

Conclusion Assume that $\sigma_A^2 \neq \sigma_B^2$, with $\alpha = 0.05$.

Now we are ready to test the difference between the means by use of case 2 procedures.

STEP 1 $H_0: \mu_A = \mu_B$ or $\mu_A - \mu_B = 0 \; (\geq)$

$H_a: \mu_A < \mu_B$ or $\mu_A - \mu_B < 0$

STEP 2 df = 10 − 1 = 9 (smaller sample is of size 10); $\alpha = 0.05$. See the accompanying figure. The critical value is $-t(9, 0.05) = -1.83$.

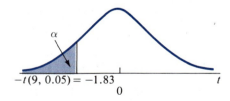

STEP 3

$$t = \frac{(5.38 - 5.92) - (0)}{\sqrt{[(1.59)^2/10] + [(0.83)^2/12]}}$$

$$= \frac{-0.54}{\sqrt{0.2528 + 0.0574}} = \frac{-0.54}{0.557} = -0.969$$

$t^* = \mathbf{-0.97}$

STEP 4 **Decision:** Fail to reject H_0 (t^* is in the acceptance region).

Conclusion There is not sufficient evidence shown by these samples to conclude that machine A dispenses less drink than does machine B. ▲▲

If confidence interval estimates are to be constructed to estimate the difference between the means of two independent small samples, the F test will again be used to determine which of the following two formulas we need to use.

CASE 1 σ's unknown but assumed equal. The $1 - \alpha$ confidence interval is given by

$$(\bar{x}_1 - \bar{x}_2) - t(\mathrm{df}, \alpha/2) \cdot s_p \cdot \sqrt{\frac{1}{n_1} + \frac{1}{n_2}} \text{ to } (\bar{x}_1 - \bar{x}_2) + t(\mathrm{df}, \alpha/2) \cdot s_p \cdot \sqrt{\frac{1}{n_1} + \frac{1}{n_2}} \quad (10\text{-}17)$$

where df $= n_1 + n_2 - 2$ and s_p is the pooled estimate for the standard deviation found by using formula (10-13).

CASE 2 σ's unknown but assumed unequal. The $1 - \alpha$ confidence interval is given

$$(\bar{x}_1 - \bar{x}_2) - t(\mathrm{df}, \alpha/2) \cdot \sqrt{\frac{s_1^2}{n_1} + \frac{s_2^2}{n_2}} \text{ to } (\bar{x}_1 - \bar{x}_2) + t(\mathrm{df}, \alpha/2) \cdot \sqrt{\frac{s_1^2}{n_1} + \frac{s_2^2}{n_2}} \quad (10\text{-}18)$$

where df is the smaller of $n_1 - 1$ or $n_2 - 1$.

▼ ILLUSTRATION 10-14

The heights of 20 randomly selected women and 30 randomly selected men were obtained from the student body of a certain college in order to estimate the difference in their mean heights. Find (a) a point estimate and (b) a 95% confidence interval estimate for the difference between the means, $\mu_m - \mu_f$.

SOLUTION The sample information is given in Table 10-7.

TABLE 10-7 Sample Results for Illustration 10-14

Sample	Number	Mean	Standard Deviation
Female (f)	20	63.8	2.18
Male (m)	30	69.8	1.92

a. The point estimate for $\mu_m - \mu_f$ is **6.0**.
b. First we must decide whether we should assume that $\sigma_f^2 = \sigma_m^2$ or $\sigma_f^2 \neq \sigma_m^2$.

$$H_0: \sigma_f^2 = \sigma_m^2 \text{ or } \frac{\sigma_f^2}{\sigma_m^2} = 1$$

$$H_a: \sigma_f^2 \neq \sigma_m^2$$

$\alpha = 0.05$; see the following figure for the test criteria. The critical value is $F(19, 29, 0.025) \approx 2.24$.

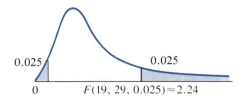

NOTE When the correct degrees of freedom cannot be found in the table, the critical value is found by using the linear interpolation technique. (See Appendix E.)

$$F^* = \frac{(2.18)^2}{(1.92)^2} = 1.289$$

So we fail to reject H_0. Therefore, we assume that $\sigma_f^2 = \sigma_m^2$. The 0.95 confidence interval is then found by using formulas (10-13) and (10-17).

$$6.0 \pm t(48, 0.025) \cdot \sqrt{\frac{19(2.18)^2 + 29(1.92)^2}{48}} \cdot \sqrt{\frac{1}{20} + \frac{1}{30}}$$

$$6.0 \pm 1.96\sqrt{4.1084}\sqrt{0.0833}$$

$$6.0 \pm (1.96)(2.0269)(0.2886)$$

$$6.0 \pm 1.1465$$

$$6.0 \pm 1.15$$

4.85 to **7.15**, the 0.95 confidence interval for $\mu_m - \mu_f$.

(No specific illustration is shown for estimation where $\sigma_1^2 \neq \sigma_2^2$. All parts of such a solution can be found in various illustrations in this chapter.) ▲▲

Case Study 10-5

An Empirical Study of Faculty Evaluation Systems: Business Faculty Perceptions

Tong and Bures report a whole series of t-test results from comparing various factors of perceptions between AACSB accredited schools and non-AACSB accredited schools. For each t-test they report a two-tailed probability (p-value). Many of them are 0.000 meaning that the p-value is less than 0.0005. The p-value for "activity in professional societies" is 0.394. What does that mean? (See Exercise 10.67, p. 531.)

Table 1 shows how respondents from American Assembly of Collegiate Schools of Business (AACSB)-accredited and non-AACSB-accredited schools rated the importance of the ten faculty evaluation factors. Clearly AACSB-accredited schools and non-AACSB-accredited schools are different, with the AACSB-accredited institutions placing greater weight on research and publication, and with the non-accredited institutions putting more emphasis on classroom teaching, campus committee work, student advising, public service, advisor to student organizations, and consultation (business, government). There is no significant difference between AACSB-accredited and non-AACSB-accredited schools in the importance given to a faculty member's activity in professional societies.

TABLE 1 *t*-test Results of Importance of Faculty Evaluation Factors, AACSB-Accredited vs. Non-AACSB-Accredited Schools

Factor	AACSB-Accredited Schools ($n = 176$)	Non-AACSB-Accredited Schools ($n = 74$)	Two-tailed *t*-test Probability
Articles in professional journals	4.49	2.87	0.000
Classroom teaching	3.34	4.31	0.000
Books as author or editor	3.23	2.45	0.000
Papers at professional meetings	3.09	2.65	0.001
Activity in professional societies	2.55	2.65	0.394
Campus committee work	2.25	3.16	0.000
Student advising	1.80	3.36	0.000
Public service	1.98	2.47	0.000
Advisor to student organizations	1.71	2.30	0.000
Consultation (business, government)	1.73	2.30	0.001

Note: A five-point scale with 1 = "not at all important" to 5 = "extremely important" was used in this analysis.

Source: Hsin-Min Tong and Allen L. Bures in *Journal of Education for Business*, April 1987. Reprinted with permission of the Helen Dwight Reid Educational Foundation. Published by Heldref Publications, 4000 Albemarle St., N.W., Washington, D.C. 20016. Copyright © 1987.

Case Study 10-6

Youthful Ideas About Old Age: An Analysis of Children's Drawings

An interesting application of two sample statistics is discussed in the following article. The "heights" and the "quality" of the children's artwork were measured and analyzed, and hypothesis tests were completed for both independent and dependent samples. The "correlated samples" mentioned are dependent samples as described in this section. Why were both independent and dependent sample techniques used to analyze the data? (See Exercise 10.68, p. 531.)

ABSTRACT Youthful ideas about old age are investigated by means of an analysis of children's drawings of young and old people. The age of the artists ranged between ten and a half years and eleven and a half years of age, and each artist drew four pictures (a young woman, an old woman, a young man, and an old man). Analyses consisted of a content analysis of drawings; the calculation of standard scores by means of the Harris-Goodenough Draw-A-Person (DAP) test scoring procedure (in order to effect comparisons of scores within each child's corpus of drawings); and measurement of the height of each drawing. The content analysis suggested a greater degree of stereotyping in boys' pictures of young women and in girls' pictures of old men than in other drawings. Pictures of old people overall were no more stereotyped than were those of young people, though they were more negative in content. Results of the DAP standard score calculation comparisons found consistent differences between children's drawings of old and young people, with pictures of old people attaining lower standard scores than those of young people. Size measurements revealed pictures of old people to be significantly smaller than those of young people.

FIGURE 1 A Young Woman (Boy Artist)

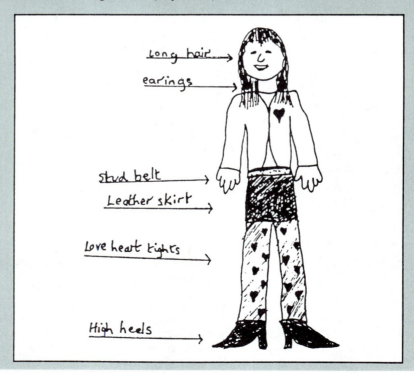

FIGURE 2 An Old Woman (Girl Artist)

TABLE 3 Mean Height of Figures Measured in cms (Sample Standard Deviations in Parentheses)

	Young Woman	Old Woman	Young Man	Old Man
Drawings by girls	14.01 (2.29)	12.46 (2.67)	12.24 (3.44)	11.69 (3.26)
Drawings by boys	11.88 (2.90)	10.36 (2.21)	9.13 (2.32)	8.29 (2.61)

HEIGHT COMPARISONS

Inspection of drawings suggested that there were regular differences in sizes of the different classes of people depicted. Measurement supported this view. Mean heights of figures in the four groups are shown in Table 3. The old people were invariably smaller than the young ones, women were larger than men, and drawings by girls were bigger than those by boys.

Raw scores were converted to standard scores, and mean standard scores were calculated for all categories of drawing. These values are shown in Table 4. Given that the higher the standard score, the more "sophisticated," detailed and complex the drawing, it appears that DAP scores echo some of the findings of the mean height calculations: namely, that drawings of young people were "better" than those of old people, and that drawings of women, on the whole, were "better" than those of men.

TABLE 4 Mean Standard Scores (Standard Deviations in Parentheses)

Gender of Artist	Subject of Drawing			
	Young Woman	Old Woman	Young Man	Old Man
Girl	110.6 (8.25)	98.9 (7.84)	104.9 (9.00)	100.2 (8.42)
Boy	113.6 (11.66)	102.4 (9.26)	108.3 (9.44)	100.1 (9.47)

TESTS FOR THE SIGNIFICANCE OF DIFFERENCES

t-tests for the significance of the difference between two means for correlated samples were carried out, in order to compare scores for drawings of young and old people and for scores of drawings of female and male people. Scores of girl and boy artists were also treated separately. Similarly, t-tests for the significance of the difference between two means for independent samples enabled comparisons of drawings of girls and boys.

In the correlated samples, all differences between drawings of old and young people were found to be significant. Differences between boys' drawings of old and young women were found to be highly significant at the .001 level ($t = 7.42$) as were those of girls ($t = 6.55$). Similarly, boys' drawings of old and young men were also significantly different at the .001 level ($t = 5.02$). Girls' drawings of young and old men differed at the .05 level ($t = 2.3225$). Tests for independent samples carried out on drawings of girls and boys in all four categories (young woman, old woman, young man, old man) produced no significant differences.

Source: Nancy Falchikov, *International Journal of Aging and Human Development*, 3 (no. 2), 1990, 79–99. Reprinted by permission.

▼▲ EXERCISES

10.49 State the null hypothesis, H_0, and the alternative hypothesis, H_a, that would be used to test the following claims:
  a. There is no difference between the two population means.
  b. There is a difference between the mean age of employees at two different large companies.

Section 10.5 ▼ INFERENCES BETWEEN THE MEANS OF TWO INDEPENDENT SAMPLES

c. The mean of population I is at least 5 units larger than the mean of population II.

10.50 Determine the test criteria that would be used to test the following hypotheses when t is used as the test statistic:
  a. $H_0: \mu_A - \mu_B = 0$ and $H_a: \mu_A - \mu_B > 0$, with $n_A = 21$, $n_B = 18$, σ's equal and $\alpha = 0.05$.
  b. $H_0: \mu_A - \mu_B = 0$ and $H_a: \mu_A - \mu_B \neq 0$, with $n_A = 12$, $n_B = 9$, σ's equal and $\alpha = 0.05$.
  c. $H_0: \mu_A - \mu_B = 1.0$ and $H_a: \mu_A - \mu_B < 1.0$, with $n_A = 21$, $n_B = 18$, σ's unequal and $\alpha = 0.01$.
  d. $H_0: \mu_A - \mu_B = 0$ and $H_a: \mu_A - \mu_B > 0$, with $n_A = 9$, $n_B = 13$, σ's unequal and $\alpha = 0.01$.

10.51 Certain "adjustments" are made between the test for the equality of two population means when the variances are unknown and judged to be equal and the test for equality of two population means when variances are unknown and judged to be unequal. (The t-distribution is used in both cases.) Explain how each of the following adjustments is accomplished:
  a. The test statistic is modified.
  b. The critical region is modified.
  c. The number of degrees of freedom is modified.

10.52 Calculate the pooled estimate for the standard error of difference between two independent means for each of the following cases:
  a. $s_1^2 = 12$, $s_2^2 = 15$, $n_1 = 16$ and $n_2 = 21$.
  b. $s_1^2 = 0.054$, $s_2^2 = 0.087$, $n_1 = 8$ and $n_2 = 10$.
  c. $s_1 = 2.8$, $s_2 = 6.4$, $n_1 = 16$ and $n_2 = 21$.

10.53 Two independent samples are drawn from normal populations, with the results as shown in the following table. Does this information provide sufficient reason to reject the null hypothesis in favor of the claim that the mean of population R is significantly larger than the mean of population S? Use $\alpha = 0.05$.

Sample	n	$\sum x$	$\sum (x - \bar{x})^2$
R	10	295	75
S	8	195	90

  a. Can we assume $\sigma_R = \sigma_S$? How do we decide? Explain.
  b. Test the null hypothesis $\sigma_R = \sigma_S$ against the alternative hypothesis $\sigma_R \neq \sigma_S$.
  c. Assume $\sigma_R = \sigma_S$. What effect does this assumption have on the hypothesis test about means?
  d. Is there sufficient reason to reject the hypothesis that $\mu_R = \mu_S$ at $\alpha = 0.05$?

10.54 If a random sample of 18 homes south of Center Street in Provo, New York, has a mean selling price of $15,000 and a variance of $2,400, and a random sample of 18 homes north of Center Street has a mean selling price of $16,000 and a variance of $4,800, can you conclude that there is a significant difference between the selling price of homes in these two areas of Provo at the 0.05 level?
   a. Solve using the classical approach.
   b. Solve using the prob-value approach.

10.55 A study was designed to test the hypothesis that men have a higher diastolic blood pressure than do women. MINITAB was used to analyze the following sample data. The subcommand POOLED is used when the variances are assumed to be equal.

   males: 76 76 74 70 80 68 90 70 90 72 76 80 68 72 96 80
   females: 76 70 82 90 68 60 62 68 80 74 60 62 72

```
MTB > SET MALE DIASTOLIC BLOOD PRESSURES IN C1
MTB > SET FEMALE DIASTOLIC BLOOD PRESSURES IN C2
MTB > TWOSAMP .99 C1 C2;
SUBC> POOLED;
SUBC> ALTERNATIVE +1.
TWOSAMPLE T FOR C1 VS C2
         N      MEAN     STDEV    SE MEAN
C1      16     77.37      8.35       2.1
C2      13     71.08      9.22       2.6
99 PCT CI FOR MU C1 - MU C2: (-2.8, 15.4)
TTEST MU C1 = MU C2 (VS GT): T= 1.93  P=0.032  DF= 27
POOLED STDEV =        8.75
```

   a. Use the standard deviations to verify that the hypothesis of equal variances is not rejected at $\alpha = 0.05$.
   b. Verify the value for t^*.
   c. Use the Student's t table to bound the p-value and compare with the MINITAB value.

10.56 The following two samples were both generated by MINITAB as simulated random samples drawn from a normal population with mean 18 and standard deviation 4.

Sample 1

```
MTB > RANDOM    28 OBSERVATIONS INTO C1;
SUBC> NORMAL    MU = 18  SIGMA = 4.
MTB > PRINT C1

C1
  26.1763    11.9731    12.5655    15.6145    18.2149    27.8638    20.9445
  19.9671    11.5700    21.9374    17.2521    13.4863    18.6078    11.7857
  15.7301    24.2869    15.5793    15.8590    12.8898    16.8578    20.2236
  21.2893    24.6022    16.0486    21.1467    18.1897    13.6682    10.6549
```

Sample 2

```
MTB > RANDOM    28 OBSERVATIONS INTO C2;
SUBC> NORMAL    MU = 18  SIGMA = 4.
MTB > PRINT C2

C2
    18.3001    19.6821    22.0718    19.2413    17.5341    12.6875    22.0327
    21.1418    16.2281    15.3774    19.9150    21.4764    19.7811    22.5094
    13.0132    22.1110    27.6955    17.9114    22.2057    19.7361    18.0574
    17.2909    13.3404    22.0134    16.3378    21.9791    16.9883    15.4049
```

a. Calculate the mean and standard deviation of each sample.
b. Is the ratio of the variances significantly different from one? Use $\alpha = 0.05$.
c. Is there a significant difference between the means of these two samples? Use $\alpha = 0.05$.

10.57 Independent samples were taken from each of two normal populations in order to compare the two population means. The sample data are summarized in the following table. Assume that $\sigma_1 \neq \sigma_2$.

a. What effect does the assumption $\sigma_1 \neq \sigma_2$ have on the situation?
b. Determine the number of degrees of freedom to use in comparing μ_1 and μ_2.
c. Is there sufficient reason to reject the hypothesis that $\mu_1 = \mu_2$, at $\alpha = 0.05$?

Sample	n	$\bar{x}$	s^2
1	10	12.3	6.8
2	15	13.9	23.5

10.58 Two samples of data were obtained to compare the means of two normal populations. The data are summarized in the following table: Is there sufficient evidence to conclude that $\mu_A > \mu_B$ at $\alpha = 0.05$? (*Hint:* Does $\sigma_A = \sigma_B$?)

Sample	n	$\sum x$	$\sum x^2$
A	10	223	6735
B	10	110	1322

10.59 MINITAB was used to complete a *t*-test of the difference of two independent means for the following two samples. The command TWOSAMP uses formula (10-12) unless the subcommand POOLED is used. Verify the results.

```
MTB > SET INTO C1
33.7 21.6 32.1 38.2 33.2 35.9 34.1 39.8 23.5 21.2 23.3
18.9 30.3
MTB > SET INTO C2
28.0 59.9 22.3 43.3 43.6 24.1 6.9 14.1 30.2 3.1 13.9
19.7 16.6 13.8 62.1 28.1
MTB > TWOSAMP .95 C1 C2
TWOSAMPLE T FOR C1 VS C2
         N       MEAN      STDEV    SE MEAN
C1   13      29.68      7.07       2.0
C2   16      26.86     17.41       4.4
95 PCT CI FOR MU C1 - MU C2:  (-7.1, 12.8)
TTEST MU C1 = MU C2 (VS NE): T= 0.59  P= 0.56  DF= 20
MTB > STOP
```

10.60 The quality of latex paint is monitored by measuring different characteristics of the paint. One characteristic of interest is the particle size. Two different types of disc centrifuges (JLDC, Joyce Loebl Disc Centrifuge, and the DPJ, Dwight P. Joyce disc) are used to measure the particle size. It is thought that these two methods yield different measurements. Thirteen readings were taken from the same batch of latex paint using both the JLDC and the DPJ discs.

JLDC	DPJ
4714	4295
4601	4271
4696	4326
4896	4530
4905	4618
4870	4779
4987	4752
5144	4744
3962	3764
4006	3797
4561	4401
4626	4339
4924	4700

Source: With permission of SCM Corporation.

a. Determine whether there is a significant difference between the readings at the 0.10 level of significance.
b. What is your estimate for the difference between the two readings?

Section 10.5 ▼ INFERENCES BETWEEN THE MEANS OF TWO INDEPENDENT SAMPLES

10.61 To compare the mathematical competency of men and women, the Beckmann-Beal test of mathematical competencies was administered to 15 men and 15 women. The results were

Males: $\bar{x} = 39.5$ $s = 2.9$
Females: $\bar{x} = 38.4$ $s = 3.4$

a. Test $H_0: \sigma_M = \sigma_F$ versus $H_a: \sigma_M \neq \sigma_F$ at $\alpha = 0.05$.
b. Test $H_0: \mu_M = \mu_F$ versus $H_a: \mu_M \neq \mu_F$ at $\alpha = 0.05$.
c. Give the p-value for the test in (b).

10.62 A study was designed to compare the attitudes of two groups of nursing students toward computers. Group 1 had previously taken a statistical methods course that involved significant computer interaction through the use of statistical packages. Group 2 had taken a statistical methods course that did not use computers. The students' attitudes were measured by administering the Computer Anxiety Index (CAIN). The results were

group 1 (with computers): $n = 10$ $\bar{x} = 60.3$ $s = 7.5$
group 2 (without computers): $n = 15$ $\bar{x} = 67.2$ $s = 2.1$

a. Test for equality of variances at $\alpha = 0.05$.
b. Do the data show that the mean score for those with computer experience was significantly less than the mean score for those without computer experience? Use $\alpha = 0.05$.

10.63 Twenty laboratory mice were randomly divided into two groups of 10. Each group was fed according to a prescribed diet. At the end of three weeks the weight gained by each animal was recorded. Do the data in the following table justify the conclusion that the mean weight gained on diet B was greater than the mean weight gained on diet A, at the $\alpha = 0.05$ level of significance?

Diet A	5	14	7	9	11	7	13	14	12	8
Diet B	5	21	16	23	4	16	13	19	9	21

10.64 The material used in making parts affects not only how long the part lasts but also how difficult it is to take apart to repair. The following measurements are for screw torque removal for a specific screw after several operations of use. The first column lists the part number, the second column lists the screw torque removal measurements for assemblies made with material A, and the third column lists the screw torque removal measurements for assemblies made with material B.

	Removal Torque (NM, Newton-meters)	
Part Number	Material A	Material B
1	16	11
2	14	14
3	13	13
4	17	13
5	18	10
6	15	15
7	17	14
8	16	12
9	14	11
10	16	14
11	15	13
12	17	12
13	14	11
14	16	13
15	15	12

Source: Problem data provided by AC Rochester Division, General Motors, Rochester, NY.

    a. Find the sample mean, variance, and standard deviation for the material A data.
    b. Find the sample mean, variance, and standard deviation for the material B data.
    c. At the 0.01 level, do these data show a significant difference in the mean torque required to remove the screws from the two different materials?

10.65 A study comparing attitudes toward death was conducted in which organ donors (individuals who had signed organ donor cards) were compared with nondonors. The study is reported in the journal *Death Studies* (Vol. 14, no. 3, 1990, pages 219–227). *Templer's Death Anxiety Scale* (DAS) was administered to both groups. On this scale, high scores indicate high anxiety concerning death. The results were reported as follows:

	n	Mean	St. Dev.	t
Organ Donors	25	5.36	2.91	
Nonorgan Donors	69	7.62	3.45	2.92

Verify the computed value of t.

10.66 A study was conducted to assess the safety and efficiency of receiving nitroglycerin from a transdermal system (i.e., a patch worn on the skin), which intermit-

Section 10.5 ▼ INFERENCES BETWEEN THE MEANS OF TWO INDEPENDENT SAMPLES

tently delivers the medication versus oral medication (pills). Twenty patients who suffer from angina (chest pain) due to physical effort were enrolled in trials. All received patches, some ($n = 8$) contained nitroglycerin, the others ($n = 12$) contained a placebo. Suppose the resulting "time to angina" data were summarized:

	Mean Time to Angina (sec)				
	Active	Placebo	Difference	SE	p Value[b]
Day 1 AM	320.00	287.00	33.00	9.68	0.0029
Day 7 PM	314.00	285.25	28.75	13.74	0.0500

[b]For treatment difference.

a. Determine the value of t for the difference between two independent means given the difference and the standard error (SE) for the day 1 AM data.
b. Verify the *p*-value.
c. Determine the value of t for the difference between two independent means given the difference and the standard error (SE) for the day 7 PM data.
d. Verify the *p*-value.

10.67 Explain the meaning of the *p*-value 0.394 reported in Case Study 10-5 (p. 520) for the two-tailed *t*-test comparing the difference between the perception of activities in professional societies.

10.68 Explain why both independent and dependent sample techniques were used to analyze the data discussed in Case Study 10-6? (p. 521)

10.69 The two independent samples shown in the following table were obtained in order to estimate the difference between the two population means. Construct the 0.98 confidence interval estimate.

Sample A	6	7	7	6	6	5	6	8	5	4
Sample B	7	2	4	3	3	5	4	6	4	2

10.70 The February 1991 issue of *Changing Times* (page 22) reported that the average starter home had a sale price of $71,100 in Atlanta, while in Dallas the average starter home sale price was $76,000. Suppose that another survey found that 50 starter homes in Atlanta had a mean price of $70,025 and a standard deviation of $5,000, and 50 starter homes in Dallas were priced such that the mean was $75,000 and the standard deviation was $6,500. Construct a 90% confidence interval estimate for the difference in the mean prices for all starter homes in Atlanta and Dallas.

10.71 The MINITAB output shown in Exercise 10.55, p. 526, includes a 99% confidence interval.
  a. Verify the confidence interval.
  b. Find the 0.95 confidence interval estimate for the difference between the two means.

10.72 The MINITAB output shown in Exercise 10.59, p. 527, includes a 95% confidence interval.
  a. Verify the confidence interval.
  b. Find the 0.99 confidence interval estimate for the difference between the two means.

10.6 ▼ Inferences Concerning the Difference Between Proportions of Two Independent Samples

We are often interested in making statistical comparisons between the proportions, percentages, or probabilities associated with two populations. Such questions as the following are frequently asked: Is the proportion of homeowners who favor a certain tax proposal different from the proportion of renters who favor it? Did a larger percentage of this semester's class than of last semester's class pass statistics? Do students' opinions about the new code of conduct differ from those of the faculty? You can probably see the many types of questions involved. In this section we will compare two population proportions by using the difference between the observed proportions, $p'_1 - p'_2$, of two independent samples.

RECALL

1. The observed probability is $p' = x/n$, where x is the number of observed successes in n trials.
2. $q' = 1 - p'$.
3. p is the probability of success on an individual trial in a binomial probability experiment of n repeated independent trials (see p. 285).

The sampling distribution of $p'_1 - p'_2$ is approximately normally distributed with a mean $\mu_{p'_1 - p'_2} = p_1 - p_2$ and with a standard error of

$$\sqrt{\frac{p_1 q_1}{n_1} + \frac{p_2 q_2}{n_2}}$$

if n_1 and n_2 are sufficiently large. Therefore, when the null hypothesis that there is no difference between two proportions is to be tested, the *test statistic* will be z. The calculated value of z will be determined by formula (10-19):

$$z = \frac{p'_1 - p'_2}{\sqrt{pq[(1/n_1) + (1/n_2)]}} \tag{10-19}$$

NOTES

1. The null hypothesis is $p_1 = p_2$, or $p_1 - p_2 = 0$.
2. The numerator of formula (10-19) written in the usual manner is $(p'_1 - p'_2) - (p_1 - p_2)$, but since the null hypothesis is assumed to be true during the test, $p_1 - p_2 = 0$ and the numerator becomes simply $p'_1 - p'_2$.

Section 10.6 ▼ INFERENCES BETWEEN PROPORTIONS OF TWO INDEPENDENT SAMPLES

3. Since the null hypothesis is $p_1 = p_2$, the standard error of $p'_1 - p'_2$ can be written as $\sqrt{pq[(1/n_1) + (1/n_2)]}$, where $p = p_1 = p_2$ and $q = 1 - p$.

▼ **ILLUSTRATION 10-15**

A salesman for a new manufacturer of walkie-talkies claims that the percentage of defective walkie-talkies found among his products will be no higher than the percentage of defectives found in a competitor's line. To test his statement, random samples were taken of each manufacturer's product. The sample summaries are given in Table 10-8. Can we reject the salesman's claim at the 0.05 level of significance?

TABLE 10-8
Sample Results for Illustration 10-15

Product	Sample Number	Number of Defectives	Number Checked
Salesman's	1	8	100
Competitor's	2	2	100

SOLUTION

STEP 1 $H_0: p_1 - p_2 = 0 \;(\leq)$ (no higher than)
$H_a: p_1 - p_2 > 0$ (higher than)

STEP 2 $\alpha = 0.05$. The test criteria are shown in the accompanying figure. The critical value is $z(0.05) = 1.65$.

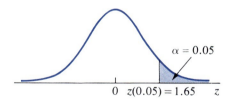

STEP 3 $p'_1 = 8/100 = 0.08$, $p'_2 = 2/100 = 0.02$.

Since the values of p_1 and p_2 are unknown but assumed equal, (H_0), the best estimate we have for p ($p = p_1 = p_2$), is obtained by pooling the two samples to obtain a **pooled observed probability**, p'_p:

pooled observed probability

$$p'_p = \frac{x_1 + x_2}{n_1 + n_2} \qquad (10\text{-}20)$$

Our pooled estimate for p, p'_p, then becomes

$$p'_p = \frac{8 + 2}{100 + 100} = \frac{10}{200} = \mathbf{0.05} \quad \text{and} \quad q'_p = 1 - p'_p = 1 - 0.05 = \mathbf{0.95}$$

The value of the test statistic z is calculated by using formulas (10-19) and (10-20) together. p'_p and q'_p are used as estimates for p and q.

$$z = \frac{p'_1 - p'_2}{\sqrt{p'_p q'_p [(1/n_1) + (1/n_2)]}}$$

$$= \frac{0.08 - 0.02}{\sqrt{(0.05)(0.95)[(1/100) + (1/100)]}}$$

$$= \frac{0.06}{\sqrt{(0.05)(0.95)(0.02)}} = \frac{0.06}{0.031} = 1.935$$

$$z^* = \mathbf{1.94}$$

See the accompanying figure.

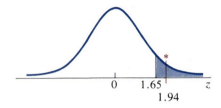

STEP 4 **Decision:** Reject H_0 (z^* is in the critical region).

Conclusion At the 0.05 level of significance, there is sufficient evidence to reject the salesman's claim. ▲▲

▼ ILLUSTRATION 10-16

Joe claimed that the probability that a commuting college student has car trouble of some type (car wouldn't start, flat tire, accident) on the way to college in the morning is greater than the probability that the student will have car trouble on the way home after class. He has several theories to explain his claim. The Sports Car Club thinks that the ideas of "before class" and "after class" have nothing to do with whether or not a student has car trouble. The club decides to challenge Joe's claim, and two large samples are gathered in order to test the theory. The resulting sample statistics are given in Table 10-9. Complete the hypothesis test with $\alpha = 0.02$.

TABLE 10-9
Sample Results for Illustration 10-16

Time of Trouble	Number Sampled (n)	Number Having Trouble
Before (b)	500	30
After (a)	600	28

SOLUTION

STEP 1 $H_0: p_b = p_a$ or $p_b - p_a = 0$ (no difference)

$H_a: p_b > p_a$ or $p_b - p_a > 0$ (Joe's claim)

STEP 2 $\alpha = 0.02$. The test criteria are shown in the accompanying figure.

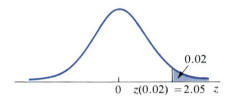

STEP 3

$$z = \frac{p_b' - p_a'}{\sqrt{p_p' q_p'[(1/n_b) + (1/n_a)]}}$$

$$p_b' = \frac{30}{500} = 0.06 \qquad p_a' = \frac{28}{600} = 0.04666 = 0.047$$

The pooled estimate for p is

$$p_p' = \frac{30 + 28}{500 + 600} = \frac{58}{1100} = 0.0527 = \mathbf{0.053}$$

and thus $q_p' = 1 - 0.053 = \mathbf{0.947}$. Therefore,

$$z = \frac{0.06 - 0.047}{\sqrt{(0.053)(0.947)(1/500 + 1/600)}}$$

$$= \frac{0.013}{\sqrt{0.05019(0.002 + 0.0017)}}$$

$$= \frac{0.013}{\sqrt{0.000186}} = \frac{0.013}{0.0136} = 0.956$$

$$z^* = \mathbf{0.96}$$

STEP 4 **Decision:** Fail to reject H_0 (z^* is in the noncritical region).

Conclusion At the 0.02 level of significance, there is no evidence to indicate that the probability of car trouble is greater before class than after class. ▲▲

To estimate the difference between the proportions of two populations, you should use the observed difference $(p_1' - p_2')$ as the point estimate. The $1 - \alpha$ *confidence interval* is given by

$$(p_1' - p_2') - z(\alpha/2) \cdot \sqrt{\frac{p_1' q_1'}{n_1} + \frac{p_2' q_2'}{n_2}} \text{ to } (p_1' - p_2') + z(\alpha/2) \cdot \sqrt{\frac{p_1' q_1'}{n_1} + \frac{p_2' q_2'}{n_2}} \qquad \text{(10-21)}$$

NOTE A pooled sample proportion is not used in the confidence interval because it is not known whether $p_1 = p_2$.

▼ ILLUSTRATION 10-17

In studying his campaign plans, Mr. Morris wishes to estimate the difference between men's and women's views regarding his appeal as a candidate. He asks his campaign manager to take two samples and find the 99% confidence interval estimate of the difference. A sample of 1000 voters was taken from each population, with 388 men and 459 women favoring Mr. Morris.

SOLUTION The campaign manager calculated the confidence interval estimate by using formula (10-21), as follows.

$$p'_m = \frac{388}{1000} = 0.388 \qquad p'_w = \frac{459}{1000} = 0.459$$

$$(0.459 - 0.388) \pm 2.58 \sqrt{\frac{(0.459)(0.541)}{1000} + \frac{(0.388)(0.612)}{1000}}$$

$$0.071 \pm 2.58\sqrt{0.000248 + 0.000237}$$

$$0.071 \pm 2.58\sqrt{0.000485}$$

$$0.071 \pm (2.58)(0.022)$$

$$0.071 \pm 0.057$$

0.014 to **0.128**, the 0.99 confidence interval for $p_w - p_m$.

With 99% confidence we can say that there is a difference of from 1.4% to 12.8% in Mr. Morris's voter appeal. That is, a larger proportion of women than men favor Mr. Morris, and the difference in proportion is between 1.47% and 12.8%.

▲▲

Case Study 10-7

Smokers Need More Time in Recovery Room

The following report says that a study involving 327 patients in Long Beach, N.J., showed that 38% of the nonsmokers spent less than one hour in recovery compared to 23% of smokers. Further, 19% of the smokers spent more than two hours, compared to 7% of the nonsmokers. Do these differences seem to be significant? Explain. (See Exercise 10.80, p. 541.)

Smokers take longer to recover from anesthesia after surgery than nonsmokers do, new research shows.

A study reported over the weekend at the American Society of Anesthesiologists meeting in Las Vegas found that smokers were nearly three times more likely than nonsmokers to spend two or more hours in the recovery room. "It's a very strong difference," says researcher Dr. David Handlin, of Monmouth Medical Center, Long Branch, N.J.

Section 10.6 ▼ INFERENCES BETWEEN PROPORTIONS OF TWO INDEPENDENT SAMPLES

Of the 327 patients studied:
▶ 38% of non-smokers spent less than one hour in the recovery room, vs. 23% of smokers.
▶ 19% of smokers spent more than two hours in recovery, vs. 7% of non-smokers.

Longer recovery room stays demand more nursing care and drive up health care costs, says Handlin. Other studies have shown that smokers tend to have a higher risk of complications from surgery. Handlin says this is the first to show that smokers take longer to recover from anesthesia.

Longer recovery stays may be due to mild lung disease or a lowered ability of the blood to carry oxygen, says Handlin. Some experts say quitting smoking weeks before surgery may cut recovery time.

Source: Copyright 1990, USA TODAY. Reprinted with permission.

Case Study 10-8

One Too Many

"One Too Many" reports that 36% (overall) admit to sometimes drinking too much. The graphic indicates that a sample of women and a sample of men were polled separately, yielding results of 46% and 26%. The "overall" statistic appears to be a pooled statistic. Explain. (See Exercise 10.79, p. 541.)

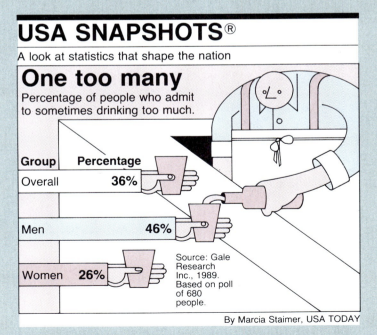

Source: Copyright 1990, USA TODAY. Reprinted with permission.

Case Study 10-9

Gallup Report: Sampling Tolerances

The Gallup Report *provides its readers with the following information and tables. The tables give the 0.95 confidence level sampling errors (maximum error of estimation) for various percentages and sample sizes. Can you verify the values in these tables? (See Exercise 10.85, a and b, p. 541.)*

In interpreting survey results, it should be borne in mind that all sample surveys are subject to sampling error, that is, the extent to which the results may differ from those that would be obtained if the whole population surveyed had been interviewed. The size of such sampling errors depends largely on the number of interviews.

In comparing survey results in two sub-samples, such as men and women, the question arises as to how large a difference between them must be before one can be reasonably sure that it reflects a real difference. In Tables 1 and 2, the number of points which must be allowed for in such comparisons is indicated.

For percentages near 20 or 80, use Table 1, for those near 50, Table 2. For percentages in between, the error factor is between that shown in the two tables.

Here is an example of how the tables should be used: Say 50 percent of men and 40 percent of women respond the same way to a question, a difference of 10 percentage points. Can it be said with any assurance that the 10 point difference reflects a real difference between men and women on the question? (Unless otherwise noted, the sample contains approximately 750 men and 750 women.)

Because the percentages are near 50, consult Table 2. Since the two samples are about 750 persons each, look for the place in the table where the column and row labeled "750" converge. The number seven appears there. This means the allowance for error should be seven points, and the conclusion that the percentage among men is somewhere between 3 and 17 points higher than the percentage among women would be wrong only about 5 percent of the time. In other words, there is a considerable likelihood that a difference exists in the direction observed and that it amounts to at least 3 percentage points.

If, in another case, male respondents amount to 22 percent, and female to 24 percent, consult Table 1 because these percentages are near 20. The column and row labeled "750" converge on the number 5. Obviously then, the 2 point difference is inconclusive.

TABLE 1 Sampling Tolerances

	Recommended Allowance for Sampling Error of the Difference			
	In Percentage Points (at 95 in 100 confidence level)*			
	Percentages Near 20 or Percentages Near 80			
	750	600	400	200
Size of the Sample				
750	5			
600	6	6		
400	7	7	7	
200	8	8	9	10

The chances are 95 in 100 that the sampling error is not larger than the figures shown.

TABLE 2 Sampling Tolerances

	Percentages Near 50			
Size of the Sample				
750	6			
600	8	8		
400	8	8	9	
200	10	11	11	13

The chances are 95 in 100 that the sampling error is not larger than the figures shown.

Source: Gallup Reports, November 1986.

▼▲ EXERCISES

10.73 State the null hypothesis, H_0, and the alternative hypothesis, H_a, that would be used to test the following claims:
  a. There is no difference between the proportions of men and women who will vote for the incumbent in next month's election.
  b. The percentage of boys who cut classes is greater than the percentage of girls who cut classes.
  c. The percentage of college students who drive old cars is higher than the percentage of noncollege people of the same age who drive old cars.

10.74 Determine the test criteria that would be used to test the following hypotheses when z is used as the test statistic:
  a. $H_0: p_1 = p_2$ and $H_a: p_1 > p_2$, with $\alpha = 0.05$.
  b. $H_0: p_A = p_B$ and $H_a: p_A \neq p_B$, with $\alpha = 0.05$.

c. $H_0: p_1 - p_2 = 0$ and $H_a: p_1 - p_2 < 0$, with $\alpha = 0.04$.
d. $H_0: p_m - p_f = 0$ and $H_a: p_m - p_f > 0$, with $\alpha = 0.01$.

10.75 The July 28, 1990 issue of *Science News* (page 61) reported that smoking boosts death risk for diabetics. The death risk is increased more for women than men. Suppose as a follow-up study we investigated the smoking rates for male and female diabetics and obtained the following data:

Gender	n	Number Who Smoke
Male	500	215
Female	500	170

a. Test the research hypothesis that the smoking rate (proportion of smokers) is higher for males than females. Calculate the prob-value.
b. What decision and conclusion would be reached at the 0.05 level of significance?

10.76 In a survey of working parents (both parents working) one of the questions asked was "Have you refused a job, promotion, or transfer because it would mean less time with your family?" Two hundred men and 200 women were asked this question. Twenty-nine percent of the men and 24% of the women responded "yes." Based on this survey, can we conclude that there is a difference in the proportion of men and women responding "yes" at the 0.05 level of significance?

10.77 Two randomly selected groups of citizens were exposed to different media campaigns that dealt with the image of a political candidate. One week later the citizen groups were surveyed to see whether they would vote for the candidate. The results were as follows:

	Exposed to Conservative Image	Exposed to Moderate Image
Number in Sample	100	100
Proportion for the Candidate	0.40	0.50

Is there sufficient evidence to show a difference in the effectiveness of the two image campaigns at the 0.05 level of significance?
a. Solve using the classical approach.
b. Solve using the prob-value approach.

10.78 U.S. military active duty personnel and their dependents are provided with free medical care. A survey was conducted to compare obstetrical care between military and civilian (pay for medical care) families ("Use of Obstetrical Care Compared Among Military and Civilian Families," *Public Health Reports*, May/June 1989, pages 309, 310). The number of women who began prenatal care by the second trimester may be reported:

Section 10.6 ▼ INFERENCES BETWEEN PROPORTIONS OF TWO INDEPENDENT SAMPLES

	Military	Civilian
Prenatal Care Began by 2nd Trimester	358	6786
Total Sample	407	7363

Is there a significant difference between the proportion of military and civilian women who begin prenatal care by the second trimester? Use $\alpha = 0.02$.

10.79 "One Too Many" (Case Study 10-8, p. 537) shows the percentage (relative frequency) of people who admit to sometimes drinking too much.
   a. What was the percentage of men who admitted drinking too much? women?
   b. In this section, a pooled observed probability, p'_p, was introduced. Explain the apparent relationship between the pooled observed probability and the overall group percentage reported.
   c. Assuming that 340 men and 340 women were polled, calculate the 0.95 confidence interval estimate for the difference in the proportion of men and women who drink too much.

10.80 "38% of nonsmokers spent less than one hour in the recovery room vs. 23% of smokers" was reported in "Smokers Need More Time in Recovery Room" (Case Study 10-7, p. 536). Is this difference significant at $\alpha = 0.05$?

10.81 According to the publication *Housing Economics* (March 1991, page 6), 10.6% of all renters have an IRA or Keogh retirement plan compared to 24.2% of all households and 32.0% of homeowners. Suppose you conducted a survey of 200 renters and 300 homeowners. If 15% of the renters and 25% of the homeowners had an IRA or Keogh plan, find a 99% confidence interval estimating the difference in proportions for the two groups.

10.82 In a random sample of 40 brown-haired individuals, 22 indicated that they used hair coloring. In another random sample of 40 blonde individuals, 26 indicated that they used hair coloring. Use a 0.92 confidence interval to estimate the difference in the proportion of these groups that use hair coloring.

10.83 The proportions of defective parts produced by two machines were compared and the following data were collected:

   Machine 1: $n = 150$; number of defective parts = 10
   Machine 2: $n = 150$; number of defective parts = 4

   Determine a 90% confidence interval estimate for $p_1 - p_2$.

10.84 In a survey of 300 people from city A, 128 preferred New Spring soap to all other brands of deodorant soap. In city B, 149 of 400 people preferred New Spring. Find the 98% confidence interval estimate for the difference in the two proportions.

10.85 Calculate the 0.95 confidence maximum error of estimation for the difference between the following proportions and compare your answers to the corresponding values in the tables in Case Study 10-9 (p. 538).
   a. p is near 0.20 and both samples are of size 750.
   b. p is near 0.50 and both samples are of size 750.

10.86 Show that the standard error of $p'_1 - p'_2$, which is $\sqrt{(p_1q_1/n_1) + (p_2q_2/n_2)}$, reduces to $\sqrt{pq[(1/n_1) + (1/n_2)]}$ when $p_1 = p_2 = p$.

IN RETROSPECT

In this chapter we began the comparisons of two populations by first distinguishing between independent and dependent samples, which are statistically very important and useful sampling procedures. We then proceeded to examine the inferences concerning the comparison of means, variances, and proportions for two populations.

The use of hypothesis testing and interval estimates can sometimes be interchanged; that is, the calculation of a confidence interval can often be used in place of a hypothesis test. For example, Illustration 10-17 called for a confidence interval estimate. Now suppose that Mr. Morris asked: "Is there a difference in my voter appeal to men voters as opposed to women voters?" To answer his question, you would not need to calculate a test statistic if you chose to test at $\alpha = 0.01$ using a two-tailed test. "No difference" would mean a difference of zero, which is not included in the interval from 0.014 to 0.128 (the interval determined in Illustration 10-17). Therefore, a null hypothesis of "no difference" would be rejected, thereby substantiating the conclusion that there is a significant difference in voter appeal between the two groups.

We are always making comparisons between two groups: we compare means and we compare proportions. In this chapter we have learned how to statistically compare two populations by making inferences about their means, proportions, or variances. For convenience, Table 10-10 identifies the formulas to use when making inferences of comparisons between two populations. The article at the beginning of this chapter talks about two subpopulations in the workforce: those who consider their work "just a job" and those who consider their work "a

TABLE 10-10
Formulas to Use for Inferences Involving Two Populations

Situation	Test Statistic	Formula to Be Used — Hypothesis Test	Formula to Be Used — Interval Estimate
Two independent means			
σ known	z	Formula (10-5)	(10-6)
σ unknown (large samples)	z	Formula (10-7)	(10-8)
σ unknown (equal and small samples)	t	Formulas (10-13), (10-14), (10-15)	(10-13), (10-14), (10-17)
σ unknown (unequal and small samples)	t	Formula (10-16)	(10-18)
Two dependent means	t	Formulas (10-1), (10-2), (10-3)	(10-1), (10-2), (10-4)
Two proportions	z	Formulas (10-19), (10-20)	(10-21)
Two variances	F	Formula (10-9)	(10-12)

career." Notice the various ways in which these two subgroups may be compared: (1) average number of hours at work per day, (2) proportions concerning how they feel about their job, (3) how their boss says "good job," and so on.

In Chapters 8 through 10 we have introduced and completed hypothesis testing and confidence interval estimation for questions that involve one or two means, proportions, and variances. There is much more to be done. However, this much inferential statistics is sufficient to answer many questions. So in succeeding chapters we are ready to look at some other useful tests for other types of questions that may occur.

CHAPTER EXERCISES

 10.87 Ten new recruits participated in a rifle-shooting competition at the end of their first day at training camp. The same ten competed again at the end of a full week of training and practice. Their resulting scores are shown in the following table:

Time of Competition	Recruit									
	1	2	3	4	5	6	7	8	9	10
First day	72	29	62	60	68	59	61	73	38	48
One week later	75	43	63	63	61	72	73	82	47	43

a. Does this set of ten pairs of data show that there was a significant amount of improvement in the recruits' shooting abilities during the week? Use $\alpha = 0.05$.

 10.88 Twelve automobiles were selected at random to test two new mixtures of unleaded gasolines. Each car was given a measured allotment of the first mixture, x, and driven; then the distance traveled was recorded. The second mixture, y, was immediately tested in the same manner. The order in which the x and y mixtures were tested was also randomly assigned. The results are given in the following table:

Mixture	Car											
	1	2	3	4	5	6	7	8	9	10	11	12
x	7.9	5.6	9.2	6.7	8.1	7.3	8.1	5.4	6.9	6.1	7.1	8.1
y	7.7	6.1	8.9	7.1	7.9	6.7	8.2	5.0	6.2	5.7	6.2	7.5

Can you conclude that there is no real difference in mileage obtained by these two gasoline mixtures at the 0.10 level of significance?
    a. Solve using the classical approach.
    b. Solve using the prob-value approach.

10.89 Using a 95% confidence interval, estimate the difference in IQ between the oldest and the youngest members (brothers and sisters) of a family based on the following random sample of IQs:

Oldest	145	133	116	128	85	100	105	150	97	110	120	130
Youngest	131	119	103	93	108	100	111	130	135	113	108	125

10.90 The diastolic blood pressures for 15 patients were determined using two techniques: the standard method used by medical personnel and a method using an electronic device with a digital readout. The results were as follows:

Patient	1	2	3	4	5	6	7	8	9	10	11	12	13	14	15
Standard method	72	80	88	80	80	75	92	77	80	65	69	96	77	75	60
Digital readout method	70	76	87	77	81	75	90	75	82	64	72	95	80	70	61

Determine the 90% confidence interval estimate for the mean difference in the two readings, where d = standard method − digital readout.

10.91 Two semesters ago a sample of 150 math students showed a mean of 41.9 on the departmental final exam. Last semester another sample of 150 math students obtained an average of 44.4 on an equivalent exam. Assume that $\sigma = 12.0$ for both sets of exam grades. Should the null hypothesis "there is no real difference between these two groups" be rejected at $\alpha = 0.05$?
  a. Solve using the classical approach.
  b. Solve using the prob-value approach.

10.92 In the article "Management of Family and Employment Responsibilities by Mexican-American and Anglo-American Women" (*Social Work,* May 1990), Marlow reports the following results:

Age	Mexican-American Women	Anglo-American Women
18–24	15	11
25–34	31	29
35–44	15	15
45–54	6	16
55–64	3	1

Source: Copyright 1990, National Association of Social Workers, Inc. *Social Work.*

CHAPTER EXERCISES

For the purposes of this exercise, use 15–24 as the class limits for the first classification in place of 18–24.

    a. Draw histograms of these two distributions.
    b. Calculate the mean and standard deviation for the Mexican-American distribution.
    c. Calculate the mean and standard deviation for the Anglo-American distribution.
    d. Is the difference between these two sample means significant at the 0.05 level?
    e. Does it appear that these two samples have the same distributions? Justify your conclusion.

10.93 Assuming that the standard deviation of the lifetime for two brands of flashlight batteries is 2.8 hours, do the sample statistics in the following table substantiate the claim that brand S has a longer life than brand K at the 0.01 level?

Brand	n	$\bar{x}$
S	25	47.6
K	30	45.2

    a. Solve using the classical approach.
    b. Solve using the prob-value approach.

10.94 A test concerning some of the fundamental facts about AIDS was administered to two groups, one consisting of college graduates and the other consisting of high school graduates. A summary of the test results follows:

    College graduates: $n = 75$ $\bar{x} = 77.5$ $s = 6.2$
    High school graduates: $n = 75$ $\bar{x} = 50.4$ $s = 9.4$

Do these data show that the college graduates, on the average, score significantly higher on the test? Use $\alpha = 0.05$.

10.95 A test that measures math anxiety was given to 50 male and 50 female students. The results were as follows.

    Males: $\bar{x} = 70.5$ $s = 13.2$
    Females: $\bar{x} = 75.7$ $s = 13.6$

Construct a 95% confidence interval estimate for the difference between the mean anxiety scores.

10.96 The same achievement test is given to soldiers selected at random from two units. The scores they obtained are summarized as follows:

    Unit 1: $n_1 = 70$ $\bar{x}_1 = 73.2$ $s_1 = 6.1$
    Unit 2: $n_2 = 60$ $\bar{x}_2 = 70.5$ $s_2 = 5.5$

Construct a 90% confidence interval estimate for the difference in the mean level of the two units.

10.97 A manufacturer designed an experiment to compare the difference between men and women with respect to the times they required to assemble a product. Fifteen men and 15 women were tested to determine the time they required, on the average, to assemble the product. The times required by the men had a standard deviation of 4.5 min, and the times required by the women had a standard deviation of 2.8 min. Do these data show that the amount of time needed by men is more variable than the time needed by women? Use $\alpha = 0.05$.

10.98 A soft-drink distributor is considering two new models of dispensing machines. Both the Harvard Company machine and the Fizzit machine can be adjusted to fill the cups to a certain mean amount. However, the variation in the amount dispensed from cup to cup is a primary concern. Ten cups dispensed from the Harvard machine showed a variance of 0.065, whereas 15 cups dispensed from the Fizzit machine showed a variance of 0.033. The factory representative from the Harvard Company maintains that his machine had no more variability than the Fizzit machine.

    a. At the 0.05 level of significance, does the sample refute the representative's assertion?

    b. Estimate the variance for the amount of fill dispensed by the Harvard machine using a 90% confidence interval.

10.99 To estimate the ratio of the variances among the weights of cans of drained peaches of the leading brand to her own brand, a store manager had samples taken with the following results:

Leading brand: $n = 16$ $s^2 = 1.968$
Her brand: $n = 25$ $s^2 = 2.834$

    a. Give a point estimate for the ratio of "leading brand" variance to "her brand" variance.

    b. Construct the 95% confidence interval estimate for the same ratio of variances.

    c. Find the 95% confidence interval for the ratio of standard deviations.

10.100 A person might use the variance, or standard deviation, of the daily change in the stock market price as a measure of stability. Suppose that you want to compare the stability of a company's stock this year to its stability last year. You are given the following results of taking random samples from the daily gains and losses over the last two years:

This year: $n = 25$ $s = 2.57$
Last year: $n = 25$ $s = 1.26$

    a. Construct the 90% confidence interval estimate for the ratio of last year's standard deviation to this year's.

    b. Is the daily gain more, less, or about the same, as far as stability is concerned, for this year as compared to last year?

10.101 Two diets will be compared. Eighty individuals are selected at random from a population of overweight musicians. Forty-five musicians are assigned diet A and the other 35 are placed on diet B. After one week the weight losses (in pounds) are recorded as shown in the following table:

Diet	Sample Size	Sample Mean (pounds)	Sample Variance
A	45	10.3	7
B	35	7.3	3.25

a. Test the hypothesis that $\sigma_A = \sigma_B$ at $\alpha = 0.10$.
b. Do the data substantiate the conclusion that the expected weight loss μ_A under diet A is greater than the expected weight loss μ_B under diet B? Test at the 0.10 level. Draw the appropriate conclusion.
c. Construct a 90% confidence interval for $\mu_A - \mu_B$.

10.102 To compare the merits of two short-range rockets, 8 of the first kind and 10 of the second kind are fired at a target. If the first kind has a mean target error of 36 ft and a standard deviation of 15 ft, while the second kind has a mean target error of 52 ft and a standard deviation of 18 ft, does this indicate that the second kind of rocket is less accurate than the first? Use $\alpha = 0.01$.

10.103 The score on a certain psychological test is used as an index of status frustration. The scale ranges from 0 (low frustration) to 10 (high frustration). The test was administered to independent random samples of seven radical rightists and eight Peace Corps volunteers.

Radical Rightists	6	10	3	8	8	7	9	
Peace Corps Volunteers	3	5	2	0	3	1	0	4

Using the 10% level of significance, test that the mean score for both groups is the same against the alternative that it is not the same.
a. Solve using the classical approach.
b. Solve using the prob-value approach.

10.104 The women's health section of *Good Housekeeping* (February 1991, page 100) mentions a study connecting a woman's athletic ability and her menstrual cycle. Researchers selected 16 runners who logged at least 35 miles a week. Half had ceased having menstrual periods and half had normal cycles. They found no significant difference in the exercise capacities between the two groups. Suppose the following data were obtained:

Group	n	Mean Exercise Capacity	St. Dev.
No period	8	30.5 min	9.5
Normal cycle	8	31.3	8.0

a. Test the research hypothesis that the athletic capacities differ and give the prob-value.

b. If the test is being completed at the 0.05 level of significance, what decision and conclusion is reached?

 10.105 Two methods were used to study the latent heat of ice fusion. Both method A (an electrical method) and method B (a method of mixtures) were conducted with the specimens cooled to $-0.72°C$. The data in the following table represent the change in total heat from $-0.72°C$ to water at $0°C$ in calories per gram of mass.

Method A	Method B
79.98	80.02
80.04	79.94
80.02	79.98
80.04	79.97
80.03	79.97
80.03	80.03
80.04	79.95
79.97	79.97
80.05	
80.03	
80.02	
80.00	
80.02	

Is there a significant difference in the mean values at the 0.05 level?
a. Solve using the classical approach.
b. Solve using the prob-value approach.

10.106 Ten soldiers were selected at random from each of two companies to participate in a rifle-shooting competition. Their scores are shown in the following table. Can you conclude that Company B has a higher mean score than Company A? Use $\alpha = 0.05$.

Company A	72	29	62	60	68	59	61	73	38	48
Company B	75	43	63	63	61	72	73	82	47	43

10.107 The following data were collected concerning waist sizes of men and women. Do these data present sufficient evidence to conclude that men have larger mean waist sizes than women at the 0.05 level of significance?

Men	33	33	30	34	34	40	35	35	32
	34	32	35	32	32	34	36	30	38
Women	22	29	27	24	28	28			
	27	26	27	26	25				

 10.108 The performance on an achievement test in a beginning computer science course was administered to two groups. One group had a previous computer science course in high school; the other group did not. The test results were as follows:

Group 1 (had high school course)	17	18	27	19	24	36	27	26	
	35	22	18	29	29	26	33		
Group 2 (no high school course)					19	25	28	27	21
	24	18	14	28	21	22	20	21	14
	29	28	25	17	20	28	31	27	

Can we conclude that the mean test score for students who had a high school course in computer science will be higher than the mean score for those students who did not have a high school course? Use $\alpha = 0.05$.

 10.109 A group of 17 students participated in an evaluation of a special training session that claimed to improve memory. The students were randomly assigned to two groups: group A, the test group, and group B, the control group. All 17 students were tested for the ability to remember certain material. Group A was given the special training; group B was not. After one month both groups were tested again, with the results as shown in the following table. Do these data support the alternative hypothesis that the special training is effective at the $\alpha = 0.01$ level of significance?

Group A

Time of Test	Student								
	1	2	3	4	5	6	7	8	9
Before	23	22	20	21	23	18	17	20	23
After	28	29	26	23	31	25	22	26	26

Group B

Time of Test	Student							
	10	11	12	13	14	15	16	17
Before	22	20	23	17	21	19	20	20
After	23	25	26	18	21	17	18	20

10.110 A study was conducted to investigate the effectiveness of two different teaching methods used in physical education: the task method and the command method. Both were used to teach tennis to college students. A group of size 30 was used for each method. The students in both groups were instructed for the same length of time and then tested on the forehand and backhand tennis strokes.

Results	Command Group	Task Group	Calculated t
Forehand	$\bar{x} = 55.6$	$\bar{x} = 59.8$	$t^* = 2.39$
Backhand	$\bar{x} = 35.87$	$\bar{x} = 41.83$	$t^* = 3.59$

  a. Is one method more effective than the other in teaching tennis to college students?
  b. If so, for which stroke and with what level of significance?

10.111 In determining the "goodness" of a test question, a teacher will often compare the percentage of better students who answer it correctly to the percentage of the poorer students who answer it correctly. One expects that the better students will answer the question correctly more frequently than the poorer students. On the last test, 35 of the students with the top 60 grades and 27 with the bottom 60 answered a certain question correctly. Did the students with the top grades do significantly better on this question? Use $\alpha = 0.05$.
  a. Solve using the classical approach.
  b. Solve using the prob-value approach.

10.112 A survey was conducted to determine the proportion of Democrats as well as Republicans who support a "get tough" policy in South America. The results of the survey were as follows:

Democrats: $n = 250$ number in support = 120
Republicans: $n = 200$ number in support = 105

Do these data show that there is a significant difference in the proportion who support this policy? Use $\alpha = 0.05$.

10.113 The December 31, 1991 issue of *Newsweek* (p. 68) summarized results obtained at the University of Utah Medical Center concerning breast cancer. One hundred and three healthy women who each had two close relatives with breast cancer were compared with 31 women with no such family history. Both sets were examined for proliferative breast disease (PBD), benign but multiplying breast lesions. The results may be summarized as follows:

Group	Number	Percent with PBD
1. Family history of breast cancer	103	35
2. No family history	31	13

  a. Calculate the test statistic for the research hypothesis that $p_1 > p_2$ and the prob-value.

b. State the decision and the conclusion if this hypothesis test is completed at the 0.02 level of significance.

10.114 The April 4, 1991 issue of *USA Today* reported results from the *New England Journal of Medicine*. In a study of 987 deaths in southern California, the average right-hander died at age 75 and the average left-hander died at age 66. In addition, it was found that 7.9% of the lefties died from accident-related injuries, excluding vehicles, versus 1.5% for the right-handers; and 5.3% of the left-handers died while driving vehicles versus 1.4% of the right-handers.

Suppose you examine 1000 randomly selected death certificates of which 100 were left-handers and 900 were right-handers. If you found that 5 of the left-handers and 18 of the right-handers died while driving a vehicle, would you have evidence to show the proportion of left-handers who die at the wheel is significantly higher than the proportion of right-handers? Calculate the prob-value and interpret its meaning.

10.115 "It's a draw, according to two Australian researchers. By age 25, up to 29% of all men and up to 34% of all women have some gray hair, but this difference is so small that it's considered insignificant" ("Silver Threads Among the Gold: Who'll Find Them First, a Man or a Woman?" *Family Circle*, June 26, 1990). If 1000 men and 1000 women were involved in this research, would the 5% difference mentioned be significant at the 0.01 level?

10.116 According to a report in *Science News* (Vol. 137, no. 8, page 125), the percentage of seniors who used an illicit drug during the previous month was 19.7% in 1989. The figure was 21.3% in 1988. The annual survey of 17,000 seniors is conducted by researchers at the University of Michigan in Ann Arbor.
  a. Set a 95% confidence interval on the true decrease in usage.
  b. Does the interval found in part (a) suggest that there has been a significant decrease in the usage of illicit drugs by seniors? Explain.

10.117 Of a random sample of 100 stocks on the New York Exchange, 32 made a gain today. A random sample of 100 stocks on the American Stock Exchange showed 27 stocks making a gain.
  a. Construct the 99% confidence interval, estimating the difference in the proportion of stocks making a gain.
  b. Does the answer to (a) suggest that there is a significant difference between the proportions of stocks making gains on the two different stock exchanges?

10.118 A consumer group compared the reliability of two comparable microcomputers from two different manufacturers. The proportion requiring service within the first year after purchase was determined for samples from each of two manufacturers.

Manufacturer	Sample Size	Proportion Needing Service
1	75	0.15
2	75	0.09

Find a 0.95 confidence interval estimate for $p_1 - p_2$.

VOCABULARY LIST

Be able to define each term. In addition, describe, in your own words, and give an example of each term. Your examples should not be ones given in class or in the textbook.

The bracketed numbers indicate the chapters in which the term previously appeared, but you should define the terms again to show increased understanding of their meaning.

calculated value of t
confidence interval estimate [8, 9]
dependent means
dependent samples
F distribution
hypothesis test [8, 9]
independent means
independent samples
large sample
mean difference
paired difference
pooled estimate for the standard deviation
pooled observed probability
ratio of standard deviation
ratio of variances
source (of data)
test statistic t
unequal variance
z-statistic

KEY CONCEPTS

confidence interval estimate [8, 9]
dependent samples
F distribution
hypothesis test [8, 9]
independent samples
paired difference

QUIZ A

Answer "True" if the statement is always true. If the statement is not always true, replace the words shown in bold with words that make the statement always true.

10.1 When the means of two unrelated samples are used to compare two populations, we are dealing with **two dependent means**.

10.2 The use of **paired data (dependent means)** often allows for the control of unmeasurable or confounding variables because each pair is subjected to these confounding effects equally.

10.3 The **chi-square distribution** is used for making inferences about the ratio of the variances of two populations.

10.4 The **F distribution** is used when two dependent means are to be compared.

10.5 In comparing two independent means when the σ's are unknown, we need to perform an **F test on their variances** to determine the proper formula to use.

10.6 The **standard normal score** is used for all inferences concerning population proportions.

10.7 The F distribution is a **symmetric** distribution.

10.8 The number of degrees of freedom for the critical value of t is equal to **the smaller of $n_1 - 1$ or $n_2 - 1$** when making inferences about the difference between two independent means for the case where the variances are unknown and assumed to be unequal, and the sample sizes are small.

10.9 A pooled estimate for the standard deviation of two combined populations is calculated in all problems dealing with the difference between two **independent means**.

10.10 A **pooled estimate** for any statistic in a problem dealing with two populations is a value arrived at by combining the two separate sample statistics so as to achieve the best possible point estimate.

QUIZ B

Answer all questions, showing all formulas, substitutions, and work.

10.1 State the null (H_0) and the alternative (H_a) hypotheses that would be used to test each of the following claims:
  a. There is no significant difference in the mean batting averages for the baseball players of the two major leagues.
  b. The standard deviation for the monthly amounts of rainfall in Monroe County is less than the standard deviation for the monthly amounts of rainfall in Orange County.
  c. There is a significant difference in the percentage of male and female college students who own their own car.

10.2 Determine the test criteria [level of significance, test statistic, critical region(s), and critical value(s)] that would be used in completing each hypothesis test using $\alpha = 0.05$.

  a. $H_0: \mu_1 - \mu_2 = 43$
     $H_a: \mu_1 - \mu_2 < 43$
     (σ's known)
  b. $H_0: \sigma_1/\sigma_2 = 0.0$
     $H_a: \sigma_1/\sigma_2 > 0.0$
     ($n_1 = 16, n_2 = 25$)
  c. $H_0: p_1 - p_2 = 0.2$
     $H_a: p_1 - p_2 \neq 0.2$
  d. $H_0: \mu_d = 12$
     $H_a: \mu_d \neq 12$
     ($n = 28$)
  e. $H_0: \mu_1 - \mu_2 = 17$
     $H_a: \mu_1 - \mu_2 > 17$
     ($n_1 = 8, n_2 = 10$)

10.3 Find each of the following:
  a. $z(0.04)$
  b. $t(15, 0.025)$
  c. $\chi^2(20, 0.01)$
  d. $t(45, 0.01)$
  e. $F(8, 12, 0.01)$
  f. $\chi^2(30, 0.99)$

10.4 Twenty college freshmen were randomly divided into two groups. The members of one group were assigned to a statistics section using programmed materials only. Members of the other group were assigned to a section in which the professor lectured. At the end of the semester, all were given the same final exam. The results were as follows:

Programmed	76	60	85	58	91	44	82	64	79	88
Lecture	81	62	87	70	86	77	90	63	85	83

At the 5% level of significance, do these data show sufficient evidence to conclude that on the average the students in the lecture sections performed significantly better on the final exam?

10.5 The weights of eight people before they stopped smoking and five weeks after they stopped smoking are as follows:

Person	1	2	3	4	5	6	7	8
Before	148	176	153	116	129	128	120	132
After	154	179	151	121	130	136	125	128

At the 0.05 level of significance, does this sample present enough evidence to justify the conclusion that weight increases if one quits smoking?

10.6 In a nationwide sample of 600 school-age boys and 500 school-age girls, 288 boys and 175 girls admitted to having committed a destruction-of-property offense. Use this sample data and construct the 0.95 confidence interval estimate for the difference between the proportion of boys and girls who have committed this offense.

QUIZ C

10.1 To compare the accuracy of two short-range missiles, 8 of the first kind and 10 of the second kind are fired at a target. Let x be the distance by which the missile missed the target. Do these two sets of data (8 distances and 10 distances) represent dependent or independent samples? Explain.

10.2 Let's assume that there are 400 students in our college taking elementary statistics this semester. Describe how you could obtain two dependent samples of size 20 from these students in order to test some precourse skill against the same skill after completing the course. Be very specific.

10.3 Student A says he "doesn't see what all the fuss about the difference between independent and dependent means is all about; the results are almost the same regardless of the method used." Professor C suggests, "Student A should compare the procedures a bit more carefully." Help student A discover that there is a substantial difference between the procedures.

10.4 Suppose you are testing $H_0: \mu_d = 0$ versus $H_a: \mu_d < 0$ and the sample paired differences are all negative. Does this mean there is sufficient evidence to reject the null hypothesis? How can it not be significant? Explain.

10.5 Truancy is very disruptive to the educational system. A group of high school teachers and counselors have developed a group counseling program that they hope will help improve the truancy situation in their school. They have selected the 80 students in their school with the worst truancy records and have randomly assigned half of them to the group counseling program. At the end of the school year, the 80 students will be rated with regard to their truancy. When the scores have been collected, they will be turned over to you for evaluation. Explain what you will do to complete the study.

10.6 Explain the role of the F distribution in testing the difference between the means of two independent samples.

10.7 You wish to estimate and compare the proportion of Catholic families whose children attend a private school to the proportion of non-Catholic families whose children attend private schools. How would you go about estimating the two proportions and the difference between them?

WORKING WITH YOUR OWN DATA

As consumers, we all purchase many bottled, boxed, canned, or packaged products. Seldom, if ever, do we question whether or not the content amount stated on the container is accurate.

Here are a few typical content listings found on various containers:

28 FL OZ (1 PT 12 OZ)

750 ml

5 FL OZ (148 ml)

32 FL OZ (1 QT) 0.95 l

NET WT 10 OZ 283 GRAMS

NET WT $3\frac{3}{4}$ OZ 106 g—48 tea bags

140 1-PLY NAPKINS

77 SQ FT—92 TWO-PLY SHEETS—11 × 11 IN.

Have you ever thought, "I wonder if I am getting the amount that I am paying for?" And if this thought did cross your mind, did you attempt to check the validity of the content claim? The following article appeared in the *Times Union* of Rochester, New York on February 16, 1972. (This kind of situation occurs frequently.)

Milk Firm Accused of Short Measure*

The processing manager of Dairylea Cooperative, Inc., has been named in a warrant charging that the cooperative is distributing cartons of milk in the Rochester area containing less than the quantity represented.

. . . an investigator found shortages in four quarts of Dairylea milk purchased Friday.

Asst. Dist. Atty. Howard R. Relin, who issued the warrant, said the shortages ranged from $1\frac{1}{8}$ to $1\frac{1}{4}$ ounces per quart. A quart of milk contains 32 fluid ounces.

. . . the state Agriculture and Markets Law . . . provides that a seller of a commodity shall not sell or deliver less of the commodity than the quantity represented to be sold.

. . . the purpose of the law under which . . . the dairy is charged is to ensure honest, accurate, and fair dealing with the public. There is no requirement that intent to violate the law be proved, he said.

This situation poses a very interesting legal problem: There is no need to show intent to "short the customer." If caught, the violators are fined automatically and the fines are often quite severe.

---

*From *The Times-Union*, Rochester, N.Y., Feb. 16, 1972.

A ▼ A High-Speed Filling Operation

A high-speed piston-type machine used to fill cans with hot tomato juice was sold to a canning company. The guarantee stated that the machine would fill 48-oz cans with a mean amount of 49.5 oz, a standard deviation of 0.072 oz, and a maximum spread of 0.282 oz while operating at a rate of filling 150 to 170 cans per minute. On August 12, 1986, a sample of 42 cans was gathered and the following weights were recorded. The weights, measured to the nearest $\frac{1}{8}$ oz, are recorded as variations from 49.5 oz.

$-\frac{1}{8}$	0	$-\frac{1}{8}$	0	0	0	$-\frac{1}{8}$	0	0	0
0	0	$-\frac{1}{8}$	0	$\frac{1}{8}$	0	$\frac{1}{8}$	0	$\frac{1}{8}$	0
0	0	$-\frac{1}{8}$	0	0	0	0	$-\frac{1}{8}$	0	0
0	0	0	0	0	0	0	0	0	0
0	0								

1. Calculate the mean $\bar{x}$, the standard deviation s, and the range of the sample data.
2. Construct a histogram picturing the sample data.
3. Does the amount of fill differ from the prescribed 49.5 oz at the $\alpha = 0.05$ level? Test the hypothesis that $\mu = 49.5$ against an appropriate alternative.
4. Does the amount of variation, as measured by the standard deviation, satisfy the guarantee at the $\alpha = 0.05$ level?
5. Assuming that the filling machine continues to fill cans with an amount of tomato juice that is distributed normally and the mean and standard deviation are equal to the values found in question 1, what is the probability that a randomly selected can will contain less than the 48 oz claimed on the label?
6. If the amount of fill per can is normally distributed and the standard deviation can be maintained, find the setting for the mean value that would allow only 1 can in every 10,000 to contain less than 48 oz.

B ▼ Your Own Investigation

Select a packaged product that has a quantity of fill per package that you can and would like to investigate.

1. Describe your selected product, including the quantity per package, and describe how you plan to obtain your data.
2. Collect your sample of data. (Consult your instructor for advice on size of sample.)
3. Calculate the mean $\bar{x}$ and standard deviation s for your sample data.
4. Construct a histogram or stem-and-leaf diagram picturing the sample data.
5. Does the mean amount of fill meet the amount prescribed on the label? Test using $\alpha = 0.05$.
6. Assume that the item you selected is filled continually. The amount of fill is normally distributed and the mean and standard deviation are equal to the values found in question 3. What is the probability that one randomly selected package contains less than the prescribed amount?

PART FOUR

More Inferential Statistics

IN PART 3 WE STUDIED the two inferential statistical techniques: hypothesis testing and confidence interval estimating, for one and two populations, and the three population parameters: mean, μ; standard deviation, σ; and proportion, p. In Part 4 we will learn how to use the hypothesis test and the interval estimation techniques in other statistical situations. Some of these situations will be extensions of previous methods. (For example, analysis of variance will be used to deal with more than two populations when the question involves the mean, and the binomial experiment will be expanded to a multinomial experiment.) Other situations will be alternatives to methods previously studied (such as the nonparametric techniques) or will deal with new inferences (such as the nonparametric methods and the regression and correlation inferential techniques).

Sir Ronald A. Fisher

Library of Congress

Sir Ronald Alymer Fisher, British statistician, was born in London on February 17, 1890. In 1912, after earning a B.A. from Gonville & Caius College in Cambridge, Fisher worked as a statistician for an investment company until 1915, at which time he became a public school teacher. In 1917, he married Ruth Eillean Gralton Guiness with whom he had eight children. Fisher received his M.A. in 1920 and his Sc.D. in 1926. Many years later, he retired to Australia, where he died at the age of 72 on July 29, 1962.

In 1919, Fisher was hired by Rothamstead Experimental Station to do statistical work with their plant breeding experiments. It was there that he pioneered the application of statistical procedures to the design of scientific experiments. During his employment at Rothamstead, Fisher introduced the principle of randomization and originated the concept of the analysis of variance (an even more important achievement). In 1925, Fisher wrote "Statistical Methods for Research," a work that remained in print for more than 50 years. Fisher has played a major role in the development and in the application of statistics.

11 Additional Applications of Chi-Square

Chapter Outline

11.1 Chi-Square Statistic
The chi-square distribution will be used to test hypotheses concerning **enumerated data**.

11.2 Inferences Concerning Multinomial Experiments
A multinomial experiment differs from a binomial experiment in that **each trial has many outcomes** rather than two outcomes.

11.3 Inferences Concerning Contingency Tables
Contingency tables are tabular representations of frequency counts for data in a **two-way classification**.

Trial By Number

Karl Pearson's chi-square test measured the fit between theory and reality, ushering in a new sort of decision making.

Our world is inundated with statistics. Every medical fear or triumph is charted by a complex analysis of chances. Think of cancer, heart disease, AIDS: The less we know, the more we hear of probabilities. This daily barrage is not a matter of mere counting but of inference and decision in the face of uncertainty. No committee changes our schools or our prisons without studies on the effects of busing or early parole. Money markets, drunken driving, family life, high energy physics, and deviant human cells are all subject to tests of significance and data analysis.

This all began in 1900 when Karl Pearson published his chi-square test of goodness of fit, a formula for measuring how well a theoretical hypothesis fits some observations. The basic idea is simple enough. Suppose that you think a die will fall equally often on each of its six faces. You roll it 600 times. It seems to come up six all too frequently. Could this simply be chance? How well does the hypothesis—that the die is fair—fit the data? The result that would best fit your theory would be that each face came up just 100 times in 600. In practice the ratios are almost always different, even with a fair die, because even for many throws there will always be the factor of chance. How different should they be to make you suspect a poor fit between your theory and your 600 observations? Pearson's chi-square test gives one measure of how well theory and data correspond.

The chi-square test can be used for hypotheses and data where observations naturally fall into discrete categories that statisticians call cells. If, for example, you are testing to find whether a certain treatment for cholera is worthless, then the patients divide among four cells: treated and recovered, treated and died, untreated and recovered, and untreated and died. If the treatment is worthless, you expect no difference in recovery rate between treated and untreated patients. But chance and uncontrollable variables dictate that there will almost always be some difference. Pearson's test takes this into account, telling you how well your hypothesis—that the treatment is worthless—fits your observations. . . .

The chi-square test was a tiny event in itself, but it was the signal for a sweeping transformation in the ways we interpret our numerical world. Now there could be a standard way in which ideas were presented to policy makers and to the general public. . . .

Statistical talk has now invaded many aspects of daily life, creating sometimes spurious impressions of objectivity. Policy makers demand numbers to measure every risk and hope. Thus the 1975 *Reactor Safety Study* of the Nuclear Regulatory Commission attached probabilities to various kinds of danger. Some critics suggest that this is just a way of escaping the responsibility of admitting ignorance. Other critics point to the increasing use of complex models generated to "fit" reams of data in some statistical sense—models that can become a fantasyland without any connection to the real world.

For better or worse, statistical inference has provided an entirely new style of reasoning. The quiet statisticians have changed our world—not by discovering new facts or technical developments but by changing the ways we reason, experiment, and form our opinions about it.

Source: Ian Hacking. Reprinted by permission from the November issue of SCIENCE '84. Copyright © 1984 by the American Association for the Advancement of Science.

Chapter Objectives

Previously we discussed and used the chi-square distribution both to test and to estimate the value of the variance (or standard deviation) of a single population. The chi-square distribution may also be used for tests in other types of situations. In this chapter we are going to investigate some of these uses. Specifically, we will look at two tests: a multinomial experiment and the contingency table. These two types of tests will be used to compare experimental results with expected results in order to determine (1) preferences, (2) independence, and (3) homogeneity.

enumerative data

The information that we will be using in these techniques will be **enumerative**; that is, the data will be placed in categories and counted. These counts become the enumerative information used.

11.1 ▼ Chi-Square Statistic

There are many problems for which the data are categorized and the results shown by way of counts. For example, a set of final exam scores can be displayed as a frequency distribution. These frequency numbers are counts, the number of data that fall in each cell. A survey asks voters whether they are registered as Republican, Democrat, or other, and whether or not they support a particular candidate. The results are usually displayed on a chart that shows the number of voters for each possible category. There were numerous illustrations of this way of presenting data throughout the previous ten chapters.

cell

Suppose that we have a number of **cells** into which n observations have been sorted. (The term *cell* is synonymous with the term *class*; the terms *class* and *frequency* were defined and first used in earlier chapters. Before continuing, a brief review of Sections 2.2, 2.3, and 2.4 might be beneficial.) The **observed frequencies** in each cell are denoted by $O_1, O_2, O_3, \ldots, O_k$ (see Table 11-1). Note that the sum of all observed frequencies is equal to $O_1 + O_2 + \cdots + O_k = n$, where n is the sample size. What we would like to do is to compare the observed frequencies with some **expected or theoretical frequencies**, denoted by $E_1, E_2, E_3, \ldots, E_k$, for each of these cells. Again, the sum of these expected frequencies must be exactly

expected frequency

$$E_1 + E_2 + \cdots + E_k = n$$

TABLE 11-1
Observed Frequencies

	\multicolumn{5}{c}{k Categories}	
	1st	2d	3d	...	kth	Total
Observed Frequency	O_1	O_2	O_3	...	O_k	n

We will then decide whether the observed frequencies seem to agree or seem to disagree with the expected frequencies. This will be accomplished by a hypothesis test using **chi-square**, χ^2.

chi-square

Section 11.2 ▼ INFERENCES CONCERNING MULTINOMIAL EXPERIMENTS

The calculated value of the test statistic will be

$$\chi^2 = \sum_{\text{all cells}} \frac{(O - E)^2}{E} \qquad \text{(11-1)}$$

This calculated value for chi-square will be the sum of several nonnegative numbers, one from each cell (or category). The numerator of each term in the formula for χ^2 is the square of the difference between the values of the observed and the expected frequencies. The closer together these values are, the smaller the value of $(O - E)^2$; the farther apart, the larger the value of $(O - E)^2$. The denominator for each cell puts the size of the numerator into perspective. That is, a difference $(O - E)$ of 10 resulting from frequencies of 110 (O) and 100 (E) seems quite different from a difference of 10 resulting from 15 (O) and 5 (E).

These ideas suggest that small values of chi-square indicate agreement between the two sets of frequencies, whereas larger values indicate disagreement. Therefore, it is customary for these tests to be one-tailed, with the critical region on the right.

In repeated sampling, the calculated value of χ^2 [formula (11-1)] will have a sampling distribution that can be approximated by the chi-square probability distribution when n is large. This approximation is generally considered adequate when all the expected frequencies are equal to or greater than 5. Recall that there is a separate chi-square distribution for each degree of freedom, df. The appropriate value of df will be described with each specific test. The critical value of chi-square, $\chi^2(\text{df}, \alpha)$, can be found in Table 7 of Appendix F.

In this chapter we will permit a certain amount of "liberalization" with respect to the null hypothesis and its testing. In previous chapters the null hypothesis has always been a statement about a population parameter (μ, σ, or p). However, there are other types of hypotheses that can be tested, such as "this die is fair" or "height and weight of individuals are independent." Notice that these hypotheses are not claims about a parameter, although sometimes they could be stated with parameter values specified.

Suppose that I claim that "this die is fair," $p = P(\text{any one number}) = \frac{1}{6}$, and you want to test it. What would you do? Was your answer something like "Roll this die many times, recording the results"? Suppose that you decide to roll the die 60 times. If the die is fair, what do you expect will happen? Each number $(1, 2, \ldots, 6)$ should appear approximately $\frac{1}{6}$ of the time (that is, 10 times). If it happens that approximately 10 of each number occur, you will certainly accept the claim of fairness ($p = \frac{1}{6}$). If it happens that the die seems to favor some numbers, you will reject the claim. (The test statistic χ^2 will have a large value in this case, as we will soon see.)

11.2 ▼ Inferences Concerning Multinomial Experiments

multinomial experiment

The preceding die problem is a good illustration of a **multinomial experiment**. Let's consider this problem again. Suppose that we want to test this die (at $\alpha = 0.05$) and decide whether to fail to reject or reject the claim "this die is fair."

Chapter 11 ▼ ADDITIONAL APPLICATIONS OF CHI-SQUARE

(The probability of each number is $\frac{1}{6}$.) The die is rolled from a cup onto a smooth flat surface 60 times, with the frequencies shown in the following table:

Number	1	2	3	4	5	6
Occurrences	7	12	10	12	8	11

The null hypothesis that the die is fair is assumed to be true. This allows us to calculate the expected frequencies. If the die was fair, we certainly would expect 10 occurrences for each number.

Now let's calculate an observed value of χ^2. These calculations are shown in Table 11-2. The calculated value is $\chi^{2*} = 2.2$.

TABLE 11-2 Computations for Calculating χ^2

Number	Observed (O)	Expected (E)	O − E	(O − E)²	(O − E)²/E
1	7	10	−3	9	0.9
2	12	10	2	4	0.4
3	10	10	0	0	0.0
4	12	10	2	4	0.4
5	8	10	−2	4	0.4
6	11	10	1	1	0.1
Total	60	60	0		**2.2**

NOTE $\Sigma(O - E)$ must equal zero since $\Sigma O = \Sigma E = n$. You can use this fact as a check, as shown in Table 11-2.

Before continuing, let's set up the hypothesis-testing format.

STEP 1 H_0: The die is fair (each $p = \frac{1}{6}$).

H_a: The die is not fair (at least one p is different).

STEP 2 $\alpha = 0.05$. See the accompanying figure for the test criteria. In the multinomial test, df = $k - 1$, where k is the number of cells. The critical value is $\chi^2(5, 0.05) = 11.1$.

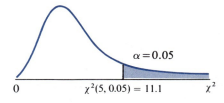

STEP 3 The test statistic was calculated in Table 11-2 and was found to be $\chi^{2*} = 2.2$.

Section 11.2 ▼ INFERENCES CONCERNING MULTINOMIAL EXPERIMENTS 565

STEP 4 **Decision:** Fail to reject H_0 (χ^{2*} is not in the critical region).

Conclusion At the 0.05 level of significance, the observed frequencies are not significantly different from those expected of a fair die. ▲▲

Before we look at other illustrations, we must define the term *multinomial experiment* and we must state the guidelines for completing the chi-square test for it.

multinomial experiment

MULTINOMIAL EXPERIMENT

An experiment with the following characteristics:

1. It consists of n identical independent trials.
2. The outcome of each trial fits into exactly one of k possible cells.
3. There is a probability associated with each particular cell, and these individual probabilities remain constant during the experiment. (It must be true that $p_1 + p_2 + \cdots + p_k = 1$.)
4. The experiment will result in a set of observed frequencies, $O_1, O_2, \ldots, O_k$, where each O_i is the number of times a trial outcome falls into that particular cell. (It must be the case that $O_1 + O_2 + \cdots + O_k = n$.)

The die illustration meets the definition of a multinomial experiment because it has all four of the characteristics described in the definition.

1. The die was rolled n (60) times in an identical fashion, and these trials were independent of each other. (The result of each trial was unaffected by the results of other trials.)
2. Each time the die was rolled, one of six numbers resulted, and each number was associated with a cell.
3. The probability associated with each cell was $\frac{1}{6}$, and this was constant from trial to trial. (Six values of $\frac{1}{6}$ sum to 1.0.)
4. When the experiment was complete, we had a list of frequencies (7, 12, 10, 12, 8, and 11) that summed to 60, indicating that each of the outcomes was taken into account.

The testing procedure for multinomial experiments is very much as it was in previous chapters. The biggest change comes with the statement of the null hypothesis. It may be a verbal statement, such as in the die illustration: "This die is fair." Often the alternative to the null hypothesis is not stated. However, in this book the alternative hypothesis will be shown, since it seems to aid in organizing and understanding the problem. However, it will not be used to determine the location of the critical region, as was the case in previous chapters. *For multinomial experiments we will always use a one-tailed critical region, and it will be the right-hand tail of the χ^2 distribution, because larger deviations from the expected values (positive or negative) lead to an increase in the calculated χ^2 value.*

The critical value will be determined by the level of significance assigned (α) and the number of degrees of freedom. The number of degrees of freedom (df) will be 1 less than the number of cells (k) into which the data are divided:

$$\text{df} = k - 1 \qquad (11\text{-}2)$$

Each expected frequency, E_i, will be determined by multiplying the corresponding probability (p_i) for that cell by the total number of trials n. That is,

$$E_i = n \cdot p_i \qquad (11\text{-}3)$$

One guideline should be met to ensure a good approximation to the chi-square distribution: Each expected frequency should be at least 5 (that is, each $E_i \geq 5$). Sometimes it is possible to combine "smaller" cells to meet this guideline. If this guideline cannot be met, corrective measures to ensure a good approximation should be used. These corrective measures are not covered in this book but are discussed in many other sources.

▼ ILLUSTRATION 11-1

College students have regularly insisted on freedom of choice when registering for courses. This semester there were seven sections of a particular mathematics course. They were scheduled to meet at various times with a variety of instructors. Table 11-3 shows the number of students who selected each of the seven sections. Do the data indicate that the students had a preference for certain sections or do they indicate that each section was equally likely to be chosen?

TABLE 11-3 Data for Illustration 11-1

	Section							
	1	2	3	4	5	6	7	Total
Number of Students	18	12	25	23	8	19	14	119

SOLUTION If no preference was shown in the selection of sections, we would expect the 119 students to be equally distributed among the seven classes. Thus if no preference was the case, then we would expect 17 students to register for each section. The test is completed as shown in Steps 1–4, at the 5% level of significance.

STEP 1 H_0: There was no preference shown (equally distributed).

H_a: There was a preference shown (not equally distributed).

STEP 2 $\alpha = 0.05$. See the accompanying figure for the test criteria.

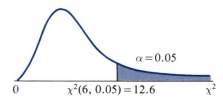

$\chi^2(6, 0.05) = 12.6$

Section 11.2 ▼ INFERENCES CONCERNING MULTINOMIAL EXPERIMENTS

STEP 3 We use formula (11-1) to calculate the test statistic.

$$\chi^2 = \frac{(18-17)^2}{17} + \frac{(12-17)^2}{17} + \frac{(25-17)^2}{17}$$
$$+ \frac{(23-17)^2}{17} + \frac{(8-17)^2}{17} + \frac{(19-17)^2}{17} + \frac{(14-17)^2}{17}$$
$$= \frac{(1)^2 + (-5)^2 + (8)^2 + (6)^2 + (-9)^2 + (2)^2 + (-3)^2}{17}$$
$$= \frac{1 + 25 + 64 + 36 + 81 + 4 + 9}{17} = \frac{220}{17}$$
$$= 12.9411$$

$\chi^{2*} = \mathbf{12.94}$

STEP 4 **Decision:** Reject H_0 (χ^{2*} falls in the critical region).

Conclusion At the 0.05 level of significance, there does seem to be a preference shown. We cannot determine, from the given information, what the preference is. It could be teacher preference, time preference, or a case of schedule conflict. Conclusions must be worded carefully to avoid suggesting conclusions that we cannot support. ▲▲

Not all multinomial experiments result in equal expected frequencies, as we will see in Illustration 11-2.

▼ ILLUSTRATION 11-2

The Mendelian theory of inheritance claims that the frequencies of round and yellow, wrinkled and yellow, round and green, and wrinkled and green peas will occur in the ratio 9:3:3:1 when two specific varieties of peas are crossed. In testing this theory, Mendel obtained frequencies of 315, 101, 108, and 32, respectively. Do these sample data provide sufficient evidence to reject this theory at the 0.05 level of significance?

SOLUTION

STEP 1 H_0: 9:3:3:1 is the ratio of inheritance.

H_a: 9:3:3:1 is not the ratio of inheritance.

STEP 2 $\alpha = 0.05$, $k = 4$, and df $= 3$. See the accompanying figure.

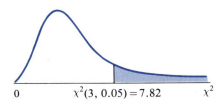

STEP 3 The ratio 9:3:3:1 indicates probabilities of $\frac{9}{16}$, $\frac{3}{16}$, $\frac{3}{16}$, and $\frac{1}{16}$. Therefore, the expected frequencies are $9n/16$, $3n/16$, $3n/16$, and $n/16$, where

$$n = \sum O_i = 315 + 101 + 108 + 32 = 556$$

The computations necessary for calculating χ^2 are given in Table 11-4. Thus $\chi^{2*} = \mathbf{0.47}$.

TABLE 11-4
Computations Needed to Calculate χ^2

O	E	O − E	(O − E)²/E
315	312.75	2.25	0.0162
101	104.25	−3.25	0.1013
108	104.25	3.75	0.1349
32	34.75	−2.75	0.2176
556	556.00	0	0.4700

STEP 4 Decision: Fail to reject H_0
Conclusion At the 0.05 level of significance, there is not sufficient evidence to reject Mendel's theory.

Case Study 11-1

The data collected from a survey of 300 adults and used to create "Why We Rearrange Furniture" is actually that of a multinomial experiment. What was one trial? What was the variable? What were the many levels of results from each trial? (See Exercise 11.1.)

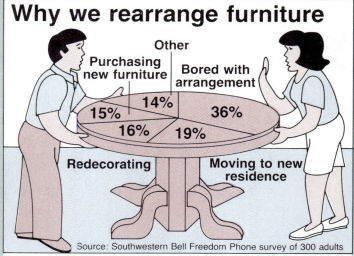

Source: Copyright 1990, USA TODAY. Reprinted with permission.

Case Study 11-2

"Methods Used to Fall Asleep" displays the results of surveying 1000 adults about what method they use to fall asleep. Why is the information shown not that of a multinomial experiment? (See Exercise 11.2.)

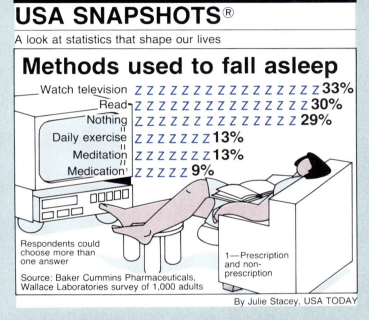

Source: Copyright 1990, USA TODAY. Reprinted with permission.

▼▲ EXERCISES

 11.1 Explain how the sample information shown in "Why We Rearrange Furniture" (Case Study 11-1) satisfies the properties of a multinomial experiment. Specifically identify: a trial, the variable, and the various levels of the variable.

 11.2 Explain why the sample information shown in "Methods Used to Fall Asleep" (Case Study 11-2) does not satisfy the properties of a multinomial experiment.

11.3 State the null hypothesis H_0 and the alternative hypothesis H_a that would be used to test the following statements:
  a. The five numbers, 1, 2, 3, 4, and 5, are equally likely to be drawn.
  b. That multiple choice question has a history of students selecting answers in the ratio of 2:3:2:1.
  c. The poll will show a distribution of 16%, 38%, 41%, and 5% for the possible ratings of excellent, good, fair, poor on that issue.

11.4 Determine the test criteria that would be used to test the null hypothesis for the following multinomial experiments:
  a. H_0: $P(1) = P(2) = P(3) = P(4) = 0.25$, with $\alpha = 0.05$.
  b. H_0: $P(I) = 0.25$, $P(II) = 0.40$, $P(III) = 0.35$, with $\alpha = 0.01$.

11.5 A manufacturer of floor polish conducted a consumer-preference experiment to determine which of five different floor polishes was the most appealing in appearance. A sample of 100 consumers viewed five patches of flooring that had each received one of the five polishes. Each consumer indicated the patch he or she preferred. The lighting and background were approximately the same for all patches. The results were as follow:

Polish	A	B	C	D	E	Total
Frequency	27	17	15	22	19	100

a. State the hypothesis for "no preference" in statistical terminology.
b. What test statistic will be used in testing this null hypothesis?
c. Complete the hypothesis test using $\alpha = 0.10$.
  (1) Solve using the classical approach.
  (2) Solve using the prob-value approach.

11.6 A certain type of flower seed will produce magenta, chartreuse, and ochre flowers in the ratio 6:3:1 (one flower per seed). A total of 100 seeds are planted and all germinate, yielding the following results:

Magenta	Chartreuse	Ochre
52	36	12

a. If the null hypothesis (6:3:1) is true, what is the expected number of magenta flowers?
b. How many degrees of freedom are associated with chi-square?
c. Complete the hypothesis test using $\alpha = 0.10$.
  (1) Solve using the classical approach.
  (2) Solve using the prob-value approach.

11.7 "Looking Up to Athletes" (*USA Today,* May 7, 1991) reported: "Here's how sports team members say athletes do as role models for children: Excellent—16%, Good—38%, Fair—41%, Poor—5%." Suppose you took a poll of 350 members within your community and obtained the following results: Excellent—44, Good—145, Fair—133, Poor—28. Do your results show that your community has a significantly different idea about athletes as role models than the sports team members?

MINITAB was used to analyze this problem. The results are shown below. Verify the calculations of expected values, cell chi-square, and total chi-square.

```
MTB > READ C1 C2
    4 ROWS READ
MTB > NAME C1 'P' C2 'OBS' C3 'EXP' C4 'CHI-SQ'
MTB > LET C3 = C1 * SUM(C2)
MTB > LET C4 = ((C2-C3)**2)/C3
MTB > SUM C1   K1
    SUM     =     1.0000
MTB > SUM C2 K2
    SUM     =     350.00
MTB > SUM C3 K3
    SUM     =     350.00
MTB > SUM C4 K4
    SUM     =     10.722
MTB > PRINT C1-C4

ROW      P       OBS      EXP      CHI-SQ

 1      0.16      44      56.0     2.57142
 2      0.38     145     133.0     1.08271
 3      0.41     133     143.5     0.76829
 4      0.05      28      17.5     6.30002

MTB > PRINT K1-K4
K1         1.00000
K2       350.000
K3       350.000
K4        10.7224
MTB > STOP
```

11.8 A large supermarket carries four qualities of ground beef. Customers are believed to purchase these four varieties with probabilities of 0.10, 0.30, 0.35, and 0.25, respectively, from the least to most expensive variety. A sample of 500 purchases resulted in sales of 46, 162, 191, and 101 of the respective qualities. Does this sample contradict the expected proportions? Use $\alpha = 0.05$. Use a computer to complete this exercise. (If you use MINITAB, see Exercise 11.7.)

11.9 A standard diet is recommended to reduce cholesterol in individuals who have elevated cholesterol levels. The expected response to the standard diet after six months may be categorized as follows:

Category	Percentage
1. Marked decrease in cholesterol	60
2. Moderate decrease in cholesterol	20
3. Slight decrease in cholesterol	10
4. No change in cholesterol	10

An experimental diet is to be compared to the standard diet. Two hundred individuals were placed on the experimental diet and the following results were obtained:

Category	1	2	3	4
Observed Cell Counts	110	30	10	50

Do the results show sufficient reason to reject the hypothesis that "the responses to the two diets are the same"? Use $\alpha = 0.05$.
   a. Solve using the classical approach.
   b. Solve using the prob-value approach.

11.10 Previous sales records show that Inboard Custom Boats sold 20% of its boats in its Northeast sales district, and 28%, 8%, 12%, and 32%, respectively, in its Southeast, North Central, South Central, and West Coast sales districts. Of the first 500 boats sold this year, 120, 128, 43, 66, and 143 were sold in each of the five districts, respectively. Does the sales distribution so far this year seem to be the same as in previous years? Use $\alpha = 0.05$.
   a. Solve using the classical approach.
   b. Solve using the prob-value approach.

11.11 A program for generating random numbers on a computer is to be tested. The program is instructed to generate 100 single-digit integers between 0 and 9. The frequencies of the observed integers were as follows:

Integer	0	1	2	3	4	5	6	7	8	9
Frequency	11	8	7	7	10	10	8	11	14	14

At the 0.05 level of significance, is there sufficient reason to believe that the integers are not being generated uniformly?

11.12 *Nursing Magazine* (March 1991, pages 50–54) reports results of a survey of over 1800 nurses across the country concerning job satisfaction and retention. Nurses from magnet hospitals (hospitals that successfully attract and retain nurses) describe the staffing situation on their units as follows:

Staffing Situation	Percent
1. Desperately short of help— patient care has suffered	12%
2. Short, but patient care hasn't suffered	32
3. Adequate	38
4. More than adequate	12
5. Excellent	6

Section 11.3 ▼ INFERENCES CONCERNING CONTINGENCY TABLES

A survey of 500 nurses from nonmagnet hospitals gave the following responses to the staffing situation:

Staffing situation	1	2	3	4	5
Number	165	140	125	50	20

Do the data indicate that the nurses from the nonmagnet hospitals have a different distribution of opinions? Use $\alpha = 0.05$.

11.3 ▼ Inferences Concerning Contingency Tables

contingency table

A **contingency table** is an arrangement of data into a two-way classification. The data are sorted into cells and the number of data in each cell is reported. The contingency table involves two factors (or variables), and the usual question concerning such tables is whether the data indicate that the two variables are independent or dependent.

TEST OF INDEPENDENCE

To illustrate the use and analysis of a contingency table, let's consider the gender of liberal arts college students and their favorite academic area.

▼ ILLUSTRATION 11-3

Each person in a group of 300 students was identified as male or female and then asked whether he or she preferred taking liberal arts courses in the area of math-science, social science, or humanities. Table 11-5 is a contingency table that shows the frequencies found for these categories. Does this sample present sufficient evidence to reject the null hypothesis "preference for math-science, social science, or humanities is independent of the sex of a college student" at the 0.05 level of significance?

TABLE 11-5 Contingency Table Showing Sample Results for Illustration 11-3

Sex of Student	Favorite Subject Area			Total
	Math-Science (M-S)	Social Science (SS)	Humanities (H)	
Male	37	41	44	122
Female	35	72	71	178
Total	72	113	115	300

SOLUTION

STEP 1 H_0: Preference for math-science, social science, or humanities is independent of the sex of a college student.
H_a: Subject preference is not independent of the sex of the student.

Chapter 11 ▼ ADDITIONAL APPLICATIONS OF CHI-SQUARE

STEP 2 To determine the critical value of chi-square, we need to know df, the number of degrees of freedom, involved. In the case of contingency tables, the number of degrees of freedom is exactly the same number as the number of cells in the table that may be filled in freely when you are given the totals. The totals in this illustration are shown in the following table:

			122
			178
72	113	115	300

Given these totals, you can fill in only two cells before the others are all determined. (The totals must, of course, be the same.) For example, once we pick two arbitrary values (say 50 and 60) for the first two cells of the first row, the other four cell values are fixed (see the following table):

50	60	C	122
D	E	F	178
72	113	115	300

They have to be $C = 12$, $D = 22$, $E = 53$, and $F = 103$. Otherwise, the totals will not be correct. Therefore, for this problem there are two free choices. Each free choice corresponds to one degree of freedom. Hence the number of degrees of freedom for our example is 2 (df = 2). Thus, if $\alpha = 0.05$ is used, the critical value is $\chi^2(2, 0.05) = 6.00$. See the accompanying figure. (Following the conclusion of this illustration, a formula for finding df will be discussed.)

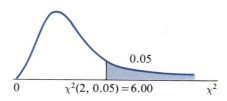

STEP 3 Before the calculated value of chi-square can be found, we need to determine the expected values E for each cell. To do this we must recall the null hypothesis, which asserts that these factors are independent. Therefore, we would expect the values to be distributed in proportion to the **marginal totals**. There are 122 males; we would expect them to be distributed among M-S, SS, and H proportionally to the 72, 113, and 115 totals. Thus the expected cell counts for males are

$$\frac{72}{300} \cdot 122 \qquad \frac{113}{300} \cdot 122 \qquad \frac{115}{300} \cdot 122$$

Similarly, we would expect for the females

$$\frac{72}{300} \cdot 178 \qquad \frac{113}{300} \cdot 178 \qquad \frac{115}{300} \cdot 178$$

Section 11.3 ▼ INFERENCES CONCERNING CONTINGENCY TABLES

Thus the expected values are as shown in Table 11-6. (Always check the new totals against the old totals.)

TABLE 11-6
Expected Values

	M-S	SS	H	Total
	29.28	45.95	46.77	122.00
	42.72	67.05	68.23	178.00
Total	72.00	113.00	115.00	300.00

NOTE We can think of the computation of the expected values in a second way. Recall that we assume the null hypothesis to be true until there is evidence to reject it. Having made this assumption in our example, in effect we are saying that the event that a student picked at random is male and the event that a student picked at random prefers math-science courses are independent. Our point estimate for the probability that a student is male is 122/300, and the point estimate for the probability that the student prefers the math-science courses is 72/300. Therefore, the probability that both events occur is the product of the probabilities. [Refer to formula (4-8a), p. 234.] Thus $(122/300) \cdot (72/300)$ is the probability of a selected student being male and preferring math-science. Therefore, the number of students out of 300 that are expected to be male and prefer math-science is found by multiplying the probability (or proportion) by the total number of students (300). Thus the expected number of males who prefer math-science is $(122/300)(72/300)(300) = (122/300)(72) = 29.28$. The other expected values can be determined in the same manner.

Typically the contingency table is written so that it contains all this information (see Table 11-7, p. 576).

The calculated chi-square is

$$\chi^2 = \sum \frac{(O - E)^2}{E}$$

$$= \frac{(37 - 29.28)^2}{29.28} + \frac{(41 - 45.95)^2}{45.95} + \frac{(44 - 46.77)^2}{46.77}$$

$$+ \frac{(35 - 42.72)^2}{42.72} + \frac{(72 - 67.05)^2}{67.05} + \frac{(71 - 68.23)^2}{68.23}$$

$$= 2.035 + 0.533 + 0.164 + 1.395 + 0.365 + 0.112$$

$$= 4.604$$

$$\chi^{2*} = 4.604$$

STEP 4 **Decision:** Fail to reject H_0 (χ^{2*} did not fall in the critical region).

Conclusion At the 0.05 level of significance, the evidence does not allow us to reject the idea of independence between the sex of a student and the student's preferred academic subject area.

▲▲

TABLE 11-7
Contingency Table Showing Sample Results and Expected Values

Six of Student	Favorite Subject Area			Total
	Math-Science (M-S)	Social Science (SS)	Humanities (H)	
Male	37 (29.28)	41 (45.95)	44 (46.77)	122
Female	35 (42.72)	72 (67.05)	71 (68.23)	178
Total	72	113	115	300

$r \times c$ contingency table

In general, the $r \times c$ **contingency table** (r is the number of rows; c is the number of columns) will be used to test the independence of the row factor and the column factor. The number of degrees of freedom will be determined by

$$\mathrm{df} = (r - 1) \cdot (c - 1) \qquad (11\text{-}4)$$

where r and c are both greater than 1. (This value for df should agree with the number of cells counted according to the general description on pages 573 and 574.)

NOTE If either r or c is 1, then df $= k - 1$, where k represents the number of cells [formula (11-2)].

The expected values for an $r \times c$ contingency table will be found by means of the formulas in each cell in Table 11-8, where $n =$ grand total. In general, the **expected value** at the intersection of the *i*th **row and the *j*th column** is given by

expected value

$$E_{i,j} = \frac{R_i \times C_j}{n} \qquad (11\text{-}5)$$

We should again observe the previously mentioned guideline: each $E_{i,j}$ should be at least 5.

NOTE The notation used in Table 11-8 and formula (11-5) may be unfamiliar to you. For convenience in referring to cells or entries in a table, $E_{i,j}$ (or E_{ij}) can be used to denote the entry in the *i*th row and the *j*th column. That is, the first letter in the subscript corresponds to the row number and the second letter corresponds to the column number. Thus $E_{1,2}$ is the entry in the first row, second column, and $E_{2,1}$ is the entry in the second row, first column. Referring to Table 11-6, $E_{1,2}$ for that table is 45.95 and $E_{2,1}$ is 42.72. The notation used in Table 11-8 is interpreted in a similar manner; that is, R_1 corresponds to the total from row 1, and C_1 corresponds to the total from column 1.

TEST OF HOMOGENEITY

homogeneity

There is another type of contingency table problem. It is called a **test of homogeneity.** This test is used when one of the two variables is controlled by the experimenter so that the row (or column) totals are predetermined.

TABLE 11-8 Expected Values for an $r \times c$ Contingency Table

Rows	1	2	...	j	...	c	Total
1	$\dfrac{R_1 \times C_1}{n}$	$\dfrac{R_1 \times C_2}{n}$		$\dfrac{R_1 \times C_j}{n}$		$\dfrac{R_1 \times C_c}{n}$	R_1
2	$\dfrac{R_2 \times C_1}{n}$	.		.		.	R_2
.	.	.		.		.	.
i	$\dfrac{R_i \times C_1}{n}$	.		$\dfrac{R_i \times C_j}{n}$		.	R_i
.	.			.			.
r	$\dfrac{R_r \times C_1}{n}$			.		$\dfrac{R_r \times C_c}{n}$	R_r
Total	C_1	C_2	...	C_j	...	C_c	n

For example, suppose that we were to poll registered voters about a piece of legislation proposed by the governor. In the poll, 200 urban, 200 suburban, and 100 rural residents will be randomly selected and asked whether they favor or oppose the governor's proposal. That is, a simple random sample is taken for each of these three groups. A total of 500 voters are to be polled. But notice that it has been predetermined (before the sample is taken) just how many are to fall within each row category, as shown in Table 11-9, and each category is sampled separately.

TABLE 11-9 Registered Voter Poll with Predetermined Row Totals

Type of Residence	Governor's Proposal		
	Favor	Oppose	Total
Urban			200
Suburban			200
Rural			100
Total			500

In a test of this nature, we are actually testing the hypothesis "the distribution of proportions within the rows is the same for all rows." That is, the distribution of proportions in row 1 is the same as in row 2, is the same as in row 3, and so on. The alternative to this is "the distribution of proportions within the rows is not the same for all rows." This type of example may be thought of as a comparison of several multinomial experiments.

Beyond this conceptual difference, the actual testing for independence and homogeneity with contingency tables is the same. Let's demonstrate this by completing the polling illustration.

ILLUSTRATION 11-4

Each person in a random sample of 500 registered voters (200 urban, 200 suburban, and 100 rural residents) was asked his or her opinion about the governor's proposed legislation. Does the sample evidence shown in Table 11-10 support the hypothesis that "voters within the different residence groups have different opinions about the governor's proposal"? Use $\alpha = 0.05$.

TABLE 11-10 Sample Results for Illustration 11-4

Type of Residence	Governor's Proposal		Total
	Favor	Oppose	
Urban	143	57	200
Suburban	98	102	200
Rural	13	87	100
Total	254	246	500

SOLUTION

STEP 1 H_0: The proportion of voters favoring the proposed legislation is the same in all three groups.

H_a: The proportion of voters favoring the proposed legislation is not the same in all three groups. (That is, in at least one group the proportions are different from the others.)

STEP 2 $\alpha = 0.05$ and df $= (3-1)(2-1) = 2$. See the accompanying figure.

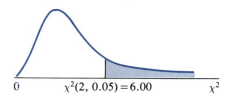

$\chi^2(2, 0.05) = 6.00$

STEP 3 The expected values are found by using formula (11-5) and are given in Table 11-11.

TABLE 11-11 Sample Results and Expected Values

Type of Residence	Governor's Proposal		Total
	Favor	Oppose	
Urban	143 (101.6)	57 (98.4)	200
Suburban	98 (101.6)	102 (98.4)	200
Rural	13 (50.8)	87 (49.2)	100
Total	254	246	500

Section 11.3 ▼ INFERENCES CONCERNING CONTINGENCY TABLES

NOTE Each expected value is used twice in the calculation of χ^{2*}; therefore, it is a good idea to keep extra decimal places while doing the calculations.

$$\chi^2 = \frac{(143 - 101.6)^2}{101.6} + \frac{(57 - 98.4)^2}{98.4} + \frac{(98 - 101.6)^2}{101.6}$$

$$+ \frac{(102 - 98.4)^2}{98.4} + \frac{(13 - 50.8)^2}{50.8} + \frac{(87 - 49.2)^2}{49.2}$$

$$= 16.87 + 17.42 + 0.13 + 0.13 + 28.13 + 29.04$$

$$= 91.72$$

$$\chi^{2*} = \mathbf{91.72}$$

STEP 4 **Decision:** Reject H_0.

Conclusion The three groups of voters do not all have the same proportions favoring the proposed legislation. ▲▲

COMPUTER-SOLUTION MINITAB Printout for Illustration 11-4

Information given to computer

```
MTB > READ TABLE INTO C1-C2
      COLUMN    C1         C2
      COUNT     3          3
ROW
  1             143.       57.
  2             98.        102.
  3             13.        87.
```

Contingency table with expected values, compare to Table 11-11

```
MTB > CHISQUARE FOR DATA IN C1 C2
Expected counts are printed below observed counts
             C1        C2      Total
  1          143       57      200
             101.60    98.40
  2          98        102     200
             101.60    98.40
  3          13        87      100
             50.80     49.20
Total        254       246     500
```

The calculated value of chi-square, cell by cell, compare to solution above

```
ChiSq =  16.870 + 17.418 +
          0.128 +  0.132 +
         28.127 + 29.041 = 91.715
```

The calculated chi-square value, χ^{2*}
The number of degrees of freedom

```
df = 2
--STOP
```

Case Study 11-3

50% Opposed to Building Nuclear Plants

The results of a poll by the National Center for Telephone Research of New York for Gannett News Service asked registered voters, "Do you favor or oppose building more nuclear power plants in New York State?" Percentages of voters who answered "yes," "no," or "not sure" for each of four geographical regions of the state as well as for the whole state are shown in the table. If the actual number of voters in each category were given, we would have a contingency table and we would be able to complete a hypothesis test about the homogeneity of the four regions. Explain why the following question could be tested using the chi-square statistic: "Is the support for building more nuclear power plants the same in all four regions of New York State?" (See Exercise 11.17, p. 584.)

NEW YORK STATE POLL

The New York Poll was conducted Dec. 15–18 by the National Center for Telephone Research of New York for Gannett News Service. Trained interviewers surveyed 1,000 registered voters. The calls were placed at random, but in proportion to past voter turnouts around the state.

Half of all New Yorkers oppose construction of more nuclear power plants in the state, the New York State Poll shows.

And women, far more than men, account for the opposition.

Exactly 50 percent of New Yorkers surveyed said they opposed more nuclear plants in the state. That matches the percentage of voters who said they were against more plants two years ago in a New York State Poll.

The latest poll, however, shows a drop since 1977 in the percentage that favors nuclear plants—down to 32 percent from 38 percent. There is a corresponding increase of 6 percent among those who say they're not sure.

Resistance to nuclear plants, regionally, ran highest in western New York, and lowest in the New York City north suburbs.

There is a great deal of uncertainty about the safety of nuclear plants, the poll found. But slightly more people seem to think they're safe than think they are unsafe.

Thus, it seems that most New Yorkers agree with Gov. Carey's decision last year during the campaign, to not allow more nuclear plants to be built until the radioactive waste problem is solved.

Here are the figures on the main question. "Do you favor or oppose building more nuclear power plants in New York State?":

	Yes	No	Not Sure
Total	32	50	18
Upstate	35	48	17
NY City	30	54	16
No. suburbs	32	45	23
Western NY	28	56	16

Source: John Omicinski, Chief, GNS Albany Bureau. Copyright 1979 by the Gannett News Service, Albany Bureau. Reprinted by permission.

Case Study 11-4

The USA Snapshots below show circle graphs that represent relative frequency distributions for six categories of responses by men and women separately. Does the distribution of responses given by men appear to be significantly different from the distribution of responses given by women? This question calls for a test of homogeneity. Can the sample information given here be expressed on a contingency table? How would significance be determined? (See Exercise 11.18, p. 584.)

Source: Copyright 1991, USA TODAY. Reprinted with permission.

Case Study 11-5

Table 2 shows the distribution (in percentages) of the level of student usage for each of 11 different drugs. The 11 distributions are shown side-by-side for comparison. A test of homogeneity asks, "Are the distributions all the same?" What do you think—are the distributions of drug usage all the same? Explain. (See Exercise 11.19, p. 584.)

TABLE 2 Percentages of Students Who Used Selected Drugs in the Last Three Months

	Frequency				
Drug	Daily	Weekly	Monthly	Less Than Once a Month	Never
Caffeine	73.8%	15.9%	4.0%	2.4%	3.9%
Tobacco	17.1	5.9	2.8	7.0	67.2
Beer or Wine	2.9	23.5	21.6	26.1	25.8
Hard Liquor	1.4	10.2	21.6	26.6	40.2
Inhaled Solvents	.7	1.1	1.1	3.2	93.9
Stimulants	1.2	2.2	5.8	9.2	81.5
Depressants	.3	.7	2.2	2.4	94.5
Marijuana	3.1	4.2	4.6	8.7	79.5
Hallucinogens	.3	.7	1.2	1.5	96.3
Opiates	.3	.3	1.0	.4	98.1
Cocaine	.7	.3	1.3	2.7	95.0

Source: Health Education, July/August 1990. Reprinted by permission.

Case Study 11-6

"Washers/Dryers by Region" shows the distribution of homes with washing machines region by region.

	West	Midwest	South	Northeast
With	92.8%	93.4%	92.6%	92.2%
Without	7.2	6.6	7.4	7.8

The graphic shows in picture form the contingency table of percentages as determined by the Census Bureau in 1987. Construct a contingency table showing the percentages of homes with and without dryers. (See Exercise 11.20, p. 584.)

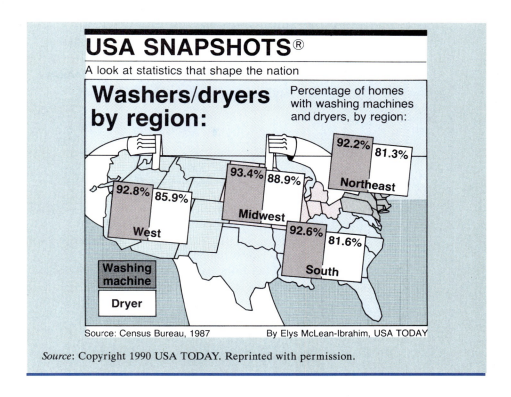

Source: Copyright 1990 USA TODAY. Reprinted with permission.

▼▲ EXERCISES

11.13 State the null hypothesis H_0 and the alternative hypothesis H_a that would be used to test the following statements:
   a. The voters expressed preferences that were not independent of their party affiliations.
   b. The distribution of opinions is the same for all three communities.
   c. The proportion of "yes" responses was the same for all categories surveyed.

11.14 The "test of independence" and the "test of homogeneity" are completed in identical fashion, using the contingency table to display and organize the calculations. Explain how these two hypothesis tests differ.

11.15 The manager of an assembly process wants to determine whether the number of defective articles manufactured depends on the day of the week the articles are produced. She collected the following information:

Day of Week	Mon.	Tues.	Wed.	Thurs.	Fri.
Nondefective	85	90	95	95	90
Defective	15	10	5	5	10

Is there sufficient evidence to reject the hypothesis that the number of defective articles is independent of the day of the week on which they are produced? Use $\alpha = 0.05$.

11.16 A random sample of 1500 registered voters in Rochester showed the following cross-tabulation for political party and religion:

	Catholic	Protestant	Jewish
Democrat	180	410	86
Republican	160	590	74

a. State the null and alternative hypotheses for a test of independence between party affiliation and religion.
b. Using $\alpha = 0.01$, determine the test criteria: level of significance, test statistic, critical region, critical value(s).
c. Using the information on the computer printout, state the decision and conclusion to the hypothesis test.

```
MTB> PRINT C1-C3
    ROW   C1    C2    C3
     1   180   410    86
     2   160   590    74
MTB> CHIS C1-C3
Expected counts are printed below observed counts
            C1       C2       C3     Total
      1    180      410       86      676
         153.23   450.67    72.11
      2    160      590       74      824
         186.77   549.33    87.89
  Total   340     1000      160     1500
  ChiSq = 4.678 + 3.670 + 2.677
          3.838 + 3.011 + 2.196 = 20.069
  df = 2
```

11.17 Refer to Case Study 11-3, p. 580.
a. Explain why the following question could be tested using the chi-square statistic: "Is the support for building more nuclear power plants the same in all four regions of New York State?"
b. Explain why this is a test of homogeneity.

 11.18 The *USA Snapshot* "Why We Don't Exercise" (Case Study 11-4, p. 581) shows the responses given by 1000 men and 1000 women who do not exercise.
a. Express the information on these circle graphs as a contingency table showing relative frequencies. Use two columns (M,W) and six rows (the six different responses).
b. Change the relative frequencies to frequencies and construct another 2 × 6 contingency table.

Section 11.3 ▼ INFERENCES CONCERNING CONTINGENCY TABLES

c. Do you think reasons for not exercising are distributed the same for men and women? Explain.
d. Complete a hypothesis test to determine if there is a significant difference between the distribution of responses given by men and women. Use $\alpha = 0.05$.
e. Suppose only 100 adult men and 100 adult women had been polled with the same relative frequencies resulting. How would the calculated value of χ^2 and the critical value of χ^2 be affected?

11.19 a. Explain why the information shown on Table 2 in Case Study 11-5, p. 582, is in fact 11 relative frequency distributions shown side-by-side.
b. Do you think the 11 distributions are all the same or does there seem to be a significant difference? Explain.

11.20 The *USA Snapshot* "Washers/Dryers by Region" (Case Study 11-6, p. 582) shows the percentage of homes with washing machines and dryers by regions of the United States.
a. In Case Study 11-6, a 2 × 4 contingency table was constructed showing the percentage of homes, by region, with and without washing machines. Does there appear to be a significant difference in the percentage of homes with washing machines from region to region? Explain.
b. Construct a 2 × 4 contingency table showing the percentage of homes with and without clothes dryers. Does there appear to be a significant difference in the percentage of homes with dryers from region to region?
c. Construct four 2 × 2 contingency tables (one for each region) with columns for washer and dryer, and rows for with and without.
d. Does there appear to be a significant difference in the percentage of homes with washing machines and with dryers in the Midwest region? in the South region? the northeast? the West? Did you answer them all the same? Explain why or why not.
e. What information would you need to complete a hypothesis test for the significance of such sample information?

11.21 Eighty-five college sophomores were classified according to their high school academic performance (HSAP) as well as their current college academic performance (CCAP). Excellent performance was coded 1, above-average was coded 2, and average or below was coded 3.

		\multicolumn{3}{c}{CCAP}		
		1	2	3
HSAP	1	10	10	5
	2	6	25	6
	3	5	8	10

The following MINITAB analysis shows the expected values as well as the calculated values for chi-square.
a. Verify the expected numbers, the chi-square values.
b. Find the *p*-value.

```
MTB > READ DATA IN C1-C3
DATA > 10 10 5
DATA > 6 25 6
DATA > 5 8 10
DATA > END DATA
      3 ROWS READ
MTB > CHISQUARE ANALYSIS ON DATA IN C1-C3
Expected counts are printed below observed counts
             C1        C2        C3     Total
    1        10        10         5       25
            6.18     12.65      6.18
    2         6        25         6       37
            9.14     18.72      9.14
    3         5         8        10       23
            5.68     11.64      5.68
Total        21        43        21       85
ChiSq =   2.367 +   0.554 +   0.224 +
          1.079 +   2.109 +   1.079 +
          0.082 +   1.136 +   3.281 = 11.911
df = 4
MTB > STOP
```

11.22 Students use many kinds of criteria when selecting courses. "Teacher who is a very easy grader" is often one criterion. Three teachers are scheduled to teach statistics next semester. A sample of the grade distributions for these three teachers is shown below.

	Professor		
Grades	#1	#2	#3
A	12	11	27
B	16	29	25
C	35	30	15
Other	27	40	23

Using the 0.01 level of significance, test the hypothesis "Student's grades are independent of the student's professor."
   a. State the null and alternative hypotheses.
   b. Determine the test criteria: level of significance, test statistic, critical region, critical value(s).
   c. Use a computer to complete the calculation of χ^{2*}.
   d. State the decision and conclusion.
   e. Which professor is the easiest grader? Explain, citing specific supporting evidence.

Section 11.3 ▼ INFERENCES CONCERNING CONTINGENCY TABLES

11.23 A paper entitled "Research Skills and the Research Environment: A Needs Assessment of Allied Health Faculty" (*Journal of Allied Health,* May 1988, pages 101–113) reported a survey of allied health faculty. One question asked was "Do you need help with statistical analysis?" Suppose the data, with the respondents, broken down by degree, were as follows:

	Degree		
	BS	Master's	Doctoral
n	513	953	525
Percentage Needing Help with Statistics	76.4	77.2	52.8

The paper gives the calculated χ^2 value as 108.94. Verify this value.

11.24 Fear of darkness is a common emotion. The following data were obtained by asking 200 individuals in each age group whether they had serious fears of darkness. At $\alpha = 0.01$, do we have sufficient evidence to reject the hypothesis that "the same proportion of each age group has serious fears of darkness"? (*Hint*: The contingency table must account for all 1000 people.)

Age Group	Elementary	Jr. High	Sr. High	College	Adult
No. Who Fear Darkness	83	72	49	36	114

a. Solve using the classical approach.
b. Solve using the prob-value approach.

11.25 A psychologist is investigating how a person reacts to a certain situation. He feels the reaction may be influenced by how ethnically pure the person's neighborhood is. He collects data on 500 people and finds the results shown in the following table:

Ethnically Pure Neighborhood	Reaction			
	Mild	Medium	Strong	Total
Yes	170	100	30	300
No	70	100	30	200
Total	240	200	60	500

Does there appear to be a relationship between neighborhood and reaction at the 0.10 level of significance?
a. Solve using the classical approach.
b. Solve using the prob-value approach.

11.26 "Doctor's Age Key in Decision on Cesareans" (*USA Today*, August 29, 1990) reports on a survey of 2213 obstetricians, "98% of doctors under 40 encouraged vaginal delivery in patients with a previous C-section vs. 84% over 55." How can a doctor's age be a key to the decision as to whether a woman has a cesarean childbirth or not? We know that people's ages affect their opinions; so why not expect a doctor's age to affect his or her opinion on method of childbirth? If you were commissioned to carry out a study regarding the effect that a doctor's age had on preference for or against C-sections, what information would you want to collect and how would you chart and analyze it?

IN RETROSPECT

In this chapter we have been concerned with tests of hypotheses using chi-square, with the cell probabilities associated with the multinominal experiment, and with the simple contingency table. In each case the basic assumptions are that a large number of observations have been made and that the resulting test statistic, $\Sigma [(O - E)^2/E]$, is approximately distributed as chi-square. In general, if n is large and the minimum allowable expected cell size is 5, this assumption is satisfied.

The article in Case Study 11-3 shows a contingency table with the results listed as relative frequencies. This is a common practice when reporting the results of public opinion polls. However, to statistically interpret the results, the actual frequencies must be used. (See Exercise 11.17.)

The contingency table can also be used to test homogeneity. The test for homogeneity and the test for independence look very similar and, in fact, are carried out in exactly the same way. The concepts being tested, however, equal distributions and independence, are quite different. The two tests are easily distinguished from one another, for the test of homogeneity has predetermined marginal totals in one direction in the table. That is, before the data are collected, the experimenter determines how many subjects will be observed in each category. The only predetermined number in the test of independence is the grand total.

A few words of caution: The correct number of degrees of freedom is critical if the test results are to be meaningful. The degrees of freedom determine, in part, the critical region, and its size is important. Like other tests of hypothesis, failure to reject H_0 does not mean outright acceptance of the null hypothesis.

CHAPTER EXERCISES

11.27 When interbreeding two strains of roses, we expect the hybrid to appear in three genetic classes in the ratio 1:3:4. If the results of an experiment yield 80 hybrids of the first type, 340 of the second type, and 380 of the third type, do we have sufficient evidence to reject the hypothesized genetic ratio at the 0.05 level of significance?
   a. Solve using the classical approach.
   b. Solve using the prob-value approach.

11.28 The psychology department at a certain college claims that the grades in its introductory course are distributed as follows: 10% A's, 20% B's, 40% C's, 20% D's, and 10% F's. In a poll of 200 randomly selected students who had completed this course, it was found that 16 had received A's, 43 B's, 65 C's, 48 D's, and 28 F's. Does this sample contradict the department's claim at the 0.05 level?

    a. Solve using the classical approach.
    b. Solve using the prob-value approach.

11.29 A sample of 200 individuals are tested for their blood type and the results are used to test the hypothesized distribution of blood types:

Blood Type	A	B	O	AB
Percent	0.41	0.09	0.46	0.04

The observed results were as follows:

Blood Type	A	B	O	AB
Number	75	20	95	10

At the 0.05 level of significance, is there sufficient evidence to show that the stated distribution is incorrect?

11.30 The weights (x) of 300 adult males were determined and used to test the hypothesis that the weights were normally distributed with a mean of 160 lb and a standard deviation of 15 lb. The data were grouped into the following classes:

Weight (x)	Observed Frequency
$x \leq 130$	7
$130 \leq x < 145$	38
$145 \leq x < 160$	100
$160 \leq x < 175$	102
$175 \leq x < 190$	40
190 and over	13

Using the normal tables, the percentages for the classes are 2.28%, 13.59% 34.13%, 34.13%, 13.59%, and 2.28%, respectively. Do the observed data show significant reason to discredit the hypothesis that the weights are normally distributed with a mean of 160 lb and a standard deviation of 15 lb? Use the prob-value approach.

11.31 Four brands of popcorn were tested for popping. One hundred kernels of each brand were popped and the number of kernels not popped was recorded in each test (see the following table). Can we reject the null hypothesis that all four brands pop equally? Test at $\alpha = 0.05$.

Brand	A	B	C	D
No. Not Popped	14	8	11	15

a. Solve using the classical approach.
b. Solve using the prob-value approach.

11.32 A survey report in the March 1991 issue of *McCall's* magazine presented findings concerning job stress affecting women. One question concerned women getting verbal abuse in offices. Suppose the results were reported as follows:

Country	Percent Who Often Get Verbal Abuse in Office
USA	11
Australia	5
Brazil	17
Germany	4
Japan	10

If there were 4400 respondents in each country, would the data above indicate a true difference in the rate of verbal office abuse in the five countries?

a. Calculate the value of chi-square for these data.
b. Complete the hypothesis test at the 0.01 level of significance.
c. Place a bound on the prob-value.

11.33 The January 2, 1991 *USA Snapshot*, "Bring on the Water Bottle," reported two relative frequency distributions for the reasons why people drink bottled water versus tap water.

	1990	1987
Concerned about tap water purity	32%	42%
Bottled tastes better	44	33
Both	20	19
Other/don't know	4	6

If we assume that 2000 people were surveyed each year, does there appear to be a significant change in the distribution of the reasons? Use $\alpha = 0.05$.

11.34 The tumor-producing potential of a new drug was tested. One hundred rats were used as a control group, 100 were exposed to a low dose of the drug, and 100 were exposed to a high dose. The results were as follows:

CHAPTER EXERCISES

	Number of Tumors		
Rat Group	None	One or More	Total
Control	93	7	100
Low dose	89	11	100
High dose	86	14	100
Total	268	32	300

Is there sufficient evidence to conclude that the dosage does, in fact, affect the occurrence of tumors? Use $\alpha = 0.05$.

11.35 In the May 1990 issue of *Social Work*, Marlow reported the following results:

Number of Children Living at Home	Mexican-American Women	Anglo-American Women
0	23	38
1	22	9
2	17	15
3	7	9
4	1	1

Source: Copyright 1990, National Association of Social Workers, Inc. *Social Work*.

  a. Calculate the expected values for each of the cells on the contingency table.
  b. Calculate the chi-square for the data above.
  c. State the null and alternative hypotheses used to test the claim, "The distribution for number of children living at home is different for Mexican-American and Anglo-American women."
  d. State the test criteria using $\alpha = 0.05$.
  e. State the decision and conclusion.
  f. Did the statistical decision above result in the same conclusion as you expressed in answering part (d) of Exercise 2.114, p. 120, for the same data?

11.36 In the May 1990 issue of *Social Work*, Marlow reported the following results:

Choice of Working	Mexican-American Women	Anglo-American Women
Yes	23	13
No, woman would not if she did not have to	21	12
No, but woman still would work	26	47

Source: Copyright 1990, National Association of Social Workers, Inc. *Social Work*.

a. Calculate the expected values for each of the cells on the contingency table.
b. Calculate the chi-square for the data above.

11.37 a. Case Study 11-3 (p. 580) reports the findings of a sample of 1000 people. Suppose that the percentages reported in this news article had come from the following contingency table. Can the null hypothesis "the distribution of yes, no, and not sure opinions is the same for all four geographic areas of New York State" be true? Use $\alpha = 0.05$.

	Yes	No	Not Sure
Upstate	96	150	54
NY City	105	144	51
No. suburbs	64	90	46
Western NY	56	112	32

b. Suppose that there had been only 400 people in the sample, 100 from each geographic location. Calculate the observed value of chi-square, χ^{2*}. Compare this value to that found in (a). Explain the difference.

11.38 "Attitudes in Conflict" (*USA Today*, March 11, 1991) reported, "We love our doctors . . . and hospitals . . . but dislike the system." Below are the ratings for each from a Gallup Poll of 1000 adults.

Rating	Physician	Hospital	Health-Care System
Excellent	53%	47%	6%
Good	39	35	32
Fair	6	12	32
Poor	1	5	24
Don't know	1	1	4
Total	100%	100%	98%[1]

[1]Doesn't total 100% due to rounding

a. Change the two 32s to 33 and convert the table of relative frequencies above to frequencies, assuming that each column represents 1000 responses.
b. Is there sufficient reason to reject the claim that "we love our doctors and hospitals" with the same distribution of attitudes? Use $\alpha = 0.05$.
c. Is there sufficient reason to claim that "our love for the doctors, the hospitals, and the system" is not all distributed the same? Use $\alpha = 0.05$.
d. Does it seem fair to conclude, as the article did, that "we love doctors and hospitals but dislike the system"? How do you justify your answer?

CHAPTER EXERCISES

 11.39 In an article in the *Journal of Geography* (January/February 1991, pages 18–26), Brothers reports a relationship between the presence or absence of beech trees and the classification of uplands (above 550 ft elevation) and bottomlands (below 550 ft elevation).

	Uplands	Bottomlands
Beech Present	18	6
Beech Absent	36	40

Use the data to test the independence of these two factors. Use $\alpha = 0.05$.

 11.40 A survey of travelers who visited the service station restrooms of a large U.S. petroleum distributor showed the following results:

	Quality of Restroom Facilities			
Sex of Respondent	Above Average	Average	Below Average	Total
Female	7	24	28	59
Male	8	26	7	41
Total	15	50	35	100

Using $\alpha = 0.05$, test to see whether the null hypothesis "Quality of responses is independent of the sex of the respondent" can be rejected.
   a. Solve using the classical approach.
   b. Solve using the prob-value approach.

 11.41 Does test failure reduce academic aspirations and thereby contribute to the decision to drop out of school? These were the concerns of a study entitled "Standards and School Dropouts: A National Study of Tests Required for High School Graduation." The table reports the responses of 283 students selected from schools with low graduation rates to the question, "Do tests required for graduation discourage some students from staying in school?"

	Urban	Suburban	Rural	Total
Yes	57	27	47	131
No	23	16	12	51
Unsure	45	25	31	101
Total	125	68	90	283

Source: American Journal of Education (November 1989), University of Chicago Press. © 1989 by The University of Chicago. All Rights Reserved.

Does there appear to be a relationship at the 0.05 level of significance between a student's response and the school's location?

11.42 The following table shows the number of reported crimes committed last year in the inner part of a large city. The crimes were classified according to type of crime and district of the inner city where it occurred. Do these data show sufficient evidence to reject the hypothesis that the type of crime and the district in which it occurred are independent? Use $\alpha = 0.01$.

	Crime				
District	Robbery	Assault	Burglary	Larceny	Stolen Vehicle
1	54	331	227	1090	41
2	42	274	220	488	71
3	50	306	206	422	83
4	48	184	148	480	42
5	31	102	94	596	56
6	10	53	92	236	45

11.43 Last year's work record for absenteeism in each of four categories for 100 randomly selected employees is compiled in the following table. Do these data provide sufficient evidence to reject the hypothesis that the rate of absenteeism is the same for all categories of employees? Use $\alpha = 0.01$ and 240 work days for the year.

	Married Male	Single Male	Married Female	Single Female
Number of Employees	40	14	16	30
Days Absent	180	110	75	135

a. Solve using the classical approach.
b. Solve using the prob-value approach.

11.44 A survey of employees at an insurance firm was concerned with worker-supervisor relationships. One statement for evaluation was "I am not sure what my supervisor expects." The results of the survey are found in the following contingency table:

Years of Employment	I Am Not Sure What My Supervisor Expects		
	True	Not True	Totals
Less than 1 year	18	13	31
1–3 years	20	8	28
3–10 years	28	9	37
10 years or more	26	8	34
Total	92	38	130

Can we reject the null hypothesis that "the responses to the statement and years of employment are independent" at the 0.10 level of significance?
  a. Solve using the classical approach.
  b. Solve using the prob-value approach.

11.45 Consider the following set of data:

	Response		
	Yes	No	Total
Group 1	75	25	100
Group 2	70	30	100
Total	145	55	200

  a. Compute the value of the test statistic z^* that would be used to test the null hypothesis that $p_1 = p_2$, where p_1 and p_2 are the proportions of "yes" responses in the respective groups.
  b. Compute the value of the test statistic χ^{2*} that would be used to test the hypothesis that "response is independent of group."
  c. Show that $\chi^2 = (z^*)^2$.

VOCABULARY LIST

Be able to define each term. In addition, describe, in your own words, and give an example of each term. Your examples should not be ones given in class or in the textbook.

The bracketed numbers indicate the chapters in which the term previously appeared, but you should define the terms again to show increased understanding of their meaning.

cell
chi-square [9]
column
contingency table
degrees of freedom [9, 10]
enumerative data
expected frequency
expected value
frequency distribution [2]

homogeneity
hypothesis test [8, 9, 10]
independence [4]
marginal totals
multinomial experiment
observed frequency [2, 4]
$r \times c$ contingency table
rows
statistic [1, 2, 8]

KEY CONCEPTS

chi-square [9]
enumerative data
homogeneity

hypothesis test [8, 9, 10]
independence [4]
multinomial experiment

QUIZ A

Answer "True" if the statement is always true. If the statement is not always true, replace the words shown in bold with words that make the statement always true.

11.1 The number of degrees of freedom for a test of a multinomial experiment is **equal to** the number of cells in the experimental data.

11.2 The **expected frequency** in a chi-square test is found by multiplying the hypothesized probability of a cell by the number of pieces of data in the sample.

11.3 The **observed** frequency of a cell is not allowed to be smaller than 5 when a chi-square test is being conducted.

11.4 In the **multinomial experiment** we have $(r - 1)$ times $(c - 1)$ degrees of freedom (r is the number of rows and c is the number of columns).

11.5 A multinomial experiment consists of **n identical, independent trials**.

11.6 A **multinomial experiment** arranges the data into a two-way classification such that the totals in one direction are predetermined.

11.7 The charts for both the multinomial experiment and the contingency table **must** be set up in such a way that each piece of data will fall into exactly one of the categories.

11.8 The test statistic $\Sigma[(O - E)^2/E]$ has a distribution that is **approximately normal**.

11.9 The data used in a chi-square multinomial test are always **enumerative** in nature.

11.10 The null hypothesis being tested by a test of **homogeneity** is that the distribution of proportions is the same for each of the subpopulations.

QUIZ B

Answer all questions. Show formulas, substitutions, and work.

11.1 State the null and alternative hypotheses that would be used to test each of the following claims:
  a. The single-digit numerals generated by a certain random number generator were not equally likely.
  b. The results of the last election in our city suggest that the votes cast were not independent of the voter's registered party.
  c. The distributions of types of crimes committed against society are the same in the four largest U.S. cities.

11.2 Find each of the following:
  a. $\chi^2(12, 0.975)$ b. $\chi^2(17, 0.005)$

11.3 Three hundred consumers were each asked to identify which one of three different items they found to be the most appealing. The table shows the number that preferred each item.

Item	1	2	3
Number	85	103	112

Do these data present sufficient evidence at the 0.05 level of significance to indicate that the three items are not equally preferred?

11.4 To study the effect of the type of soil on the amount of growth attained by a new hybrid plant, saplings were planted in three different types of soil and their subsequent amounts of growth classified into three categories:

	Soil Type		
Growth	Clay	Sand	Loam
Poor	16	8	14
Average	31	16	21
Good	18	36	25
Total	65	60	60

Does the quality of growth appear to be distributed differently for the tested soil types at the 0.05 level?
    a. State the null and alternative hypotheses.
    b. Find the test criteria [level of significance, test statistic, its distribution, critical region, and critical value(s)].
    c. Find the expected value for the cell containing 36.
    d. Calculate the value of chi-square for these data.
    e. State the decision and the conclusion for this hypothesis test.

QUIZ C

11.1 Explain how a multinomial experiment and a binomial experiment are similar and also how they are different.

11.2 Explain the distinction between a test for independence and a test for homogeneity.

11.3 Student A says that the tests for independence and homogeneity are the same and student B says that they are not at all alike because they are tests of different concepts. Both students are partially right and partially wrong. Explain.

11.4 You are interpreting the results of an opinion poll on the role of recycling in your town. A random sample of 400 people were asked to respond strongly in favor, slightly in favor, neutral, slightly against, or strongly against on each of several questions. There are four key questions that concern you and you plan to analyze their results.
    a. How do you calculate the expected probabilities for each answer?
    b. How would you decide if the four questions were answered the same?

12 ANALYSIS OF VARIANCE

Chapter Outline

12.1 Introduction to the Analysis of Variance Technique
Analysis of variance (ANOVA) is used to test a hypothesis about **several population means**.

12.2 The Logic Behind ANOVA
Between-sample variance and **within-sample variation** are compared in an ANOVA test.

12.3 Applications of Single-Factor ANOVA
Notational considerations and a **mathematical model** explaining the composition of each piece of data are presented.

The Behavioral Sciences and Management: An Evaluation of Relevant Journals

One hundred eighty-nine academics rated 54 journals concerned with the behavioral aspects of management. Journals were evaluated in regard to the quality of research they published. Results of respondents' ratings are compared to the Social Science Citation Index *and earlier studies evaluating managerial journals. In addition, biases hypothesized to affect these ratings are analyzed. Although findings are generally consistent with earlier research, this survey is unique in its focus on the behavioral aspects of management, the large number of rated journals, and its analysis of differences in ratings.*

Although academic journals are used primarily to disseminate new knowledge and research findings to members of a discipline, they also serve as arbiters of the quality of scholarly activity. Individuals needing to judge the quality of a research article, but unable to judge its merit directly for any number of reasons, rely on the journal in which the article is published to serve as an indicator of quality.

DIFFERENCES AMONG JOURNALS: RESEARCH PROPOSITIONS

Characteristics of the journals themselves might also influence the evaluation process. A high rejection rate or the perception of stringent review practices might lead a journal to be more highly regarded. Research supports the position that groups that impose a strenuous or difficult entrance requirement are more likely to be viewed favorably—apart from the actual value of the group (Aronson & Mills, 1959). Mahoney (1985) contends that prestige attributions for journals are both cause and consequence of low acceptance rates, and low acceptance rates tend to be associated with journals publishing high quality articles. Coe and Weinstock (1983) found a significant correlation between journal ratings and perceived acceptance rates ($r = -.96$). Thus, apart from the quality of the research appearing in a particular publication, a high rejection rate might lead evaluators to rate such journals more favorably.

Hypothesis 3: Ratings of Journal Quality Will Be Positively Associated with Journal Rejection Rates

Journals are designed to appeal to different audiences. In management, journals typically are designed to appeal to either academics or practitioners. Stahl, Leap, and Wei (1988) found significantly different patterns of publication between journals whose chief audiences are academics and journals whose chief audiences are practitioners. Overall, academics tended to publish more in scholarly journals than in those aimed at practitioners. Journals designed for practitioners typically contain less technical jargon and fewer details of scientific method and statistical analyses. Because of this non-technical approach, they may be thought of as lower in quality than journals designed for an academic audience.

Hypothesis 4: Journals Designed to Appeal to a Practitioner Audience Will Be Rated Lower in Quality by Academic Readers than Those Journals Designed for an Academic Readership

METHODS

Analysis. Analysis of the data was done in two parts. In the first part, descriptive statistics regarding each journal's quality were analyzed and compared to previous ratings of that journal. Second, the four hypotheses were tested by performing the following analyses:

Hypothesis 3: Cabell (1985) lists submission rejection rates for most of the journals in this study; rejection rate information for several other journals was obtained from editors' notes. Because Cabell (1985) supplies this information in six categories (5% or less, 6–10%, 11–20%, 21–30%, 31–50% and over 50%), all data were converted to this format. One-way analysis of variance was used to examine differences in average journal quality ratings across the six categories.

Hypothesis 4: Cabell (1985) reports editors' classifications of their journal's readership in one of three categories: academic, businessperson, or a combination of the two. One-way analysis of variance was used to examine differences in average journal ratings across the three categories.

RESULTS

Hypothesis 3: Acceptance rate. Are journals that have low acceptance rates perceived as having higher quality? A one-way analysis of variance of average journal rating across the six acceptance rate categories, as defined by Cabell (1985), revealed some support for this negative association ($F = 2.57$, significance $p < .05$). A subsequent Duncan multiple range test (significance $p < .05$)

revealed that respondents perceived journals in category 2 (acceptance rates of 6–10%) to be of significantly higher quality than those in both categories 4 (21–30%) and 6 (> 50%), respectively.

Hypothesis 4: Audience. Does the readership categorization of a journal influence its ratings? A one-way analysis of variance of average journal rating across the three readership categories (academic, businessperson, both) revealed some support for this association ($F = 7.95$, significance $p < .01$). A subsequent Scheffe test revealed that academic journals were consistently rated higher than practitioner and mixed audience journals; practitioner and mixed audience journals ratings were not significantly different.

Source: Marian M. Extejt and Jonathan E. Smith, "The Behavioral Sciences and Management: An Evaluation of Relevant Journals," *Journal of Management*, 16 (no. 3), 1990. Reprinted by permission.

Chapter Objectives

analysis of variance

Previously, we have tested hypotheses about two means. In this chapter we are concerned with testing a hypothesis about several means. The **analysis of variance technique (ANOVA)**, which we are about to explore, will be used to test a hypothesis about several means, for example,

$$H_0: \mu_1 = \mu_2 = \mu_3 = \mu_4 = \mu_5$$

By using our former technique for hypotheses about two means, we could test several hypotheses if each stated a comparison of two means. For example, we could test

$H_1: \mu_1 = \mu_2$ $H_2: \mu_1 = \mu_3$
$H_3: \mu_1 = \mu_4$ $H_4: \mu_1 = \mu_5$
$H_5: \mu_2 = \mu_3$ $H_6: \mu_2 = \mu_4$
$H_7: \mu_2 = \mu_5$ $H_8: \mu_3 = \mu_4$
$H_9: \mu_3 = \mu_5$ $H_{10}: \mu_4 = \mu_5$

In order to test the null hypothesis, H_0, that all five means are equal, we would have to test each of these ten hypotheses using our former technique for two means. Rejection of any one of the ten hypotheses about two means would cause us to reject the null hypotheses that all five means are equal. If we failed to reject all ten hypotheses about the means, we would fail to reject the main null hypothesis. Suppose we tested a null hypothesis that dealt with several means by testing all the possible pairs of two means; the overall type-I error rate would become much larger than the value of α associated with a single test. The ANOVA techniques allow us to test the null hypothesis (all means are equal) against the alternative hypothesis (at least one mean value is different) with a specified value of α.

In this chapter we will introduce ANOVA. ANOVA experiments can be very complex, depending on the situation. We will restrict our discussion to the most basic experimental design, the single-factor ANOVA.

12.1 ▼ Introduction to the Analysis of Variance Technique

We will begin our discussion of the analysis of variance technique by looking at an illustration.

▼ ILLUSTRATION 12-1

The temperature at which a plant is maintained is believed to affect the rate of production in the plant. The data in Table 12-1 are the number, x, of units produced in one hour for randomly selected one-hour periods when the production process in the plant was operating at each of three temperature **levels**. The data values from repeated samplings are called **replicates**. Four replicates, or data values, were obtained for two of the temperatures and five were obtained for the third temperature. Do these data suggest that temperature has a significant effect on the production level at the 0.05 level?

TABLE 12-1
Sample Results for Illustration 12-1

	Temperature Levels		
	Sample from 68°F ($i = 1$)	Sample from 72°F ($i = 2$)	Sample from 76°F ($i = 3$)
	10	7	3
	12	6	3
	10	7	5
	9	8	4
		7	
Column Totals	$C_1 = 41$ $\bar{x}_1 = 10.25$	$C_2 = 35$ $\bar{x}_2 = 7.0$	$C_3 = 15$ $\bar{x}_3 = 3.75$

The level of production is measured by the mean value; $\bar{x}_i$ indicates the observed production mean at level i, where $i = 1, 2,$ and 3 corresponds to temperatures of 68°F, 72°F, and 76°F, respectively. There is a certain amount of variation among these means. Since sample means do not necessarily repeat when repeated samples are taken from a population, some variation can be expected, even if all three population means are equal. We will next pursue the question "Is this variation among the $\bar{x}$'s due to chance or is it due to the effect that temperature has on the production rate?"

SOLUTION The null hypothesis that we will test is

$$H_0: \mu_{68} = \mu_{72} = \mu_{76}$$

That is, the true production mean is the same at each temperature level tested. In other words, the temperature does not have a significant effect on the production rate. The alternative to the null hypothesis is

$$H_a: \text{not all temperature level means are equal}$$

Thus we will want to reject the null hypothesis if the data show that one or more of the means are significantly different from the others.

We will make the decision to reject H_0 or fail to reject H_0 by using the F distribution and the F test statistic. Recall from Chapter 10 that the calculated value of F is the ratio of two variances. The analysis of variance procedure will separate the variation among the entire set of data into two categories. To accomplish this separation, we first work with the numerator of the fraction used to define sample variance, formula (2-7):

$$s^2 = \frac{\sum (x - \bar{x})^2}{n - 1}$$

sum of squares

The numerator of this fraction is called the ***sum of squares***:

$$\text{sum of squares} = \sum (x - \bar{x})^2 \tag{12-1}$$

total sum of squares, SS(total)

We calculate the **total sum of squares, SS(total)**, for the total set of data by using a formula that is equivalent to formula (12-1) but does not require the use of $\bar{x}$. This equivalent formula is

$$\text{SS(total)} = \sum (x^2) - \frac{(\sum x)^2}{n} \tag{12-2}$$

Now we can find the SS(total) for our illustration by using formula (12-2):

$$\sum (x^2) = 10^2 + 12^2 + 10^2 + 9^2 + 7^2 + 6^2 + 7^2$$
$$+ 8^2 + 7^2 + 3^2 + 3^2 + 5^2 + 4^2$$
$$= 731$$

$$\sum x = 10 + 12 + 10 + 9 + 7 + 6 + 7$$
$$+ 8 + 7 + 3 + 3 + 5 + 4$$
$$= 91$$

$$\text{SS(total)} = 731 - \frac{(91)^2}{13}$$
$$= 731 - 637 = \mathbf{94}$$

partitioning

Next, 94, the SS(total), must be separated into two parts: the sum of squares, SS(temperature), due to temperature levels; and the sum of squares, SS(error), due to experimental error of replication. This splitting is often called **partitioning**, since SS(temperature) + SS(error) = SS(total); that is, in our illustration SS(temperature) + SS(error) = 94.

variation between the factor levels

The sum of squares, SS(factor) [SS(temperature) for our illustration] that measures the **variation between the factor levels** (temperatures) is found by using formula (12-3):

$$\text{SS(factor)} = \left(\frac{C_1^2}{k_1} + \frac{C_2^2}{k_2} + \frac{C_3^2}{k_3} + \cdots \right) - \frac{(\sum x)^2}{n} \tag{12-3}$$

where C_i represents the column total, k_i represents the number of replicates at each level of the factor, and n represents the total sample size ($n = \Sigma k_i$).

NOTE The data have been arranged so that each column represents a different level of the factor being tested.

Now we can find the SS(temperature) for our illustration by using formula (12-3):

$$\text{SS(temperature)} = \left(\frac{41^2}{4} + \frac{35^2}{5} + \frac{15^2}{4}\right) - \frac{(91)^2}{13}$$
$$= (420.25 + 245.00 + 56.25) - 637.0$$
$$= 721.5 - 637.0 = \mathbf{84.5}$$

variation within rows

The sum of squares *SS(error)* that measures the **variation within the rows** is found by using formula (12-4):

$$\text{SS(error)} = \Sigma (x^2) - \left(\frac{C_1^2}{k_1} + \frac{C_2^2}{k_2} + \frac{C_3^2}{k_3} + \cdots\right) \quad \textbf{(12-4)}$$

The SS(error) for our illustration can now be found by using formula (12-4):

$$\Sigma (x^2) = 731 \text{ (found previously)}$$
$$\left(\frac{C_1^2}{k_1} + \frac{C_2^2}{k_2} + \frac{C_3^2}{k_3}\right) = 721.5 \text{ (found previously)}$$
$$\text{SS(error)} = 731.0 - 721.5 = 9.5$$

NOTE SS(total) = SS(factor) + SS(error). Inspection of formulas (12-2), (12-3), and (12-4) will verify this.

For convenience we will use an ANOVA table to record the sums of squares and to organize the rest of the calculations. The format of an ANOVA table is shown in Table 12-2.

TABLE 12-2
Format for
ANOVA Table

Source	SS	df	MS
Factor			
Error			
Total			

We have calculated the three sums of squares for our illustration. The degrees of freedom, df, associated with each of the three sources are determined as follows:

1. df(factor) is one less than the number of levels (columns) for which the factor is tested:

$$\text{df(factor)} = c - 1 \quad \textbf{(12-5)}$$

where c represents the number of levels for which the factor is being tested (number of columns on the data table).

2. df(total) is one less than the total number of data:

$$\text{df(total)} = n - 1 \qquad (12\text{-}6)$$

where n represents the number of data in the total sample (that is, $n = k_1 + k_2 + k_3 + \cdots$, where k_i is the number of replicates at each level tested).

3. df(error) is the sum of the degrees of freedom for all the levels tested (columns in the data table). Each column has $k_i - 1$ degrees of freedom; therefore,

$$\text{df(error)} = (k_1 - 1) + (k_2 - 1) + (k_3 - 1) + \cdots$$

or

$$\text{df(error)} = n - c \qquad (12\text{-}7)$$

The degrees of freedom for our illustration are

$$\text{df(temperature)} = c - 1 = 3 - 1 = \mathbf{2}$$
$$\text{df(total)} = n - 1 = 13 - 1 = \mathbf{12}$$
$$\text{df(error)} = n - c = 13 - 3 = \mathbf{10}$$

The sums of squares and the degrees of freedom must check. That is,

$$\text{SS(factor)} + \text{SS(error)} = \text{SS(total)} \qquad (12\text{-}8)$$
$$\text{df(factor)} + \text{df(error)} = \text{df(total)} \qquad (12\text{-}9)$$

mean square, MS(factor), MS(error)

The **mean square** for the factor being tested, the **MS(factor)**, and for error, the **MS(error)**, will be obtained by dividing the sum-of-square value by the corresponding number of degrees of freedom. That is,

$$\text{MS(factor)} = \frac{\text{SS(factor)}}{\text{df(factor)}} \qquad (12\text{-}10)$$

$$\text{MS(error)} = \frac{\text{SS(error)}}{\text{df(error)}} \qquad (12\text{-}11)$$

The mean squares for our illustration are

$$\text{MS(temperature)} = \frac{\text{SS(temperature)}}{\text{df(temperature)}} = \frac{84.5}{2}$$
$$= \mathbf{42.25}$$

$$\text{MS(error)} = \frac{\text{SS(error)}}{\text{df(error)}} = \frac{9.5}{10}$$
$$= \mathbf{0.95}$$

The completed ANOVA table appears as shown in Table 12-3. The hypothesis test is now completed using the two mean squares as measures of variance. The

TABLE 12-3
ANOVA Table for
Illustration 12-1

Source	SS	df	MS
Temperature	84.5	2	42.25
Error	9.5	10	0.95
Total	94.0	12	

test statistic, F^*

calculated value of the **test statistic, F^***, is found by dividing the MS(factor) by the MS(error):

$$F = \frac{\text{MS(factor)}}{\text{MS(error)}} \quad (12\text{-}12)$$

The calculated value of F for our illustration is found by using formula (12-12):

$$F = \frac{\text{MS(temperature)}}{\text{MS(error)}} = \frac{42.25}{0.95}$$

$$F^* = \mathbf{44.47}$$

The decision to reject H_0 or fail to reject H_0 will be made by comparing the calculated value of F, $F^* = 44.47$, to a one-tailed critical value of F obtained from Table 8a of Appendix F. See the accompanying figure. The critical value is $F(2, 10, 0.05) = 4.10$.

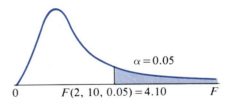

NOTE Since the calculated value of F, F^*, is found by dividing MS(temperature) by MS(error), the number of degrees of freedom for the numerator is df(temperature) $= 2$ and the number of degrees of freedom for the denominator is df(error) $= 10$.

We rejected H_0 because the value F^* fell in the critical region. Therefore we conclude that at least one of the room temperatures does have a significant effect on the production rate. The differences in the mean production rates at the tested temperature levels were found to be significant. The mean at 68°F is certainly different from the mean at 76°F, since the sample means for these levels are the largest and smallest, respectively. Whether any other pairs of means are significantly different cannot be determined from the ANOVA procedure alone. ▲▲

In this section we saw how the ANOVA technique separated the variance among the sample data into two measures of variance: (1) MS(factor), the measure of variance between the levels tested, and (2) MS(error), the measure of variance within the levels being tested. Then these measures of variance can be compared. For our illustration, the between-level variance was found to be significantly larger than the within-level variance (experimental error). This led us to the conclusion that temperature did have a significant effect on the variable x, the number of units of production completed per hour.

In the next section we will use several illustrations to demonstrate the logic of the analysis of variance technique.

12.2 ▼ The Logic Behind ANOVA

Many experiments are conducted to determine the effect that different levels of some test factor have on a response variable. The test factor may be temperature (as in Illustration 12-1), the manufacturer of a product, the day of the week, or any number of other things. In this chapter we are investigating the *single-factor* analysis of variance. Basically, the design for the single-factor ANOVA is to obtain independent random samples at each of the several levels of the factor being tested. We will then make a statistical decision concerning the effect that the levels of the test factors have on the response (observed) variable.

Illustrations 12-2 and 12-3 demonstrate the logic of the analysis of variance technique. Briefly, the reasoning behind the technique proceeds like this: In order to compare the means of the levels of the test factor, a measure of the **variation between the levels** (between columns on the data table), the **MS(factor)**, will be compared to a measure of the **variation within the levels** (within the columns on the data table), the **MS(error)**. If the MS(factor) is significantly larger than the MS(error), then we will conclude that the means for each of the factor levels being tested are not all the same. This implies that the factor being tested does have a significant effect on the response variable. If, however, the MS(factor) is not significantly larger than the MS(error), we will not be able to reject the null hypothesis that all means are equal.

variation between levels

variation within levels

▼ ILLUSTRATION 12-2

Do the data in Table 12-4 show sufficient evidence to conclude that there is a difference in the three population means μ_F, μ_G, and μ_H?

TABLE 12-4
Sample Results for Illustration 12-2

	Factor Levels		
	Sample from Level F	Sample from Level G	Sample from Level H
	3	5	8
	2	6	7
	3	5	7
	4	5	8
Column Totals	$C_F = 12$ $\bar{x} = 3.00$	$C_G = 21$ $\bar{x} = 5.25$	$C_H = 30$ $\bar{x} = 7.50$

Section 12.2 ▼ THE LOGIC BEHIND ANOVA

FIGURE 12-1

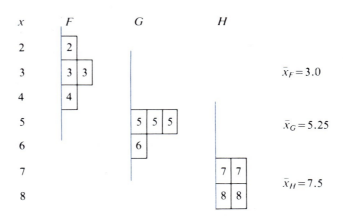

Figure 12-1 shows the relative relationship among the three samples. A quick look at the figure suggests that the three sample means are different from each other, implying that the sampled populations have different mean values. ▲▲

within-sample and between-sample variation

These three samples demonstrate relatively little **within-sample variation**, although there is a relatively large amount of **between-sample variation**. Let's look at another illustration.

▼ **ILLUSTRATION 12-3**

Do the data in Table 12-5 show sufficient evidence to conclude that there is a difference in the three population means μ_J, μ_K, and μ_L?

TABLE 12-5
Sample Results for Illustration 12-3

	Factor Levels		
	Sample from Level J	Sample from Level K	Sample from Level L
	3	5	6
	8	4	2
	6	3	7
	4	7	5
Column Totals	$C_J = 21$ $\bar{x}_J = 5.25$	$C_K = 19$ $\bar{x}_K = 4.75$	$C_L = 20$ $\bar{x}_L = 5.00$

Figure 12-2 (p. 608) shows the relative relationship among the three samples. A quick look at the figure *does not suggest* that the three sample means are different from each other.

FIGURE 12-2

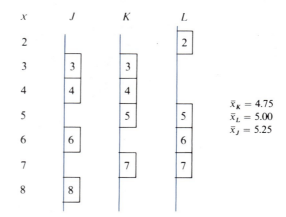

There is little between-sample variation for these three samples (that is, the sample means are relatively close in value), whereas the within-sample variation is relatively large (that is, the data values within each sample cover a relatively wide range of values).

To complete a hypothesis test for analysis of variance, we must agree on some ground rules, or assumptions. In this chapter we will use the following three basic assumptions:

1. Our goal is to investigate the effect that various levels of the factor under test have on the response variable. Typically, we want to find the level that yields the most advantageous values of the response variable. This, of course, means that we will probably want to reject the null hypothesis in favor of the alternative. Then a follow-up study could determine the "best" level of the factor.
2. We must assume that the effects due to chance and due to untested factors are normally distributed and that the variance caused by these effects is constant throughout the experiment.
3. We must assume independence among all observations of the experiment. (Recall that independence means that the results of one observation of the experiment do not affect the results of any other observation.) We will usually conduct the tests in a randomly assigned order to ensure independence. This technique also helps to avoid data contamination.

Case Study 12-1

Waiting for a Verdict

This USA SNAPSHOT *reports the average amount of time it takes for each of several types of felony charges to move through trial and to a verdict. Does the type of felony appear to have an effect on the average amount of time required? What additional information would be needed in order to determine whether the type of felony has a significant effect on the waiting time? (See Exercise 12.8, p. 620.)*

Section 12.3 ▼ APPLICATIONS OF SINGLE-FACTOR ANOVA

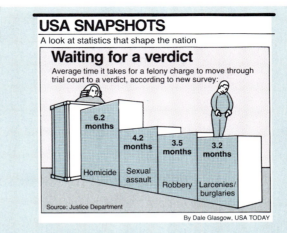

Source: Copyright 1986, USA TODAY. Reprinted with permission.

Case Study 12-2

Quality of Faculty Life

The number of hours per week that faculty spend preparing for teaching (one variable discussed) varies from professor to professor as well as from institution to institution. This article subdivides colleges into four partitions (the factor), very large, large, medium, and small; and the diagram depicts the mean value for each level of the factor.

Is it true that "faculty from smaller colleges spent significantly more time on teaching-related activities"? If you had the data to test this statement, how would you organize it and how would you analyze it? (See Exercise 12.9, p. 620.)

Faculty at larger institutions may take home more money and bask in the glow of greater academic prestige, but faculty at smaller colleges feel better about what they are doing, a recent report shows.

The Carnegie Foundation for the Advancement of Teaching analyzed data gathered through its 1989 National Survey of Faculty. It broke institutional size into four categories representing average faculty size (very large = 1,607 faculty; large = 822 faculty; medium = 407 faculty; small = 91 faculty). The survey asked faculty how many hours a week they spent on teaching, advising students and doing service with co-corricular student activities and how they felt about their work.

Not surprisingly, faculty from smaller colleges spent significantly more time on teaching-related activities (see chart), but they gave equally high marks to the importance of their academic discipline in their work. Regardless of institutional size, 98 percent of all faculty say their discipline is "very" or "fairly" important to them.

The report, which was published in the November/December 1990 issue of *Change* magazine, concludes that large institutions are paying a higher price in human terms than smaller colleges. "At smaller institutions, the faculty interact more with undergraduates and are more positive about that educational enterprise; they are more active as campus citizens; they are less alienated and less critical of the administration and their departmental colleagues; and they attach greater importance to their academic departments and the institution as a whole."

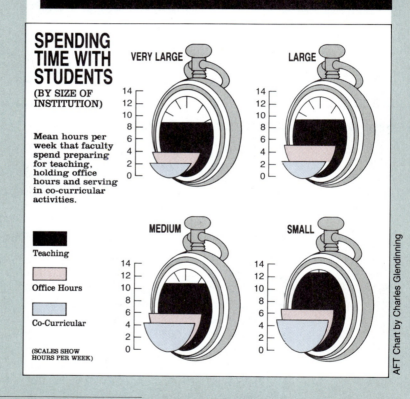

Source: *On Campus*, May/June 1991. Reprinted by permission.

Case Study 12-3

The average spent per week by teenagers in each age group seems to increase substantially as the teenager gets older. Data on the amounts spent by each of 2110 teenagers were collected and divided by age group. This is similar to how one-way analysis of variance treats data of this type. Explain. (See Exercise 12.10, p. 620.)

Section 12.3 ▼ APPLICATIONS OF SINGLE-FACTOR ANOVA

Source: Copyright 1990, USA TODAY. Reprinted with permission.

12.3 ▼ Applications of Single-Factor ANOVA

Before continuing our ANOVA discussion, let's identify the notation, particularly the subscripts that are used (see Table 12-6). Notice that each piece of data has two subscripts; the first subscript indicates the column number (test factor level) and the second subscript identifies the replicate (row) number. The column totals, C_i, are listed across the bottom of the table. The grand total, T, is equal to the sum of all x's and is found by adding the column totals. Row totals can be used as a cross-check but serve no other purpose.

TABLE 12-6
Notation Used in ANOVA

		Factor Levels			
Replication	Sample from Level 1	Sample from Level 2	Sample from Level 3	...	Sample from Level C
$k = 1$	$x_{1,1}$	$x_{2,1}$	$x_{3,1}$		$x_{c,1}$
$k = 2$	$x_{1,2}$	$x_{2,2}$	$x_{3,2}$		$x_{c,2}$
$k = 3$	$x_{1,3}$	$x_{2,3}$	$x_{3,3}$		$x_{c,3}$
⋮					
Column Totals	C_1	C_2	C_3	...	C_c T

T = grand total = sum of all x's = $\sum x = \sum C_i$

A mathematical model (equation) is often used to express a particular situation. In Chapter 3 we used a mathematical model to help explain the relationship between the values of bivariate data. The equation $\hat{y} = b_0 + b_1 x$ served as the model when we believed that a straight-line relationship existed. The probability functions studied in Chapter 5 are also examples of mathematical models. For the single-factor ANOVA, the **mathematical model**, formula (12-13), is an expression of the composition of each piece of data entered in our data table.

mathematical model

$$x_{c,k} = \mu + F_c + \varepsilon_{k(c)} \tag{12-13}$$

We interpret each term of this model as follows:

1. μ is the mean value for all the data without respect to the test factor.
2. F_c is the effect that the factor being tested has on the response variable at each different level c.
3. $\varepsilon_{k(c)}$ (ε is the lowercase Greek letter epsilon) is the **experimental error** that occurs among the k replicates in each of the c columns.

experimental error

Let's look at another hypothesis test concerning an analysis of variance.

▼ ILLUSTRATION 12-4

A rifle club performed an experiment on a randomly selected group of beginning shooters. The purpose of the experiment was to determine whether shooting accuracy is affected by the method of sighting used: only the right eye open, only the left eye open, or both eyes open. Fifteen beginning shooters were selected and split into three groups. Each group experienced the same training and practicing procedures with one exception: the method of sighting used. After completing training, each student was given the same number of rounds and asked to shoot at a target. Their scores appear in Table 12-7.

TABLE 12-7
Sample Results for Illustration 12-4

	Method of Sighting	
Right Eye	Left Eye	Both Eyes
12	10	16
10	17	14
18	16	16
12	13	11
14		20
		21

At the 0.05 level of significance, is there sufficient evidence to reject the claim that these methods of sighting are equally effective?

SOLUTION In this experiment the factor is "method of sighting" and the levels are the three different methods of sighting (right eye, left eye, and both eyes open). The replicates are the scores received by the students in each group. The null hypothesis to be tested is "the three methods of sighting are equally effective," or "the mean scores attained using each of the three methods are the same."

STEP 1 H_0: $\mu_R = \mu_L = \mu_B$
H_a: The means are not all equal (that is, at least one mean is different).

STEP 2 $\alpha = 0.05$. The test criteria are shown in the accompanying figure.

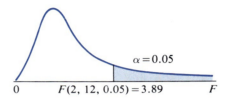

$$df(\text{numerator}) = df(\text{method}) = 3 - 1 = 2 \quad [\text{using formula (12-5)}]$$
$$df(\text{denominator}) = df(\text{error}) = 15 - 3 = 12 \quad [\text{using formula (12-7)}]$$

The critical value is $F(2, 12, 0.05) = 3.89$.

STEP 3 Calculate the test statistic F^*. Table 12-8 is used to find column totals.

TABLE 12-8
Sample Results for Illustration 12-4

Replicates	Factor Levels: Method of Sighting		
	Right Eye	Left Eye	Both Eyes
$k = 1$	12	10	16
$k = 2$	10	17	14
$k = 3$	18	16	16
$k = 4$	12	13	11
$k = 5$	14		20
$k = 6$			21
Totals	$C_R = 66$	$C_L = 56$	$C_B = 98$

First, the summations Σx and Σx^2 need to be calculated:
$$\Sigma x = 12 + 10 + 18 + 12 + 14 + 10 + 17 + \cdots + 21 = \mathbf{220}$$
$$\Sigma x^2 = 12^2 + 10^2 + 18^2 + 12^2 + 14^2 + 10^2 + \cdots + 21^2 = \mathbf{3392}$$

Using formula (12-2), we find
$$SS(\text{total}) = 3392 - \frac{(220)^2}{15} = 3392 - 3226.67 = \mathbf{165.33}$$

Using formula (12-3), we find

$$SS(\text{method}) = \left(\frac{66^2}{5} + \frac{56^2}{4} + \frac{98^2}{6}\right) - 3226.67$$

$$= 3255.87 - 3226.67 = \mathbf{29.20}$$

Using formula (12-4), we find

$$SS(\text{error}) = 3392 - 3255.87 = \mathbf{136.13}$$

Use formula (12-8) to check the sum of squares.

$$SS(\text{method}) + SS(\text{error}) = SS(\text{total})$$
$$29.20 \quad + \quad 136.13 \quad = \quad 165.33$$

The number of degrees of freedom is found using formulas (12-5), (12-6), and (12-7):

$$df(\text{method}) = 3 - 1 = \mathbf{2}$$
$$df(\text{total}) = 15 - 1 = \mathbf{14}$$
$$df(\text{error}) = 15 - 3 = \mathbf{12}$$

Using formulas (12-10) and (12-11), we find

$$MS(\text{method}) = \frac{29.20}{2} = \mathbf{14.60}$$

$$MS(\text{error}) = \frac{136.13}{12} = \mathbf{11.34}$$

The results of these computations are combined in the ANOVA table shown in Table 12-9.

TABLE 12-9
ANOVA Table for Illustration 12-4

Source	SS	df	MS
Method	29.20	2	14.60
Error	136.13	12	11.34
Total	165.33	14	

The calculated value of the test statistic is then found using formula (12-12).

$$F^* = \frac{14.60}{11.34} = \mathbf{1.287}$$

STEP 4 The decision to fail to reject or reject the null hypothesis is made by comparing F^* to the critical value shown in the test criteria (Step 2).

Section 12.3 ▼ APPLICATIONS OF SINGLE-FACTOR ANOVA

Decision Fail to reject H_0.

Conclusion The data show no evidence that would give reason to reject the null hypothesis that the three methods are equally effective. ▲▲

NOTE MINITAB's DOTPLOT command produces side-by-side dot plots that are very useful in visualizing the within-sample variation, the between-sample variation, and the relationship between them.

COMPUTER SOLUTION   MINITAB Printout for Illustration 12-4

Information given to computer

```
MTB > READ DATA INTO C1, LEVELS INTO C2
      COLUMN   C1         C2
      COUNT    15         15
      ROW
      1        12.        1.
      2        10.        1.
      3        18.        1.
      4        12.        1.
      5        14.        1.
      6        10.        2.
      7        17.        2.
      8        16.        2.
      9        13.        2.
      10       16.        3.
      11       14.        3.
      12       16.        3.
      13       11.        3.
      14       20.        3.
      15       21.        3.
MTB > ONEWAY ANALYSIS OF VARIANCE, DATA IN C1, LEVELS IN C2
              ANALYSIS OF VARIANCE
```

The ANOVA table → compare to Table 12-9

SOURCE	DF	SS	MS	F	P
FACTOR	2	29.2	14.6	1.29	0.312
ERROR	12	136.1	11.3		
TOTAL	14	165.3			

The calculated value of F, F^*

LEVEL	N	MEAN	ST.DEV.
1	5	13.200	3.033
2	4	14.000	3.162
3	6	16.333	3.724

(continued)

```
MTB  > DOTPLOT C1;
SUBC > BY C2.
```

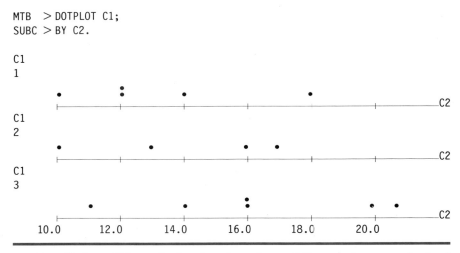

Recall the null hypothesis, that there is no difference between the levels of the factor being tested. A "fail to reject H_0" decision must be interpreted as the conclusion that there is no evidence of a difference due to the levels of the tested factor, whereas the rejection of H_0 implies that there is a difference between the levels. That is, at least one level is different from the others. If there is a difference, the next problem is to locate the level or levels that are different. Locating this difference may be the main object of the analysis. In order to find the difference, the only method that is appropriate at this stage is to inspect the data. It may be obvious which level(s) caused the rejection of H_0. In Illustration 12-1 it seems quite obvious that at least one of the levels [level 1 (68°F) or level 3 (76°F), because they have the largest and smallest sample means] is different from the other two. If the higher values are more desirable for finding the "best" level to use, we would choose that corresponding level of the factor.

Thus far we have discussed analysis of variance for data dealing with one factor. It is not unusual for problems to have several factors of concern. The ANOVA techniques presented in this chapter can be developed further and applied to more complex cases.

Case Study 12-4

Tillage Test Plots Revisited

The 18 plot yields can be analyzed using the one-way analysis of variance technique with 3 replicates at each of 6 different treatment levels (tillage method). Draw a dot plot showing these 6 sets of data side by side on the scale. Is there much variation in the data? Is more of the variation between or within the levels? Does the method of tillage seem to have an effect on the mean yield? (See Exercise 12.19, p. 625.)

All tillage tools are not created equal although the weather can certainly make them look that way. That's our analysis after putting six of your favorite primary tillage tools through the paces last season.

We staked out 18 plots on a gently sloping field with a Chalmers silty clay loam soil type.

How good is a tie. The yields from individual tillage treatments showed less variability than we had hoped, probably due to the late planting. That's why university researchers conduct these studies over many years. The only yield differences worth noting are the 7.4 bu/A spreads between the standard chisel and light disk at 115.2 bu. and the moldboard plow at 122.6 bu. The coulter-chisel and heavy disk, and surprisingly enough the V-chisel, yielded virtually the same as the moldboard plow. However, we think there are other factors that add even more merit to heavy-duty conservation tillage.

Tillage Plot Yields Bu A @ 15.5% Moisture

Tillage	Rep			Tillage Average
	I	II	III	
Plow	118.3	125.6	123.8	122.6
V chisel	115.8	122.5	118.9	119.1
Coulter chisel	124.1	118.5	113.3	118.6
Std. chisel	109.2	114.0	122.5	115.2
Hvy. disk	118.1	117.5	121.4	119.0
Lt. disk	118.3	113.7	113.7	115.2

Source: Larry Reichenberger, Associate Machine Editor. Reprinted with permission from *Successful Farming*, February 1979. Copyright 1979 by Meredith Corporation. All rights reserved.

Case Study 12-5

The Effect of the Class Evaluation Method on Learning in Certain Mathematics Courses

In the following article, Larry Stephens discusses how the statistical techniques of one-way analysis of variance can be used to evaluate three teaching methods. The percentages in Table 1 are the students' test scores. Verify the results shown. (See Exercise 12.20, p. 625.)

Much has been written concerning the merits of different teaching methods and their effectiveness in different disciplines. The purpose of this paper is to investigate the effects of different evaluation methods on the learning process. . . .

To determine whether different methods of evaluation influence the learning in such a course, three different methods were used. Method I can be described as follows. Homework was collected weekly and mid-term and comprehensive final exams were administered. Method II consisted of four tests administered at the end of each four weeks covering the material of the past four weeks. Method III consisted of 30-minute weekly tests and a comprehensive final. Homework was assigned but not collected in methods II and III. . . .

The same text and instructor were used in all three methods, and the three groups were representative of the junior and senior students from the school....

The response variable measured was a percentage of test points obtained by each student....

TABLE 1 Percentages, Means and Standard Deviations for the Three Methods

Method I	91, 62, 77, 84, 52, 67, 58, 78, 88, 72 87, 75, 93, 62, 63, 56, 83, 97, 72, 68 72, 85, 66, 94, 63, 60, 75, 79, 87	$n_1 = 29$ $\bar{x} = 74.7, s = 12.5$
Method II	85, 91, 79, 84, 73, 47, 92, 91, 91, 62 64, 73, 67, 75, 87, 77, 92, 83	$n_2 = 18$ $\bar{x} = 78.5, s = 12.6$
Method III	79, 85, 74, 89, 79, 80, 94, 91, 79, 71 73, 86, 70, 71, 71	$n_3 = 15$ $\bar{x} = 79.5, s = 8.0$

Table 1 shows that the weekly test with homework not taken up produced the largest average of the three methods. To answer the question of whether these sample means are statistically different requires an analysis of variance. In other words, if μ_1, μ_2 and μ_3 are the average scores made by all students taking such a course and being evaluated by methods I, II and III, respectively, what can be inferred about μ_1, μ_2 and μ_3? To test the hypothesis that $\mu_1 = \mu_2 = \mu_3$, i.e., that there is no difference in the three methods, the sample results given in Table 1 are subjected to an analysis of variance.

TABLE 2 One-Way Analysis of Variance for the Three Methods

Source	df	SS	MS	F
Total	61	8257		
Between methods	2	286.4	143.2	
Within methods	59	7970.6	135.1	1.06

Table 2 shows the results of a one-way analysis of variance for the three methods. The F-value is very nonsignificant, indicating that, statistically speaking, $\mu_1 = \mu_2 = \mu_3$.

The results of this experiment indicate that collecting and grading homework does not produce significantly higher test scores. Also, the frequency of testing does not seem to produce any different results.

Source: Larry J. Stephens, *International Journal of Math, Education, Science, and Technology*, 1977, Vol. 8., no. 4., pp. 477–479.

Section 12.3 ▼ APPLICATIONS OF SINGLE-FACTOR ANOVA

▼▲ EXERCISES

12.1 State the null hypothesis H_0 and the alternative hypothesis H_a that would be used to test the following statements:
  a. The mean value of x is the same at all five levels of the experiment.
  b. The scores are the same at all four locations.
  c. The four levels of the test factor do not significantly affect the data.
  d. The three different methods of treatment do affect the variable.

12.2 Determine the test criteria that would be used to test the null hypothesis for the following multinomial experiments:
  a. $H_0: \mu_1 = \mu_2 = \mu_3 = \mu_4$ with $n = 18$, $\alpha = 0.05$.
  b. $H_0: \mu_1 = \mu_2 = \mu_3 = \mu_4 = \mu_5$ with $n = 15$, $\alpha = 0.01$.
  c. $H_0: \mu_1 = \mu_2 = \mu_3$ with $n = 25$, $\alpha = 0.05$.

12.3 Suppose that an F test (as described in this chapter) has a critical value of 2.2, as shown in the following figure:

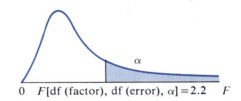

  a. What is the interpretation of a calculated value of F larger than 2.2?
  b. What is the interpretation of a calculated value of F smaller than 2.2?
  c. What is the interpretation if the calculated F were 0.1? 0.01?

12.4 Why does df(factor), the number of degrees of freedom, always appear first in the critical value notation F [df(factor), df(error), α]?

12.5 Consider the following table for a single-factor ANOVA. Find the following:
  a. $x_{1,2}$ b. $x_{2,4}$ c. C_1 d. Σx e. $\Sigma (C_i)^2$

Replicates	Level of Factor		
	1	2	3
1	3	2	7
2	0	5	4
3	1	4	5
4	5	3	6
5	7	6	2

12.6 The following table of data is to be used for single-factor ANOVA. Find each of the following:
  a. $x_{3,2}$ b. $x_{4,3}$ c. c_3 d. Σx e. $\Sigma (c_i)^2$

	Level of Factor			
Replicates	1	2	3	4
1	13	12	16	14
2	17	8	18	11
3	9	15	10	19

12.7
a. State the null hypothesis, in a general form, for the one-way ANOVA.
b. State the alternative hypothesis, in a general form, for the one-way ANOVA.
c. What must happen in order to "reject H_0"?
d. How would a decision of "reject H_0" be interpreted?
e. What must happen in order to "fail to reject H_0"?
f. How would a decision of "reject H_0" be interpreted?

12.8 Refer to Case Study 12-1, p. 608.
a. Does the type of felony charge appear to have an effect on the average amount of time it takes to move through the trial to a verdict?
b. What additional information would be needed in order to determine whether the type of felony has a significant effect on the waiting time?

12.9 The "Quality of Faculty Life" (Case Study 12-2, p. 609) presents information about how faculty members spend their time. An inference was made about the relationship between the size of the college and the mean number of hours faculty spend on teaching-related duties: "faculty from smaller colleges spent significantly more time on teaching-related activities." If you had the data to test this statement, describe what the data would be, how you would organize the data, and how you would analyze the data.

12.10 The amount of money spent varies from teenager to teenager, just as it varies from person to person at all ages. "Teens Are Big Spenders" (Case Study 12-3, p. 610) suggests that the average amount spent is affected by the age category the teenager belongs to. Describe how the data presented are organized in the same way as data are organized for one-way analysis of variance.

12.11 A new operator was recently assigned to a crew of workers who perform a certain job. From the records of the number of units of work completed by each worker each day last month, a sample of size five was randomly selected for each of the two experienced workers and the new worker. At the 0.05 level of significance, does the evidence provide sufficient reason to reject the claim that there is no difference in the amount of work done by the three workers?

	Workers		
	New	A	B
Units of Work (replicates)	8	11	10
	10	12	13
	9	10	9
	11	12	12
	8	13	13

Section 12.3 ▼ APPLICATIONS OF SINGLE-FACTOR ANOVA

12.12 An employment agency wants to see which of three types of ads in the help-wanted section of local newspapers is the most effective. Three types of ads (big headline, straightforward, and bold print) were randomly alternated over a period of weeks and the number of people responding to the ads was noted each week. Do these data support the null hypothesis that there is no difference in the effectiveness of the ads, as measured by the mean number responding, at the 0.01 level of significance?

	Type of Advertisement		
	Big Headline	Straightforward	Bold Print
Number of Responses (replicates)	23	19	28
	42	31	33
	36	18	46
	48	24	29
	33	26	34
	26		34

12.13 What is the average number of years people expect their home furnishings or appliances to last? Does the kind of furnishing or appliance affect the length of usable life that can be expected? The following set of data was collected in order to investigate these questions. x = number of years of service owner had from item.

Refrig.	Washer	Dryer	Stereo	Couch	Mattress
14	10	11	9	11	10
9	9	13	6	7	11
18	13	14	12	9	8
16	7	9	11	7	10
	9	12		6	12

a. Construct a graph picturing these data and showing the relationship between the 6 sets of data.
b. Does the kind of furnishing/appliance have a significant effect on the number of years? Use $\alpha = 0.05$.

12.14 The medical section of *Health* magazine (February 1991, page 22) describes an experiment in which the cholesterol levels of four groups of walkers were compared. The four groups were

Group	Description
1	none
2	walks 0.5 to 2.0 hr per week
3	walks 2.5 to 4.0 hr per week
4	walks 4.5 or more hr per week

The study included 3621 individuals. Subjects in the moderate and high exercise group had less than half the prevalence of high cholesterol levels of those who did not exercise.

Consider a small-scale study with five in each group. The cholesterol readings for each subject are given.

	Group		
1	2	3	4
240	250	220	190
190	200	200	180
230	215	180	200
215	190	170	150
200	185	150	160

Complete an ANOVA table and test the hypothesis, "The mean cholesterol reading is the same for all test groups." Use $\alpha = 0.05$.

12.15 Students with three different high school academic backgrounds were compared with respect to their aptitude in computer science. The students were classified as having excellent, above-average, or average or below-average high school academic backgrounds. Each student was given the KSW computer-science aptitude test and the score was recorded. The results were as follows:

	High School Academic Performance	
Excellent	Above Average	Average or Below
16	21	4
22	19	20
15	16	13
20	17	18
23	5	8
16	20	6
21	18	11
	19	
	14	
	22	
	13	

Refer to the following MINITAB analysis of the experiment and verify the analysis of variance table and test for the null hypothesis $\mu_E = \mu_{AA} = \mu_{AB}$ at $\alpha = 0.05$.

```
MTB  > READ DATA IN C1, LEVELS IN C2
DATA > 16 1
DATA > 22 1
DATA > 15 1
DATA > 20 1
DATA > 23 1
DATA > 16 1
DATA > 21 1
DATA > 21 2
DATA > 19 2
DATA > 16 2
DATA > 17 2
DATA > 5 2
DATA > 20 2
DATA > 18 2
DATA > 19 2
DATA > 14 2
DATA > 22 2
DATA > 13 2
DATA > 4 3
DATA > 20 3
DATA > 13 3
DATA > 18 3
DATA > 8 3
DATA > 6 3
DATA > 11 3
DATA > END DATA
      25 ROWS READ
MTB  > ONEWAY ANOVA, DATA IN C1, LEVELS IN C2

ANALYSIS OF VARIANCE ON C1
SOURCE    DF      SS       MS      F       P
CS         2    214.7    107.4   4.65   0.021
ERROR     22    507.9     23.1
TOTAL     24    722.6
```

12.16 A study was conducted to assess the effectiveness of treating vertigo (motion sickness) with the Transdermal Therapeutic System (TTS—patch worn on skin).

Two other treatments, both oral (one pill containing a drug and one a placebo) were used. The ages and the gender of the patients for each treatment are listed below.*

TTS		Antivert		Placebo	
47-f	53-m	51-f	43-f	67-f	38-m
41-f	58-f	53-f	56-f	52-m	59-m
63-m	62-f	27-m	48-m	47-m	33-f
59-f	34-f	29-f	52-f	35-f	32-f
62-f	47-f	31-f	19-f	37-f	26-f
24-m	35-f	25-f	31-f	40-f	37-m
43-m	34-f	52-f	48-f	31-f	49-f
20-m	63-m	55-f	53-m	45-f	49-m
55-f	46-f	32-f	63-m	41-f	38-f
		51-f	54-m	49-m	
		21-f			

Is there a significant difference between the mean age of the three test groups? Use $\alpha = 0.05$. Use a computer to complete this exercise. (If you use MINITAB, see Exercise 12.15.)

12.17 The article "An Investigation of High School Preparation As Predictors of the Cultural Literacy of Developmental, Nondevelopmental and ESL College Students" (*RTDE*, Fall 1990) reported on a study that examined the cultural literacy of developmental, nondevelopmental, and ESL (English as a Second Language) college freshmen.

a. How many student scores were in the samples?
b. The students were divided into how many groups?
c. Given the sum of square (SS) and degree of freedom (df) values, verify the mean square (MS), the calculated F value and the P value for each figure.

Analysis of Variance by Group for Total Score

Source	df	SS	MS	F	P
Group	2	4062.06	2031.03	14.49	0.0001
Error	117	16394.53	140.12		
Total	119	20456.59			

Analysis of Variance by Group for Foreign Language Preparation

Source	df	SS	MS	F	P
Group	2	0.95	0.475	1.93	0.1493
Error	117	28.75	0.246		
Total	119	29.70			

d. Do the statistics in the first table show that the total scores were different for the groups involved? Explain.
   e. Do the statistics in the second table show that the foreign language preparation scores were different for the groups involved? Explain.

12.18 An article entitled "The Effectiveness of Biofeedback and Home Relaxation Training on Reduction of Borderline Hypertension" (*Health Education,* Oct./Nov. 1988, pages 19–22) compared different methods of reducing blood pressure. Biofeedback ($n = 13$ subjects), Biofeedback/Relaxation ($n = 15$), and Relaxation ($n = 14$) were three methods compared. There were no differences among the three groups on pretest diastolic or systolic blood pressures. There was a significant posttest difference between groups on the systolic measure, $F(2, 39) = 4.14$, $p < 0.02$, and diastolic measure, $F(2, 39) = 5.56$, $p < 0.008$.
   a. Verify that df(method) = 2 and df(error) = 39.
   b. Use Tables 8a, 8b, and 8c in Appendix F to verify that for systolic, $p < 0.025$ and for diastolic, $p < 0.01$.

12.19 "Tillage Test Plots Revisited" (Case Study 12-4, p. 616) presents the yield per plot obtained in an experiment designed to compare six different methods of tilling the ground.
   a. Construct a dot plot showing the six samples separately and side-by-side. Using one-way ANOVA, test the claim that "all tillage tools are not created equal."
   b. State the null and alternative hypotheses and describe the meaning of each.
   c. Complete the hypothesis test using $\alpha = 0.05$.

12.20 Refer to Case Study 12-5, p. 617.
   a. State the null hypothesis being tested.
   b. Verify the $n, \bar{x}$, and s reported for method I.
   c. What was the value of the calculated F?
   d. Approximate the critical value of F if $\alpha = 0.05$.
   e. Why did Mr. Stephens say, "The F-value is very nonsignificant"?

IN RETROSPECT

In this chapter we have presented an introduction to the statistical techniques known as analysis of variance. The techniques studied here were restricted to the test of a hypothesis that dealt with questions about means from several populations. We were restricted to normal populations and populations with homogeneous (equal) variances. The test of multiple means is accomplished by partitioning the sum of squares into two segments: (1) the sum of squares due to variation between the levels of the factor being tested, and (2) the sum of squares due to the variation between the replicates within each level. The null hypothesis about means is then tested by using the appropriate variance measurements.

Note that we restricted our development to one-factor experiments. This one-factor technique represents only a beginning to the study of analysis of variance techniques.

Refer back to Case Study 12-5, p. 617, and you will see that the data given use the method of tillage (plow, and so on) as the factor and the yield from each

Chapter 12 ▼ ANALYSIS OF VARIANCE

field plot as the replicate of a one-factor analysis of variance experiment. (See Exercise 12.19.)

CHAPTER EXERCISES

12.21 Samples of peanut butter produced by three different manufacturers were tested for salt content with the following results:

Brand 1:	2.5	8.3	3.1	4.7	7.5	6.3
Brand 2:	4.5	3.8	5.6	7.2	3.2	2.7
Brand 3:	5.3	3.5	2.4	6.8	4.2	3.0

Is there a significant difference in the mean amount of salt in these samples? Use $\alpha = 0.05$.
   a. State the null and alternative hypotheses.
   b. Determine the test criteria: level of significance, test statistic, critical region, critical value(s).
   c. Using the information on the computer printout below, state the decision and conclusion to the hypothesis test.
   d. What does the p-value tell you? Explain.

Each level of data is entered into a separate column.

```
MTB > SET C1
2.5 8.3 3.1 4.7 7.5 6.3
MTB > SET C2
4.5 3.8 5.6 7.2 3.2 2.7
MTB > SET C3
5.3 3.5 2.4 6.8 4.2 3.0
MTB > AOVONEWAY C1 C2 C3

ANALYSIS OF VARIANCE
SOURCE    DF      SS      MS      F       P
FACTOR     2    4.68    2.34    0.64   0.541
ERROR     15   54.88    3.66
TOTAL     17   59.56
                                INDIVIDUAL 95 PCT CI'S FOR MEAN
                                BASED ON POOLED STDEV
LEVEL    N     MEAN    STDEV  ----+---------+---------+---------+--
C1       6    5.400    2.359                 (----------*----------)
C2       6    4.500    1.669      (----------*----------)
C3       6    4.200    1.621   (----------*----------)
                               ----+---------+---------+---------+--
POOLED STDEV =  1.913            3.0       4.5       6.0       7.5
```

12.22 A new all-purpose cleaner is being test marketed by placing sales displays in three different locations within various supermarkets. The number of bottles sold from each location within each of the supermarkets tested is reported below:

Locations		
I	II	III
40	32	45
35	38	48
44	30	50
38	35	52

a. State the null and alternative hypotheses for testing "the location of the sales display had no effect on the number of bottles sold."
b. Using $\alpha = 0.01$, determine the test criteria: level of significance, test statistic, critical region, critical value(s).
c. Using the information on the computer printout below, state the decision and conclusion to the hypothesis test.
d. What does the p value tell you? Explain.

Data entered into C1, location identifier entered into C2.

```
MTB > READ C1 C2
40 1
32 2
45 3
35 1
38 2
48 3
44 1
30 2
50 3
38 1
35 2
52 3
MTB > ONEWAY C1 C2

ANALYSIS OF VARIANCE ON C1
SOURCE    DF       SS        MS       F        P
C2         2     460.7     230.3    19.51    0.001
ERROR      9     106.2      11.8
TOTAL     11     566.9
                                 INDIVIDUAL 95 PCT CI'S FOR MEAN
                                 BASED ON POOLED STDEV
LEVEL      N      MEAN    STDEV   --------+---------+---------+--------
  1        4    39.250    3.775                 (----*-----)
  2        4    33.750    3.500      (----*----)
  3        4    48.750    2.986                              (-----*----)
                                 --------+---------+---------+--------
POOLED STDEV = 3.436                   35.0      42.0      49.0
```

12.23 For the following data, find SS(error) and show that
$$SS(error) = [(k_1 - 1)s_1^2 + (k_2 - 1)s_2^2 + (k_3 - 1)s_3^2]$$
where s_i is the variance for the ith factor level.

Factor Level		
1	2	3
8	6	10
4	6	12
2	4	14

12.24 For the following data, show that
$$SS(factor) = k_1(\bar{x}_1 - \bar{x})^2 + k_2(\bar{x}_2 - \bar{x})^2 + k_3(\bar{x}_3 - \bar{x})^2$$
where $\bar{x}_1, \bar{x}_2, \bar{x}_3$ are the means for the three factor levels and $\bar{x}$ is the overall mean.

Factor Level		
1	2	3
6	13	9
8	12	11
10	14	7

12.25 An article in the *Journal of Pharmaceutical Sciences* (December 1987, pages 880–885) discusses the change of plasma protein binding of Diazepam at various concentrations of Imipramine. Suppose the results were reported as follows:

Diazepam Alone (1.25 mg/ml)	Diazepam with Imipramine		
	1.25	2.50	5.00
97.99	97.68	96.29	93.92

The values given represent mean plasma protein binding and $n = 8$ for each of the four groups. Find the sum of squares among the four groups.

12.26 A study reported in the *Journal of Research and Development in Education* (Summer 1989, pages 1–9) evaluates the effectiveness of social skills training and cross-age tutoring for improving academic skills and social communication behaviors among boys with learning disabilities. Twenty boys were divided into three groups and their scores on the *Test of Written Spelling* (TWS) may be summarized as follows:

Group	n	TWS Mean	St. Dev.
Social skills training and tutoring components	7	21.43	9.48
Social skills training only	7	20.00	8.91
Neither component	6	20.83	9.06

Calculate the entries of the ANOVA table that follows from these results.

12.27 An experiment was designed to compare the lengths of time that four different drugs provided pain relief following surgery. The results (in hours) are shown in the following table:

	Drug		
A	B	C	D
8	6	8	4
6	6	10	4
4	4	10	2
2	4	10	
		12	

Is there enough evidence to reject the null hypothesis that there is no significant difference in the length of pain relief for the four drugs at $\alpha = 0.05$?

12.28 The distance required to stop a vehicle on wet pavement was measured to compare the stopping power of four major brands of tires. A tire of each brand was tested on the same vehicle on a controlled wet pavement. The resulting distances are shown in the following table. At $\alpha = 0.05$, is there sufficient evidence to conclude that there is a difference in the mean stopping distance?

	Brand of Tire			
	A	B	C	D
Distance (replicate)	37	37	33	41
	34	40	34	41
	38	37	38	40
	36	42	35	39
	40	38	42	41
	32		34	43

12.29 A certain vending company's soft-drink dispensing machines are supposed to serve six ounces of beverage. Various machines were sampled and the resulting amounts of dispensed drink were recorded, as shown in the following table. Does this sample evidence provide sufficient reason to reject the null hypothesis that all five machines dispense the same average amount of soft drink? Use $\alpha = 0.01$.

	\multicolumn{5}{c}{Machines}				
	A	B	C	D	E
Amounts of Soft Drink Dispensed	3.8	6.8	4.4	6.5	6.2
	4.2	7.1	4.1	6.4	4.5
	4.1	6.7	3.9	6.2	5.3
	4.4		4.5		5.8

12.30 Suburbs, each with its own attributes, are located around every metropolitan area. There is always the "rich" one (the most expensive one), the least expensive one, and so on. Does the suburb affect the transfer value of its homes? x = transfer value, the amount on which county transfer taxes are paid.

Suburb A	Suburb B	Suburb C	Suburb D	Suburb E
105	101	95	74	79
114	88	107	135	89
85	105	101	165	140
177	100	92	114	114
104	161	91	80	80
135	113	89	115	86
	94			94
				102

a. Do the sample data show sufficient evidence to conclude that the suburbs represented do have a significant effect on the transfer value of their homes? Use $\alpha = 0.01$.
b. Construct a graph that demonstrates the conclusion reached in (a).

12.31 An experiment compared typing speeds for clerk-typists on a standard electric typewriter and on a Teletype Model 43 computer terminal. Twelve typists were randomly assigned to type on the two machines. The scores are shown in the following table:

Standard Electric	Terminal
62	52
78	60
48	47
63	48
55	52
	40
	51

Is there sufficient evidence to conclude that there is a difference in the population means for the two types of machines? Use $\alpha = 0.05$.

 12.32 To compare the effectiveness of three different methods of teaching reading, 26 children of equal reading aptitude were divided into three groups. Each group was instructed for a given period of time using one of the three methods. After completing the instruction period, all students were tested. The test results are shown in the following table. Is the evidence sufficient to reject the hypothesis that all three instruction methods are equally effective? Use $\alpha = 0.05$.

	Method I	Method II	Method III
	45	45	44
	51	44	50
	48	46	45
	50	44	55
Test Scores (replicates)	46	41	51
	48	43	51
	45	46	45
	48	49	47
	47	44	

 12.33 The following table shows the number of arrests made last year for violations of the narcotic drug laws in 24 communities. The data given are rates of arrest per 10,000 inhabitants. At $\alpha = 0.05$, is there sufficient evidence to reject the hypothesis that the mean rates of arrests are the same in all four sizes of communities?

Cities (over 250,000)	Cities (under 250,000)	Suburban Communities	Rural Communities
45	23	25	8
34	18	17	16
41	27	19	14
42	21	28	17
37	26	31	10
28	34	37	23

12.34 Seven golf balls from each of six manufacturers were randomly selected and tested for durability. Each ball was hit 300 times or until failure occurred, whichever came first. Do these sample data show sufficient reason to reject the null hypothesis that the six different brands tested withstood the durability test equally well? Use $\alpha = 0.05$.

	Manufacturer Brand					
	A	B	C	D	E	F
Number of Hits (replicate)	300	190	228	276	162	264
	300	164	300	296	175	168
	300	238	268	62	157	254
	260	200	280	300	262	216
	300	221	300	230	200	257
	261	132	300	175	256	183
	300	156	300	211	92	93

VOCABULARY LIST

Be able to define each term. In addition, describe in your own words and give an example of each term. Your examples should not be ones given in class or in the textbook.

The bracketed numbers indicate the chapters in which the term previously appeared, but you should define the terms again to show increased understanding of their meaning.

analysis of variance (ANOVA)
between-sample variation
degrees of freedom [9, 10, 11]
experimental error
levels of the tested factor
mathematical model
mean square, MS(factor), MS(error)
partitioning
randomize [2]

replicate
response variable [1]
sum of squares
test statistic, F^*
total sum of squares, SS(total)
variance [2, 9, 10]
variation between levels, MS(factor)
variation within a level, MS(error)
within-sample variation

KEY CONCEPTS

analysis of variance
experimental error
hypothesis test [8, 9, 10, 11]
mean square
replicate
response variable [1]
variation between levels
variation within a level

QUIZ A

Answer "True" if the statement is always true. If the statement is not always true, replace the words shown in bold with words that make the statement always true.

12.1 To partition the sum of squares for the total is to separate the numerical value of SS(total) into two values such that the **sum** of these two values is equal to SS(total).

12.2 A **sum of squares** is actually a measure of variance.

12.3 **Experimental error** is the name given to the variability that takes place between the levels of the test factor.

12.4 **Experimental error** is the name given to the variability that takes place among the replicates of an experiment as it is repeated under constant conditions.

12.5 **Fail to reject H_0** is the desired decision when the means for the levels of the factor being tested are all different.

12.6 The **mathematical model** for a particular problem is an equational statement showing the anticipated makeup of an individual piece of data.

12.7 The degrees of freedom for the factor are equal to **the number** of factors tested.

12.8 The measure of a specific level of a factor being tested in an ANOVA is the **variance** of that factor level.

12.9 We **need not** assume that the observations are independent to do analysis of variance.

12.10 The rejection of H_0 **indicates** that you have identified the level(s) of the factor that is (are) different from the others.

QUIZ B

12.1 Determine the truth (T/F) for each statement with regard to the one-factor analysis of variance technique.
 ____ a. The mean squares are measures of variance.
 ____ b. "There is no difference between the mean values of the random variable at the various levels of the test factor" is a possible interpretation of the null hypothesis.
 ____ c. "The factor being tested has no effect on the random variable x" is a possible interpretation of the alternative hypothesis.
 ____ d. "There is no variance among the mean values of x for each of the different factor levels" is a possible interpretation of the null hypothesis.
 ____ e. The "partitioning" of the variance occurs when the SS(total) is separated into SS(factor) and SS(error).
 ____ f. We will want to reject the null hypothesis and conclude that the factor has an effect on the variable when the amount of variance assigned to factor is significantly larger than the variance assigned to error.
 ____ g. In order to apply the F-test, the sample size from each factor level must be the same.
 ____ h. In order to apply the F-test, the sample standard deviation from each factor level must be the same.

_____ **i.** If 20 is subtracted from every data value, then the calculated value of the F statistic is also reduced by 20.

When the calculated value of F, F^*, is greater than the table value for F,
_____ **j.** the decision will be: "Fail to reject H_0."
_____ **k.** the conclusion will be: "The factor being tested does have an effect on the variable."

Independent samples were collected in order to test the effect a factor had on a variable. The data are summarized in the following ANOVA table:

	SS	df
Factor	810	2
Error	720	8
Total	1530	10

Is there sufficient evidence to reject the null hypothesis that all levels of the test factor have the same effect on the variable?
_____ **l.** The null hypothesis could be: $\mu_A = \mu_B = \mu_C = \mu_D$.
_____ **m.** The calculated value of F is 1.125.
_____ **n.** The critical value of F for $\alpha = 0.05$ is 6.06.
_____ **o.** The null hypothesis can be rejected at $\alpha = 0.05$.

12.2 Determine the values A, B, C, D, and E missing in the table.

	SS	df	MS	F*
Factor	A	4	18	E
Error	B	18	D	
Total	144	C		

Find the values of
    a. A **b.** B **c.** C **d.** D **e.** E

QUIZ C

12.1 In 50 words or less, explain what a single-factor ANOVA experiment is.

12.2 A state environmental agency tested three different scrubbers used to reduce the resulting air pollution in the generation of electricity. The primary concern was the emission of particulate matter. Several trials were run with each scrubber. The amount of particulate emission was recorded for each trial.

		Replicates Amounts of Emission					
Scrubbers Tested	I	11	10	12	9	13	12
	II	12	10	12	8	9	
	III	9	11	10	7	8	

a. State the mathematical model for this experiment.
b. State the null and alternative hypotheses.
c. Calculate and form the ANOVA table.
d. Complete the testing of H_0 using 0.05 level of significance. State the decision and conclusion clearly.
e. Construct a graph representing the data that is helpful in picturing the results of the hypothesis test.

13 LINEAR CORRELATION AND REGRESSION ANALYSIS

Chapter Outline

13.1 Linear Correlation Analysis
Analysis of linear dependency uses two measures: **covariance** and the **coefficient of linear correlation**.

13.2 Inferences About the Linear Correlation Coefficient
We **question and interpret** the correlation coefficient once we have obtained it.

13.3 Linear Regression Analysis
To analyze two related variables, we use a **line of best fit**.

13.4 Inferences Concerning the Slope of the Regression Line
We judge the **usefulness of the equation of best fit** before predicting the value of one variable given the value of another variable.

13.5 Confidence Interval Estimates for Regression
If the line of best fit is usable, we can establish confidence interval estimates.

13.6 Understanding the Relationship Between Correlation and Regression
Linear correlation analysis tells us whether or not we are dealing with a **linear relationship**; regression analysis tells us **what the relationship is**.

Developmental Factors Associated with Self-Perceptions of Mentoring Competence and Mentoring Needs

This research examined the psychosocial development of undergraduate students in relationship to their perceived competency, readiness, focus, and choices regarding participation in mentoring relationships.

The specific research questions were the following:

1. Is there a significant relationship between students' psychosocial developmental levels as indicated by scores on measures of Autonomy, Interpersonal Relationships, and Purpose and their self-perceptions of their competence in serving as mentors?

2. Is there a significant relationship between students' psychosocial developmental levels as indicated by scores on measures of Autonomy, Interpersonal Relationships, and Purpose and their self-perceived readiness to participate in mentoring relationships as mentees?

RESULTS

The means and standard deviations for the sample on the Autonomy, Interpersonal relationships, purpose, mentor competence, and mentee readiness scores are reported in Table 1. No significant differences were found between these sample means and the mean scores of the normative sample (Hood, 1986).

Relationship Between Developmental Status and Perceived Competence as a Mentor

Pearson r correlation coefficients between the developmental measures scores and the perceived competence as a mentor scores were statistically significant ($p < .001$). Autonomy ($r = -.45$), Interpersonal Relationships ($r = -.49$), and Purpose ($r = -.41$) correlated significantly with perceived competence.

Relationships Between Developmental Levels and Perceived Readiness To Be a Mentee

Pearson r correlation coefficients between the developmental measures and self-perceptions of readiness to be a mentee were also statistically significant, $p < .001$. The coefficients were smaller, but the pattern was similar to that of the relationships with the mentor competence scores. Autonomy ($r = -.36$), Purpose ($r = -.33$), and Interpersonal Relationships ($r = -.31$) correlated significantly with mentee readiness scores.

TABLE 1 Means and Standard Deviations for Perceived Developmental Levels and Mentor Competence and Mentee Readiness

Variable	N	M	SD
Autonomy	142	157.78	14.30
Interpersonal Relationships	142	124.68	14.64
Purpose	142	104.89	12.90
Mentor	139	26.07	5.93
Mentee	141	19.53	5.39

Source: Mary Beth Rice and Robert D. Brown. Reprinted from the *Journal of College Student Development*, Vol. 31, July 1990, p. 295. © AACD. Reprinted with permission. No further reproduction authorized without written permission of American Association for Counseling and Development.

Chapter 13 ▼ LINEAR CORRELATION AND REGRESSION ANALYSIS

Chapter Objectives

In Chapter 3 the basic ideas of regression and linear correlation analysis were introduced. (If these concepts are not fresh in your mind, review Chapter 3 before beginning this chapter.) Chapter 3 was only a first look—a presentation of the basic graphic and descriptive statistical aspects of linear correlation and regression analysis. In this chapter we will take a second, more detailed look at linear correlation and regression analysis.

Previously we used the linear correlation coefficient to measure the strength of the linear relationship between two variables. Now we will determine whether there is a linear relationship by using a hypothesis test in which the probability of a type I error is fixed by the value assigned to α. In Chapter 3 we introduced a set of formulas for finding the equation of the straight line of best fit. Now we wish to ask, "Is the linear equation of any real use?" Previously we used the equation of the line of best fit to make point predictions. Now we will make confidence interval estimations. In short, this second look at linear correlation and regression analysis will be a much more complete presentation than that in Chapter 3.

bivariate data

Recall that **bivariate data** are ordered pairs of numerical values. The two values are each response variables. They are paired with each other as a result of a common bond (see p. 145).

13.1 ▼ Linear Correlation Analysis

In Chapter 3 the linear correlation coefficient was presented as a quantity that measures the strength of a linear relationship (dependency). Now let's take a second look at this concept and see how r, the coefficient of linear correlation, works. Intuitively, we want to think about how to measure the mathematical linear dependency of one variable on another. As x increases, does y tend to increase (decrease)? How strong (consistent) is this tendency? We are going to use two measures of dependence—covariance and the coefficient of linear correlation—to measure the relationship between two variables. We'll begin our discussion by examining a set of bivariate data and identifying some related facts as we prepare to define covariance.

▼ ILLUSTRATION 13-1

Let's consider the following sample of six pieces of bivariate data: $(2, 1), (3, 5), (6, 3), (8, 2), (11, 6), (12, 1)$. (See Figure 13-1.) The mean of the six x values $(2, 3, 6, 8, 11, 12)$ is $\bar{x} = 7$. The mean of the six y values $(1, 5, 3, 2, 6, 1)$ is $\bar{y} = 3$.

The point $(\bar{x}, \bar{y})$, which is $(7, 3)$, is located as shown on the graph of the sample points in Figure 13-2. The point $(\bar{x}, \bar{y})$ is called the **centroid** of the data. If a vertical and a horizontal line are drawn through the centroid, the graph is di-

centroid

Section 13.1 ▼ LINEAR CORRELATION ANALYSIS 639

FIGURE 13-1
Graph of Data for
Illustration 13-1

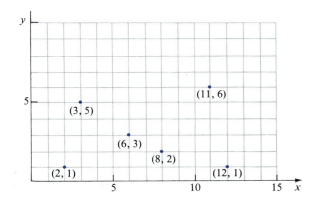

FIGURE 13-2 The Point (7, 3) Is the Centroid

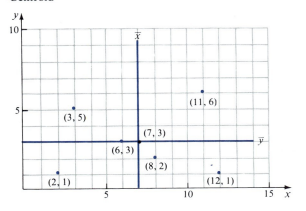

FIGURE 13-3 Measuring the Distance of Each Data Point from the Centroid

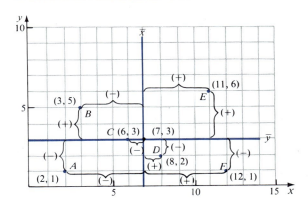

vided into four sections, as shown in Figure 13-2. Each point (x, y) lies a certain distance from each of these two lines. $(x - \bar{x})$ is the horizontal distance from (x, y) to the vertical line passing through the centroid. $(y - \bar{y})$ is the vertical distance from (x, y) to the horizontal line passing through the centroid. Both the horizontal and vertical distances of each data point from the centroid can be measured, as shown in Figure 13-3. The distances may be positive, negative, or zero, depending on the position of the point (x, y) in reference to $(\bar{x}, \bar{y})$. $[(x - \bar{x})$ and $(y - \bar{y})$ are represented by means of braces, with positive or negative signs, as shown in Figure 13-3.] ▲▲

covariance

One measure of linear dependency is the covariance. The **covariance of x and**

y is defined as the sum of the products of the distances of all values of x and y from the centroid, $\Sigma[(x - \bar{x})(y - \bar{y})]$, divided by $n - 1$:

$$\text{covar}(x, y) = \frac{\sum_{i=1}^{n}(x_i - \bar{x})(y_i - \bar{y})}{n - 1} \qquad (13\text{-}1)$$

The covariance for the data given in Illustration 13-1 is calculated in Table 13-1. The covariance, written as covar(x, y), of the data is $+\frac{3}{5} = 0.6$.

TABLE 13-1 Calculations for Finding covar(x, y) for the Data of Illustration 13-1

Points	$x - \bar{x}$	$y - \bar{y}$	$(x - \bar{x})(y - \bar{y})$
$A(2, 1)$	-5	-2	10
$B(3, 5)$	-4	2	-8
$C(6, 3)$	-1	0	0
$D(8, 2)$	1	-1	-1
$E(11, 6)$	4	3	12
$F(12, 1)$	5	-2	-10
Total	0	0	3

NOTE $\Sigma(x - \bar{x}) = 0$ and $\Sigma(y - \bar{y}) = 0$. This will always happen. Why? (See p. 81.)

The covariance is positive if the graph is dominated by points to the upper right and to the lower left of the centroid. The products of $(x - \bar{x})$ and $(y - \bar{y})$ are positive in these two sections. If the majority of the points are in the upper-left and lower-right sections relative to the centroid, the sum of the products is negative. Figure 13-4 shows data that represent a positive dependency (a), a negative dependency (b), and little or no dependency (c). The covariances for these three situations would definitely be positive in part (a), negative in (b), and near

FIGURE 13-4 Data and Covariance

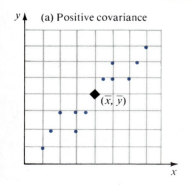
(a) Positive covariance

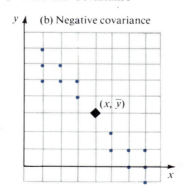
(b) Negative covariance

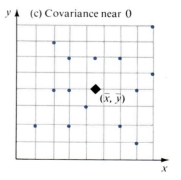
(c) Covariance near 0

Section 13.1 ▼ LINEAR CORRELATION ANALYSIS

zero in (c). (The sign of the covariance is always the same as the sign of the slope of the regression line.)

The biggest disadvantage of covariance as a measure of linear dependency is that it does not have a standardized unit of measure. One reason for this is that the spread of the data is a strong factor in the size of the covariance. For example, if we were to multiply each data point in Illustration 13-1 by 10, we would have (20, 10), (30, 50), (60, 30), (80, 20), (110, 60), and (120, 10). The relationship of the points to each other would be changed only in that they would be much more spread out. However, the covariance for this new set of data is 60. (This calculation is assigned in Exercise 13.9, p. 645.) Does this mean that the amount of dependency between the x and y variables is stronger than in the original case? No, it does not; the relationship is the same even though each data point was multiplied by 10. This is the trouble with covariance as a measure. We must find some way to eliminate the effect of the spread of the data when we measure dependency.

If we standardize x and y by dividing the distance of each from the respective mean by the respective standard deviation,

$$x' = \frac{x - \bar{x}}{s_x} \quad \text{and} \quad y' = \frac{y - \bar{y}}{s_y}$$

and then compute the covariance of x' and y', we will have a covariance that is *not* affected by the spread of the data. This is exactly what is accomplished by the linear correlation coefficient. It divides the covariance of x and y by a measure of the spread of x and by a measure of the spread of y (the standard deviations of x and of y are used as measures of spread). Therefore, by definition, the **coefficient of linear correlation** is

coefficient of linear correlation

$$r = \text{covar}(x', y') = \frac{\text{covar}(x, y)}{s_x \cdot s_y} \quad (13\text{-}2)$$

The coefficient of linear correlation standardizes the measure of dependency and allows us to compare the relative strengths of dependency of different sets of data. [Formula (13-2) for linear correlation is also commonly referred to as **Pearson's product moment, r**.]

Pearson's product moment

The value of r, the coefficient of linear correlation, for the data in Illustration 13-1 can be found by calculating the two standard deviations and then dividing:

$$s_x = 4.099 \quad \text{and} \quad s_y = 2.098$$

$$r = \frac{0.6}{(4.099)(2.098)} = \mathbf{0.07}$$

Finding the correlation coefficient by using formula (13-2) can be a very tedious arithmetic process. The formula can be written in a more workable form, as it was in Chapter 3:

$$r = \frac{\text{covar}(x, y)}{s_x \cdot s_y} = \frac{\sum[(x - \bar{x}) \cdot (y - \bar{y})]/(n - 1)}{s_x \cdot s_y}$$

$$= \frac{SS(xy)}{\sqrt{SS(x) \cdot SS(y)}} \quad (13\text{-}3)$$

▼▲ EXERCISES

13.1 Explain why $\Sigma(x - \bar{x}) = 0$ and $\Sigma(y - \bar{y}) = 0$.

13.2 Consider a set of paired bivariate data. Describe the relationship of the ordered pairs that will cause $\Sigma[(x - \bar{x}) \cdot (y - \bar{y})]$ to be
    a. positive   **b.** negative   **c.** near zero

13.3 **a.** Construct a scatter diagram of the following bivariate data:

	Point									
	A	B	C	D	E	F	G	H	I	J
x	1	1	3	3	5	5	7	7	9	9
y	1	2	2	3	3	4	4	5	5	6

    b. Calculate the covariance.
    c. Calculate s_x and s_y.
    d. Calculate r using formula (13-2).
    e. Calculate r using formula (13-3).

13.4 Consider the accompanying bivariate data:

	Point									
	A	B	C	D	E	F	G	H	I	J
x	0	1	1	2	3	4	5	6	6	7
y	6	6	7	4	5	2	3	0	1	1

    a. Draw a scatter diagram for the data.
    b. Calculate the covariance.
    c. Calculate s_x and s_y.
    d. Calculate r by formula (13-2).
    e. Calculate r by formula (13-3).

13.5 MINITAB was used to form the extensions table, calculate the summations $\Sigma x, \Sigma y, \Sigma x^2, \Sigma xy, \Sigma y^2$, and find the SS($x$), SS($y$), SS($xy$) for the following set of bivariate data. Verify the results.

Section 13.1 ▼ LINEAR CORRELATION ANALYSIS

x	45	52	49	60	67	61
y	22	26	21	28	33	32

```
MTB > READ C1 C2
      6 ROWS READ
MTB > NAME C1 'X',  C2 'Y',  C3 'XSQ',  C4 'XY',  C5 'YSQ'
MTB > LET C3 = C1*C1
MTB > LET C4 = C1*C2
MTB > LET C5 = C2*C2
MTB > SUM C1 C6
    SUM    =       334.00
MTB > SUM C2 C7
    SUM    =       162.00
MTB > SUM C3 C8
    SUM    =       18940
MTB > SUM C4 C9
    SUM    =       9214.0
MTB > SUM C5 C10
    SUM    =       4498.0
MTB > PRINT C1-C5

  ROW     X     Y     XSQ      XY     YSQ

    1    45    22    2025     990     484
    2    52    26    2704    1352     676
    3    49    21    2401    1029     441
    4    60    28    3600    1680     784
    5    67    33    4489    2211    1089
    6    61    32    3721    1952    1024

MTB > PRINT C6-C10

  ROW     C6     C7      C8      C9     C10

    1    334    162   18940    9214    4498

MTB > # K1 = SS(X)
MTB > LET K1 = SUM(C3)-((SUM(C1)**2)/COUNT(C1))
MTB > PRINT K1
K1       347.333
MTB > # K2 = SS(Y)
MTB > LET K2 = SUM(C5)-((SUM(C2)**2)/COUNT(C2))
MTB > PRINT K2
K2       124.000

MTB > #K3 = SS(XY)
MTB > LET K3 = SUM(C4)-((SUM(C1)*SUM(C2))/COUNT(C1))
MTB > PRINT K3
K3       196.000
```

13.6 Use a computer to form the extensions table, calculate the summations $\Sigma x, \Sigma y,$ $\Sigma x^2, \Sigma xy, \Sigma y^2,$ and find the $SS(x), SS(y), SS(xy)$ for the following set of bivariate data. (If you use MINITAB, see Exercise 13.5 for the necessary commands.)

x	11.4	9.4	6.5	7.3	7.9	9.0	9.3	10.6
y	8.1	8.2	5.8	6.4	5.9	6.5	7.1	7.8

13.7 The weight, x, and the waist size, y, were determined for 11 women. The data were as follows:

x	110	143	120	127	143	111	137	154	123	104	128
y	22	29	27	26	27	24	28	28	26	25	27

Verify the correlation coefficient in the following MINITAB output:

```
MTB  > READ X IN C1, Y IN C2
DATA > 110 22
DATA > 143 29
DATA > 120 27
DATA > 127 26
DATA > 143 27
DATA > 111 24
DATA > 137 28
DATA > 154 28
DATA > 123 26
DATA > 104 25
DATA > 128 27
DATA > END DATA
      11 ROWS READ
MTB  > CORRELATION C1, C2

Correlation of C1 and C2 = 0.805

MTB  > STOP
```

13.8 An article in *Physical Therapy* (March 1990, pages 150–156) describes two different methods for measuring the differences in leg lengths in a group of ten physical therapy patients. One method (TMM) uses a tape measure and the other utilizes a radiographic technique. The following data are taken from the paper:

Leg-Length Differences (cm)	
TMM	Radiographic
1.30	1.20
−0.10	−0.15
1.50	1.55
0.70	1.15
−2.70	−2.00
2.50	1.15
−0.40	−0.10
−0.70	−0.50
−0.60	−0.05
−1.00	−1.10

a. Calculate the linear correlation coefficient.
b. What conclusions might you draw from your answer in (a)?

13.9 a. Calculate the covariance of the set of data (20, 10), (30, 50), (60, 30), (80, 20), (110, 60), and (120, 10).
b. Calculate the standard deviation of the six x values and the standard deviation of the six y values.
c. Calculate r, the coefficient of linear correlation, for the data in part (a).
d. Compare these results to those found in the text for Illustration 13-1, p. 638.

13.10 A formula that is sometimes given for computing the correlation coefficient is

$$r = \frac{n(\sum xy) - (\sum x)(\sum y)}{\sqrt{n(\sum x^2) - (\sum x)^2} \sqrt{n(\sum y^2) - (\sum y)^2}}$$

Use this expression as well as the formula

$$r = \frac{SS(xy)}{\sqrt{SS(x) \cdot SS(y)}}$$

to compute r for the data in the following table:

x	2	4	3	4	0
y	6	7	5	6	3

13.2 ▼ Inferences About the Linear Correlation Coefficient

After the linear correlation coefficient, r, has been calculated for the sample data, it seems necessary to ask this question: Does the value of r indicate that there is a linear dependency between the two variables in the population from which the sample was drawn? To answer this question we can perform a hypothesis test. The null hypothesis is "the two variables are linearly unrelated" ($\rho = 0$), where **ρ** (the lowercase Greek letter rho) is the **linear correlation coefficient for the population**. The alternative hypothesis may be either one-tailed or two-tailed. Most frequently it is two-tailed. However, when we suspect that there is only a positive or only a negative correlation, we should use a one-tailed test. The alternative hypothesis of a one-tailed test is $\rho > 0$ or $\rho < 0$.

ρ (rho)

The critical region for the test is on the right when a positive correlation is expected and on the left when a negative correlation is expected. The test statistic used to test the null hypothesis is the calculated value of r from the sample. Critical values for r are found in Table 9 of Appendix F at the intersection of the column identified by the appropriate value of α and the row identified by the degrees of freedom. The number of degrees of freedom for the r statistic is 2 less than the sample size, $df = n - 2$.

The rejection of the null hypothesis means that there is evidence of a linear dependency between the two variables in the population. Failure to reject the null hypothesis is interpreted as meaning that linear dependency between the two variables in the population has not been shown.

> **CAUTION**
>
> The sample evidence may say only that the pattern of behavior of the two variables is related in that one can be used effectively to predict the other. *This does not mean that you have established a cause-and-effect relationship.*

Now let's look at such a hypothesis test.

▼ ILLUSTRATION 13-2

For Illustration 13-1, where $n = 6$, we found $r = 0.07$. Is this significantly different from zero at the 0.02 level of significance?

SOLUTION

STEP 1 $H_0: \rho = 0$.
$H_a: \rho \neq 0$.

STEP 2 $\alpha = 0.02$, $df = n - 2 = 6 - 2 = 4$. See the accompanying figure.

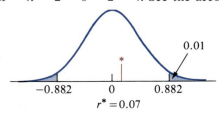

Section 13.2 ▼ INFERENCES ABOUT THE LINEAR CORRELATION COEFFICIENT

The critical values (-0.882 and 0.882) were obtained from Table 9 of Appendix F.

STEP 3 The calculated value of r, r^*, was found earlier. It is $r^* = 0.07$.

STEP 4 **Decision:** Fail to reject H_0.

Conclusion At the 0.02 level of significance, we have failed to show that x and y are correlated.
▲▲

As in other problems, a confidence interval estimate of the population correlation coefficient is sometimes required. It is possible to estimate the value of ρ, the linear correlation coefficient of the population. Usually this is accomplished by using a table showing **confidence belts**. Table 10 in Appendix F gives confidence belts for 95% confidence interval estimates. This table is a bit tricky to read, so be extra careful when you use it. The next illustration demonstrates the procedure for estimating ρ.

▼ ILLUSTRATION 13-3

A sample of 15 ordered pairs of data have a calculated r value of 0.35. Find the 95% confidence interval estimate for ρ, the population linear correlation coefficient.

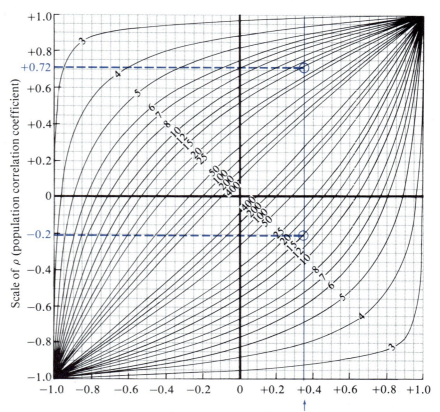

FIGURE 13-5
Using Table 10 of Appendix F, Confidence Belts for the Correlation Coefficient

SOLUTION Find $r = 0.35$ at the bottom of Table 10. (See the arrow on Figure 13-5.) Visualize a vertical line through that point. Find the two points where the belts marked for the correct sample size cross the vertical line. The sample size is 15. These two points are circled in Figure 13-5. Now look horizontally from the two circled points to the vertical scale on the left and read the confidence interval. The values are 0.72 and -0.20. Thus the 95% confidence interval estimate for ρ, the population coefficient of linear correlation, is **-0.20 to 0.72.**

▲▲

Case Study 13-1 — Personality Characteristics of Police Applicants

Pearson's linear correlation coefficient is often used in the study of bivariate data. Bruce N. Carpenter and Susan M. Raza report correlation coefficients for several age-related variables (and their p-values) in their study of personality characteristics of police applicants. The following excerpt also contains the conclusions Carpenter and Raza reached as a result of the correlation analysis. Notice that when the age of male applicants was correlated with the MMPI score, older applicants obtained significantly lower k scores. How is this demonstrated by the information in the article? (See Exercise 13.21, p. 650.)

The Minnesota Multiphasic Personality Inventory (MMPI) continues to be a very widely used measure in the selection of police officers and other law enforcement officials (for example, Parisher, Rios, and Reilley 1979). The characterization of police applicants with the MMPI is therefore important both for better understanding how the measure works as a selection device and for gaining increased understanding of applicant characteristics, the latter of which is examined in this article.

    The age of the male applicants was correlated with their MMPI scale scores to reveal any age-related differences. Older male applicants obtained significantly lower K scores, $r(237) = .11, p < .05$. However, they obtained significantly higher scores on scale 1, $r(237) = .27, p < .001$; scale 2, $r(237) = .22, p < .001$; scale 3, $r(237) = .15, p < .01$; and scale 0, $r(237) = .14, p < .05$ than younger male applicants. This suggests that older applicants tend to be less satisfied, to have more physical complaints, to be more likely to develop physical symptoms in response to stress, and to be more conservative than their younger counterparts. With the exception of developing physical symptoms in response to stress, these personality characteristics have been found to be related to age in a random sample of United States males (Colligan *et al.* 1984).

Source: Bruce N. Carpenter and Susan M. Raza, reproduced from the *Journal of Police Science and Administration,* Vol. 15, No. 1, pp. 10–17, with permission of the International Association of Chiefs of Police, P.O. Box 6010, 13 Firstfield Road, Gaithersburg, Maryland 20878.

Case Study 13-2

DNA Quantitation by Image Cytometry of Touch Preparations from Fresh and Frozen Tissue

The linear relationship between DNA indices from Frozen Touch preps and Fresh Touch preps appears to be very strong. The reported Pearson correlation coefficient is 0.94. Is that significant? Explain. (See Exercise 13.22, p. 650.)

Quantitation of DNA content by flow cytometry of formalin-fixed, paraffin-embedded tissues can provide satisfactory results but has several disadvantages, including sacrifice of tissue blocks and relatively poor resolution of DNA histograms as a result of cell fragments. Satisfactory results with this method may also depend on optimum fixation time, dehydration, temperature, and embedding. DNA analysis by microspectrophotometry of Feulgen-stained tissue sections taken from paraffin blocks has similar limitations in addition to stereologic problems associated with section thickness and operation of a microspectrophotometer.

DNA quantitation from Feulgen-stained touch preparations correlates well with results of flow cytometry and may be more sensitive in detecting some aneuploid populations. The ability to measure DNA from touch preparations of frozen blocks would facilitate retrospective analysis of tumors for which frozen tissue is available and would provide an alternative method of specimen transport for reference laboratories. . . .

A scatter plot comparing the DNA indices obtained from fresh and frozen tissue samples is shown in Figure 2. In 54 of 59 cases (91.5%) there was excellent agreement between the two methods (Spearman $r = 0.91$; Pearson $r = 0.94$).

FIGURE 2 Scatter Plot Showing Correlation of DNA Indices of Fresh and Frozen Touch Preparations

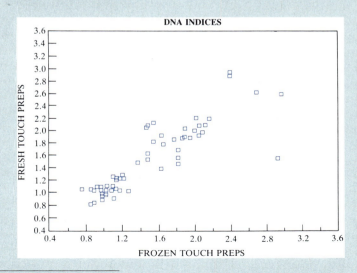

Source: Paula F. Suit, M.D., and Thomas W. Bauer, M.D., Ph.D., *American Journal of Clinical Pathology*, July 1990. Reprinted by permission.

▼▲ EXERCISES

13.11 Using graphs to illustrate, explain the meaning of a correlation coefficient whose value is (a) -1, (b) 0, (c) $+1$.

13.12 Using graphs to illustrate, explain the meaning of a correlation coefficient whose value is (a) $+0.5$, (b) -0.6.

13.13 State the null hypothesis, H_0, and the alternative hypothesis, H_a, that would be used to test the following statements:
    a. The linear correlation coefficient is positive.
    b. There is no linear correlation.
    c. There is evidence of negative correlation.
    d. There is a positive linear relationship.

13.14 Determine the test criteria that would be used in testing each of the following null hypotheses:
    a. $H_0: \rho = 0$ vs. $H_a: \rho \neq 0$, with $n = 18, \alpha = 0.05$.
    b. $H_0: \rho = 0$ vs. $H_a: \rho > 0$, with $n = 32, \alpha = 0.01$.
    c. $H_0: \rho = 0$ vs. $H_a: \rho < 0$, with $n = 16, \alpha = 0.05$.

13.15 A sample of 20 pieces of bivariate data has a linear correlation coefficient of $r = 0.43$. Does this provide sufficient evidence to reject the null hypothesis that $\rho = 0$ in favor of a two-sided alternative? Use $\alpha = 0.10$.

13.16 If a sample of size 18 has a linear correlation coefficient of -0.50, is there significant reason to conclude that the linear correlation coefficient of the population is negative? Use $\alpha = 0.01$.

13.17 A sample of size 10 produced $r = -0.67$. Is this sufficient evidence to conclude that ρ is different from zero at the 0.05 level of significance?

13.18 Is a value of $r = +0.24$ significant in trying to show that ρ is greater than zero for a sample of 62 data at the 0.05 level of significance?

13.19 Use Table 10 of Appendix F to determine a 0.95 confidence interval estimate for the true population linear correlation coefficient based on the following sample statistics:
    a. $n = 8, r = 0.20$ **b.** $n = 100, r = -0.40$
    c. $n = 25, r = +0.65$ **d.** $n = 15, r = -0.23$

13.20 Carpenter and Raza report several correlation coefficients in this article in Case Study 13-1, p. 648. Two of them are $r = -0.11$ and $r = 0.27$. Use Table 10 in Appendix F to determine a 0.95 confidence interval estimate given that $n = 237$.

13.21 In reference to Case Study 13-1 on p. 648,
    a. what does it mean to say "older applicants obtained significantly lower scores"?
    b. how is this demonstrated using correlation?
    c. how is the significance shown?
    d. how can it happen that "they obtained a higher score on another scale"? How is the significance shown?

13.22 In reference to Case Study 13-2 on p. 649, the linear relationship between DNA indices from Frozen Touch preps and Fresh Touch preps appears to be very

Section 13.2 ▼ INFERENCES ABOUT THE LINEAR CORRELATION COEFFICIENT

strong. The reported Pearson correlation coefficient is 0.94. Is that significant? Explain.

 13.23 The population (in millions) and the violent crime rate (per 1000) were recorded for ten metropolitan areas. The data are shown in the following table:

Population	10.0	1.3	2.1	7.0	4.4	0.3	0.3	0.2	0.2	0.4
Crime Rate	12.0	9.5	9.2	8.4	8.2	7.3	7.1	7.0	6.9	6.9

Do these data provide evidence to reject the null hypothesis that $\rho = 0$ in favor of $\rho \neq 0$ at $\alpha = 0.05$?

 13.24 In a study involving 24 coastal drainage basins in the Mendocino triple junction region of northern California (*Geological Society of America Bulletin*, November 1989, pages 1373–1388), it is reported that the Pearson correlation coefficient between uplift rate and the length of the drainage basin equals 0.16942. Use Table 9 in Appendix F to determine whether these data provide evidence sufficient to reject $H_0: \rho = 0$ in favor of $H_a: \rho \neq 0$ at $\alpha = 0.05$.

 13.25 The Test-Retest Method is one way of establishing the reliability of a test. The test is administered and then, at a later date, the same test is readministered to the same individuals. The correlation coefficient is computed between the two sets of scores. The following test scores were obtained in a Test-Retest situation.

First Score	75	87	60	75	98	80	68	84	47	72
Second Score	72	90	52	75	94	78	72	80	53	70

Find r and set a 95 percent confidence interval for ρ.

 13.26 An article in the *Journal for Research in Mathematics Education* (Vol. 19, no. 5, 1988, pages 439–448) determined the regression equation connecting the SDMT score, x, and the summative test score, y, for a group of third-graders. The SDMT score is for the Stanford Diagnostic Mathematics Test. The summative test consists of 30 items dealing with problems concerning place value, greatest and smallest numbers, and the use of greater than, less than, and equal to symbols. Suppose in another study, the data for 10 third-graders were

x	15	20	25	17	20	19	26	25	20	26
y	18	22	25	21	24	18	24	25	21	24

Find the correlation coefficient and find the 0.95 confidence interval estimate for ρ.

13.3 ▼ Linear Regression Analysis

line of best fit

Recall that the **line of best fit** results from an analysis of two or more related variables. (We will restrict our work to two variables. However, on occasion more than two will be mentioned to clarify the analysis.) When two variables are studied jointly, we often would like to control one variable by means of controlling the other. Or we might want to predict the value of a variable based on knowledge about another variable. In both cases we want to find the line of best fit, provided one exists, that will best predict the value of the dependent, or output, variable. Recall that the variable we know or can control is called the *independent*, or *input*, *variable*; the variable resulting from using the equation of the line of best fit is called the *dependent*, or *predicted*, *variable*.

Recall that in Chapter 3 the *method of least squares* was developed. From this concept formulas (3-6) and (3-7) were obtained and are used to calculate b_0 (the y-intercept) and b_1 (the slope of the line of best fit).

$$b_0 = \frac{1}{n}\left(\sum y - b_1 \cdot \sum x\right) \tag{3-6}$$

$$b_1 = \frac{SS(xy)}{SS(x)} \tag{3-7}$$

Then these two coefficients are used to write the equation for the line of best fit in the form

$$\hat{y} = b_0 + b_1 x$$

When the line of best fit is plotted, it does more than just show us a pictorial representation. It tells us two things: (1) whether or not there really is a functional (equational) relationship between the two variables and (2) the quantitative relationship between the two variables. Recall that the line of best fit is of no use when a change in the input variable does not seem to have a definite effect on the output variable. When there is no relationship between the variables, a horizontal line of best fit will result. A horizontal line has a slope of zero, which implies that the value of the input variable has no effect on the output variable. (This idea will be amplified later in this chapter.)

The result of regression analysis is the mathematical equation that is the equation of the line of best fit. We will, as mentioned before, restrict our work to the simple linear case, that is, one input variable and one output variable where the line of best fit is straight. However, you should be aware that not all relationships are of this nature. If the scatter diagram suggests something other than a straight line, we have **curvilinear regression.** In cases of this type we must introduce terms to higher powers, x^2, x^3, and so on; or other functions, e^x, $\log x$, and so on; or we must introduce other input variables. Maybe two or three input variables would improve the usefulness of our regression equation. These possibilities are examples of curvilinear regression and **multiple regression.**

curvilinear regression

multiple regression

The linear model used to explain the behavior of linear bivariate data *in the population* is

$$y = \beta_0 + \beta_1 x + \varepsilon \tag{13-4}$$

intercept (b_0 or β_0)
slope (b_1 or β_1)

This equation represents the linear relationship between the two variables in a population. **β_0 is the y-intercept and β_1 is the slope.** ε (lowercase Greek letter

Section 13.3 ▼ LINEAR REGRESSION ANALYSIS

experimental error (ε or e)

regression line

epsilon) **is the random experimental error** in the observed value of y at a given value of x.

The **regression line** from the sample data gives us b_0, which is **our estimate of β_0**, and b_1, which is **our estimate of β_1**. The *error ε is approximated by $e = y - \hat{y}$*, the difference between the observed value of y and the predicted value of y, $\hat{y}$, at a given value of x.

$$e = y - \hat{y} \tag{13-5}$$

The random variable e is positive when the observed value of y is larger than the predicted value, $\hat{y}$; e is negative when y is smaller than $\hat{y}$. The sum of the errors for the different values of y for a given value of x is exactly zero. (This is part of the least squares criteria.) Thus the mean value of the experimental error is zero; its variance is σ_ε^2. Our next goal is to estimate this variance of the experimental error.

Before we estimate the variance of ε, let's try to understand exactly what the error represents. ε is the amount of error in our observed value of y. That is, it is the difference between the observed value of y and the mean value of y at that particular value of x. Since we do not know the mean value of y, we will use the regression equation and estimate it with $\hat{y}$, the predicted value of y at this same value of x. Thus the best estimate that we have for ε is $e = (y - \hat{y})$, as shown in Figure 13-6.

FIGURE 13-6
The Error e Is $y - \hat{y}$

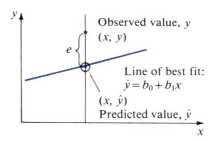

NOTE e is the observed error in measuring y at a specified value of x.

If we were to observe several values of y at a given value of x, we could plot a distribution of y values about the line of best fit (about $\hat{y}$, in particular). Figure 13-7 shows a sample of bivariate values for which the value of x is the

FIGURE 13-7
Sample of y Values at a Given x

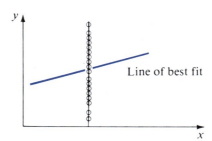

FIGURE 13-8

Theoretical Distribution of y Values for a Given x

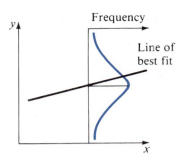

same. Figure 13-8 shows the theoretical distribution of all possible y values at a given x value. A similar distribution occurs at each different value of x. The mean of the observed y's at a given value of x varies, but it can be estimated by $\hat{y}$.

Before we can make any inferences about a regression line, we must assume that the distribution of y's is approximately normal and that the variances of the distributions of y at all values of x are the same. That is, the standard deviation of the distribution of y about $\hat{y}$ is the same for all values of x, as shown in Figure 13-9.

FIGURE 13-9

Standard Deviation of the Distribution of y Values for All x Is the Same

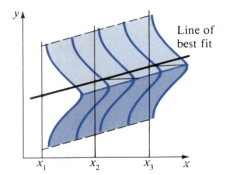

Before looking at the variance of e, let's review the definition of sample variance. The sample variance, s^2, is defined as $\Sigma(x - \bar{x})^2/(n - 1)$, the sum of the squares of each deviation divided by the number of degrees of freedom, $n - 1$, associated with a sample of size n. The variance of y involves an additional complication: There is a different mean for y at each value of x. (Notice the many distributions in Figure 13-9.) However, each of these "means" is actually the predicted value, $\hat{y}$, that corresponds to the x that fixes the distribution. So the **variance of the error e** is estimated by the formula

variance (s^2 or σ^2)

$$s_e^2 = \frac{\Sigma(y - \hat{y})^2}{n - 2} \qquad (13\text{-}6)$$

where $n - 2$ is the number of degrees of freedom.

NOTE The variance of y about the line of best fit is the same as the variance of the error e. Recall that $e = y - \hat{y}$.

Section 13.3 ▼ LINEAR REGRESSION ANALYSIS

Formula (13-6) can be rewritten by substituting $b_0 + b_1x$ for $\hat{y}$. Since $\hat{y} = b_0 + b_1x$, then s_e^2 becomes

$$s_e^2 = \frac{\sum(y - b_0 - b_1x)^2}{n - 2} \tag{13-7}$$

With some algebra and some patience, this formula can be rewritten once again into a more workable form. The form we will use is

$$s_e^2 = \frac{(\sum y^2) - (b_0)(\sum y) - (b_1)(\sum xy)}{n - 2} \tag{13-8}$$

sum of squares for error (SSE)

For ease of discussion let's agree to call the numerator of formulas (13-6), (13-7), and (13-8) the **sum of squares for error (SSE)**.

Now let's see how all this information can be used.

▼ ILLUSTRATION 13-4

Suppose that you move into a new city and take a job. You will, of course, be concerned about the problems you will face commuting to and from work. For example, you would like to know how long (in minutes) it will take you to drive to work each morning. Let's use "one-way distance to work" as a measure of where you live. You live x miles away from work and want to know how long it will take you to commute each day. Your new employer, foreseeing this question, has already collected a random sample of data to be used in answering your question. Fifteen of your co-workers were asked to give their one-way travel time and distance to work. The resulting data are shown in Table 13-2. (For convenience the data have been arranged so that the x values are in numerical order.) Find the line of best fit and the variance of y about the line of best fit, s_e^2.

TABLE 13-2
Data for Illustration 13-4

Co-worker	Miles (x)	Minutes (y)	x^2	xy	y^2
1	3	7	9	21	49
2	5	20	25	100	400
3	7	20	49	140	400
4	8	15	64	120	225
5	10	25	100	250	625
6	11	17	121	187	289
7	12	20	144	240	400
8	12	35	144	420	1,225
9	13	26	169	338	676
10	15	25	225	375	625
11	15	35	225	525	1,225
12	16	32	256	512	1,024
13	18	44	324	792	1,936
14	19	37	361	703	1,369
15	20	45	400	900	2,025
Total	184	403	2,616	5,623	12,493

SOLUTION The extensions and summations needed for this problem are shown in Table 13-2. The line of best fit, using formulas (2-9), (3-4), (3-7), and (3-6), can now be calculated.

Using formula (2-9),

$$SS(x) = 2616 - \frac{(184)^2}{15} = 358.9333$$

Using formula (3-4),

$$SS(xy) = 5623 - \frac{(184)(403)}{15} = 679.5333$$

Using formula (3-7),

$$b_1 = \frac{679.5333}{358.9333} = 1.893202 = 1.89$$

Using formula (3-6),

$$b_0 = \frac{1}{15}[403 - (1.893202)(184)] = 3.643387$$

$$= 3.64$$

Therefore, the equation for the line of best fit is

$$\hat{y} = 3.64 + 1.89x$$

The variance of y about the regression line is calculated by using formula (13-8).

$$s_e^2 = \frac{SSE}{n-2} = \frac{(\sum y^2) - (b_0)(\sum y) - (b_1)(\sum xy)}{n-2}$$

$$= \frac{12{,}493 - (3.643387)(403) - (1.893202)(5{,}623)}{15-2}$$

$$= \frac{379.2402}{13} = 29.1723$$

$$= 29.17$$

NOTE Extra decimal places are often needed for this type of calculation. Notice that b_1 (1.893202) was multiplied by 5,623. If 1.89 had been used instead, that one product would have changed the numerator by approximately 18. That, in turn, would have changed the final answer by almost 1.4, and that is a sizable round-off error.

$s_e^2 = 29.17$ is the variance of the 15 e's. In Figure 13-10 the 15 e's are shown as vertical line segments.   ▲▲

In the sections that follow, the variance of e will be used in much the same way as the variance of x (as calculated in Chapter 2) was used in Chapters 8, 9, and 10 to complete the statistical inferences studied there.

Section 13.3 ▼ LINEAR REGRESSION ANALYSIS

FIGURE 13-10
The 15 Random Errors as Line Segments

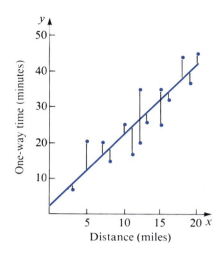

▼▲ EXERCISES

13.27 Ten salespeople were surveyed and the average number of client contacts per month, x, and the sales volume, y (in thousands), were recorded for each:

x	12	14	16	20	23	46	50	48	50	55
y	15	25	30	30	30	80	90	95	110	130

Refer to the following MINITAB output and verify that the equation of the line of best fit is $\hat{y} = -13.4 + 2.3x$ and that $s_e = 10.17$:

```
        READ C1, C2
DATA >  12  15
DATA >  14  25
DATA >  16  30
DATA >  20  30
DATA >  23  30
DATA >  46  80
DATA >  50  90
DATA >  48  95
DATA >  50  110
DATA >  55  130
DATA >  END DATA
        10 ROWS READ
MTB  >  REGRESS Y IN C2 ON 1 PRED IN C1
The regression equation is
C2 = -13.4 + 2.30 C1
Predictor      Coef
Constant      -13.414
C1              2.3028
s = 10.17
```

13.28 A paper entitled "Blue Grama Response to Zinc Source and Rates" by E. M. White (*Journal of Range Management*, January 1991) described five different experiments involving herbage yield and zinc application. Some of the data in experiment number two were

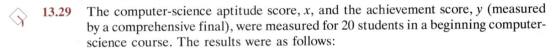

Grams of Zinc per kg Soil	0.0	0.1	0.2	0.4	0.8
Grams of Herbage	3.2	2.8	2.6	2.0	0.1

The paper quotes the correlation coefficient as equal to -0.99 and the line of best fit as $\hat{y} = -3.825x + 3.29$, where x = grams of zinc and y = grams of herbage.
 a. Verify the value for r.
 b. Verify the equation of the line of best fit.

13.29 The computer-science aptitude score, x, and the achievement score, y (measured by a comprehensive final), were measured for 20 students in a beginning computer-science course. The results were as follows:

x	4	16	20	13	22	21	15	20	19	16	18	17	8	6	5	20	18	11	19	14
y	19	19	24	36	27	26	25	28	17	27	21	24	18	18	14	28	21	22	20	21

Find the equation of the line of best fit and s_e^2.

13.30 In a regression problem where x = the number of days on a diet, and y = the number of pounds lost, we obtain the following:

$$n = 4 \quad \sum xy = 132$$

$$\sum x = 20 \quad \sum x^2 = 110$$

$$\sum y = 24 \quad \sum y^2 = 162$$

 a. Find the equation of the line of best fit.
 b. How much is the average weight loss in pounds for each additional day on the diet?

13.31 a. Using the ten points shown in the following table, find the equation of the line of best fit, $\hat{y} = b_0 + b_1 x$, and graph it on a scatter diagram:

	Point									
	A	B	C	D	E	F	G	H	I	J
x	1	1	3	3	5	5	7	7	9	9
y	1	2	2	3	3	4	4	5	5	6

b. Find the ordinates $\hat{y}$ for the points on the line of best fit whose abscissas are $x = 1, 3, 5, 7$, and 9.
c. Find the value of e for each of the points in the given data ($e = y - \hat{y}$).
d. Find the variance s_e^2 of those points about the line of best fit by using formula (13-6).
e. Find the variance s_e^2 by using formula (13-8). [Answers to (d) and (e) should be the same.]

13.32 The following data show the number of hours studied for an exam, x, and the grade received on the exam, y (y is measured in 10's; that is, $y = 8$ means that the grade, rounded to the nearest 10 points, is 80):

x	2	3	3	4	4	5	5	6	6	6	7	7	7	8	8
y	5	5	7	5	7	7	8	6	9	8	7	9	10	8	9

a. Draw a scatter diagram of the data.
b. Find the equation of the line of best fit and graph it on the scatter diagram.
c. Find the ordinates, $\hat{y}$, that correspond to $x = 2, 3, 4, 5, 6, 7$, and 8.
d. Find the five values of e that are associated with the points where $x = 3$ and $x = 6$.
e. Find the variances s_e^2 of all the points about the line of best fit.

13.4 ▼ Inferences Concerning the Slope of the Regression Line

Now that the equation of the line of best fit has been determined and the linear model has been verified (by inspection of the scatter diagram), we are ready to determine whether we can use the equation to predict y. We will test the null hypothesis "the equation of the line of best fit is of no value in predicting y given x." That is, the null hypothesis to be tested in "β_1 (the slope of the relationship in the population) is zero." If $\beta_1 = 0$, then the linear equation will be of no real use in predicting y. To test this hypothesis we will use a t test.

Before we look at the hypothesis test, let's discuss the sampling distribution of the slope. If random samples of size n are repeatedly taken from a bivariate population, the calculated slopes, the b_1's, would form a sampling distribution that is approximately normally distributed with a mean of β_1, the population value of the slope, and with a variance of $\sigma_{b_1}^2$, where

$$\sigma_{b_1}^2 = \frac{\sigma_\varepsilon^2}{\sum(x - \bar{x})^2} \qquad (13\text{-}9)$$

provided there is no lack of fit. An appropriate estimator for $\sigma_{b_1}^2$ is obtained by replacing σ_ε^2 by s_e^2, the estimate of the variance of the error about the regression line:

$$s_{b_1}^2 = \frac{s_e^2}{\sum(x - \bar{x})^2} \qquad (13\text{-}10)$$

This formula may be rewritten in the following, more manageable form:

$$s_{b_1}^2 = \frac{s_e^2}{SS(x)} = \frac{s_e^2}{\sum x^2 - \left[\left(\sum x\right)^2 / n\right]} \qquad (13\text{-}11)$$

NOTE The "standard error of ___" is the standard deviation of the sampling distribution of ___. Therefore, the standard error of regression (slope) is σ_{b_1} and is estimated by s_{b_1}.

We are now ready to *test the hypothesis* $\beta_1 = 0$. Let's use the line of best fit determined in Illustration 13-4: $\hat{y} = 3.64 + 1.89x$. That is, we want to determine whether this equation is of any use in predicting travel time y. In this type of hypothesis test, the null hypothesis is always H_0: $\beta_1 = 0$.

STEP 1 H_0: $\beta_1 = 0$. (This implies that x is of no use in predicting y; that is, that $\hat{y} = \bar{y}$ would be as effective.)

The alternative hypothesis can be either one-tailed or two-tailed. If we suspect that the slope is positive, as in Illustration 13-4 (we would expect travel time, y, to increase as the distance, x, increased), a one-tailed test is appropriate:

$$H_a\text{: } \beta_1 > 0.$$

STEP 2 The test statistic is t. The number of degrees of freedom for this test is $n - 2$, df $= n - 2$. Thus for our example, df $= 15 - 2 = 13$. If we use $\alpha = 0.05$, the critical value of t is $t(13, 0.05) = 1.77$, as found in Table 6 of Appendix F (see the accompanying figure).

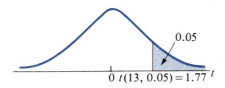

The formula used to calculate the value of the *test statistic t* for inferences about the slope is

$$t = \frac{b_1 - \beta_1}{s_{b_1}} \qquad (13\text{-}12)$$

STEP 3 In our illustration of travel times and distances, the variance among the b_1's is estimated by use of formula (13-11):

$$s_{b_1}^2 = \frac{29.1723}{358.9333} = 0.081275 = 0.0813$$

Using formula (13-12), the observed value of t becomes

$$t = \frac{b_1 - \beta_1}{s_{b_1}} = \frac{1.89 - 0}{\sqrt{0.0813}} = 6.629$$

$$t^* = 6.63$$

STEP 4 Decision: Reject H_0 (t^* is in the critical region; see the figure that accompanies Step 2).

Conclusion At the 0.05 level of significance, we conclude that the slope of the line of best fit in the population is greater than zero. The evidence indicates that there is a linear relationship and that the one-way distance (x) is useful in predicting the travel time to work (y).

The slope β_1 of the regression line of the population can be estimated by means of a confidence interval. The confidence interval is given by

$$b_1 \pm t(n-2, \alpha/2) \cdot s_{b_1} \qquad (13\text{-}13)$$

The 95% confidence interval for the estimate of the population's slope, β_1, for Illustration 13-4 is

$$1.89 \pm (2.16)(\sqrt{0.0813})$$
$$1.89 \pm 0.6159$$
$$1.89 \pm 0.62$$

1.27 to **2.51**, the 0.95 confidence interval for β_1.

That is, we can say that the slope of the line of best fit of the population from which the sample was drawn is between 1.27 and 2.51 with 95% confidence.

Case Study 13-3

Reexamining the Use of Seriousness Weights in an Index of Crime

Figure 1 is the scatter diagram of a strong linear relationship, $S_t = -3953.85 + 3.13 A_t$. The vertical scale is drawn at $A_t = 12{,}700$, and the line of best fit appears to intersect the vertical scale at approximately 35,750. Verify the coordinates of this point of intersection. Verify the 95% confidence interval for the slope of the line of best fit. (See Exercise 13.41, p. 667.)

ABSTRACT The index of crime has become one of the most important social measurements for political jurisdictions in the United States. To characterize their crime problems, national, state, and local governments rely on a single index of crime, which is invariably constructed from crime statistics reported to the FBI known as Uniform Crime Reports (UCRs). The UCR index, composed of seven general crime categories, is often criticized for failing to account for the relative seriousness of its components. Blumstein (1974) examined whether the national UCR index could be improved by adding crime seriousness weights but found that the weighted index contributed no further information to national crime trends. This study replicated that research using recent Arizona UCRs to address criticisms of Blumstein's study. It also considered the appropriateness of a single index of crime, the UCRs, and how they might best be used. The findings support the conclusions of the original study.

Regression and Correlational Measures Between Indices

Regression of the Arizona UCR index on the average seriousness index produces the linear relationship depicted in figure 1. Also shown is the ninety-five percent confidence interval (3.001, 3.262), which is based upon a standard error of .065 on the estimate of the slope. The regression equation for this relationship is

$$S_t = -3953.85 + 3.13A_t.$$

FIGURE 1 Regression of Average Seriousness and Arizona UCR Indices

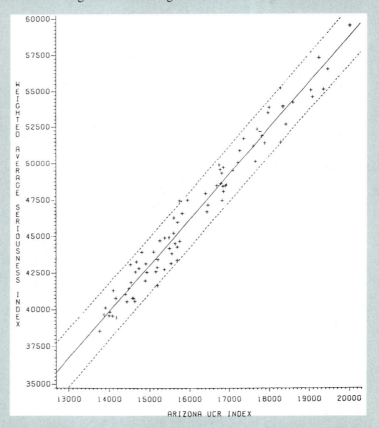

Source: Reprinted with permission from the *Journal of Criminal Justice,* Volume 17, Thomas Epperlein and Barbara C. Nienstedt, "Reexamining the Use of Seriousness Weights in an Index of Crime," 1989, Pergamon Press, Inc.

Case Study 13-4

Charting the Pounds

Height and weight charts, similar to the one below, have been around for many years. They are used in research, by insurance companies, by doctors, and by each of us to determine the "ideal" weight for our height. Draw a scatter diagram showing the information for medium-frame women. (Use height as the input variable.) Do the variables appear to have a linear correlation? One inch in height seems to add how many pounds to the ideal weight? (See Exercise 13.42, p. 666.)

This is the standard chart used in a major new study of women's weight and heart attacks. The study found that women weighing at least 6 percent below these "ideals" had 23 percent less of a risk of heart attacks.

Height	Small Frame	Women's Weight (lbs.) Medium Frame	Large Frame
4'10"	102–111	109–121	118–131
4'11"	103–113	111–123	120–134
5'0"	104–115	113–126	122–137
5'1"	106–118	115–129	125–140
5'2"	108–121	118–132	128–143
5'3"	111–124	121–135	131–147
5'4"	114–127	124–138	134–151
5'5"	117–130	127–141	137–155
5'6"	120–133	130–144	140–159
5'7"	123–136	133–147	143–163
5'8"	126–139	136–150	146–167
5'9"	129–142	139–153	149–170
5'10"	132–145	142–156	152–173
5'11"	135–148	145–159	155–176
6'0"	138–151	148–162	158–179

Source: Metropolitan Life Insurance Co.; figures for adult women ages 25-59, in 3 pounds of clothing, wearing shoes with one-inch heels

By Suzy Parker, USA TODAY

Source: Copyright 1990, USA TODAY. Reprinted with permission.

▼▲ EXERCISES

13.33 State the null hypothesis, H_0, and the alternative hypothesis, H_a, that would be used to test the following statements:
   a. The slope for the line of best fit is positive.
   b. There is no regression.
   c. There is evidence of negative regression.

13.34 Determine the test criteria that would be used in testing each of the following null hypotheses:
   a. $H_0: \beta_1 = 0$ vs. $H_a: \beta_1 \neq 0$, with $n = 18$, $\alpha = 0.05$.
   b. $H_0: \beta_1 = 0$ vs. $H_a: \beta_1 > 0$, with $n = 28$, $\alpha = 0.01$.
   c. $H_0: \beta_1 = 0$ vs. $H_a: \beta_1 < 0$, with $n = 16$, $\alpha = 0.05$.

13.35 Calculate the estimated standard error of regression, s_{b_1}, for the computer-science aptitude score–achievement score relationship in Exercise 13.29.

13.36 Calculate the estimated standard error of regression, s_{b_1}, for the number of hours studied–exam grade relationship in Exercise 13.32.

13.37 The undergraduate grade point average (GPA) and the composite score on the graduate record exam (GRE) were recorded for 15 graduates in a given department at a university. The data are shown in the following table:

CPA (x)	2.30	3.65	3.00	2.75	3.10	2.55	2.50	2.30	2.90	3.15	3.25	2.00	2.75	2.65	3.13
GRE (y)	925	1300	1150	1400	900	825	950	1050	1200	1200	1100	700	850	990	1000

Verify that the value for testing $\beta_1 = 0$ versus $\beta_1 \neq 0$ is $t^* = 2.70$, as shown on the following MINITAB printout. Complete the test using $\alpha = 0.05$.

```
         READ X IN C1, Y IN C2
DATA> 2.30 925
DATA> 3.65 1300
DATA> 3.00 1150
DATA> 2.75 1400
DATA> 3.10 900
DATA> 2.55 825
DATA> 2.50 950
DATA> 2.30 1050
DATA> 2.90 1200
DATA> 3.15 1200
DATA> 3.25 1100
DATA> 2.00 700
DATA> 2.75 850
DATA> 2.65 990
DATA> 3.13 1000
DATA> END DATA
```

```
               15 ROWS READ
MTB > REGRESS Y IN C2 ON 1 PRED IN C1

The regression equation is
C2 = 297 + 264 C1

Predictor      Coef      Stdev     t-ratio        P
Constant      296.9
C1            264.09     97.66      2.70       0.018

s = 158.0
```

 13.38 The relationship between the diameter of a spot weld, x, and the shear strength of the weld, y, is very useful. The diameter of the spot weld can be measured after the weld is completed. The shear strength of the weld can only be measured by applying force to the weld until it breaks. Thus it would be very useful to be able to predict the shear strength based only on the diameter. The following data were obtained from several sample welds:

x, Dia. of Weld (.001 in.)	190	215	200	230	209	250	215	265	215	250
y, Shear Strength (lb)	680	1025	800	1100	780	1030	885	1175	975	1300

Complete these questions with the aid of a computer. (If you use MINITAB, see Exercise 13.37 for the necessary commands.)
  a. Draw a scatter diagram.
  b. Find the equation for the line of best fit.
  c. Is the value of b_1 significantly greater than zero at the 0.05 level?
  d. Find the 0.95 confidence interval estimate for β_1.

 13.39 A sample of 10 students were asked for the distance and the time required to commute to college yesterday. The data collected are shown in the following table:

Distance (x)	1	3	5	5	7	7	8	10	10	12
Time (y)	5	10	15	20	15	25	20	25	35	35

  a. Draw a scatter diagram of these data.
  b. Find the equation that describes the regression line for these data.
  c. Does the value of b_1 show sufficient strength to conclude that β_1 is greater than zero at the $\alpha = 0.05$ level?
  d. Find the 0.98 confidence interval for the estimation of β_1. (Retain these answers for use in Exercise 13.43.)

13.40 Interest rates are alleged to have an effect on the level of employment. The data in the following table show, by quarters, the bank interest rates on short-term loans and the unemployment rate:

Interest Rate	12.27	12.34	12.31	15.81	15.67	17.75	11.56	15.71	19.91	19.99	21.11
Unemployment Rate	5.9	5.6	5.9	5.9	6.2	7.6	7.5	7.3	7.6	7.2	8.3

    a. Calculate the equation of the line of best fit.
    b. Does this sample present sufficient evidence at the 0.05 level of significance to reject the null hypothesis (slope is zero) in favor of the alternative hypothesis that the slope is positive?

13.41 **a.** Figure 1 in Case Study 13-3, p. 661, is a scatter diagram showing a very strong linear relationship. The vertical scale is drawn at $A_t = 12{,}700$, and the line of best fit appears to intersect the vertical scale at approximately 35,750. Verify the coordinates of this point of intersection.
    b. The article also gives an interval estimate of (3.001, 3.262). Verify this 95% interval using the information given in the article.

13.42 Case Study 13-4, p. 663, contains the Metropolitan Life Insurance Company weight chart for women. Draw a scatter diagram showing the intervals for medium-frame women. Use height as the input variable, x.
    a. Do the variables height and weight seem to have a linear relationship? Why is it so strong?
    b. One inch of height adds how many pounds to the ideal weight? How is the number of pounds per inch related to the line of best fit?

13.5 ▼ Confidence Interval Estimates for Regression

Once the equation for the line of best fit has been obtained and determined usable, we are ready to use the equation to make predictions. There are two different quantities that we can estimate: (1) the mean of the population y values at a given value of x, written $\mu_{y|x_0}$, and (2) the individual y value selected at random that will occur at a given value of x, written y_{x_0}. The best **point estimate**, or pre-

predicted value of y ($\hat{y}$)

diction, for both $\mu_{y|x_0}$ and y_{x_0} is $\hat{y}$. This is the y value obtained when an x value is substituted into the equation of the line of best fit. Like other point estimates it is seldom correct. The actual values for both $\mu_{y|x_0}$ and y_{x_0} will vary above and below the calculated value of $\hat{y}$.

Before developing confidence intervals for $\mu_{y|x_0}$ and y_{x_0}, recall the development of confidence intervals for the population mean μ in Chapter 8 when the variance was known, and in Chapter 9 when the variance was estimated. The sample mean, $\bar{x}$, was the best point estimate of μ. We used the fact that $\bar{x}$ is normally distributed with a standard deviation of $\sigma/\sqrt{n}$ to construct formula (8-2) for the confidence interval for μ. When σ had to be estimated, we used formula (9-2) for the confidence interval.

Section 13.5 ▼ CONFIDENCE INTERVAL ESTIMATES FOR REGRESSION

confidence interval estimate

The **confidence intervals** for $\mu_{y|x_0}$ and y_{x_0} are constructed in a similar fashion. $\hat{y}$ replaces $\bar{x}$ as our point estimate. If we were to take random samples from the population, construct the line of best fit for each sample, calculate $\hat{y}$ for a given x using each regression line, and plot the various $\hat{y}$ values (they would vary since each sample would yield a slightly different regression line), we would find that the $\hat{y}$ values form a normal distribution. That is, the sampling distribution of $\hat{y}$ is normal, just as the sampling distribution of $\bar{x}$ is normal. What about the appropriate standard deviation of $\hat{y}$? The standard deviation in both cases ($\mu_{y|x_0}$ and y_{x_0}) is calculated by multiplying the square root of the variance of the error by an appropriate correction factor. Recall that the variance of the error, s_e^2, is calculated by means of formula (13-8).

Before looking at the correction factors for the two cases, let's see why they are necessary. Recall that the line of best fit passes through the point $(\bar{x}, \bar{y})$, the centroid. In Section 13.3 we formed a confidence interval estimate for the slope β_1 (see Illustration 13-4) by using formula (13-13). If we draw lines with slopes equal to the extremes of that confidence interval, 1.27 to 2.51, through the point $(\bar{x}, \bar{y})$ [which is (12.3, 26.9)] on the scatter diagram, we will see that the value for $\hat{y}$ fluctuates considerably for different values of x (Figure 13-11). Therefore, we should suspect a need for a wider confidence interval as we select values of x that are further away from $\bar{x}$. Hence we need a correction factor to adjust for the distance between x_0 and $\bar{x}$. This factor must also adjust for the variation of the y values about $\hat{y}$.

First, let's estimate the *mean value of y* at a given value of x, $\mu_{y|x_0}$. The confidence interval estimate formula is

$$\hat{y} \pm t(n - 2, \alpha/2) \cdot s_e \cdot \sqrt{\frac{1}{n} + \frac{(x_0 - \bar{x})^2}{\sum (x - \bar{x})^2}} \qquad (13\text{-}14)$$

NOTE The numerator of the second term under the radical sign is the square of the distance of x_0 from $\bar{x}$. The denominator is closely related to the variance of x and has a "standardizing" effect on this term.

FIGURE 13-11
Lines Representing the Confidence Interval for Slope

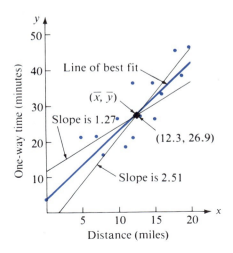

Formula (13-14) can be modified to avoid having $\bar{x}$ in the denominator. The new form is

$$\hat{y} \pm t(n-2, \alpha/2) \cdot s_e \cdot \sqrt{\frac{1}{n} + \frac{(x_0 - \bar{x})^2}{SS(x)}} \qquad (13\text{-}15)$$

Let's compare formula (13-14) with formula (9-2). $\hat{y}$ replaces $\bar{x}$, and

$$s_e \sqrt{\frac{1}{n} + \frac{(x_0 - \bar{x})^2}{\sum (x - \bar{x})^2}} \qquad \text{(the standard error of } \hat{y}\text{)}$$

the estimated standard deviation of $\hat{y}$ in predicting $\mu_{y|x_0}$, replaces $s/\sqrt{n}$, the standard deviation of $\bar{x}$. The degrees of freedom are now $n-2$ instead of $n-1$ as before.

These ideas are explored in the next illustration.

▼ ILLUSTRATION 13-5

Construct a 95% confidence interval estimate for the mean travel time for the co-workers who travel seven miles to work (refer to Illustration 13-4).

SOLUTION

STEP 1 Find $\hat{y}_{x_0}$ when $x_0 = 7$:

$$\hat{y} = 3.64 + 1.89x = 3.64 + (1.89)(7) = \mathbf{16.87}$$

STEP 2 Find s_e:

$$s_e^2 = 29.17 \qquad \text{(found in Illustration 13-4)}$$

$$s_e = \sqrt{29.17} = \mathbf{5.40}$$

STEP 3 Find $t(13, 0.025) = 2.16$ (from Table 6 in Appendix F).

STEP 4 Use formula (13-15):

$$16.87 \pm (2.16) \cdot (5.40) \cdot \sqrt{\frac{1}{15} + \frac{(7 - 12.27)^2}{358.9333}}$$

$$16.87 \pm (2.16)(5.40)\sqrt{0.06667 + 0.07738}$$

$$16.87 \pm (2.16)(5.40)\sqrt{0.14405}$$

$$16.87 \pm (2.16)(5.40)(0.38)$$

$$16.87 \pm 4.43$$

12.44 to **21.30**, 0.95 confidence interval for $\mu_{y|x_0=7}$

This confidence interval estimate is shown in Figure 13-12 by the heavy vertical line. The **confidence interval belt** showing the upper and lower boundaries of all interval estimates at 95% confidence is also shown. Notice that the boundary lines for x values far away from $\bar{x}$ become close to the two lines that represent the equations having slopes equal to the extreme values of the 95% confidence interval estimate for the slope (see Figure 13-11).   ▲▲

Often when making a prediction, we want to predict the value of an individual y. For example, you live seven miles from your place of business and you are interested in an estimate of how long it will take you to get to work. You are somewhat less interested in the average time for all those who live seven miles away. The formula for the interval estimate of the value of a *single randomly selected y* is

$$\hat{y} \pm t(n-2, \alpha/2) \cdot s_e \cdot \sqrt{1 + \frac{1}{n} + \frac{(x_0 - \bar{x})^2}{SS(x)}} \qquad (13\text{-}16)$$

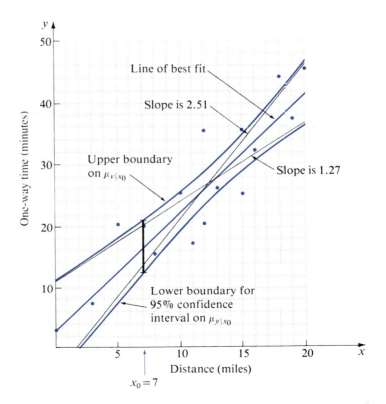

FIGURE 13-12
Confidence Belts for $\mu_{y|x_0}$

▼ **ILLUSTRATION 13-6**

What is the 95% confidence interval prediction for the time it will take you to commute to work if you live seven miles away?

SOLUTION

STEP 1 $\hat{y}$ at $x_0 = 7$ is 16.87, as found in Illustration 13-5.

STEP 2 $s_e = 5.40$, as found in Illustration 13-5.

STEP 3 $t(13, 0.025) = 2.16$.

STEP 4 Use formula (13-16):

$$16.87 \pm (2.16)(5.40)\sqrt{1 + 0.06667 + 0.07738}$$

$$16.87 \pm (2.16)(5.40)(1.0696)$$

$$16.87 \pm 12.48$$

4.39 to **29.35**, 0.95 confidence interval for $y_{x_0} = 7$

The confidence interval is shown in Figure 13-13 as the vertical line segment at $x_0 = 7$. Notice that it is much longer than the confidence interval for $\mu_{y|x_0=7}$. The dashed lines represent the upper and lower boundaries of the confidence intervals for individual y values for all given x values.

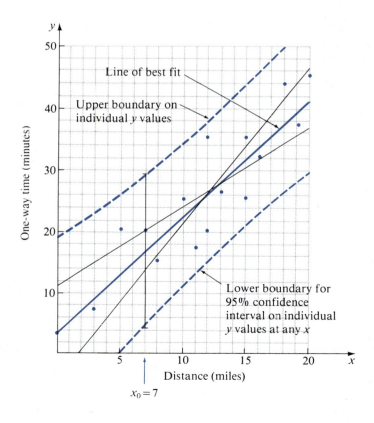

FIGURE 13-13
Confidence Belts for y_{x_0}

FIGURE 13-14

Confidence Belts for Individual y's and for the Mean Value of y

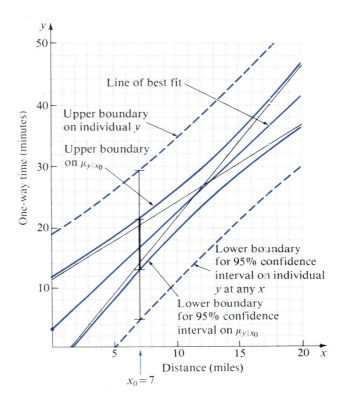

Can you justify the fact that the confidence interval for individual values of y is wider than the confidence interval for the mean values? Think about "individual values" and "mean values" and study Figure 13-14.

The techniques of coding (for large numbers) and the techniques of frequency distributions (for large sets of numbers) are adaptable to the calculations presented in this chapter. Information about these techniques and about models other than linear models may be found in many other textbooks. It is worth noting here that if the data suggest a logarithmic or an exponential functional relationship, a simple change of variable will allow for the use of linear analysis.

There are three basic precautions that you need to be aware of as you work with regression analysis:

1. Remember that the regression equation is meaningful only in the domain of the x variable studied. Prediction outside this domain is extremely dangerous; it requires that we know or assume that the relationship between x and y remains the same outside the domain of the sample data. For example, Joe says that he lives 75 miles from work, and he wants to know how long it will take him to commute. We certainly can use $x = 75$ in all the formulas, but we do not expect the answers to carry the confidence or validity of the values of x between 3 and 20, which were in the sample. The

75 miles may represent a distance to the heart of a nearby major city. Do you think the estimated times, which were based on local distances of 3 to 20 miles, would be good predictors in this situation? Also, at $x = 0$ the equation has no real meaning. However, although projections outside the interval may be somewhat dangerous, they may be the best predictors available.

2. Don't get caught by the common fallacy of applying the regression results inappropriately. For example, this fallacy would include applying the results of Illustration 13-4 to another company. But suppose that the second company had a city location, whereas the first company had a rural location, or vice versa. Do you think the results for a rural location would also be valid for a city location? Basically, the results of one sample should not be used to make inferences about a population other than the one from which the sample was drawn.

3. Don't jump to the conclusion that the results of the regression prove that x causes y to change. (This is perhaps the most common fallacy.) Regressions only measure movement between x and y; they *never prove causation*. A judgment of causation can be made only when it is based on theory or

COMPUTER SOLUTION MINITAB Printout
for Parts of Illustrations 13-4, 13-5, and 13-6

Information Given to Computer

```
MTB > READ DATA, X-VALUES INTO C1, Y-VALUES INTO C2
   COLUMN  C1              C2
   COUNT   15              15
   ROW
   1        3               7
   2        5              20
   3        7              20
   4        8              15
   5       10              25
   6       11              17
   7       12              20
   8       12              35
   9       13              26
  10       15              25
  11       15              35
  12       16              32
  13       18              44
  14       19              37
  15       20              45
MTB > PLOT Y-VALUES IN C2 VS X-VALUES IN C1
```

Scatter Diagram of Given Data

(continued on pp. 673 and 674)

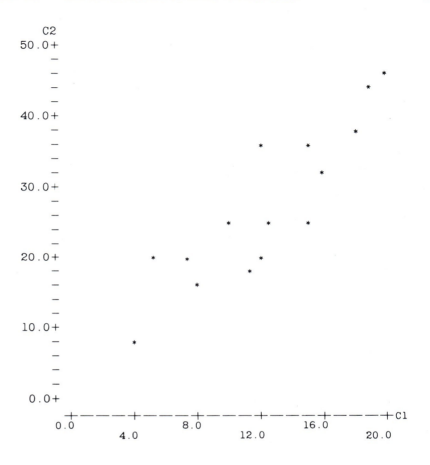

knowledge of the relationship separate from the regression results. The most common difficulty in this regard occurs because of what is called the *missing variable,* or third-variable, *effect.* That is, we observe a relationship between x and y because a third variable, one that is not in the regression, affects both x and y.

▼▲ EXERCISES

  13.43 Use the data and the answers found in Exercise 13.39, p. 666 to make the following estimates:
  a. Give a point estimate for the mean time required to commute four miles.
  b. Give a 0.90 confidence interval estimate for the mean travel time required to commute four miles.

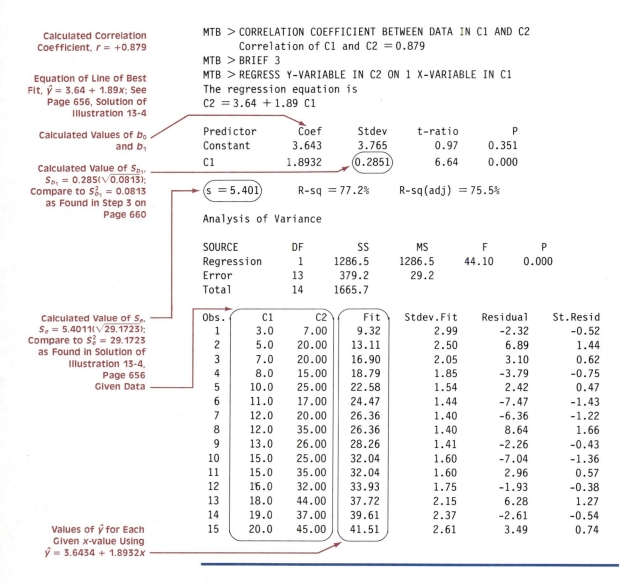

c. Give a 0.90 confidence interval estimate for the travel time required for one person to commute the four miles.

d. Answer (a), (b), and (c) for $x = 9$.

13.44 Use the data and the answers found in Exercise 13.40, p. 666, to make the following estimates:

a. Give a point estimate for the mean unemployment rate given that the interest rate is 14.2.

Section 13.5 ▼ CONFIDENCE INTERVAL ESTIMATES FOR REGRESSION

b. Give a 0.95 confidence interval estimate for the mean unemployment rate when the interest is 14.2.

c. Give a 0.95 confidence interval estimate for the unemployment rate when the interest rate is 14.2.

13.45 An experiment was conducted to study the effect of a new drug in lowering the heart rate in adults. The data collected are shown in the following table:

Drug Dose in mg (x)	0.5	0.75	1.00	1.25	1.50	1.75	2.00	2.25	2.50	2.75
Heart-rate Reduction (y)	10	7	15	12	15	14	20	20	18	21

a. Find the 95% confidence interval estimate for the mean heart-rate reduction for a dose of 2.00 mg.

b. Find the 95% confidence prediction interval for the heart-rate reduction expected for an individual receiving a dose of 2.00 mg.

13.46 The relationship between the "strength" and "fineness" of cotton fibers was the subject of a study that produced the following data:

x, Strength	76	69	71	76	83	72	78	74	80	82	90	81	78	80	81	78
y, Fineness	4.4	4.6	4.6	4.1	4.0	4.1	4.9	4.8	4.2	4.4	3.8	4.1	3.8	4.2	3.8	4.2

a. Draw a scatter diagram.

b. Find the 99% confidence interval estimate for the mean measurement of fineness for fibers with a strength of 80.

c. Find the 99% confidence estimate for an individual measurement of fineness for fibers with a strength of 75.

13.47 A study in *Physical Therapy* (April 1991, pages 294–300) reports on seven different methods to determine crutch length plus two new techniques utilizing linear regression. One of the regression techniques uses the patient's reported height. One hundred and seven individuals were in the study. The mean of the self-reported heights was 68.84 in. The regression equation determined was $y = 0.68x + 4.8$, where $y =$ crutch length and $x =$ self-reported height. The MSE (s_e^2) was reported to be 0.50. In addition, the standard deviation of the self-reported heights was 7.35 in. Use this information to determine a 95% confidence interval estimate for the mean crutch length for individuals who say they are 70 in. tall.

13.48 People not only live longer today but they are living independently longer. The May/June 1989 issue of *Public Health Reports* (pages 222–225) included an article entitled "A Multistate Analysis of Active Life Expectancy." Two of the variables studied were people's ages at which they became dependent and the number of independent years they had remaining. Suppose the data were as follows:

Age When Became Dependent, x	65	66	67	68	70	72	74	76	78	80	83	85
Independent Years Remaining, y	11.1	10.0	10.4	9.3	8.2	6.8	6.8	4.4	5.4	2.5	2.7	0.9

a. Draw a scatter diagram.
b. Calculate the equation for the equation of best fit.
c. Draw the line of best fit on the scatter diagram.
d. For a person who becomes dependent at age 80, how many years of independent living can be expected to remain? Find the answer two different ways: use equation (b) and use the line on the scatter diagram (c).
e. Construct a 99% confidence interval estimating the number of years of independent living remaining for a person who becomes dependent at age 80.
f. Draw a vertical line segment on the scatter diagram representing the interval found in (e).

13.49 Explain why a 95% confidence interval estimate for the mean value of y at a particular x is much narrower than a 95% confidence interval for an individual y value at the same value of x.

13.50 When $x_0 = \bar{x}$, is the formula for the standard error of $\hat{y}_{x_0}$ what you might have expected it to be, $s \cdot \sqrt{1/n}$?

13.6 ▼ Understanding the Relationship Between Correlation and Regression

Now that we have taken a closer look at both correlation and regression analysis, it is necessary to decide when to use them. Do you see any duplication of work?

The primary use of the linear correlation coefficient is in answering the question, "Are these two variables linearly related?" There are other words that may be used to ask this basic question. For example, "Is there a linear correlation between the annual consumption of alcoholic beverages and the salary paid to firemen?"

The linear correlation coefficient can be used to indicate the usefulness of x as a predictor of y in the case where the linear model is appropriate. The test concerning the slope of the regression line (H_0: $\beta_1 = 0$) also tests this same basic concept. Either one of the two is sufficient to determine the answer to this query.

Although the "lack-of-fit" test can be statistically determined, it is beyond the scope of this text. However, we are very likely to carry out this test at a subjective level when we view the scatter diagram. If the scatter diagram is carefully constructed, this subjective decision will determine the mathematical model for the regression line that you believe will fit the data.

The concepts of linear correlation and regression are quite different, because each measures different characteristics. It is possible to have data that yield a strong linear correlation coefficient and have the wrong model. For example, the straight line can be used to approximate almost any curved line if the interval of domain is restricted sufficiently. In such a case the linear correlation coefficient can become quite high but the curve will still not be a straight line. Figure 13-15 suggests one such interval where r could be significant, but the scatter diagram does not suggest a straight line.

Regression analysis should be used to answer questions about the relationship between two variables. Such questions as "What is the relationship?" "How are two variables related?" and so on, require this regression analysis.

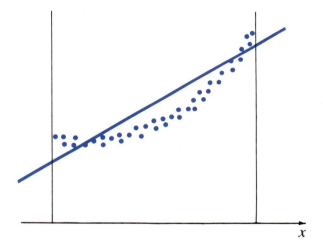

FIGURE 13-15
The Value of r Is High but the Relationship Is Not Linear

IN RETROSPECT

In this chapter we have made a more thorough inspection of the linear relationship between two variables. Although the curvilinear and multiple regression situations were only mentioned in passing, the basic techniques and concepts have been explored. We would only have to modify our mathematical model and our formulas if we wanted to deal with these other relationships.

Although it was not directly emphasized, we have applied many of the topics of earlier chapters in this chapter. The ideas of hypothesis testing and confidence interval estimates were applied to the regression problem. Reference was made to the sampling distribution of the sample slope b_1. This allowed us to make inferences about β_1, the slope of the population from which the sample was drawn. We estimated the mean value of y at a fixed value of x by pooling the variance for the slope with the variance of the *y*'s. This was allowable since they are independent. You should recall that in Chapter 10 we presented formulas for combining the variances of independent samples. The idea here is much the same. Finally, we added a measure of variance for individual values of *y* and made estimates for these individual values of *y* at fixed values of *x*.

Case Study 13-3 presents the results of regression analysis on data collected to compare two crime-reporting indices. (Take another look at Case Study 13-3, p. 661.) The scatter diagram very convincingly shows that the two crime indices being compared are related to each other in a very strong and predictable pattern. Thus, as stated, "the weighted index contributed no further information to national trends." Even though the numerical values of the two indices are quite different, the information coming from them is basically the same. Thus, the introduction of the weighted index seems unnecessary since the Uniform Crime Reports index is a recognized standard.

As this chapter ends you should be aware of the basic concepts of regression analysis and correlation analysis. You should now be able to collect the data for and do a complete analysis on any two-variable linear relationship.

CHAPTER EXERCISES

13.51 Answer the following as "sometimes," "always," or "never." Explain each "never" and "sometimes" response.
   a. The correlation coefficient has the same sign as the slope of the least squares line fitted to the same data.
   b. A correlation coefficient of 0.99 indicates a strong causal relationship between the variables under consideration.
   c. An *r* value greater than zero indicates that ordered pairs with high *x*-values will have low *y*-values.
   d. The two coefficients for the line of best fit have the same sign.
   e. If *x* and *y* are independent, then the population correlation coefficient equals zero.

13.52 Suppose that you calculated the correlation coefficient between two variables and found it to be -1.50. What conclusion would you reach?

13.53 A study in the *Journal of Range Management* (September 1990, pages 407–411) examines the relationships between elements in Russian wild rye. The correlation coefficient between magnesium and calcium was reported to be 0.69 for a sample of size 45. Is there a significant correlation between magnesium and calcium in Russian wild rye (i.e., is $\rho = 0$)?

13.54 A study concerning the plasma concentration of the drug Ranitidine was reported in the *Journal of Pharmaceutical Sciences* (December 1989, pages 990–993). The drug was administered (coded I) and the plasma concentration of Ranitidine was followed for twelve hours. The time to the first peak in concentration was called T_{max1}. The same experiment was repeated one week later (coded II). Twelve subjects participated in the study. The correlation coefficient between T_{max1}, I and T_{max1}, II was reported to be 0.818. Use Table 9 in Appendix F to determine bounds on the p-value for the hypothesis test of $H_0: \rho = 0$ versus $H_a: \rho \neq 0$.

13.55 When two dice are rolled simultaneously, it is expected that the result on each roll will be independent of the result on the other. To test this, roll a pair of dice 12 times. Identify one die as x and the other as y. Record the number observed on each of the two dice.

  a. Draw the scatter diagram of x versus y.
  b. If the two do behave independently, what value of r can be expected?
  c. Calculate r.
  d. Test for independence of these dice at $\alpha = 0.10$.

13.56 The coding scheme used for a variable can determine whether a negative or a positive correlation coefficient is obtained between two variables.

  a. Suppose that high school academic performance (HSAP) is coded as: 1 = excellent, 2 = above average, and 3 = average or below. The score made on a final exam in a college algebra course is also recorded. The resulting data are shown in the following table:

HSAP	1	2	3	1	3	2	1	3	3	2
Final Exam Score	79	80	65	89	70	75	90	65	70	80

Compute r for these data.

  b. Now suppose that HSAP is coded as: 3 = excellent, 2 = above average, and 1 = average or below. The data would now appear as follow.

HSAP	3	2	1	3	1	2	3	1	1	2
Final Exam Score	79	80	65	89	70	75	90	65	70	80

Compute r for these data.

  c. What connection do you see between the two resulting correlation coefficients?

13.57 One would usually expect that the number of hours studied, x, in preparation for a particular examination would have direct (positive) correlation with the grade, y, attained on that exam. The hours studied and the grades obtained by ten students selected at random from a large class are shown in the following table:

	Student									
	1	2	3	4	5	6	7	8	9	10
Hours Studied (x)	10	6	15	11	7	19	17	3	13	17
Exam Grade (y)	51	36	67	63	44	89	80	26	50	85

a. Draw the scatter diagram and estimate r.
b. Calculate r. (How close was your estimate?)
c. Is there significant positive correlation shown by these data? Test at $\alpha = 0.01$.
d. Find the 95% confidence interval estimate for the true population value of ρ.

13.58 When buying most items, it is advantageous to buy in as large a quantity as possible. The unit price is usually less for the larger quantities. The following data were obtained to test this theory:

Number of Units (x)	1	3	5	10	12	15	24
Cost per Unit (y)	55	52	48	36	32	30	25

a. Calculate r.
b. Does the number of units ordered have an effect on the unit cost? Test at $\alpha = 0.05$.
c. Find the 95% confidence interval estimate for the true value of ρ.

13.59 The use of electrical stimulation (ES) to increase muscular strength is discussed in the *Journal of Orthopaedic and Sports Physical Therapy* (September 1990, pages 101–104). Seventeen healthy volunteers were used in the experiment. Muscular strength, Y, was measured as a torque in foot-pounds, and electrical stimulation, X, was measured in mA (micro-amps). The equation for the line of best fit is given as $Y = 1.8X - 28.7$, and the Pearson correlation coefficient as 0.61.
a. Was the correlation coefficient significantly different from zero. Use $\alpha = 0.05$.
b. Predict the torque for a current equal to 50 mA.

13.60 An article in the *New Zealand Geographer* (Vol. 45, no. 1, 1989) discusses the assumptions and limitations of regression analysis. One of the limitations is that regression models may only be used for predicting values within the range of the combinations of the values for the independent variables (see page 29 of the article). Consider, for example, an experiment in which yield of tomatoes (in pounds per plot), y, and the amount of fertilizer (in ounces), x, are recorded. Suppose the data were as follows:

x	1.1	1.5	1.9	2.1	2.5
y	20.1	22.4	22.0	25.0	30.0

a. Determine the regression equation and predict the yield for $x = 4.0$.
b. If the actual yield were 23.5, how might this be explained?

CHAPTER EXERCISES

13.61 An article in *Geology* (September 1989, pages 837–841) gives the following equation relating pressure, P, and total aluminum content, AL, for 12 Horn blende rims: $P = -3.46(+0.24) + 4.23(+0.13)AL$. The quantities shown in parenthesis are standard errors for the y-intercept and slope estimates. Find a 95% confidence interval for the slope, β_1.

13.62 The following data show the gain in reading speed, y, and the number of weeks in a speed-reading program, x, for eight students:

x	3	5	2	8	6	9	3	4
y	80	110	50	190	160	230	70	110

For these data, find
  a. the equation of the line of best fit.
  b. s_e.
  c. the standard error of b_1.

13.63 The following data resulted from an experiment performed for the purpose of regression analysis. The input variable, x, was set at five different levels, and observations were made at each level.

x	0.5	1.0	2.0	3.0	4.0
y	3.8	3.2	2.9	2.4	2.3
	3.5	3.4	2.6	2.5	2.2
	3.8	3.3	2.7	2.7	2.3
		3.6	3.2	2.3	

  a. Draw a scatter diagram of the data.
  b. Draw the regression line by eye.
  c. Place an asterisk, *, at each level approximately where the mean of the observed y values is located. Does your regression line look like the line of best fit for these five mean values?
  d. Calculate the equation of the regression line.
  e. Find the standard deviation of y about the regression line.
  f. Construct a 95% confidence interval estimate for the true value of β_1.
  g. Construct a 95% confidence interval estimate for the mean value of y at $x = 3.0$. at $x = 3.5$.
  h. Construct a 95% confidence interval prediction for an individual value of y at $x = 3.0$. at $x = 3.5$.

Chapter 13 ▼ LINEAR CORRELATION AND REGRESSION ANALYSIS

13.64 The following set of 25 scores was randomly selected from a teacher's class list. Let x be the prefinal average and y the final examination score. (The final examination had a maximum of 75 points.)

Student	x	y	Student	x	y
1	75	64	14	73	62
2	86	65	15	78	66
3	68	57	16	71	62
4	83	59	17	86	71
5	57	63	18	71	55
6	66	61	19	96	72
7	55	48	20	96	75
8	84	67	21	59	49
9	61	59	22	81	71
10	68	56	23	58	58
11	64	52	24	90	67
12	76	63	25	92	75
13	71	66			

a. Draw a scatter diagram for these data.
b. Draw the regression line (by eye) and estimate its equation.
c. Estimate the value of the coefficient of linear correlation.
d. Calculate the equation of the line of best fit.
e. Draw the line of best fit on your graph. How does it compare with your estimate?
f. Calculate the linear correlation coefficient. How does it compare with your estimate?
g. Test the significance of r at $\alpha = 0.10$.
h. Find the 95% confidence interval estimate for the true value of ρ.
i. Find the standard deviation of the y values about the regression line.
j. Calculate a 95% confidence interval estimate for the true value of the slope, β_1.
k. Test the significance of the slope at $\alpha = 0.05$.
l. Estimate the mean final exam grade that all students with an 85 prefinal average will obtain (95% confidence interval).
m. Using the 95% confidence interval, predict the grade that John Henry will receive on his final, knowing that his prefinal average is 78.

13.65 Twenty-one mature flowers of a particular species were dissected and the number of stamens and carpels present in each flower were counted. See the following table:

Stamens (x)	Carpels (y)	Stamens (x)	Carpels (y)
52	20	74	29
68	31	38	28
70	28	35	25
38	20	45	27
61	19	72	21
51	29	59	35
56	30	60	27
65	30	73	33
43	19	76	35
37	25	68	34
36	22		

a. Is there sufficient evidence to claim a linear relationship between these two variables at $\alpha = 0.05$?
b. What is the relationship between the number of stamens and carpels in this variety of flower?
c. Is the slope of the regression line significant at $\alpha = 0.05$?
d. Give the 0.95 confidence interval prediction for the number of carpels that one would expect to find in a mature flower of this variety if the number of stamens were 64.

 13.66 It is believed that the amount of nitrogen fertilizer used per acre has a direct effect on the amount of wheat produced. The following data show the amount of nitrogen fertilizer used per test plot and the amount of wheat harvested per test plot:

Pounds of Fertilizer (x)	100 Pounds of Wheat (y)	Pounds of Fertilizer (x)	100 Pounds of Wheat (y)
30	5	70	19
30	9	70	23
30	14	70	31
40	6	80	24
40	14	80	32
40	18	80	35
50	12	90	27
50	14	90	32
50	23	90	38
60	18	100	34
60	24	100	35
60	28	100	39

a. Is there sufficient reason to conclude that the use of more fertilizer results in a higher yield? Use $\alpha = 0.05$.
b. Estimate, with a 0.98 confidence interval, the mean yield that could be expected if 50 pounds of fertilizer were used per plot.
c. Estimate, with a 0.98 confidence interval, the mean yield that could be expected if 75 pounds of fertilizer were used per plot.

13.67 The correlation coefficient, r, is related to the slope of best fit, b_1, by the equation

$$r = b_1 \sqrt{\frac{SS(x)}{SS(y)}}$$

Verify this equation using the following data:

x	1	2	3	4	6
y	4	6	7	9	12

13.68 The following equation is known to be true for any set of data: $\Sigma(y - \bar{y})^2 = \Sigma(y - \hat{y})^2 + \Sigma(\hat{y} - \bar{y})^2$. Verify this equation with the following data:

x	0	1	2
y	1	3	2

VOCABULARY LIST

Be able to define each term. In addition, describe in your own words, and give an example of each term. Your examples should not be ones given in class or in the textbook.

The bracketed numbers indicate the chapters in which the term previously appeared, but you should define the terms again to show increased understanding of their meaning.

bivariate data [3]
centroid
coefficient of linear correlation [3]
confidence belts
confidence interval estimate [8, 9, 10]
covariance
curvilinear regression
experimental error (ε or e)
hypothesis tests [8, 9, 10, 11, 12]
intercept (b_0 or β_0) [3]

line of best fit [3]
linear correlation
linear regression [3]
multiple regression
Pearson's product moment, r
predicted value of μ_y
predicted value of y ($\hat{y}$)
regression line [3]
rho (ρ)
sampling distribution [7, 10]

scatter diagram [3]
slope (b_1 or β_1) [3]
standard error [7, 8, 9]

sum of squares for error (SSE) [1, 2]
variance (s^2 or σ^2) [2, 8, 9, 10, 12]

KEY CONCEPTS

bivariate data [3]
confidence interval estimate [8, 9, 10]
experimental error (ε or e)
hypothesis test [8, 9, 10, 11, 12]
linear correlation [3]

linear regression [3]
Pearson's product moment r
slope (b_1 or β_1) [3]
standard error [7, 8, 9]

QUIZ A

Answer "True" if the statement is always true. If the statement is not always true, replace the words shown in bold with words that make the statement always true.

13.1 The error **must be** normally distributed if inferences are to be made.

13.2 x and y **must both be** normally distributed.

13.3 A high correlation between x and y **proves** that x causes y.

13.4 The values of the input variable **must be** randomly selected to achieve valid results.

13.5 The output variable must be **normally distributed** about the regression line for each value of x.

13.6 **Covariance** measures the strength of the linear relationship and is a standardized measure.

13.7 The **sum of squares for error** is the name given to the numerator portion of the formula for the calculation of the variance of y about the line of regression.

13.8 **Correlation** analysis attempts to find the equation of the line of best fit for two variables.

13.9 There are $n - 3$ degrees of freedom involved with the inferences about the regression line.

13.10 $\hat{y}$ serves as the **point estimate** for both $\mu_{y|x_0}$ and y_{x_0}.

QUIZ B

Answer all questions, showing formulas and work.

It is believed that the amount of nitrogen fertilizer used per acre has a direct effect on the amount of wheat produced. The data show the amount of nitrogen fertilizer used per test plot and the amount of wheat harvested per test plot. All test plots were of the same size.

Pounds of Fertilizer, x	100 Pounds of Wheat, y
30	9
30	11
30	14
50	12
50	14
50	23
70	19
70	22
70	31
90	29
90	33
90	35

13.1 Draw a scatter diagram of the data (use graph paper and a straight edge). Be sure to label completely.

13.2 Complete an extensions table.

13.3 Calculate $SS(x), SS(xy), SS(y)$.

13.4 Calculate the linear correlation coefficient, r.

13.5 Determine the 0.95 confidence interval estimate for the population linear correlation coefficient.

13.6 Calculate the equation for the line of best fit.

13.7 Draw the line of best fit on the scatter diagram (in red ink).

13.8 Calculate the standard deviation of the y values about the line of best fit.

13.9 Does the value of b_1 show strength significant enough to conclude that the slope is greater than zero at the 0.05 level?

13.10 Determine the 0.95 confidence interval estimate for the mean yield when 85 pounds of fertilizer is used per plot.

13.11 Draw a line on the scatter diagram representing the 0.95 confidence interval found in (j) (in blue ink).

QUIZ C

13.1 "There is a high correlation between how frequently skiers have their bindings tested and the incidence of lower-leg injuries, according to researchers at the Rochester Institute of Technology. To make sure your bindings release properly when you begin to fall, you should have them serviced by a ski mechanic every 15 to 30 ski days or at least at the start of each ski season." (University of California, Berkeley, "Wellness Letter," February 1991) Explain what two variables are being discussed in this statement and interpret the "high correlation" mentioned.

13.2 Describe why the method used to define the correlation coefficient is referred to as "a product moment."

13.3 If you know the value of r is very close to zero, what value would you anticipate for b_1? Explain why.

13.4 Describe why the method used to find the line of best fit is referred to as "the method of least squares."

13.5 Explain the difference between the standard deviation about the line of best fit, the standard error for the mean value of y at x_0, and the standard error of estimate for an individual y at x_0.

13.6 You wish to study the relationship between the amount of sugar contained in a child's breakfast and the child's hyperactivity in school during the four hours after breakfast. You ask 200 mothers of fifth-grade children to keep a careful record of what the child eats and drinks each morning. The parent's report is analyzed and the sugar consumption is determined. During the same time period, data on hyperactivity are collected at school. What statistic will measure the strength and kind of relationship that exists between the amount of sugar and the amount of hyperactivity? Explain why the statistic you selected is appropriate and what value you expect this statistic might have.

13.7 You are interested in studying the relationship between the length of time a person has been supported by welfare and self-esteem. You believe that the longer a person is supported, the lower the self-esteem. What data would you need to collect and what statistics would you calculate if you wish to predict a person's level of self-esteem after having been on welfare for a certain period of time? Explain in detail.

14 ELEMENTS OF NONPARAMETRIC STATISTICS

Chapter Outline

14.1 Nonparametric Statistics
Distribution-free, or **nonparametric**, methods provide test statistics for an unspecified distribution.

14.2 The Sign Test
A simple **quantitative count** of plus and minus signs tells us whether or not to reject the null hypothesis.

14.3 The Mann-Whitney U Test
The **rank number** for each piece of data is used to compare **two independent samples**.

14.4 The Runs Test
A **sequence of data** that possesses a common property, a **run**, is used to test the question of randomness.

14.5 Rank Correlation
The linear correlation coefficient's **nonparametric alternative**, rank correlation, uses only rankings to determine whether or not to reject the null hypothesis.

14.6 Comparing Statistical Tests
When choosing between parametric and nonparametric tests, we are interested primarily in the **control of error**, the relative **power** of the test, and **efficiency**.

Study Finds That Money Can't Buy Job Satisfaction

When it comes to getting workers to produce—do their level best—money is less than everything. Feeling appreciated—having a sense of being recognized—is more important.

That is the thesis of Kenneth A. Kovach, of George Mason University....

Kovach reproduces a consequential survey of 30 years ago in which workers and supervisors gave their opinions of "what workers want from their jobs and what management thinks they want." See the accompanying table.

Managements today are more "behavioral" in their approach to managing. Nevertheless, Kovach contends that a wide gap still exists, a gap that is suggested by what workers rank as No. 1 in importance to them ("appreciation") and what supervisors consider No. 1 to workers ("good wages").

	Worker Ranking	Boss Ranking
Full appreciation of work done	1	8
Feeling of being in on things	2	10
Sympathetic help on personal problems	3	9
Job security	4	2
Good wages	5	1
Interesting work	6	5
Promotion and growth in the organization	7	3
Personal loyalty to employees	8	6
Good working conditions	9	4
Tactful disciplining	10	7

Source: Reprinted by permission of the *Philadelphia Inquirer*, 29 December 1976.

Chapter Objectives

Nonparametric methods of statistics dominate the success story of statistics in recent years. Unlike their parametric counterparts, many of the best-known nonparametric tests, also known as *distribution-free* tests, are founded on a basis of elementary probability theory. The derivation of most of these tests is well within the grasp of the student who is competent in high school algebra and understands binomial probability. Thus the nonmathematical statistics user is much more at ease with the nonparametric techniques.

This chapter is intended to give you a feeling for basic concepts involved in nonparametric techniques and to show you that nonparametric methods are extremely versatile and easy to use once a table of critical values is developed for a particular application. The selection of the nonparametric methods presented here includes only a few of the common tests and applications. You will learn about the sign test, the Mann-Whitney U test, the runs test, and Spearman's rank correlation test. These will be used to make inferences corresponding to both one- and two-sample situations.

14.1 ▼ Nonparametric Statistics

parametric method

Most of the statistical procedures that we have studied in this book are known as **parametric methods**. For a statistical procedure to be parametric, we either assume that the parent population is at least approximately normally distributed or we rely on the central limit theorem to give us a normal approximation. This is particularly true of the statistical methods studied in Chapters 8, 9, and 10.

nonparametric, or distribution-free, methods

The **nonparametric methods**, or **distribution-free methods**, as they are also known, do not depend on the distribution of the population being sampled. The nonparametric statistics are usually subject to much less confining restrictions than are their parametric counterparts. Some, for example, require only that the parent population be continuous.

The recent popularity of nonparametric statistics can be attributed to the following characteristics:

1. Nonparametric methods require few assumptions about the parent population.
2. Nonparametric methods are generally easier to apply than their parametric counterparts.
3. Nonparametric methods are relatively easy to understand.
4. Nonparametric methods can be used in situations where the normality assumptions cannot be made.
5. Nonparametric methods appear to be wasteful of information in that they sacrifice the value of the variable for only a sign or a rank number. However, nonparametric statistics are generally only slightly less efficient than their parametric counterparts.

14.2 ▼ The Sign Test

sign test

The ordinary **sign test** is a versatile and exceptionally easy-to-apply nonparametric method that *uses only plus and minus signs.* (There are several specific techniques.) The sign test is useful in two situations: (1) a hypothesis test concerning the value of the median for one population and (2) a hypothesis test concerning the median difference (paired difference) for two dependent samples. Both tests are carried out using the same basic procedures and are nonparametric alternatives to the *t* tests used with one mean (Section 9.1) and the difference between two dependent means (Section 10.2).

SINGLE-SAMPLE HYPOTHESIS TEST PROCEDURE

The sign test can be used when a random sample is drawn from a population with an unknown **median**, *M*, and the population is assumed to be continuous in the vicinity of *M*. The null hypothesis to be tested concerns the value of the population median *M*. The test may be either one- or two-tailed. This test procedure is presented in the following illustration.

▼ ILLUSTRATION 14-1

A random sample of 75 students was selected and each student was asked to carefully measure the amount of time required to commute from his or her front door to the college parking lot. The data collected were used to test the hypothesis "the median time required for students to commute is 15 minutes" against the alternative that the median is unequal to 15 minutes. The 75 pieces of data were summarized as follows:

```
Under 15:  18
15:        15
Over 15:   42
```

Use the sign test to test the null hypothesis against the alternative hypothesis.

SOLUTION The data are converted to (+) and (−) signs. A plus sign will be assigned to each piece of data larger than 15, a minus sign to each piece of data smaller than 15, and a zero to those data equal to 15. The sign test uses only the plus and minus signs; therefore, the zeros are discarded and the usable sample size becomes 60. That is, $n(+) = 42$, $n(-) = 18$, and $n = n(+) + n(-) = 42 + 18 = 60$.

STEP 1 $H_0: M = 15$

$H_a: M \neq 15$

STEP 2 $\alpha = 0.05$ for a two-tailed test. The **test statistic** that will be used is the **number of the less frequent sign**, the smaller of $n(+)$ and $n(-)$, which is $n(-)$ for our illustration. We will want to reject the null hypothesis whenever the number of the less-frequent sign is extremely small. Table 11 of Appendix F gives the maximum allowable number of the less-frequent sign, k, that will allow us to reject the null hypothesis. That is, if the number of the less-frequent sign is less than or equal to the critical value in the table, we will reject H_0. If the observed value of the less-frequent sign is larger than the table value, we will fail to reject H_0. In the table, n is the total number of signs, not including zeros.

For our illustration, $n = 60$ and the critical value from the table is 21. See the accompanying figure.

Reject H_0	Fail to reject H_0
0 21	22

Number of less frequent sign

STEP 3 The observed statistic is $x = n(-) = 18$, and it falls in the critical region.

STEP 4 Decision: Reject H_0.

Conclusion The sample shows sufficient evidence at the 0.05 level to reject the claim that the median is 15 minutes. ▲▲

TWO-SAMPLE HYPOTHESIS TEST PROCEDURE

The sign test may also be applied to a hypothesis test dealing with the *median difference* between paired data that result from *two dependent samples*. A familiar application is the use of before-and-after testing to determine the effectiveness of some activity. In a test of this nature, the signs of the differences are used to carry out the test. Again, zeros are disregarded. The following illustration shows this procedure.

▼ ## ILLUSTRATION 14-2

A new no-exercise, no-starve weight-reducing plan has been developed and advertised. To test the claim that "you will lose weight within two weeks or . . . ," a local statistician obtained the before-and-after weights of 18 people who had used this plan. Table 14-1 lists the people, their weights, a minus (−) for those who lost weight during the two weeks, a 0 for those who remained the same, and a plus (+) for those who actually gained weight.

The claim being tested is that people are able to lose weight. The null hypothesis that will be tested is that "there is no weight loss (or the median weight loss is zero)," meaning that only a rejection of the null hypothesis will allow us to conclude in favor of the advertised claim. Actually we will be testing to see whether there are significantly more minus signs than plus signs. If the weight-reducing plan is of absolutely no value, we would expect to find an equal number of plus and minus signs. If it works, there should be significantly more minus signs than plus signs. Thus the test performed here will be a one-tailed test. (We will want to reject the null hypothesis in favor of the advertised claim if there are "many" minus signs.)

TABLE 14-1
Sample Results for Illustration 14-2

Person	Weight Before	Weight After	Sign of Difference, Before to After
Mrs. Smith	146	142	−
Mr. Brown	175	178	+
Mrs. White	150	147	−
Mr. Collins	190	187	−
Mr. Gray	220	212	−
Miss Collins	157	160	+
Mrs. Allen	136	135	−
Mrs. Noss	146	138	−
Miss Wagner	128	132	+
Mr. Carroll	187	187	0
Mrs. Black	172	171	−
Mrs. McDonald	138	135	−
Miss Henry	150	151	+
Miss Greene	124	126	+
Mr. Tyler	210	208	−
Mrs. Williams	148	148	0
Mrs. Moore	141	138	−
Mrs. Sweeney	164	159	−

SOLUTION

STEP 1 $H_0: M = 0$ (no weight loss).

$H_a: M < 0$ (weight loss).

STEP 2 Use $\alpha = 0.05$, $n(+) = 5$, $n(-) = 11$, and $n = 16$. The critical value from Table 11 shows $k = 4$ as the maximum allowable number. (You must use the $\alpha = 0.10$ column, because the table is set up for two-tailed tests.)

Reject H_0	Fail to reject H_0
0 1 2 3 4	5

Number of less frequent sign

STEP 3 $x = n(+) = 5$

STEP 4 Decision: Fail to reject H_0 (we have too many plus signs).

Conclusion The evidence observed is not sufficient to allow us to reject the no-weight-loss null hypothesis at the 0.05 level of significance. ▲▲

normal approximation

The sign test may be carried out by means of a **normal approximation** using the standard normal variable z. The normal approximation will be used if

Table 11 does not show the particular levels of significance desired or if n is large. z will be calculated by using the formula

$$z = \frac{x' - n/2}{(\frac{1}{2})\sqrt{n}} \qquad (14\text{-}1)$$

(See Note 3 with regard to x'.)

NOTES

1. x may be the number of the less-frequent sign or the most-frequent sign. You will have to determine this in such a way that the direction is consistent with the interpretation of the situation.
2. x is really a binomial random variable, where $p = \frac{1}{2}$. The sign test statistic satisfies the properties of a binomial experiment (see p. 285). Each sign is the result of an independent trial. There are n trials, and each trial has two possible outcomes ($+$ or $-$). Since the median is used, the probabilities for each outcome are both $\frac{1}{2}$. Therefore, the mean, μ_x, is equal to

$$n/2 \quad [\mu = np = n \cdot \tfrac{1}{2} = n/2]$$

and the standard deviation, σ_x, is equal to

$$\tfrac{1}{2}\sqrt{n} \quad [\sigma = \sqrt{npq} = \sqrt{n \cdot \tfrac{1}{2} \cdot \tfrac{1}{2}} = \tfrac{1}{2}\sqrt{n}]$$

3. x is a discrete variable. But recall that the normal distribution must be used only with continuous variables. However, although the binomial random variable is discrete, it does become approximately normally distributed for large n. Nevertheless, when using the normal distribution for testing, we should make an adjustment in the variable so that the approximation is more accurate. (See Section 6.5, p. 331, on the normal approximation.) This adjustment is illustrated in Figure 14-1 and is called a **continuity correction**. For this discrete variable the area that represents the probability is a rectangular bar. Its width is 1 unit wide, from $\frac{1}{2}$ unit below to $\frac{1}{2}$ unit above the value of interest. Therefore, when z is to be used, we will need to make a $\frac{1}{2}$-unit adjustment before calculating the observed value of z. x'

FIGURE 14-1
Continuity Correction

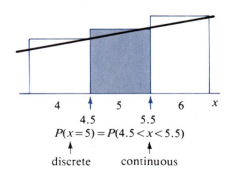

will be adjusted value for x. If x is larger than $n/2$, $x' = x - \frac{1}{2}$. If x is smaller than $n/2$, $x' = x + \frac{1}{2}$. The test is then completed by the usual procedure.

▼ ILLUSTRATION 14-3

Use the sign test to test the hypothesis that the median number of hours, M, worked by students of a certain college is at least 15 hours per week. A survey of 120 students was taken; a plus sign was recorded if the number of hours the student worked last week was equal to or greater than 15, and a minus sign was recorded if the number of hours was less than 15. Totals showed 80 minus signs and 40 plus signs.

SOLUTION

STEP 1 H_0: $M = 15$ ($\geq$) (at least as many plus signs as minus signs).

H_a: $M < 15$ (fewer plus signs than minus signs).

STEP 2 $\alpha = 0.05$; x is the number of plus signs. The critical value is shown in the accompanying figure.

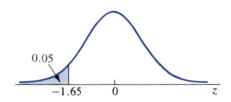

STEP 3

$$z = \frac{x' - n/2}{(\frac{1}{2})\sqrt{n}} = \frac{40.5 - 60}{(\frac{1}{2})\sqrt{120}} = \frac{-19.5}{(\frac{1}{2})(10.95)}$$

$$= \frac{-19.5}{5.475} = -3.562$$

$z^* = -3.56$

STEP 4 **Decision:** Reject H_0.

Conclusion At the 0.05 level, there are significantly more minus signs than plus signs, thereby implying that the median is less than the claimed 15 hours. ▲▲

CONFIDENCE INTERVAL PROCEDURE

The sign test techniques can be applied to obtain a confidence interval estimate for the unknown population median, M. To accomplish this we will need to arrange the data in ascending order (smallest to largest). The data are identified as

x_1 (smallest), $x_2, x_3, \ldots, x_n$ (largest). The critical value, k, for the maximum allowable number of signs, obtained from Table 11, Appendix F, tells us the number of positions to be dropped from each end of the ordered data. The remaining extreme values become the bounds of the $1 - \alpha$ confidence interval. That is, the lower boundary for the confidence interval is x_{k+1}, the $(k + 1)$th piece of data; the upper boundary is x_{n-k}, the $(n - k)$th piece of data. The following illustration will clarify this procedure.

▼ **ILLUSTRATION 14-4**

Suppose that we have 12 pieces of data in ascending order $(x_1, x_2, x_3, \ldots, x_{12})$ and we wish to form a 0.95 confidence interval for the population median. Table 11 shows a critical value of 2 ($k = 2$) for $n = 12$ and $\alpha = 0.05$ for a hypothesis test. This means that the critical region for a two-tailed hypothesis test would contain the last two values on each end (x_1 and x_2 on the left; x_{11} and x_{12} on the right). The noncritical region is then from x_3 to x_{10}, inclusive. The 0.95 confidence interval is then x_3 to x_{10}, expressed as

x_3 to x_{10} 0.95 confidence interval for M ▲▲

In general, the two pieces of data that bound the confidence interval will occupy positions $k + 1$ and $n - k$, where k is the critical value read from Table 11. Thus

x_{k+1} to x_{n-k} $1 - \alpha$ confidence interval for M

If the normal approximation is to be used (including the continuity correction), the position numbers become

$$(\tfrac{1}{2})n \pm (\tfrac{1}{2} + \tfrac{1}{2} \cdot z(\alpha/2) \cdot \sqrt{n}) \qquad (14\text{-}2)$$

The interval is

x_L to x_U $1 - \alpha$ confidence interval for M

where

$$L = \frac{n}{2} - \frac{1}{2} - \frac{z(\alpha/2)}{2} \cdot \sqrt{n}$$

$$U = \frac{n}{2} + \frac{1}{2} + \frac{z(\alpha/2)}{2} \cdot \sqrt{n}$$

(L should be rounded down and U should be rounded up to be sure that the level of confidence is at least $1 - \alpha$.)

▼ **ILLUSTRATION 14-5**

Estimate the population median with a 0.95 confidence interval for a given set of 60 pieces of data: $x_1, x_2, x_3, \ldots, x_{59}, x_{60}$.

SOLUTION
When we use formula (14-2), the position numbers L and U are

$$(\tfrac{1}{2})(60) \pm [\tfrac{1}{2} + \tfrac{1}{2}(1.96)\sqrt{60}]$$
$$30 \pm [0.50 + 7.59]$$
$$30 \pm 8.09$$

Thus,

$$L = 30 - 8.09 = 21.91 \quad \text{(21st piece of data)}$$
$$U = 30 + 8.09 = 38.09 \quad \text{(39th piece of data)}$$

Therefore,

$$x_{21} \text{ to } x_{39} \quad 0.95 \text{ confidence interval for } M$$

▼▲ EXERCISES

14.1 State the null hypothesis, H_0, and the alternative hypothesis, H_a, that would be used to test the following statements:
  a. The median value is at least 32.
  b. People prefer the taste of the bread made with the new recipe.
  c. There is no change in weight from weigh-in until after two weeks of the diet.

14.2 Determine the test criteria that would be used to test the null hypothesis for the following multinomial experiments:
  a. $H_0: P(+) = 0.5$ vs. $H_a: P(+) \neq 0.5$, with $n = 18$, $\alpha = 0.05$.
  b. $H_0: P(+) = 0.5$ vs. $H_a: P(+) > 0.5$, with $n = 78$, $\alpha = 0.05$.
  c. $H_0: P(+) = 0.5$ vs. $H_a: P(+) < 0.5$, with $n = 38$, $\alpha = 0.005$.
  d. $H_0: P(+) = 0.5$ vs. $H_a: P(+) \neq 0.5$, with $n = 148$, $\alpha = 0.05$.

14.3 Use the sign test to test the hypothesis that the median daily high temperature in the city of Rochester, New York, during the month of December is 48 degrees. The following daily highs were recorded on 20 randomly selected days during December:

| 47 | 46 | 40 | 40 | 46 | 35 | 34 | 59 | 54 | 33 |
| 65 | 39 | 48 | 47 | 46 | 46 | 42 | 36 | 45 | 38 |

  a. State the null hypothesis for the test.
  b. State specifically what it is that you are actually testing when using the sign test.
  c. Complete the test for $\alpha = 0.05$, and carefully state your findings.

14.4 *USA Today* reported on May 8, 1991, "Teacher's pay increases 5.4%." The following sample of average teacher salary increases was taken from the annual report of the National Education Association:

1.6	11.4	4.2	8.7	4.5	6.4	5.9	3.7	4.5
4.6	4.4	4.1	7.6	5.5	10.6	2.8	6.8	3.0
3.5	7.9	8.0	4.3	2.4	5.0			

Do the sample data suggest that the claim "median pay increase is 5.4%" be rejected at the 0.05 level of significance?

 14.5 In order to test the null hypothesis that there is no difference in the ages of husbands and wives, the following data were collected:

Husbands	28	45	40	37	25	42	21	22	54	47	35	62	29	44	45	38	59
Wives	26	36	40	42	28	40	20	24	50	54	37	60	25	40	34	42	49

Does the sign test show that there is a significant difference in the ages of husbands and wives at $\alpha = 0.05$?

 14.6 Sixteen students were given an elementary statistics test on a hot day. Eight randomly selected students took the test in a room that was not air-conditioned. Then, after a short break, they completed a similar test in an air-conditioned room. The procedure was reversed for the other eight students.

Student	1	2	3	4	5	6	7	8	9	10	11	12	13	14	15	16
Not Air-Conditioned	52	90	63	74	87	77	92	72	77	94	67	86	78	84	57	55
Air-Conditioned	49	94	60	78	93	77	93	74	78	93	78	89	92	83	49	68

Does the sample provide sufficient reason to conclude that the use of air conditioning on a hot day has an effect on test grades? Use $\alpha = 0.05$.

 14.7 A blind taste test was used to determine people's preference for the taste of the old cola and new cola. The results were

645 preferred the new
583 preferred the old
272 had no preference

Is the preference for the taste of the new cola significantly greater than one-half? Use $\alpha = 0.01$.

14.8 A taste test was conducted with a regular beef pizza. Each of 133 individuals was given two pieces of pizza, one with a whole wheat crust and the other with a white crust. Each person was then asked whether he or she preferred whole wheat or white crust. The results were

65 preferred whole wheat to white crust
53 preferred white to whole wheat crust
15 had no preference

Is there sufficient evidence to verify the hypothesis that whole wheat crust is preferred to white crust at the $\alpha = 0.05$ level of significance?

14.9 Determine the 0.95 confidence interval for the medium daily high temperature in Rochester during December based on the sample given in Exercise 14.3.

14.10 Find the 0.75 confidence interval estimate for the median swim time for a swimmer whose recorded times are

| 24.7 | 24.7 | 24.6 | 25.5 | 25.7 | 25.8 | 26.5 | 24.5 | 25.3 |
| 26.2 | 25.5 | 26.3 | 24.2 | 25.3 | 24.3 | 24.2 | 24.2 | |

14.11 Use the data in Exercise 14.6 to find a 0.95 confidence interval estimate for the median difference in scores attained in the air-conditioned room and in the not-air-conditioned room.

14.12 A sample of the daily rental-car rates for a compact car was collected in order to estimate the average daily cost of renting a compact car.

33.93	35.00	36.99	32.99	36.93	29.00	34.95	23.99
43.93	44.95	28.95	22.99	37.93	37.00	35.99	36.99
30.93	28.95	29.99	25.99	39.93	40.50	28.90	23.80
26.93	23.70	26.99	21.94	47.93	40.00	29.94	28.99
23.93	22.70	26.99	25.48	31.93	31.90	31.92	29.99

Find the 0.99 confidence interval estimate for the median daily rental cost.

14.13 According to an article in a *Newsweek* special issue (Fall/Winter 1990, pages 16–22), 51.1% of 17-year-olds answered the following question correctly:

If $7X + 4 = 5X + 8$, then $X =$
☐ 1   ☐ 2   ☐ 4   ☐ 6

Suppose we wished to test the null hypothesis that one-half of all 17-year-olds could solve the problem above against the alternative hypothesis, "the proportion who can solve differs from one-half." Furthermore, suppose we asked 75 randomly selected 17-year-olds to solve the problem. Let + represent a correct solution and − represent an incorrect solution. If we obtain 25 + signs and 50 − signs, do we have sufficient evidence to show the proportion who can solve is different than one-half? Use $\alpha = 0.05$.

14.14 According to an article in *USA Today*, June 7, 1991, "only 46% of high school seniors can solve problems involving fractions, decimals and percentages." Suppose we wish to test the null hypothesis, "one-half of all seniors can solve problems involving fractions, decimals and percentages," against an alternative that the proportion who can solve differs from one-half. Let + represent passed and − represent failed the test on fractions, decimals, and percentages. If a random sample of 1500 students is tested, what value of x, the number of the least-frequent sign, will be the critical value at the 0.05 level of significance?

14.3 ▼ The Mann-Whitney *U* Test

Mann-Whitney *U* test

The **Mann-Whitney *U* test** is a nonparametric alternative for the *t* test for the **difference between two independent means**. It can be applied when we have two independent random samples (independent within each sample as well as between

Chapter 14 ▼ ELEMENTS OF NONPARAMETRIC STATISTICS

samples) in which the random variable is continuous. This test is often applied in situations in which the two samples are drawn from the same population but different "treatments" are used on each set. We will demonstrate the procedure in the following illustration.

▼ ILLUSTRATION 14-6

In a large lecture class, when the instructor gives a one-hour exam, she gives two "equivalent" examinations. Students in even-numbered seats take exam A and those in the odd-numbered seats take exam B. It is reasonable to ask, "Are these two different exam forms equivalent?" Assuming that the odd- or even-numbered seats had no effect, we would want to test the hypothesis "the test forms yielded scores that had identical distributions."

To test this hypothesis the following two random samples were taken:

A	52	78	56	90	65	86	64	90	49	78
B	72	62	91	88	90	74	98	80	81	71

The size of the individual samples will be called n_a and n_b; actually, it makes no difference which way these are assigned. In our illustration they both have the value 10.

The first thing that must be done with the entire sample (all $n_a + n_b$ pieces of data) is to order it into one sample, smallest to largest:

49 52 56 62 64 65 71 72 74 78
78 80 81 86 88 90 90 90 91 98

rank

Each piece of data is then assigned a **rank number**. The smallest (49) is assigned rank 1, the next smallest (52) is assigned rank 2, and so on, up to the largest, which is assigned rank $n_a + n_b$ (20). Ties are handled by assigning to each of the tied observations the mean rank of those rank positions that they occupy. For example, in our illustration there are two 78s; they are the 10th and 11th pieces of data. The mean rank for each is then $(10 + 11)/2 = 10.5$. In the case of the three 90s, the 16th, 17th, and 18th pieces of data, each is assigned 17, since $(16 + 17 + 18)/3 = 17$. The rankings are shown in Table 14-2.

The calculation of the test statistic U is a two-step procedure. We first determine the sum of the ranks for each of the two samples. Then, using the two sums of ranks, we calculate two U scores for each sample. The **smaller U score is the test statistic**.

test statistic

The sum of ranks R_a for sample A is computed as

$$R_a = 1 + 2 + 3 + 5 + 6 + 10.5 + 10.5 + 14 + 17 + 17 = 86$$

The sum of ranks R_b for sample B is

$$R_b = 4 + 7 + 8 + 9 + 12 + 13 + 15 + 17 + 19 + 20 = 124$$

Section 14.3 ▼ THE MANN-WHITNEY U TEST

TABLE 14-2
Ranked Data for Illustration 14-6

Ranked Data	Rank	Source	Ranked Data	Rank	Source
49	1	A	78	10.5	A
52	2	A	80	12	B
56	3	A	81	13	B
62	4	B	86	14	A
64	5	A	88	15	B
65	6	A	90	17	A
71	7	B	90	17	A
72	8	B	90	17	B
74	9	B	91	19	B
78	10.5	A	98	20	B

The U score for each sample is obtained by using the following pair of formulas:

$$U_a = n_a \cdot n_b + \frac{(n_b)(n_b + 1)}{2} - R_b \qquad (14\text{-}3)$$

$$U_b = n_a \cdot n_b + \frac{(n_a)(n_a + 1)}{2} - R_a \qquad (14\text{-}4)$$

U, the test statistic, will be the smaller of U_a and U_b.

For our illustration, we obtain

$$U_a = (10)(10) + \frac{(10)(10 + 1)}{2} - 124 = 31$$

$$U_b = (10)(10) + \frac{(10)(10 + 1)}{2} - 86 = 69$$

Therefore,

$$U = 31$$

Before we carry out the test for this illustration, let's try to understand some of the underlying possibilities. Recall that the null hypothesis is that the distributions are the same and that we will most likely want to conclude from this that the means are approximately equal. Suppose for a moment that they are indeed quite different—say all of one sample comes before the smallest piece of data in the second sample when they are ranked together. This would certainly mean that we want to reject the null hypothesis. What kind of a value can we expect for U in this case? Suppose, in Illustration 14-6, that the ten A values had ranks 1 through 10 and the ten B values had ranks 11 through 20. Then we would obtain

$$R_a = 55 \qquad R_b = 155$$

$$U_a = (10)(10) + \frac{(10)(10 + 1)}{2} - 155 = 0$$

$$U_b = (10)(10) + \frac{(10)(10 + 1)}{2} - 55 = 100$$

Therefore,
$$U = 0$$

Suppose, on the other hand, that both samples were perfectly matched, that is, a score in each set identical to one in the other. Now what would happen?

54	54	62	62	71	71	72	72	...
A	B	A	B	A	B	A	B	...
1.5	1.5	3.5	3.5	5.5	5.5	7.5	7.5	...

$$R_a = R_b = 105$$

$$U_a = U_b = (10)(10) + \frac{(10)(10+1)}{2} - 105 = 50$$

Therefore,
$$U = 50$$

If this were the case, we certainly would want to reach the decision "fail to reject the null hypothesis."

Note that the sum of the two U's ($U_a + U_b$) will always be equal to the product of the two sample sizes ($n_a \cdot n_b$). For this reason we need to concern ourselves with only one of them, the smaller one.

Now let's return to the solution of Illustration 14-6. In order to complete our hypothesis test, we need to be able to determine a critical value for U. Table 12 in Appendix F gives us the critical value for some of the more common testing situations as long as both sample sizes are reasonably small. Table 12 shows only the critical region in the left-hand tail, and the null hypothesis will be rejected if the observed value for U is less than or equal to the value read from the table. For our example, $n_a = n_b = 10$; at $\alpha = 0.05$ in a two-tailed test, the critical value is 23. We observed a value of 31 and therefore we make the decision "fail to reject the null hypothesis." This means that we do not have sufficient evidence to reject the "equivalent" hypothesis. ▲▲

If the *samples are larger than size 20*, we may make the test decision with the aid of the standard normal variable, z. This is possible since the distribution of U is approximately normal with a mean

$$\mu_U = \frac{n_a \cdot n_b}{2} \qquad (14\text{-}5)$$

and a standard deviation

$$\sigma_U = \sqrt{\frac{n_a n_b (n_a + n_b + 1)}{12}} \qquad (14\text{-}6)$$

The null hypothesis is then tested by using

$$z = \frac{U - \mu_U}{\sigma_U} \qquad (14\text{-}7)$$

Section 14.3 ▼ THE MANN-WHITNEY U TEST

in the typical fashion. The test statistic z may be used whenever n_a and n_b are both greater than 10.

▼ **ILLUSTRATION 14-7**

A dog obedience trainer is training 27 dogs to obey a certain command. The trainer is using two different training techniques, the reward-and-encouragement method (I) and the no-reward method (II). The following table shows the number of obedience sessions that were necessary before the dogs would obey the command. Does the trainer have sufficient evidence to claim that the reward method will, on the average, require less training time ($\alpha = 0.05$)?

I	29	27	32	25	27	28	23	31	37	28	22	24	28	31	34
II	40	44	33	26	31	29	34	31	38	33	42	35			

SOLUTION

STEP 1 H_0: The average amount of training time required is the same for both methods.
H_a: The reward method requires less time on the average.

STEP 2 $\alpha = 0.05$; the critical value is shown in the accompanying figure.

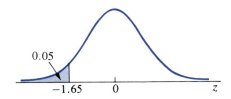

STEP 3 The two sets of data are ranked jointly and ranks are assigned, as shown in Table 14-3, p. 704. Then the sums are

$$R_I = 1 + 2 + 3 + 4 + 6.5 + \cdots + 20.5 + 23 = 151.0$$
$$R_{II} = 5 + 11.5 + \cdots + 26 + 27 = 227.0$$

Now by using formulas (14-3) and (14-4), we can obtain the value of the U scores:

$$U_I = (15)(12) + \frac{(12)(12 + 1)}{2} - 227 = 180 + 78 - 227 = 31$$

$$U_{II} = (15)(12) + \frac{(15)(15 + 1)}{2} - 151 = 180 + 120 - 151 = 149$$

Therefore,

$$U = 31$$

TABLE 14-3 Rankings for Training Methods

Number of Sessions	Group	Rank		Number of Sessions	Group	Rank	
22	I	1		31	II	15	14.5
23	I	2		31	II	16	14.5
24	I	3		32	I	17	
25	I	4		33	II	18	18.5
26	II	5		33	II	19	18.5
27	I	6	6.5	34	I	20	20.5
27	I	7	6.5	34	II	21	20.5
28	I	8	9	35	II	22	
28	I	9	9	37	I	23	
28	I	10	9	38	II	24	
29	I	11	11.5	40	II	25	
29	II	12	11.5	42	II	26	
31	I	13	14.5	44	II	27	
31	I	14	14.5				

Now we use formulas (14-5), (14-6), and (14-7) to determine the z statistic.

$$\mu_U = \frac{12 \cdot 15}{2} = 90$$

$$\sigma_U = \sqrt{\frac{(12)(15)(12 + 15 + 1)}{12}} = \sqrt{\frac{(180)(28)}{12}} = \sqrt{420} = 20.49$$

$$z = \frac{31 - 90}{20.49} = \frac{-59}{20.49} = -2.879$$

$$z^\ast = -2.88$$

STEP 4 Decision: Reject H_0.

Conclusion At the 0.05 level of significance, the data show sufficient evidence to conclude that the reward method does, on the average, require less training time. ▲▲

Case Study 14-1

Health Beliefs and Practices of Runners Versus Nonrunners

Valerie Walsh used the Mann-Whitney U statistic to conclude that runners place a greater value on personal health than do nonrunners.

Returns of mailed questionnaires from 77 runner and 63 nonrunner respondents showed that runners placed a statistically higher value on health and performed greater numbers of health-related behaviors. Major differences were found in nutrition, exercise, and medical awareness and self-care. No major differences were found in addictive substance use, stress management, or safety practices. A number of concerns regarding runners' health practices were identified, including running while ill or in pain, incidence of injuries, negative feelings when unable to run, neglect of a conscious cool-down period, low weight levels, and a tendency to increase workouts following perceived dietary indiscretions....

The purpose of this investigation was to explore the differences between runners and nonrunners in terms of specific health beliefs and behaviors. It was hypothesized that there is a difference between runners and nonrunners in the relative value placed on personal health....

RESULTS

The first hypothesis, that there is a difference between runners and nonrunners in the relative value placed on personal health, was tested using the Mann-Whitney U, with alpha set at .05, and was accepted, $p < .019$. The value of U was found to be 1876.5; of U, 4990.5. Greater value was placed on personal health by the runners than by the nonrunners....

DISCUSSION

Runners placed a higher value on health and performed more health-related behaviors than did nonrunners. Although these results might have been anticipated in light of findings from other studies (Dawber, 1980; Paffenbarger et al., 1977), the information derived from them regarding health behaviors of runners was limited and inferential at best. The findings from this investigation, however, are congruent with those of Blair et al. (1981), in that runners exert tighter control over the types of nutrients they consume than do nonrunners.

Source: Valerie R. Walsh, *Nursing Research*, November/December 1985, Vol. 34, no. 6.

▼▲ EXERCISES

14.15 State the null hypothesis, H_0, and the alternative hypothesis, H_a, that would be used to test the following statements:
  a. There is a difference in the value of the variable between the two groups of subjects.
  b. The average value is not the same for both groups.
  c. The blood pressure for group A is higher than for group B.

14.16 Determine the test criteria that would be used to test the following hypotheses for experiments involving two independent samples:
  a. H_0: Average(A) = Average(B)
  H_a: Average(A) > Average(B),
  with $n_A = 18$, $n_B = 15$, and $\alpha = 0.05$.
  b. H_0: The average score is the same for both groups.
  H_a: Group I average score are less than Group II,
  with $n_I = 78$, $n_{II} = 45$, and $\alpha = 0.05$.

14.17 A study entitled "Textbook Pictures and First-Grade Children's Perception of Mathematical Relationships" by Patricia Campbell (*Journal for Research in Mathematics Education*, November 1978) investigated the influence of artistic style and the number of pictures on first-grade children's perception of mathematics textbook pictures. Analysis with the Mann-Whitney U test indicated that students who initially viewed and described sequences had significantly higher story-response scores than those students who viewed only single pictures. Consider the following data from two such groups. Group 1 is the group who viewed sequences of pictures, and Group 2 is the group who viewed only single pictures.

Group 1	30	35	40	42	40	45	36
Group 2	25	32	27	39	30		

Using the Mann-Whitney U test, determine if the Group 1 scores are significantly higher than the Group 2 scores. Use $\alpha = 0.05$.

14.18 The reaction times for an emergency situation were recorded for individuals from two different age groups. The data are shown in the following table:

Age Group I (20–30)	5.1	6.2	6.0	5.5	4.9	5.0	5.5	6.0	5.7
Age Group II (50–65)	7.0	6.5	6.0	7.2	7.2	5.9			

Do these samples show a significant difference in the reaction times for the two groups at the $\alpha = 0.05$ level of significance?

14.19 Pulse rates were recorded for 16 men and 13 women. The results are shown in the following table:

Males	61	73	58	64	70	64	72	60
	65	80	55	72	56	56	74	65
Females	83	58	70	56	76	64	80	
	68	78	108	76	70	97		

These data were used to test the hypothesis that the distribution of pulse rates differs for men and women. The following MINITAB output printed out the sum of ranks for males (W = 192.0) and the *p*-value for z^* of 0.0373. Verify these two values.

Section 14.3 ▼ THE MANN-WHITNEY U TEST

```
MTB > SET MALE PULSE RATES IN C1
DATA> 61 73 58 64 70 64 72 60 74 65 65 80 55 72 56 56
DATA> END DATA
MTB > SET FEMALE PULSE RATES IN C2
DATA> 83 58 70 56 76 64 80 76 70 97 68 78 108
DATA> END DATA
MTB > MANN-WHITNEY ON DATA IN C1, C2

Mann-Whitney Confidence Interval and Test

C1           N = 16       MEDIAN =         64.50
C2           N = 13       MEDIAN =         76.00
POINT ESTIMATE FOR ETA1-ETA2 IS          -9.00
95.4  PCT C.I. FOR ETA1-ETA2 IS (    -18.00,       0.00)
W = 192.0
TEST OF ETA1 = ETA2 VS. ETA1 N.E. ETA2 IS SIGNIFICANT AT 0.0373
```

14.20 The following set of data represents the ages of drivers involved in automobile accidents. Do these data present sufficient evidence to conclude that there is a difference in the average age of men and women drivers involved in accidents? Use a two-tailed test at $\alpha = 0.05$.

Men	70	60	77	39	36	28	19	40
	23	23	63	31	36	55	24	76
Women	62	46	43	28	21	22	27	42
	21	46	33	29	44	29	56	70

a. State the null hypothesis that is being tested.
b. Complete the test using a computer. See Exercise 14.19 for the necessary commands.

14.21 In a heart study the systolic blood pressure was measured for 24 men of age 25 and for 30 men of age 40. (For your convenience, the data are ranked.) Do these data show sufficient evidence to conclude that the older men have a higher systolic blood pressure at the 0.02 level of significance?

25-year-olds	95	100	100	105	106	108	110	110		
	115	118	120	122	124	125	130	130		
	130	132	136	138	140	148	150	156		
40-year-olds	108	110	110	114	114	116	118	120	122	124
	126	126	128	130	130	132	136	136	136	140
	142	142	146	148	150	152	154	160	164	176

14.22 To determine whether there is a difference in the breaking strength of lightweight monofilament fishing line produced by two companies, the following test data (pounds of force required to break the line) were obtained. Use the Mann-Whitney U test and $\alpha = 0.05$ to determine whether there is a difference.

A	12.4	11.9	11.8	13.5	11.6	12.0	12.9	11.3	13.8	12.3		
B	12.7	10.7	13.2	11.8	11.8	12.1	12.5	11.9	12.8	11.2	11.5	11.3

14.4 ▼ The Runs Test

runs test
randomness
run

The **runs test** is most frequently used to test the **randomness of data** (or lack of randomness). A **run** is a sequence of data that possesses a common property. One run ends and another starts when a piece of data does not display the property in question. The random variable that will be used in this test is V, the number of runs observed.

▼ **ILLUSTRATION 14-8**

To illustrate the idea of runs, let's draw a sample of ten single-digit numbers from the telephone book, using the next-to-last digit from each of the selected telephone numbers.

Sample: 2, 3, 1, 1, 4, 2, 6, 6, 6, 7

Let's consider the property of "odd" (*o*) or "even" (*e*). The sample, as it was drawn, becomes *e, o, o, o, e, e, e, e, e, o*, which displays four runs.

<u>e</u> <u>o o o</u> <u>e e e e e</u> <u>o</u>

Thus $V = 4$.

In Illustration 14-8, if the sample contained no randomness, there would be only two runs—all the evens, then all the odds, or the other way around. We would also not expect to see them alternate—odd, even, odd, even. The maximum number of possible runs would be $n_1 + n_2$, or less (provided n_1 and n_2 are not equal), where n_1 and n_2 are the number of data that have each of the two properties being identified.

We will often want to interpret the maximum number of runs as a rejection of a null hypothesis of randomness, since we often want to test randomness of the data in reference to how they were obtained. For example, if the data alternated all the way down the line, we might suspect that the data had been tampered with. There are many aspects to the concept of randomness. The occurrence of odd and even as discussed in Illustration 14-8 is one aspect. Another aspect of randomness that we might wish to check is the ordering of fluctuations of the data above (*a*) or below (*b*) the mean or median of the sample.

▼ ILLUSTRATION 14-9

Consider the sequence that results from determining whether each of the data points in the sample of Illustration 14-8 is above or below the median value. Test the null hypothesis that this sequence is random. Use $\alpha = 0.05$.

SOLUTION

STEP 1 H_0: The numbers in the sample form a random sequence with respect to the two properties "above" and "below" the median value.

H_a: The sequence is not random.

Sample: 2,   3,   1,   1,   4,   2,   6,   6,   6,   7

First we must rank the data and find the median. The ranked data are $1, 1, 2, 2, 3, 4, 6, 6, 6, 7$. Since there are ten pieces of data, the median is at the $i = 5.5$ position. Thus $\tilde{x} = (3 + 4)/2 = 3.5$. By comparing each number in the original sample to the value of the median, we obtain the following sequence of a's (above) and b's (below):

b   b   b   b   a   b   a   a   a   a

We observe $n_a = 5$, $n_b = 5$, and 4 runs. So $V = 4$.

If n_1 and n_2 are both less than or equal to 20 and a two-tailed test at $\alpha = 0.05$ is desired, then Table 13 of Appendix F will give us the two critical values for the test. For our illustration with $n_a = 5$ and $n_b = 5$, Table 13 shows critical values of 2 and 10. This means that if 2 or fewer, or 10 or more, runs are observed, the null hypothesis will be rejected. If between 3 and 9 runs are observed, we will fail to reject the null hypothesis.

STEP 2 $\alpha = 0.05$ and a two-tailed test is used. The critical values for V are found in Table 13 (see the accompanying figure). The critical values are 2 and 10.

Reject H_0	Fail to reject H_0	Reject H_0
2	3 9	10

V, number of runs

STEP 3   Four runs were observed; $V^* = 4$.

STEP 4   **Decision:** Fail to reject H_0.

Conclusion   We are unable to reject the hypothesis of randomness at the 0.05 level of significance.   ▲▲

To complete the hypothesis test about randomness when n_1 and n_2 is larger than 20 or when α is other than 0.05, we will use z, the standard normal random variable. V is approximately normally distributed with a mean of μ_V and a standard deviation of σ_V. The formulas are as follows:

$$\mu_V = \frac{2n_1 \cdot n_2}{n_1 + n_2} + 1 \tag{14-8}$$

$$\sigma_V = \sqrt{\frac{(2n_1 \cdot n_2)(2n_1 \cdot n_2 - n_1 - n_2)}{(n_1 + n_2)^2(n_1 + n_2 - 1)}} \tag{14-9}$$

$$z = \frac{V - \mu_V}{\sigma_V} \tag{14-10}$$

▼ ILLUSTRATION 14-10

Test the null hypothesis that the sequence that results from classifying the sample data in Illustration 14-8 as "odd" or "even" is a random sequence. Use $\alpha = 0.10$.

SOLUTION

STEP 1 H_0: The sequence of odd and even occurrences is a random sequence.
           H_a: The sequence is not random.

STEP 2 $\alpha = 0.10$ and a two-tailed test is to be used. The test criteria are shown in the following figure:

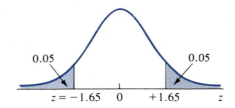

STEP 3   The sample and the sequence of odd and even properties are shown in Illustration 14-8. $n_1 = n(\text{even}) = 6$ and $n_2 = n(\text{odd}) = 4$. There are 4 runs, so $V = 4$.

$$\mu_V = \frac{2n_1 n_2}{n_1 + n_2} + 1 = \frac{2 \cdot 6 \cdot 4}{6 + 4} + 1 = \frac{48}{10} + 1 = 5.8$$

$$\sigma_V = \sqrt{\frac{(2n_1 n_2)(2n_1 n_2 - n_1 - n_2)}{(n_1 + n_2)^2(n_1 + n_2 - 1)}}$$

$$= \sqrt{\frac{(2 \cdot 6 \cdot 4)(2 \cdot 6 \cdot 4 - 6 - 4)}{(6 + 4)^2(6 + 4 - 1)}}$$

$$= \sqrt{\frac{(48)(38)}{(10)^2(9)}} = \sqrt{\frac{1824}{900}} = \sqrt{2.027} = 1.42$$

$$z = \frac{V - \mu_V}{\sigma_V} = \frac{4.0 - 5.8}{1.42} = \frac{-1.8}{1.42} = -1.268$$

$$z^* = -1.27$$

Section 14.4 ▼ THE RUNS TEST

STEP 4 **Decision:** Fail to reject H_0.

Conclusion We are unable to reject the hypothesis of randomness at the 0.10 level of significance.

▼▲ EXERCISES

14.23 State the null hypothesis, H_0, and the alternative hypothesis, H_a, that would be used to test the following statements:
   a. The data did not occur in a random order about the median.
   b. The sequence of odd and even is not random.
   c. The gender of customers entering a grocery store was recorded; the order of entry is not random in order.

14.24 Determine the test criteria that would be used to test the null hypothesis for the following multinomial experiments:
   a. H_0: The results collected occurred in random order above and below the median.
   H_a: The results were not random, with $n(A) = 14$, $n(B) = 15$, and $\alpha = 0.05$.
   b. H_0: The two properties alternated randomly.
   H_a: The two properties didn't occur in random fashion, with $n(I) = 78$, $n(II) = 45$, and $\alpha = 0.05$.

14.25 A manufacturing firm hires both men and women. The following shows the sex of the last 20 individuals hired (M = male, F = female):

M	M	F	M	F	F	M	M	M	M
M	M	F	M	M	F	M	M	M	M

At the $\alpha = 0.05$ level of significance, are we correct in concluding that this sequence is not random?

14.26 A student was asked to perform an experiment that involved tossing a coin 25 times. After each toss the student recorded the results. The following data were reported (H = heads, T = tails):

H	T	H	T	H	T	H	T	H	H	T	T	H
H	T	T	H	T	H	T	H	T	H	T	H	

Use the runs test at a 5% level of significance to test the student's claim that the results reported are random.

14.27 The following are 24 consecutive downtimes (in minutes) of a particular machine:

| 20 | 33 | 33 | 35 | 36 | 36 | 22 | 22 | 25 | 27 | 30 | 30 |
| 30 | 31 | 31 | 32 | 32 | 36 | 40 | 40 | 50 | 45 | 45 | 40 |

The null hypothesis of randomness is to be tested against the alternative that there is a trend. A MINITAB analysis of the number of runs above and below the median follows. Confirm that the number of runs is 4 and compute the value of z^*. Would you reject the hypothesis of randomness?

```
MTB > SET THE FOLLOWING TIMES IN C1
DATA> 20 33 33 35 36 36 22 22 25 27 30 30
DATA> 30 31 31 32 32 36 40 40 50 45 45 40
DATA> END DATA
MTB > RUNS ABOVE OR BELOW 32.5 FOR DATA IN C1

   C1

   K = 32.5000

   THE OBSERVED NO. OF RUNS =    4
   THE EXPECTED NO. OF RUNS = 13.0000
   12 OBSERVATIONS ABOVE K     12 BELOW
              THE TEST IS SIGNIFICANT AT 0.0002
```

14.28 The article "Water Boosts Hemoglobin's Love for Oxygen" (*Science News,* March 30, 1991, page 198) discusses the ability of the iron-rich protein pigment hemoglobin to carry oxygen throughout the body. The article states that the protein's conversion to an oxygen-loving state involves between 60 and 80 water molecules. Suppose 20 different determinations of the number of water molecules needed resulted in the following data:

| 79 | 75 | 69 | 70 | 70 | 65 | 75 | 75 | 65 | 70 |
| 60 | 62 | 63 | 63 | 67 | 65 | 70 | 60 | 65 | 62 |

a. Determine the median and the number of runs above and below the median.
b. Use the runs test to test these data for randomness about the median.
c. State your conclusion.

14.29 The following data were collected in an attempt to show that the number of minutes the city bus is late is steadily growing larger. The data are in order of occurrence.

| 6 | 1 | 3 | 9 | 10 | 10 | 2 | 5 | 5 | 6 |
| 12 | 3 | 7 | 8 | 9 | 4 | 5 | 8 | 11 | 14 |

At $\alpha = 0.05$, do these data show sufficient lack of randomness to support the claim?

14.30 Mrs. Brown attended a special reading course that claimed to improve reading speed and comprehension. She took a pretest (on which she scored 106) and a test at the end of each of 15 sessions. Her test scores are shown in the following table. Do these data support the claim of improvement at the $\alpha = 0.025$ level?

Session	1	2	3	4	5	6	7	8	9	10	11	12	13	14	15
Score	109	108	110	112	108	111	112	113	110	112	115	114	116	116	118

Section 14.5 ▼ RANK CORRELATION

14.31 The number of absences recorded at a lecture that met at 8 A.M. on Mondays and Thursdays last semester were (in order of occurrence)

| 5 | 16 | 6 | 9 | 18 | 11 | 16 | 21 | 14 | 17 | 12 | 14 | 10 |
| 6 | 8 | 12 | 13 | 4 | 5 | 5 | 6 | 1 | 7 | 18 | 26 | 6 |

Do these data show a randomness about the median value at $\alpha = 0.05$? Complete this test by using (1) critical values from Table 13 in Appendix F, and (2) the standard normal distribution.

14.32 In an attempt to answer the question, "Does the husband (h) or wife (w) do the family banking?" the results of a sample of 28 married customers doing the family banking show the following sequence of arrivals at the bank:

| w | w | w | w | h | w | h | h | h | h | w | w | w | w |
| w | h | h | w | w | w | h | h | h | h | w | h | h | w |

Do these data show lack of randomness with regard to whether the husband or wife does the family banking? Use $\alpha = 0.05$.

14.5 ▼ Rank Correlation

The rank correlation coefficient was developed by C. Spearman in the early 1900s. It is a nonparametric alternative to the linear correlation coefficient that was discussed in Chapters 3 and 13. Only rankings are used in the calculation of this coefficient. If the data are quantitative, each of the two variables must be ranked separately.

Spearman rank correlation coefficient

The **Spearman rank correlation coefficient**, r_s, is found by using the formula

$$r_s = 1 - \frac{6 \sum (d_i)^2}{n(n^2 - 1)} \qquad (14\text{-}11)$$

where d_i is the difference in the rankings and n is the number of pairs of data. The value of r_s will range from -1 to $+1$ and will be used in much the same manner as the linear correlation coefficient was used previously.

The null hypothesis that we will be testing is "there is no correlation between the two rankings." The alternative hypothesis may be either two-tailed, "there is correlation," or one-tailed, if we anticipate the existence of either positive or negative correlation. The critical region will be on the side(s) corresponding to the specific alternative that is expected. For example, if we suspect negative correlation, the critical region will be in the left-hand tail.

▼ ILLUSTRATION 14-11

Let's consider a hypothetical situation in which four judges rank five contestants in a contest. Let's identify the judges as A, B, C, and D and the contestants as a, b, c, d, and e. Table 14-4 lists the awarded rankings.

TABLE 14-4
Rankings for Five Contestants

Contestant	Judge			
	A	B	C	D
a	1	5	1	5
b	2	4	2	2
c	3	3	3	1
d	4	2	4	4
e	5	1	5	3

When we compare judges A and B, we see that they ranked the contestants in exactly the opposite order—perfect disagreement (see Table 14-5). From our previous work with correlation, we expect the calculated value for r_s to be exactly -1 for these data.

$$r_s = 1 - \frac{6\left[\sum (d_i)^2\right]}{n(n^2 - 1)} = 1 - \frac{(6)(40)}{(5)(5^2 - 1)} = 1 - \frac{240}{120} = \mathbf{-1}$$

TABLE 14-5
Rankings of A and B

Contestant	A	B	$d_i = A - B$	$(d)^2$
a	1	5	-4	16
b	2	4	-2	4
c	3	3	0	0
d	4	2	2	4
e	5	1	4	16
				40

When judges A and C are compared, we see that their rankings of the contestants are identical (see Table 14-6). We would expect to find a calculated correlation coefficient of $+1$ for these data.

$$r_s = 1 - \frac{(6)(0)}{(5)(5^2 - 1)} = 1 - 0 = \mathbf{1}$$

TABLE 14-6
Rankings of A and C

Contestant	A	C	$d_i = A - C$	$(d)^2$
a	1	1	0	0
b	2	2	0	0
c	3	3	0	0
d	4	4	0	0
e	5	5	0	0
				0

By comparing the rankings of judge A with those of judge B and then with those of judge C, we have seen the extremes: total agreement and total disagreement. Now let's compare the rankings of judge A with those of judge D (see Table 14-7). There seems to be no real agreement or disagreement here. Let's compute r_s:

$$r_s = 1 - \frac{(6)(24)}{(5)(5^2 - 1)} = 1 - \frac{144}{120} = 1 - 1.2 = -0.2$$

TABLE 14-7 Rankings of A and D

Contestant	A	D	$d_i = A - D$	$(d)^2$
a	1	5	−4	16
b	2	2	0	0
c	3	1	2	4
d	4	4	0	0
e	5	3	2	4
				24

This is fairly close to zero, which is what we should have suspected since there was no real agreement or disagreement.

The test of significance will result in a failure to reject the null hypothesis when r_s is close to zero and will result in a rejection of the null hypothesis in cases where r_s is found to be close to +1 or −1. The critical values found in Table 14 of Appendix F are positive critical values only. Since the null hypothesis is "the population correlation coefficient is zero" (that is, $\rho_s = 0$), we have a symmetric test statistic. Hence we need only add a plus or minus sign to the value found in the table, as appropriate. This will be determined by the specific alternative that we have in mind.

In our illustration, the critical values for a two-tailed test at $\alpha = 0.10$ are ±0.900. (Remember that n represents the number of pairs.) If the calculated value for r_s is between 0.9 and 1.0 or between −0.9 and −1.0, we will reject the null hypothesis in favor of the alternative "there is a correlation." ▲▲

The Spearman rank coefficient is defined by formula (3-1) [Pearson's product moment, r], with rankings used in place of the quantitative x and y values. If there are no ties in the rankings, formula (14-11) is equivalent to formula (3-1). Formula (14-11) provides us with a much easier procedure to use for calculating the r_s statistic. When there are only a few ties, it is common practice to use formula (14-11). The resulting value of r_s is not exactly the same; however, it is generally considered to be an acceptable estimate. Illustration 14-12 shows the procedure for handling ties and uses formula (14-11) for the calculation of r_s. For comparison, Exercise 14.43, p. 721, asks you to calculate r_s using formula (3-1).

If ties occur in either set of the ordered pairs of rankings, assign each tied observation the mean of the ranks that would have been assigned had there been no ties, as was done in the Mann-Whitney U test.

ILLUSTRATION 14-12

Students who finish exams more quickly than the rest of the class are often thought to be smarter. The following set of data shows the score and order of finish for 12 students on a recent one-hour exam. At the 0.05 level, do these data support the alternative hypothesis that the first students to complete an exam have higher grades?

Order of Finish	1	2	3	4	5	6	7	8	9	10	11	12
Exam Score	90	74	76	60	68	86	92	60	78	70	78	64

SOLUTION

STEP 1 H_0: Order of finish has no relationship to exam score.

H_a: First to finish tend to have higher grades.

STEP 2 $\alpha = 0.05$ and $n = 12$; the critical region is shown in the following diagram:

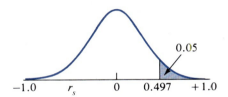

STEP 3   Rank the scores from higher to lowest, assigning the highest score the rank number 1, as shown.

92	90	86	78	78	76	74	70	68	64	60	60
1	2	3	4	5	6	7	8	9	10	11	12
			4.5	4.5						11.5	11.5

The rankings and the preliminary calculations are shown in Table 14-8. Using formula (14-11), we obtain

$$r_s = 1 - \frac{(6)(235.0)}{(12)(143)} = 1 - \frac{1410}{1716} = 1 - 0.822$$

$$= 0.178$$

STEP 4   **Decision:** Fail to reject H_0.

Conclusion   There is not sufficient evidence presented by these sample data to enable us to conclude that the first students to finish have the higher grades at the 0.05 level of significance.

TABLE 14-8
Rankings for Test Scores and Differences

Test Score Rank	Order of Finish	Difference (d_i)	$(d_i)^2$
1	7	−6	36.00
2	1	1	1.00
3	6	−3	9.00
4.5	9	−4.5	20.25
4.5	11	−6.5	42.25
6	3	3	9.00
7	2	5	25.00
8	10	−2	4.00
9	5	4	16.00
10	12	−2	4.00
11.5	4	7.5	56.25
11.5	8	3.5	12.25
			235.00

Case Study 14-2

Effects of Therapeutic Touch on Tension Headache Pain

Elizabeth Keller and Virginia Bzdek made use of the Spearman rank correlation coefficient to determine the influence of the subjects' education, age, and several other attributes on the results of therapeutic touch. They reported several observed values and the associated prob-values.

Therapeutic touch (TT) is a modern derivative of the laying on of hands that involves touching with the intent to help or heal. This study investigated the effects of TT on tension headache pain in comparison with a placebo simulation of TT. Sixty volunteer subjects with tension headaches were randomly divided into treatment and placebo groups. The McGill-Melzack Pain Questionnaire was used to measure headache pain levels before each intervention, immediately afterward, and 4 hours later....

HYPOTHESES

I. Tension headache pain will be reduced following therapeutic touch, and the initial reduction will be maintained for a 4-hour period.

II. Subjects who receive therapeutic touch will experience greater tension headache pain reduction than subjects receiving a placebo simulation of therapeutic touch.

III. Subjects who receive therapeutic touch will maintain greater tension headache pain reduction than subjects receiving a placebo simulation of therapeutic touch 4 hours following the intervention....

Findings

Spearman correlation coefficients were calculated to determine the influence of the subject's education, age, sex, practice of meditation, religion,

> and level of initial skepticism toward TT. The differences in the MMPQ pain scores were not significantly correlated with any belief system or demographic variable in either group. The one exception was the posttest difference on the PPI scale in the placebo group, which was inversely correlated with years of education, $r = -.53, p = .002$.
>
> The MMPQ was internally consistent in this study to a statistically significant degree between all three subscales on the pretest, posttest, and delayed posttest. The highest correlation was between the PRI and the NWC, $r = .79, p < .0001$. The lowest correlation was between NWC and the PPI, $r = .30, p < .02$. When one subscale reported a significant difference between the groups, the other two scales also reported significant differences in the same direction. Based on these observations of internal consistency, the MMPQ was considered to be a reliable instrument in this study.
>
> *Source*: Elizabeth Keller and Virginia M. Bzdek, *Nursing Research,* March/April 1986, Vol. 35, no. 2.

▼▲ EXERCISES

14.33 State the null hypothesis, H_0, and the alternative hypothesis, H_a, that would be used to test the following statements:
   a. There is no relationship between the two rankings.
   b. The two variables are unrelated.
   c. There is a positive correlation between the two variables.
   d. Age has a decreasing effect on monetary value.

14.34 Determine the test criteria that would be used to test the null hypothesis for the following multinomial experiments:
   a. H_0: No relationship between the two variables.
   H_a: There is a relationship,
   with $n = 14$ and $\alpha = 0.05$.
   b. H_0: No correlation.
   H_a: Positively correlated,
   with $n = 27$ and $\alpha = 0.05$.
   c. H_0: Variable A has no effect on variable B.
   H_a: Variable B decreases as A increases,
   with $n = 18$ and $\alpha = 0.01$.

14.35 The following data involving molecular weight and percent absorption in humans for 13 compounds are extracted from an article about nasal absorption of drugs (*Journal of Pharmaceutical Sciences,* July 1987, pages 535–540):

Molecular Wt.	362	1007	1056	1069	1069	1182	1182	1182	1209	1297	1337	4800	5000
% Absorption	40	10	10	15	40	1	2	2.5	15	2.5	5	2	1

The article states that the Spearman rank correlation coefficient equals −0.67. Verify this result.

14.36 An advertising company wants to determine whether the number of television commercials (x) is related to the number of sales (y) of a certain product. The following data were obtained:

x	10	12	15	5	7	5	5	15	8	1	13	15
y	30	60	50	12	10	25	10	60	25	10	75	95

a. State the null and alternative hypotheses.
b. Compute the Spearman rank correlation coefficient, r_s.
c. At the 0.05 level of significance, is there sufficient reason to conclude that a positive correlation exists?

14.37 The following data represent the ages of 12 subjects and the mineral concentration (in parts per million) in their tissue samples:

Age (x)	82	83	64	53	47	50	70	62	34	27	75	28
Mineral Concentration (y)	170	40	64	5	15	5	48	34	3	7	50	10

Refer to the following MINITAB output and verify that the Spearman rank correlation coefficient equals 0.753.

```
MTB > READ AGE IN C1, PPM IN C2
DATA> 82 170
DATA> 83 40
DATA> 64 64
DATA> 53 5
DATA> 47 15
DATA> 50 5
DATA> 70 48
DATA> 62 34
DATA>34 3
DATA> 27 7
DATA> 75 50
DATA> 28 10
DATA> END DATA
      12 ROWS READ
MTB > RANK C1,PUT RANKS IN C3
MTB > RANK C2, PUT RANKS IN C4
MTN > CORRELATION C3,C4

Correlation of C3 and C4 = 0.753
```

Chapter 14 ▼ ELEMENTS OF NONPARAMETRIC STATISTICS

14.38 The following sets of air pollution readings were taken from a local newspaper on ten separate days:

	Day									
Pollution	1	2	3	4	5	6	7	8	9	10
Sulfur Dioxide	0.07	0.01	0.01	0.02	0.07	0.02	0.00	0.01	0.05	0.04
Carbon Monoxide	0.9	2.0	0.8	0.8	0.9	1.3	3.8	2.0	1.3	2.5
Soiling or Dust	0.3	0.5	0.3	0.3	0.3	0.5	0.3	0.5	0.3	0.5

Use a computer to complete this exercise. See Exercises 3.15 and 14.37 for the necessary commands.
   a. Draw a scatter diagram showing the sulfur dioxide and carbon monoxide readings as paired data. Use Spearman's rank correlation coefficient to test the correlation between the sulfur dioxide and the carbon monoxide readings. Complete a two-tailed test at $\alpha = 0.10$.
      (1) State the null hypothesis being tested.
      (2) Complete the test and carefully state the conclusion reached.
   b. Repeat (a) using the sulfur dioxide and the soiling readings ($\alpha = 0.05$).
   c. Repeat (a) using the carbon monoxide and the soiling readings ($\alpha = 0.01$).

 14.39 An article in *Self* magazine (February 1991, page 52) discusses the relationship between the pace of life and the coronary-heart-disease death rate. New York City, for example, was ranked as only the third-fastest-paced city but number one for deadly heart attacks. Suppose the data from another such study involving eight cities were as follows:

City	Rank for Pace-of-Life	Rank for Heart-Disease Death Rate
Salt Lake City	4	7
Buffalo	2	2
Columbus	5	8
Worcester	6	4
Boston	1	6
Providence	8	3
New York	3	1
Paterson	7	5

Find the Spearman rank correlation coefficient.

 14.40 A computer-science aptitude test score, x, and a mathematical competency score, y, were determined for 15 students. The results are shown in the following table:

x	4	16	20	13	22	21	15	20	19	16	18	17	8	6	5
y	28	35	42	41	44	42	36	44	39	36	40	40	33	27	44

a. Find r_s.
b. Is there sufficient evidence to conclude that the two test scores are correlated? Use $\alpha = 0.05$.

14.41 The Spearman rank correlation coefficient between urine volume and salicylic acid excretion was reported to be 0.539 in a study involving 129 healthy volunteers (*Xenobiotica*, 1986, Vol. 16, no. 2, 239–249). Assuming there were no or few ties in the rankings, find Σd^2.

14.42 "The highest correlation was between the PRI and the NWC, $r = .79, p < .0001$. The lowest correlation was between NWC and the PPI, $r = .30, p < .02$." These statements are from the article in Case Study 14-2 on page 717. Explain the meaning of this quote.

14.43 Using formula (3-2), calculate the Spearman rank correlation coefficient for the data in Illustration 14-12, p. 716. Recall that formula (3-2) is equivalent to the definition formula (3-1) and that rank numbers must be used with this formula in order for the resulting statistic to be Spearman's r_s.

14.44 Refer to the bivariate data shown in the following table:

x	−2	−1	1	2
y	4	1	1	4

a. Construct a scatter diagram.
b. Calculate Spearman's correlation coefficient, r_s [formula (14-11)].
c. Calculate Pearson's correlation coefficient, r [formula (3-2)].
d. Compare the two results from (b) and (c). Do the two measures of correlation measure the same thing?

14.6 ▼ Comparing Statistical Tests

Only a few nonparametric tests have been presented in this chapter. Many more nonparametric tests can be found in other books. In addition, most of these other nonparametric tests can be used in place of certain parametric tests. The question is, then, which test do we use, the parametric or the nonparametric? Furthermore, when there is more than one nonparametric test, which one do we use?

The decision about which test to use must be based on the answer to the question "Which test will do the job best?" First, let's agree that we are dealing with two or more tests that are equally qualified to be used. That is, each test has a set of assumptions that must be satisfied before it can be applied. From this starting point we will attempt to define "best" to mean the test that is best able to control the risks of error and at the same time keep the size of the sample to a number that is reasonable to work with. (Sample size means cost, cost to you or your employer.)

Let's look first at the ability to *control the risk of error*. The risk associated with a type I error is controlled directly by the level of significance α. Recall that P (type I error) $= \alpha$ and P (type II error) $= \beta$. Therefore, it is β that we must control. Statisticians like to talk about power (as do others) and the **power of a statistical test** is defined to be $1 - \beta$. Thus the power of a test, $1 - \beta$, is the probability that we reject the null hypothesis when we should have rejected it. If two tests with the same α are equal candidates for use, the one with the greater power is the one you would want to choose.

The other factor is the *sample size required* to do a job. Suppose that you set the levels of risk that you can tolerate, α and β, and then are able to determine the sample size that it would take to meet your specified challenge. The test that required the smaller sample size would then seem to have the edge. Statisticians usually use the term *efficiency* to talk about this concept. **Efficiency** is defined to be the ratio of the sample size of the best parametric test to the sample size of the best nonparametric test when compared under a fixed set of risk values. For example, the efficiency rating for the sign test is approximately 0.63. This means that a sample of size 63 with a parametric test will do the same job as a sample of size 100 will do with the sign test.

The power and the efficiency of a test cannot be used alone to determine the choice of test. Sometimes you will be forced to use a certain test because of the data you are given. When there is a decision to be made, the final decision rests in a trade-off of three factors: (1) the power of the test, (2) the efficiency of the test, and (3) the data (and the number of data) available. Table 14-9 shows how the nonparametric tests discussed in this chapter compare with the parametric tests covered in previous chapters.

TABLE 14-9 Comparison of Parametric and Nonparametric Tests

Test Situation	Parametric Test	Nonparametric Test	Efficiency of Nonparametric Test
One mean	t test (page 439)	Sign test (page 691)	0.63
Two independent means	t test (page 515)	U test (page 699)	0.95
Two dependent means	t test (page 489)	Sign test (page 692)	0.63
Correlation	Pearson's (page 646)	Spearman test (page 713)	0.91
Randomness		Runs test (page 708)	Not meaningful; there is no parametric test for comparison

IN RETROSPECT

In this chapter you have become acquainted with some of the basic concepts of nonparametric statistics. While learning about the use of nonparametric methods and specific nonparametric tests of significance, you should have also come to realize and understand some of the basic assumptions that are needed when the parametric techniques of the earlier chapters are encountered. You now have seen a variety of tests, many of which somewhat duplicate the job done by others. What you must keep in mind is that you should use the best test for your particular needs. The power of the test and the cost of sampling, as related to size and availability of the desired response variable, will play important roles in determining the specific test to be used.

The article at the beginning of this chapter illustrates only one of many situations in which a nonparametric test can be used. Spearman's rank correlation test may be used to statistically compare the two rankings. They appear to be quite different, but are they significantly different? The answer to this is left for you to determine in Exercise 14.60.

CHAPTER EXERCISES

14.45 Is the absentee rate in the 8 A.M. statistics class the same as in the 11 A.M. statistics class? The following sample of the daily number of absences was taken from the attendance records of the two classes:

	Day											
Class	1	2	3	4	5	6	7	8	9	10	11	12
8 A.M.	0	1	3	1	0	2	4	1	3	5	3	2
11 A.M.	1	0	1	0	1	2	3	0	1	3	2	1

Is there sufficient reason to conclude that there is more absence at the 8 A.M. class? Use $\alpha = 0.05$.

14.46 Track coaches, runners, and fans talk a lot about the "speed of the track." The surface of the track is believed to have a direct effect on the amount of time that it takes a runner to cover the required distance. To test this effect, ten runners were asked to run a 220-yard sprint on each of two tracks. Track A is a cinder track and track B is made of a new synthetic material. The running times are given in the following table. Test the claim that the surface on track B is conducive to faster running times.

	Runner									
Track	1	2	3	4	5	6	7	8	9	10
A	27.7	26.8	27.0	25.5	26.6	27.4	27.2	27.4	25.8	25.1
B	27.0	26.7	25.3	26.0	26.1	25.3	26.7	27.1	24.8	27.1

a. State the null and alternative hypotheses being tested. Complete the test by using the sign test with $\alpha = 0.05$.
b. State your conclusions.

14.47 A test that measures computer anxiety was administered to students in a statistics course that uses statistical packages on the computer. The test was given at the beginning and end of the course. The hypothesis being studied was that computer anxiety would be reduced as the students used computers in the course. (A high score on the test indicates high computer anxiety.) Can we conclude that computer anxiety was reduced during this course? Use the sign test and $\alpha = 0.05$.

Beginning	68	79	49	34	60	87	68	70	60	67	39	59	42	40	62	80	62	63	102	42	68	70	44	63	55
End	77	78	32	40	60	68	87	88	62	68	40	60	31	39	62	63	58	59	110	49	54	67	43	54	41

14.48 A candy company has developed two new chocolate-covered candy bars. Six randomly selected people all preferred candy bar I. Is this statistical evidence, at $\alpha = 0.05$, that the general public will prefer candy bar I?

14.49 While trying to decide on the best time to harvest his crop, a commercial apple farmer recorded the day on which the first apple on the top half and the first apple on the bottom half of 20 randomly selected trees were ripe. The variable x was assigned a value of 1 on the first day that the first ripe apple appeared on 1 of the 20 trees. The days were then numbered sequentially. The observed data are shown in the following table. Do these data provide convincing evidence that the apples on the top of the trees start to ripen before the apples on the bottom half? Use $\alpha = 0.05$.

	Tree																				
Position	1	2	3	4	5	6	7	8	9	10	11	12	13	14	15	16	17	18	19	20
Top	5	6	1	4	5	3	6	7	8	5	8	6	4	7	8	10	3	2	9	7
Bottom	6	5	5	7	3	6	6	8	9	4	10	7	5	11	6	11	5	6	9	8

14.50 A sample of 32 students received the following grades on an exam:

41	42	48	46	50	54	51	42
51	50	45	42	32	45	43	56
55	47	45	51	60	44	57	57
47	28	41	42	54	48	47	32

a. Does this sample show that the median score for the exam differs from 50? Use $\alpha = 0.05$.
b. Does this sample show that the median score for the exam is less than 50? Use $\alpha = 0.05$.

CHAPTER EXERCISES

14.51 An article in the journal *Sedimentary Geology* (Vol. 57, 1988, pages 119–129) compares a measure called the *roughness coefficient* for translucent and opaque quartz sand grains. If you measured the roughness coefficient for 20 sand grains of each type (translucent and opaque), for what values of the Mann-Whitney U statistic would you reject the null hypothesis in a two-tailed test with alpha equal to 0.05?

14.52 Twenty students were randomly divided into two equal groups. Group 1 was taught an anatomy course using a standard lecture approach. Group 2 was taught using a computer-assisted approach. The test scores on a comprehensive final exam were as follows:

Group 1	75	83	60	89	77	92	88	90	55	70
Group 2	77	92	90	85	72	59	65	92	90	79

Use the Mann-Whitney U test to test the claim that a computer-assisted approach produces higher achievement (as measured by final exam scores) in anatomy courses than does a lecture approach. Use $\alpha = 0.05$.

14.53 The use of nuclear magnetic resonance (NMR) spectroscopy for detection of malignancy is discussed in the journal *Clinical Chemistry* (Vol. 34, no. 3, 1988, pages 505–511). The line width at the half height of peaks in the NMR spectra is measured. The spectra is produced from assaying plasma from an individual. Suppose the following line widths were obtained from a normal group and a group known to have malignancies:

Normal Group	35.1	32.9	30.6	30.5	30.9
Malignancy Group	28.5	29.5	30.7	27.5	28.0

a. Compute the Mann-Whitney U statistic.
b. Would you reject a two-tailed research hypothesis at the 0.05 level of significance?

14.54 A firm is currently testing two different procedures for adjusting the cutting machines used in the production of greeting cards. The results of two samples show the following recorded adjustment times:

Method 1	17	15	14	18	16	15	17	18	15	14	14	16	15		
Method 2	14	14	13	13	15	12	16	14	16	13	14	13	12	15	17	13

Is there sufficient reason to conclude that method 2 requires less time (on the average) than method 1 at the 0.05 level of significance?

14.55 The fixed interest rates charged on auto loans varies between banks and other lending institutions. Some of the rates charged by several banks and credit unions that were listed in the *Democrat and Chronicle*, May 6, 1991, were

Banks	12.50	12.50	11.10	12.00	12.25
	11.25	12.25	10.99	11.25	11.25
	12.90	11.40	12.25	11.25	11.98
Credit Unions	9.75	10.25	11.25	9.90	10.40
	11.00	10.25	9.95	10.80	8.90
	10.90	11.75			

Do these data present sufficient evidence to conclude that the credit unions charge a lower fixed auto-loan rate than the banks charge? Use $\alpha = 0.05$.

14.56 Two table tennis ball manufacturers have agreed that the quality of their products can be measured by the height to which the balls rebound. A test is arranged, the balls are dropped from a constant height, and the rebound heights are measured. The results (in inches) are shown in the following table. Manufacturer A claims, "The results show my product to be superior." Manufacturer B replies, "I know of no statistical test that supports this claim." Can you find a test that supports A's claim?

A	14.0	12.5	11.5	12.2	12.4	12.3	11.8	11.9	13.7	13.2
B	12.0	12.5	11.6	13.3	13.0	13.0	12.1	12.8	12.2	12.6

a. Does the appropriate parametric test show that A's product is superior? [What parametric test (or tests) is appropriate and what exactly does it show?]

b. Does the appropriate nonparametric test show that A's product is superior?

14.57 Consider the following sequence of defective parts (d) and nondefective parts (n) produced by a machine:

```
n  n  n  d  n  n  n  n  n  d  n  n
n  n  n  n  n  d  n  d  n  n  n  n
```

Can we reject the hypothesis of randomness at $\alpha = 0.05$?

14.58 A patient was given two different types of vitamin pills, one containing iron and one iron-free. The patient was instructed to take the pills on alternate days. To free himself from remembering which pill he needed to take, he mixed all of the pills together in a large bottle. Each morning he took the first pill that came out of the bottle. To see whether this was a random process, for 25 days he recorded an "I" each morning that he took a vitamin with iron and an "N" for no iron.

Day	1	2	3	4	5	6	7	8	9	10	11	12	13
Type	I	I	N	I	I	N	N	I	N	N	N	N	N
Day	14	15	16	17	18	19	20	21	22	23	24	25	
Type	I	I	I	N	I	I	I	I	N	I	I	N	

Is there sufficient reason to reject the null hypothesis that the vitamins were taken in a random order at the 0.05 level of significance?

 14.59 Can today's high temperature be effectively predicted using yesterday's high? Pairs of yesterday's and today's high temperatures were randomly selected. The results are shown in the following table. Do the data present sufficient evidence to justify the statement "Today's high temperature tends to correlate with yesterday's high temperature"? Use $\alpha = 0.05$.

	Pairs of Days																	
Reading	1	2	3	4	5	6	7	8	9	10	11	12	13	14	15	16	17	18
Yesterday's	40	58	46	33	40	51	55	81	85	83	89	64	73	63	46	58	28	69
Today's	40	56	34	59	46	51	74	77	83	84	85	68	65	60	54	62	34	66

 14.60 The article at the beginning of this chapter shows the rankings assigned to ten components of job satisfaction. Do the rankings assigned by the workers and the boss show a significant difference in what each thinks is important? Test by using $\alpha = 0.05$.

 14.61 In a study to see whether spouses are consistent in their preferences for television programs, a market research firm asked several married couples to rank a list of 12 programs (1 represents the highest score; 12 represents the lowest). The average ranks for the programs, rounded to the nearest integer, were as follows:

	Program											
Rank	1	2	3	4	5	6	7	8	9	10	11	12
Husbands	12	2	6	10	3	11	7	1	9	5	8	4
Wives	5	4	1	9	3	12	2	8	6	10	7	11

Is there significant evidence of negative correlation at the 0.01 level of significance?

14.62 Nonparametric tests are also called *distribution-free* tests. However, the normal and the chi-square distributions are used in the inference-making procedures.
    a. To what does the *distribution-free* term apply? (The population? The sample? The sampling distribution?) Explain.
    b. What is it that has the normal or the chi-square distribution? Explain.

VOCABULARY LIST

Be able to define each term. In addition, describe in your own words and give an example of each term. Your examples should not be ones given in class or in the textbook.

The bracketed numbers indicate the chapters in which the term previously appeared, but you should define the terms again to show increased understanding of their meaning.

binomial random variable [5, 9]
continuity correction [6]
correlation [3, 13]
distribution [2, 5, 6, 7]
distribution-free test
efficiency
independent sample [10]
Mann-Whitney U test
median, M [2]
nonparametric test
normal approximation [6]

paired data [3, 10, 13]
parametric test
power
randomness [2, 7]
rank
run
runs test
sign test
Spearman rank correlation coefficient
test statistic

KEY CONCEPTS

distribution-free test
Mann-Whitney U test
median [2]
randomness [2]
rank

runs test
sign test
Spearman rank correlation coefficient

QUIZ A

Answer "True" if the statement is always true. If the statement is not always true, replace the words shown in bold with words that make the statement always true.

14.1 One of the advantages that the nonparametric tests have is the necessity for **less restrictive** assumptions.

14.2 The sign test is a possible replacement for the **F test**.

14.3 The **sign test** can be used to test the randomness of a set of data.

14.4 If a tie occurs in a set of ranked data, the data that form the tie are **removed from the set**.

14.5 Two dependent **means** can be compared nonparametrically by using the sign test.

14.6 The sign test is a possible alternative to the Student's t test for **one mean value**.

14.7 The **runs test** is a nonparametric alternative to the difference between two independent means.

14.8 The **confidence level** of a statistical hypothesis test is measured by $1 - \beta$.

14.9 Spearman's rank correlation coefficient is an alternative to using the **linear correlation coefficient**.

14.10 The **efficiency** of a nonparametric test is the probability that a false null hypothesis is rejected.

QUIZ B

14.1 The weights of nine people before they stopped smoking and five weeks after they stopped smoking are as follows:

Person	1	2	3	4	5	6	7	8	9
Before	148	176	153	116	128	129	120	132	154
After	155	178	151	120	130	136	126	128	158

Find the 0.95 confidence interval estimate for the average weight change.

14.2 The following data show the weight gains for 20 laboratory mice, half of which were fed one diet and half a different diet. Test to determine if the difference in weight gain is significant at $\alpha = 0.05$.

Diet A	41	40	36	43	36	43	39	36	24	41
Diet B	35	34	27	39	31	41	37	34	42	38

14.3 A large textbook publishing company hired nine new sales representatives three years ago. At the time of hire, the nine were rated according to their potential. Now three years later the company president wants to know how well their potential rates correlate with their sales totals for the three years.

Sales Rep.	a	b	c	d	e	f	g	h	i
Potential	2	5	6	1	4	3	9	8	7
Sales Tot.	450	410	350	345	330	400	250	310	270

Is there significant correlation at the 0.05 level?

14.4 The new school principal thought there might be a pattern to the order in which discipline problems arrived at his office. He had his secretary record the grade level of the students as they arrived.

9 10 11   9 12 11   9 10 10 11 10 11 10 10 11
12 12   9   9 11 12 10   9 12 10 11 12 11 10 10

At the 0.05 level, is there significant evidence of randomness?

QUIZ C

14.1 What advantages do nonparametric statistics have over parametric methods?

14.2 Explain how the sign test is based on the binomial distribution and is often approximated by the normal distribution.

14.3 Why does the sign test use a null hypothesis about the median instead of the mean like a *t* test uses?

14.4 Explain why a nonparametric test is not as sensitive to an extreme datum as a parametric test might be.

14.5 A restaurant has collected data on which of two seating arrangements its customers prefer. In a sign test to determine if one seating arrangement is significantly preferred, the null hypothesis would be
   a. $\mu = 0$
   b. $\mu = 0.5$
   c. $p = 0$
   d. $p = 0.5$

Explain your choice.

WORKING WITH YOUR OWN DATA

Many variables in everyday life can be treated as bivariate. Often two such variables have a mathematical relationship that can be approximated by means of a straight line. The following illustration demonstrates such a situation.

A ▼ The Age and Value of Peggy's Car

Peggy would like to sell her 1988 Corvette, and she needs to know what price to ask for it in order to write a newspaper advertisement. The car is in average condition and Peggy expects to get an average price for it. ("Average for a Corvette!") She must answer the question "What is an average selling price for a 1988 Corvette?"

Inspection of many classified sections of newspapers turned up only four advertisements for 1988 Corvettes. The prices listed varied a great deal. Peggy finally decided that, in order to determine an accurate selling price, she would define two variables and collect several pairs of values based on the following definitions:

POPULATION Used Chevrolet Corvettes advertised for sale by individual owners, dealers not included.

INDEPENDENT VARIABLE, x The age of the car as measured in years and defined by

$$x = \text{(present calendar year)} - \text{(year of manufacture)} + 1.$$

Example During 1991, Peggy's 1988 Corvette is considered to be 4 years old.

$$x = (1991 - 1988) + 1 = 3 + 1 = 4$$

DEPENDENT VARIABLE, y The advertised asking price.

The table on the next page lists the data collected in September 1991.

Year of Manufacture:			1989	1990	1985	1980	1987	1983
Asking Price:			$22,900	$27,500	$19,500	$13,500	$18,900	$15,750
Year:	1985	1986	1981	1984	1985	1989	1986	1987
Price:	$15,500	$16,700	10,900	14,500	13,900	27,500	19,900	23,500
Year:	1988	1987	1989	1980	1990	1986	1989	1987
Price:	23,900	18,500	21,900	10,500	24,900	17,900	26,250	19,950
Year:	1988	1984	1988	1986	1983	1988	1986	1990
Price:	25,900	16,900	20,500	20,500	12,000	24,500	21,500	29,250
Year:	1982	1987						
Price:	14,000	21,500						

1. Construct and label a scatter diagram of Peggy's data.
2. Determine the equation for the line of best fit.
3. Draw the line of best fit on the scatter diagram.
4. Test the equation of the line of best fit to see whether the linear model is appropriate for the data. Use $\alpha = 0.05$.
5. Construct a 0.95 confidence interval estimate for the mean advertised price for 1983 Corvettes.
6. Draw a line segment on the scatter diagram that represents the interval estimate found for question 5.
7. What does the value of the slope, b_1, represent? Explain.
8. What does the value of the y-intercept, b_0, represent? Explain.

B ▼ Your Own Investigation

Identify a situation of interest to you that can be investigated statistically using bivariate data. (Consult your instructor for specific guidance.)

1. Define the population, the independent variable, the dependent variable, and the purpose for studying these two variables as a regression analysis.
2. Collect 15 to 20 ordered pairs of data.
3. Construct and label a scatter diagram of your data.
4. Determine the equation for the line of best fit.
5. Draw the line of best fit on the scatter diagram.
6. Test the equation of the line of best fit to see whether the linear model is appropriate for the data. Use $\alpha = 0.05$.
7. Construct a 0.95 confidence interval estimate for the mean value of the dependent variable at the following value of x: Let x be equal to one-third the sum of the lowest value of x in your sample and twice the largest value. That is,

$$x = \frac{L + 2H}{3}$$

8. Draw a line segment on the scatter diagram that represents the interval estimate found for question 7.
9. What does the value of the slope, b_1, represent? Explain.
10. What does the value of the y-intercept, b_0, represent? Explain.

APPENDIXES

A	Summation Notation	A-1
B	Using the Random Number Table	B-1
C	Round-off Procedure	C-1
D	Basic Principles of Counting	D-1
E	Interpolation Procedure for F Distribution	E-1
F	Tables	F-1
1.	Random Numbers	F-1
2.	Factorials	F-3
3.	Binomial Coefficients	F-4
4.	Binomial Probabilities $\left[\binom{n}{x} \cdot p^x q^{n-x} \right]$	F-5
5.	Areas of the Standard Normal Distribution	F-8
6.	Critical Values of Student's t-Distribution	F-9
7.	Critical Values of the χ^2 Distribution	F-10
8a.	Critical Values of the F Distribution ($\alpha = 0.05$)	F-11
8b.	Critical Values of the F Distribution ($\alpha = 0.025$)	F-13
8c.	Critical Values of the F Distribution ($\alpha = 0.01$)	F-15
9.	Critical Values of r when $\rho = 0$	F-17
10.	Confidence Belts for the Correlation Coefficient ($1 - \alpha = 0.95$)	F-18
11.	Critical Values for the Sign Test	F-19
12.	Critical Values of U in the Mann-Whitney Test	F-20
13.	Critical Values for Total Number of Runs (V)	F-21
14.	Critical Values of Spearman's Rank Correlation Coefficient	F-22

APPENDIX A

SUMMATION NOTATION

The Greek capital letter sigma (Σ) is used in mathematics to indicate the summation of a set of addends. Each of these addends must be of the form of the variable following Σ. For example:

1. Σx means sum the variable x.
2. $\Sigma(x - 5)$ means sum the set of addends that are each 5 less than the values of each x.

When large quantities of data are collected, it is usually convenient to index the response variable so that at a future time its source will be known. This indexing is shown on the notation by using i (or j or k) and affixing the index of the first and last addend at the bottom and top of the Σ. For example,

$$\sum_{i=1}^{3} x_i$$

means to add all the consecutive values of x's starting with source number 1 and proceeding to source number 3.

▼ **ILLUSTRATION A-1**

Consider the inventory in the following table concerning the number of defective stereo tapes per lot of 100.

Lot Number (i)	1	2	3	4	5	6	7	8	9	10
Number of Defective Tapes per Lot (x)	2	3	2	4	5	6	4	3	3	2

a. Find $\sum_{i=1}^{10} x_i$. **b.** Find $\sum_{i=4}^{8} x_i$.

SOLUTION

a. $\sum_{i=1}^{10} x_i = x_1 + x_2 + x_3 + x_4 + \cdots + x_{10}$
$= 2 + 3 + 2 + 4 + 5 + 6 + 4 + 3 + 3 + 2 = \mathbf{34}$

b. $\sum_{i=4}^{8} x_i = x_4 + x_5 + x_6 + x_7 + x_8 = 4 + 5 + 6 + 4 + 3 = \mathbf{22}$ ▲▲

This index system must be used whenever only part of the available information is to be used. In statistics, however, we will usually use all the available information, and to simplify the formulas we will make an adjustment. This adjustment is actually an agreement that allows us to do away with the index system in situations where all values are used. Thus in our previous illustration, $\sum_{i=1}^{10} x_i$ could have been written simply as Σx.

NOTE The lack of the index indicates that all data are being used.

▼ ILLUSTRATION A-2

Given the following six values for x, 1, 3, 7, 2, 4, 5, find Σx.

SOLUTION

$$\Sigma x = 1 + 3 + 7 + 2 + 4 + 5 = \mathbf{22}$$

▲▲

Throughout the study and use of statistics you will find many formulas that use the Σ symbol. Care must be taken so that the formulas are not misread. Symbols like Σx^2 and $(\Sigma x)^2$ are quite different. Σx^2 means "square each x value and then add up the squares," while $(\Sigma x)^2$ means "sum the x values and then square the sum."

▼ ILLUSTRATION A-3

Find (a) Σx^2 and (b) $(\Sigma x)^2$ for the sample in Illustration A-2.

SOLUTION

a.

x	1	3	7	2	4	5
x^2	1	9	49	4	16	25

$$\Sigma x^2 = 1 + 9 + 49 + 4 + 16 + 25 = \mathbf{104}$$

b. $\Sigma x = 22$, as found in Illustration A-2. Thus,

$$(\Sigma x)^2 = (22)^2 = \mathbf{484}$$

As you can see, there is quite a difference between Σx^2 and $(\Sigma x)^2$.

▲▲

Likewise, Σxy and $\Sigma x \Sigma y$ are different. These forms will appear only when there are paired data, as shown in the following illustration.

▼ ILLUSTRATION A-4

Given the five pairs of data shown in the following table, find (a) Σxy and (b) $\Sigma x \Sigma y$.

x	1	6	9	3	4
y	7	8	2	5	10

SOLUTION

a. Σxy means to sum the products of the corresponding x and y values. Therefore, we have

x	1	6	9	3	4
y	7	8	2	5	10
xy	7	48	18	15	40

$$\Sigma xy = 7 + 48 + 18 + 15 + 40 = \mathbf{128}$$

b. $\Sigma x \Sigma y$ means the product of the two summations, Σx and Σy. Therefore, we have

$$\Sigma x = 1 + 6 + 9 + 3 + 4 = 23$$
$$\Sigma y = 7 + 8 + 2 + 5 + 10 = 32$$
$$\Sigma x \Sigma y = (23)(32) = \mathbf{736}$$

▲▲

There are three basic rules for algebraic manipulation of the Σ notation.

NOTE c represents any constant value.

RULE 1

$$\sum_{i=1}^{n} c = nc$$

To prove this rule, we need only write down the meaning of $\sum_{i=1}^{n} c$:

$$\sum_{i=1}^{n} c = \underbrace{c + c + c + \cdots + c}_{n \text{ addends}}$$

Therefore,

$$\sum_{i=1}^{n} c = n \cdot c$$

Appendix A ▼ SUMMATION NOTATION

▼ **ILLUSTRATION A-5**

Show that $\sum_{i=1}^{5} 4 = (5)(4) = 20$.

SOLUTION

$$\sum_{i=1}^{5} 4 = \underbrace{4_{(\text{when }i=1)} + 4_{(\text{when }i=2)} + 4_{(i=3)} + 4_{(i=4)} + 4_{(i=5)}}_{\text{five 4s added together}}$$

$$= (5)(4) = \mathbf{20}$$

▲▲

RULE 2

$$\sum_{i=1}^{n} cx_i = c \cdot \sum_{i=1}^{n} x_i$$

To demonstrate the truth of Rule 2, we will need to expand the term $\sum_{i=1}^{n} cx_i$ and then factor out the common term c.

$$\sum_{i=1}^{n} cx_i = cx_1 + cx_2 + cx_3 + \cdots + cx_n$$

$$= c(x_1 + x_2 + x_3 + \cdots + x_n)$$

Therefore,

$$\sum_{i=1}^{n} cx_i = c \cdot \sum_{i=1}^{n} x_i$$

RULE 3

$$\sum_{i=1}^{n} (x_i + y_i) = \sum_{i=1}^{n} x_i + \sum_{i=1}^{n} y_i$$

The expansion and regrouping of $\sum_{i=1}^{n} (x_i + y_i)$ is all that is needed to show this rule.

$$\sum_{i=1}^{n} (x_i + y_i) = (x_1 + y_1) + (x_2 + y_2) + \cdots + (x_n + y_n)$$

$$= (x_1 + x_2 + \cdots + x_n) + (y_1 + y_2 + \cdots + y_n)$$

Therefore,

$$\sum_{i=1}^{n} (x_i + y_i) = \sum_{i=1}^{n} x_i + \sum_{i=1}^{n} y_i$$

▼ **ILLUSTRATION A-6**

Show that $\sum_{i=1}^{3}(2x_i + 6) = 2 \cdot \sum_{i=1}^{3} x_i + 18$.

SOLUTION

$$\sum_{i=1}^{3}(2x_i + 6) = (2x_1 + 6) + (2x_2 + 6) + (2x_3 + 6)$$
$$= (2x_1 + 2x_2 + 2x_3) + (6 + 6 + 6)$$
$$= (2)(x_1 + x_2 + x_3) + (3)(6)$$
$$= 2\sum_{i=1}^{3} x_i + 18$$ ▲▲

▼ **ILLUSTRATION A-7**

Let $x_1 = 2, x_2 = 4, x_3 = 6, f_1 = 3, f_2 = 4,$ and $f_3 = 2$. Find $\sum_{i=1}^{3} x_i \cdot \sum_{i=1}^{3} f_i$.

SOLUTION

$$\sum_{i=1}^{3} x_i \cdot \sum_{i=1}^{3} f_i = (x_1 + x_2 + x_3) \cdot (f_1 + f_2 + f_3)$$
$$= (2 + 4 + 6) \cdot (3 + 4 + 2)$$
$$= (12)(9) = \mathbf{108}$$ ▲▲

▼ **ILLUSTRATION A-8**

Using the same values for the x's and f's as in Illustration A-7, find $\Sigma(xf)$.

SOLUTION Recall that the use of no index numbers means "use all data."

$$\Sigma(xf) = \sum_{i=1}^{3}(x_i f_i) = (x_1 f_1) + (x_2 f_2) + (x_3 f_3)$$
$$= (2 \cdot 3) + (4 \cdot 4) + (6 \cdot 2) = 6 + 16 + 12 = \mathbf{34}$$ ▲▲

▼▲ **EXERCISES**

A.1 Write each of the following in expanded form (without the summation sign):

a. $\sum_{i=1}^{4} x_i$ b. $\sum_{i=1}^{3}(x_i)^2$ c. $\sum_{i=1}^{5}(x_i + y_i)$

d. $\sum_{i=1}^{5}(x_i + 4)$ e. $\sum_{i=1}^{8} x_i y_i$ f. $\sum_{i=1}^{4} x_i^2 f_i$

A.2 Write each of the following expressions as summations, showing the subscripts and the limits of summation:
   a. $x_1 + x_2 + x_3 + x_4 + x_5 + x_6$
   b. $x_1 y_1 + x_2 y_2 + x_3 y_3 + \cdots + x_7 y_7$
   c. $x_1^2 + x_2^2 + \cdots + x_9^2$
   d. $(x_1 - 3) + (x_2 - 3) + \cdots + (x_n - 3)$

A.3 Show each of the following to be true:
   a. $\sum_{i=1}^{4} (5x_i + 6) = 5 \cdot \sum_{i=1}^{4} x_i + 24$
   b. $\sum_{i=1}^{n} (x_i - y_i) = \sum_{i=1}^{n} x_i - \sum_{i=1}^{n} y_i$

A.4 Given $x_1 = 2$, $x_2 = 7$, $x_3 = -3$, $x_4 = 2$, $x_5 = -1$, and $x_6 = 1$, find each of the following:
   a. $\sum_{i=1}^{6} x_i$ b. $\sum_{i=1}^{6} x_i^2$ c. $\left(\sum_{i=1}^{6} x_i\right)^2$

A.5 Given $x_1 = 4$, $x_2 = -1$, $x_3 = 5$, $f_1 = 4$, $f_2 = 6$, $f_3 = 2$, $y_1 = -3$, $y_2 = 5$, and $y_3 = 2$, find each of the following:
   a. $\sum x$ b. $\sum y$ c. $\sum f$ d. $\sum (x - y)$
   e. $\sum x^2$ f. $\left(\sum x\right)^2$ g. $\sum xy$ h. $\sum x \cdot \sum y$
   i. $\sum xf$ j. $\sum x^2 f$ k. $\left(\sum xf\right)^2$

A.6 Suppose that you take out a $12,000 small-business loan. The terms of the loan are that each month for 10 years (120 months) you will pay back $100 plus accrued interest. The accrued interest is calculated by multiplying 0.005 (6 percent/12) times the amount of the loan still outstanding. That is, the first month you pay $12,000 × 0.005 in accrued interest, the second month ($12,000 − 100) × 0.005 in interest, the third month [$12,000 − (2)(100)] × 0.005, and so forth. Express the total amount of interest paid over the life of the loan by using summation notation.

The answers to these exercises can be found in the answer section.

APPENDIX B

USING THE RANDOM NUMBER TABLE

The random number table is a collection of "random" digits. The term *random* means that each of the ten digits $(0, 1, 2, 3, \ldots, 9)$ has an equal chance of occurrence. The digits in Table 1 of Appendix F can be thought of as single-digit numbers (0–9), as two-digit numbers (00–99), as three-digit numbers (000–999), or as numbers of any desired size. The digits presented in Table 1 are arranged in pairs and grouped into blocks of five rows and five columns. This format is used for convenience. Tables in other books may be arranged differently.

Random numbers are used primarily for one of two reasons: (1) to identify the source element of a population (the source of data) or (2) to simulate an experiment.

▼ ILLUSTRATION B-1

A simple random sample of 10 people is to be drawn from a population of 7564 people. Each person will be assigned a number, using the numbers from 0001 to 7564. We will view Table 1 as a collection of four-digit numbers (two columns used together), where the numbers $0001, 0002, 0003, \ldots, 7564$ identify the 7564 people. The numbers $0000, 7565, 7566, \ldots, 9999$ represent no one in our population; that is, they will be discarded if selected.

Now we are ready to select our 10 people. Turn to Table 1 (p. F-1). We need to select a starting point and a "path" to be followed. Perhaps the most common way to locate a starting point is to look away and arbitrarily point to a starting point. The number we located this way was 3909. (It is located in the upper left corner of the block that is in the fourth large block from the left and the second large block down.) From here we will proceed down the column, then go to the top of the next set of columns, if necessary. The person identified by number 3909 is the first source of data selected. Proceeding down the column, we find 8869 next. This number is discarded. The number 2501 is next. Therefore, the person identified by 2501 is the second source of data to be selected. Continuing down this column, our sample will be obtained from those people identified by the numbers 3909, 2501, 7485, 0545, 5252, 5612, 0997, 3230, 1051, 2712. (The numbers 8869, 8338, and 9187 were discarded.) ▲▲

▼ ILLUSTRATION B-2

Let's use the random number table and simulate 100 tosses of a coin. The simulation is accomplished by assigning numbers to each of the possible outcomes of a particular experiment. The assignment must be done in such a way as to preserve the probabilities. Perhaps the simplest way to make the assignment for the coin toss is to let the even digits $(0, 2, 4, 6, 8)$ represent heads and the odd digits $(1, 3, 5, 7, 9)$ represent tails. The correct probabilities are maintained: $P(H) = P(0, 2, 4, 6, 8) = \frac{5}{10} = 0.5$ and $P(T) = P(1, 3, 5, 7, 9) = \frac{5}{10} = 0.5$. Once this assignment is complete, we are ready to obtain our sample.

Since the question asked for 100 tosses and there are 50 digits to a "block" in Table 1, let's select two blocks as our sample of random one-digit numbers (instead of a column 100 lines long). Let's look away and point to one block on p. F-1 and then do the same to select one block from p. F-2. We picked the sixth block down in the first column of blocks on p. F-1 (24 even and 26 odd numbers) and the sixth block down in the third column of blocks on p. F-2 (23 even and 27 odd numbers). Thus we obtain a sample of 47 heads and 53 tails for our 100 simulated tosses. ▲▲

There are, of course, many ways to use the random number table. You must use your good sense in assigning the numbers to be used and in choosing the "path" to be followed through the table. One bit of advice is to make the assignments in as simple and easy a method as possible to avoid errors.

▼▲ EXERCISES

B.1 A random sample of size 8 is to be selected from a population that contains 75 elements. Describe how the random sample of the 8 objects could be made with the aid of the random number table.

B.2 A coin-tossing experiment is to be simulated. Two coins are to be tossed simultaneously and the number of heads appearing is to be recorded for each toss. Ten such tosses are to be observed. Describe two ways to use the random number table to simulate this experiment.

B.3 Simulate five rolls of three dice by using the random number table.

The answers to these exercises can be found in the answer section.

APPENDIX C

ROUND-OFF PROCEDURE

When rounding off a number, we use the following procedure.

STEP 1 Identify the position where the round-off is to occur. This is shown by using a vertical line that separates the part of the number to be kept from the part to be discarded. For example,

    125.267   to the nearest tenth is written as   125.2|67

    7.8890   to the nearest hundredth is written as   7.88|90

STEP 2 Step 1 has separated all numbers into one of four cases. (X's will be used as placeholders for number values in front of the vertical line. These X's can represent any number value.)

    Case I: $X\,X\,X\,X\,|\,000\ldots$

    Case II: $X\,X\,X\,X\,|\,$---- (any value from $000\ldots1$ to $499\ldots9$)

    Case III: $X\,X\,X\,X\,|\,5000\ldots0$

    Case IV: $X\,X\,X\,X\,|\,$---- (any value from $5000\ldots1$ to $999\ldots9$)

STEP 3 Perform the rounding off.

    Case I requires no round-off. It is exactly $X\,X\,X\,X$.

▼ ILLUSTRATION C-1

Round 3.5000 to the nearest tenth.

$$3.5|000 \quad \text{becomes} \quad \mathbf{3.5}$$ ▲▲

Case II requires rounding. We will round down for this case. That is, just drop the part of the number that is behind the vertical line.

▼ ILLUSTRATION C-2

Round 37.6124 to the nearest hundredth.

$$37.61|24 \quad \text{becomes} \quad \mathbf{37.61}$$ ▲▲

Case III requires rounding. This is the case that requires special attention. **When a 5 (exactly a 5) is to be rounded off, round to the even digit.** In the long run, half of the time the 5 will be preceded by an even digit (0, 2, 4, 6, 8) and you will round down, while the other half of the time the 5 will be preceded by an odd digit (1, 3, 5, 7, 9) and you will round up.

▼ ILLUSTRATION C-3

Round 87.35 to the nearest tenth.

$$87.3|5 \quad \text{becomes} \quad \textbf{87.4}$$

Round 93.445 to the nearest hundredth.

$$93.44|5 \quad \text{becomes} \quad \textbf{93.44}$$

(Note: 87.35 is 87.35000... and 93.445 is 93.445000....) ▲▲

Case IV requires rounding. We will round up for this case. That is, we will drop the part of the number that is behind the vertical line and we will increase the last digit in front of the vertical line by one.

▼ ILLUSTRATION C-4

Round 7.889 to the nearest tenth.

$$7.8|89 \quad \text{becomes} \quad \textbf{7.9}$$

▲▲

NOTE Cases I, II, and IV describe what is commonly done. Our guidelines for Case III are the only ones that are different typical procedure.

If the typical round-off rule (0, 1, 2, 3, 4 are dropped; 5, 6, 7, 8, 9 are rounded up) is followed, then $(n + 1)/(2n + 1)$ of the situations are rounded up. (n is the number of different sequences of digits that fall into each of Case II and Case IV.) That is more than half. You (as many others have) may say, "So what?" In today's world that tiny, seemingly insignificant amount becomes very significant when applied repeatedly to large numbers.

▼▲ EXERCISES

C.1 Round each of the following to the nearest integer:
    a. 12.94   **b.** 8.762   **c.** 9.05   **d.** 156.49
    e. 45.5   **f.** 42.5   **g.** 102.51   **h.** 16.5001

C.2 Round each of the following to the nearest tenth:
    a. 8.67   **b.** 42.333   **c.** 49.666   **d.** 10.25   **e.** 10.35
    f. 8.4501   **g.** 27.35001   **h.** 5.65   **i.** 3.05   **j.** $\frac{1}{4}$

C.3 Round each of the following to the nearest hundredth:
    a. 17.6666   **b.** 4.444   **c.** 54.5454   **d.** 102.055   **e.** 93.225
    f. 18.005   **g.** 18.015   **h.** 5.555   **i.** 44.7450   **j.** $\frac{2}{3}$

The answers to these exercises can be found in the answer section.

APPENDIX D

BASIC PRINCIPLES OF COUNTING

In order to find the probability of many events, it is necessary to determine the number of possible outcomes for the experiment involved. This requires us to enumerate (obtain a "count" of) the possibilities. This "count" can be obtained by using one of two methods: (1) list all the possibilities and then proceed to count them $(1, 2, 3, \ldots)$; or (2) since it is often not necessary to delineate (obtain a representation of) all possibilities, the count can be determined by calculating its numerical value. In this section, we are going to learn three commonly used methods for obtaining the count by calculation: the fundamental technique and two specific techniques.

▼ ILLUSTRATION D-1

An automobile dealer offers one of its small sporty models with two transmission options (standard or automatic) and in one of three colors (black, red, or white). How many different choices of transmission and color combinations are there for the customer?

SOLUTION The number of choices available can easily be found by listing and counting them. There are six.

Standard, black   Automatic, black
Standard, red   Automatic, red
Standard, white   Automatic, white

The possible choices can also be demonstrated by use of a tree diagram.

Trans. Opt.	Color Opt.	Possible Choices
standard	black	standard, black
	red	standard, red
	white	standard, white
automatic	black	automatic, black
	red	automatic, red
	white	automatic, white

▲▲

NOTE More information and additional illustrations of tree diagrams can be found in Chapter 4 and in Lesson I-5 of the Study Guide.

Each of the two transmission choices can be paired with any one of three colors; thus there are 2 × 3 or six different possible choices. This suggests the following rule:

FUNDAMENTAL COUNTING RULE

If an experiment is composed of two trials, where one of the trials (single action or choice) has m possible outcomes (results) and the other trial has n possible outcomes, then when the two trials are performed together, there are

$$m \times n \qquad \text{(D-1)}$$

possible outcomes for the experiment.

In Illustration D-1, $m = 2$ (the number of transmission choices) and $n = 3$ (the number of color choices). Using the Fundamental Counting Rule (formula D-1), the number of possible choices available to a customer is

$$m \times n = 2 \times 3 = 6$$

This fundamental counting rule may be extended to include experiments that have more than two trials.

GENERAL COUNTING RULE

If an experiment is composed of k trials performed in a definite order, where the first trial has n_1 possible outcomes, the second trial has n_2 possible outcomes, the third trial has n_3 outcomes, and so on, then the number of possible outcomes for the experiment is

$$n_1 \times n_2 \times n_3 \times \cdots \times n_k. \qquad \text{(D-2)}$$

▼ ILLUSTRATION D-2

In many states, automobile license plates use three letters followed by three numerals to make up the "license plate number." (There are other combinations of letters and numerals used; however, let's focus only on this six-character "number" for now.) If we assume that any one of the 26 letters may be used for each of the first three characters and that any one of the 10 numerals 0 through 9 can be used for each of the last three characters, how many different license plate numbers are possible?

SOLUTION There are 26 possible choices for the first letter ($n_1 = 26$), 26 possible choices for the second letter ($n_2 = 26$), and 26 possible choices for the third letter ($n_3 = 26$). In similar fashion, there are 10 choices for the numeral to be used for each of the fourth ($n_4 = 10$), fifth ($n_5 = 10$), and sixth ($n_6 = 10$) characters. Therefore, using the General Counting Rule (formula D-2), we find there are

$$26 \times 26 \times 26 \times 10 \times 10 \times 10 = \mathbf{17{,}576{,}000}$$

different "license plate numbers" using this six-character scheme. ▲▲

▼ ILLUSTRATION D-3

How many different "license plate numbers" are possible if the non-zero numerals are used for the three leading characters, letters are used for the three trailing characters, and the letters are not allowed to repeat?

SOLUTION There are 9 possible choices for each of the first three characters (since only 1 thru 9 may be used). Thus, $n_1 = 9$, $n_2 = 9$ and $n_3 = 9$. The fourth character may be chosen from any one of the 26 letters ($n_4 = 26$). However, the fifth character must be chosen from any one of the 25 letters not previously used ($n_5 = 25$), and the sixth character must be chosen from the 24 letters not previously used ($n_6 = 24$). Applying the General Counting Rule (D-2), we find there are

$$9 \times 9 \times 9 \times 26 \times 25 \times 24 = \mathbf{11{,}372{,}400}$$

different "license plate numbers" using this second six-character scheme. ▲▲

We are now ready to investigate two additional concepts commonly encountered when enumerating possibilities: *permutations* and *combinations*. A permutation is a collection of distinct objects arranged in a specific order, while a combination is a collection of distinct objects without any specific order.

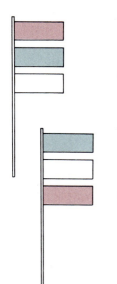

▼ ILLUSTRATION D-4

There are four flags of different colors (one each of red, white, blue, and green) in a box, and you are asked to select any three. If you select {red, white, green} you have the same combination of colors as {green, red, white}. This question does not require or distinguish between different "orders" or arrangements; thus each set of flags is one combination. ▲▲

▼ ILLUSTRATION D-5

There are four flags of different colors (one each of red, white, blue, and green) in a box, and you are asked to select any three of them and make a "signal" by hanging the three different flags, one above the other, on a flagpole. Since red over green over white is different from green over white over red, order is important and each possible signal is one permutation. ▲▲

Permutations

▼ ILLUSTRATION D-6

Select four different letters from the English alphabet and arrange them in any specific order. As a result of following these instructions, Barbara created the "four-letter word" BSJT. Rob created the word EOST. Steve selected KOCM. How many different "four-letter words" can be created?

Each of these "words" is a *permutation* of four letters selected from the set of 26 different letters forming the alphabet.

PERMUTATION

An ordered arrangement of a set of distinct objects. That is, there is a first object, a second object, a third object, and so on; and each object is distinctly different from the others.

The number of permutations that can be formed is calculated using an adaption of the General Counting Rule.

PERMUTATION FORMULA

The number of permutations that can be formed using r different objects selected from a set of n distinct objects (symbolized by $_nP_r$ and read "the number of permutations of n objects selected r at a time") is

$$_nP_r = n \times (n-1) \times (n-2) \times \cdots \times (n-r+1) \quad \textbf{(D-3)}$$

or, in factorial notation,

$$_nP_r = \frac{n!}{(n-r)!} \quad \textbf{(D-4)}$$

NOTE More information and additional illustrations of factorial notation can be found in Chapter 5 and in Lesson I-8 of the Study Guide. Remember: $0! = 1$.

Let's continue with the solution of Illustration D-6. Since the 4 letters were selected from the 26 letters of the alphabet, the value of $r = 4$ (the number of selections) and $n = 26$ (the number of objects available for selection). Using formula D-3,

$$_{26}P_4 = 26 \times 25 \times \cdots \times (26 - 4 + 1)$$
$$= 26 \times 25 \times 24 \times 23 = \textbf{358,800}$$

Or using formula D-4,

$$_{26}P_4 = \frac{26!}{(26-4)!} = \frac{26!}{22!}$$

$$= \frac{26 \times 25 \times 24 \times 23 \times 22 \times 21 \times \cdots \times 1}{22 \times 21 \times \cdots \times 1}$$

$$= \frac{26 \times 25 \times 24 \times 23 \times (22 \times 21 \times \cdots \times 1)}{(22 \times 21 \times \cdots \times 1)}$$

$$= 26 \times 25 \times 24 \times 23 = \textbf{358,800}$$

ILLUSTRATION D-7

A group of eight finalists in a ceramic art competition are to be awarded five prizes—first, second and so on. How many different ways are there to award these five prizes?

SOLUTION Since the prizes are ordered, this is a permutation of $n = 8$ different people, taken 5 at a time (only five prizes, and each prize is distinctly different from the others). Using formula (D-3), we find there are

$$_8P_5 = 8 \times 7 \times \cdots \times (8 - 5 + 1)$$
$$= 8 \times 7 \times 6 \times 5 \times 4 = \mathbf{6{,}720}$$

different possible ways of awarding these five prizes.

Combinations

ILLUSTRATION D-8

Select a set of four different letters from the English alphabet. As a result of following this instruction, Kevin selected A, E, R, and T. Karen selected D, E, N, and Q. Sue selected R, E, A, and T. Notice that Kevin and Sue selected the same set of letters, even though they selected them in different orders. These three people have selected two different sets of four letters. How many different sets of four letters can be selected?

SOLUTION Each of these "sets" of four letters represents a *combination* of $r = 4$ objects having been selected from a set of $n = 26$ distinct objects.

> **COMBINATION**
>
> *A set of distinct objects without regard to an arrangement or an order.* That is, the membership of the set is all that matters.

The number of combinations that can be selected is related to the number of permutations. In Illustration D-6, we found that there were 358,800 permutations of four letters possible. Many permutations were "words" formed from the same set of four letters. For example, the set of four letters A, B, C, and D can be used to form many permutations ("words"):

ABCD	ABDC	ACBD	ACDB	ADBC	ADCB
BACD	BADC	BCAD	BCDA	BDAC	BDCA
CBAD	CBDA	CABD	CADB	CDBA	CDAB
DBCA	DBAC	DCBA	DCAB	DABC	DACB

There are 4! ($4 \times 3 \times 2 \times 1$) or 24 different permutations for this set of four letters. Every other set of four letters can also be used to form 24 permutations.

Therefore, if we divide the number of permutations possible (358,800) by the number of permutations each set has (24), the quotient will be the number of different sets (combinations) possible. That is, there are 14,950 (358,800/24) combinations of four letters possible. This concept is generalized in the following formula:

> **COMBINATION FORMULA**
>
> The number of combinations of r objects that can be selected from a set of n distinct objects (symbolized by $_nC_r$, and read "the number of combinations of n things taken r at a time") is
>
> $$_nC_r = \frac{n(n-1)(n-2)\cdots(n-r+1)}{r!} \quad \text{(D-5)}$$
>
> or, in factorial notation,
>
> $$_nC_r = \frac{n!}{(n-r)! \times r!} \quad \text{(D-6)}$$

Let's continue with the solution of Illustration D-8 using these new formulas. First using formula (D-5),

$$_nC_r = \frac{n(n-1)(n-2)\cdots(n-r+1)}{r!}$$

$$_{26}C_4 = \frac{26(25)(24)\cdots(26-4+1)}{4!} = \frac{26 \times 25 \times 24 \times 23}{4 \times 3 \times 2 \times 1}$$

$$= \frac{358{,}800}{24} = \mathbf{14{,}950}$$

Using formula (D-6),

$$_nC_r = \frac{n!}{(n-r)! \times r!}$$

$$_{26}C_4 = \frac{26!}{(26-4)! \times 4!} = \frac{26!}{22! \times 4!}$$

$$= \frac{26 \times 25 \times 24 \times 23 \times 22 \times 21 \times \cdots \times 2 \times 1}{(22 \times 21 \times \cdots \times 2 \times 1)(4 \times 3 \times 2 \times 1)}$$

$$= \frac{26 \times 25 \times 24 \times 23 \times (22 \times 21 \times \cdots \times 2 \times 1)}{(22 \times 21 \times \cdots \times 2 \times 1)(4 \times 3 \times 2 \times 1)}$$

$$= \frac{358{,}800}{24} = \mathbf{14{,}950}$$

ILLUSTRATION D-9

A department has 30 members and a committee of 5 people is needed to carry out a task. How many different possible committees are there?

SOLUTION As stated, there is no specific assignment or order to the members of the committee; therefore, each possible committee is a combination and $n = 30$ (the number of people eligible to be selected), and $r = 5$ (the number to be selected).

$$_nC_r = \frac{n!}{(n - r)! \times r!}$$

$$_{30}C_5 = \frac{30!}{(30 - 5)! \times 5!} = \frac{30!}{25! \times 5!}$$

$$= \frac{30 \times 29 \times 28 \times 27 \times 26 \times 25 \times 24 \times \cdots \times 2 \times 1}{(25 \times 24 \times \cdots \times 2 \times 1)(5 \times 4 \times 3 \times 2 \times 1)}$$

$$= \frac{30 \times 29 \times 28 \times 27 \times 26 \times (25 \times 24 \times \cdots \times 2 \times 1)}{(25 \times 24 \times \cdots \times 2 \times 1)(5 \times 4 \times 3 \times 2 \times 1)}$$

$$= \frac{17,100,720}{120} = \mathbf{142{,}506}$$

It is possible to select 142,506 different committees of 5 people from this department of 30 people. ▲▲

NOTE The number of combinations $_nC_r$ and the binomial coefficient $\binom{n}{r}$ or $\binom{n}{x}$ are numerically equivalent.

The three "counting" formulas described above in this section (formulas D-2, D-4, and D-6) can be and often are used together to solve problems.

ILLUSTRATION D-10

A department has 30 members and a committee is needed to carry out a task. The committee is to be composed of a chairperson and four members. How many different possible committees are there?

SOLUTION This problem is solved by treating it in two parts: consider the chairperson position and the committee members as two separate parts to be combined using the Fundamental Counting Rule ($m \times n$). Let m be the number of possible choices for the chairperson. Since any one of the 30 department members could serve as the chair, $m = 30$. Let n be the number of four-person committees that can be selected from the remaining 29 department members. Since these four have no specific assignment, the number of possibilities, n, is the number of combinations of 29 things taken 4 at a time.

$$n(\text{committees}) = m \times n$$
$$= 30 \times {}_{29}C_4$$
$$= 30 \times \frac{29 \times 28 \times 27 \times 26 \times (25 \times 24 \times \cdots \times 2 \times 1)}{(25 \times 24 \times \cdots \times 2 \times 1) \times 4 \times 3 \times 2 \times 1}$$
$$= 30 \times 23{,}751 = \mathbf{712{,}530}$$

There are 712,530 different committees possible of size 5 with an assigned chairperson. ▲▲

▼ ILLUSTRATION D-11

A department has 30 members and a committee is needed to carry out a task. The committee is to be composed of two co-chairpersons and three members. How many different possible committees are there?

SOLUTION This problem is solved by treating it in two parts: consider the selecting of two co-chairpersons and then the remaining committee members as two separate parts to be combined using the Fundamental Counting Rule ($m \times n$). Let m be the number of possible choices for the co-chairpersons. This is like a committee of two, since there is no further distinction between them; therefore $m = {}_{30}C_2$, since any two of the 30 department members could serve as the co-chairs. Let n be the number of three-person committees that can be selected from the remaining 28 department members. Since these three have no specific assignment, the number of possibilities, n, is the number of combinations of 28 things taken 3 at a time.

$$n(\text{committees}) = m \times n$$
$$= {}_{30}C_2 \times {}_{28}C_3$$
$$= \frac{30!}{28! \times 2!} \times \frac{28!}{25! \times 3!}$$
$$= \frac{(30 \times 29 \times 28 \times \cdots \times 1) \times (28 \times 27 \times 26 \times \cdots \times 1)}{(28 \times 27 \times \cdots \times 1) \times (2 \times 1) \times (25 \times \cdots \times 1) \times (3 \times 2 \times 1)}$$
$$= 15 \times 29 \times 14 \times 9 \times 26$$
$$= \mathbf{1{,}425{,}060}$$

There are 1,425,060 different committees of size 5 with assigned co-chairpersons possible. ▲▲

▼▲ EXERCISES

D.1 A long weekend of three days is being planned. The three days are to be spent taking scenic drives through the countryside and ending each day at a motel where reservations have been previously made for Friday and Saturday nights (will

Appendix D ▼ BASIC PRINCIPLES OF COUNTING

be home Sunday night). There are three scenic routes that may be traveled on Friday, two choices for Saturday, and three scenic route choices for Sunday's return trip. How many different trips are possible if
- a. all the scenic options are considered?
- b. one of the Friday routes has been previously driven, and is not a choice?
- c. on Sunday, it is decided to drive straight home and not take one of the scenic routes?

D.2
- a. Show that formulas (D-3) and (D-4) are equivalent.
- b. Show that formulas (D-5) and (D-6) are equivalent.

D.3 Explain why each of the following pairs of "counts" are equal:
- a. $_nP_n$ and $_nP_{n-1}$
- b. $_nP_1$ and $_nC_1$
- c. $_nC_r$ and $_nC_{n-r}$
- d. $_nC_r$ and the binomial coefficient $\binom{n}{r}$.

D.4 A department of 30 people is to select a committee of 5 persons. How many different committees are possible if the committee is composed of
- a. a chairperson, a secretary, and three others?
- b. two co-chairs and three others?
- c. two co-chairs, a secretary, and two others?

D.5 License plates are to be "numbered" using a combination or letters and numerals. How many different "numbers" are possible if each of the following sets of restrictions is used?
- a. Six characters using any combination or arrangement of the 26 letters and 10 single-digit numerals.
- b. Six characters using any combination or arrangement of letters and single-digit numerals, except that "zero" and "one" are not to be used because they are hard to distinguish from "o" and "i."
- c. Six characters using letters for the two leading characters and the 10 single-digit numerals for the four trailing characters.
- d. Six characters using the 10 single-digit numerals for the four leading characters and letters for the two trailing characters.
- e. Six characters using the 10 single-digit numerals for the four leading characters and letters for the two trailing characters, except that "zero" cannot be the leading character.

D.6 Mathew has six shirts, four pairs of pants, and five pairs of socks clean and ready to wear. How many different "outfits" can he assemble, if
- a. he wears one item from each category?
- b. he wears one specific shirt and one item each from the other two categories?
- c. he only wears two of the shirts with one specific pair of pants and no socks, but the rest are worn in any complete combination?

D.7 Five cards are to be randomly selected from a standard bridge deck of 52 cards.
- a. How many different "hands" of five cards are possible?
- b. How many different "hands" of five cards are possible if the first card drawn is an ace?

c. How many different "hands" of five cards are possible if the first card drawn is an ace and the remaining four are not aces?
d. How many different "hands" of five cards are possible if the first card drawn is a club and the remaining four are not clubs?
e. How many different "hands" of five cards are possible if the first card drawn is an ace and the remaining four are not clubs?

APPENDIX E

INTERPOLATION PROCEDURE FOR F DISTRIBUTION

ONE df VALUE IN TABLE

▼ ILLUSTRATION E-1

Find the critical value $F(22, 11, 0.05)$.

SOLUTION Critical values of F are found in Table 8a of Appendix F, p. F-11. Turn to Table 8a and you will see that there is no column for $df_n = 22$. There is a row for $df_d = 11$. Therefore the values closest to $F(22, 11, 0.05)$ are $F(20, 11, 0.05) = 2.65$ and $F(24, 11, 0.05) = 2.61$. Since df = 22 is halfway between df = 20 and df = 24, it seems logical to assume that a good estimate for $F(22, 11, 0.05)$ is 2.63, the value halfway between 2.65 and 2.61. ▲▲

Illustration E-1 is a simple illustration of linear interpolation. The method of **linear interpolation** assumes that if a number of degrees of freedom lies between two of those listed in the table of critical values, then the F-value for that number of degrees of freedom is a value with the same proportional difference between the F-values listed, and vice versa. In Illustration E-1 the proportion was one-half. The following illustration shows the general procedure used to interpolate when one of the two degrees of freedom is in the table:

▼ ILLUSTRATION E-2

Find the value of $F(75, 15, 0.05)$.

SOLUTION In Table 8a there is no column for $df_n = 75$, but there is a row for $df_d = 15$. Therefore we have two nearby values: $F(60, 15, 0.05) = 2.16$ and $F(120, 15, 0.05) = 2.11$. Since degrees of freedom for the denominator is 15 for both, we need to interpolate between $df_n = 60$ and $df_n = 120$ in order to find $df_n = 75$. The following schematic arrangement shows the information we have and the proportional differences that form the proportion that we must solve to find d, the proportional difference.

$$60 \begin{array}{|c|} \hline 15 \\ \hline \end{array} \begin{array}{l} F(60, 15, 0.05) = 2.16 \\ F(75, 15, 0.05) = ? \\ F(120, 15, 0.05) = 2.11 \end{array} \begin{array}{|c|} \hline d \\ \hline \end{array} 0.05$$

The corresponding differences are in proportion; therefore:

$$\frac{15}{60} = \frac{d}{0.05}$$

$$d = \frac{(15)(0.05)}{60}$$

$$d = 0.0125$$

Notice that the value of F decreased as the degrees of freedom increased; therefore the value of d is subtracted from 2.16:

$$F(75, 15, 0.05) = 2.16 - 0.0125 = 2.1475 \quad \text{or} \quad \mathbf{2.15}$$

to the nearet hundredth. ▲▲

NEITHER df VALUE IN TABLE

When neither of the two degrees of freedom are in the table we will need to interpolate **three** times in order to find the desired value. This interpolation process will follow an "H" pattern as demonstrated by the following illustration:

▼ ILLUSTRATION E-3

Find the value of $F(28, 72, 0.05)$.

SOLUTION Table 8a has no column for $df_n = 28$ and no row for $df_d = 72$. Therefore we will use the two columns and the two rows that have df values immediately smaller and immediately larger than 28 and 72, respectively. We find the following four F-values:

$$F(24, 60, 0.05) = 1.70 \qquad F(30, 60, 0.05) = 1.65$$

$$Ⓐ \quad F(28, 72, 0.05) = ? \quad Ⓑ$$

$$F(24, 120, 0.05) = 1.61 \qquad F(30, 120, 0.05) = 1.55$$

The first two interpolations will be to find F-values in positions A $[F(24, 72, 0.05)]$ and B $[F(30, 72, 0.05)]$.

For A:

$$\frac{12}{60} = \frac{d}{0.09}$$

$$d = \frac{(12)(0.09)}{60}$$

$$d = 0.018$$

For B:

$$\frac{12}{60} = \frac{d}{0.10}$$

$$d = \frac{(12)(0.10)}{60}$$

$$d = 0.020$$

Therefore,

$$A = F(24, 72, 0.05) \qquad\qquad B = F(30, 72, 0.05)$$
$$= 1.70 - 0.018 \qquad\qquad\quad = 1.65 - 0.020$$
$$= \mathbf{1.682} \qquad\qquad\qquad\quad = \mathbf{1.630}$$

The final step is to interpolate between A and B to find our answer.

$$\begin{array}{c} F(24, 72, 0.05) = 1.682 \\ 6\ \boxed{4\ \ F(28, 72, 0.05) = ?\ \ \boxed{d}\ \ 0.052} \\ F(30, 72, 0.05) = 1.630 \end{array}$$

The corresponding differences are in proportion, therefore,

$$\frac{4}{6} = \frac{d}{0.052}$$

$$d = \frac{(4)(0.052)}{6}$$

$$d = 0.03467 = \mathbf{0.035}$$

$F(28, 72, 0.05) = 1.682 - 0.035 = 1.647$ or **1.65** to the nearest hundredth.

APPENDIX F

TABLES

TABLE 1
Random Numbers

10	09	73	25	33	76	52	01	35	86	34	67	35	48	76	80	95	90	91	17	39	29	27	49	45
37	54	20	48	05	64	89	47	42	96	24	80	52	40	37	20	63	61	04	02	00	82	29	16	65
08	42	26	89	53	19	64	50	93	03	23	20	90	25	60	15	95	33	47	64	35	08	03	36	06
99	01	90	25	29	09	37	67	07	15	38	31	13	11	65	88	67	67	43	97	04	43	62	76	59
12	80	79	99	70	80	15	73	61	47	64	03	23	66	53	98	95	11	68	77	12	17	17	68	33
66	06	57	47	17	34	07	27	68	50	36	69	73	61	70	65	81	33	98	85	11	19	92	91	70
31	06	01	08	05	45	57	18	24	06	35	30	34	26	14	86	79	90	74	39	23	40	30	97	32
85	26	97	76	02	02	05	16	56	92	68	66	57	48	18	73	05	38	52	47	18	62	38	85	79
63	57	33	21	35	05	32	54	70	48	90	55	35	75	48	28	46	82	87	09	83	49	12	56	24
73	79	64	57	53	03	52	96	47	78	35	80	83	42	82	60	93	52	03	44	35	27	38	84	35
98	52	01	77	67	14	90	56	86	07	22	10	94	05	58	60	97	09	34	33	50	50	07	39	98
11	80	50	54	31	39	80	82	77	32	50	72	56	82	48	29	40	52	42	01	52	77	56	78	51
83	45	29	96	34	06	28	89	80	83	13	74	67	00	78	18	47	54	06	10	68	71	17	78	17
88	68	54	02	00	86	50	75	84	01	36	76	66	79	51	90	36	47	64	93	29	60	91	10	62
99	59	46	73	48	87	51	76	49	69	91	82	60	89	28	93	78	56	13	68	23	47	83	41	13
65	48	11	76	74	17	46	85	09	50	58	04	77	69	74	73	03	95	71	86	40	21	81	65	44
80	12	43	56	35	17	72	70	80	15	45	31	82	23	74	21	11	57	82	53	14	38	55	37	63
74	35	09	98	17	77	40	27	72	14	43	23	60	02	10	45	52	16	42	37	96	28	60	26	55
69	91	62	68	03	66	25	22	91	48	36	93	68	72	03	76	62	11	39	90	94	40	05	64	18
09	89	32	05	05	14	22	56	85	14	46	42	75	67	88	96	29	77	88	22	54	38	21	45	98
91	49	91	45	23	68	47	92	76	86	46	16	28	35	54	94	75	08	99	23	37	08	92	00	48
80	33	69	45	98	26	94	03	68	58	70	29	73	41	35	54	14	03	33	40	42	05	08	23	41
44	10	48	19	49	85	15	74	79	54	32	97	92	65	75	57	60	04	08	81	22	22	20	64	13
12	55	07	37	42	11	10	00	20	40	12	86	07	46	97	96	64	48	94	39	28	70	72	58	15
63	60	64	93	29	16	50	53	44	84	40	21	95	25	63	43	65	17	70	82	07	20	73	17	90
61	19	69	04	46	26	45	74	77	74	51	92	43	37	29	65	39	45	95	93	42	58	26	05	27
15	47	44	52	66	95	27	07	99	53	59	36	78	38	48	82	39	61	01	18	33	21	15	94	66
94	55	72	85	73	67	89	75	43	87	54	62	24	44	31	91	19	04	25	92	92	92	74	59	73
42	48	11	62	13	97	34	40	87	21	16	86	84	87	67	03	07	11	20	59	25	70	14	66	70
23	52	37	83	17	73	20	88	98	37	68	93	59	14	16	26	25	22	96	63	05	52	28	25	62
04	49	35	24	94	75	24	63	38	24	45	86	25	10	25	61	96	27	93	35	65	33	71	24	72
00	54	99	76	54	64	05	18	81	59	96	11	96	38	96	54	69	28	23	91	23	28	72	95	29
35	96	31	53	07	26	89	80	93	54	33	35	13	54	62	77	97	45	00	24	90	10	33	93	33
59	80	80	83	91	45	42	72	68	42	83	60	94	97	00	13	02	12	48	92	78	56	52	01	06
46	05	88	52	36	01	39	09	22	86	77	28	14	40	77	93	91	08	36	47	70	61	74	29	41

*For specific details on the use of this table, see Appendix B.

TABLE 1 (Continued)

32	17	90	05	97	87	37	92	52	41	05	56	70	70	07	86	74	31	71	57	85	39	41	18	38
69	23	46	14	06	20	11	74	52	04	15	95	66	00	00	18	74	39	24	23	97	11	89	63	38
19	56	54	14	30	01	75	87	53	79	40	41	92	15	85	66	67	43	68	06	84	96	28	52	07
45	15	51	49	38	19	47	60	72	46	43	66	79	45	43	59	04	79	00	33	20	82	66	95	41
94	86	43	19	94	36	16	81	08	51	34	88	88	15	53	01	54	03	54	56	05	01	45	11	76
98	08	62	48	26	45	24	02	84	04	44	99	90	88	96	39	09	47	34	07	35	44	13	18	80
33	18	51	62	32	41	94	15	09	49	89	43	54	85	81	88	69	54	19	94	37	54	87	30	43
80	95	10	04	06	96	38	27	07	74	20	15	12	33	87	25	01	62	52	98	94	62	46	11	71
79	75	24	91	40	71	96	12	82	96	69	86	10	25	91	74	85	22	05	39	00	38	75	95	79
18	63	33	25	37	98	14	50	65	71	31	01	02	46	74	05	45	56	14	27	77	93	89	19	36
74	02	94	39	02	77	55	73	22	70	97	79	01	71	19	52	52	75	80	21	80	81	45	17	48
54	17	84	56	11	80	99	33	71	43	05	33	51	29	69	56	12	71	92	55	36	04	09	03	24
11	66	44	98	83	52	07	98	48	27	59	38	17	15	39	09	97	33	34	40	88	46	12	33	56
48	32	47	79	28	31	24	96	47	10	02	29	53	68	70	32	30	75	75	46	15	02	00	99	94
69	07	49	41	38	87	63	79	19	76	35	58	40	44	01	10	51	82	16	15	01	84	87	69	38
09	18	82	00	97	32	82	53	95	27	04	22	08	63	04	83	38	98	73	74	64	27	85	80	44
90	04	58	54	97	51	98	15	06	54	94	93	88	19	97	91	87	07	61	50	68	47	66	46	59
73	18	95	02	07	47	67	72	62	69	62	29	06	44	64	27	12	46	70	18	41	36	18	27	60
75	76	87	64	90	20	97	18	17	49	90	42	91	22	72	95	37	50	58	71	93	82	34	31	78
54	01	64	40	56	66	28	13	10	03	00	68	22	73	98	20	71	45	32	95	07	70	61	78	13
08	35	86	99	10	78	54	24	27	85	13	66	15	88	73	04	61	89	75	53	31	22	30	84	20
28	30	60	32	64	81	33	31	05	91	40	51	00	78	93	32	60	46	04	75	94	11	90	18	40
53	84	08	62	33	81	59	41	36	28	51	21	59	02	90	28	46	66	87	95	77	76	22	07	91
91	75	75	37	41	61	61	36	22	69	50	26	39	02	12	55	78	17	65	14	83	48	34	70	55
89	41	59	26	94	00	39	75	83	91	12	60	71	76	46	48	94	97	23	06	94	54	13	74	08
77	51	30	38	20	86	83	42	99	01	68	41	48	27	74	51	90	81	39	80	72	89	35	55	07
19	50	23	71	74	69	97	92	02	88	55	21	02	97	73	74	28	77	52	51	65	34	46	74	15
21	81	85	93	13	93	27	88	17	57	05	68	67	31	56	07	08	28	50	46	31	85	33	84	52
51	47	46	64	99	68	10	72	36	21	94	04	99	13	45	42	83	60	91	91	08	00	74	54	49
99	55	96	83	31	62	53	52	41	70	69	77	71	28	30	74	81	97	81	42	43	86	07	28	34
33	71	34	80	07	93	58	47	28	69	51	92	66	47	21	58	30	32	98	22	93	17	49	39	72
85	27	48	68	93	11	30	32	92	70	28	83	43	41	37	73	51	59	04	00	71	14	84	36	43
84	13	38	96	40	44	03	55	21	66	73	85	27	00	91	61	22	26	05	61	62	32	71	84	23
56	73	21	62	34	17	39	59	61	31	10	12	39	16	22	85	49	65	75	60	81	60	41	88	80
65	13	85	68	06	87	60	88	52	61	34	31	36	58	61	45	87	52	10	69	85	64	44	72	77
38	00	10	21	76	81	71	91	17	11	71	60	29	29	37	74	21	96	40	49	65	58	44	96	98
37	40	29	63	97	01	30	47	75	86	56	27	11	00	86	47	32	46	26	05	40	03	03	74	38
97	12	54	03	48	87	08	33	14	17	21	81	53	92	50	75	23	76	20	47	15	50	12	95	78
21	82	64	11	34	47	14	33	40	72	64	63	88	59	02	49	13	90	64	41	03	85	65	45	52
73	13	54	27	42	95	71	90	90	35	85	79	47	42	96	08	78	98	81	56	64	69	11	92	02
07	63	87	79	29	03	06	11	80	72	96	20	74	41	56	23	82	19	95	38	04	71	36	69	94
60	52	88	34	41	07	95	41	98	14	59	17	52	06	95	05	53	35	21	39	61	21	20	64	55
83	59	63	56	55	06	95	89	29	83	05	12	80	97	19	77	43	35	37	83	92	30	15	04	98
10	85	06	27	46	99	59	91	05	07	13	49	90	63	19	53	07	57	18	39	06	41	01	93	62
39	82	09	89	52	43	62	26	31	47	64	42	18	08	14	43	80	00	93	51	31	02	47	31	67
59	58	00	64	78	75	56	97	88	00	88	83	55	44	86	23	76	80	61	56	04	11	10	84	08
38	50	80	73	41	23	79	34	87	63	90	82	29	70	22	17	71	90	42	07	95	95	44	99	53
30	69	27	06	68	94	68	81	61	27	56	19	68	00	91	82	06	76	34	00	05	46	26	92	00
65	44	39	56	59	18	28	82	74	37	49	63	22	40	41	08	33	76	56	76	96	29	99	08	36
27	26	75	02	64	13	19	27	22	94	07	47	74	46	06	17	98	54	89	11	97	34	13	03	58
91	30	70	69	91	19	07	22	42	10	36	69	95	37	28	28	82	53	57	93	28	97	66	62	52
68	43	49	46	88	84	47	31	36	22	62	12	69	84	08	12	84	38	25	90	09	81	59	31	46
48	90	81	58	77	54	74	52	45	91	35	70	00	47	54	83	82	45	26	92	54	13	05	51	60
06	91	34	51	97	42	67	27	86	01	11	88	30	95	28	63	01	19	89	01	14	97	44	03	44
10	45	51	60	19	14	21	03	37	12	91	34	23	78	21	88	32	58	08	51	43	66	77	08	83
12	88	39	73	43	65	02	76	11	84	04	28	50	13	92	17	97	41	50	77	90	71	22	67	69
21	77	83	09	76	38	80	73	69	61	31	64	94	20	96	63	28	10	20	23	08	81	64	74	49
19	52	35	95	15	65	12	25	96	59	86	28	36	82	58	69	57	21	37	98	16	43	59	15	29
67	24	55	26	70	35	58	31	65	63	79	24	68	66	86	76	46	33	42	22	26	65	59	08	02
60	58	44	73	77	07	50	03	79	92	45	13	42	65	29	26	76	08	36	37	41	32	64	43	44
53	85	34	13	77	36	06	69	48	50	58	83	87	38	59	49	36	47	33	31	96	24	04	36	42
24	63	73	97	36	74	38	48	93	42	52	62	30	79	92	12	36	91	86	01	03	74	28	38	73
83	08	01	24	51	38	99	22	28	15	07	75	95	17	77	97	37	72	75	85	51	97	23	78	67
16	44	42	43	34	36	15	19	90	73	27	49	37	09	39	85	13	03	25	52	54	84	65	47	59
60	79	01	81	57	57	17	86	57	62	11	16	17	85	76	45	81	95	29	79	65	13	00	48	60

From tables of the RAND Corporation. Reprinted from Wilfred J. Dixon and Frank J. Massey, Jr., *Introduction to Statistical Analysis*. 3rd ed. (New York: McGraw-Hill, 1969), pp. 446–447. Reprinted by permission of the RAND Corporation.

Appendix F ▼ TABLES

TABLE 2
Factorials

n	$n!$
0	1
1	1
2	2
3	6
4	24
5	120
6	720
7	5,040
8	40,320
9	362,880
10	3,628,800
11	39,916,800
12	479,001,600
13	6,227,020,800
14	87,178,291,200
15	1,307,674,368,000
16	20,922,789,888,000
17	355,687,428,096,000
18	6,402,373,705,728,000
19	121,645,100,408,832,000
20	2,432,902,008,176,640,000

*For specific details on the use of this table, see p. 287.

TABLE 3   Binomial Coefficients

n	$\binom{n}{0}$	$\binom{n}{1}$	$\binom{n}{2}$	$\binom{n}{3}$	$\binom{n}{4}$	$\binom{n}{5}$	$\binom{n}{6}$	$\binom{n}{7}$	$\binom{n}{8}$	$\binom{n}{9}$	$\binom{n}{10}$
0	1										
1	1	1									
2	1	2	1								
3	1	3	3	1							
4	1	4	6	4	1						
5	1	5	10	10	5	1					
6	1	6	15	20	15	6	1				
7	1	7	21	35	35	21	7	1			
8	1	8	28	56	70	56	28	8	1		
9	1	9	36	84	126	126	84	36	9	1	
10	1	10	45	120	210	252	210	120	45	10	1
11	1	11	55	165	330	462	462	330	165	55	11
12	1	12	66	220	495	792	924	792	495	220	66
13	1	13	78	286	715	1287	1716	1716	1287	715	286
14	1	14	91	364	1001	2002	3003	3432	3003	2002	1001
15	1	15	105	455	1365	3003	5005	6435	6435	5005	3003
16	1	16	120	560	1820	4368	8008	11440	12870	11440	8008
17	1	17	136	680	2380	6188	12376	19448	24310	24310	19448
18	1	18	153	816	3060	8568	18564	31824	43758	48620	43758
19	1	19	171	969	3876	11628	27132	50388	75582	92378	92378
20	1	20	190	1140	4845	15504	38760	77520	125970	167960	184756

If necessary, use the identity $\binom{n}{k} = \binom{n}{n-k}$

From John E. Freund, *Statistics, A First Course*, Prentice-Hall, Inc., Englewood Cliffs, N. J., 1970, p. 313. Reprinted by permission.
*For specific details on the use of this table, see p. 287.

TABLE 4 Binomial Probabilities $\left[\binom{n}{x} \cdot p^x q^{n-x}\right]$

								p							
n	x	0.01	0.05	0.10	0.20	0.30	0.40	0.50	0.60	0.70	0.80	0.90	0.95	0.99	x
2	0	980	902	810	640	490	360	250	160	090	040	010	002	0+	0
	1	020	095	180	320	420	480	500	480	420	320	180	095	020	1
	2	0+	002	010	040	090	160	250	360	490	640	810	902	980	2
3	0	970	857	729	512	343	216	125	064	027	008	001	0+	0+	0
	1	029	135	243	384	441	432	375	288	189	096	027	007	0+	1
	2	0+	007	027	096	189	288	375	432	441	384	243	135	029	2
	3	0+	0+	001	008	027	064	125	216	343	512	729	857	970	3
4	0	961	815	656	410	240	130	062	026	008	002	0+	0+	0+	0
	1	039	171	292	410	412	346	250	154	076	026	004	0+	0+	1
	2	001	014	049	154	265	346	375	346	265	154	049	014	001	2
	3	0+	0+	004	026	076	154	250	346	412	410	292	171	039	3
	4	0+	0+	0+	002	008	026	062	130	240	410	656	815	961	4
5	0	951	774	590	328	168	078	031	010	002	0+	0+	0+	0+	0
	1	048	204	328	410	360	259	156	077	028	006	0+	0+	0+	1
	2	001	021	073	205	309	346	312	230	132	051	008	001	0+	2
	3	0+	001	008	051	132	230	312	346	309	205	073	021	001	3
	4	0+	0+	0+	006	028	077	156	259	360	410	328	204	048	4
	5	0+	0+	0+	0+	002	010	031	078	168	328	590	774	951	5
6	0	941	735	531	262	118	047	016	004	001	0+	0+	0+	0+	0
	1	057	232	354	393	303	187	094	037	010	002	0+	0+	0+	1
	2	001	031	098	246	324	311	234	138	060	015	001	0+	0+	2
	3	0+	002	015	082	185	276	312	276	185	082	015	002	0+	3
	4	0+	0+	001	015	060	138	234	311	324	246	098	031	001	4
	5	0+	0+	0+	002	010	037	094	187	303	393	354	232	057	5
	6	0+	0+	0+	0+	001	004	016	047	118	262	531	735	941	6
7	0	932	698	478	210	082	028	008	002	0+	0+	0+	0+	0+	0
	1	066	257	372	367	247	131	055	017	004	0+	0+	0+	0+	1
	2	002	041	124	275	318	261	164	077	025	004	0+	0+	0+	2
	3	0+	004	023	115	227	290	273	194	097	029	003	0+	0+	3
	4	0+	0+	003	029	097	194	273	290	227	115	023	004	0+	4
	5	0+	0+	0+	004	025	077	164	261	318	275	124	041	002	5
	6	0+	0+	0+	0+	004	017	055	131	247	367	372	257	066	6
	7	0+	0+	0+	0+	0+	002	008	028	082	210	478	698	932	7
8	0	923	663	430	168	058	017	004	001	0+	0+	0+	0+	0+	0
	1	075	279	383	336	198	090	031	008	001	0+	0+	0+	0+	1
	2	003	051	149	294	296	209	109	041	010	001	0+	0+	0+	2
	3	0+	005	033	147	254	279	219	124	047	009	0+	0+	0+	3
	4	0+	0+	005	046	136	232	273	232	136	046	005	0+	0+	4
	5	0+	0+	0+	009	047	124	219	279	254	147	033	005	0+	5
	6	0+	0+	0+	001	010	041	109	209	296	294	149	051	003	6
	7	0+	0+	0+	0+	001	008	031	090	198	336	383	279	075	7
	8	0+	0+	0+	0+	0+	001	004	017	058	168	430	663	923	8

*For specific details on the use of this table, see p. 290.
Source: Frederick B. Mosteller, Robert E. K. Rourke, and George B. Thomas, Jr. *Probability with Statistical Applications*, Second Edition. © 1970 by Addison-Wesley Publishing Company, Inc. Reprinted with permission of the publisher.

TABLE 4 (Continued)

n	x	0.01	0.05	0.10	0.20	0.30	0.40	0.50	0.60	0.70	0.80	0.90	0.95	0.99	x
9	0	914	630	387	134	040	010	002	0+	0+	0+	0+	0+	0+	0
	1	083	299	387	302	156	060	018	004	0+	0+	0+	0+	0+	1
	2	003	063	172	302	267	161	070	021	004	0+	0+	0+	0+	2
	3	0+	008	045	176	267	251	164	074	021	003	0+	0+	0+	3
	4	0+	001	007	066	172	251	246	167	074	017	001	0+	0+	4
9	5	0+	0+	001	017	074	167	246	251	172	066	007	001	0+	5
	6	0+	0+	0+	003	021	074	164	251	267	176	045	008	0+	6
	7	0+	0+	0+	0+	004	021	070	161	267	302	172	063	003	7
	8	0+	0+	0+	0+	0+	004	018	060	156	302	387	299	083	8
	9	0+	0+	0+	0+	0+	0+	002	010	040	134	387	630	914	9
10	0	904	599	349	107	028	006	001	0+	0+	0+	0+	0+	0+	0
	1	091	315	387	268	121	040	010	002	0+	0+	0+	0+	0+	1
	2	004	075	194	302	233	121	044	011	001	0+	0+	0+	0+	2
	3	0+	010	057	201	267	215	117	042	009	001	0+	0+	0+	3
	4	0+	001	011	088	200	251	205	111	037	006	0+	0+	0+	4
	5	0+	0+	001	026	103	201	246	201	103	026	001	0+	0+	5
	6	0+	0+	0+	006	037	111	205	251	200	088	011	001	0+	6
	7	0+	0+	0+	001	009	042	117	215	267	201	057	010	0+	7
	8	0+	0+	0+	0+	001	011	044	121	233	302	194	075	004	8
	9	0+	0+	0+	0+	0+	002	010	040	121	268	387	315	091	9
	10	0+	0+	0+	0+	0+	0+	001	006	028	107	349	599	904	10
11	0	895	569	314	086	020	004	0+	0+	0+	0+	0+	0+	0+	0
	1	099	329	384	236	093	027	005	001	0+	0+	0+	0+	0+	1
	2	005	087	213	295	200	089	027	005	001	0+	0+	0+	0+	2
	3	0+	014	071	221	257	177	081	023	004	0+	0+	0+	0+	3
	4	0+	001	016	111	220	236	161	070	017	002	0+	0+	0+	4
	5	0+	0+	002	039	132	221	226	147	057	010	0+	0+	0+	5
	6	0+	0+	0+	010	057	147	226	221	132	039	002	0+	0+	6
	7	0+	0+	0+	002	017	070	161	236	220	111	016	001	0+	7
	8	0+	0+	0+	0+	004	023	081	177	257	221	071	014	0+	8
	9	0+	0+	0+	0+	001	005	027	089	200	295	213	087	005	9
	10	0+	0+	0+	0+	0+	001	005	027	093	236	384	329	099	10
	11	0+	0+	0+	0+	0+	0+	0+	004	020	086	314	569	895	11
12	0	886	540	282	069	014	002	0+	0+	0+	0+	0+	0+	0+	0
	1	107	341	377	206	071	017	003	0+	0+	0+	0+	0+	0+	1
	2	006	099	230	283	168	064	016	002	0+	0+	0+	0+	0+	2
	3	0+	017	085	236	240	142	054	012	001	0+	0+	0+	0+	3
	4	0+	002	021	133	231	213	121	042	008	001	0+	0+	0+	4
	5	0+	0+	004	053	158	227	193	101	029	003	0+	0+	0+	5
	6	0+	0+	0+	016	079	177	226	177	079	016	0+	0+	0+	6
	7	0+	0+	0+	003	029	101	193	227	158	053	004	0+	0+	7
	8	0+	0+	0+	001	008	042	121	213	231	133	021	002	0+	8
	9	0+	0+	0+	0+	001	012	054	142	240	236	085	017	0+	9
	10	0+	0+	0+	0+	0+	002	016	064	168	283	230	099	006	10
	11	0+	0+	0+	0+	0+	0+	003	017	071	206	377	341	107	11
	12	0+	0+	0+	0+	0+	0+	0+	002	014	069	282	540	886	12

TABLE 4 (Continued)

n	x	0.01	0.05	0.10	0.20	0.30	0.40	p 0.50	0.60	0.70	0.80	0.90	0.95	0.99	x
13	0	878	513	254	055	010	001	0+	0+	0+	0+	0+	0+	0+	0
	1	115	351	367	179	054	011	002	0+	0+	0+	0+	0+	0+	1
	2	007	111	245	268	139	045	010	001	0+	0+	0+	0+	0+	2
	3	0+	021	100	246	218	111	035	006	001	0+	0+	0+	0+	3
	4	0+	003	028	154	234	184	087	024	003	0+	0+	0+	0+	4
	5	0+	0+	006	069	180	221	157	066	014	001	0+	0+	0+	5
	6	0+	0+	001	023	103	197	209	131	044	006	0+	0+	0+	6
	7	0+	0+	0+	006	044	131	209	197	103	023	001	0+	0+	7
	8	0+	0+	0+	001	014	066	157	221	180	069	006	0+	0+	8
	9	0+	0+	0+	0+	003	024	087	184	234	154	028	003	0+	9
	10	0+	0+	0+	0+	001	006	035	111	218	246	100	021	0+	10
	11	0+	0+	0+	0+	0+	001	010	045	139	268	245	111	007	11
	12	0+	0+	0+	0+	0+	0+	002	011	054	179	367	351	115	12
	13	0+	0+	0+	0+	0+	0+	0+	001	010	055	254	513	878	13
14	0	869	488	229	044	007	001	0+	0+	0+	0+	0+	0+	0+	0
	1	123	359	356	154	041	007	001	0+	0+	0+	0+	0+	0+	1
	2	008	123	257	250	113	032	006	001	0+	0+	0+	0+	0+	2
	3	0+	026	114	250	194	085	022	003	0+	0+	0+	0+	0+	3
	4	0+	004	035	172	229	155	061	014	001	0+	0+	0+	0+	4
	5	0+	0+	008	086	196	207	122	041	007	0+	0+	0+	0+	5
	6	0+	0+	001	032	126	207	183	092	023	002	0+	0+	0+	6
	7	0+	0+	0+	009	062	157	209	157	062	009	0+	0+	0+	7
	8	0+	0+	0+	002	023	092	183	207	126	032	001	0+	0+	8
	9	0+	0+	0+	0+	007	041	122	207	196	086	008	0+	0+	9
	10	0+	0+	0+	0+	001	014	061	155	229	172	035	004	0+	10
	11	0+	0+	0+	0+	0+	003	022	085	194	250	114	026	0+	11
	12	0+	0+	0+	0+	0+	001	006	032	113	250	257	123	008	12
	13	0+	0+	0+	0+	0+	0+	001	007	041	154	356	359	123	13
	14	0+	0+	0+	0+	0+	0+	0+	001	007	044	229	488	869	14
15	0	860	463	206	035	005	0+	0+	0+	0+	0+	0+	0+	0+	0
	1	130	366	343	132	031	005	0+	0+	0+	0+	0+	0+	0+	1
	2	009	135	267	231	092	022	003	0+	0+	0+	0+	0+	0+	2
	3	0+	031	129	250	170	063	014	002	0+	0+	0+	0+	0+	3
	4	0+	005	043	188	219	127	042	007	001	0+	0+	0+	0+	4
	5	0+	001	010	103	206	186	092	024	003	0+	0+	0+	0+	5
	6	0+	0+	002	043	147	207	153	061	012	001	0+	0+	0+	6
	7	0+	0+	0+	014	081	177	196	118	035	003	0+	0+	0+	7
	8	0+	0+	0+	003	035	118	196	177	081	014	0+	0+	0+	8
	9	0+	0+	0+	001	012	061	153	207	147	043	002	0+	0+	9
	10	0+	0+	0+	0+	003	024	092	186	206	103	010	001	0+	10
	11	0+	0+	0+	0+	001	007	042	127	219	188	043	005	0+	11
	12	0+	0+	0+	0+	0+	002	014	063	170	250	129	031	0+	12
	13	0+	0+	0+	0+	0+	0+	003	022	092	231	267	135	009	13
	14	0+	0+	0+	0+	0+	0+	0+	005	031	132	343	366	130	14
	15	0+	0+	0+	0+	0+	0+	0+	0+	005	035	206	463	860	15

TABLE 5
Areas of the Standard Normal Distribution

The entries in this table are the probabilities that a random variable having the standard normal distribution assumes a value between 0 and z; the probability is represented by the area under the curve shaded in the accompanying figure. Areas for negative values of z are obtained by symmetry.

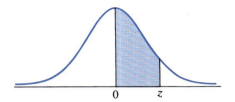

	Second Decimal Place in z									
z	0.00	0.01	0.02	0.03	0.04	0.05	0.06	0.07	0.08	0.09
0.0	0.0000	0.0040	0.0080	0.0120	0.0160	0.0199	0.0239	0.0279	0.0319	0.0359
0.1	0.0398	0.0438	0.0478	0.0517	0.0557	0.0596	0.0636	0.0675	0.0714	0.0753
0.2	0.0793	0.0832	0.0871	0.0910	0.0948	0.0987	0.1026	0.1064	0.1103	0.1141
0.3	0.1179	0.1217	0.1255	0.1293	0.1331	0.1368	0.1406	0.1443	0.1480	0.1517
0.4	0.1554	0.1591	0.1628	0.1664	0.1700	0.1736	0.1772	0.1808	0.1844	0.1879
0.5	0.1915	0.1950	0.1985	0.2019	0.2054	0.2088	0.2123	0.2157	0.2190	0.2224
0.6	0.2257	0.2291	0.2324	0.2357	0.2389	0.2422	0.2454	0.2486	0.2517	0.2549
0.7	0.2580	0.2611	0.2642	0.2673	0.2704	0.2734	0.2764	0.2794	0.2823	0.2852
0.8	0.2881	0.2910	0.2939	0.2967	0.2995	0.3023	0.3051	0.3078	0.3106	0.3133
0.9	0.3159	0.3186	0.3212	0.3238	0.3264	0.3289	0.3315	0.3340	0.3365	0.3389
1.0	0.3413	0.3438	0.3461	0.3485	0.3508	0.3531	0.3554	0.3577	0.3599	0.3621
1.1	0.3643	0.3665	0.3686	0.3708	0.3729	0.3749	0.3770	0.3790	0.3810	0.3830
1.2	0.3849	0.3869	0.3888	0.3907	0.3925	0.3944	0.3962	0.3980	0.3997	0.4015
1.3	0.4032	0.4049	0.4066	0.4082	0.4099	0.4115	0.4131	0.4147	0.4162	0.4177
1.4	0.4192	0.4207	0.4222	0.4236	0.4251	0.4265	0.4279	0.4292	0.4306	0.4319
1.5	0.4332	0.4345	0.4357	0.4370	0.4382	0.4394	0.4406	0.4418	0.4429	0.4441
1.6	0.4452	0.4463	0.4474	0.4484	0.4495	0.4505	0.4515	0.4525	0.4535	0.4545
1.7	0.4554	0.4564	0.4573	0.4582	0.4591	0.4599	0.4608	0.4616	0.4625	0.4633
1.8	0.4641	0.4649	0.4656	0.4664	0.4671	0.4678	0.4686	0.4693	0.4699	0.4706
1.9	0.4713	0.4719	0.4726	0.4732	0.4738	0.4744	0.4750	0.4756	0.4761	0.4767
2.0	0.4772	0.4778	0.4783	0.4788	0.4793	0.4798	0.4803	0.4808	0.4812	0.4817
2.1	0.4821	0.4826	0.4830	0.4834	0.4838	0.4842	0.4846	0.4850	0.4854	0.4857
2.2	0.4861	0.4864	0.4868	0.4871	0.4875	0.4878	0.4881	0.4884	0.4887	0.4890
2.3	0.4893	0.4896	0.4898	0.4901	0.4904	0.4906	0.4909	0.4911	0.4913	0.4916
2.4	0.4918	0.4920	0.4922	0.4925	0.4927	0.4929	0.4931	0.4932	0.4934	0.4936
2.5	0.4938	0.4940	0.4941	0.4943	0.4945	0.4946	0.4948	0.4949	0.4951	0.4952
2.6	0.4953	0.4955	0.4956	0.4957	0.4959	0.4960	0.4961	0.4962	0.4963	0.4964
2.7	0.4965	0.4966	0.4967	0.4968	0.4969	0.4970	0.4971	0.4972	0.4973	0.4974
2.8	0.4974	0.4975	0.4976	0.4977	0.4977	0.4978	0.4979	0.4979	0.4980	0.4981
2.9	0.4981	0.4982	0.4982	0.4983	0.4984	0.4984	0.4985	0.4985	0.4986	0.4986
3.0	0.4987	0.4987	0.4987	0.4988	0.4988	0.4989	0.4989	0.4989	0.4990	0.4990
3.1	0.4990	0.4991	0.4991	0.4991	0.4992	0.4992	0.4992	0.4992	0.4993	0.4993
3.2	0.4993	0.4993	0.4994	0.4994	0.4994	0.4994	0.4994	0.4995	0.4995	0.4995
3.3	0.4995	0.4995	0.4995	0.4996	0.4996	0.4996	0.4996	0.4996	0.4996	0.4997
3.4	0.4997	0.4997	0.4997	0.4997	0.4997	0.4997	0.4997	0.4997	0.4997	0.4998
3.5	0.4998									
4.0	0.49997									
4.5	0.499997									
5.0	0.4999997									

Reprinted with permission from *Standard Mathematical Tables*, 15th ed. Copyright The Chemical Rubber, Co., CRC Press, Inc.

*For specific details on the use of this table, see p. 310.

TABLE 6
Critical Values of Student's t-Distribution

The entries in this table are the critical values for Student's t for an area of α in the right-hand tail. Critical values for the left-hand tail are found by symmetry.

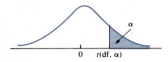

df	Amount of α in One-tail					
	0.25	0.10	0.05	0.025	0.01	0.005
1	1.000	3.08	6.31	12.7	31.8	63.7
2	0.816	1.89	2.92	4.30	6.97	9.92
3	0.765	1.64	2.35	3.18	4.54	5.84
4	0.741	1.53	2.13	2.78	3.75	4.60
5	0.727	1.48	2.02	2.57	3.37	4.03
6	0.718	1.44	1.94	2.45	3.14	3.71
7	0.711	1.42	1.89	2.36	3.00	3.50
8	0.706	1.40	1.86	2.31	2.90	3.36
9	0.703	1.38	1.83	2.26	2.82	3.25
10	0.700	1.37	1.81	2.23	2.76	3.17
11	0.697	1.36	1.80	2.20	2.72	3.11
12	0.695	1.36	1.78	2.18	2.68	3.05
13	0.694	1.35	1.77	2.16	2.65	3.01
14	0.692	1.35	1.76	2.14	2.62	2.98
15	0.691	1.34	1.75	2.13	2.60	2.95
16	0.690	1.34	1.75	2.12	2.58	2.92
17	0.689	1.33	1.74	2.11	2.57	2.90
18	0.688	1.33	1.73	2.10	2.55	2.88
19	0.688	1.33	1.73	2.09	2.54	2.86
20	0.687	1.33	1.72	2.09	2.53	2.85
21	0.686	1.32	1.72	2.08	2.52	2.83
22	0.686	1.32	1.72	2.07	2.51	2.82
23	0.685	1.32	1.71	2.07	2.50	2.81
24	0.685	1.32	1.71	2.06	2.49	2.80
25	0.684	1.32	1.71	2.06	2.49	2.79
26	0.684	1.32	1.71	2.06	2.48	2.78
27	0.684	1.31	1.70	2.05	2.47	2.77
28	0.683	1.31	1.70	2.05	2.47	2.76
29	0.683	1.31	1.70	2.05	2.46	2.76
z	0.674	1.28	1.65	1.96	2.33	2.58

NOTE: For df $\geq$ 30, the critical value $t(df, \alpha)$ is approximated by $z(\alpha)$, given in the bottom row of table.

Adapted from E. S. Pearson and H. O. Hartley, *Biometrika Tables for Statisticians*, vol. I (1966), p. 146. Reprinted by permission of the Biometrika Trustees. The two columns headed "0.10" and "0.01" are taken from Table III (adapted) on p. 46 of Fisher and Yates, *Statistical Tables for Biological, Agricultural and Medical Research*, 6th ed., published by Longman Group Ltd., London, 1974 (previously published by Oliver and Boyd, Edinburgh), and by permission of the authors and publishers.

*For specific details on the use of this table, see p. 440.

TABLE 7
Critical Values of the χ^2 Distribution

The entries in this table are the critical values for chi square for which the area to the right under the curve is equal to α.

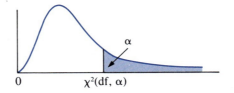

df	\multicolumn{9}{c}{Amount of α in Right-hand Tail}
	0.995	0.990	0.975	0.950	0.900	0.100	0.050	0.025	0.010	0.005
1	0.0000393	0.000157	0.000982	0.00393	0.0158	2.71	3.84	5.02	6.65	7.88
2	0.0100	0.0201	0.0506	0.103	0.211	4.61	6.00	7.38	9.21	10.6
3	0.0717	0.115	0.216	0.352	0.584	6.25	7.82	9.35	11.4	12.9
4	0.207	0.297	0.484	0.711	1.0636	7.78	9.50	11.1	13.3	14.9
5	0.412	0.554	0.831	1.15	1.61	9.24	11.1	12.8	15.1	16.8
6	0.676	0.872	1.24	1.64	2.20	10.6	12.6	14.5	16.8	18.6
7	0.990	1.24	1.69	2.17	2.83	12.0	14.1	16.0	18.5	20.3
8	1.34	1.65	2.18	2.73	3.49	13.4	15.5	17.5	20.1	22.0
9	1.73	2.09	2.70	3.33	4.17	14.7	17.0	19.0	21.7	23.6
10	2.16	2.56	3.25	3.94	4.87	16.0	18.3	20.5	23.2	25.2
11	2.60	3.05	3.82	4.58	5.58	17.2	19.7	21.9	24.7	26.8
12	3.07	3.57	4.40	5.23	6.30	18.6	21.0	23.3	26.2	28.3
13	3.57	4.11	5.01	5.90	7.04	19.8	22.4	24.7	27.7	29.8
14	4.07	4.66	5.63	6.57	7.79	21.1	23.7	26.1	29.1	31.3
15	4.60	5.23	6.26	7.26	8.55	22.3	25.0	27.5	30.6	32.8
16	5.14	5.81	6.91	7.96	9.31	23.5	26.3	28.9	32.0	34.3
17	5.70	6.41	7.56	8.67	10.1	24.8	27.6	30.2	33.4	35.7
18	6.26	7.01	8.23	9.39	10.9	26.0	28.9	31.5	34.8	37.2
19	6.84	7.63	8.91	10.1	11.7	27.2	30.1	32.9	36.2	38.6
20	7.43	8.26	9.59	10.9	12.4	28.4	31.4	34.2	37.6	40.0
21	8.03	8.90	10.3	11.6	13.2	29.6	32.7	35.5	39.0	41.4
22	8.64	9.54	11.0	12.3	14.0	30.8	33.9	36.8	40.3	42.8
23	9.26	10.2	11.0	13.1	14.9	32.0	35.2	38.1	41.6	44.2
24	9.89	10.9	12.4	13.9	15.7	33.2	36.4	39.4	43.0	45.6
25	10.5	11.5	13.1	14.6	16.5	34.4	37.7	40.7	44.3	46.9
26	11.2	12.2	13.8	15.4	17.3	35.6	38.9	41.9	45.6	48.3
27	11.8	12.9	14.6	16.2	18.1	36.7	40.1	43.2	47.0	49
28	12.5	13.6	15.3	16.9	18.9	37.9	41.3	44.5	48.3	51.0
29	13.1	14.3	16.1	17.7	19.8	39.1	42.6	45.7	49.6	52.3
30	13.8	15.0	16.8	18.5	20.6	40.3	43.8	47.0	50.9	53.7
40	20.7	22.2	24.4	26.5	29.1	51.8	55.8	59.3	63.7	66.8
50	28.0	29.7	32.4	34.8	37.7	63.2	67.5	71.4	76.2	79.5
60	35.5	37.5	40.5	43.2	46.5	74.4	79.1	83.3	88.4	92.0
70	43.3	45.4	48.8	51.8	55.3	85.5	90.5	95.0	100.0	104.0
80	51.2	53.5	57.2	60.4	64.3	96.6	102.0	107.0	112.0	116.0
90	59.2	61.8	65.7	69.1	73.3	108.0	113.0	118.0	124.0	128.0
100	67.3	70.1	74.2	77.9	82.4	114.0	124.0	130.0	136.0	140.0

Adapted from E. S. Pearson and H. O. Hartley, *Biometrika Tables for Statisticians*, vol. I (1962), pp. 130–131. Reprinted by permission of the Biometrika Trustees.

*For specific details on the use of this table, see p. 464.

TABLE 8a
Critical Values of the
F Distribution
($\alpha = 0.05$)

The entries in this table are critical values of F for which the area under the curve to the right is equal to 0.05.

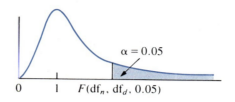

		\multicolumn{10}{c}{Degrees of Freedom for Numerator}
		1	2	3	4	5	6	7	8	9	10
Degrees of Freedom for Denominator	1	161	200	216	225	230	234	237	239	241	242
	2	18.5	19.0	19.2	19.2	19.3	19.3	19.4	19.4	19.4	19.4
	3	10.1	9.55	9.28	9.12	9.01	8.94	8.89	8.85	8.81	8.79
	4	7.71	6.94	6.59	6.39	6.26	6.16	6.09	6.04	6.00	5.96
	5	6.61	5.79	5.41	5.19	5.05	4.95	4.88	4.82	4.77	4.74
	6	5.99	5.14	4.76	4.53	4.39	4.28	4.21	4.15	4.10	4.06
	7	5.59	4.74	4.35	4.12	3.97	3.87	3.79	3.73	3.68	3.64
	8	5.32	4.46	4.07	3.84	3.69	3.58	3.50	3.44	3.39	3.35
	9	5.12	4.26	3.86	3.63	3.48	3.37	3.29	3.23	3.18	3.14
	10	4.96	4.10	3.71	3.48	3.33	3.22	3.14	3.07	3.02	2.98
	11	4.84	3.98	3.59	3.36	3.20	3.09	3.01	2.95	2.90	2.85
	12	4.75	3.89	3.49	3.26	3.11	3.00	2.91	2.85	2.80	2.75
	13	4.67	3.81	3.41	3.18	3.03	2.92	2.83	2.77	2.71	2.67
	14	4.60	3.74	3.34	3.11	2.96	2.85	2.76	2.70	2.65	2.60
	15	4.54	3.68	3.29	3.06	2.90	2.79	2.71	2.64	2.59	2.54
	16	4.49	3.63	3.24	3.01	2.85	2.74	2.66	2.59	2.54	2.49
	17	4.45	3.59	3.20	2.96	2.81	2.70	2.61	2.55	2.49	2.45
	18	4.41	3.55	3.16	2.93	2.77	2.66	2.58	2.51	2.46	2.41
	19	4.38	3.52	3.13	2.90	2.74	2.63	2.54	2.48	2.42	2.38
	20	4.35	3.49	3.10	2.87	2.71	2.60	2.51	2.45	2.39	2.35
	21	4.32	3.47	3.07	2.84	2.68	2.57	2.49	2.42	2.37	2.32
	22	4.30	3.44	3.05	2.82	2.66	2.55	2.46	2.40	2.34	2.30
	23	4.28	3.42	3.03	2.80	2.64	2.53	2.44	2.37	2.32	2.27
	24	4.26	3.40	3.01	2.78	2.62	2.51	2.42	2.36	2.30	2.25
	25	4.24	3.39	2.99	2.76	2.60	2.49	2.40	2.34	2.28	2.24
	30	4.17	3.32	2.92	2.69	2.53	2.42	2.33	2.27	2.21	2.16
	40	4.08	3.23	2.84	2.61	2.45	2.34	2.25	2.18	2.12	2.08
	60	4.00	3.15	2.76	2.53	2.37	2.25	2.17	2.10	2.04	1.99
	120	3.92	3.07	2.68	2.45	2.29	2.18	2.09	2.02	1.96	1.91
	∞	3.84	3.00	2.60	2.37	2.21	2.10	2.01	1.94	1.88	1.83

*For specific details on the use of this table, see p. 506.

TABLE 8a (Continued)

		\multicolumn{9}{c}{Degrees of Freedom for Numerator}
		12	15	20	24	30	40	60	120	∞
Degrees of Freedom for Denominator	1	244	246	248	249	250	251	252	253	254
	2	19.4	19.4	19.4	19.5	19.5	19.5	19.5	19.5	19.5
	3	8.74	8.70	8.66	8.64	8.62	8.59	8.57	8.55	8.53
	4	5.91	5.86	5.80	5.77	5.75	5.72	5.69	5.66	5.63
	5	4.68	4.62	4.56	4.53	4.50	4.46	4.43	4.40	4.37
	6	4.00	3.94	3.87	3.84	3.81	3.77	3.74	3.70	3.67
	7	3.57	3.51	3.44	3.41	3.38	3.34	3.30	3.27	3.23
	8	3.28	3.22	3.15	3.12	3.08	3.04	3.01	2.97	2.93
	9	3.07	3.01	2.94	2.90	2.86	2.83	2.79	2.75	2.71
	10	2.91	2.85	2.77	2.74	2.70	2.66	2.62	2.58	2.54
	11	2.79	2.72	2.65	2.61	2.57	2.53	2.49	2.45	2.40
	12	2.69	2.62	2.54	2.51	2.47	2.43	2.38	2.34	2.30
	13	2.60	2.53	2.46	2.42	2.38	2.34	2.30	2.25	2.21
	14	2.53	2.46	2.39	2.35	2.31	2.27	2.22	2.18	2.13
	15	2.48	2.40	2.33	2.29	2.25	2.20	2.16	2.11	2.07
	16	2.42	2.35	2.28	2.24	2.19	2.15	2.11	2.06	2.01
	17	2.38	2.31	2.23	2.19	2.15	2.10	2.06	2.01	1.96
	18	2.34	2.27	2.19	2.15	2.11	2.06	2.02	1.97	1.92
	19	2.31	2.23	2.16	2.11	2.07	2.03	1.98	1.93	1.88
	20	2.28	2.20	2.12	2.08	2.04	1.99	1.95	1.90	1.84
	21	2.25	2.18	2.10	2.05	2.01	1.96	1.92	1.87	1.81
	22	2.23	2.15	2.07	2.03	1.98	1.94	1.89	1.84	1.78
	23	2.20	2.13	2.05	2.01	1.96	1.91	1.86	1.81	1.76
	24	2.18	2.11	2.03	1.98	1.94	1.89	1.84	1.79	1.73
	25	2.16	2.09	2.01	1.96	1.92	1.87	1.82	1.77	1.71
	30	2.09	2.01	1.93	1.89	1.84	1.79	1.74	1.68	1.62
	40	2.00	1.92	1.84	1.79	1.74	1.69	1.64	1.58	1.51
	60	1.92	1.84	1.75	1.70	1.65	1.59	1.53	1.47	1.39
	120	1.83	1.75	1.66	1.61	1.55	1.50	1.43	1.35	1.25
	∞	1.75	1.67	1.57	1.52	1.46	1.39	1.32	1.22	1.00

From E. S. Pearson and H. O. Hartley, *Biometrika Tables for Statisticians*, vol. I (1958), pp. 159–163. Reprinted by permission of the Biometrika Trustees.

TABLE 8b
Critical Values of the
F Distribution
($\alpha = 0.025$)

The entries in this table are critical values of F for which the area under the curve to the right is equal to 0.025.

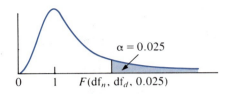

		Degrees of Freedom for Numerator									
		1	2	3	4	5	6	7	8	9	10
Degrees of Freedom for Denominator	1	648	800	864	900	922	937	948	957	963	969
	2	38.5	39.0	39.2	39.2	39.3	39.3	39.4	39.4	39.4	39.4
	3	17.4	16.0	15.4	15.1	14.9	14.7	14.6	14.5	14.5	14.4
	4	12.2	10.6	9.98	9.60	9.36	9.20	9.07	8.98	8.90	8.84
	5	10.0	8.43	7.76	7.39	7.15	6.98	6.85	6.76	6.68	6.62
	6	8.81	7.26	6.60	6.23	5.99	5.82	5.70	5.60	5.52	5.46
	7	8.07	6.54	5.89	5.52	5.29	5.12	4.99	4.90	4.82	4.76
	8	7.57	6.06	5.42	5.05	4.82	4.65	4.53	4.43	4.36	4.30
	9	7.21	5.71	5.08	4.72	4.48	4.32	4.20	4.10	4.03	3.96
	10	6.94	5.46	4.83	4.47	4.24	4.07	3.95	3.85	3.78	3.72
	11	6.72	5.26	4.63	4.28	4.04	3.88	3.76	3.66	3.59	3.53
	12	6.55	5.10	4.47	4.12	3.89	3.73	3.61	3.51	3.44	3.37
	13	6.41	4.97	4.35	4.00	3.77	3.60	3.48	3.39	3.31	3.25
	14	6.30	4.86	4.24	3.89	3.66	3.50	3.38	3.28	3.21	3.15
	15	6.20	4.77	4.15	3.80	3.58	3.41	3.29	3.20	3.12	3.06
	16	6.12	4.69	4.08	3.73	3.50	3.34	3.22	3.12	3.05	2.99
	17	6.04	4.62	4.01	3.66	3.44	3.28	3.16	3.06	2.98	2.92
	18	5.98	4.56	3.95	3.61	3.38	3.22	3.10	3.01	2.93	2.87
	19	5.92	4.51	3.90	3.56	3.33	3.17	3.05	2.96	2.88	2.82
	20	5.87	4.46	3.86	3.51	3.29	3.13	3.01	2.91	2.84	2.77
	21	5.83	4.42	3.82	3.48	3.25	3.09	2.97	2.87	2.80	2.73
	22	5.79	4.38	3.78	3.44	3.22	3.05	2.93	2.84	2.76	2.70
	23	5.75	4.35	3.75	3.41	3.18	3.02	2.90	2.81	2.73	2.67
	24	5.72	4.32	3.72	3.38	3.15	2.99	2.87	2.78	2.70	2.64
	25	5.69	4.29	3.69	3.35	3.13	2.97	2.85	2.75	2.68	2.61
	30	5.57	4.18	3.59	3.25	3.03	2.87	2.75	2.65	2.57	2.51
	40	5.42	4.05	3.46	3.13	2.90	2.74	2.62	2.53	2.45	2.39
	60	5.29	3.93	3.34	3.01	2.79	2.63	2.51	2.41	2.33	2.27
	120	5.15	3.80	3.23	2.89	2.67	2.52	2.39	2.30	2.22	2.16
	∞	5.02	3.69	3.12	2.79	2.57	2.41	2.29	2.19	2.11	2.05

*For specific details on the use of this table, see page 506.

TABLE 8b (Continued)

		Degrees of Freedom for Numerator								
		12	15	20	24	30	40	60	120	∞
Degrees of Freedom for Denominator	1	977	985	993	997	1,001	1,006	1,010	1,014	1,018
	2	39.4	39.4	39.4	39.5	39.5	39.5	39.5	39.5	39.5
	3	14.3	14.3	14.2	14.1	14.1	14.0	14.0	13.9	13.9
	4	8.75	8.66	8.56	8.51	8.46	8.41	8.36	8.31	8.26
	5	6.52	6.43	6.33	6.28	6.23	6.18	6.12	6.07	6.02
	6	5.37	5.27	5.17	5.12	5.07	5.01	4.96	4.90	4.85
	7	4.67	4.57	4.47	4.42	4.36	4.31	4.25	4.20	4.14
	8	4.20	4.10	4.00	3.95	3.89	3.84	3.78	3.73	3.67
	9	3.87	3.77	3.67	3.61	3.56	3.51	3.45	3.39	3.33
	10	3.62	3.52	3.42	3.37	3.31	3.26	3.20	3.14	3.08
	11	3.43	3.33	3.23	3.17	3.12	3.06	3.00	2.94	2.88
	12	3.28	3.18	3.07	3.02	2.96	2.91	2.85	2.79	2.72
	13	3.15	3.05	2.95	2.89	2.84	2.78	2.72	2.66	2.60
	14	3.05	2.95	2.84	2.79	2.73	2.67	2.61	2.55	2.49
	15	2.96	2.86	2.76	2.70	2.64	2.59	2.52	2.46	2.40
	16	2.89	2.79	2.68	2.63	2.57	2.51	2.45	2.38	2.32
	17	2.82	2.72	2.62	2.56	2.50	2.44	2.38	2.32	2.25
	18	2.77	2.67	2.56	2.50	2.44	2.38	2.32	2.26	2.19
	19	2.72	2.62	2.51	2.45	2.39	2.33	2.27	2.20	2.13
	20	2.68	2.57	2.46	2.41	2.35	2.29	2.22	2.16	2.09
	21	2.64	2.53	2.42	2.37	2.31	2.25	2.18	2.11	2.04
	22	2.60	2.50	2.39	2.33	2.27	2.21	2.14	2.08	2.00
	23	2.57	2.47	2.36	2.30	2.24	2.18	2.11	2.04	1.97
	24	2.54	2.44	2.33	2.27	2.21	2.15	2.08	2.01	1.94
	25	2.51	2.41	2.30	2.24	2.18	2.12	2.05	1.98	1.91
	30	2.41	2.31	2.20	2.14	2.07	2.01	1.94	1.87	1.79
	40	2.29	2.18	2.07	2.01	1.94	1.88	1.80	1.72	1.64
	60	2.17	2.06	1.94	1.88	1.82	1.74	1.67	1.58	1.48
	120	2.05	1.95	1.82	1.76	1.69	1.61	1.53	1.43	1.31
	∞	1.94	1.83	1.71	1.64	1.57	1.48	1.39	1.27	1.00

From E. S. Pearson and H. O. Hartley, *Biometrika Tables for Statisticians*, vol. I (1958), pp. 159–163. Reprinted by permission of the Biometrika Trustees.

TABLE 8C
Critical Values of the F Distribution ($\alpha = 0.01$)

The entries in the table are critical values of F for which the area under the curve to the right is equal to 0.01.

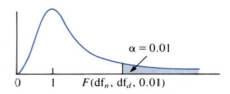

$F(df_n, df_d, 0.01)$

		Degrees of Freedom for Numerator								
		1	2	3	4	5	6	7	8	9	10
Degrees of Freedom for Denominator	1	4,052	5,000	5,403	5,625	5,764	5,859	5,928	5,982	6,023	6,056
	2	98.5	99.0	99.2	99.2	99.3	99.3	99.4	99.4	99.4	99.4
	3	34.1	30.8	29.5	28.7	28.2	27.9	27.7	27.5	27.3	27.2
	4	21.2	18.0	16.7	16.0	15.5	15.2	15.0	14.8	14.7	14.5
	5	16.3	13.3	12.1	11.4	11.0	10.7	10.5	10.3	10.2	10.1
	6	13.7	10.9	9.78	9.15	8.75	8.47	8.26	8.10	7.98	7.87
	7	12.2	9.55	8.45	7.85	7.46	7.19	6.99	6.84	6.72	6.62
	8	11.3	8.65	7.59	7.01	6.63	6.37	6.18	6.03	5.91	5.81
	9	10.6	8.02	6.99	6.42	6.06	5.80	5.61	5.47	5.35	5.26
	10	10.0	7.56	6.55	5.99	5.64	5.39	5.20	5.06	4.94	4.85
	11	9.65	7.21	6.22	5.67	5.32	5.07	4.89	4.74	4.63	4.54
	12	9.33	6.93	5.95	5.41	5.06	4.82	4.64	4.50	4.39	4.30
	13	9.07	6.70	5.74	5.21	4.86	4.62	4.44	4.30	4.19	4.10
	14	8.86	6.51	5.56	5.04	4.70	4.46	4.28	4.14	4.03	3.94
	15	8.68	6.36	5.42	4.89	4.56	4.32	4.14	4.00	3.89	3.80
	16	8.53	6.23	5.29	4.77	4.44	4.20	4.03	3.89	3.78	3.69
	17	8.40	6.11	5.19	4.67	4.34	4.10	3.93	3.79	3.68	3.59
	18	8.29	6.01	5.09	4.58	4.25	4.01	3.84	3.71	3.60	3.51
	19	8.19	5.93	5.01	4.50	4.17	3.94	3.77	3.63	3.52	3.43
	20	8.10	5.85	4.94	4.43	4.10	3.87	3.70	3.56	3.46	3.37
	21	8.02	5.78	4.87	4.37	4.04	3.81	3.64	3.51	3.40	3.31
	22	7.95	5.72	4.82	4.31	3.99	3.76	3.59	3.45	3.35	3.26
	23	7.88	5.66	4.76	4.26	3.94	3.71	3.54	3.41	3.30	3.21
	24	7.82	5.61	4.72	4.22	3.90	3.67	3.50	3.36	3.26	3.17
	25	7.77	5.57	4.68	4.18	3.86	3.63	3.46	3.32	3.22	3.13
	30	7.56	5.39	4.51	4.02	3.70	3.47	3.30	3.17	3.07	2.98
	40	7.31	5.18	4.31	3.83	3.51	3.29	3.12	2.99	2.89	2.80
	60	7.08	4.98	4.13	3.65	3.34	3.12	2.95	2.82	2.72	2.63
	120	6.85	4.79	3.95	3.48	3.17	2.96	2.79	2.66	2.56	2.47
	∞	6.63	4.61	3.78	3.32	3.02	2.80	2.64	2.51	2.41	2.32

TABLE 8C (Continued)

		Degrees of Freedom for Numerator								
		12	15	20	24	30	40	60	120	∞
Degrees of Freedom for Denominator	1	6,106	6,157	6,209	6,235	6,261	6,287	6,313	6,339	6,366
	2	99.4	99.4	99.4	99.5	99.5	99.5	99.5	99.5	99.5
	3	27.1	26.9	26.7	26.6	26.5	26.4	26.3	26.2	26.1
	4	14.4	14.2	14.0	13.9	13.8	13.7	13.7	13.6	13.5
	5	9.89	9.72	9.55	9.47	9.38	9.29	9.20	9.11	9.02
	6	7.72	7.56	7.40	7.31	7.23	7.14	7.06	6.97	6.88
	7	6.47	6.31	6.16	6.07	5.99	5.91	5.82	5.74	5.65
	8	5.67	5.52	5.36	5.28	5.20	5.12	5.03	4.95	4.86
	9	5.11	4.96	4.81	4.73	4.65	4.57	4.48	4.40	4.31
	10	4.71	4.56	4.41	4.33	4.25	4.17	4.08	4.00	3.91
	11	4.40	4.25	4.10	4.02	3.94	3.86	3.78	3.69	3.60
	12	4.16	4.01	3.86	3.78	3.70	3.62	3.54	3.45	3.36
	13	3.96	3.82	3.66	3.59	3.51	3.43	3.34	3.25	3.17
	14	3.80	3.66	3.51	3.43	3.35	3.27	3.18	3.09	3.00
	15	3.67	3.52	3.37	3.29	3.21	3.13	3.05	2.96	2.87
	16	3.55	3.41	3.26	3.18	3.10	3.02	2.93	2.84	2.75
	17	3.46	3.31	3.16	3.08	3.00	2.92	2.83	2.75	2.65
	18	3.37	3.23	3.08	3.00	2.92	2.84	2.75	2.66	2.57
	19	3.30	3.15	3.00	2.92	2.84	2.76	2.67	2.58	2.49
	20	3.23	3.09	2.94	2.86	2.78	2.69	2.61	2.52	2.42
	21	3.17	3.03	2.88	2.80	2.72	2.64	2.55	2.46	2.36
	22	3.12	2.98	2.83	2.75	2.67	2.58	2.50	2.40	2.31
	23	3.07	2.93	2.78	2.70	2.62	2.54	2.45	2.35	2.26
	24	3.03	2.89	2.74	2.66	2.58	2.49	2.40	2.31	2.21
	25	2.99	2.85	2.70	2.62	2.53	2.45	2.36	2.27	2.17
	30	2.84	2.70	2.55	2.47	2.39	2.30	2.21	2.11	2.01
	40	2.66	2.52	2.37	2.29	2.20	2.11	2.02	1.92	1.80
	60	2.50	2.35	2.20	2.12	2.03	1.94	1.84	1.73	1.60
	120	2.34	2.19	2.03	1.95	1.86	1.76	1.66	1.53	1.38
	∞	2.18	2.04	1.88	1.79	1.70	1.59	1.47	1.32	1.00

From E. S. Pearson and H. O. Hartley, *Biometrika Tables for Statisticians*, vol. I (1958), pp. 159–163. Reprinted by permission of the Biometrika Trustees.

TABLE 9
Critical Values of r When $\rho = 0$

The entries in this table are the critical values of r for a two-tailed test at α.

For simple correlation, df $= n - 2$, where n is the number of pairs of data in the sample. For a one-tailed test, the value of α shown at the top of the table is double the value of α being used in the hypothesis test.

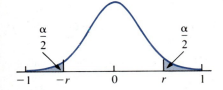

df	α = 0.10	0.05	0.02	0.01
1	0.988	0.997	1.000	1.000
2	0.900	0.950	0.980	0.990
3	0.805	0.878	0.934	0.959
4	0.729	0.811	0.882	0.917
5	0.669	0.754	0.833	0.874
6	0.662	0.707	0.789	0.834
7	0.582	0.666	0.750	0.798
8	0.549	0.632	0.716	0.765
9	0.521	0.602	0.685	0.735
10	0.497	0.576	0.658	0.708
11	0.476	0.553	0.634	0.684
12	0.458	0.532	0.612	0.661
13	0.441	0.514	0.592	0.641
14	0.426	0.497	0.574	0.623
15	0.412	0.482	0.558	0.606
16	0.400	0.468	0.542	0.590
17	0.389	0.456	0.528	0.575
18	0.378	0.444	0.516	0.561
19	0.369	0.433	0.503	0.549
20	0.360	0.423	0.492	0.537
25	0.323	0.381	0.445	0.487
30	0.296	0.349	0.409	0.449
35	0.275	0.325	0.381	0.418
40	0.257	0.304	0.358	0.393
45	0.243	0.288	0.338	0.372
50	0.231	0.273	0.322	0.354
60	0.211	0.250	0.295	0.325
70	0.195	0.232	0.274	0.302
80	0.183	0.217	0.256	0.283
90	0.173	0.205	0.242	0.267
100	0.164	0.195	0.230	0.254

From E. S. Pearson and H. O. Hartley, *Biometrika Tables for Statisticians*, vol. I (1962), p. 138. Reprinted by permission of the Biometrika Trustees.

*For specific details on the use of this table, see p. 646.

TABLE 10
Confidence Belts for the Correlation Coefficient $(1 - \alpha) = 0.95$

The numbers on the curves are sample sizes.

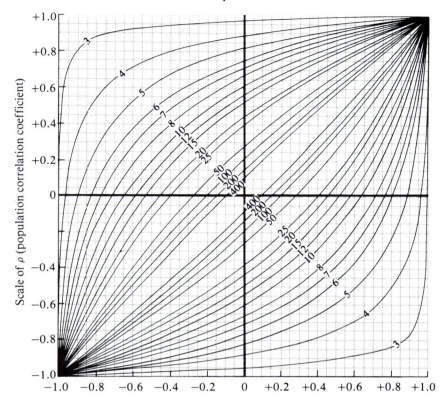

Scale of r (sample correlation coefficient)

*For specific details on the use of this table, see p. 647.

TABLE 11
Critical Values of the Sign Test

The entries in this table are the critical values for the number of the least frequent sign for a two-tailed test at α for the binomial $p = 0.5$. For a one-tailed test, the value of α shown at the top of the table is double the value of α being used in the hypothesis test.

	α					α			
n	0.01	0.05	0.10	0.25	n	0.01	0.05	0.10	0.25
1					51	15	18	19	20
2					52	16	18	19	21
3				0	53	16	18	20	21
4				0	54	17	19	20	22
5			0	0	55	17	19	20	22
6		0	0	1	56	17	20	21	23
7		0	0	1	57	18	20	21	23
8	0	0	1	1	58	18	21	22	24
9	0	1	1	2	59	19	21	22	24
10	0	1	1	2	60	19	21	23	25
11	0	1	2	3	61	20	22	23	25
12	1	2	2	3	62	20	22	24	25
13	1	2	3	3	63	20	23	24	26
14	1	2	3	4	64	21	23	24	26
15	2	3	3	4	65	21	24	25	27
16	2	3	4	5	66	22	24	25	27
17	2	4	4	5	67	22	25	26	28
18	3	4	5	6	68	22	25	26	28
19	3	4	5	6	69	23	25	27	29
20	3	5	5	6	70	23	26	27	29
21	4	5	6	7	71	24	26	28	30
22	4	5	6	7	72	24	27	28	30
23	4	6	7	8	73	25	27	28	31
24	5	6	7	8	74	25	28	29	31
25	5	7	7	9	75	25	28	29	32
26	6	7	8	9	76	26	28	30	32
27	6	7	8	10	77	26	29	30	32
28	6	8	9	10	78	27	29	31	33
29	7	8	9	10	79	27	30	31	33
30	7	9	10	11	80	28	30	32	34
31	7	9	10	11	81	28	31	32	34
32	8	9	10	12	82	28	31	33	35
33	8	10	11	12	83	29	32	33	35
34	9	10	11	13	84	29	32	33	36
35	9	11	12	13	85	30	32	34	36
36	9	11	12	14	86	30	33	34	37
37	10	12	13	14	87	31	33	35	37
38	10	12	13	14	88	31	34	35	38
39	11	12	13	15	89	31	34	36	38
40	11	13	14	15	90	32	35	36	39
41	11	13	14	16	91	32	35	37	39
42	12	14	15	16	92	33	36	37	39
43	12	14	15	17	93	33	36	38	40
44	13	15	16	17	94	34	37	38	40
45	13	15	16	18	95	34	37	38	41
46	13	15	16	18	96	34	37	39	41
47	14	16	17	19	97	35	38	39	42
48	14	16	17	19	98	35	38	40	42
49	15	17	18	19	99	36	39	40	43
50	15	17	18	20	100	36	39	41	43

*For specific details on the use of this table, see pp. 691–696.

From Wilfred J. Dixon and Frank J. Massey, Jr., *Introduction to Statistical Analysis*, 3d ed., (New York: McGraw-Hill, 1969), p. 509. Reprinted by permission.

TABLE 12
Critical Values of U in the Mann-Whitney Test

A. The entries are the critical values of U for a one-tailed test at 0.025 or for a two-tailed test at 0.05.

n_2 \ n_1	1	2	3	4	5	6	7	8	9	10	11	12	13	14	15	16	17	18	19	20
1																				
2								0	0	0	0	1	1	1	1	1	2	2	2	2
3					0	1	1	2	2	3	3	4	4	5	5	6	6	7	7	8
4				0	1	2	3	4	4	5	6	7	8	9	10	11	11	12	13	13
5			0	1	2	3	5	6	7	8	9	11	12	13	14	15	17	18	19	20
6			1	2	3	5	6	8	10	11	13	14	16	17	19	21	22	24	25	27
7			1	3	5	6	8	10	12	14	16	18	20	22	24	26	28	30	32	34
8		0	2	4	6	8	10	13	15	17	19	22	24	26	29	21	34	36	38	41
9		0	2	4	7	10	12	15	17	20	23	26	28	31	34	37	39	42	45	48
10		0	3	5	8	11	14	17	20	23	26	29	33	36	39	42	45	48	52	55
11		0	3	6	9	13	16	19	23	26	30	33	37	40	44	47	51	55	58	62
12		1	4	7	11	14	18	22	26	29	33	37	41	45	49	53	57	61	65	69
13		1	4	8	12	16	20	24	28	33	37	41	45	50	54	59	63	67	72	76
14		1	5	9	13	17	22	26	31	36	40	45	50	55	59	64	67	74	78	83
15		1	5	10	14	19	24	29	34	39	44	49	54	59	64	70	75	80	85	90
16		1	6	11	15	21	26	31	37	42	47	53	59	64	70	75	81	86	92	98
17		2	6	11	17	22	28	34	39	45	51	57	63	67	75	81	87	93	99	105
18		2	7	12	18	24	30	36	42	48	55	61	67	74	80	86	93	99	106	112
19		2	7	13	19	25	32	38	45	52	58	65	72	78	85	92	99	106	113	119
20		2	8	13	20	27	34	41	48	55	62	69	76	83	90	98	105	112	119	127

B. The entries are the critical values of U for a one-tailed test at 0.05 or for a two-tailed test at 0.10.

n_2 \ n_1	1	2	3	4	5	6	7	8	9	10	11	12	13	14	15	16	17	18	19	20
1																			0	0
2					0	0	0	1	1	1	1	2	2	2	3	3	3	4	4	4
3			0	0	1	2	2	3	3	4	5	5	6	7	7	8	9	9	10	11
4			0	1	2	3	4	5	6	7	8	9	10	11	12	14	15	16	17	18
5		0	1	2	4	5	6	8	9	11	12	13	15	16	18	19	20	22	23	25
6		0	2	3	5	7	8	10	12	14	16	17	19	21	23	25	26	28	30	32
7		0	2	4	6	8	11	13	15	17	19	21	24	26	28	30	33	35	37	39
8		1	3	5	8	10	13	15	18	20	23	26	28	31	33	36	39	41	44	47
9		1	3	6	9	12	15	18	21	24	27	30	33	36	39	42	45	48	51	54
10		1	4	7	11	14	17	20	24	27	31	34	37	41	44	48	51	55	58	62
11		1	5	8	12	16	19	23	27	31	34	38	42	46	50	54	57	61	65	69
12		2	5	9	13	17	21	26	30	34	38	42	47	51	55	60	64	68	72	77
13		2	6	10	15	19	24	28	33	37	42	47	51	56	61	65	70	75	80	84
14		2	7	11	16	21	26	31	36	41	46	51	56	61	66	71	77	82	87	92
15		3	7	12	18	23	28	33	39	44	50	55	61	66	72	77	83	88	94	100
16		3	8	14	19	25	30	36	42	48	54	60	65	71	77	83	89	95	101	107
17		3	9	15	20	26	33	39	45	51	57	64	70	77	83	89	96	102	109	115
18		4	9	16	22	28	35	41	48	55	61	68	75	82	88	95	102	109	116	123
19	0	4	10	17	23	30	37	44	51	58	65	72	80	87	94	101	109	116	123	130
20	0	4	11	18	25	32	39	47	54	62	69	77	84	92	100	107	115	123	130	138

*For specific details on the use of this table, see p. 702.
Reproduced from the *Bulletin of the Institute of Educational Research at Indiana University*, vol. 1, no. 2; with the permission of the author and the publisher.

TABLE 13
Critical Values for Total Number of Runs (V)

The entries in this table are the critical values* for a two-tailed test using $\alpha = 0.05$. For a one-tailed test, $\alpha = 0.025$ and use only one of the critical values: the smaller critical value for a left-hand critical region, the larger for a right-hand critical region.

		The larger of n_1 and n_2														
The smaller of n_1 and n_2		5	6	7	8	9	10	11	12	13	14	15	16	17	18	19	20
2								2/6	2/6	2/6	2/6	2/6	2/6	2/6	2/6	2/6
3			2/8	2/8	2/8	2/8	2/8	2/8	2/8	2/8	3/8	3/8	3/8	3/8	3/8	3/8
4		2/9	2/9	2/10	3/10	3/10	3/10	3/10	3/10	3/10	3/10	3/10	4/10	4/10	4/10	4/10	4/10
5		2/10	3/10	3/11	3/11	3/12	4/12	4/12	4/12	4/12	4/12	4/12	4/12	5/12	5/12	5/12	5/12
6			3/11	3/12	3/12	4/13	4/13	4/13	4/13	5/14	5/14	5/14	5/14	5/14	5/14	6/14	6/14
7				3/13	4/13	4/14	5/14	5/14	5/14	5/15	5/15	6/15	6/16	6/16	6/16	6/16	6/16
8					4/14	5/14	5/15	5/15	6/16	6/16	6/16	6/16	7/17	7/17	7/17	7/17	7/17
9						5/15	5/16	6/16	6/16	6/17	7/17	7/18	7/18	7/18	8/18	8/18	8/18
10							6/16	6/17	7/17	7/18	7/18	7/18	8/19	8/19	8/19	8/20	9/20
11								7/17	7/18	7/19	8/19	8/20	8/20	9/20	9/20	9/21	9/21
12									7/19	8/19	8/20	8/20	9/21	9/21	9/21	10/22	10/22
13										8/20	9/20	9/21	9/21	10/22	10/22	10/23	10/23
14											9/21	9/22	10/22	10/23	10/23	11/23	11/24
15												10/22	10/23	11/23	11/24	11/24	12/25
16													11/23	11/24	11/25	12/25	12/25
17														11/25	12/25	12/26	13/26
18															12/26	13/26	13/27
19																13/27	13/27
20																	14/28

*See p. 709 in regard to critical values.

From C. Eisenhart and F. Swed, "Tables for testing randomness of grouping in a sequence of alternatives," *The Annals of Statistics*, vol. 14 (1943): 66–87. Reprinted by permission.

For $n_1 > 20$ or $n_2 > 20$, treat V as a normal variable with a mean and a standard deviation of

$$\mu_v = \frac{2n_1 n_2}{n_1 + n_2} + 1 \qquad \sigma_v = \sqrt{\frac{2n_1 n_2 (2n_1 n_2 - n_1 - n_2)}{(n_1 + n_2)^2 (n_1 + n_2 - 1)}}$$

TABLE 14
Critical Values of Spearman's Rank Correlation Coefficient

The entries in this table are the critical values of r_s for a two-tailed test at α. For a one-tailed test, the value of α shown at the top of the table is double the value of α being used in the hypothesis test.

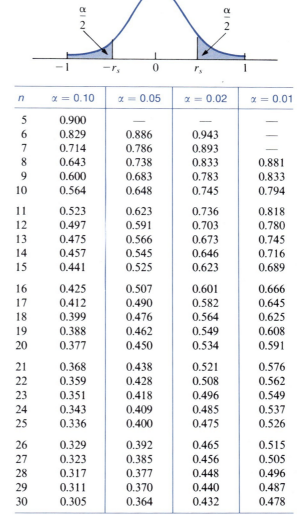

n	$\alpha = 0.10$	$\alpha = 0.05$	$\alpha = 0.02$	$\alpha = 0.01$
5	0.900	—	—	—
6	0.829	0.886	0.943	—
7	0.714	0.786	0.893	—
8	0.643	0.738	0.833	0.881
9	0.600	0.683	0.783	0.833
10	0.564	0.648	0.745	0.794
11	0.523	0.623	0.736	0.818
12	0.497	0.591	0.703	0.780
13	0.475	0.566	0.673	0.745
14	0.457	0.545	0.646	0.716
15	0.441	0.525	0.623	0.689
16	0.425	0.507	0.601	0.666
17	0.412	0.490	0.582	0.645
18	0.399	0.476	0.564	0.625
19	0.388	0.462	0.549	0.608
20	0.377	0.450	0.534	0.591
21	0.368	0.438	0.521	0.576
22	0.359	0.428	0.508	0.562
23	0.351	0.418	0.496	0.549
24	0.343	0.409	0.485	0.537
25	0.336	0.400	0.475	0.526
26	0.329	0.392	0.465	0.515
27	0.323	0.385	0.456	0.505
28	0.317	0.377	0.448	0.496
29	0.311	0.370	0.440	0.487
30	0.305	0.364	0.432	0.478

From E. G. Olds, "Distribution of sums of squares of rank differences for small numbers of individuals," *Annals of Statistics*, vol. 9 (1938), pp. 138–148, and amended, vol. 20 (1949), pp. 117–118. Reprinted by permission.

*For specific details on the use of this table, see p. 715.

ANSWERS TO SELECTED EXERCISES

CHAPTER 1

1.1 a. college students
c. What is in for Fall? What's out?
e. Each student could have more than *one* answer.

1.3 a. all U.S. motorcyclists
c. where in the U.S. the motorcyclist lives.

1.5 a. descriptive
c. descriptive

1.7 a. those individuals who have hypertension
c. the proportion of the population for which the drug is effective
e. No, but it is estimated to be approximately 80%.

1.9 a. all results, past, present and future, from the first ace experiment
c. the results obtained from the 5 trials performed for the exercise

1.11 a. U.S. adults who spend time in bed
c. the 1013 adults surveyed
e. (1) number of hours spent sleeping
(2) number of hours spent listening to music
(3) number of hours spent watching TV
(4) number of hours spent romancing

1.13 a. continuous c. discrete e. discrete

1.15 a. average cost of textbooks for the semester for one student
c. the cost of textbooks for the semester for one student
e. the average cost of textbooks for one student for the semester; add all 100 values and divide the total by 100

1.17 a. The population contains all objects of interest, while the sample contains only those actually sampled.

1.19 group 2, the football players because their weights cover a wider range of values, probably 175 to 300+, while the cheerleaders probably all weigh between 110 and 150

1.21 A lack of variability would indicate all students attained very similar scores, even when they do not all know the same amount.

1.23 a. the set (list) from which the sample was actually drawn

1.25 systematic sampling

1.27 A simple random sample would be very difficult to obtain from an extremely large or spread-out population.

1.29 a. No. Often the people who respond to this kind of sampling are either very strongly for or very strongly opposed; that is why they take the time to respond.

1.31 a. statistics c. statistics

1.33 Several large comprehensive computer programs (called statistical packages) have been developed that perform many of the computations and tests you will study in this text. In order to have the statistical package perform the computations, you simply enter the data into the computer and the computer does the rest on command.

1.35 Each student's answers will be different. Here are a few possibilities:
a. color of hair, major, gender, marital status
c. height, weight, distance from hometown to college, cost of textbooks

1.37 a. the proportion of registered voters who will vote for the candidate
c. No. If the sample is representative of the population, the candidate will not win.

1.39 a. T = 3 is a piece of data—a value from one person.
c. What is the average number of times per week that people go shopping?

1.41 Cluster sampling. The blocks are clusters, or strata. Some, but not all, are sampled. In this example, cluster sampling would be less costly than simple random sampling.

1.43 Each will have different examples.

1.45 Each will have different examples.

CHAPTER 2

2.1 a. "I enjoy using computers."

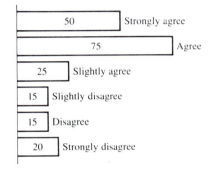

c. The circle graph is based on a 100% total; therefore, it is more informative to most people.

2.3 a. "What's in U.S. landfills?"

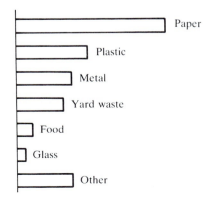

2.5 region of residence

2.7 points scored per game

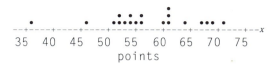

2.9 MINITAB verify

2.11 one-way travel time

```
0 | 5 5
1 | 5 5 5 0
2 | 0 0 5 0 5 5 5 5 5 0 5 0 0 0 0 5 0 0 0
3 | 0 5 0
4 | 0 5
```

2.13 MINITAB verify

2.15 MINITAB verify

2.17 Number of children living at home

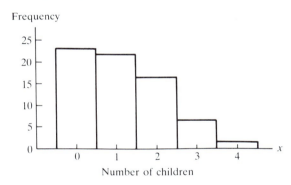

2.19 a. 1, 3, 16, 10, 12, 5, 1, 2
c. 0.02, 0.08, 0.40, 0.60, 0.84, 0.94, 0.96, 1.00

e.

2.21 a. −0.5, 3.5, 7.5, 11.5, 15.5, 19.5, 23.5, 27.5
c. 4

Chapter 2 ▼ ANSWERS TO SELECTED EXERCISES

2.23 Salaries for club managers

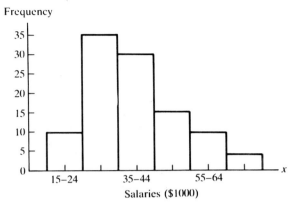

2.25 a. 50—twice, 150—once, open interval—once, undefined interval—once
b. unequal class intervals and the fifth class

2.27 a. 4, 14, 19, 7, 5, 3, 2, 1
c. 22, 20, 24.5

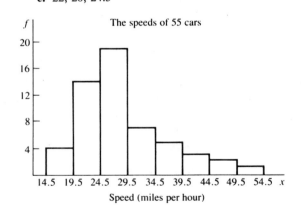

2.29 a. f: 4, 14, 11, 7, 3, 1
c. 2.505; 2.01 and 3.00; 2.005 and 3.005

2.31 MINITAB verify—no answer needed

2.33 a. 6
c. no mode

2.35 a. 7.2
c. 7

2.37 a. 30.05, 30, 30, 29.5

2.39 a. 162,470.40
c. 347,175.00

2.41 a. 5,192,000
c. 423,000

2.43 a. 2.3
c. 2

2.45 a. 10.7
c. 10

2.47 28.5

2.49 58.50

2.51 MINITAB verify

2.53 MINITAB verify

2.55 a. 7
c. 2.9

2.57 a. 8.2
c. 2.9

2.59 $\bar{x}_1 = 2.000$ and $s_1 = 0.0027$
$\bar{x}_2 = 2.000$ and $s_2 = 0.0074$
Even though both machines produced shafts of the same mean diameter, machine 2 had a standard deviation that was over 2.5 times the standard deviation of machine 1. Machine 2 will produce more shafts that do not meet specifications.

2.61 1.17, 1.1

2.63 12.2, 3.5

2.65 2.37

2.67 7.9

2.69 a. 22,153.6

2.71 Set 1: 0, 14, 54, 9
Set 2: 0, 46, 668, 35

2.73 MINITAB verify

2.75 MINITAB verify

2.77 formula verification

2.79 a. 70
c. 95

2.81 a. 8.3
c. 10.85
e. 7.1, 8.3, 9.25, 10.85, 15.5

2.83 a. closing price, variability between the high and low price of the day; 2910, 2870, 2880

2.85 a. 1.0
c. 2.25

2.87 680

2.89 a. 152 is one and one-half standard deviations above the mean.
c. the number of standard deviations from the mean

2.91 A has the higher relative position.

2.93 a. Range should be approximately equal to 6 times the standard deviation.

2.95 a. at most 11%

2.97 a. 9.10 to 37.18

2.99 a. frequencies are: 7, 10, 22, 8, 7, 2, 3, 0, 1
  c. 1.7, 5.1
  e. 0.0, 6.8
  g. −1.7, 8.5
  i. 93% is at least 75%, 98.3% is at least 89%; both agree with Chebyshev's theorem.

2.101 Misrepresentative in that the 1990 bar looks to be three times longer than 1985 bar thus suggesting that the number of bankruptcies has tripled. Not true, 1461 to 2869 is about double. Drawn this way probably to save space.

2.103 a. Mean increased; sum increased.
  c. Mode is unchanged.
  e. Range increased; difference between high and low values increased.
  g. Standard deviation increased; data are now more spread out.

2.105 a. 18.62
  c. 6.81

2.107 a. 4.56
  c. Yes, they seem to average more than 4%.

2.109 a. median annual salaries

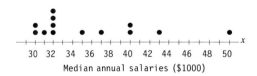

  c. 31,696 and 39,893

2.111 a. 44.6

2.113 a. April: 7.08, 1.49 and August: 7.13, 1.50

2.115 a. freq: 1, 2, 3, 4, 5, 5, 6, 7, 8, 9, 6, 2
  c. number of litters, 58

2.117 a. 14.42
  c. 3.65

2.119 79.88, 12.4

2.121 a. 15.45
  c. 16

2.123 a.

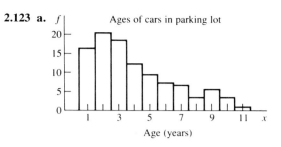

  c. 2, 5.5
  e. 10, 6.7, 2.6

2.125 a. freq: 6, 4, 10, 5, 4, 4
  c. 218.50 and 3.21

2.127 a. first class: 300–399
  freq: 4, 4, 3, 4, 3, 4, 4, 1
  c. 664.3 and 219.6

2.129 21.84, 4.83

2.131 a. 4.05, 6.25, 8.45, 10.65, 12.85
  c. cum. freq.: 7, 19, 23, 26, 27

2.133 a.

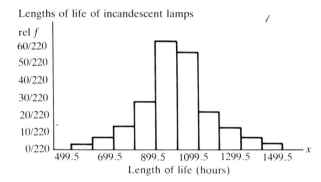

  c. 169.2

2.135 a. P_{98}

2.137 a. 0.71 and 0.62

2.139 58, 0, 9.8, 6.9, 181

2.141 a. at least 75%

2.143 The empirical rule states that 99.7% of a normal distribution is between the z-scores of −3 and +3.

2.145 a. 3978.1
  c. 3570 to 4386

CHAPTER 3

3.1 a.

		Gender		Marginal Total
		Male	Female	
Made a Purchase	yes	10	14	24
	no	17	19	36
Marginal Total		27	33	60

c.

		Gender		Marginal Total
		Male	Female	
Made a Purchase	yes	37%	42%	40%
	no	63	58	60
Marginal Total		100%	100%	100%

3.3 a. 3000
c. 50%

3.5 a. fixed rates at banks

c. 14.975 and 14.00
e. 3.80 and 7.50
g. The credit unions tend to have lower rates, with one exception.

3.7 MINITAB verify
3.9 MINITAB verify
3.11

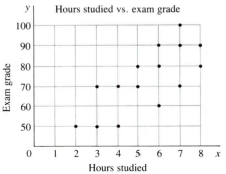

3.13

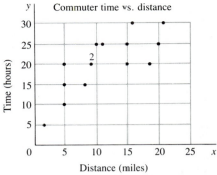

3.15 MINITAB verify
3.17 As the batch number increases, the box plot shifts so as to cover a range of smaller values. The high end of the box plot for batch 10 is approximately even with the low end of batch 1, and there seems to be a gradual shift in between.
3.19 Impossible. The correlation coefficient must be a numerical value between -1 and $+1$. There must be a calculation error.
3.21 a. 7.449
c. -79.09
3.23 a. estimate r to be near 0.75
3.25 0.926
3.27 MINITAB verify
3.29 a.

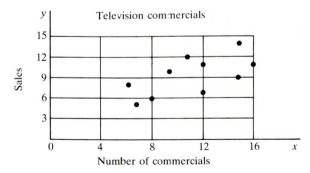

c. 0.661

3.31

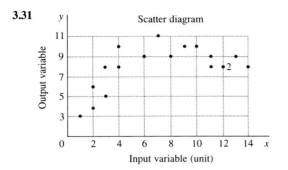

This scatter diagram does not suggest a linear relationship between the two variables. Therefore, we would not be justified in using techniques of linear regression. If we were to calculate the line of best fit, we would get an equation that would be worthless.

3.33 68,100

3.35 1.158

3.37 a.

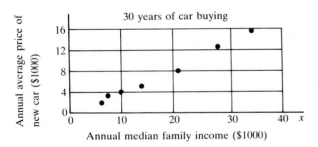

3.39 Verify

3.41 MINITAB verify

3.43 a.

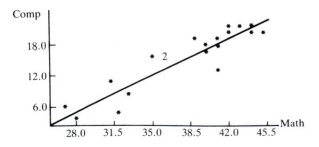

c. Line on graph in (a).

3.45 a. Column totals: 105, 78, 68, 94, 9
Row totals: 184, 170
Grand total: 354

c.

	Parents	Relatives	Friends	Combination	Others
Rural	71%	59%	35%	39%	22%
Nonrural	29%	41%	65%	61%	78%

e.

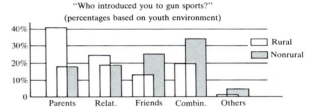

3.47 a. Column totals: 1640, 688, 772
Row totals: 900, 1000, 1200
Grand total: 3100
c. 46%   21%   33%
  57%   21%   22%
  54%   24%   22%
e.

3.49 a, c and e. correlation

3.51

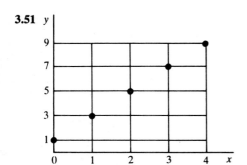

$r = 1.0, \hat{y} = 1.0 + 2.0x$

3.53 0.937

3.55 a.

3.57 a.

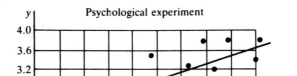

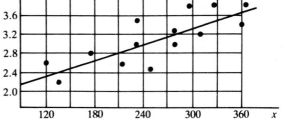

c. Yes, there appears to be correlation.
e. See line on graph in (a).

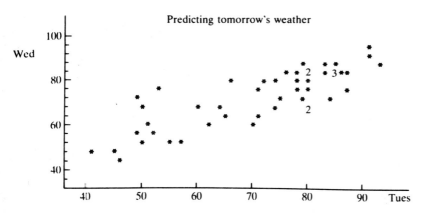

c. They seem to be strongly related.

3.59 a.

3.61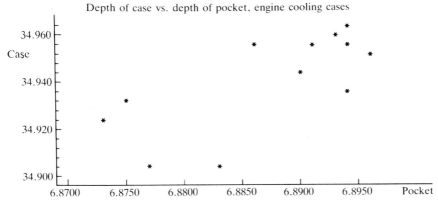

c. $\hat{y} = 22.7 + 1.78x$

3.63 Numerator of formula (3-1):

$$\text{Numerator} = \sum (x - \bar{x})(y - \bar{y})$$

$$= \sum (xy - \bar{x}y - x\bar{y} + \bar{x}\bar{y})$$

$$= \sum xy - \bar{x}\sum y - \bar{y}\sum x + n\bar{x}\bar{y}$$

$$= \sum xy - \left(\frac{\sum x}{n}\right)\sum y - \left(\frac{\sum y}{n}\right)\sum x + n\left(\frac{\sum x}{n}\right)\left(\frac{\sum y}{n}\right)$$

$$= \sum xy - \frac{\sum x \sum y}{n}$$

$$= SS(xy)$$

Denominator of formula (3-1):

$$\text{Denominator} = (n-1)s_x s_y$$

$$= (n-1)\sqrt{\frac{SS(x)}{n-1}}\sqrt{\frac{SS(y)}{n-1}}$$

$$= \sqrt{SS(x)SS(y)}$$

Therefore, formula (3-1) is equivalent to formula (3-2).

CHAPTER 4

4.1 If your results are 6 heads and 4 tails;
a. 0.6

4.3 Results will vary.

4.5 MINITAB verify

4.7 You can expect a 1 to occur approximately $\frac{1}{6}$th of the time when you roll a die repeatedly.

4.9 0.18

4.11 Results will vary.

4.13 Let J = jack, Q = queen, K = king, H = heart, C = club, D = diamond, S = spade.
Sample space = {JH, JC, JD, JS, QH, QC, QD, QS, KH, KC, KD, KS}.

4.15 Let MM = malignant melanoma, BC = basal-cell carcinoma, SC = squamous-cell carcinoma, O = other.
Sample space = {(MM, MM), (MM, BC), (MM, SC), (MM, O), (BC, MM), (BC, BC), (BC, SC), (BC, O), (SC, MM), (SC, BC), (SC, SC), (SC, O), (O, MM), (O, BC), (O, SC), (O, O)}

4.17 S = {H1, H2, H3, H4, H5, H6, T1, T2, T3, T4, T5, T6}

4.19 S = {HH, HT, T1, T2, T3, T4, T5, T6}

4.21 a. 0.50   **c.** 0.06

4.23 $P(R) = X, P(Y) = X, P(G) = 2X$, and $X + X + 2X = 1, 4X = 1$ and $X = \frac{1}{4}$; therefore $P(R) = \frac{1}{4}, P(Y) = \frac{1}{4}$, and $P(G) = \frac{2}{4} = \frac{1}{2}$.

4.25 $P(A) = X, P(B) = 2X, P(C) = 4X$, and $X + 2X + 4X = 1, 7X = 1$, and $X = \frac{1}{7}$; therefore $P(A) = \frac{1}{7}, P(B) = \frac{2}{7}$, and $P(C) = \frac{4}{7}$.

4.27 The three success ratings (highly successful, successful, and not successful) appear to be nonintersecting and their union appears to be the sample space. If this is true, none of the three sets of probabilities is appropriate.
a. A has a total probability of 1.2. The total must be exactly 1.0.
b. B has a negative probability of -0.1. All probability numbers are values between 0 and 1.
c. C has a total probability of 0.9. The total must be exactly 1.0.

4.29 a. 0.55

4.31 0.89

4.33 a, c. Not mutually exclusive

4.35 If two events are mutually exclusive then there is no intersection. The event "A and B" is the intersection. If no intersection, then $P(A$ and $B) = 0.0$.

4.37 a. 0.7
c. 0.7

4.39 No. "Female" students can be "working" students. Further, if the probabilities are correct, there must be an intersection; otherwise, the total probability would be more than 1.0.

4.41 0.04

4.43 The percentages total more than 1.0.

4.45 a, c. Independent
e. Not independent (if you do not own a car, how can your car have a flat tire?)

4.47 a. 0.12
c. 0.3

4.49 a. 0.5
c. no

4.51 a. independent
c. dependent

4.53 a. 0.042

4.55 a. 0.36
c. 0.48

4.57 a. $\frac{3}{5}$

4.59 4/28,224,105, or approximately one chance in 7 million.

4.61 a, c. Verify

4.63 0.7

4.65 a. 0.15
c. 0.7
e. 0.70

4.67 a. $\frac{20}{56}$
c. $\frac{6}{56}$

4.69 a. $\frac{1}{2}, \frac{1}{4}, \frac{1}{8}$

4.71 a. 0.2
c. 0.5

4.73 $\frac{8}{30}$

4.75 a. 0.9
c. $\frac{1}{40}$ or 0.025

4.77 **a.** 0.625
  c. dependent
4.79 0.462 and 0.538
4.81 **a.** True; law of large numbers.
  c. False; law of large numbers says expect about one-half of the 100 million to be H's and the other half to be T's.
4.83 **a.** 0.30
  c. Not independent; $(0.3)(0.4) = 0.12$ not $\frac{70}{300}$.
4.85 **a.** False; if mutually exclusive, $P(R \text{ or } S)$ is found by adding 0.2 and 0.5.
  c. False; if mutually exclusive, $P(R \text{ and } S)$ must be equal to zero—there is no intersection.
4.87 **a.** S = {GGG, GGR, GRG, GRR, RGG, RGR, RRG, RRR}
  c. $\frac{7}{8}$
4.89 **a.** 0.3168
  c. No, $P(A) \neq P(A:B)$
  e. It would mean that the two events "candidate wants job" and "RJB wants candidate" could both not happen.
4.91 0.080
4.93 **a.** 0.22
  c. 0.0; based on the appearance of the list, these appear to be mutually exclusive categories.
4.95 **a.** Surgeon: 0.0000023 to 0.0000238
  Dentist: 0.0000003 to 0.0000038
4.97 **a.** 0.5
  c. 0.167
4.99 0.300
4.101 **a.** 0.30
  c. 0.10
  e. 0.333
4.103 0.56
4.105 **a.**

1st Drawing	2nd Drawing
writes	writes
	does not write
does not write	writes
	does not write

  c. 0.8
  e. 0.3778

  g. 0.2
  i. 0.622
  k. 0.178
  m. 1.0
  o. No; $P(A \text{ and } B) = 0.622$
  q. No; $P(A) \times P(B)$ not equal to $P(A \text{ and } B)$.
4.107 **a.** 0.531
  c. 0.047
4.109 **a.**

	Coin A	Coin B	(A, B)	Probability
	0.6 H	0.5 H	HH	0.30
		0.5 T	HT	0.30
	0.4 T	0.5 H	TH	0.20
		0.5 T	TT	0.20

  c. 0.50
  e. 0.50
  g. 0.60
4.111 0.592
4.113 independent
4.115 0.769, 0.154, 0.077
4.117 **a.** 0.9986 and 0.10

CHAPTER 5

5.1 number of children per family; $x = 0, 1, 2, 3, \ldots$.
5.3 $x = n(\text{bull's-eye shots}); x = 0, 1, 2, \ldots, 8$.
5.5

x	0	1	2	3
$P(x)$	0.20	0.30	0.40	0.10

5.7

x	$P(x)$
1	$\frac{4}{10}$
2	$\frac{3}{10}$
3	$\frac{2}{10}$
4	$\frac{1}{10}$
sum	$\frac{10}{10} = 1.0$

It is a probability function: (a) Each $P(x)$ is a value between zero and one and (b) the sum of the $P(x)$s is one.

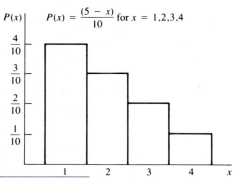

5.9

x	S(x)
2	1/36
3	2/36
4	3/36
5	4/36
6	5/36
7	6/36
8	5/36
9	4/36
10	3/36
11	2/36
12	1/36
sum	1.0

It is a probability function: (a) each $S(x)$ is a value between zero and one; (b) the sum of the $S(x)$s is 1.

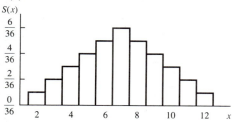

5.11 a.

x	0	1	2	3	4	5	6	7	8	9
P(x)	0.05	0.12	0.15	0.25	0.21	0.11	0.05	0.03	0.02	0.01

c. The variable is discrete since x cannot take on any fractional values. There are no fractional trips. Thus, the distribution is discrete in nature.

5.13 MINITAB verify—no answer needed

5.15 2.0, 1.0

5.17 2.0, 1.1

5.19 0.9

5.21 $\sigma^2 = \sum (x - \mu)^2 P(x)$

$\sum (x^2 - 2x + \mu^2) P(x)$

$\sum x^2 P(x) - 2\mu \sum x P(x) + \mu^2 \sum P(x)$

$\sum x^2 P(x) - 2\mu^2 + \mu^2$

$\sigma^2 = \sum x^2 P(x) - \mu^2$ or

$\sum x^2 P(x) - \left\{\sum x P(x)\right\}^2$

5.23 a. 24
c. 1
e. 10
g. 0.0081
i. 10
k. 0.4096

5.25 $n = 100$ trials (shirts); two outcomes (first quality or irregular): $p = P(\text{irregular})$, $x = n(\text{irregular})$

5.27 a. x is not a binomial random variable because the trials are not independent. The probability of success (get an ace) changes from trial to trial. On the first trial it is $\frac{4}{52}$. The probability of an ace on the second trial can be either $\frac{4}{51}$ or $\frac{3}{51}$, depending on whether or not the first trial was a success. The probability of success on the third and fourth trials depends on what has occurred on the previous trials. The probability of success does not remain constant throughout the experiment because we do not have independence of trials.

5.29 a.

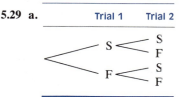

$2^2 = (2)(2) = 4$. The tree diagram has 4 branches, also.

c. 32, 1024

5.31 a.

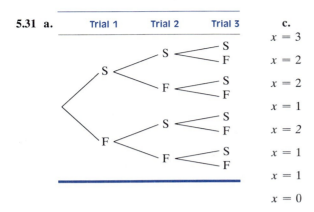

c.
$x = 3$
$x = 2$
$x = 2$
$x = 1$
$x = 2$
$x = 1$
$x = 1$
$x = 0$

e. $P(x) = \binom{3}{x} p^x q^{3-x}$ for $x = 0, 1, 2, 3$

5.33 a. 0.4116   **c.** 0.5625   **e.** 0.375

5.35

x	P(x)
0	1/32
1	5/32
2	10/32
3	10/32
4	5/32
5	1/32
sum	32/32 = 1.0

It is a probability function: (a) each $P(x)$ is a value between zero and 1, (b) the sum of the $P(x)$s is 1.
$P(x) = \binom{5}{x}\left(\frac{1}{2}\right)^x\left(\frac{1}{2}\right)^{5-x}$, for $x = 0, 1, 2, 3, 4, 5$

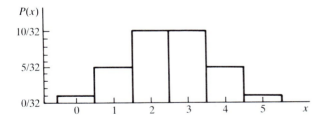

Yes, it is binomial.

5.37 a. Yes. $n = 5$, $p = P(H) = 0.5$, $q = P(T) = 0.5$; each toss is a trial.
c. $\frac{1}{32}, \frac{5}{32}, \frac{10}{32}, \frac{10}{32}, \frac{5}{32}, \frac{1}{32}$

5.39 a. 0.590

5.41 0.0011

5.43 a. 0.463

5.45 The number of defective items should be fairly small and therefore easier to count.

5.47 a. 2.5, 1.1
   c. the same

5.49 a. 7.7, 2.7
   c. 44.0, 2.3

5.51 400, 0.5

5.53 a. If assume x is binomial (approximately):
$P(x) = \binom{5}{x}(0.75)^x(0.25)^{5-x}$, for $x = 0, 1, \ldots, 5$.

x	P(x)
0	0.00098
1	0.01465
2	0.08789
3	0.26367
4	0.39551
5	0.23730
sum	1.00000

c. 3.75, 0.97

5.55 a. $p^3 + 3p^2q$
   c. 0.896
   e. 0, 0.5, 1.0

5.57 (i) Each probability, $P(x)$, is a value between zero and 1.
(ii) The sum of all the $P(x)$ is exactly 1.

5.59

x	f(x)
0	(3/4)/(0!)(3!) = 0.125
1	(3/4)/(1!)(2!) = 0.375
2	(3/4)/(2!)(1!) = 0.375
3	(3/4)/(3!)(0!) = 0.125

It is a probability function since both properties stated in answer to Exercise 5.57 hold.

5.61 a. 0.1
   c. 0.6

5.63 a.

Grade	Probability
F	21/101
D	20/101
C	22/101
B	25/101
A	13/101
sum	101/101

5.65

x	P(x)	xP(x)
0	6/12	0/12
1	1/12	1/12
2	1/12	2/12
3	1/12	3/12
4	1/12	4/12
5	1/12	5/12
6	1/12	6/12
sum	12/12	21/12

$\mu = \sum xP(x) = 21/12 = 1.75$

5.67 2600

5.69 0.984

5.71 0.1035

5.73 0.930

5.75 0.668

5.77 a. 0.930

5.79 a. 0.116

5.81 two-engine: 0.997; four-engine: 1.000

5.83 Let success be the event that an item is defective on a given draw. The probability of success on the first draw is $\frac{3}{10}$, or 0.3. The probability of success on the second draw is dependent on the results of the first draw. If the first was defective, then the probability is $\frac{2}{9}$. If the first is not defective, then the probability is $\frac{3}{9}$. Since the trials are not independent, x is not a binomial random variable.

5.85 a. 6

5.87

x	1	2	3
P(x)	0.23	0.41	0.36

5.89 a. 28
c. $\frac{1}{28}$
e. 9.0, 3.0

5.91 $\mu = \sum xP(x)$

$= (1)\left(\frac{1}{n}\right) + (2)\left(\frac{1}{n}\right) + \cdots + (n)\left(\frac{1}{n}\right)$

$= \left(\frac{1}{n}\right)(1 + 2 + \cdots + n)$

$= \left(\frac{1}{n}\right)\left(\frac{(n)(n+1)}{2}\right) = \frac{n+1}{2}$

CHAPTER 6

6.1 A bell-shaped distribution with a mean of zero and a standard deviation of 1.

6.3 a. 0.4032   **c.** 0.4993

6.5 a. 0.4394   **c.** 0.9394

6.7 a. 0.7737   **c.** 0.8983

6.9 a. 0.5000   **c.** 0.9893   **e.** 0.0548

6.11 a. 0.4906   **c.** 0.4483

6.13 a. 1.14   **c.** 1.66   **e.** 1.74

6.15 a. 1.65   **c.** 2.33

6.17 $+1.28$ or -1.28

6.19 -0.67 and $+0.67$

6.21 a. 1.28   **c.** 2.33

6.23 a. 0.5000   **c.** 0.6072   **e.** 0.9502

6.25 a. 0.0038 or 0.38%

6.27 a. 89.6   **c.** 57.3

6.29 a. 0.2857

6.31 a. 0.1131   **c.** 4.64 min.

6.33 20.264

6.35 a. 5.32   **c.** 70.778

6.37 MINITAB verify

6.39 a. $z(0.03)$ **c.** $z(0.75)$ **e.** $z(0.91)$

6.41 a. 1.96   **c.** 2.33

6.43 a. 1.28, 1.65, 1.96, 2.05, 2.33, 2.58

6.45 a. A is an area. z is 0.10 and the area to the right of $z = 0.10$ is $0.5000 - 0.0398 = 0.4602$.
c. C is an area. z is -0.05 and the area to the right of $z = -0.05$ is $0.5000 + 0.0199 = 0.5199$.

6.47 a. $np = 3$ and $nq = 7$; therefore, the approximation is not appropriate, since $np < 5$.
c. $np = 50$ and $nq = 450$; therefore, the approximation is appropriate, since both $np > 5$ and $nq > 5$.

6.49 0.1822

6.51 normal approx.: 0.9429, and binomial: 0.943

6.53 a. 5.0, 2.18   **c.** 0.1251

6.55 0.0087

6.57 0.0287

6.59 a. $P(x \text{ is 90 or fewer}) = P(x = 0, 1, 2, 3, \ldots, 90)$

$$P(x = 0) = \binom{500}{0}(0.2^0)(0.8^{500})$$

$$P(x = 1) = \binom{500}{1}(0.2^1)(0.8^{499})$$

$$P(x = 2) = \binom{500}{2}(0.2^2)(0.8^{498})$$

on so on until $x = 90$.

$$P(x = 90) = \binom{500}{90}(0.2^{90})(0.8^{410})$$

$P(x \text{ is 90 or fewer})$ is the sum of the above.

6.61 The range from $z = -2$ to $z = +2$ represents two standard deviations on either side of the mean. According to Chebyshev's theorem, there shall be at least $\frac{3}{4}$ or at least 0.75 of a distribution in this interval.
   The actual area under a normal curve is $2(0.4772)$ or 0.9544.

6.63 a.

x	-7	-6	-5	-4	-3	-2	-1	0	1	2	3	4	5	6	7	8	9	10	11	12	13	14	15	16	17	18	19	20	21
f	4	2	2	4	2	9	18	24	39	55	48	52	62	46	40	34	22	22	9	13	9	1	2	6	2	2	0	2	1

c. Empirical rule: 68%, 95%, 99.7%
This data has: 74.8%, 94.7%, 99.1%. Not very close.
e. The histogram looks normal; however, the percentages are not very close to those of the empirical rule.

6.65 a. −0.92   **c.** −0.74

6.67 a. 0.0930

6.69 a. 1.175 or 1.18
c. −1.04

6.71 18.8

6.73 10.9

6.75 a. 0.0158

6.77 10.033

6.79 a. The normal approximation is reasonable since both $np = 7.5$ and $nq = 17.5$ are greater than 5.

6.81 MINITAB verify

6.83 a. 0.001
c. 0.995

6.85 a. 0.1170
c. 0.0008

6.87 a. 1.0
c. 0.4772

CHAPTER 7

7.1 a. A sampling distribution of sample means is the distribution formed by the means from all possible samples of a fixed size that can be taken from a population.
c. There were 25 equally likely samples. Therefore, each has a probability of $\frac{1}{25}$ or 0.04.

7.3 a.

11	31	51	71	91
13	33	53	73	93
15	35	55	75	95
17	37	57	77	97
19	39	59	79	99

c.

R	0	2	4	6	8
P(R)	0.20	0.32	0.24	0.16	0.08

7.5 a. Each company reported the average age of aircraft in its fleet.
c. Since the fleets are not of equal size, the averages reported are not from samples of equal size.

7.7 Every student will have different results; however, they should be at least somewhat similar to these:

Samples	997 870 725 745 603 073 508 562 683 510
	325 629 343 273 750 611 826 361 744 879
(a)	8.3 5.0 4.7 5.3 3.0 3.3 4.3 4.3 5.7 2.0
	3.3 5.7 3.3 4.0 4.0 2.7 5.3 3.3 5.0 8.0

7.9 MINITAB verify

7.11 a. one

7.13 a. 500   **c.** approximately normal

7.15 MINITAB verify

7.17 a. approximately normal   **c.** $\frac{10}{\sqrt{36}} = 1.667$
e. 0.8849

7.19 a. 0.3830   **c.** 0.3085

7.21 0.9992

7.23 MINITAB verify

7.25 a. 0.6826   **c.** 0.9974

7.27 0.7888

7.29 a.

Samples			$\bar{x}$	Samples			$\bar{x}$
2	2	2	6/3	4	4	4	12/3
2	2	4	8/3	4	4	6	14/3
2	2	6	10/3	4	6	2	12/3
2	4	2	8/3	4	6	4	14/3
2	4	4	10/3	4	6	6	16/3
2	4	6	12/3	6	2	2	10/3
2	6	2	10/3	6	2	4	12/3
2	6	4	12/3	6	2	6	14/3
2	6	6	14/3	6	4	2	12/3
4	2	2	8/3	6	4	4	14/3
4	2	4	10/3	6	4	6	16/3
4	2	6	12/3	6	6	2	14/3
4	4	2	10/3	6	6	4	16/3
				6	6	6	18/3

c.

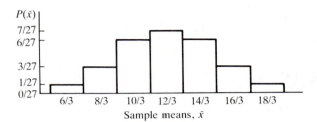

e. 0.9428

7.31 a. Individual score x: the distribution of x's is normal, with a mean of 720 and a standard deviation of 60.
c. 0.5359

7.33 0.49

7.35 0.9544

7.37 a. 0.1498

7.39 0.0009

7.41 0.0228

7.43 0.0004

7.45 a. 0.9759

7.47 a. within 3 standard deviations: 40.5 to 49.5
c. 4.5

CHAPTER 8

8.1 a. Type A correct decision: The accused is indeed innocent and is acquitted. Type I error: The accused is in fact innocent but is convicted. Type II error: The accused is actually guilty but is acquitted. Type B correct decision: The accused is guilty and is convicted.
c. no

8.3 a. Type A correct decision: The parachute will open and the inspector okayed its use. Type I error: The parachute will open and the inspector did not okay its use. Type II error: The parachute will not open and the inspector okayed its use. Type B correct decision: The parachute will not open and the inspector did not okay its use.
c. Set(3), keep β as small as possible.

8.5 A type I error occurs if the company concludes that the additive increases average coverage when in fact it does not. A type II error occurs if the company concludes that the additive does not increase the coverage when in fact it does.

8.7 a. type I   **c.** type I

8.9 We are willing to allow the type I error to occur with a probability of: **a.** 0.001   **c.** 0.10

8.11 The null hypothesis states that the teaching techniques have no effect (change is equal to zero), while the alternative (what they want to show) is the statement that the teaching techniques do have an effect on students' exam scores. They are looking for improvement, and are therefore using a one-tail test.

8.13 a. Commercial is not effective.

8.15 a. 0.05

8.17 a. "Reject H_0" or "Fail to reject H_0"

8.19 "alpha" and "beta" are interrelated; if one is reduced, the other one becomes larger.

8.21 a. $H_0: \mu = 26$ yrs $(<, =)$; $H_a: \mu > 26$
    c. $H_0: \mu = 1600$ hrs $(>, =)$; $H_a: \mu < 1600$
    e. $H_0: \mu = 4.7$ mi $(>, =)$; $H_a: \mu < 4.7$
    g. $H_0: \mu = 20$ deg $(>, =)$; $H_a: \mu < 20$

8.23 (a) A type I error would be committed if a decision of reject H_0 was interpreted as mean hourly wage is less than $50.00 per hour when in fact it was true that the mean wage was at least $50.00 per hour.

8.25 a.

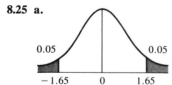

c.

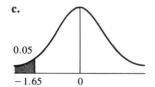

e.

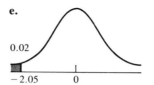

8.27 a. 3

8.29 MINITAB verify—no answer needed

8.31 a. +1.65

8.33 $z^* = 1.643$, fail to reject H_0

8.35 $H_a: \mu < 36.7$, $z^* = -2.59$, reject H_0

8.37 $H_a: \mu \neq 65$, $z^* = -6.48$, reject H_0

8.39 a. (1) Mean wear is less than in past (longer life).
       (2) Mean wear is different than in past.
       (3) Mean wear is greater than in past (shorter life).

8.41 a. 0.0694   **c.** 0.2420   **e.** 0.3524

8.43 a. 1.57   **c.** 2.87

8.45 a. fail to reject H_0

8.47 MINITAB verify

8.49 a. $H_0: \mu = 525$ vs. $H_a: \mu < 525$
    c. Verification

8.51 0.0548

8.53 $z^* = 10.5$, p-value is very near zero.

8.55 Width $= (\bar{x} + z(\alpha/2)(\sigma/\sqrt{n}))$
            $- (\bar{x} - z(\alpha/2)(\sigma/\sqrt{n}))$
           $= 2z(\alpha/2)\sigma/\sqrt{n}$

    a. The higher the level of confidence, the larger the value of $z(\alpha/2)$, and the greater the width. Therefore, as level of confidence increases, the width of the confidence interval increases.
    c. The more variable the characteristic measured is, the larger the standard deviation will be, the larger the standard error will be, and the greater the width. Therefore, as the standard deviation increases, the width of the confidence interval increases.

8.57 MINITAB verify

8.59 a. 7280.0   **c.** 6764 to 7796

8.61 a. 25.3   **c.** 23.97 to 26.63

8.63 959

8.65 a. 97
8.67 39.5 to 41.9
8.69 a. H_0: $\mu = 100$ **c.** 0.01   **e.** 100
     g. 12   **i.** -2.35 **k.** $\alpha = 0.01$

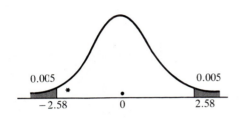

8.71 a. 1.40, 39.7 to 41.7
8.73 a. H_0: $\mu = 300$ vs. H_a: $\mu < 300$ **c.** 0.0808
8.75 a. H_a: $\mu > 42$, $z^* = 1.20$, fail to reject H_0
8.77 a. H_a: $\mu \neq 15{,}650$, $z^* = 1.41$, fail to reject H_0
8.79 a. Alpha determines the size of the critical region.
8.81 MINITAB verify
8.83 MINITAB verify
8.85 a. 69.89 to 75.31
8.87 a. 9.75 to 9.99
     c. The increased confidence widened the interval.
8.89 $z^* = 2.89$, p-value $= 0.002$
8.91 19.1 to 20.9
8.93 97
8.95 273

CHAPTER 9

9.1 a. H_0: $\mu = 11$ vs. H_a: $\mu < 11$
   c. H_0: $\mu = 75$ vs. H_a: $\mu \neq 75$
9.3 a. 1.71   **c.** 2.60
   e. -1.72 **g.** -2.47
9.5 7
9.7 a. -2.49 **c.** -0.685
9.9 a. symmetric about mean, mean is zero
9.11 a. H_a: $\mu < 25$, $t^* = -3.50$, reject H_0
9.13 a. H_a: $\mu \neq 35$, $t^* = 1.02$, fail to reject H_0
9.15 MINITAB verify
9.17 51.5 to 56.5
9.19 79.15 to 92.85
9.21 MINITAB verify
9.23 a. $0.025 < p < 0.05$ **c.** Yes
9.25 a. H_0: $p = P(\text{work}) = 0.60$ vs. H_a: $p > 0.60$
   c. H_0: $p = P(\text{interested in quitting}) = \frac{1}{3}$ vs. H_a: $p > \frac{1}{3}$
   e. H_0: $p = P(\text{vote for}) = 0.50$ vs. H_a: $p > 0.50$
   g. H_0: $p = P(\text{fair}) = 0.50$ vs. H_a: $p \neq 0.50$
9.27 a. 0.017   **c.** 0.101
9.29 a. 0.090   **c.** 0.007
9.31 a. correctly fail to reject H_0
   c. commit a type II error
9.33 a. H_a: $p = P(\text{claim settled within 30 days}) < 0.90$, $z^* = -4.82$, reject H_0
9.35 a. H_a: $p = P(\text{voter will support her}) < 0.60$, $z^* = -2.04$, reject H_0
9.37 H_a: $p = P(\text{worried}) < 0.27$, $z^* = -4.47$, reject H_0
9.39 a. 44% is the proportion from the sample.
   c. high school student leaders
9.41 a. 0.21
9.43 0.206 to 0.528
9.45 0.015
9.47 a. 0.162, 0.438
   c. 0.24, 0.76
   e. 0.474, 0.526
   g. As sample size increased, interval became narrower.
9.49 522
9.51 1068
9.53 a. H_0: $\sigma = 24$ vs. H_a: $\sigma > 24$
   c. H_0: $\sigma = 10$ vs. H_a: $\sigma \neq 10$
   e. H_0: $\sigma^2 = 0.025$ vs. H_a: $\sigma^2 \neq 0.025$
9.55 a. 34.8   **c.** 13.4   **e.** 12.3   **g.** 37.7
9.57 a. 11.1   **c.** 9.24
9.59 0.94
9.61 H_a: $\sigma \neq 8$, $\chi^{2*} = 29.3$, reject H_0
9.63 a. H_a: $\sigma \neq 3.50$, $\chi^{2*} = 19.95$, fail to reject H_0
9.65 30.2 to 80.8
9.67 a. 9.49   **c.** 2.47 to 4.14
9.69 a. H_a: $\mu > 4$, $t^* = 1.18$, fail to reject H_0
9.71 a. H_a: $\mu < 3.8$, $t^* = -1.89$, reject H_0
9.73 a. H_a: $\mu \neq 80$, $t^* = -4.47$, reject H_0

9.75 a. $0.025 < P < 0.05$ **c.** $0.05 < P < 0.10$
9.77 1480 to 1620
9.79 67,756 to 82,244
9.81 60
9.83 a. $H_a: p = P(\text{like}) \neq 0.60, z^* = -2.60$, reject H_0
9.85 0.0401
9.87 a. $n = 800$, trial = one person surveyed, success = "doctor should not be prosecuted." $p = P(\text{"should not be..."}), x = n(\text{"should not be..."})$.
9.89 0.088 to 0.232
9.91

p	0.1	0.2	0.3	0.4	0.5	0.6	0.7	0.8	0.9
q	0.9	0.8	0.7	0.6	0.5	0.4	0.3	0.2	0.1
pq	0.09	0.16	0.21	0.24	0.25	0.24	0.21	0.16	0.09

9.93 a. 0.027   **c.** 2213
9.95 $H_a: \sigma \neq 12, \chi^{2*} = 41.979$, fail to reject H_0
9.97 0.79 to 1.17
9.99 a. $H_a: \sigma > 0.5, \chi^{2*} = 87.8$, reject H_0
9.101
Formula (9-5): $E = z(\alpha/2)\sqrt{\dfrac{pq}{n}}$

$$E = \dfrac{z(\alpha/2)\sqrt{pq}}{\sqrt{n}}$$

$$E\sqrt{n} = z(\alpha/2)\sqrt{pq}$$

$$\sqrt{n} = \dfrac{z(\alpha/2)\sqrt{pq}}{E}$$

$$n = \left\{\dfrac{z(\alpha/2)\sqrt{pq}}{E}\right\}^2$$

Formula (9-6): $n = \dfrac{\{z(\alpha/2)\}^2 pq}{E^2}$

CHAPTER 10

10.1 Independent samples. Each sample was randomly selected from the entire population of 4000 trees.
10.3 Independent samples. The two samples are from two separate unrelated sets of rats.
10.5 a. Independent samples will result if they split the 25 cars into two groups, and send one group to each garage.

10.7 Obtain a measurement of this skill from 25 randomly selected students before the course begins. Then obtain another sample of 25 from those completing the course.
10.9 a. $H_0: \mu_d = 10$ vs. $H_a: \mu_d > 10$
   c. $H_0: \mu_d = 12$ vs. $H_a: \mu_d < 12$
10.11 a. $H_a: \mu_d > 0, t^* = 3.06$, reject H_0
10.13 MINITAB verify—no answer needed
10.15 a. $H_a: \mu_d \neq 0, t^* = 3.30$, reject H_0
10.17 $-1.03, 8.53$
10.19 a. $H_0: \mu_1 = \mu_2$ vs. $H_a: \mu_1 \neq \mu_2$
   c. $H_0: \mu_1 - \mu_2 = 10$ vs. $H_a: \mu_1 - \mu_2 > 10$
10.21 a. 4340, or 9960 with MBA
10.23 a. $H_a: \mu_I - \mu_{II} \neq 0, z^* = -1.31$, fail to reject H_0
10.25 a. $H_a: \mu_C - \mu_E < 0, z^* = -10.0$, reject H_0
10.27 a. $H_a: \mu_m - \mu_f > 2.5, z^* = 2.63$, reject H_0
10.29 -0.77 to -0.43, for $\mu_A - \mu_B$
10.31 59.19 to 94.67, for $\mu_w - \mu_r$
10.33 a. $H_0: \sigma_A^2/\sigma_B^2 = 1$ vs. $H_a: \sigma_A^2/\sigma_B^2 \neq 1$
   c. $H_0: \sigma_A^2/\sigma_B^2 = 1$ vs. $H_a: \sigma_A^2/\sigma_B^2 \neq 1$
10.35 a. $F(9, 11, 0.025)$ **c.** $F(8, 15, 0.01)$
10.37 a. 2.51   **c.** 2.91   **e.** 2.67   **g.** 1.79
10.39 a. $H_0: \sigma_{\text{pop}}^2 = \sigma_{\text{app}}^2$ vs. $H_a: \sigma_{\text{pop}}^2 > \sigma_{\text{app}}^2$
   c. The prob-value is less than 0.005; the probability of committing a type I error by rejecting the null hypothesis is less than 0.005.
10.41 a. $H_a: \sigma_k^2 \neq \sigma_m^2, F^* = 1.33$, fail to reject H_0
10.43 MINITAB verify—no answer needed
10.45 0.47 to 2.41
10.47 0.10
10.49 a. $H_0: \mu_1 - \mu_2 = 0$ vs. $H_a: \mu_1 - \mu_2 \neq 0$
   c. $H_0: \mu_1 - \mu_2 = 5$ vs. $H_a: \mu_1 - \mu_2 > 5$
10.51 a. Formula (10-15) is used when it is assumed that $\sigma_1 = \sigma_2$, and formula (10-16) is used when it is assumed that the standard deviations are not equal. The difference is the formula used to estimate the standard error.
   c. The number of degrees of freedom is smaller in the case where it is assumed that the standard deviations are not equal.

10.53 **a.** An F-test for the ratio of variances must be used to make that decision.
c. It determines the formula for the standard error and the number of degrees of freedom.

10.55 **a.** MINITAB verify

10.57 **a.** It determines the formula for calculating the standard error and number of degrees of freedom.
c. H_a: $\mu_2 - \mu_1 \neq 0$, $t^* = 1.07$, fail to reject H_0

10.59 MINITAB verify

10.61 **a.** $F^* = 1.37$, assume $\sigma_m = \sigma_f$.
c. $0.20 < P < 0.50$

10.63 1st: H_a: $\sigma_A^2/\sigma_B^2 \neq 1$, $F^* = 4.407$, assume $\sigma_A = \sigma_B$
H_a: $\mu_A - \mu_B < 0$, $t^* = -1.98$, reject H_0

10.65 Verify

10.67 $P = 0.394$ means that if the null hypothesis is rejected, there is a high probability (0.39) of committing the type I error.

10.69 1st: H_a: $\sigma_A^2/\sigma_B^2 \neq 1$, $F^* = 2.0$, assume $\sigma_A = \sigma_B$
0.39 to 3.61

10.71 MINITAB verify

10.73 **a.** H_0: $p_m - p_w = 0$ vs. H_a: $p_m - p_w \neq 0$
c. H_0: $p_c - p_{nc} = 0$ vs. H_a: $p_c - p_{nc} > 0$

10.75 H_a: $p_m - p_w > 0$, $z^* = 2.92$, p-value $= 0.0018$

10.77 **a.** H_a: $p_c - p_m \neq 0$, $z^* = -1.42$, fail to reject H_0

10.79 **a.** 46%, 26%   **c.** 0.129 to 0.271

10.81 0.01 to 0.19

10.83 0.000 to 0.080

10.85 **a.** 0.0405

10.87 H_a: $\mu_d > 0$, $t^* = 2.22$, reject H_0

10.89 -8.85 to 16.02

10.91 **a.** H_a: $\mu_t - \mu_l \neq 0$, $z^* = -1.80$, fail to reject H_0

10.93 **a.** H_a: $\mu_S - \mu_K > 0$, $z^* = 3.17$, reject H_0

10.95 -0.05 to 10.45

10.97 H_a: $\sigma_m/\sigma_f > 1$, $F^* = 2.58$, reject H_0

10.99 **a.** 0.6944   **c.** 0.533 to 1.369

10.101 **a.** H_a: $\sigma_A \neq \sigma_B$, $F^* = 2.15$, assume $\sigma_A = \sigma_B$
c. 2.18 to 3.82

10.103 1st: H_a: $\sigma_R \neq \sigma_P$, $F^* = 1.57$, assume σ's equal
H_a: $\mu_R - \mu_P \neq 0$, $t^* = 4.73$, reject H_0

10.105 1st: H_a: $\sigma_A \neq \sigma_B$, $F^* = 1.713$, assume σ's equal
H_a: $\mu_A - \mu_B \neq 0$, $t^* = 3.475$, reject H_0

10.107 1st: H_a: $\sigma_m \neq \sigma_f$, $F^* = 1.59$, assume σ's equal
H_a: $\mu_m - \mu_f \neq 0$, $t^* = 8.4$, reject H_0

10.109 1st: H_a: $\sigma_A \neq \sigma_B$, $F^* = 1.49$, assume σ's equal
H_a: $\mu_A - \mu_B > 0$, $t^* = 4.48$, reject H_0

10.111 H_a: $p_b - p_p > 0$, $z^* = 1.46$, fail to reject H_0

10.113 **a.** $z^* = 2.35$, p-value $= 0.0094$

10.115 yes [H_a: $p_w - p_m > 0$, $z^* = 2.40$, p-value $= 0.0082$]

10.117 **a.** -0.116 to 0.216

CHAPTER 11

11.1 trial: each adult surveyed, variable: reason why we rearrange furniture, multiple answers: bored, moving, new furniture, etc.

11.3 **a.** H_0: $P(1) = P(2) = P(3) = P(4) = P(5) = 0.2$
H_a: The numbers are not equally likely.
c. H_0: $P(E) = 0.16$, $P(G) = 0.38$, $P(F) = 0.41$, $P(P) = 0.05$
H_a: The percentages are different than specified in H_0.

11.5 **a.** H_0: $P(A) = P(B) = P(C) = P(D) = P(E) = 0.2$
c. $\chi^{2*} = 4.40$, fail to reject H_0

11.7 MINITAB verify

11.9 H_0: $P(1) = 0.6$, $P(2) = 0.2$, $P(3) = 0.1$, $P(4) = 0.1$
$\chi^{2*} = 53.33$, reject H_0

11.11 H_0: $P(0) = P(1) = P(2) = \cdots = P(9) = 0.1$
$\chi^{2*} = 6.00$, fail to reject H_0

11.13 **a.** H_0: Voter's preference and voter's party affiliation are independent
c. H_0: The proportion of yesses is the same in all categories sampled.

11.15 H_0: The number of defective items is independent of the day of the week.
$\chi^{2*} = 8.548$, fail to reject H_0

11.17 **a.** The information compares several distributions, a distribution for each region of New York State.

11.19 a. The 11 rows total to 100% (except for round-off error).

11.21 MINITAB verify

11.23 Verify

11.25 H_0: The distribution of reactions is the same for both groups.
$\chi^{2*} = 22.56$, reject H_0

11.27 H_0: 1:3:4 proportions
$\chi^{2*} = 10.33$, reject H_0

11.29 H_0: $P(A) = 0.41$, $P(B) = 0.09$, $P(O) = 0.46$, $P(AB) = 0.04$
$\chi^{2*} = 1.418$, fail to reject H_0

11.31 H_0: Proportion of popcorn that popped is the same for all brands.
$\chi^{2*} = 2.839$, fail to reject H_0

11.33 H_0: The reasons are distributed with the same proportions for both years.
$\chi^{2*} = 66.96$, reject H_0

11.35 a. Column 1: 30.07, 15.28, 15.77, 8.87
Column 2: 30.93, 15.72, 16.23, 9.13
  c. H_0: The number of children is distributed the same for both groups of women.
H_a: The number of children is distributed differently for the two groups.
  e. Reject H_0; the distributions are significantly different.

11.37 a. H_0: The distribution is the same for all regions.
$\chi^{2*} = 7.788$, fail to reject H_0

11.39 H_0: Independence between presence of beech trees and classification of lands
$\chi^{2*} = 5.61$, reject H_0

11.41 H_0: The response and the school's location are independent.
$\chi^{2*} = 3.651$, fail to reject H_0

11.43 H_0: Rate of absenteeism is the same for all groups.
$\chi^{2*} = 27.24$, reject H_0

11.45 a. 0.7918   **c.** Both are 0.6269.

CHAPTER 12

12.1 a. H_0: $\mu_1 = \mu_2 = \mu_3 = \mu_4 = \mu_5$
H_a: Means not all equal
  c. H_0: $\mu_1 = \mu_2 = \mu_3 = \mu_4$
H_a: Means not all equal

12.3 a. The mean levels of the test factor are not all equal.
  c. The mean levels of the test factor are not significantly different.

12.5 a. 0   **c.** 16   **e.** 1232

12.7 a. The test factor has no effect on the mean at the tested levels.
  c. The calculated value of F must fall in the critical region; that is, the variance between factors must be significantly larger than variance within rows.
  e. The calculated value of F must fall in the noncritical region; that is, the variance between factors must not be significantly larger than variance within rows.

12.9 Data needed: the number of hours spent on teaching-related duties per week for faculty members at many different colleges, and the size of the college of each of those professors. Data organization: The data would be grouped according to the size of the college of each respondent. Analyze: The data would be analyzed using ANOVA.

12.11 H_0: The mean values for workers are all equal.

Source	SS	df	MS	
Workers	17.73	2	8.87	$F^* = 4.22$
Error	25.20	12	2.10	
Total	42.93	14		

Reject H_0.

12.13 a. $x =$ years of appliance's useful life

```
Refrig.  -+--+--+--!--+--+--+--!--+--!--+--!--x
         6   8   10  12  14  16  18

Washer   -+--!--+--!-!-+--+--!--+--+--+--+--+-x
         6   8   10  12  14  16  18

Dryer    -+--+--!--+--!-!-!-!-+--+--+--+--x
         6   8   10  12  14  16  18

Stereo   -!--+--!-+-!-!-+--+--+--+--+--+-x
         6   8   10  12  14  16  18

Couch    -!-!--+-!--+--!--+--+--+--+--+--+-x
         6   8   10  12  14  16  18

Mattress -+--!--+-!-!-!-+--+--+--+--+--x
         6   8   10  12  14  16  18
```

12.15 MINITAB verify

12.17 a. 120
  c. Verify
  e. No, the p-value is too large.

12.19 a. $x =$ tillage plot yield

```
Plow
                         P        P  P
-+--+--+--+--+--+--+--+--+--+--+--+--+-x
 109  112  115  118  121  124
V chisel
           V        V        V
-+--+--+--+--+--+--+--+--+--+--+--+-x
 109  112  115  118  121  124
C chisel
       C            C            C
-+--+--+--+--+--+--+--+--+--+--+--+-x
 109  112  115  118  121  124
S chisel
     S       S                 S
-+--+--+--+--+--+--+--+--+--+--+--+-x
 109  112  115  118  121  124
Hvy disk
                   h h       h
-+--+--+--+--+--+--+--+--+--+--+--+-x
 109  112  115  118  121  124
Lt disk
            l            l
-+--+--+--+--+--+--+--+--+--+--+--+-x
 109  112  115  118  121  124
```

c. $F^* = 1.23$, fail to reject H_0

12.21 a. H_0: The mean amount of salt is the same in all tested brands of peanut butter.
H_a: The mean amounts are not all equal.
c. Fail to reject H_0. There appears to be no significant difference in the mean amounts of salt.

12.23 29.3333

12.25 82.4738

12.27 H_0: The mean amount of relief time is the same for all four drugs.
$F^* = 12.5$, reject H_0

12.29 H_0: The mean amounts dispensed by the machines are all equal.
$F^* = 31.6$, reject H_0

12.31 H_0: The mean typing speeds are equal on the two types of machines.
$F^* = 5.08$, reject H_0

12.33 H_0: The mean rates of arrests are the same in all four sizes of communities.
$F^* = 14.02$, reject H_0

CHAPTER 13

13.1 x and y are the mean values of the two variables. In Chapter 2 we learned that the summation of the deviations about the mean was zero.

13.3 a.

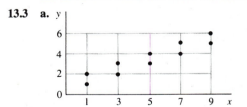

c. 2.981, 1.581   **e.** 0.943

13.5 MINITAB verify—no answer needed

13.7 MINITAB verify—no answer needed

13.9 a. 60   **c.** 0.07

13.11 (a) $r = -1$ (c) $r = +1$

13.13 a. $H_0: \rho = 0$ vs. $H_a: \rho > 0$
c. $H_0: \rho = 0$ vs. $H_a: \rho < 0$
13.15 $H_a: \rho \neq 0$, $r^* = 0.43$, reject H_0
13.17 $H_a: \rho \neq 0$, $r^* = -0.67$, reject H_0
13.19 a. -0.55 to 0.75 **c.** 0.32 to 0.82
13.21 a. It was reported that $r = -0.11$, $n = 237$, and the p-value < 0.05. Therefore if $\alpha = 0.05$ or more, the reported value of r is significant.
c. The p-value < 0.05.
13.23 $H_a: \rho \neq 0$, $r^* = 0.798$, reject H_0
13.25 0.955, 0.8 to 1.0
13.27 MINITAB verify
13.29 $\hat{y} = 16.34 + 0.4245x$, 21.088
13.31 a. $\hat{y} = 1.0 + 0.5x$
c. $-.5\ .5\ -.5\ .5\ -.5\ .5\ -.5\ .5\ -.5\ .5$
e. 0.3125
13.33 a. $H_0: \beta_1 = 0$ vs. $H_a: \beta_1 > 0$
c. $H_0: \beta_1 = 0$ vs. $H_a: \beta_1 < 0$
13.35 0.1894
13.37 MINITAB verify—no answer needed
13.39 a.

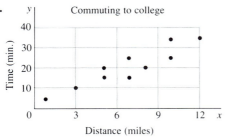

c. $H_a: \beta_1 > 0.0$, $t^* = 6.54$, reject H_0
13.41 a. Verify
13.43 a. 13.02   **c.** 4.67 to 21.37
13.45 a. 15.4 to 19.1
13.47 52.3 to 52.5
13.49 The standard error for $\bar{x}$'s is much smaller than the standard deviation for individual x's (CLT). Thus the confidence interval will be narrower in accordance.
13.51 a. always   **c.** sometimes   **e.** always
13.53 Yes, $H_a: \rho > 0$, $r^* = 0.69$, reject H_0
13.55 a. Results will vary; however, most of the time the data will indicate no relationship (independence).

13.57 a.

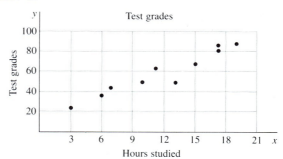

c. $H_a: \rho > 0.0$, $r^* = 0.961$, reject H_0
13.59 Yes, $H_a: \rho = 0$, $r^* = 0.61$, reject H_0
13.61 3.94 to 4.52
13.63 a.

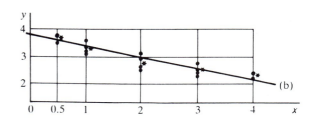

e. 0.1236
g. 2.483 to 2.637, 2.202 to 2.518
13.65 a. $H_a: \rho \neq 0$, $r^* = 0.513$, reject H_0
c. $H_a: \beta_1 > 0$, $t^* = 2.60$, reject H_0
13.67 $r = 0.9973$ using both formulas.

CHAPTER 14

14.1 a. H_0: median $= 32$ vs. H_a: median < 32
c. $H_0: P(+) = 0.5$ vs. $H_a: P(+) \neq 0.5$ where $(+) =$ weight gain
14.3 a. H_0: median $= 48$
c. H_a: median $\neq 48$, $x = 3$, reject H_0
14.5 H_a: There is a difference in their ages; $(+) =$ husband older, $x = 6$; crit. val. $= 3$; fail to reject H_0.
14.7 H_a: There is a preference for the new; $(+) =$ prefer new, $z^* = 1.74$, fail to reject H_0.

14.9 39 to 47

14.11 -1 to $+6$, where $d =$ air $-$ not air

14.13 $H_a: P(+) \neq 0.5$, $x = 25$, crit. val. $= 28$, reject H_0

14.15 a. H_0: The average value is the same for both groups.
H_a: The average value is different for the groups.
c. H_0: The average blod pressure is the same for both groups.
H_a: The average blood pressure for group A is higher than for group B.

14.17 H_a: Group 1 scores are higher, $U_{0.05} = 6$, $U^* = 4.5$, reject H_0.

14.19 MINITAB verify

14.21 H_0: No difference between the average systolic blood pressure of men of age 40 and men of age 25, $U = 480$, $z^* = 2.09$, reject H_0.

14.23 a. H_0: The data did occur in a random order.
H_a: The data did not occur in a random order.
c. H_0: The order of entry by gender was random.
H_a: The order of entry was not random.

14.25 H_0: The hiring sequence is random, crit. vals. are 4 and 12, $V^* = 9$, fail to reject H_0

14.27 Yes, reject the hypothesis of random runs above and below the-median.

14.29 H_0: Random order of increase and decrease in value from previous value, crit. vals. are 3 and 10, $V^* = 8$; fail to reject H_0

14.31 H_0: Randomness in number of absences (about median), crit. vals. are 8 and 20, $V^* = 9$, fail to reject H_0

14.33 a. H_0: There is a no relationship between the two rankings vs. H_a: There is a relationship between the two rankings.
c. H_0: The is no correlation between the two variables vs. H_a: There is correlation.

14.35 Verify

14.37 MINITAB verify

14.39 0.048

14.41 164,927.36

14.43 0.175

14.45 H_a: There is a difference in the absentee rates, $(+) =$ more at 8 AM, $x = 2$, crit. val. $= 2$, reject H_0.

14.47 H_a: computer anxiety was reduced, $d =$ before $-$ after, $x = 10$, crit. val. $= 7$, fail to reject H_0.

14.49 H_a: top ripens first, $+ =$ top ripened first, crit. val. $= 5$, reject H_0.

14.51 Reject for $U \leq 127$.

14.53 a. $U = 2$

14.55 H_a: Credit unions charge lower fixed rates than banks, $U_{0.05} = 55$, $U^* = 12$, reject H_0.

14.57 H_a: lack of randomness; crit. vals. $V = 4$ and 10, $V^* = 9$, fail to reject H_0

14.59 H_a: positive correlation, crit. val. $= 0.399$, $r_s^* = 0.880$, reject H_0

14.61 H_a: negative correlation, crit. val. $= -0.703$, $r_s^* = 0.168$; fail to reject H_0

ANSWERS TO CHAPTER QUIZZES

QUIZ A ANSWERS

Only the replacement for the word(s) in boldface type is given. (If the statement is true, no answer is shown. If the statement is false, a replacement is given.)

CHAPTER 1 QUIZ A, PAGE 34

1.1 descriptive
1.2 inferential
1.4 sample
1.5 population
1.6 attribute data
1.7 discrete
1.8 continuous
1.9 random

CHAPTER 2 QUIZ A, PAGE 132

2.1 median
2.2 dispersion
2.3 never
2.5 zero
2.6 higher than

CHAPTER 3 QUIZ A, PAGE 191

3.1 regression
3.2 strength of the
3.3 $+1$ or -1
3.5 positive
3.7 positive
3.8 -1 and $+1$
3.9 output or predicted value

CHAPTER 4 QUIZ A, PAGE 263

4.1 any number value between 0 and 1, inclusive
4.4 simple
4.5 seldom
4.6 sum to 1.0
4.7 dependent
4.8 complementary
4.9 mutually exclusive or dependent
4.10 multiplication rule

CHAPTER 5 QUIZ A, PAGE 303

5.1 continuous
5.3 one
5.5 exactly two
5.6 binomial
5.7 one success occurring on 1 trial
5.8 population
5.9 population parameters

CHAPTER 6 QUIZ A, PAGE 341

6.1 its mean
6.4 one standard deviation
6.6 right
6.7 zero, 1
6.8 some (many)
6.9 mutually exclusive events
6.10 normal

CHAPTER 7 QUIZ A, PAGE 372

7.1 is not
7.2 some (many)
7.3 population
7.4 divided by $\sqrt{n}$
7.5 decreases
7.6 approximately normal
7.7 sampling
7.8 means
7.9 random

CHAPTER 8 QUIZ A, PAGE 433

8.1 alpha
8.2 alpha
8.3 sample distribution of the mean
8.7 type II error
8.8 beta
8.9 correct decision
8.10 critical (rejection) region

CHAPTER 9 QUIZ A, PAGE 479

9.2 Student's t
9.3 chi-square
9.4 to be rejected
9.6 t score
9.7 $n - 1$
9.9 $\sqrt{pq/n}$
9.10 z(normal)

CHAPTER 10 QUIZ A, PAGE 552

10.1 two independent means
10.3 F distribution
10.4 Student's t distribution
10.7 is not

CHAPTER 11 QUIZ A, PAGE 596

11.1 one less than
11.3 expected
11.4 contingency table
11.6 test of homogeneity
11.8 approximated by chi-square

CHAPTER 12 QUIZ A, PAGE 633

12.2 mean square
12.3 SS(factor) or MS(factor)
12.5 reject H_0
12.7 the number of factors less one
12.8 mean
12.9 need to
12.10 does not indicate

CHAPTER 13 QUIZ A, PAGE 685

13.2 need not be
13.3 does not prove
13.4 need not be
13.6 the linear correlation coefficient
13.8 regression
13.9 $n - 2$

CHAPTER 14 QUIZ A, PAGE 728

14.2 t-test
14.3 runs test
14.4 assigned equal ranks
14.7 Mann-Whitney U test
14.8 power
14.10 power

QUIZ B ANSWERS

CHAPTER 1 QUIZ B, PAGE 35

1. **a.** A **b.** D **c.** C **d.** B **e.** D
2. c, g, h, b, e, a, d, f

CHAPTER 2 QUIZ B, PAGE 132

1. **a.** 30 **b.** 45.5 **c.** 90.5 **d.** 15
   e. 1 **f.** 61 **g.** 75 **h.** 75.5
   i. 90.5 **j.** 105.5
2. **a.** two items purchased
   b. Nine people purchased 3 items each.
   c. 40 **d.** 120 **e.** 5 **f.** 2
   g. 3 **h.** 3 **i.** 3.0 **j.** 1.795
   k. 1.34

3. a. 6.7 **b.** 7 **c.** 8 **d.** 6.5
e. 5 **f.** 6 **g.** 3.0 **h.** 1.7
i. 5
4. a. −1.5 **b.** 153

CHAPTER 3 QUIZ B, PAGE 191

1. a. B, D, A, C **b.** 12 **c.** 10
d. 175 **e.** N **f.** (125, 13) **g.** N
h. P
2. Someone made a mistake in arithmetic, r must be between −1 and +1.
3. a. 12 **b.** 10 **c.** 8 **d.** 0.73
e. 0.67 **f.** 4.33 **g.** $\hat{y} = 4.33 + 0.67x$

CHAPTER 4 QUIZ B, PAGE 264

1. a. $\frac{4}{8}$ **b.** $\frac{4}{8}$ **c.** $\frac{2}{8}$
d. $\frac{6}{8}$ **e.** $\frac{2}{8}$ **f.** $\frac{6}{8}$
g. 0 **h.** $\frac{6}{8}$ **i.** $\frac{1}{8}$
j. $\frac{5}{8}$ **k.** $\frac{2}{4}$ **l.** 0
m. $\frac{1}{2}$ **n.** no (e) **o.** yes (g)
p. no (i) **q.** yes (a, k) **r.** no (b, l)
s. yes (a, m)
2. a. 0 **b.** 0.7 **c.** 0 **d.** no (c)
3. a. 0.14 **b.** 0.76 **c.** 0.2 **d.** no (a)
4. a. 0.7 **b.** 0.5 **c.** no, $P(E \text{ and } F) = 0.2$
d. yes, $P(E) = P(E|F)$
5. a. 0.4 **b.** 0.5 **c.** no, $P(G \text{ and } H) = 0.1$
d. no, $P(G)$ not equal to $P(G|H)$
6. 0.51

CHAPTER 5 QUIZ B, PAGE 304

1. a. Each $P(x)$ is between zero and 1, and the sum of all $P(x)$ is exactly one.
b. 0.2 **c.** 0 **d.** 0.8 **e.** 3.2
f. 1.25
2. a. 0.230 **b.** 0.085 **c.** 1.2 **d.** 1.04

CHAPTER 6 QUIZ B, PAGE 342

1. a. 0.4922 **b.** 0.9162 **c.** 0.1020
d. 0.9082

2. a. 0.63 **b.** −0.95 **c.** 1.75
3. a. $z(0.8100)$ **b.** $z(0.2830)$
4. 0.7910
5. 28.03
6. a. 0.0569 **b.** 0.9890 **c.** 537
d. 417 **e.** 605

CHAPTER 7 QUIZ B, PAGE 373

1. a. 0.4364 **b.** 0.2643
2. a. 0.0918 **b.** 0.9525
3. 0.6247

CHAPTER 8 QUIZ B, PAGE 433

1. a. $H_0: \mu = 245, H_a: \mu > 245$
b. $H_0: \mu = 4.5, H_a: \mu < 4.5$
c. $H_0: \mu = 35, H_a: \mu \neq 35$
2. a. 0.05, $z, z \leq -1.65$
b. 0.05, $z, z \geq +1.65$
c. 0.05, $z, z \leq -1.96$ or $z \geq +1.96$
3. a. 1.65 **b.** 2.33 **c.** 1.18
d. −1.65 **e.** −2.05 **f.** −0.67
4. $H_0: \mu = 1520$ vs. $H_a: \mu < 1520$, crit. reg. $z \leq -2.33$, $z^* = -1.61$, fail to reject H_0
5. a. $z^* = 2.50$ **b.** 0.0062
6. 4.72 to 5.88

CHAPTER 9 QUIZ B, PAGE 480

1. a. $H_0: \mu = 225, H_a: \mu > 225$
b. $H_0: \sigma = 3.7, H_a: \sigma < 3.7$
c. $H_0: p = 0.40, H_a: p \neq 0.40$
2. a. 0.05, $z, z \leq -1.65$
b. 0.05, $z, z \geq +1.65$
c. 0.05, $t, t \leq -2.08$ or $t \geq +2.08$
d. 0.05, $\chi^2, \chi^2 \leq 14.6$ or $\chi^2 \geq 43.2$
3. a. 2.05 **b.** −1.73 **c.** 14.6
4. a. 28.6 **b.** 1.44 **c.** 27.16 to 30.04
5. 0.528 to 0.752
6. $H_0: \mu = 26$ vs. $H_a: \mu < 26$, crit. reg. $t \leq -1.71$, $t^* = -1.86$, reject H_0

7. H_0: $\sigma = 0.1$ vs. H_a: $\sigma > 0.1$, crit. reg. $\chi^2 \geq 21.1$, $\chi^{2*} = 23.66$, reject H_0
8. H_0: $p = 0.50$ vs. H_a: $p > 0.50$, crit. reg. $z \geq 2.05$, $z^* = 1.29$, fail to reject H_0

CHAPTER 10 QUIZ B, PAGE 553

1. a. H_0: $\mu_N - \mu_A = 0$, H_a: $\mu_N - \mu_A \neq 0$
   b. H_0: $\sigma_o/\sigma_m = 1.0$, H_a: $\sigma_o/\sigma_m > 1.0$
   c. H_0: $p_m - p_f = 0$, H_a: $p_m - p_f \neq 0$
2. a. 0.05, z, $z \leq -1.65$
   b. 0.05, F, $F \geq 2.11$
   c. 0.05, z, $z \leq -1.96$ or $z \geq +1.96$
   d. 0.05, t, $t \leq -2.05$ or $t \geq +2.05$
   e. 0.05, t, if $\sigma_1 \neq \sigma_2$, then $df = 7$ and $t \geq 1.89$ or if $\sigma_1 = \sigma_2$, then $df = 16$ and $t \geq 1.75$
3. a. 1.75 b. 2.13 c. 37.6 d. 2.33
   e. 4.50 f. 15.0
4. 1st: $F^* = 2.329$, assume $\sigma_P = \sigma_L$
   H_0: $\mu_L - \mu_P = 0$ vs. H_a: $\mu_L - \mu_P > 0$, crit. reg. $t \geq +1.73$, $t^* = 0.979$, fail to reject H_0
5. H_0: $\mu_d = 0$ vs. H_a: $\mu_d > 0$, crit. reg. $t \geq 1.89$, $t^* = 1.88$, fail to reject H_0
6. 0.072 to 0.188

CHAPTER 11 QUIZ B, PAGE 596

1. a. H_0: Digits generated occur with equal probability.
   H_a: Digits do not occur with equal probability.
   b. H_0: Votes were cast independently of party affiliation.
   H_a: Votes were not cast independently of party affiliation.
   c. H_0: The crimes distributions are the same for all four cities.
   H_a: The crimes distribution are not all the same.
2. a. 4.40 b. 35.7
3. H_0: $P(1) = P(2) = P(3) = \frac{1}{3}$
   H_a: preferences not all equal, crit. val.
   $\chi^2(2, 0.05) = 6.00$, $\chi^{2*} = 3.78$, fail to reject H_0
4. a. H_0: The distribution is the same for all types of soil.
   H_a: The distributions are not all the same.
   b. Test crit.: 0.05, χ^2, $\chi^2 \geq 9.50$
   c. 25.622

   d. 13.746
   e. Reject H_0. There is sufficient evidence to show that the growth distribution is different for at least one of the three soil types.

CHAPTER 12 QUIZ B, PAGE 633

1. a. T   b. T   c. F   d. T
   e. T   f. T   g. F   h. F
   i. F   j. F   k. T   l. F
   m. F   n. F   o. T
2. a. 72 b. 72 c. 22 d. 4
   e. 4.5

CHAPTER 13 QUIZ B, PAGE 685

1.

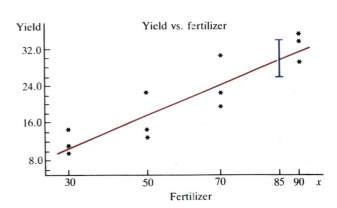

2. $\Sigma x = 720$, $\Sigma y = 252$, $\Sigma x^2 = 49{,}200$, $\Sigma xy = 17{,}240$, $\Sigma y^2 = 6{,}228$
3. $SS(x) = 6000$, $SS(y) = 936$, $SS(xy) = 2120$
4. 0.895
5. 0.65 to 0.97
6. $\hat{y} = -0.20 + 0.353x$
7. See (a) in figure above.
8. 4.324
9. Yes; H_0: $\beta_1 = 0$ vs H_a: $\beta_1 > 0$, $t^* = 6.33$, reject H_0
10. 25.63 to 33.98
11. See (a) in figure above.

CHAPTER 14 QUIZ B, PAGE 729

1. -2 to $+7$
2. H_0: No difference in weight gain.
   H_a: There is a difference in weight gain, crit. val.: 23, $U^* = 32.5$, fail to reject H_0
3. H_0: no correlation
   H_a: correlated, crit. val.: ± 0.683, $r_s^* = -0.70$, reject H_0. Yes, there is significant correlation.
4. $(+)$ = higher grade level than previous problem
   $(-)$ = lower grade level than previous problem
   H_0: $P(+) = 0.5$
   H_a: $P(+) = 0.5$, crit. val.: 7, $x = 11$, fail to reject H_0. This sample does not show a significant pattern.

QUIZ C ANSWERS

CHAPTER 1 QUIZ C, PAGE 35

1. See definitions; examples will vary. Note: *population* is set of ALL possible, while *sample* is the actual set of subjects studied.
2. See definitions; examples will vary. Note: *variable* is the idea of interest, while *data* are the actual values obtained.
3. See definitions; examples will vary. Note: *data* is the value describing one source, the *statistic* is a value (usually calculated) describing all the data in the sample, the *parameter* is a value describing the entire population (usually unknown).
4. Every element of the population has an equal chance of being selected.

CHAPTER 2 QUIZ C, PAGE 134

1. a. 98   b. 50   c. 121   d. 100
2. a. $32,000, $26,500, $20,000, $50,000
   b.
   ```
             :  :  .:  .        .                              .
           --+---+---+---+---+---+---+---+---+---+---+---+---+--
           mode median mean        midrange
             *      *      *           *
           --+---+---+---+---+---+---+---+---+---+---+---+---+--
             20    30    40    50    60    70    80
                              Salary ($1000)
   ```
   c. Mr. VanCott—midrange; business manager—mean; foreman—median; new worker—mode
   d. The distribution is *J*-shaped.

3. There is more than one possible answer for these.
   a. 12, 12, 12
   b. 15, 20, 25
   c. 12, 15, 15, 18
   d. 12, 15, 16, 25, 25
   e. 12, 12, 15, 16, 17
   f. 20, 25, 30, 32, 32, 80
4. A is right; B is wrong; standard deviation will not change.
5. B is correct. For example, if standard deviation is $5, then the variance, (standard deviation)^2, is "25 dollars squared." Who knows what "dollars squared" are?

CHAPTER 3 QUIZ C, PAGE 193

1. Young children have small feet and probably tend to have less mathematics ability, while adults have larger feet and would tend to have more ability.
2. Student B is correct. -1.78 can occur only as a result of faulty arithmetic.
3. These answers will vary, but should somehow include the basic thought:
   a. strong negative   b. strong positive
   c. no correlation   d. no correlation
   e. impossible value, bad arithmetic
4. There is more than one possible answer for these.
   a. $(1, 1), (2, 1), (3, 1)$
   b. $(1, 1), (3, 3), (5, 5)$
   c. $(1, 5), (3, 3), (5, 1)$
   d. $(1, 1), (5, 1), (1, 5), (5, 5)$

CHAPTER 4 QUIZ C, PAGE 265

1. Check the weather reports for a long period of time and determine the relative frequency with which each occurs.
2. Student B is right. *Mutually exclusive* means not a common occurrence, while *independence* means one event does not affect the probability of the other.
3. These answers will vary, but should somehow include the basic thought:
   a. no common occurrence
   b. either event has no effect on the probability of the other
   c. the relative frequency with which the event occurs
   d. probability that an event will occur even though the conditional event has previously occurred

CHAPTER 5 QUIZ C, PAGE 304

1. n independent repeated trials of two outcomes; the two outcomes are "success" and "failure"; $p = P(\text{success})$ and $q = P(\text{failure})$ and $p + q = 1$; $x = n(\text{success}) = 0, 1, 2, \ldots, n$.
2. Student B is correct. The sample mean and standard deviation are statistics found using formulas studied in Chapter 2. The probability distributions studied in Chapter 5 are theoretical populations and their means and standard deviations are parameters.
3. Student B is correct. There are no restrictions on the values of the variable x.

CHAPTER 6 QUIZ C, PAGE 343

1. This answer will vary but should somehow include the basic properties: bell-shaped, mean of 0, standard deviation of 1.
2. This answer will vary but should somehow include the basic ideas: it is a z-score, α represents the area under the curve and to the right of z.
3. All normal distributions have the same shape and probabilities relative to the z-score.

CHAPTER 7 QUIZ C, PAGE 374

1. In this case each head produced one piece of data, the estimated length of the line. The CLT assures us that the mean value of a sample is far less variable than individual values of the variable x.
2. All samples must be of one fixed size.
3. Student A is correct. A population distribution is a distribution formed by all x values that make up the entire population.
4. Student A is correct. The standard error is found by dividing the standard deviation by the square root of the *sample size*.

CHAPTER 8 QUIZ C, PAGE 434

1. **a.** H_0 – (a), H_a – (b)
   b. 4   **c.** 2
   d. $P(\text{type I error})$ is alpha, decreases; $P(\text{type II error})$ increases

CHAPTER 9 QUIZ C, PAGE 481

1. If the distribution is normal, six standard deviations is approximately equal to the range.
2. B
3. They are both correct.
4. When the sample size, n, is larger than 30, the critical value of t is estimated by using the critical value from the standard normal distribution of z.
5. Student A
6. Student B is right. It is significant at the 0.01 level of significance.
7. Student A is correct.
8. It depends on what it means to improve the confidence interval. For most purposes, an increased sample size would be the best improvement.

CHAPTER 10 QUIZ C, PAGE 554

1. independent
2. One possibility: Test all students before the course starts, then randomly select 20 of those who finish the course and test them afterwards.
3. For starters, if the two independent samples are of different sizes, the techniques for dependent samples could not be completed. They are testing very different concepts, the "mean of the differences of paired data" and the "difference between two mean values."
4. It is only significant if the calculated t-score is in the critical region. The variation among the data and their relative size will play a role.
5. The 80 scores actually are two independent samples of size 40. A test to compare the mean scores of the two groups could be completed.
6. When the samples are small and the standard deviations are unknown, the F-test is used to determine which case of two independent means is being tested.
7. A fairly large sample of both Catholic and non-Catholic families would need to be taken, and the number of each whose children attended private schools would need to be obtained. The difference between two proportions could then be estimated.

CHAPTER 11 QUIZ C, PAGE 597

1. Similar in that there are n repeated independent trials. Different in that the binomial has two possible outcomes, while the multinomial has several. Each possible outcome has a probability and these probabilities sum to 1 for each different experiment, both for binomial and multinomial.

2. The test of homogeneity compares several distributions in a side-by-side comparison, while the test for independence tests the independence of the two factors that create the rows and columns of the contingency table.

3. Student A is right in that the calculations are completed in the same manner. Student B is incorrect in that they have similarities, they are tabled in a very similar manner, etc.

4. a. If a chi-square test is to be used, the results of the four questions would be pooled to estimate the expected probability.
   b. Use a chi-square test for homogeneity.

CHAPTER 12 QUIZ C, PAGE 634

1. This answer will vary but should somehow include the basic ideas: It is the comparison of several mean values that result from testing some statistical population by measuring a variable repeatedly at each of the several levels for which the factor is being tested.

2. a. $x_{r,k} = \mu + F$ scrubber $+ \varepsilon_{k(r)}$
   b. H_0: The mean amount of emissions is the same for all three scrubbers tested.
   H_a: The mean amounts are not all equal.

   c.
Source	df	SS	MS
Scrubber	2	12.80	6.40
Error	13	33.63	2.59
Total	15	46.44	

   d. $F(2, 13, 0.05) = 3.81$, $F^* = 2.47$, fail to reject H_0. The difference in the mean values for the scrubbers is not significant.

   e. I
   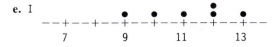

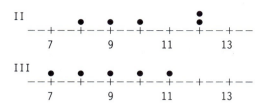

CHAPTER 13 QUIZ C, PAGE 686

1. Variable 1: The frequency of skiers having their bindings tested
Variable 2: The incidence of lower-leg injury
The statement implies that as the frequency with which the bindings are tested increases, the frequency of lower-leg injury decreases; thus the strong correlation must be negative for these variables.

2. A "moment" is the distance from the mean, and the product of both the horizontal moment and the vertical moment is summed in calculating the correlation coefficient.

3. A value close to zero, also. The formulas used to calculate both values have the same numerator, namely $SS(xy)$.

4. The vertical distance from a potential line of best fit to the data point is measured by $(y - \hat{y})$. The line of best fit is defined to be the line that results in the smallest possible total when the squared values of $(y - \hat{y})$ are totaled. Thus "the method of least squares."

5. The strength of the linear relationship could be measured with the correlation coefficient.

6. A random sample will be needed from the population of interest. The data collected need to be for the variables length of time on welfare and the measure of current level of self-esteem.

CHAPTER 14 QUIZ C, PAGE 730

1. The nonparametric statistics do not require assumptions about the distribution of the variable.

2. The sign test is a binomial experiment of n trials (the n data observations) with two outcomes for each data [$(+)$ or $(-)$], and $p = (+) = 0.5$. The variable x is the number of the least frequent sign.

3. The median is the middle value such that 50% of the distribution is larger in value and 50% is smaller in value.
4. The extreme value in a set of data can have a sizeable effect on the mean and standard deviation in the parametric methods. The nonparametric methods typically use rank numbers. The extreme value with ranks is either 1 or n, and neither change if the value is more extreme.
5. d, $p = P(+) = P(\text{prefer seating arrangement A}) = 0.5$; no preference

APPENDIX A

EXERCISES, PAGE A-5

A-1 (a) $x_1 + x_2 + x_3 + x_4$
  (b) $x_1^2 + x_2^2 + x_3^2$
  (c) $(x_1 + y_1) + (x_2 + y_2) + (x_3 + y_3) + (x_4 + y_4) + (x_5 + y_5)$
  (d) $(x_1 + 4) + (x_2 + 4) + (x_3 + 4) + (x_4 + 4) + (x_5 + 4)$
  (e) $x_1y_1 + x_2y_2 + x_3y_3 + x_4y_4 + x_5y_5 + x_6y_6 + x_7y_7 + x_8y_8$
  (f) $x_1^2 f_1 + x_2^2 f_2 + x_3^2 f_3 + x_4^2 f_4$

A-2 (a) $\sum_{i=1}^{6} x_i$ (b) $\sum_{i=1}^{7} x_i y_i$ (c) $\sum_{i=1}^{9} (x_i)^2$
  (d) $\sum_{i=1}^{n} (x_i - 3)$

A-4 (a) 8 (b) 68 (c) 64
A-5 (a) 8 (b) 4 (c) 12
  (d) 4 (e) 42 (f) 64
  (g) −7 (h) 32 (i) 20
  (j) 120 (k) 400

A-6 $\sum_{i=1}^{120} [0.005(12{,}000 - (i-1)100)]$

APPENDIX B

EXERCISES, PAGE B-2

B-2 (a) Use a two-digit number to represent the results obtained. Let the first digit represent one of the coins and the second digit represent the other coin. Let an even digit indicate heads and an odd digit, tails. Observe 10 two-digit numbers from the table. If a 16 is observed, it represents tails and heads on the two coins. One head was therefore observed. The probabilities have been preserved.
  (b) A second way to simulate this experiment is to find the probabilities associated with the various possible results. The number of heads that can be seen on two coins is 0, 1, or 2. (HH, HT, TH, TT is the sample space.) $P(\text{no heads}) = \frac{1}{4}$; $P(\text{one head}) = \frac{1}{2}$; $P(\text{two heads}) = \frac{1}{4}$. Using two-digit numbers, let the numbers 00 to 24 stand for no head appeared, 25 to 74 stand for one head appeared, and 75 to 99 stand for two heads appeared. The probabilities have again been preserved. Observe 10 two-digit numbers.

APPENDIX C

EXERCISES, PAGE C-2

C-1 (a) 13 (b) 9 (c) 9
  (d) 156 (e) 46 (f) 42
  (g) 103 (h) 17

C-2 (a) 8.7 (b) 42.3 (c) 49.7
  (d) 10.2 (e) 10.4 (f) 8.5
  (g) 27.4 (h) 5.6 (i) 3.0
  (j) 0.2

C-3 (a) 17.67 (b) 4.44 (c) 54.55
  (d) 102.06 (e) 93.22 (f) 18.00
  (g) 18.02 (h) 5.56 (i) 44.74
  (j) 0.67

APPENDIX D

EXERCISES, PAGE D-8

D-1 (a) 18
  (c) 6

D-3 (a) $_nP_n = n!$ and $_nP_{n-1} = n \times n-1 \times n-2 \times \cdots \times 2$. $_nP_{n-1}$ is the same product as $_nP_n$, only the last factor of 1 is missing.

(c) $_nC_r = \dfrac{n!}{r!(n-r)!}$ and

$_nC_{n-r} = \dfrac{n!}{(n-r)!(n-(n-r))!} = \dfrac{n!}{(n-r)!\,r!}$

D-5 (a) 2,176,782,336 assuming zero and 0 can be used as first characters.

(c) 6,760,000
(e) 6,084,000

D-7 (a) $_{52}C_5 = 2{,}598{,}960$
(c) $_4C_1 \times {_{48}}C_4 = 4 \times 194{,}580 = 778{,}320$
(e) $_1C_1 \times {_{39}}C_4 + {_3}C_1 \times {_{38}}C_4 = 303{,}696$

INDEX

Acceptance region, 388
Addition rule, 225
All-inclusive events, 211
Alpha, 384, 402, 407, 410, 415
Alternate hypothesis, 383, 393
Analysis of variance, 600
Asterisk, use of, 388, 396
Attribute data, 13
Averages, 38, 68, 74
   long term, 206

Bar graph, 39
Bayes, Reverend Thomas, 249
Beta, 384
   intercept, 652
   slope, 652, 659
   type II error, 384
Bimodal distribution, 57
Binomial coefficient, 287, F-4
Binomial experiment
   definition, 285
   properties, 285
Binomial parameters, 451
Binomial probability
   by calculation, 287
   normal approximation, 331
   by table, 290, F-5
Binomial probability distribution, 282
   mean, 294
   standard deviation, 294
Binomial probability function, 287
Bivariate data, 139, 638
   both qualitative, 139
   both quantitative, 145
   one qualitative, one quantitative, 143
Box-and-whiskers display, 98
Box plots, 144

Cause-and-effect relation, 646, 672
Cells, 562, 573, 574
Census, 23
Central limit theorem, 354, 385, 416
   application, 362, 385, 416

Centroid, 638
Chebyshev's Theorem, 106
Chi-square distributions
   calculation of, 463, 563, 575
   critical value, 464, F-10
   distribution, properties, 464
   tests of enumerative data, 562
   tests of standard deviation, 463
   tests of variance, 463
Circle graphs, 39
Class, 52
Classical hypothesis test, model, 394
Classification of data
   basic guideline, 52
   class bounds, 53
   class frequency, 52
   class limits, 53
   class marks, 54
   class width, 53
Collection of data, 21
Combinations, D-5
Comparing statistical hypothesis tests, 721
Complementary events, 219
Compound events, 222
Computer, 30
Conclusions (hypothesis tests), 388
Conditional probability, 231, 232
Confidence coefficient, 415
Confidence interval, 416
Confidence-interval belt, 647, 669, F-18
Confidence-interval estimations
   level of confidence, 415
   lower limit, 418
   maximum error, 417
   point estimate, 415
   standard error, 417
   upper limit, 418
Constant function, 271
Contingency table, 139, 573
Continuity correction factor, 332, 694
Continuous variable, 14, 308
Correct decision
   type A, 384

type B, 384
Correlation
 negative, 159
 no, 159
 perfect, 159
 positive, 159
Correlation analysis, 158, 638
Correlation coefficient (*see* linear
  correlation coefficient)
Covariance
 calculation, 640
 defined, 639
Critical region, 385, 394
Critical value, 385, 387
Critical-value notation
 chi-square, 464
 F, 506
 t, 441
 z, 325
Cross-tabulation, 139
Cumulative frequency distribution, 58
Curvilinear regression, 652

Data, 12, 13
 bivariate, 139, 638
 enumerative, 562
 single variable, 38
Data collection (*see* sample)
Decision
 fail to reject H_0, 384
 reject H_0, 384
Decision making (*see* hypothesis tests)
Decision rule, 388, 410
Decisions, types of, 384
Degrees of freedom, 440
 ANOVA, 601
 chi-square, 464
 F, 506, 603
 linear correlation, 646
 number of, 440
 regression, 660
 Student's t, 440, 490, 515, 660, 668
Dependent events, 231
Dependent means, 488
Dependent samples, 484, 692
Dependent variable, 145, 158, 652
Depth, 72, 95
Descriptive statistics, defined, 4
Determination of sample size, confidence
  intervals, 422, 455, 722

Deviation
 about mean, 81
 standard, 84
Discrete variable, 14, 268, 331
Dispersion, measure of, 38, 81
Dispersion about the mean, 81
Disraeli, 112
Distribution
 bell-shaped, 107, 308, 309
 bimodal, 57
 binomial, 282
 frequency, 51
 Gaussian, 308
 J-shaped, 57
 normal, 57, 107, 308
 probability, 270
 rectangular, 57
 sampling, 346
 skewed, 57
 standard normal, 310
 symmetrical, 57
 uniform, 57
Distribution-free tests (*see* nonparametric
  tests)
Dot plot display, 39
Dot plots, 144

Efficiency, 722
Empirical rule, 107
Enumerative data, 562
Errors
 probability of, 384
 type I, 384
 type II, 384
Estimation (*see also* confidence interval),
  382, 415
 correlation coefficient (ρ), 161, 647
 difference between two dependent
  means, 492
 difference between two independent
  means, 498, 519
 interval estimate, 415
 means
  one, sigma known, 415
  one, sigma unknown, 445
 median (sign test), 695
 proportion
  one, 453
  two, 535
 regression line, 175

individual value of y, 669
mean value of y, 667
slope, 174, 659
variance (standard deviation),
one, 469
two, 510
Event, 211
Events
all-inclusive, 211
complementary, 219
compound, 222
dependent, 231
equally likely, 217
"failure", 285
independent, 231, 285
intersection of, 223
mutually exclusive, 211, 222, 226
"success", 285
Expected value, 205
Expected value, (E)
contingency table, 574
multinomial experiment, 562
Experiment
defined, 12, 210
first-ace, 18
Experimental error, 612, 653
variance of, 653
Experimentation (*see* statistical experimentation)
Extension(s), 70, 86, 160, 278

Factorial notation, 287
Factorials, D-4, F-3
Factors (of ANOVA), 600, 602
Fail to reject, 388
Failure, 285
F distribution, F-11
calculated value, 506, 605
critical-value notation, 506
defined, 506
test of hypothesis
ANOVA, 600
two variances, 505
Finite population, 11
First-ace experiment, 18
First quartile, 94, 98
Five-number summary, 98
Frequency, 51, 562
Frequency distributions
calculations using, 70, 86
cumulative, 58
cumulative relative, 58
grouped, 52, 54
relative, 55
ungrouped, 51
Frequency histogram, 55
Function
constant, 271
probability, 271

Galton, Sir Francis, 1
Gaussian distribution, 308
Gosset, William, 379, 439
Grand total, 141
Graphic presentations, 39, 145
Greek letters, use of, 12, 278 (*see also* inside front cover)
Grouped frequency distribution, 52, 71, 88

Highest score (H), 53, 73
Hinge, 98
Histogram, 55
line, 272
probability, 273
shape of
bimodal, 57
j-shaped, 57
rectangular, 57
skewed, 57
symmetrical, 57
Homogeneity, 576
Hypothesis, 383
Hypothesis test, 382
acceptance region, 388
alternative hypothesis, 383, 393
ANOVA, 600
calculated value, 396
classical approach, 392, 394
classical model, 394
conclusion, 397
correlation coefficient, 646
critical value, 387, 394, 396
decision, 388, 396
decision rule, 410
homogeneity (contingency table), 576
independence (contingency table), 573
line of best fit, slope, 659

mean, one
  sigma unknown, 439
  sigma known, 392
multinomial experiment, 563
nature of, 382
noncritical region, 388
nonparametric
  Mann-Whitney U, 699
  rank correlation, 713
  runs, 708
  sign, 691
null hypothesis, 383, 393
one-tailed, 410
prob-value, 407, 410
  approach, 407
  model, 407, 410
proportion
  one, 451
  two, 532
  two or more, 577
significance level, 385
test
  criteria, 385
  statistic, 385, 396
two
  dependent means, 489
  independent means, 496, 515
two-tailed, 410
variance (standard deviation)
  one, 463
  two, 505

Independence tested by chi-square, 573
Independent
  events, 231
  means, 496, 515, 699
  samples, 484
  trials, 285, 286, 565
  variable, 145
Independent variable, 145, 158, 652
Inferences, types of (*see* estimation and hypothesis tests)
Inferential statistics, defined, 4
Input variable, 145, 158, 652
Intercept, 169, 652
Interpolation, D-11
Interquartile range, 98
Intersection, 224
Interval estimate (*see* estimation)

Lack of fit, 676
Law of large numbers, 206
Laws of probability
  additive, 226
  multiplicative, 233
Leaf, 41
Least-frequent sign, number of, 691
Least-squares criterion, 168, 652
Level of confidence, 415, 421
Level of significance, 385, 394, 395
Levels of test factor, 601
Limits
  class, 53
  confidence, 418
Linear correlation coefficient
  calculation, 160
  confidence belts, 647, 775
  critical value, 641, 647, 715, F-17, F-22
  defined, 159, 641
  described, 158
  estimation, 161
  formula, 159, 641
Linear model, 652
Line of best fit
  approximation, 174
  calculation, 169
  estimation, 174
  method of least squares, 168
  point $(\bar{x},\bar{y})$, 171, 667
  predictions, 171
  slope, 169, 652
  standard deviation, 654
  y intercept, 169, 652
Lower class limit, 53
Lower confidence limit, 418
Lowest score (L), 53, 73

Mann-Whitney U
  normal approximation, 702
  test statistic, 700, F-20
Marginal totals, 140, 574
Mathematical model, 612, 652
Maximum error of estimation, 417, 454
Mean absolute deviation, 82
Mean difference, 490
Mean square (MS), 604
Mean
  defined, 69
  formula, 69, 70
  inferences (*see* estimation, hypothesis test)
  population, 278

sample, 69, 278
Measurability, 21
Measures
  of central tendency, 38, 68
  of dispersion, 38, 81
  of position, 38, 94
  of spread, 38
Median, 72, 74, 75, 94, 97, 98, 691
  defined, 72
  position of, 72
Median difference, 692
Method of least squares, 168, 652
Midquartile, defined, 97
Midrange, 73
  formula, 73
Misuses of statistics, 6, 112
Modal class, 58
Mode, 58, 73
  defined, 73
Mu (μ), 278
Multinomial experiment, defined, 565
Multiple regression, 652
Multiplication rule, 233
Mutually exclusive events, 211, 222, 226

Negative correlation, 159
No correlation, 159
Noncritical region, 388
Nonparametric tests, 690
  characteristics of, 690
  Mann-Whitney, 699
  rank correlation, 713
  runs, 708
  sign, 691
Normal approximation
  to binomial, 331
  sign test, 693
Normal distribution, 57, 107, 308
  area under, 309
  standard, 310, F-8
Normal probabilities by table, 310
Notation, critical value
  chi-square, 464
  F, 506
  t, 441
  z, 325
Null hypothesis, 383, 393

Observed frequency, 562
Ogive, 58, 108, 109
One-tailed test, 399

Ordered pairs, 145, 210, 638, 713
Outcome, 210
Output variable, 145, 158, 652

p-value (*see* prob-value)
Paired data, 145, 489, 638, 713
Paired differences, 489, 692
Parameter, 12, 278
Parametric methods, 690
Pearson's product moment r, 160, 641
Pearson, Egon S., 199
Percentages
  based on column totals, 142
  based on grand total, 141
  based on row totals, 142
Percentiles, 94
Perfect correlation, 159
Permutations, D-4
Pie charts, 39, 40
Pie graphs, 39, 40
Piece of data, 12
Point estimate, 415
Population, 11, 278
  defined, 11
  finite, 11
  infinite, 11
  mean, 278
  parameter, 12, 278
  standard deviation, 278, 280
Positive correlation, 159
Power of statistical test, 722
Prediction equation, 167
Prediction for y given x, 166, 171, 666
Prediction interval of
  mean given x, 667
  y given x, 669
Prob-value
  calculation, 410, 443, 467
  decision rule, 410
  defined, 407
Probability, 29, 202, 309
  additive law, 226
  Bayes's rule, 249
  binomial, 285, 451, 532
  conditional, 232
  defined, 204, 218
  empirical, 205
  experimental, 205
  function, 271
  multiplicative law, 233
  normal, 310

observed, 205, 451, 533
properties of, 218, 271
subjective, 218
theoretical, 217
Probability distribution
  continuous, 309
  defined, 270
  discrete, 268
  function, 271, 308
  mean of, 278
  properties of, 271
  variance of, 279
Probability of an event, 205
Probability paper, 108, 109
Proportion, pooled estimate, 533

Qualitative data, 13
Quantitative data, 13
Quartiles, 94

Random, 24, B-1
Random error, 652
Random numbers, table of, F-1
  uses of, B-1
Random sample, 24, 349
Random sampling, 349
Random variable, 268
  binomial, 285
Range, 53, 81
Rank correlation
  coefficient, 713
    critical values, 715, F-22
    hypothesis test, 714
Ranked data, 72, 95, 700, 714
Ranks, 700
Reduced sample space, 232
Regression analysis, 166, 652
Regression line (*see* line of best fit)
Regression models, 167, 652
Rejection region, 386, 399
Relative frequency, 55, 203
Relative frequency histogram, 55
Repeated sampling, 346
Replicate, 601
Representative sampling frame, 23
Response variable, 11
Rho, 646, 715
Risks, 384, 722
Round-off rule, 84, C-1
Run, defined, 708
Runs test, 708, F-21
  normal approximation, 709

Sample, 5, 11
  design, 23
  mean, 69, 278
  standard deviation, 84, 278
  statistics, 12, 278
  variance, 82
Sample points, 211
Sample size (n), 69
Sample-size determination, 422, 454, 722
Sample space, 210
  listing, 211
  reduced, 232
  tree diagram (*see also* Study Guide), 211
Samples
  cluster, 25
  dependent, 484
  independent, 484
  judgment, 23
  probability, 24
  random, 24, 349
  repeated, 346
  stratified, 25
  systematic, 25
Sampling distribution, 346, 347, 451, 497
Sampling frame, 23
Sampling plan, 23
Scatter diagram, 145
Second quartile, 94
Short-cut formula (variance), 85
Sigma
  (Σ), 19, 69, A-1
  (σ), 278, 463
Sign test, 691
  confidence-interval procedure, 695
  critical values, 692, F-19
  hypothesis test, 691
  normal approximation, 693
  zeros, 691
Significance level, 385
Simulation of experiments, B-1
Slope, 167, 659
Sort, 41
Spearman, C., 713
Spearman rank correlation coefficient, 713, F-22
Standard deviation
  calculation of, 84, 281, 294
  defined, 84
  pooled estimate, 515
Standard error
  defined, 355

difference between means, 490, 497, 499, 515
difference between proportions, 532
means, 355
proportion, 451
Standard normal distribution, 310
  application of, 317
Standard score, 101, 310
  notation, 325
Standardized variable, 310
Statistic
  defined, 12
  sample, 278
Statistical deception, 112
Statistical experiment, 12
Statistical tables, F-1
Statistics, 29, 30
  defined, 4
  descriptive, 4
  inferential, 4
  misuse of, 6, 112
  use of, 6
Stem, 41
Stem-and-leaf display, 41, 56
Strata, 25
Student, 379, 439
Student's t distribution, 418, F-9
  properties, 418
Success, 285
Sum of squares, 84, 160
  for error, 603, 655
  for factor, 602
  for total, 602
Summation notation, 19, 69, A-1

t
  critical value notation, 441
  critical values of, F-9
  standard score, 418
  test statistic, 418, 439, 490, 515, 660
Tallies, 53
Test criteria, 385
Test for independence, 573
Test for normality, 109
Test of homogeneity, 576
Test of hypothesis (*see* hypothesis test)
Test statistic
  asterisk notations, 388, 396
  calculated value of, 387
  critical value of, 387

  defined, 385
Testing for independence, 234
Third quartile, 94, 98
Tree diagram (*see also* Study Guide), 211
Trial (binomial), 285
Trials
  number of (n), 285
  repeated independent, 285, 565
Trichotomy law, 383
Two-tailed test, 399
type I error, 384
type II error, 384

U score, 700
Ungrouped frequency distribution, 51
Upper class boundaries, 53, 54, 59
Upper class limits, 53
Upper confidence limit, 418

Variable (data), 11
  attribute, 13
  continuous, 14, 332
  dependent, 145, 158, 652
  discrete, 14, 268, 332
  independent, 145, 158, 652
  input, 145, 158, 652
  numerical, 13
  output, 145, 158, 652
  predicted, 652
  qualitative, 13
  quantitative, 13
  random, 268
  response, 11
  standardized, 310
Variability, 21
Variance, population (σ^2), 279
Variance, sample (s^2), 82, 87, 654
  defined, 82
  formula, 82
  short-cut formula, 85
Venn diagram (*see also* Study Guide), 224

"x-bar $(\bar{x})$," 69

"y-hat $(\hat{y})$," 168

z
  notation, 325
  standard score, 101, 310
  test statistic, 396, 451, 497, 499, 532, 693, 702, 709